# GLOSSARY OF SYMBOLS

| Symbol | Meaning* |
|---|---|
| $A$ | An event (3.3) |
| $A$ | Intercept of population (or true) regression line (11.4) |
| $A$ and $B$ | Both event $A$ and event $B$ (3.4) |
| $A$ or $B$ | Either event $A$ or event $B$ or both (3.4) |
| $A\|B$ | Event $A$ occurs, given that event $B$ occurs (3.5) |
| $a$ | Intercept of sample regression line (11.5) |
| $\alpha$ | Probability of Type I error (8.4) |
| $A_n$ | Critical limit in sequential sampling plan (16.7) |
| $AQL$ | Acceptable quality level (page 332) |
| $B$ | An event (3.3) |
| $B$ | Slope of population (or true) regression line (11.4) |
| $B_1, B_2$ | True regression coefficients of $X_1$ and $X_2$ (12.2) |
| $b$ | Slope of sample regression line (11.5) |
| $b_1, b_2$ | Sample regression coefficients of $X_1$ and $X_2$ (12.4) |
| $\beta$ | Probability of Type II error (8.4) |
| $BSS$ | Between-group sum of squares (10.6) or treatment sum of squares (10.8) |
| $C$ | Cyclical variation (13.2) |
| $c$ | Number of measurements in class intervals below the one containing the median (2.4) |
| $c$ | Number of columns (9.5) |
| $\chi^2$ | Distribution of sum of squares of $v$ independent standard normal variables (9.3) |
| $\chi^2_\alpha$ | Value of $\chi^2$ variable exceeded with a probability of $\alpha$ (9.3) |
| $d$ | Durbin-Watson test statistic (12.11) |
| $d_L, d_u$ | Critical values of Durbin-Watson test statistic (12.11) |

| Symbol | Meaning* |
|---|---|
| $\Delta$ | Length of time interval (5.11) |
| $\delta$ | Measure of desired accuracy of sample mean or proportion (7.8) |
| $E_1, \ldots, E_n$ | $n$ events (3.4) |
| $E(X)$ | Expected value of random variable $X$ (4.4) |
| $E_L$ | Mean length of waiting line (Appendix 5.1) |
| $E_r$ | Expected number of runs (9.12) |
| $E_u$ | Expected value of $U$ in Mann-Whitney test (9.11) |
| $ESS$ | Error sum of squares (10.6 and 10.8) |
| $e$ | Expected or theoretical frequency (9.4) |
| $e_i$ | Error term in $i$th observation (11.4) |
| $\hat{e}_i$ | Residual for $i$th observation (that is, difference between $Y_i$ and $\hat{Y}_i$) (12.11) |
| $F$ | Distribution of the ratio of two independent $\chi^2$ variables, each divided by its degrees of freedom (10.4) |
| $F_\alpha$ | Value of $F$ variable that is exceeded with a probability of $\alpha$ (10.5) |
| $F$ | Variance ratio (10.5) |
| $F$ | Ratio of treatment mean square to error mean square (10.8) |
| $F$ | Ratio of block mean square to error mean square (10.8) |
| $F$ | Ratio of explained mean square to unexplained mean square (12.8) |
| $f$ | Actual frequency (9.4) |
| $f_j$ | Number of observations in $j$th class interval (2.4) |
| $f_m$ | Number of observations in class interval containing the median (2.4) |
| $H_o$ | Null hypothesis (8.2) |

(Continued on next page)

*Number in parenthesis indicates the section where the symbol is introduced.

| Symbol | Meaning* | Symbol | Meaning* |
|---|---|---|---|
| $H_1$ | Alternative hypothesis (8.2) | $\mu_{y.x}$ | Mean of $Y$, given the value of $X$ (11.4) |
| $H_i$ | $i$th hypothesis (3.6) | | |
| $h$ | Number of parameters estimated from sample (9.7) | $\bar{\mu}$ | Bayesian estimate of the population mean (Appendix 7.1) |
| $I$ | Irregular variation (13.2) | $N$ | Number of observations in the population (2.4) |
| $I_1$ | Price index for year 1 (14.2) or quantity index for year 1 (14.8) | $n$ | Number of observations in the sample (2.4) |
| $k$ | Number of independent variables in a regression (12.5) | $n$ | Number of Bernoulli trials (4.7) |
| $k$ | Number of population means being compared (10.6) | $n$ | Number of blocks (10.8) |
| | | $n_1, n_2$ | Sample sizes from populations 1 and 2 (7.7) |
| $k$ | Number of treatments (10.8) | $n_1, n_2$ | Numbers of each type in runs test (9.12) |
| $l$ | Width of class interval containing the median (2.4) | $v$ | Number of degrees of freedom of the $\chi^2$ distribution (9.3) |
| $L_m$ | Lower limit of class interval containing the median (2.4) | $v_1, v_2$ | Numbers of degrees of freedom of the $F$ distribution (10.4) |
| $\lambda$ | Mean number of events (arrivals in Appendix 5.1) per unit of time (5.11) | $\Pi$ | Population proportion (7.4) |
| $M$ | Median (2.4) | $\Pi$ | Probability of success on each Bernoulli trial (4.7) |
| $M$ | Number of times an experiment occurs (3.3) | $\Pi_0$ | Population proportion if null hypothesis is true (8.5) |
| $M$ | Monetary gain (15.6) | | |
| $M_0$ | Population median if null hypothesis is true (9.10) | $\Pi_1, \Pi_2$ | Population proportions in populations 1 and 2 (7.7) |
| $M_i$ | Dummy variable for $i$th month (13.10) | $\hat{\Pi}$ | Estimate of population proportion (7.8) |
| $m$ | Number of times a particular outcome occurs (3.3) | $P(A)$ | Probability of event $A$ (3.4) |
| $m$ | Mean number of customers that can be serviced per unit of time (Appendix 5.1) | $P(A \text{ or } B)$ | Probability of event $A$, or event $B$, or both (3.4) |
| $m$ | Number of commodities in price index (14.2) | $P(A \text{ and } B)$ | Probability of both event $A$ and event $B$ (3.4) |
| $\mu$ | Population mean (2.4) | $P(A|B)$ | Probability of event $A$, given that event $B$ occurs (3.5) |
| $\mu_o$ | Population mean if null hypothesis is true (8.4) | $P(x)$ | Probability that random variable $X$ equals $x$ (4.3) |
| $\mu_1, \mu_2$ | Means of populations 1 and 2 (7.7) | $P(\text{not } A)$ | Probability that event $A$ does not occur (3.4) |
| $\mu_i$ | Mean of $i$th population (10.5) | | |
| $\mu_p$ | Mean of prior distribution of the mean (Appendix 7.1) | | |

(Continued on inside back cover)

*Number in parenthesis indicates the section where the symbol is introduced.

# STATISTICS FOR BUSINESS AND ECONOMICS

Methods and Applications

*Third Edition*

*To accompany the text:*

Statistics for Business and Economics: Problems, Exercises, and Case Studies, *Third Edition*

Statistics for Business and Economics: Readings and Cases

Using Minitab with Statistics for Business and Economics
by Mark D. Soskin and Nariman Behravesh

# EDWIN MANSFIELD

DIRECTOR, CENTER FOR ECONOMICS AND TECHNOLOGY
UNIVERSITY OF PENNSYLVANIA

# STATISTICS FOR BUSINESS AND ECONOMICS

## METHODS AND APPLICATIONS

### THIRD EDITION

W · W · NORTON & COMPANY · NEW YORK · LONDON

To Edward D. Gruen and the late Frank A. Hanna,
who encouraged and helped me to enter this field.

Cover artwork by Floyd Rollefstad, Laser Fantasy Productions

ISBN 0-393-95569-9

W. W. Norton & Company, Inc., 500 Fifth Avenue, New York, N.Y. 10110
W. W. Norton & Company Ltd., 37 Great Russell Street, London WC1B 3NU

2 3 4 5 6 7 8 9 0

# Contents

# PART II

## Probability, Probability Distributions, and Sampling          *73*

**6 SAMPLE DESIGNS AND SAMPLING DISTRIBUTIONS** *198*

# PART III

## Estimation and Hypothesis Testing  *241*

**7 STATISTICAL ESTIMATION**  *243*

**8 HYPOTHESIS TESTING**  *287*

# PART IV

## Chi-Square Tests, Nonparametric Techniques, and the Analysis of Variance                                    *345*

# PART V
## Regression and Correlation    *433*

# PART VI
## Time Series and Index Numbers
*541*

# PART VII
## Statistical Decision Theory
*623*

# Preface

Three objectives guided the writing of this text. First, to explain statistical methods used in business and by economists in the clearest possible way. Second, to draw case material from real-world situations, in order to demonstrate the practical applications of those methods. Third, to surround both theory and cases with an abundance of problems, based on real data whenever possible, and graded in level of difficulty.

In this new edition I have further strengthened the applied approach of the book, adding detailed cross-chapter cases that help students master the tools covered in each part of the book. I also include more material on the role of the computer in statistics, and I have streamlined and enhanced the presentation of material throughout, aided by the suggestions of teachers from across the country. In addition, a new design aids students in study and review.

Looking at specific changes, there are eight important ways in which this third edition differs from its predecessor.

*1. The treatment of hypothesis testing and confidence intervals has been simplified and made more consistent.* In Chapter 8, the discussion of hypothesis testing now stresses the similarities among various test procedures and the steps to be carried out. The test statistic that is used is now $Z$ in all relevant cases. Much more emphasis is put on $P$-values. Since many instructors prefer "do not reject the null hypothesis" to "accept the null hypothesis," we now use this terminology throughout the book. The presentation of confidence intervals has been streamlined.

*2. Seven cross-chapter cases, entitled "Statistics in Context," are presented as major integrating devices to help the student see the connections between various chapters.* Appearing near the end of each part (other than the introduction), each case (2–4 pages long) is based on actual data and industrial experience. To deal effectively with each case, the student must bring together and draw on what he or she has learned in the several chapters that comprise the relevant part of the book.

*3. Thirty-two "Basic Ideas" are highlighted in the text, in an effort to call attention to the fundamental concepts that the student should under-*

stand to comprehend the statistical forest, as distinct from the trees. Too frequently statistics books tend to be recitations of one specialized technique after another, and the basic ideas—which are all that can be expected to survive in the student's mind a year or two after graduation—tend to get lost in the shuffle.

*4. Much more emphasis has been put on computer techniques.* In most chapters, there is a new type of insert on "Using the Computer," which explains in a general way how computer packages like Minitab can be used and the nature and interpretation of the output of such packages. Also, a limited number of exercises involving computer output have been added. At the same time, I have taken pains to ensure that this material can easily be skipped in classes where little or no time can be devoted to computers.

For instructors who wish to spend more time on computer techniques, there is also a new workbook, *Using Minitab with Statistics for Business and Economics,* by Mark D. Soskin (State College at Potsdam, New York) and Nariman Behravesh (Wharton Econometric Forecasting Associates). This workbook introduces Minitab commands for each chapter of the text, using both examples from the text and new data drawn from studies in the field. Of special interest are datasets provided by WEFA—used in case studies in Parts, I, III, V, and VI of the workbook, and central to the chapters on regression analysis and time series—which allow students to study a simplified version of the Wharton model as well as the data. All data used in the workbook are available in computer-readable form for adopters of the text, and are suitable for use with Minitab or any other statistical package.

*5. A group of review exercises has been added to the end of each chapter.* Since these exercises refer to all of the topics of the chapter, the student, in working these exercises, must figure out which of the techniques discussed in the chapter is relevant for each exercise. This is a very useful form of review.

*6. The examples and cases have been updated and sometimes replaced entirely.* Whenever fresher or better empirical material could be found, it was used. For example, new material concerning the CEA's forecasts, the profitability of oil firms, the application of statistical techniques by industry—and even the probability of failure of the booster rockets of the U.S. space shuttle—are included.

*7. The discussion of many basic statistical techniques has been improved.* For example, the coefficient of variation is described more fully, the material on probability theory in Chapter 3 has been simplified (and shortened somewhat), an appendix on the stem-and-leaf diagram has been added, and some criticisms of $R^2$ are included.

*8. Some relatively advanced materials—such as the treatment of two-action problems in Chapter 15 and of sequential decision making under uncertainty in Chapter 16–have been moved to chapter appendices.* For those instructors who find that time limitations prevent their inclusion, they are easier now to omit.

This book is written for first courses in statistics for undergraduate and graduate students in business administration, public administration, and economics. During the past twenty years, I have taught such courses at the Graduate School of Industrial Administration at Carnegie-Mellon University, at Harvard University, and at the University of Pennsylvania. Some of these courses have been designed primarily for business students, while others have been primarily for economics majors. Despite the fact that many good books exist in this field, I felt the need in these courses for a text that contained a fuller account of how statistical methods are used in the real world—a text that was broader in scope and more flexible than those currently available. Gradually, building on materials used in the classroom, the present volume evolved.

There are several features of this book that, in my opinion, differentiate it from others.

*1. There is a continual emphasis on the practical applications of statistics.* In each chapter, there are case studies indicating how statistical techniques have been used by business firms and government agencies in solving actual problems. Also, in each chapter attention is focused on a case based on the actual experience of a particular firm or government agency. These cases, ranging from NASA's problems with the reliability of the Apollo space mission to Cutler-Hammer's decision concerning the purchase of a license, are presented for solution by the student. In addition, there are inserts asking the student to spot errors in statistical procedures. In my judgment, the case studies, the cases, and the inserts add significantly to the effectiveness of the text because students are motivated to learn material much more thoroughly when they are convinced that it has important practical uses.

*2. This book provides a broader menu of topics than is included in most other texts at this level.* I have included optional sections and appendixes dealing with basic aspects of econometrics, queuing theory, simulation, serial correlation, the analysis of residuals, and other topics frequently neglected in books of this sort. These are areas in which statistics has found important applications which many instructors would like to touch on. Also, more attention is paid to the role of the computer here than in most other books at this level. An attempt is made as well to include a fuller treatment of modern decision theory than in most other texts.

*3. If used in a one-semester (or one-quarter) statistics course this book provides more flexibility than most other texts.* It is designed so that several quite different types of one-semester courses can be taught from it: (1) for instructors who want to provide a relatively broad survey of the basic techniques of statistical inference, the outline in Table 1 may be useful; (2) for those who want to emphasize decision theory, the outline in Table 2 may be preferred; (3) for a course emphasizing regression and economic statistics, the outline in Table 3 may be more appropriate.[1] Moreover, the outlines in Tables 1, 2, and 3 are by no means the only choices. I have placed many topics in optional sections and ap-

---

[1] As indicated in Table 3, instructors who wish to skip the Poisson distribution can easily do so.

Table 1    Suggested Outline for a One-Semester Course Providing a
Relatively Broad Survey of the Basic Techniques of
Statistical Inference

Chapter 1
Chapter 2 (including Appendix 2.1)
Chapter 3
Chapter 4 (including Appendix 4.1)
Chapter 5
Chapter 6
Chapter 7
Chapter 8
Chapter 9 (probably omitting some sections)
Chapter 10 (probably omitting some sections)
Chapter 11
Chapter 15 (optional)

Table 2    Suggested Outline for a One-Semester Course
Emphasizing Decision Theory and Management Science

Chapter 1
Chapter 2 (including Appendix 2.1)
Chapter 3
Chapter 4 (including Appendix 4.1)
Chapter 5 (including Appendix 5.1)
Chapter 6 (including Appendix 6.1)
Chapter 7 (including Appendix 7.1)
Chapter 8
Chapter 11
Chapter 15
Chapter 16

Table 3    Suggested Outline for a One-Semester Course
Emphasizing Regression and Economic Statistics

Chapter 1
Chapter 2 (including Appendix 2.1)
Chapter 3
Chapter 4 (including Appendix 4.1)
Chapter 5 (omitting Sections 5.10–5.12)
Chapter 6
Chapter 7
Chapter 8
Chapter 11
Chapter 12
Chapter 13 (optional)
Chapter 14
Chapter 15 (optional)

pendixes, thus enabling the instructor to put together a course that fits his or her special need.[2]

*4. I have tried to assemble an unusually abundant and varied selection of problems, problem sets, and supplementary materials for use by both the student and the instructor.* In addition to the case studies, cases, and inserts, each chapter contains approximately 30 to 40 problems and problem sets. As indicated above, this revised edition contains a set of chapter review exercises at the end of each chapter. Further, *Statistics for Business and Economics: Problems, Exercises, and Case Studies*, third edition, which accompanies this text, includes numerous additional problems and questions as well as a detailed and realistic case study pertaining to each chapter. (There is also a *Solutions Manual* for instructors, which includes solutions to all problems and cases in the text.) Finally, *Statistics for Business and Economics: Readings and Cases* has been designed to supplement the basic text. These readings illustrate and help to illuminate the various topics in the text.

No mathematical background beyond high school algebra is required for an understanding of *Statistics for Business and Economics*. Mathematical derivations are generally not included in the text. The emphasis here is on providing students with solid and effective evidence concerning the power and applicability of modern statistical methods, on making sure that they can use these techniques, and on indicating the assumptions underlying these techniques, as well as their limitations. To accomplish these purposes, it is neither necessary nor appropriate to deluge students with mathematics.

In writing this book I have benefited from the comments and suggestions of many colleagues and students. Particular thanks go to the following teachers who have commented in detail on all or most of the manuscript: Gordon Antelmen, University of Chicago; Wallace Blischke, University of Southern California; Warren Boe, University of Iowa; Judd Hammock, California State University at Los Angeles; Arthur Hoerl, University of Delaware; D. L. Marx, Louisiana State University; Robert B. Miller, University of Wisconsin; Don R. Robinson, Illinois State University; Stanley Steinkamp, University of Illinois; Robert J. Thornton, Lehigh University; Bruch Vavrichek, University of Maryland; Albert Wolinsky, California State University at Los Angeles; Gordon P. Wright, Tuck School, Dartmouth College; and Arnold Barnett, Massachusetts Institute of Technology. For valuable comments on the second and third editions, particular thanks go to Pat Donnelly, Drexel University; John Mattila, Wayne State University; Robert Sandy, Indiana University—Purdue University; Robert Schiller of Yale University; Donna Stroup, University of Texas at Austin; Robert W. Hoyer, University of Michigan; Ronald P. LeBlanc, Idaho State University; John Mamer, University of California at Los

---

[2] In addition, some instructors prefer to take up basic aspects of simple regression and correlation relatively early in the course. This can easily be done (see footnote 9 of Chapter 11). Also some instructors prefer to take up decision theory early in the course. This, too, can easily be done.

Angeles; Robert B. Miller, University of Wisconsin at Madison; Pam Peterson, Florida State University; William Rettig, Indiana University at Pennsylvania; John F. Schank, George Mason University; Jay Weinroth, Kent State University; and Mary Sue Younger, University of Tennessee. Professor Boe supervised the checking of many of the solutions to the exercises. I also thank John Hawkins and Donald S. Lamm, who did their usual good job with the editorial and publishing aspects of the work.

I am indebted to the Biometrika Trustees for allowing me to reproduce material from the *Biometrika Tables for Statisticians;* to Professors J. Durbin and G. S. Watson for permission to reproduce their tables; and to the Rand Corporation for permission to reproduce materials from its tables of random numbers. Further, I am grateful to the literary executor of the late Sir Ronald A. Fisher, F.R.S.; to Dr. Frank Yates, F.R.S.; and to Longman Group Ltd., London, for permission to reprint part of Table 3 from their book *Statistical Tables for Biological, Agricultural, and Medical Research* (6th edition, 1974). Finally, special thanks go to my wife, Lucile, and to my two children who helped in countless ways with the preparation of this book.

Philadelphia, 1986                                             E.M.

PART I

# Descriptive Statistics

# Introduction to Statistics

## 1.1 The Field of Statistics

According to John Ruskin, the nineteenth-century English author, "Life being short, and the quiet hours of it few, we ought to waste none of them in reading valueless books." Why spend time reading a book on statistics? Because statistics is of basic importance in business and economics. It is no more possible for a business executive or an economist to function without a knowledge of statistics than for a physicist to function without a knowledge of mathematics.

Basically, the field of statistics has two aspects: descriptive statistics and analytical statistics. *Descriptive statistics* is concerned with summarizing and describing a given set of data. For example, every ten years, the Bureau of the Census gathers basic data concerning the number, age distribution, and occupational and educational composition of the American people. Since the amount of raw data gathered from the American people by the Census Bureau is immense, it is necessary to condense and interpret this information to make it useful. Because this process may distort the results if not carefully handled, it is very important that it be done properly. Otherwise, the entire census may result in misleading and incorrect conclusions.

*Analytical statistics* consists of a host of techniques that help decision makers to arrive at rational decisions under uncertainty. Generally, the data available to a business executive, economist, government official, or other such decision maker are incomplete. For example, firms continually have to test whether the materials they receive from suppliers conform to specifications of quality and performance. Since it is frequently too expensive or even impossible (in cases where the only effective tests destroy the materials) to test all incoming materials, firms must base their decisions on testing only a sample. A central function of analytical statistics is to specify ways in which decisions based on incomplete information of this sort can be made as effectively as possible.

It is important to note at the outset that statistics is concerned with *the ways in which* data should be gathered (and *whether* data should be gathered at all), as well as with *how* a particular set of data should be analyzed once it has been collected. *What one can legitimately conclude from a particular set of data depends on how the data have been collected.* If a market-research firm were to collect data concerning the tippling habits of the American people by sampling the inhabitants of nursing homes run by the Temperance Union, you would pay no attention to the results—and rightly so. But as we shall see, far less apparent—and thus more dangerous—mistakes can be made in data gathering with unfortunate and costly results both to those who collect and analyze the data and those who use them.

## 1.2  The Design of Sample Surveys

**Population vs. Sample**

Two terms, *population* and *sample,* must now be defined. *The **population** consists of the total collection of observations or measurements that are of interest to the statistician or decision maker.* It is difficult to give a more specific definition at this point, but in subsequent sections we shall provide many concrete examples of populations. *A **sample** is a subset of measurements taken from the population in which the statistician is interested.* Firms, government agencies, and other organizations continually draw samples to obtain needed information because it would be too expensive and time-consuming to try to obtain complete data concerning all relevant units. Thus, the federal government's unemployment statistics are based on a sample of persons rather than on a complete count of all workers without jobs and looking for work. Television ratings, which have so strong an influence over which programs survive and which are canceled, are based on the TV-viewing decisions of a sample of families. It is easy to find examples of the use of sampling in all areas of social and government activity.

To illustrate in greater detail the use of sampling techniques by business firms, consider the case of the Exxon Corporation, a huge U.S.-based multinational oil firm, which has a scientific and engineering affiliate, the Exxon Research and Engineering Company. This affiliate maintains a storehouse containing supplies necessary for its operations, and each year it must adjust its accounting figure concerning the total value of all these supplies to bring it into agreement with the actual contents of the storehouse. Originally, a 100 percent annual inventory was taken in which the amount of each type of item (scissors, paint brushes, and so forth) was counted, and the total value of all types of items in the storehouse was compared with what the accounting records said it should be. Then the accounting figure for the total value of the items was adjusted to make it accord with reality.

In recent years, Exxon has replaced this 100 percent annual inventory with a sampling procedure of the following sort. First, it has divided all of the types of items in the storehouse into five groups: (1) those with unit cost under $1; (2) those with unit cost of $1–$4.99; (3) those with unit cost of $5–$19.99; (4) those with unit cost of $20 and over; and (5) those that are particularly likely to be lost (based on past experience), such as paint brushes. Then, Exxon has chosen a *sample* of the types of items in each group and has compared the total value of each of these types of items with what the accounting records have said they should be. Based on this sample, it has estimated the difference between the *total* value of *all* types of items in the storehouse and what the accounting records say it should be.

How accurate has this sampling procedure been? Based on the principles that will be explained in detail in later chapters, Exxon has estimated that this statistical procedure seldom results in an error of more than $1,300. (In other words, the adjustment in the total value of all types of items seldom is in error by more than $1,300.) Since the total value is about $185,000, this is less than a 1 percent error and is acceptable to Exxon, which, as shown in Table 1.1, has saved about

| | |
|---|---|
| Total number of types of items | 3,894 |
| Number of types of items sampled | 726 |
| Manhours for 100 percent inventory | 580 |
| Manhours for sample | 260 |
| Approximate saving (dollars) due to sampling* | 3,200 |

**Table 1.1**
Characteristics and Accuracy of Exxon's Sample of Items in Storehouse

*Source:* R. Obrock, "A Case Study of Statistical Sampling," in E. Mansfield, *Statistics for Business and Economics: Readings and Cases* (New York: Norton, 1980).

* Based on an assumed wage rate of $10 per hour.

$3,200 per year by sampling rather than carrying out a 100 percent inventory. In this case, the savings due to the use of sampling techniques were of moderate, but by no means trivial, size; in other cases, the savings have turned out to be much greater.[1]

## TYPES OF POPULATIONS AND SAMPLES

At this point, let's return to the concept of the population. What was the population in the Exxon case? For each type of item in the storehouse, there is a difference (sometimes zero) between the total value of this type of item and its total value according to the accounting records. The population in this case is the entire set of these differences. For the sake of simplicity, suppose that there are only 20 types of items in the storehouse and that the difference for each type of item between its actual total value and its total value according to the accounting records is as shown in Table 1.2. Then the population consists of the 20 numbers in Table 1.2. Why? Because this set of numbers represents the difference between the actual and book value of the inventory, which is what Exxon is interested in. To calculate the adjustment that must be made in the accounting figure for the total value of all types of items in the storehouse, it must sum up these 20 numbers.

[1] For further description of this case, see the paper referred to in Table 1.1.

**Table 1.2**
Hypothetical Population of Differences (Dollars) between the Actual Total Value of a Type of Item and Its Total Value According to Accounting Records, 20 Types of Items

| | | | | |
|---|---|---|---|---|
| −3 | +20 | 0 | −18 | −98 |
| −80 | −51 | −1 | −26 | +42 |
| +14 | −32 | +11 | 0 | −103 |
| −8 | 0 | −36 | +4 | +8 |

Using a quite different example, suppose that the U.S. Department of Labor wants to estimate the percentage of unemployed individuals in the labor force 16 years old and over in a particular small town. In this case, what is the population? It is a set of observations indicating whether each person in the labor force 16 years old or over does or does not have a job. To simplify, suppose that there are in this town only 25 individuals in the labor force of age 16 or over, and that the employment status of each one is as shown in Table 1.3. (That is, the first person's employment status is given by the first word in the first column, the second person's employment status is given by the second word in the first column, and so on, until the last person's employment status is given by the last word in the fifth column.) Then the observations in Table 1.3 would constitute the population.

**Table 1.3**
Hypothetical Population of Whether or Not Employed, 25 People

| | | | | |
|---|---|---|---|---|
| Unemployed | Employed | Employed | Employed | Employed |
| Employed | Employed | Unemployed | Employed | Employed |
| Employed | Employed | Employed | Employed | Employed |
| Employed | Employed | Employed | Employed | Employed |
| Employed | Employed | Employed | Employed | Employed |

When carrying out sample surveys, *a listing of all the elements or units in the population is often called a **frame.*** In the example of the Exxon Corporation, the frame is the list of types of items (scissors, paint brushes, and so forth). In the case of the unemployment rate, a listing of all persons in the labor force 16 years old or over would be a frame. *A **census** is a survey that attempts to include all the elements or units in the frame.* For example, the Census Bureau's decennial Census of Population attempts to include each person in the United States. If the frame is very large, it is generally impossible to include all the elements or units; nevertheless, surveys that do so with reasonable success are called censuses.

Populations are of various types. For example, some consist of *quantitative* information, whereas others consist of *qualitative* information. In the Exxon example, the population consists of quantitative measurements; that is, ones that can be expressed numerically. Each observation or measurement in this population is the difference between the total value of a particular type of item and its total value according to the accounting records. Each such difference is a number (which can be positive, negative, or zero). In the unemployment case, the population consists of qualitative (nonnumerical) information.

*Quantitative vs.*
*Qualitative Data*

Each observation or measurement in this population is a statement of whether or not the person in question has a job. Other qualitative characteristics of a person include his or her sex and level of education.

Populations can also be *finite* or *infinite*. If a population contains a finite number of members, it is called a *finite population*. Some finite populations, such as the hypothetical ones in Tables 1.2 and 1.3, contain relatively few members; others have a very large number of members. If a population contains an unlimited number of members, it is called an *infinite population*. Typically, infinite populations are conceptual constructs. In flipping a coin again and again, the sequence of heads and tails is an infinite population if the flipping is conceived of as continuing indefinitely. In some cases, as we shall see in subsequent chapters, it is convenient to treat a finite population as if it were infinite if the size of the sample is a small proportion of the size of the population.

As we have already stated, a sample is a subset of measurements taken from the population.[2] For example, suppose that we choose the types of items in the first column of Table 1.2 as our sample; then the sample consists of −$3, −$80, +$14, and −$8. Or suppose that we choose as a sample the individuals in the third column in Table 1.3; then the sample consists of "Employed," "Unemployed," "Employed," "Employed," and "Employed." A sample can be drawn in a variety of ways. Among those discussed in subsequent chapters are simple random sampling, stratified random sampling, systematic sampling, and cluster sampling. In order to use sample data intelligently it is essential that you have some familiarity with each of these methods.

## 1.3 The Design of Experiments

Unlike the case of the Exxon Corporation, many statistical investigations are not aimed simply at estimating an unknown total or an unknown percentage. Instead, their purpose is to estimate the effect of one or more factors on some dependent variable. For example, the IBM Corporation may be interested in estimating how much effect a 1 percent reduction in price will have on the quantity demanded of its computers. Or the Federal Reserve Board may be interested in estimating the effect on the prime interest rate of a 3 percent increase in the money supply over a particular period of time. In many cases, it is impossible for firms or agencies to carry out experiments to help estimate these effects, either because the factors are beyond their control or because it would be too expensive or risky to experiment in this way. In such cases, statisticians have devised techniques for estimating these effects as accurately as possible from historical data.

One important limitation of historical studies of this kind is the

---

[2] Even if a subset of measurements is taken from a population in which the statistician or decision maker is *not* interested, it is a sample; but it is a sample from the wrong population.

difficulty in controlling the effects of factors other than those in which the investigator is interested. For example, the relationship between the price and the quantity demanded of, say, gasoline is likely to depend, among other things, on the level of consumer income and the prices of competing and complementary products (all of which are continually changing). Unless the effects of these other factors can be held constant, there is no meaningful way to estimate the price-demand relationship. As we shall see in later chapters, statisticians have devised techniques to help reduce—and in some cases, completely overcome—this limitation, but it should be noted that special problems are encountered in analyzing nonexperimental data of this sort.

In recent years, there have been increasing uses of direct and conscious experimentation in business and economics, ranging from experiments investigating consumers' responses to advertising to experiments investigating the response of the poor to cash subsidies. Imperial Chemical Industries (ICI), the huge British chemical firm, for instance, carried out the following experiment to estimate the effect of a chlorinating agent on the abrasion resistance of a certain type of rubber.[3] Each of 10 pieces of this rubber was cut in half, and one half of each piece was treated with the chlorinating agent, while the other half was untreated. Then the abrasion resistance of each half-piece was evaluated on a machine, and the difference between the abrasion resistance of the treated half-piece and the untreated half-piece was computed. Table 1.4 shows the 10 differences (one corresponding to each of the pieces in the sample).

These 10 differences can be viewed as a sample of 10 from the infinite population of differences which would result if pieces of rubber were subjected to this test indefinitely. Viewed in this way, the sample

**Table 1.4**
Difference in Abrasion Resistance (Treated Part Minus Untreated Part), 10 Pieces of Rubber

| Piece | Difference* |
|-------|-------------|
| 1 | 2.6 |
| 2 | 3.1 |
| 3 | −0.2 |
| 4 | 1.7 |
| 5 | 0.6 |
| 6 | 1.2 |
| 7 | 2.2 |
| 8 | 1.1 |
| 9 | −0.2 |
| 10 | 0.6 |

*Source:* O. Davies, *The Design and Analysis of Industrial Experiments* (London: Oliver and Boyd, 1956), p. 13.
* The units in which these differences are expressed are given in the source and need not concern us here.

[3] See O. Davies, *The Design and Analysis of Industrial Experiments* (London: Oliver and Boyd, 1956), p. 13.

enabled ICI to estimate the average effect of the chlorinating agent on the abrasion resistance of the rubber. Also, it enabled ICI to *test certain hypotheses* about this effect. For example, some of ICI's personnel were interested in testing the hypothesis that on the average, this chlorinating agent had no effect on the abrasion resistance of the rubber. As we shall see in subsequent chapters, statistical methods can be used to test this hypothesis, based on the sample in Table 1.4.

Just as it is not easy to design a sample survey effectively, so it is not easy to create the proper design for an experiment. The time to worry about how an experiment should be designed is before, not after, the experiment is carried out. Too often, an investigator finds that he or she cannot draw useful conclusions from an experiment because the experiment was improperly designed. *Before carrying out an experiment (or a sample survey) it is essential that the objectives of the experiment (or sample survey) be defined precisely.* Without such a statement of objectives, it is impossible to formulate a design that will obtain the desired information at reasonable cost. With such a statement of objectives, the statistician can provide useful and time-tested guidance for conducting the experiment and obtaining the desired information at minimum or close-to-minimum cost.

## 1.4 The Role of Probability in Statistics

As we have seen, a central feature of analytical statistics is its use of information concerning a sample to make inferences about the nature of the population from which the sample is drawn. For example, the Exxon Corporation used sample data to make an inference concerning the difference between the total value of all types of items in the storehouse and what the accounting records say it should be. And Imperial Chemical Industries used sample data to make an inference concerning the true effect of a chlorinating agent on the abrasion resistance of rubber. Any conclusion based on a sample must be subject to a certain amount of uncertainty. Thus, even if ICI finds that the chlorinating agent increases abrasion resistance for 150 pieces of rubber, there is always a chance that these results are somehow a fluke and that, if the sample size were increased to 1,000 or to 10,000, the results would be reversed.

How much certainty can you attach to a particular statement about a population that is based on the results of a sample? Intuitively, you are likely to feel that the answer depends on the size of the sample. The bigger the sample, the more confidence you are likely to have in the sample results; the smaller the sample, the less confidence you are likely to have in the sample results. (Subsequent chapters will show that the sample size is indeed one determinant of how much confidence you can put in a sample result, but it is by no means the only determinant.) Why do you feel that increases in sample size increase the amount of confidence you should have in the sample results? Because, as the sample size increases, the **probability** that the sample result departs greatly   *Probability* from the corresponding result for the entire population becomes smaller and smaller. Thus, as the Exxon Corporation takes a bigger and bigger sample of types of items, it seems logical that we can place more and more confidence in the sample estimate of the difference between the total value of all types of items in the storehouse and what the accounting records say it should be.

But what exactly do we mean by a probability? And how can we measure a particular probability? These questions are important to the statistician and to the user of statistics alike. To go beyond vague, intuitive notions about the degree of accuracy of a particular sample result we must draw on probability theory, a branch of mathematics distinct from, but closely related to, statistics. (To delve deeply into probability theory, one needs a considerable mathematical background. However, no mathematics beyond elementary high-school algebra is needed to understand the elements of probability theory required for an introductory course in statistics.) Until we present a more adequate definition of a probability in Chapter 3, we shall treat probability intuitively. That is, if a particular event has a 50-50 chance of occurring, we shall say that the probability of its occurring is 0.5. Similarly, if a particular event has one chance in four of occurring, we shall say that the probability of its occurring is 0.25. Or if a particular event has one chance in five of occurring, we shall say that the probability of its occurring is 0.2.

## 1.5 Decision Making under Uncertainty

As pointed out previously, statistics is meant to promote more rational decision making under uncertainty. To see how important and far-reaching this objective is, recall that practically all decisions are made under conditions of uncertainty, because it is seldom possible for the decision maker to forecast accurately the consequence of each alternative course of action. If you must decide whether to accept a job with a New York bank or a Los Angeles oil firm, there is no way for you to forecast accurately the ultimate consequences of accepting either offer. This is a decision you must make under uncertainty; and as with all decisions that must be made under uncertainty, you will be forced to gamble. Under conditions of uncertainty, even if an individual makes the best possible decisions, some will turn out to be wrong when judged with the advantages of hindsight. (The lively traffic through the divorce courts bears witness to this fact.)

A decision made by Maxwell House, the nation's largest producer of coffee, illustrates the sorts of decisions that modern statistical techniques can help to analyze. Maxwell House, together with the American Can Company, developed a new kind of keyless coffee container based on the tear-strip opening principle. One important decision that had to be made before introducing this new can was whether or not to raise the price of coffee in it by 2 cents per pound. (The quick-strip can was expected to cost an average of 0.7 cents more than the older container.) According to one analyst who studied this decision,[4] if Maxwell House raised its price by 2 cents per pound, it might have been reasonable to expect that each of the following consequences was equally likely to occur: (1) its market share would decline by about 1.5 percentage points; (2) its market share would remain constant; (3) its market share would increase by 1.0 percentage points; and (4) its market share would increase by 2.5 percentage points. The change in Maxwell House's profits corresponding to each of these changes in its market share is given in Table 1.5.

| Price held constant | | Price increased by 2 cents | |
|---|---|---|---|
| *Change in market share (percentage points)* | *Change in profit (thousands of dollars)* | *Change in market share (percentage points)* | *Change in profit (thousands of dollars)* |
| +2.8 | +4,104 | +2.5 | +11,939 |
| +1.0 | −591 | +1.0 | +6,489 |
| 0 | −840 | 0 | +2,856 |
| −0.6 | −1,218 | −1.5 | −1,050 |

**Table 1.5**
Changes in Maxwell House's Profit Corresponding to Selected Changes in Its Market Share

*Source:* J. Newman, *Management Applications of Decision Theory* (New York: Harper and Row, 1971).

According to the same analyst, if Maxwell House did not raise its price, it might have been reasonable to expect that (1) the probability was 0.1 that its market share would decline by 0.6 percentage points; (2) the probability was 0.2 that its market share would remain constant; (3) the probability was 0.5 that its market share would increase by 1.0 percentage points; and (4) the probability was 0.2 that its market share would increase by 2.8 percentage points. The change in Maxwell House's profits corresponding to each of these market-share changes is provided in Table 1.5.

Based on this welter of facts and estimates, what should Maxwell House have done? Should it have increased the coffee price or not? Given the numbers in the two previous paragraphs and in Table 1.5, you probably feel uncomfortable for at least two reasons. First, you probably wonder how such probabilities can be derived. Second, you probably wonder how data of this sort can be used to help guide decision making. Both questions will be answered in due time. (Indeed, this case will be taken up in detail in a later chapter.) For now, we simply want to introduce you to this class of problem because it is an important type that statistical methods can help to solve.

---

[4] See J. Newman, *Management Applications of Decision Theory* (New York: Harper and Row, 1971).

## 1.6 Bias and Error

*Error*

Because the field of statistics attempts to make inferences from a sample concerning a population, statistics must necessarily be concerned with **error.** Any result based on a sample is likely to depart in some measure from the corresponding result for the population. For example, although Exxon Corporation's sample result is unlikely to be in error by more than $1,300, Exxon's statisticians would not be surprised if it were off by $1,000 or so. Similarly, the results of ICI's sample (shown in Table 1.4) are almost certainly in error to some extent. That is, if ICI were to test 10,000 pieces of rubber rather than 10 pieces, the results would probably differ from those in Table 1.4. To repeat, all results based on a sample are likely to be in error, since the sample does not contain information concerning all items in the population.

The error in a particular sample result consists of two parts: ***experimental or sampling error*** (sometimes called ***random error***) and ***bias.*** Experimental or sampling errors occur because of a large number of uncontrolled factors, which we can subsume under the term *chance.* For example, if Exxon draws a sample of 726 types of items at random[5] from the population of 3,894 types of items (see Table 1.1), the results will differ from one sample to another. Why? Because of the luck of the draw. Or in the case of ICI's experiment, the average difference in abrasion resistance between treated and untreated half-pieces will differ from one sample of 10 pieces of rubber to another sample of 10 pieces of rubber. Why? Because the instruments that measure abrasion resistance contain small errors, because human beings make mistakes in reading these instruments, because the pieces of rubber sometimes are not homogeneous, and so on. Errors of this kind are experimental or sampling errors.

*Bias*

*The essential characteristic of experimental or sampling errors is that they can reasonably be expected to cancel each other out over a period of time or over a large number of experiments or samples.* Many uncontrolled factors operate to cause the results of a sample to be in error, sometimes on the high side and sometimes on the low side. However, if the sample or the experiment is repeated a large number of times, the errors tend to offset each other when the results are averaged. In contrast, **bias** *consists of a systematic and persistent type of error which will not tend to cancel out.* For example, Exxon, instead of choosing types of items at random, might select its samples in such a way that types of items that are likely to be lost or stolen have a higher chance of being picked than types of items that are unlikely to be lost or stolen. If this were the case, there would be a systematic and persistent tendency for the sample to exaggerate the difference between the total value of all types of

---

[5] The concept of drawing a sample at random is more sophisticated and technical than can be appreciated or explained adequately at this point. For present purposes, it is sufficient to regard *at random* as meaning that each unit in the population has the same chance of being drawn in the sample. Much more will be said about this in later chapters.

items in the storehouse and what the accounting records say it should be. Moreover, this error would not tend to cancel out if the sample were repeated many times and the results were averaged.

To summarize, an important reason for distinguishing between experimental or sampling error and bias is that *increases in a sample's size tend to reduce experimental or sampling errors, while such increases do not reduce bias.* Thus, although (as pointed out previously) the accuracy of the results of a sample tends to increase with the sample's size, not all errors can be eliminated by increasing sample size. If there are serious biases, the results may be considerably in error even if the sample is huge.

---

*To reduce experimental or sampling errors, one can increase the sample size. But increases in the sample size will not reduce bias.*　　　*Basic Idea #1*

---

## TWO MAJOR CAUSES OF BIAS

Bias can be particularly dangerous when, like termites or dry rot, its presence is undetected. Two types of methodological errors are commonly causes of serious bias. The first arises when *a sample is taken from a population that differs in an important way from the population that the statistician or decision maker is really interested in.* For example, suppose that you want to estimate the proportion of people in Chicago who are college graduates. To do so, you go to three Chicago suburbs and pick a sample of their residents. Such a sample is likely to result in substantial bias because you are sampling from the wrong population. Rather than sampling from the population of all Chicago residents, you are sampling from the population of Chicago suburban residents. And since the educational level is likely to be higher in the suburbs than in the inner city, the result is likely to be biased.[6]

A second methodological mistake which can be the source of serious bias arises when *the effect of the variable one wants to measure is mixed up inextricably with the effect of some other factor.* In this case, if one finds that the variable in question seems to have a certain effect, this estimated effect may be biased because it also reflects the effect of another factor. Table 1.6 presents part of the results of a nationwide test of the effectiveness of the Salk antipolio vaccine.[7] These data seem perfectly adequate to measure the effects of the vaccine on the incidence of polio in children, but in fact they contain a major bias because only those second-graders who received their parents' permission were given the

---

[6] Of course, one should not be overly dogmatic about this. Sometimes it is not possible to obtain an adequate frame for the population one is interested in, and the only possibility is to do the best one can with a frame which is only approximately what one would like to use. If this frame is close enough to what is really needed, little harm will result. Sometimes, however, it is not close enough, even though the analyst believes it to be.

[7] See W. S. Youden, "Chance, Uncertainty, and Truth in Science," *Journal of Quality Technology,* 1972.

**Table 1.6**
Incidence of Polio in
Two Groups of
Children

| School grade | Treatment | Number of children | Number afflicted with polio | Number afflicted per 100,000 children |
|---|---|---|---|---|
| Second | Salk vaccine | 222,000 | 38 | 17 |
| First and third | No vaccine | 725,000 | 330 | 46 |

*Source:* W. S. Youden, "Chance, Uncertainty, and Truth in Science," *Journal of Quality Technology,* 1972.

vaccine, whereas all first- and third-graders were used to indicate what the incidence of polio would be without the vaccine. The bias arises because the incidence of polio was substantially lower among nonvaccinated children who did not receive permission than among those who did. Why? Because children of higher-income parents were more likely than children of lower-income parents to receive permission—and they were also more likely to contract polio. (Surprising as it may seem, children who grow up in less hygienic conditions are less likely to contract this disease!)

**Table 1.7**
Modified Experiment
with Salk Vaccine

| Permission given | Treatment | Number of children | Number afflicted with polio | Number afflicted per 100,000 children |
|---|---|---|---|---|
| Yes | Salk vaccine | 201,000 | 33 | 16 |
| Yes | Placebo | 201,000 | 115 | 57 |
| No | None | 339,000 | 121 | 36 |

*Source:* See Table 1.6.

Table 1.7 shows the results of a more adequate experimental design in which this bias is eliminated. In this sample, children who received permission are differentiated from those who did not. Those with permission were assigned at random to either the group receiving the vaccine or the group receiving the placebo (something similar in appearance to the vaccine, but of no medical significance). As you can see, the results indicate that the reduction in the number afflicted with polio (per 100,000 children) due to the vaccine is much larger than in Table 1.6. Biases of the sort contained in Table 1.6 and in the Chicago educational survey can result in distorted information which can lead business executives and government officials to make costly and embarrassing mistakes. It is worth taking some trouble to avoid the methodological errors that result in these biases—unless, of course, you don't mind making serious mistakes.

## EXERCISES

**1.1** A major oil company wants to determine the percentage of its service stations that stay open 24 hours a day.
   (a) What is the population?
   (b) How can a frame be obtained?
   (c) Is the population finite or infinite?
   (d) Does the population consist of qualitative or quantitative information?

**1.2** To obtain an estimate of how well his students understand the material, an instructor asks each of the students in the first row of his class a question. Six out of the seven students in the first row come up with the right answer. Do you think that such a sample contains a bias? If so, what is it?

**1.3** The public school teachers in Philadelphia went on strike in the fall of 1981. Suppose that a newspaper asked each of the teachers on picket lines whether or not the strike would be settled on terms unfavorable to the union.
   (a) Would the results be a sample of the opinions of all public school teachers in Philadelphia?
   (b) Would the population be finite or infinite?
   (c) Would the population contain quantitative or qualitative information?
   (d) What possible biases are likely to exist in the results?

**1.4** The mayor of a seaside vacation town must decide whether or not to support an effort to close down one of the town's elementary schools. To determine the extent to which public opinion favors such an action, she hires several interviewers to ask individuals selected at random on the town's boardwalk from 10 A.M. until noon each day during the second week in August whether they feel that an elementary school with less than 200 students should remain in operation.
   (a) What is the population from which this sample is drawn?
   (b) Is this population finite or infinite?
   (c) Does this population consist of qualitative or quantitative measurements?
   (d) What deficiencies can you see in this sample design? What possible biases are likely to exist in the results?
   (e) What improvements in the sample design would you suggest?

**1.5** Another seaside vacation resort is the scene of considerable controversy over whether or not bars should be allowed to stay open after midnight. The local newspaper, which favors the existing arrangements whereby bars must close at midnight, points out that when a neighboring community allowed bars to stay open after midnight the crime rate increased.
   (a) What are the weaknesses in the newspaper's argument?
   (b) Do you think that an experiment could be run to resolve this type of controversy? If so, what sort of an experiment should it be?
   (c) Do you think that an analysis of data concerning past changes in crime rates in this and other communities might help resolve this controversy? If so, what kind of data should be examined, and how might they be analyzed?

**1.6** A major private university mails a questionnaire to a sample of 10 percent of its alumni. This questionnaire asks whether the recipient of the questionnaire is for or against the university's policies with respect to curriculum, faculty hiring, coed dormitories, and a variety of other topics. About

30 percent of the questionnaires are filled out and returned. Do you think that there is likely to be bias as well as sampling error in the results? Explain.

## 1.7 The Frequency Distribution

At the beginning of this chapter, we pointed out that the field of statistics can be divided into two parts: descriptive statistics and analytical statistics. One of the central tools of descriptive statistics is the frequency distribution. In this section, we describe the nature of a frequency distribution, how it can be represented graphically, and some guidelines that may prove useful in constructing frequency distributions.

### CLASS INTERVALS

Although one seldom is provided with the entire population in which he or she is interested, let's start by assuming that you undertake a study for which you do have the entire population. Specifically, you are interested in determining the profitability of the 17 petroleum firms in Table 1.8 in 1984. The numbers in the table constitute the population. Given these numbers, how can we summarize and describe them? (That is, how can we put them into a more compact and comprehensible form?) The need to condense and summarize the data is not as

Table 1.8
Population of Profit
Rates of 17 Major
Petroleum Companies,
1984

| Firm | Profit rate* (percent) |
|---|---|
| Exxon | 19.2 |
| Texaco | 2.3 |
| Mobil | 9.3 |
| Chevron | 10.4 |
| Standard Oil of Indiana | 17.4 |
| Shell | 14.2 |
| Atlantic Richfield | 5.7 |
| Occidental | 11.3 |
| Phillips | 12.2 |
| Unocal | 12.3 |
| Sun | 10.2 |
| Amerada Hess | 6.6 |
| Diamond Shamrock | 8.9 |
| Coastal | 17.1 |
| Standard Oil of Ohio | 17.7 |
| Pennzoil | 11.7 |
| Kerr McGee | 3.7 |

*Source: Fortune*, 1985.
* *Profit rate* is defined here as net income as a percent of stockholders' equity.

| Profit rate (percent) | Number of firms |
|---|---|
| 0.0 and under 3.0 | 1 |
| 3.0 and under 6.0 | 2 |
| 6.0 and under 9.0 | 2 |
| 9.0 and under 12.0 | 5 |
| 12.0 and under 15.0 | 3 |
| 15.0 and under 18.0 | 3 |
| 18.0 and under 21.0 | 1 |
| Total | 17 |

**Table 1.9**
Frequency Distribution of Population of 1984 Profit Rates of 17 Major Petroleum Firms

great in this example as in many other cases, since there are only 17 numbers, in contrast to the hundreds or thousands of numbers in some other populations. However, this example can illustrate the essential principles on which we shall focus.

As a first step toward summarizing these data, it is convenient to establish certain *class intervals,* which we define as *classes or ranges of values that the observations or measurements can assume.* For example, Table 1.9 shows seven class intervals for these profit rates, the first being 0.0–3.0 percent, the second 3.0–6.0 percent, and so on. In other words, all values greater than or equal to 0.0 percent and less than 3.0 percent are included in the first class interval, and all values greater than or equal to 3.0 percent and less than 6.0 percent are included in the second class interval, and so on. Each class interval must have a *lower limit* and an *upper limit.* An observation or measurement falls in a particular class interval if it is greater than or equal to this class interval's lower limit and less than its upper limit. The lower limit of the first class interval is 0.0, the lower limit of the second class interval is 3.0, and so on. The upper limit of the first class interval is 3.0, the upper limit of the second class interval is 6.0, and so on. The *width* of a class interval is equal to its upper limit minus its lower limit.

Given these class intervals, the next step is to determine how many observations or measurements in the population fall into each class interval. Using the data in Table 1.8, we find that one firm falls into the first class interval in Table 1.9, two fall into the second class interval, two fall into the third class interval, and so on. The resulting data, shown in Table 1.9, are called a *frequency distribution. A frequency distribution shows the number of observations or measurements that are included in each of the class intervals.* Thus, a frequency distribution, if properly formulated, can summarize a set of numbers very effectively. For example, the frequency distribution in Table 1.9 makes it much easier to see how profitable the 17 petroleum firms were in 1984. Clearly, only a few of them (four) had 1984 profit rates greater than or equal to 15.0 percent, and only a few (three) had 1984 profit rates of less than 6.0 percent. The bulk of the firms had profit rates between 6.0 and 15.0 percent. This frequency distribution shows the salient features of the raw data in Table 1.8.

*Frequency Distribution*

**Table 1.10**
Frequency Distribution
of Employment Status
of People in the
Civilian Labor Force,
United States,
December 1984

| Employment status | Number of persons |
|---|---|
| Employed | 106,273,000 |
| Unemployed | 8,191,000 |
| Total | 114,464,000 |

Source: *Economic Report of the President,* January 1985.

If the population consists of qualitative measurements, the classes in the frequency distribution are not ranges of numerical values, but possible qualitative observations. For example, if the population consists of observations concerning whether various people in a certain area are employed or unemployed, the frequency distribution shows the number of persons in the relevant area who are employed and the number unemployed. Table 1.10 constitutes such a frequency distribution for the United States in December 1984. Often, frequency distributions of this sort are presented graphically in the form of *bar charts* like that in Figure 1.1. Each bar in Figure 1.1 has a length—and thus an area, since the width of each bar is the same—representing the number of people employed or unemployed.

**Figure 1.1**
Bar Chart Showing
Frequency Distribution
of Employment Status
of People in the
Civilian Labor Force,
December 1984

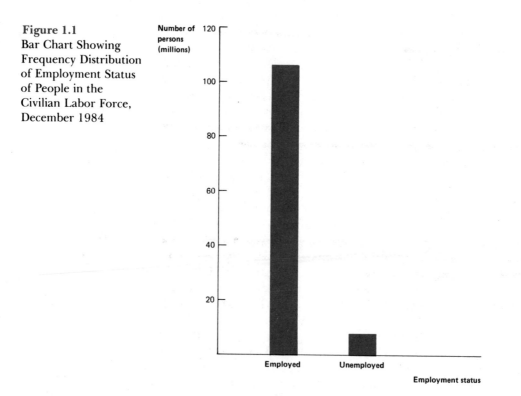

## GRAPHICAL REPRESENTATIONS OF FREQUENCY DISTRIBUTIONS

When a population consists of quantitative measurements (such as the profit rates of the 17 oil firms), a useful way to represent the population's frequency distribution is to construct a *histogram*. The histogram has a horizontal axis and a vertical axis, as shown in Figure 1.2. The horizontal axis is scaled to display all values of the measurements in the population (in this case the profit rates of the 17 oil firms). The horizontal axis is divided into segments corresponding to the class intervals, and a vertical bar is erected on each of these segments. In Figure 1.2 the segment from 0.0 percent to 3.0 percent on the horizontal axis is the first class interval, and on this segment a vertical bar is erected. Similarly, a vertical bar is erected on the segment of the horizontal axis corresponding to each of the other class intervals.

A histogram's vertical axis shows the number of observations (measurements) in the population which fall into any particular class interval. Thus, the height of the vertical bar for each class interval on the horizontal axis equals the number of observations (measurements) in the population that fall within this class interval. For example, in Figure 1.2 the height of the vertical bar for the first class interval (from 0.0 percent to 3.0 percent on the horizontal axis) is 1 unit because the

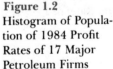
*Histogram*

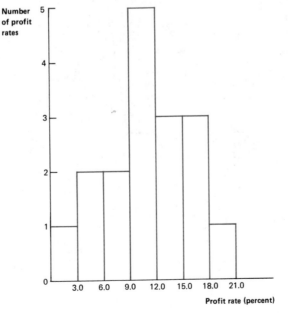

Figure 1.2
Histogram of Population of 1984 Profit Rates of 17 Major Petroleum Firms

profit rate of 1 out of 17 firms falls into this class interval. Clearly, the histogram in Figure 1.2 is a simple and effective way to present the frequency distribution in Table 1.9 in graphical form. (If the class intervals in a frequency distribution differ in their widths, the above instructions for constructing a histogram should be modified somewhat, as we shall see in Chapter 2.)

*Frequency Polygon*

An alternative way of representing a frequency distribution graphically is to construct a *frequency polygon*. Like the histogram, the frequency polygon has a horizontal axis showing possible values of the measurements in the population and a vertical axis showing the number of observations or measurements for each particular class interval. To construct a frequency polygon, we first find the *class mark* for each class interval. (The class mark is the point in a particular class interval midway between the interval's upper and lower limit.) We then plot the number of observations or measurements within a particular class interval against the class mark for that interval. In Figure 1.3, if we plot the number of oil firms with profit rates in the first class interval against the class mark for this interval (1.5 percent), we get point A. If we plot the number of oil firms with profit rates in the second class interval against the class mark for this interval (4.5 percent), we get point B. When we connect all resulting points, the resulting geometric figure is a frequency polygon. (Since there are zero firms in the −3.0– 0.0 percent class interval and in the 21.0–24.0 percent class interval, the points for these class intervals lie on the horizontal axis.) Like the histogram, the frequency polygon is useful for portraying a frequency distribution graphically.

**Figure 1.3**
Frequency Polygon of Population of 1984 Profit Rates of 17 Major Petroleum Firms

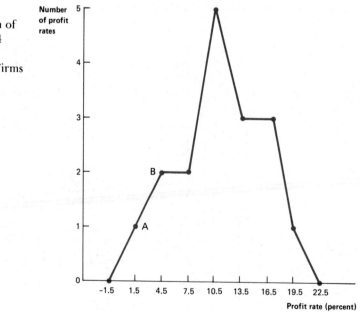

## Obtaining a Histogram

Modern computer technology has had an important impact on statistics, as well as on other fields. A variety of computer packages are available to carry out the calculations described in this book. Four of the most popular packages are: Minitab, SAS, SPSS, and BMDP. It is relatively simple to use these packages, and they take a great deal of the drudgery out of statistical calculation.[8]

To illustrate how these packages can be used, let's consider a fascinating economic activity, crime. Suppose that we want to construct a histogram showing the frequency distribution of states by crime rate (number of offenses known to the police per 1,000 population). The crime rate in each state is as follows:

| | | | | | |
|---|---|---|---|---|---|
| Maine | 4.2 | North Dakota | 3.0 | Louisiana | 5.3 |
| New Hampshire | 4.3 | South Dakota | 3.0 | Oklahoma | 4.8 |
| Vermont | 5.1 | Nebraska | 4.2 | Texas | 6.0 |
| Massachusetts | 5.8 | Kansas | 5.4 | Montana | 5.0 |
| Rhode Island | 5.9 | Delaware | 6.7 | Idaho | 4.5 |
| Connecticut | 5.8 | Maryland | 6.6 | Wyoming | 5.1 |
| New York | 6.9 | Virginia | 4.7 | Colorado | 7.4 |
| New Jersey | 6.2 | W. Virginia | 2.6 | N. Mexico | 6.2 |
| Pennsylvania | 3.7 | N. Carolina | 4.5 | Arizona | 7.6 |
| Ohio | 5.4 | S. Carolina | 5.3 | Utah | 5.8 |
| Indiana | 4.5 | Georgia | 5.6 | Nevada | 8.6 |
| Illinois | 5.0 | Florida | 8.0 | Washington | 6.7 |
| Michigan | 6.9 | Kentucky | 3.5 | Oregon | 7.0 |
| Wisconsin | 4.8 | Tennessee | 4.3 | California | 7.6 |
| Minnesota | 4.7 | Alabama | 4.9 | Alaska | 6.6 |
| Iowa | 4.7 | Mississippi | 3.5 | Hawaii | 6.5 |
| Missouri | 5.4 | Arkansas | 3.8 | | |

To use Minitab to construct such a histogram, all that one has to do (after logging on to the computer) is to type the following:

```
MTB > set in cl
DATA> 4.2 4.3 5.1 5.8 5.9 5.8 6.9 6.2 3.7 5.4 4.5
DATA> 5.0 6.9 4.8 4.7 4.7 5.4 3.0 3.0 4.2 5.4 6.7
DATA> 6.6 4.7 2.6 4.5 5.3 5.6 8.0 3.5 4.3 4.9 3.5
DATA> 3.8 5.3 4.8 6.0 5.0 4.5 5.1 7.4 6.2 7.6 5.8
DATA> 8.6 6.7 7.0 7.6 6.6 6.5
DATA> end
MTB > histogram cl
```

[8] For a description of each of these packages, see J. Lefkowitz, *Introduction to Statistical Computer Packages* (Boston: Duxbury Press, 1985). For a manual instructing students how to use Minitab, see T. Ryan, B. Joiner, and B. Ryan, *Minitab Student Handbook,* second edition (Boston: Duxbury Press, 1985).

The first line says that the numbers should be put in column C1 of the worksheet that Minitab maintains in the computer. The next several lines contain the numbers themselves. The last line commands the computer to formulate a histogram.

After you type the above lines, the computer automatically prints out the frequency distribution and the histogram, which are shown below:

```
Histogram of C1  N = 50

Midpoint  Count
    2.5      1  *
    3.0      2  **
    3.5      3  ***
    4.0      3  ***
    4.5      8  ********
    5.0      7  *******
    5.5      6  ******
    6.0      7  *******
    6.5      5  *****
    7.0      3  ***
    7.5      3  ***
    8.0      1  *
    8.5      1  *
```

Note that the histogram is printed sideways; that is, the bars (consisting of asterisks) are horizontal, not vertical.

Our aim here is not to teach you how to use these packages, but to acquaint you with their general nature and usefulness. Specialized manuals are available to teach you the details. The important point for present purposes is that computer packages exist which enable you to do statistical calculations simply, rapidly, and cheaply.

GUIDELINES FOR CONSTRUCTING FREQUENCY DISTRIBUTIONS

The purpose of constructing a frequency distribution is *to condense a body of data into a table that is more readily appraised and comprehended.* To do this, the observations in the population are grouped into class intervals, as we have seen. There are no simple, cut-and-dried rules to insure that one will construct a proper frequency distribution. Moreover, there is no single frequency distribution which necessarily is the best representation of a given set of data; ordinarily, a number of different frequency distributions will do well enough. But in constructing a frequency distribution, one would be wise to consider the following guidelines.

DEFINITION OF CLASS INTERVALS. In defining class intervals, it is essential that each measurement in the population fall into some class interval. Thus, it would be wrong to define class intervals in Table 1.9

so that profit rates above 18 percent fell into no class interval. Such a definition of class intervals would result in a frequency distribution which distorts, at least to some extent, the nature of the population. Equally important, none of the class intervals should overlap. Thus, in Table 1.9 it would be wrong to define one class interval as 9.0–12.0 percent and another as 11.0–14.0 percent. If class intervals were defined in this way, it would be possible for a given measurement in the population to fall into more than one class interval.

WIDTH OF CLASS INTERVALS. In general, statisticians prefer to construct class intervals of *equal* widths because this makes it easier to compare the number of observations in different class intervals and to carry out a variety of computations which are described in the next chapter. However, it is sometimes not practical to use class intervals of equal widths. For example, in constructing a frequency distribution of the annual incomes of American families, we might want the width of class intervals to be $4,000 up to an income level of $40,000, and to be $10,000 from an income level of $40,000 to $100,000. Families with incomes of $100,000 or more might be included in a single class interval with no finite upper limit. (Such a class interval is called *open-ended*.) In this case the reason for preferring class intervals of unequal widths is that the bulk of the nation's families have incomes of under $40,000, and it is important in this income range to have the width of class intervals small enough so that the frequency distribution portrays the distribution of income with a reasonable degree of accuracy. However, it would be inappropriate to use the same width of class interval for the very affluent, since so many class intervals would result that the frequency distribution would be very unwieldy. For example, if $4,000 were used as the width of each class interval for incomes ranging from zero to $1 million, there would be 250 intervals!

NUMBER OF CLASS INTERVALS. Although there are no hard-and-fast rules for how many class intervals to use, it is generally recommended that there be no fewer than 5 and no more than 20. If data are lumped into too few class intervals, there is a major loss of information because much of the variation cannot be determined from the frequency distribution. However, if too many class intervals are used, the frequency distribution does little more than reproduce the array of data. (If there are enough class intervals, each interval may contain only one or zero observations.) What is needed is something in between, and the answer is inevitably a matter of judgment.

RELATIONSHIP BETWEEN NUMBER AND WIDTH OF CLASS INTERVALS. When deciding on the number and width of class intervals to use, it is important to realize that interval width is related to the number of class intervals one picks. If all class intervals are to have equal widths, the following formula provides a reasonable estimate of how wide they should be:

$$\text{Width of class interval} = \frac{\text{Largest value} - \text{Smallest value}}{\text{Number of class intervals}}$$

where "Largest value" is the largest value of an observation in the population, and "Smallest value" is the smallest value of an observation in the population. What this formula does is indicate how wide each class interval must be in order to cover the distance between the largest and smallest values, given the chosen number of class intervals. Thus, if we want nine class intervals of equal widths for the data in Table 1.8,

$$\text{Width of class interval} = \frac{19.2 - 2.3}{9} = 1.9,$$

since the largest profit rate is 19.2 percent and the smallest is 2.3 percent. According to this result, the width of the class interval should be about 1.9 percentage points.

POSITION OF CLASS INTERVALS. If possible, one should try to set up class intervals so that the midpoint or class mark of each interval is close to the average of the observations included in the class interval. For example, suppose that the data to be summarized are monetary amounts that tend to concentrate on multiples of $1 ($1; $2; $3; and so on). Then it would be better to use class intervals with multiples of $1 as class marks (class intervals of $0.50 and under $1.50; $1.50 and under $2.50; and so on) than ones with class marks that are quite different from the actual concentration of observations in the class intervals. The reason is that, in calculating averages and other summary measures from frequency distributions, we assume that the class mark is representative of the values in each class interval. (More will be said about this in the next chapter.)

## THE CUMULATIVE FREQUENCY DISTRIBUTION

In some cases, statisticians and decision makers are interested in the number of measurements in the population that lie below or above a certain value. Thus, an analyst in the oil industry might want to know how many firms in Table 1.8 had profit rates below 15.0 percent in 1984. In cases of this sort, it is useful to construct a cumulative frequency distribution, showing *the number of measurements in the population that are less than particular values.* Table 1.11 shows the cumulative frequency distribution for the profit rates of the oil firms in Table 1.8. The difference between the *cumulative* frequency distribution in Table 1.11 and the *ordinary* frequency distribution in Table 1.9 is that the cumulative frequency distribution shows the number of oil firms with profit rates *less than* those given in Table 1.11, whereas the ordinary frequency distribution in Table 1.9 shows the number of oil firms in *each* of the class intervals.

If one has the frequency distribution for a set of data, it is easy to construct a cumulative frequency distribution. To construct a cumulative frequency distribution for the profit rates of oil firms in Table 1.9, we proceed as follows. The number of firms with profit rates of less than 3.0 percent is one (the figure in the lowest class interval in Table 1.9); the number of firms with profit rates of less than 6.0 percent is

| Profit rate (percent) | Number of firms |
|---|---|
| Less than 0.0 | 0 |
| Less than 3.0 | 1 |
| Less than 6.0 | 3 |
| Less than 9.0 | 5 |
| Less than 12.0 | 10 |
| Less than 15.0 | 13 |
| Less than 18.0 | 16 |
| Less than 21.0 | 17 |

**Table 1.11**
Cumulative Frequency
Distribution of
Population of 1984
Profit Rates of 17
Major Petroleum Firms

three (the sum of the figures in the lowest two class intervals); the number of firms with profit rates of less than 9.0 percent is five (the sum of the figures in the lowest three class intervals); and so on. To obtain the number of firms with profit rates of less than the upper limit of a particular class interval, we add up (or *cumulate*, which accounts for the name *cumulative frequency distribution*) the number of firms in this and all lower class intervals.

Just as the ordinary frequency distribution can be portrayed graphically, so can the cumulative frequency distribution. To construct such a graph, one plots the number of observations in the population with values less than a certain number against the number itself. For example, since there is one oil firm with a profit rate less than 3.0 percent, we would plot 1 on the vertical axis against 3.0 percent on the horizontal axis. Since there are three oil firms with profit rates less than 6.0 percent, we would plot 3 on the vertical axis against 6.0 percent on the horizontal axis. The results of plotting and connecting all such points with straight lines are shown in Figure 1.4. The resulting curve

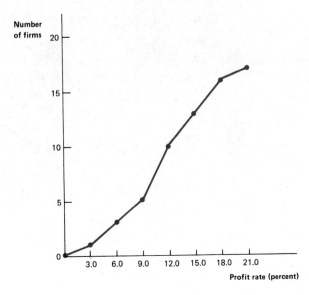

**Figure 1.4**
Ogive Showing
Cumulative Frequency
Distribution of
Population of 1984
Profit Rates of 17
Major Petroleum Firms

*Ogive*

is called an *ogive*. Such curves are encountered frequently in business and economic statistics.[9]

[9] In this section we have considered "less than" cumulative frequency distributions and ogives. It is also possible to construct "greater than" cumulative distributions and ogives, which show the number of measurements that exceed particular values. To obtain the number of firms with profit rates *greater than* the lower limit of a particular class interval, we add up the number of firms in this and all higher class intervals.

**GETTING DOWN TO CASES**

DOES THE PRODUCTION PROCESS MEET THE TOLERANCES?

A manufacturing firm (which allowed the actual data below to be published in the *Harvard Business Review*) wanted very much to control the distance between two holes stamped in a piece of metal. For a sample of 49 die stampings, it was found that the distances (in inches) between the two hole centers were as follows:

| | | | | |
|---|---|---|---|---|
| 3.008 | 3.007 | 3.007 | 3.006 | 3.006 |
| 3.006 | 3.006 | 3.006 | 3.006 | 3.006 |
| 3.005 | 3.005 | 3.005 | 3.005 | 3.005 |
| 3.004 | 3.004 | 3.004 | 3.004 | 3.004 |
| 3.004 | 3.003 | 3.002 | 3.002 | 3.000 |
| 3.004 | 3.003 | 3.003 | 3.003 | 3.002 |
| 3.002 | 3.001 | 3.001 | 3.001 | 3.001 |
| 3.001 | 3.001 | 3.001 | 3.000 | 3.000 |
| 3.000 | 3.000 | 2.999 | 2.999 | 2.999 |
| 2.998 | 2.997 | 2.997 | 2.996 | |

1. The tolerances stated that the distance between the two hole centers should be 3.000 ± .004 inches (that is, not less than 2.996 or more than 3.004 inches).

   (a) Construct a frequency distribution of the distances between the hole centers for the sample.

   (b) Use the frequency distribution to determine the percent of the sample not meeting the tolerances.

   (c) What possible reasons can you give for the unsatisfactory performance of this production process? Can you determine the actual reasons?

2. After some research by the firm's production engineers, it was found that the 25 distances shown in the first five horizontal rows of the array of data (in the paragraph before last) were stampings based on the use of die A, while the 24 distances in the last five horizontal rows were stampings based on the use of die B.

   (a) Does this information shed any light on the unsatisfactory performance of this production process?

   (b) Based on this information, what advice would you give the firm?

FREQUENCY DISTRIBUTIONS FOR SAMPLE DATA

Finally, in setting up an actual frequency distribution, we usually do not have all the measurements in the population. If, as is usually the case, we have data concerning only a sample from the population, a frequency distribution (constructed according to the principles discussed above) is a useful way to describe and summarize the data. For example, take the case of the profit rates of the 17 oil firms. Regardless of whether the data in Table 1.8 are viewed as an entire population or as a sample from a larger population, it is useful to formulate a frequency distribution to summarize the data, and the same principles apply. Thus, our discussion in previous parts of this section applies to both sample and population data.

However, two important points should be recognized. First, *the frequency distribution of a sample is different from the frequency distribution of the population from which the sample is drawn.* Because the sample contains only a portion of the measurements in the population, the two frequency distributions are different and should not be confused. Second, *although we ordinarily do not have the data to construct a frequency distribution for the entire population, such a frequency distribution exists or can be imagined. It is important that we be able to visualize and think about this frequency distribution, since the questions that statistical investigations attempt to answer often are questions about this frequency distribution.*

EXERCISES

**1.7** The McNair Company makes screws. The diameter of its screws varies from 0.1810 to 0.1849 inches. Determine upper and lower limits and the class mark for the first and last class intervals. (Assume that there are 10 class intervals.)

**1.8** The math SAT score of the entering freshmen at Old Ironsides University varies from 520 to 719. Determine upper and lower limits and the class mark for the first and last class intervals. (Assume that there are 10 class intervals.)

**1.9** The amount of life insurance carried by 40-year-old males in a Midwestern town varies from zero to $400,000. Construct a table with about ten classes into which these data might be grouped. What is the class mark in each interval?

**1.10** The monthly salary of secretaries at the McNair Company varies from $930 to $1,975. Construct a table with seven classes of equal width into which these data might be grouped. What is the class mark in each interval?

**1.11** A group of 40 railroad clerks takes an examination to test manual dexterity. Their scores are as follows:

| | | | | | | | |
|---|---|---|---|---|---|---|---|
| 81 | 62 | 76 | 81 | 61 | 80 | 42 | 53 |
| 83 | 93 | 78 | 86 | 75 | 82 | 76 | 60 |
| 78 | 98 | 92 | 74 | 73 | 81 | 78 | 65 |
| 76 | 51 | 63 | 79 | 71 | 43 | 71 | 95 |
| 74 | 75 | 50 | 71 | 69 | 58 | 72 | 98 |

(a) Group these numbers into a frequency distribution.
(b) Construct the corresponding histogram.

**1.12** A small trucking company's common stock is held by 30 people. The number of shares each person owns is as follows:

| | | | | | |
|---|---|---|---|---|---|
| 114 | 100 | 40 | 65 | 130 | 110 |
| 110 | 80 | 100 | 140 | 150 | 100 |
| 83 | 70 | 100 | 120 | 180 | 80 |
| 25 | 50 | 50 | 100 | 170 | 160 |
| 40 | 82 | 30 | 120 | 110 | 190 |

(a) Group these numbers into a frequency distribution.
(b) Construct the corresponding frequency polygon.

**1.13** (a) Given the following frequency distribution of the weights of 2,000 soldiers, construct the cumulative frequency distribution.

| Weight (pounds) | 120 and under 140 | 140 and under 160 | 160 and under 180 | 180 and under 200 | 200 and under 220 | 220 and under 240 |
|---|---|---|---|---|---|---|
| Number of soldiers | 205 | 371 | 403 | 523 | 312 | 186 |

(b) Draw the ogive.

**1.14** The frequency distribution of the ages of secretaries at the McNair Company is as follows:

| Age (years) | 18 and under 24 | 24 and under 30 | 30 and under 36 | 36 and under 42 | 42 and under 48 | 48 and under 54 | 54 and under 60 |
|---|---|---|---|---|---|---|---|
| Number of secretaries | 2 | 1 | 4 | 5 | 3 | 2 | 1 |

(a) Construct the cumulative frequency distribution.
(b) Draw the ogive.

**1.15** There are 1,712 male students and 1,024 female students at Old Ironsides University. Among the males, 785 come from the eastern U.S., the rest from the western U.S. Among the females, 302 come from the eastern U.S., the rest from the western U.S.
(a) Draw a bar chart showing the distribution of students by sex.
(b) Draw a bar chart showing the distribution of students by region.

**1.16** The McNair Company sends 112 packages by express mail to New York City during 1987. Four of these packages do not arrive on the next day (as promised by the express mail service).
(a) Draw a bar chart showing the distribution of packages by whether or not they were late.
(b) One-half of the packages that arrived on the next day (and none of those that did not arrive on the next day) did *not* contain checks. Draw a bar chart showing the distribution of packages containing checks by whether or not they were late.

# Chapter Review

1. The field of statistics consists of two parts, descriptive statistics and analytical statistics. *Descriptive statistics* is concerned with summarizing and describing a set of data. *Analytical statistics* consists of techniques which help decision makers come to rational decisions under uncertainty. Statistics is concerned with whether data should be gathered at all, with how data should be gathered, and with how a particular set of data should be analyzed once it has been collected.

2. Firms, government agencies, and other organizations are continually engaged in *sampling* in order to obtain needed information, because it would be too expensive and time-consuming to try to obtain complete data concerning all relevant units. A *population* consists of the total collection of observations or measurements that are of interest to the statistician or decision maker in solving a particular problem. (A listing of all the elements or units in the population is called a *frame*.) A population can consist of quantitative or qualitative information, and it may be finite or infinite. A *sample* is a subset of measurements taken from the population.

3. In recent years, there has been increasing utilization of direct and conscious experimentation in business and economics. Just as statistics is useful in designing a sample survey, it is also useful in designing an experiment. In both a survey and an experiment the objective should be to obtain the desired information at minimum cost. To promote this objective, it is important that the purposes of the experiment or sample survey be defined precisely before data are collected.

4. Modern statistical techniques are useful in promoting more rational decisions under uncertainty. Almost all decisions are made under uncertainty because it is seldom possible for the decision maker to forecast accurately the consequence of each alternative course of action. Statisticians use the concept of *probability* to measure the amount of confidence that one can have in various sample results.

5. Because statistics attempts to make inferences from a sample concerning a population, it must be concerned with error. After all, any sample result is likely to depart in some measure from the corresponding result for the total population. The error in any particular sample result is composed of two parts: *experimental or sampling error*, and *bias*. Experimental or sampling error is due to a large number of uncontrolled factors which we subsume under the shorthand expression *chance*. *Bias* consists of a persistent, systematic sort of error. Increases in a sample's size tend to reduce experimental or sampling error, but not bias.

6. To summarize a body of data, it is useful to construct a *frequency distribution*, which is a table showing the number of measurements or observations that fall into each of a number of class intervals. To establish a frequency distribution, one must set up certain well-defined class intervals, each interval being defined by a lower limit and an upper limit. Frequency distributions are often presented in graphical as well as tabular form. A *histogram* is composed of a series of bars or rectangles; the bottom of each bar is the line segment on the horizontal axis corresponding to the interval from the class interval's lower limit to its upper limit. The area of each bar is proportional to the number of cases in the class. A *frequency polygon* is another type of graphical representation of a frequency distribution.

# Chapter Review Exercises

**1.17** Based on the information given in Section 1.5, what was the probability that Maxwell House's profit would increase by $11,939,000 if it increased the price of coffee by 2 cents per pound? What was the probability that its profit would decrease by $1,050,000 under these circumstances?

**1.18** To determine the relative safety of today's aircraft and those of 40 years ago, a researcher compares the number of aircraft accidents in 1986 with that in 1940. Finding that the number of accidents has increased, the researcher concludes that today's airplanes are not as safe as the earlier ones. Do you agree? Why, or why not?

**1.19** A frequency distribution of bank deposits on a particular day at a small branch bank contains 10 class intervals, each with a width of $100. Does it follow that the largest bank deposit is $1,000 (that is, 10 times $100) greater than the smallest bank deposit? Why or why not?

**1.20** How can one calculate the class mark of an open-ended class interval? Is it possible?

**1.21** According to the Federal Trade Commission (FTC), the profit after taxes (as a percent of sales) of each manufacturing industry in the United States in 1974 was as follows:

| Industry | Profit (as percent of sales) |
|---|---|
| Transportation equipment | 3.9 |
| Electrical equipment | 4.3 |
| Machinery (nonelectrical) | 5.6 |
| Fabricated metal products | 4.0 |
| Primary iron and steel | 4.1 |
| Primary nonferrous metals | 5.4 |
| Stone, clay, and glass | 4.8 |
| Instruments | 8.4 |
| Other durable goods | 5.0 |
| Food and kindred products | 2.6 |
| Tobacco | 5.8 |
| Textiles | 2.8 |
| Paper | 5.4 |
| Printing | 4.8 |
| Chemicals | 6.8 |
| Petroleum and coal | 7.6 |
| Rubber | 4.0 |
| Other nondurable products | 2.1 |

(a) Construct a frequency distribution of the profit rates of all manufacturing industries.
(b) Construct a histogram based on these data.
(c) Construct a frequency polygon based on these data.

**1.22** According to the Department of Agriculture, the total net income per farm in each of 10 Midwestern states in 1973 was:

| State | Income per farm (dollars) |
|---|---|
| Illinois | 13,224 |
| Indiana | 11,282 |
| Iowa | 19,685 |
| Kansas | 17,018 |
| Michigan | 6,171 |
| Minnesota | 19,456 |
| Nebraska | 17,790 |
| North Dakota | 35,631 |
| Ohio | 5,212 |
| South Dakota | 22,928 |

(a) Are there too few states to construct a frequency distribution of their incomes per farm?
(b) If not, what frequency distribution would you suggest?

**1.23** A supermarket inspects the work of one of its clerks at the checkout counter. For each customer, it determines what the customer's bill should be and what this clerk calculates it to be. For 50 customers the frequency distribution of the difference between the latter and the former is as follows:

| Error | Number of customers |
|---|---|
| −$1.00 and under −$0.75 | 1 |
| −$0.75 and under −$0.50 | 2 |
| −$0.50 and under −$0.25 | 4 |
| −$0.25 and under −$0.00 | 30 |
| $0.00 and under $0.25 | 6 |
| $0.25 and under $0.50 | 2 |
| $0.50 and under $0.75 | 2 |
| $0.75 and under $1.00 | 2 |
| $1.00 and under $1.25 | 1 |

(a) What are the class intervals in this frequency distribution?
(b) Are the class intervals of equal width? (That is, is the difference between the upper and lower limit the same for each class interval?)
(c) If the error in a particular customer's bill equals −$0.25, into which class interval would this item fall?
(d) Construct a cumulative frequency distribution for the data.
(e) Plot the ogive for the cumulative frequency distribution for the data.

# Appendix 1.1

## THE STEM-AND-LEAF DIAGRAM

Another way of summarizing a set of data is to construct a stem-and-leaf diagram, which is a simple graphical technique to show the range of the data, where the data are concentrated, and whether there are any extremely high or low values. The guidelines for constructing a frequency distribution set forth on pages 16–24 do not apply to the construction of a stem-and-leaf diagram. Instead, each observation is characterized by its *stem* and its *leaf.*

For the two-digit state crime rates on page 21, we can use the first or leading digit as the *stem* and the second or trailing digit as the *leaf*. The leading digit or stem determines the row in which a crime rate is put. The first observation on page 21 (which is 4.2) is put in the third row of Figure 1.5, since its leading digit or stem is 4 (the figure to the left of the horizontal bar). The trailing digit or leaf (which is 2) is written in this row to the right of the vertical bar. This procedure is followed for all of the observations on page 21, the result being shown in Figure 1.5.

**Figure 1.5**
Stem-and-Leaf Diagram
for Crime Rates of 50
States

```
2 | 6
3 | 7 0 0 5 5 8
4 | 2 3 5 8 7 7 2 7 5 3 9 8 5
5 | 1 8 9 8 4 0 4 4 3 6 3 0 1 8
6 | 9 2 9 7 6 0 2 7 6 5
7 | 4 6 0 6
8 | 0 6
```

Clearly, this stem-and-leaf diagram is a useful representation of the data. It can be constructed quickly, and it has the advantage over the histogram that one loses no information concerning the value of each observation. To make the diagram somewhat neater, one can order the data within a row from lowest to highest, as shown in Figure 1.6. Also, if there seem to be too few rows, one can use two lines per stem; for example, leaf digits 0, 1, 2, 3, and 4 can be put on the first line of the stem, and leaf digits 5, 6, 7, 8, and 9 can be put on the second line. Thus, in Figure 1.6, we could use two lines for each stem. To illustrate, take the case where the stem is 5. The first line for this stem would include 0, 0, 1, 1, 3, 3, 4, 4, and 4. The second line would include 6, 8, 8, 8, and 9.

**Figure 1.6**
Stem-and-Leaf Diagram
for Crime Rates of 50
States, Within-Row
Data Ordered from
Lowest to Highest

```
2 | 6
3 | 0 0 5 5 7 8
4 | 2 2 3 3 5 5 5 7 7 7 8 8 9
5 | 0 0 1 1 3 3 4 4 4 6 8 8 8 9
6 | 0 2 2 5 6 6 7 7 9 9
7 | 0 4 6 6
8 | 0 6
```

If the data consist of three-digit numbers, the first digit can be used as the stem and the next two digits as the leaf. If the data consist of numbers with more than three digits, we might use the first two digits as the stem and the second two digits as the leaf. The remaining digits might be ignored. Thus, the stem of 305,620 might be 30, and the leaf might be 56.

# Summary and Description of Data

2

## 2.1 Introduction

As pointed out in Chapter 1, descriptive statistics is concerned with the summary and description of a body of data. While this concise definition may make descriptive statistics seem cut and dried and perhaps a bit dull, this is by no means true. The proper summary of a body of data involves much more than arithmetic. It entails avoiding a variety of pitfalls (many of which are discussed in this chapter) that can lead the unwary analyst or decision maker to false conclusions. Data can also be distorted intentionally by unscrupulous individuals and firms to mislead others. To avoid being misled, a knowledge of descriptive statistics is essential.

## 2.2 Types of Summary Measures

In the previous chapter, we showed how a frequency distribution can be constructed to summarize and describe a set of data. However, in many situations where a frequency distribution would be too detailed and cumbersome, a few summary measures can present concisely the salient features of the data. Although summary measures provide much less information than the frequency distribution, in many situations the lack of a certain amount of information is not crucial and the greater conciseness of the summary measures makes them more useful than the frequency distribution. In general, two types of summary measures are most frequently used: measures of central tendency and measures of dispersion.

MEASURES OF CENTRAL TENDENCY. Often one wants a single number to represent the "average level" of a set of data. In other

**Table 2.1**
Population of Profit
Rates of 17 Major
Petroleum Companies,
1984

| Firm | Profit rate* (percent) |
|---|---|
| Exxon | 19.2 |
| Texaco | 2.3 |
| Mobil | 9.3 |
| Chevron | 10.4 |
| Standard Oil of Indiana | 17.4 |
| Shell | 14.2 |
| Atlantic Richfield | 5.7 |
| Occidental | 11.3 |
| Phillips | 12.2 |
| Unocal | 12.3 |
| Sun | 10.2 |
| Amerada Hess | 6.6 |
| Diamond Shamrock | 8.9 |
| Coastal | 17.1 |
| Standard Oil of Ohio | 17.7 |
| Pennzoil | 11.7 |
| Kerr McGee | 3.7 |

*Source: Fortune,* 1985.
* *Profit rate* is defined here as net income as a percent of stockholders' equity.

words, a number is needed that will indicate where the frequency distribution is centered. This number should tell us what a "typical" value of the measurements might be. To illustrate, let's return to the data concerning the 1984 profit rates of 17 major oil firms, which are shown in Table 2.1. If these data were collected to determine whether the oil companies were making much higher profits than firms in other industries, it might be sensible to ask, What is the typical level of the 1984 profit rates of these 17 oil firms? As we shall see, there are several types of averages or measures of central tendency that can be used to help answer this question; and the choice among these depends on the purposes of the investigator and the nature of the data.

MEASURES OF DISPERSION. In addition to knowing the "average level" of a set of data, it is important to know the degree to which the individual measurements vary about this average. In other words, we need to know whether a frequency distribution is tightly packed around its average or whether there is a great deal of scatter about it. In the case of the oil firms in Table 2.1, an important question is: Regardless of what the average level of the 1984 profit rates of these firms may be, to what extent do their profit rates vary? Statisticians have devised a number of measures of dispersion, which will be described in subsequent sections. As in the case of measures of central tendency, the choice among measures of dispersion depends on the purposes of the investigator and the nature of the data.

## 2.3 Parameters and Statistics

If we have all the measurements in a given *population,* we can calculate summary measures for that population as a whole. Such summary measures are called *parameters.* For example, if we calculate a particular kind of average of the profit rates in Table 2.1, the resulting average is a parameter since it is calculated from all the measurements in the relevant population. (Recall from Chapter 1 that these 17 observations are regarded as the population.) Or if we calculate a particular kind of measure of dispersion (again using the profit rates in Table 2.1), the result is a parameter since the calculation is based on all the measurements in the relevant population. As has been pointed out, we seldom have all the measurements in an entire population, but this does not mean we are not interested in the parameters of the population. On the contrary, *much of analytical statistics is designed to draw inferences from a sample concerning the value of a population parameter.*

If we have only *sample* data, we can calculate summary measures for the sample; such summary measures are called *statistics.* For example, if the Exxon Corporation's statisticians calculated, from the sample of items described in Chapter 1, a particular kind of average of the differences between actual and recorded value, this resulting average is a statistic since it is calculated from the *measurements in a sample.* Or if Exxon's statisticians calculate a particular kind of measure of dispersion, using the differences between actual and recorded values for the types of items in the same sample, the result is also a statistic since the calculation is based on the *measurements in a sample.* As we shall see in subsequent chapters, a statistic from a sample is often used to estimate the analogous parameter of the entire population from which the sample is drawn. Thus, the average difference between actual and recorded values for the types of items in the sample was used by Exxon to estimate the average difference between actual and recorded values in the population of *all* types of items.

## 2.4 Measures of Central Tendency

### THE ARITHMETIC MEAN

There are several important types of measures of central tendency. The one used most frequently is the arithmetic mean. Like Molière's character in *Le Bourgeois Gentilhomme,* who was surprised to learn that he had been speaking prose all his life, you may be surprised to learn that you have been using the arithmetic mean for a long time (although you probably have not been calling it by that name). *The arithmetic mean is the sum of the numbers included in the relevant set of data divided by the number of such numbers.* Let $N$ denote how many numbers there are in a population; thus in Table 2.1, $N = 17$. If we order these numbers from 1 to $N$, $X_1$ being the first number, $X_2$ being the second number, and so on up to $X_N$, which is the $N$th number, then the population mean is

*Formula for $\mu$*

$$\mu = \frac{X_1 + X_2 + X_3 + \cdots + X_N}{N},$$    (2.1)

where $\mu$ is the Greek letter *mu*. In particular, if we order the numbers in Table 2.1 from the top down, $X_1$ being 19.2, $X_2$ being 2.3, and so forth, with $X_{17}$ being 3.7[1], then in the case of this population

$$\mu = \frac{19.2 + 2.3 + 9.3 + 10.4 + \cdots + 11.7 + 3.7}{17}$$

$$= \frac{190.2}{17}$$

$$= 11.19.$$

Consequently, the arithmetic mean of the population turns out to be 11.19 percent.

If the arithmetic mean is calculated for a sample rather than for the whole population, it is designated as $\overline{X}$ rather than $\mu$. Whereas $N$ stands for the number of measurements in the population, $n$ stands for the number of measurements in the sample. Thus, the sample mean is defined as

*Formula for $\overline{X}$*

$$\overline{X} = \frac{X_1 + X_2 + X_3 + \cdots + X_n}{n}.$$    (2.2)

This expression is sometimes written in the following form:

$$\overline{X} = \frac{\sum\limits_{i=1}^{n} X_i}{n},$$

*Summation Sign*

where $\Sigma$ is the mathematical *summation sign*. What does $\Sigma X_i$ mean? It means that the numbers to the right of the summation sign (that is, the values of $X_i$) should be summed from the lower limit on $i$ (which is given below the $\Sigma$ sign) to the upper limit on $i$ (which is given above the $\Sigma$ sign). Thus, in this case it means that $X_i$ is to be summed from $i = 1$ to $i = n$. In other words, $\Sigma X_i$ means the same as $X_1 + X_2 + \cdots + X_n$. (For further discussion of the summation sign and its uses, see Appendix 2.1.)

Whether the set of data is a sample or a whole population, it is

[1] It is important to note that this is only one of many orderings that could be used. For example, the firms (whose profit rates are included in Table 2.1) could be arranged in alphabetical order. This would do just as well.

**Table 2.2**
Frequency Distribution of Sample of 15 Profit Rates of Major Chemical Firms, 1984

| Profit rate* (percent) | Number of firms |
|---|---|
| 3 and under 6 | 2 |
| 6 and under 9 | 3 |
| 9 and under 12 | 3 |
| 12 and under 15 | 4 |
| 15 and under 18 | 3 |
| Total | 15 |

Source: *Fortune*, 1985.

* As in Table 2.1, the *profit rate* of a firm is defined as its net income as a percentage of stockholders' equity.

sometimes necessary to calculate the arithmetic mean from grouped data—that is, from a frequency distribution. For example, the 1984 profit rates of 15 chemical firms presented in Table 2.2 can be regarded as a sample from the population of 1984 profit rates of all American chemical firms. (Why are they a sample, whereas the profit rates of the 17 oil firms are a population? Because they are a subset of the profit rates of chemical firms in which we are interested, whereas in the oil industry we are interested only in the profit rates of the 17 firms.) Since the profit rates of individual firms are not given in Table 2.2, we cannot use equation (2.2) to calculate $\overline{X}$. However, we can approximate the sum of the $X_i$ by assuming that *the midpoint (class mark) of each class interval can be used to represent the value of the measurements in that class interval.* Thus, the sample mean can be approximated by

$$\overline{X} = \frac{f_1 X_1' + f_2 X_2' + \cdots + f_k X_k'}{n} = \frac{\sum_{j=1}^{k} f_j X_j'}{n}, \tag{2.3}$$

where $f_1$ is the number of measurements in the first class interval, $X_1'$ is the midpoint of the first class interval, $f_2$ is the number of measurements in the second class interval, $X_2'$ is the midpoint of the second class interval, and so on.

Applying equation (2.3) to the data in Table 2.2, we find that the sample mean can be approximated by

$$\overline{X} = \frac{2(4.5) + 3(7.5) + 3(10.5) + 4(13.5) + 3(16.5)}{15}$$

$$= \frac{166.5}{15} = 11.10.$$

Thus the sample mean is about 11.1 percent. If the mean of the measurements in each class interval is close to the midpoint of the class interval, this approximation should entail only a small amount of error.[2] Even if data are available for all the measurements, calculating the mean from a frequency distribution of the data may be easier and less expensive.

### THE WEIGHTED ARITHMETIC MEAN

In some cases the measurements in a sample or a population should not be weighted equally, as in equations (2.1) and (2.2). Since some chemical or oil firms are much bigger than others, it might be argued that a firm's profit rate should be weighted according to its size in determining the average level of profit rates. If $w_i$ is the weight attached to the $i$th measurement in a sample, the weighted arithmetic mean is

$$\overline{X}_w = \frac{\sum_{i=1}^{n} w_i X_i}{\sum_{i=1}^{n} w_i}.$$

(2.4)

For example, suppose that we have a sample of three firms' profit rates: 10 percent, 12 percent, and 15 percent. The firm with the 10 percent profit rate has assets of $2 billion, whereas the other two firms have assets of $1 billion each. If a firm's assets are used to weight its profit rate, the weighted arithmetic mean of the profit rates of these three firms is

$$\overline{X}_w = \frac{2(10) + 1(12) + 1(15)}{2 + 1 + 1} = \frac{47}{4} = 11.75.$$

Thus, the weighted mean is 11.75 percent.

Comparing equation (2.3) with equation (2.4) shows that the arithmetic mean based on grouped data is a type of weighted arithmetic mean: It is a weighted mean of the midpoints of the class intervals, the weight attached to each particular midpoint being the number of measurements falling within that class interval. In business and economics many types of weighted means are encountered and used. The Consumer Price Index, a prominent measure of the rate of inflation, is a weighted arithmetic mean of the relative changes in the prices of various goods and services, as we shall see in detail in Chapter 14.

### THE MEDIAN

Other than the mean, the most widely used measure of central tendency is the median, which is defined as the *middle value* of the rele-

---

[2] At this point, you should understand more fully why we recommended in the previous chapter that class intervals be constructed so that the midpoint of each class interval is close to the average of the observations included in it. If this is done, the error in using equation (2.3) will be very small.

vant set of data. In other words, the median is the *value that divides the set of data in half, 50 percent of the measurements being above (or equal to) it and 50 percent being below (or equal to) it.* Let's consider the profit rates of the petroleum firms in Table 2.1 again. If we list these 17 firms in the order of their 1984 profit rates (from lowest to highest) the profit rate of the ninth firm must be the middle value. In other words, as many firms have profit rates exceeding this value as have profit rates falling below (or equal to) it. According to this method of listing we find that the *median* profit rate of the oil firms is 11.3 percent.

If there is an even number of observations in a frequency distribution, *none* of the observations can be the middle value. For example, among the four numbers 2, 4, 6, and 8 there can be no "middle" number or median because the middle lies between two of the numbers—specifically, between 4 and 6. To resolve this difficulty, convention dictates that *if there is an even number of observations, the mean of the middle pair of observations is regarded as the median.* (In this case the median would be the mean of 4 and 6, or 5.)

Just as the mean frequently must be approximated from a frequency distribution, so the median also must often be approximated in this way. The first step in calculating the median from a frequency distribution is to find the class interval that contains the median. To do this, we start with the lowest class interval, cumulate the number of measurements in one, two, three, and subsequent class intervals, stopping with the interval where the cumulated number of measurements first exceeds or equals $n/2$ if the measurements are a sample or $N/2$ if they are the whole population. This particular class interval contains the median.

To calculate the median from the frequency distribution of the 15 chemical firms in Table 2.2, we cumulate the number of firms with profit rates less than the upper limit of each class interval. Thus, as shown in the third column of Table 2.3, the number of firms with profit rates under 6 percent is two, the number with profit rates under 9 percent is five, and so on. As specified in the previous paragraph, we must keep on cumulating until we reach the first class interval where the cumulated number of measurements exceeds $n/2$ (7.5 in this case, because $n = 15$). Since the cumulated number is 5 for the second class interval and 8 for the third, it is clear that the third class interval (9 and

| Profit rate (percent) | Number of firms | Cumulated number of firms* |
|---|---|---|
| 3 and under 6 | 2 | 2 |
| 6 and under 9 | 3 | 5 |
| 9 and under 12 | 3 | 8 |
| 12 and under 15 | 4 | 12 |
| 15 and under 18 | 3 | 15 |

**Table 2.3**
Finding the Class Interval in Which the Median Profit Rate Is Situated, 15 Chemical Firms

* This is the number of firms with profit rates *less than* the upper limit of the relevant class interval.

under 12 percent) is the first where the cumulated number exceeds 7.5. Thus, this is the class interval in which the median is located.

To estimate where in this class interval the median is situated, we assume that the interval's measurements are spaced evenly along its width of 3 percentage points. If so, the three measurements are 9.50; 10.50; and 11.50. The median must be the eighth lowest measurement, since there are 15 measurements in the sample. Because there are five measurements in lower class intervals, the median must therefore be the highest value in this class interval. Thus, the median must be 11.50.

A general expression for finding the median in grouped data of this sort is

$$M = \left(\frac{n/2 - c}{f_m}\right) l + L_m, \tag{2.5}$$

where $c$ = the number of measurements in class intervals below the one containing the median; $f_m$ = the number of measurements in the class interval containing the median; $l$ = the width of the class interval containing the median; and $L_m$ = the lower limit of the class interval containing the median. Since $n = 15$, $c = 5$, $f_m = 3$, $l = 3$, and $L_m = 9$,

$$M = \left(\frac{7.5 - 5}{3}\right) 3 + 9$$

$$= \left(\frac{2.5}{3}\right) 3 + 9$$

$$= 11.50.$$

Thus, this formula results in precisely the same answer (11.50) as was obtained in the previous paragraph.[3] This will always be true.

### USES OF THE MEAN AND THE MEDIAN

Both the mean and the median are important and useful measures of central tendency (that is, of the "average level" of a set of data). In some circumstances the mean is a better measure than the median, and in others the reverse is true. The following factors are among the most important determinants of whether the mean or the median should be used.

SENSITIVITY TO EXTREME OBSERVATIONS. The median is often preferred over the mean when the latter can be influenced strongly by

---

[3] It is clear from Table 2.3 that the number of measurements in class intervals below the one containing the median is five, since the "9 and under 12 percent" class interval includes the median. The width of this class interval equals 3 percent. The number of observations in this class interval is three, and the lower limit of this class interval is 9. Thus, $c = 5$, $l = 3$, $f_m = 3$, and $L_m = 9$.

extreme observations. For example, how do we go about computing the average income of the families in an apartment building containing 19 families, 8 of which earn $10,000 per year, 10 of which earn $12,000 per year, and 1 of which earns $1 million per year? (The latter presumably has the penthouse.) The mean income of the 19 families equals

$$\frac{8(\$10,000) + 10(\$12,000) + 1(\$1,000,000)}{19} = \frac{\$1,200,000}{19}$$

$$= \$63,158.$$

However, this figure is not a very good description of the yearly income level of the majority of the families in the building. A better measure might be the median, which in this case is $12,000 per year. The median is much less affected by the one extreme point (the millionaire), which raised the mean very considerably.

OPEN-ENDED CLASS INTERVALS. As you will recall from our discussion of class intervals in Chapter 1, it is not unusual for frequency distributions to have open-ended class intervals—that is, class intervals with no finite upper or lower limits. For example, in a frequency distribution of the annual income of American families, two class intervals might be "less than $1,000" and "$30,000 and more." Each of these class intervals is open-ended.[4] If one needs to calculate an average from a frequency distribution with one or more open-ended class intervals, there may be no alternative but to use the median, since calculation of the mean requires a knowledge of the sum of the measurements in the open-ended classes. Unless knowledge of this sort is available (and it frequently is not), the median is often preferable.

MATHEMATICAL CONVENIENCE. The mean rather than the median is often the preferred measure of central tendency because it possesses convenient mathematical properties that the median lacks. For example, the mean of two combined populations or samples is a weighted mean of the means of the individual populations or samples. On the other hand, given the medians of two populations or samples, there is no way to determine what the median of the two populations combined or two samples combined would be.

EXTENT OF SAMPLING VARIATION. As pointed out earlier, sample statistics such as the sample mean or the sample median are often used to estimate the population mean. A major reason for preferring the mean to the median is that the sample mean tends to be more reliable than the sample median in estimating the population mean. In other words, the sample mean is less likely than the sample median to depart considerably from the population mean. This is a very important con-

---

[4] At first glance, you may think that the "less than $1,000" class interval has a finite lower limit—namely, zero. This is incorrect because some people have negative incomes, and there is no limit on how large losses of this sort can be.

sideration which will be more fully appreciated in Chapter 7, where we shall cover this topic in greater detail.

### THE MODE

Another often-used measure of central tendency is the *mode*, which is defined as the *most frequently observed value of the measurements in the relevant set of data.* For example, if a breakfast food manufacturer were to ask people to indicate which of several colors they preferred on a particular cereal carton, and if 100 people preferred red, 50 preferred green, and 20 preferred yellow, then the mode would be at the color red. Or if there were 19 families in your apartment building, 8 of which earned $10,000 per year, 10 of which earned $12,000 per year, and 1 of which earned $1 million per year, the mode would be at $12,000. Why? Because $12,000 is the most frequently observed value of family income in the building.

When data are presented in the form of a frequency distribution, the mode can be estimated as *the midpoint or class mark of the class interval containing the largest number of measurements.* The mode of the profit rates of the chemical firms in Table 2.2 is 13.5 percent. Why? Because the largest number of measurements is contained in the class interval "12 and under 15 percent," and the midpoint of this class interval is 13.5 percent. The class interval containing the largest number of measurements is called the ***modal class.*** Thus, the modal class in Table 2.2 is "12 and under 15 percent."

Based on a graphical representation of a frequency distribution such as a frequency polygon, it is easy to find the mode of a body of data. We need only find the value along the horizontal axis where the frequency polygon achieves its *maximum* vertical height. For example, the mode of the frequency distribution portrayed in Figure 2.1 equals

**Figure 2.1**
Mode of a Frequency Polygon

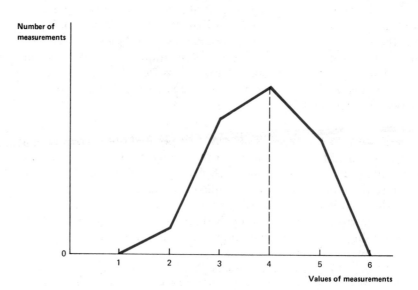

Number of measurements

0

1    2    3    4    5    6

Values of measurements

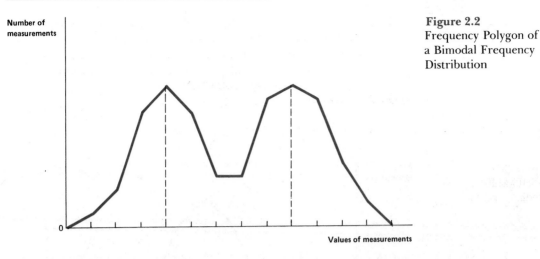

**Figure 2.2**
Frequency Polygon of
a Bimodal Frequency
Distribution

4. Some frequency distributions (like the one shown in Figure 2.2) have more than one mode. If a frequency distribution has more than one mode, it is called *multimodal;* if it has two modes, it is called *bimodal.* Frequency distributions with more than one mode often arise because two or more quite different types of measurements or observations are included. If we were to form a frequency distribution of the heights of American adults, we might find two modes, one at the modal height for men and one at the modal height for women. Great care must be exercised in constructing and interpreting measures of central tendency for multimodal distributions since *measures like the mean or the median may fall between the modes and be unrepresentative of the bulk of the measurements lying near the separate modes.* In cases where a multimodal frequency distribution arises because two or more quite different types of measurements or observations are included, it often is wise to construct a *separate* frequency distribution for each type rather than combine them. In the example above, one frequency distribution might be constructed for men's heights and another frequency distribution for the heights of women.

> *There are many commonly used averages such as the mean, median, and mode. When someone quotes an average, you would do well to determine what type of average it is, since one type can differ greatly from another.*
>
> *Basic Idea #3*

### RELATIONSHIPS AMONG THE MEAN, MEDIAN, AND MODE

Having discussed the three principal measures of central tendency (the mean, median, and mode), we must describe how these three measures are related to one another. If the frequency distribution of a set of data has a single mode and is *symmetrical,* as in panel A of Figure 2.3, the mean, median, and mode coincide. (Symmetrical means that if we were to "fold" the distribution at its mean, the part of the distribution to the left of the mean would be a perfect match for the part to the right of the mean.) Many frequency distributions are not symmetrical, but are ***skewed to the right*** (as in panel B of Figure 2.3) or ***skewed to the left*** (as in panel C of Figure 2.3). A frequency distribution

*Skewness*

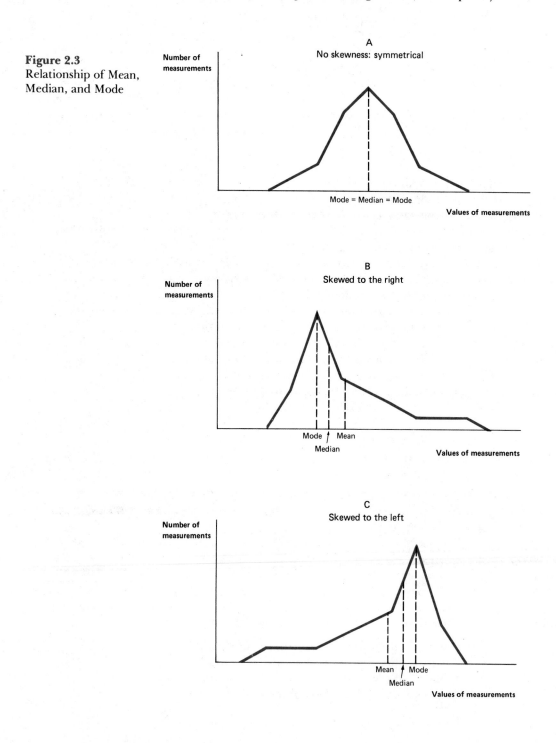

**Figure 2.3**
Relationship of Mean, Median, and Mode

that is skewed to the right has a long tail to the right, whereas one that is skewed to the left has a long tail to the left. As shown in Figure 2.3, if the frequency distribution is skewed to the right, the mean generally exceeds the median, which in turn exceeds the mode. If the frequency distribution is skewed to the left, the mode generally exceeds the median, which in turn exceeds the mean (also shown in Figure 2.3).

## EXERCISES

**2.1** For each of 18 chemical firms, the ratio of current assets to current liabilities is:

| | | | | | | | | |
|---|---|---|---|---|---|---|---|---|
| 1.14 | 1.33 | 1.04 | 1.51 | 1.68 | 1.72 | 2.04 | 1.34 | 1.77 |
| 1.21 | 1.46 | 1.85 | 1.73 | 1.46 | 1.77 | 1.83 | 1.86 | 1.56 |

   (a) Determine the mean ratio of current assets to current liabilities of these firms.
   (b) Determine the median ratio.
   (c) Is the above set of numbers a sample or a population?
   (d) Are the mean and median you determined in (a) and (b) parameters or statistics?

**2.2** An electronics firm wants to determine the average age of its sales engineers. It chooses ten such engineers (out of the 289 that work for the firm), and finds that their ages (in years) are:

| | | | | | | | | | |
|---|---|---|---|---|---|---|---|---|---|
| 46 | 49 | 32 | 30 | 27 | 49 | 62 | 53 | 37 | 39 |

   (a) Find the mean age.
   (b) Find the median age.
   (c) Is the above set of numbers a sample or a population?
   (d) Are the mean and median you determined in (a) and (b) parameters or statistics?

**2.3** A drug firm has two divisions, one with 4,000 employees, the other with 8,000 employees. The mean hourly wage of the employees is $6.05 in the first division and $7.39 in the second division. What is the mean hourly wage of all the firm's employees?

**2.4** A group of 400 female employees and 600 male employees were given an IQ test. The mean IQ for all employees was 115, and the mean IQ for the female employees was 117. What was the mean IQ for the male employees?

**2.5** The frequency distribution of the daily cost (in dollars) of commuting back and forth to work by 100 employees of a steel plant is:

| Cost | 0 and under $1 | $1 and under $3 | $3 and under $5 | $5 and under $7 |
|---|---|---|---|---|
| Number of employees | 29 | 32 | 29 | 10 |

   (a) Find the mean daily cost of commuting.
   (b) Find the median daily cost of commuting.

**2.6** The frequency distribution of the number of years of schooling of 100 workers at a hospital is:

| Number of years | 0 and under 8 | 8 and under 12 | 12 and under 16 | 16 and over |
|---|---|---|---|---|
| Number of workers | 1 | 12 | 49 | 38 |

(a) Find the mean years of schooling (if possible).
(b) Find the median years of schooling (if possible).
(c) What are the relative advantages here of the mean and median?

**2.7** A market research firm goes to 16 stores and determines how much (in cents) each charges for an identical tube of toothpaste, with the following results:

| 88 | 89 | 89 | 98 | 98 | 90 | 92 | 95 |
|---|---|---|---|---|---|---|---|
| 98 | 98 | 98 | 91 | 93 | 94 | 96 | 97 |

(a) What is the mode?
(b) Is the frequency distribution skewed? If so, is it skewed to the left or to the right?
(c) Which is largest: the mean, median, or mode?

**2.8** An Amtrak official obtains data on a particular day concerning the length of time (in minutes) that the metroliners leaving New York take to reach Philadelphia, with the following results:

| 93 | 90 | 91 | 93 | 89 | 90 | 94 | 115 | 88 | 88 |
|---|---|---|---|---|---|---|---|---|---|

(a) What is the mean?
(b) What is the median?
(c) What are the advantages here of the median over the mean?

**2.9** According to data presented in the previous chapter, the profit after taxes (as a percent of sales) of each manufacturing industry in the United States in 1974 was as follows:

| Durable goods (industry) | Profit (percent of sales) | Nondurable goods (industry) | Profit (percent of sales) |
|---|---|---|---|
| Transportation equipment | 3.9 | Food and kindred products | 2.6 |
| Electrical equipment | 4.3 | Tobacco | 5.8 |
| Machinery (except electrical) | 5.6 | Textiles | 2.8 |
| Fabricated metal products | 4.0 | Paper | 5.4 |
| Primary iron and steel | 4.1 | Printing | 4.8 |
| Primary nonferrous metals | 5.4 | Chemicals | 6.8 |
| Stone, clay, and glass | 4.8 | Petroleum and coal | 7.6 |
| Instruments | 8.4 | Rubber | 4.0 |
| Other durable goods | 5.0 | Other nondurable products | 2.1 |

(a) Calculate the mean profit rate of the durable manufacturing industries (left-hand column) in 1974.
(b) Calculate the mean profit rate of the nondurable manufacturing industries (right-hand column) in 1974.
(c) Interpret and compare the results of (a) and (b) of this question.
(d) Calculate the median profit rate of nondurable manufacturing industries in 1974. Calculate the median profit rate of durable manufacturing industries in 1974. How big is the difference between these two medians? Interpret your results.

**2.10** In a township in Virginia, all lots are 1/4 acre, 1/2 acre, 1 acre, or 2 acres. The frequency distribution of lot sizes for all residential property in this township is:

| Size of lot (acres) | Number of lots |
|---|---|
| 1/4 | 100 |
| 1/2 | 500 |
| 1 | 50 |
| 2 | 20 |

What is the mode of this frequency distribution? Is the mode bigger than the mean size of lot? Is it bigger than the median size of lot?

## 2.5 Measures of Dispersion

IMPORTANCE OF DISPERSION

In the previous section we saw that measures of central tendency provide useful summary information concerning the general level or average value of a body of data. However, this obviously does not mean that such measures alone can provide a complete or adequate description of the data. A case that is close to home can illustrate the limitations of such measures. Suppose that you have taken an examination and that the instructor, after grading the exam, announces to the class that the mean grade is 75. Then he or she hands back the exams, and you find that your paper has received a grade of 80. Clearly, it is hard to interpret this grade on the basis of information concerning only the average grade. The variability about the average is very important, too. If the grades are highly variable (as in panel A of Figure 2.4), then a large number of your classmates may have received a grade higher than yours. On the other hand, if there is little variability in the grades (as in panel B of Figure 2.4), then you may have received close to the highest grade.

If there is enough variability about an average, the average may not mean much. If Mr. Rich, one of the world's wealthiest men, is driven to the airport by his chauffeur, what is the average income level of the two occupants of the car? Mr. Rich's income is $10 million a year and his chauffeur's income is $10,000 a year; thus, the mean income of the two is $5,005,000 per year. This is a misleading figure, however, since it vastly overstates the chauffeur's income and vastly understates the income of Mr. Rich. In cases of this sort, then, we see that an average can be quite misleading. (More will be said about this later in the chapter.)

**Figure 2.4**
Histogram of Examination Grades

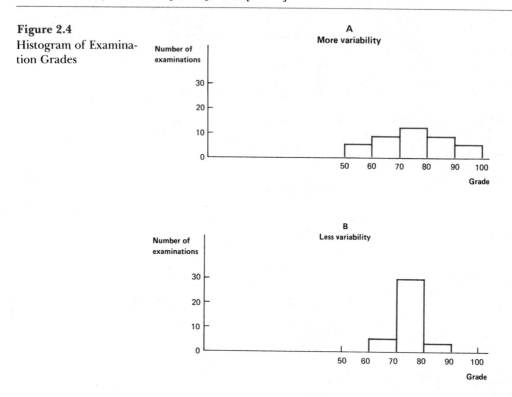

DISTANCE MEASURES OF DISPERSION

Both summary measures of dispersion and summary measures of central tendency are important in describing a body of data, but neither alone will suffice. There are two types of summary measures of dispersion: distance measures and measures of average deviation. ***Distance measures** describe the variation in the data in terms of the distance between selected measurements.* The most frequently used distance measures are the range and the interquartile range.

RANGE. Perhaps the simplest measure of variability is the *range*, which is *the difference between the highest and lowest values in the body of data.* In the example of the exam grades discussed above, if the lowest grade in the class is 50 and the highest is 99, then the range is $99 - 50 = 49$. Although the range is a popular measure of variability, particularly because it is so easy to compute, it has the important disadvantage of being unaffected by the values of all observations other than the highest and the lowest. For example, the range of the exam grades would be the same (49) if the grades were distributed evenly from 50 to 99, or if all grades other than the highest or lowest fell between 70 and 80. Yet the variability certainly is less in the latter case.

*Quartiles*

INTERQUARTILE RANGE. Another common measure of variability is the *interquartile range,* which is defined as *the difference between the third quartile and the first quartile.* The *third quartile* is the value such that 75 percent of the observations lie below it; the *first quartile* is the value such that 25 percent of the observations lie below it. Thus, the interquartile range measures *the spread bounding the middle 50 percent of the values of the observations.* (Note that the *second quartile,* which is the value such that 50 percent of the observations lie below it, is another name for the median.) One advantage of the interquartile range is that it can be calculated from frequency distributions with open-ended class intervals, while the range cannot.

MEASURES BASED ON PERCENTILES. In addition to the range and the interquartile range, other distance measures of dispersion can be based on percentiles. The *Xth percentile* is defined as the value that exceeds $X$ percent of the observations. Thus, the 90th percentile is the value such that 90 percent of the observations lie below it. The difference between the 90th percentile and the 10th percentile is a possible measure of dispersion; so is the difference between the 99th percentile and the 1st percentile. The former measures the spread bounding the middle 80 percent of the values of the observations. The latter measures the spread bounding the middle 98 percent of them.

*Percentiles*

## THE VARIANCE AND STANDARD DEVIATION

Although the distance measures of dispersion are sometimes used, they are not as important as measures of average deviation, the most significant of which are the variance and the standard deviation. The *variance* of the measurements in a population, denoted by $\sigma^2$ (sigma squared), is defined as *the arithmetic mean of the squared deviations of the measurements from their mean.* Thus if $X_1, X_2, \ldots, X_N$ are the measurements in the population,

$$\sigma^2 = \frac{(X_1 - \mu)^2 + (X_2 - \mu)^2 + \cdots + (X_N - \mu)^2}{N}$$

$$= \frac{\sum_{i=1}^{N} (X_i - \mu)^2}{N}.$$

(2.6)  *Formula for $\sigma^2$*

The variance is a measure of dispersion, but it is expressed in units of squared deviations or squares of the values of the measurements rather than in the same units as the measurements. The *standard deviation* is a measure of dispersion which is expressed in the same units as the measurements. The *standard deviation* of the measurements in a population is denoted by $\sigma$ (sigma), which is the positive square root of the variance. In other words, the standard deviation is

$$\sigma = \sqrt{\frac{\sum_{i=1}^{N} (X_i - \mu)^2}{N}}$$

(2.7)  *Formula for $\sigma$*

Since the standard deviation is so important in statistics, its definition is worth discussing in greater detail. To begin with, note that if $X_i$ is the $i$th measurement in the population, then the difference between this measurement and the population mean equals $(X_i - \mu)$. This is the deviation of the $i$th measurement from the population mean, and the square of this deviation obviously equals $(X_i - \mu)^2$. Next, let's find the mean of these squared deviations. Since there are $N$ of these squared deviations, this mean equals

$$\frac{\sum_{i=1}^{N}(X_i - \mu)^2}{N}.$$

Recall from our earlier discussion that $\Sigma$ is the summation sign, which means that the numbers to its right—that is, $(X_i - \mu)^2$—should be summed from the lower limit on $i$ (given below the summation sign) to the upper limit on $i$ (given above the summation sign). Thus, in this case, $(X_i - \mu)^2$ is to be summed from $i = 1$ to $i = N$. (In other words, the squared deviations are to be summed for all observations.) Then, to obtain the mean of the squared deviations, the resulting sum is divided by $N$. Finally, we must find the square root of this mean, the result being the standard deviation. Equation (2.7) shows this complete procedure.

Intuitively, it seems clear that *the more dispersion there is in a body of data the bigger the standard deviation will be.* If there is no dispersion at all, every observation will equal the population mean, with the result that every one of the deviations will equal zero. (Thus, the standard deviation will equal zero.) As the dispersion in the data increases, the deviations of the observations from the population mean will tend to increase as well, and so will the mean of the squared deviations. (Thus, the standard deviation will also increase.) For this reason, if one knows that the standard deviation of the measurements in one population is higher than the standard deviation of the measurements in another population, this indicates that there is more dispersion in the former population than in the latter.[5]

If the body of data is a sample rather than a population, the formulas for the variance and the standard deviation are somewhat different from those used for the entire population. Specifically, the sample variance, denoted by $s^2$, is defined as

*Formula for $s^2$*

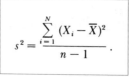

$$s^2 = \frac{\sum_{i=1}^{N}(X_i - \overline{X})^2}{n - 1}.$$

(2.8)

[5] However, if the means of the two populations differ, a better procedure for comparing their dispersion may be to calculate the coefficient of variation in each population. The coefficient of variation is defined and discussed below.

And the sample standard deviation, denoted by $s$, is defined as

$$s = \sqrt{\frac{\sum_{i=1}^{n} (X_i - \overline{X})^2}{n - 1}}.$$

(2.9) *Formula for $s$*

There are three differences between these formulas and those given for the population variance and standard deviation. First, the *sample* mean $\overline{X}$ is substituted for the *population* mean $\mu$. Second, the squared deviations from the mean are summed over all measurements in the *sample*, not the *population*. Third, the sum of the squared deviations from the mean is divided by $(n - 1)$, not by $N$.

Generations of statistics students have been puzzled over why the denominator of the sample variance is $(n - 1)$, while the denominator of the population variance is $N$. Basically, the reason is that the sample variance, if its denominator were $n$, would tend to underestimate the population variance. (More will be said about this in later chapters.)

## CALCULATING THE STANDARD DEVIATION

To illustrate the computations involved in calculating the population standard deviation, let's take two cases in which we have all the data in the population and thus are able to compute the population standard deviation. The first case is the population of incomes in a house where there are two families, one with an annual income of $20,000, the other with an annual income of $30,000. Since there are only two numbers in this population ($20,000 and $30,000), the population mean is $25,000 and the deviations from the mean (in thousands of dollars) are

$$(X_1 - \mu) = 20 - 25 = -5,$$

$$(X_2 - \mu) = 30 - 25 = 5.$$

Thus, the mean of the squared deviations (the variance) equals

$$\frac{\sum_{i=1}^{2} (X_i - \mu)^2}{2} = \frac{(-5)^2 + (5)^2}{2}$$

$$= 25,$$

and the standard deviation is[6]

$$\sqrt{\frac{\sum_{i=1}^{2} (X_i - \mu)^2}{2}} = \sqrt{25} = 5 \text{(thousands of dollars)}.$$

---

[6] In this very simple case, the standard deviation equals the deviation of each observation from the mean. For populations containing more than two observations, this generally will not be the case.

**Table 2.4**
Calculation of
$\Sigma(X_i - \mu)^2$ for the
Population of 1984
Profit Rates of 17
Major Petroleum Firms

| $X_i$ | $(X_i - \mu)$ | $(X_i - \mu)^2$ |
|---|---|---|
| 19.2 | 8.01 | 64.1601 |
| 2.3 | −8.89 | 79.0321 |
| 9.3 | −1.89 | 3.5721 |
| 10.4 | −0.79 | 0.6241 |
| 17.4 | 6.21 | 38.5641 |
| 14.2 | 3.01 | 9.0601 |
| 5.7 | −5.49 | 30.1401 |
| 11.3 | 0.11 | 0.0121 |
| 12.2 | 1.01 | 1.0201 |
| 12.3 | 1.11 | 1.2321 |
| 10.2 | −0.99 | 0.9801 |
| 6.6 | −4.59 | 21.0681 |
| 8.9 | −2.29 | 5.2441 |
| 17.1 | 5.91 | 34.9281 |
| 17.7 | 6.51 | 42.3801 |
| 11.7 | 0.51 | 0.2601 |
| 3.7 | −7.49 | 56.1001 |
| Total | | 388.3777 |

Next, let's take a somewhat more complicated case—that of the 1984 profit rates of petroleum firms in Table 2.1. What is the standard deviation of these profit rates? The second column of Table 2.4 shows the deviation from the mean (that is, $X_i - \mu$) for each of the firms. (Recall from an earlier section that $\mu = 11.19$.) In the third column, the square of this deviation—that is, $(X_i - \mu)^2$—is calculated for each firm. The sum of these squared deviations is shown to equal 388.3777:

$$\sum_{i=1}^{17} (X_i - \mu)^2 = 388.3777.$$

Thus,

$$\frac{\sum_{i=1}^{17} (X_i - \mu)^2}{N} = \frac{388.3777}{17} = 22.8457$$

and

$$\sigma = \sqrt{\frac{\sum_{i=1}^{17} (X_i - \mu)^2}{N}} = \sqrt{22.8457} = 4.8.$$

Consequently, the standard deviation of the 1984 profit rates of the petroleum firms in Table 2.1 is 4.8 percentage points.

Having calculated the standard deviation for the population in two cases, let's turn now to the calculation of the sample standard deviation; at the same time, we will illustrate how the standard deviation can be calculated from a frequency distribution. Specifically, we will estimate the standard deviation of the profit rates of the sample of 15 chemical firms in Table 2.2. As in the calculation of the arithmetic mean from a frequency distribution in equation (2.3), we assume that *the midpoint of each class interval (that is, the class mark) can be used to represent each value of the measurements in that class interval.* This means that we assume that the profit rates of the firms in the "3 and under 6 percent" class interval can be approximated by 4.5 percent, that the profit rates of the firms in the "6 and under 9 percent" class interval can be approximated by 7.5 percent, and so on. Thus the sample standard deviation can be approximated by

$$ s = \sqrt{\frac{\sum\limits_{j=1}^{k} f_j(X_j' - \overline{X})^2}{n - 1}}, \qquad (2.10) $$

where $f_j$ is the number of measurements in the $j$th class interval, $X_j'$ is the midpoint of the $j$th class interval, and $k$ is the number of the class intervals.

Table 2.5 shows how the formula for the sample standard deviation in equation (2.10) can be applied to the frequency distribution of profit rates of chemical firms. The first column of Table 2.5 provides the class intervals of this frequency distribution. The second column gives the number of measurements in each class interval (the values of $f_j$). The third column shows the midpoint of each class interval (the values of $X_j'$). Since we know from our earlier discussion (on page 37) that the sample mean is 11.1 percent, the values of $(X_j' - \overline{X})$ are as shown in the fourth column, and the values of $(X_j' - \overline{X})^2$ are as shown in the fifth column. Finally, the sixth column shows the product of $f_j$ and $(X_j' - \overline{X})^2$. Summing up the figures in the sixth column, we have the sum of the squared deviations from the sample mean, which is 237.60. Dividing this sum by 14, which is $(n - 1)$, we get 16.97, the sample variance. Taking the square root of 16.97, we get 4.1, the sample standard deviation.

**Table 2.5**
Calculation of the Standard Deviation of the Sample of 15 Profit Rates of Chemical Firms

| Profit rate (percent) | Number ($f_j$) of firms | Midpoint ($X_j'$) of class interval | $X_j' - \overline{X}$ | $(X_j' - \overline{X})^2$ | $f_j(X_j' - \overline{X})^2$ |
|---|---|---|---|---|---|
| 3 and under 6 | 2 | 4.5 | −6.6 | 43.56 | 87.12 |
| 6 and under 9 | 3 | 7.5 | −3.6 | 12.96 | 38.88 |
| 9 and under 12 | 3 | 10.5 | −0.6 | 0.36 | 1.08 |
| 12 and under 15 | 4 | 13.5 | 2.4 | 5.76 | 23.04 |
| 15 and under 18 | 3 | 16.5 | 5.4 | 29.16 | 87.48 |
| | | | | | 237.60 |

Using equation (2.10), we find that

$$s = \sqrt{\frac{\sum_{j=1}^{k} f_j (X_j' - \overline{X})^2}{n - 1}} = \sqrt{\frac{237.60}{14}} = 4.1.$$

### INTERPRETATION OF THE STANDARD DEVIATION

The standard deviation is the most important summary measure of dispersion. If the frequency distribution of a population conforms to the so-called normal distribution (to be discussed in detail in Chapter 5), then we know the percentage of measurements in the population that fall within 1, 2, or 3 standard deviations of the population mean. Specifically, 68.3 percent of the measurements lie within $\pm$ 1 standard deviation of the mean, 95.4 percent of the measurements lie within $\pm$ 2 standard deviations of the mean, and 99.7 percent of the measurements lie within $\pm$ 3 standard deviations of the mean.

Thus, if we know that the population of diameters of pieces of pipe produced by a firm conforms to the normal distribution with a mean of 4 inches and a standard deviation of 0.05 inches, it follows that 68.3 percent of the pieces produced by this firm will have diameters of 3.95 to 4.05 inches, that 95.4 percent will have diameters of 3.90 to 4.10 inches, and that 99.7 percent will have diameters of 3.85 to 4.15 inches. This is useful information. For example, if the diameters of the pieces must be between 3.90 and 4.10 inches to meet specifications, it follows that 4.6 percent of the pieces will be unacceptable.

Of course, it is essential to realize that results such as these pertain only to populations that conform to the normal distribution. However, even if a population does not conform to the normal distribution, we can still make statements (based on Chebyshev's inequality, to be discussed in Chapter 4) about the percentage of any population that will be within a certain number of standard deviations of the population mean. Much more will be said on this subject in later chapters.

### COEFFICIENT OF VARIATION

The standard deviation is a measure of *absolute* dispersion or variability, and it is affected by the units of measurement. Thus, in the case of the diameters of the pieces of pipe in the previous section, the standard deviation is 0.05 if the diameters are measured in inches, and 0.0042 ($= .05 \div 12$) if they are measured in feet. To obtain a measure of dispersion or variability that does not depend on the units of measurement, statisticians often use a measure of *relative* variability such as the coefficient of variation.

The coefficient of variation equals the standard deviation as a percentage of the mean. In other words, for a population it equals

$$V = \frac{\sigma}{\mu} \cdot 100,$$

and for a sample it equals

$$V = \frac{s}{\bar{x}} \cdot 100.$$

Since $\sigma$ equals 0.05 inches and $\mu$ equals 4 inches, the coefficient of variation for the pipe diameters is $\frac{.05}{4} \cdot 100 = 1.25$ percent.

### EXERCISES

**2.11** A sample is composed of the following five weights (in pounds): 1.8, 1.9, 2.1, 2.3, 2.0. Calculate the variance and standard deviation.

**2.12** A finite population consists of the seven integers: 3, 4, 5, 6, 7, 8, 9. Compute the variance and standard deviation.

**2.13** A sample of five tires is chosen, the diameters of the tires being 30.01, 30.02, 30.03, 30.03, and 30.02 inches.
(a) Calculate the variance and standard deviation.
(b) Subtract 30 inches from each observation in the sample, and calculate the variance and standard deviation.
(c) Explain why your results in (a) and (b) are or are not equal.

**2.14** Use the shortcut formulas in equations (2.11) and (2.12) to compute the variance and standard deviation of the ages in Exercise 2.2. (This is for students who cover Appendix 2.2.)

**2.15** Based on the frequency distribution in Exercise 2.5, use the shortcut formula in equation (2.13) to calculate the standard deviation of the daily cost of commuting, assuming that the 100 workers are a sample. If the 100 workers were the population, how would equation (2.13) have to be altered, and what would be the standard deviation? (This is for students who cover Appendix 2.2.)

**2.16** Based on the data in Exercise 2.1, what is the range of the ratio of current assets to current liabilities of the 18 chemical firms?

**2.17** Firm A is in an industry where the mean rate of return of firms is 10 percent, the standard deviation being 5 percent. Firm B is in an industry where the mean rate of return of firms is 12 percent, the standard deviation being 6 percent. If Firm A's rate of return is 16 percent and Firm B's rate of return is 18 percent, which of the two is more profitable compared to its industry?

**2.18** The mean and standard deviation of the prices charged by movie theaters in Tucson and Seattle for admission to a particular movie are:

| City | Mean | Standard deviation |
|------|------|-------------------|
| Tucson | $4.96 | $0.38 |
| Seattle | 5.22 | 0.45 |

The Bijou Theater in Tucson charges $5.50, and the Biloxi Theater in Seattle charges $5.75. Which is more expensive relative to the other theaters in the same city?

**2.19** Which of the following statements is true?
   (a) The median equals one-half of the sum of the first and third quartiles.
   (b) The first quartile equals the 25th percentile.
   (c) The third quartile equals the 75th percentile.

**2.20** In an Ohio township, the frequency distribution of the sizes of all residential lots is given below:

| Size of lot (acres) | Number of lots |
|---|---|
| 1/4 | 300 |
| 1/2 | 400 |
| 1 | 200 |
| 2 | 100 |

   (a) What is the standard deviation of the lot sizes?
   (b) What is the variance of the lot sizes?
   (c) What is the range of the lot sizes?

## 2.6  Misuses of Descriptive Statistics

The famous British prime minister Disraeli once said that there were lies, damned lies, and statistics. You are no doubt already on guard against many kinds of misuses of statistics. However, it is important for you to develop as much skill in rooting out statistical fallacies and chicanery as possible, since the world is full of pitfalls for the statistically unwary. This section covers five kinds of errors that frequently occur in descriptive statistics. These are errors that can result in costly mistakes, and even trained statisticians occasionally fall prey to them.

INAPPROPRIATE COMPARISONS. Suppose you read in the newspaper that the crime rate in your area is 2 percent lower than last year. Before accepting this conclusion, it would be wise to question whether the *definition* of crime on which these statistics were based has remained the same from last year to the present. In other words, might the apparent decrease in the crime rate be due to the fact that some types of behavior were defined as crimes last year which are no longer classified as such? Also, do the figures pertain to the same kind of population during the two periods, or could the apparent drop in the crime rate be due to the fact that the figures for the most recent period pertain to a different set of people than the earlier figures? For example, the later figures might represent all individuals, including young children, whereas the earlier ones might pertain only to individuals over 10 years old. Furthermore, are the later and earlier periods really comparable? Perhaps the later figures pertain to a portion of the year when crime is always relatively low, whereas the earlier figures pertain to a full year. Unless you can be reasonably certain that each of these possible discrepancies is nonexistent or relatively minor, it is difficult, if not impossible, to interpret this kind of statistical item in the newspaper.

*Statistics in Context/Part One*

# Japan, W. E. Deming, and the Use of Histograms

Japan's economic success in recent years is attributable in part to its emphasis on achieving and maintaining high quality levels for its products; and, according to many reports, these quality levels have been promoted by the widespread use of relatively simple statistical tools. Firms like Nissan Motor Corporation or Toshiba, a television manufacturer, use many of the statistical techniques described in this book to determine how defects occur, who is responsible for them, and how they can be eliminated. Based on what both the Japanese and their U.S. rivals say, these techniques have been very effective.

As an illustration, consider a machinery manufacturer that tested the hardness of the sheet-metal panels it received from a supplier and plotted the results in the following histogram:

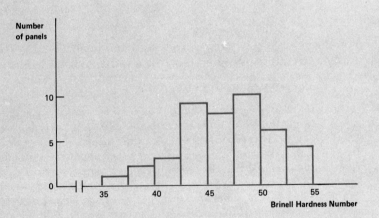

This histogram indicated that the hardness of the panels was surprisingly variable. Upon investigation, the machinery manufacturer determined that its supplier was buying metal from two firms, each using different standards, a finding which enabled the machinery manufacturer to develop ways to control the quality of its product more effectively.[7]

One indication of how seriously Japanese firms take statistical quality control is the vigorous competition among them to win the Deming Award, which is bestowed each year on a Japanese firm for the effectiveness of its quality control techniques. The awards were named after W. E. Deming, a leading American statistician, who in 1950 im-

[7] R. Wood, "Right from the Start," *Technology*, November 1981. The figures used here are hypothetical, but the case is real.

pressed Japanese scientists and engineers with the need to use statistical methods to overcome the notion that "made in Japan" meant "shoddy goods." The use of statistical quality controls helped the Japanese achieve a new reputation.

To illustrate how simple statistical tools can be used to improve productivity, Deming has described a case where a firm produced steel rods that were supposed to have diameters of at least 1 centimeter. The results of the inspection of 500 rods produced by this firm were as shown by the following histogram:

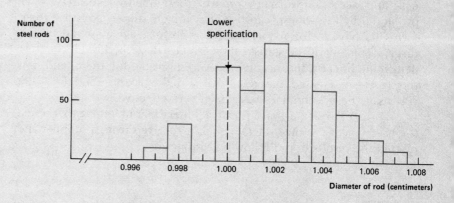

Based on this histogram, 40 of the rods were too small. (That is, they were in the class intervals centered on .997 and .998 centimeters.)

As Deming points out, this histogram

is trying to tell us something. The peak at just 1 cm. with a gap at 0.999 seems strange. It looks as if the inspectors were passing parts that were barely below the lower specification, recording them in the interval centered at 1.000. When the inspectors were asked about this possibility, they readily admitted that they were passing parts that were barely defective. They were unaware of the importance of their job, and unaware of the trouble that an undersized diameter would cause later on.

This simple chart thus detected a special cause of trouble. The inspectors themselves could correct the fault. When the inspectors in the future recorded their results more faithfully, the gap at 0.999 filled up. The number of defective rods turned out to be much bigger, 105 in the next 500, instead of the false figure of 10 + 30 + 0 = 40 in [the above histogram].

The results of inspection, when corrected, led to recognition of a fundamental fault in production: the setting of the machine was wrong. It was producing an inordinate number of rods of diameter below the lower specification limit. When the setting was corrected and the inspection carried out properly, most of the trouble disappeared.[8]

[8] W. E. Deming, "Making Things Right," in E. Mansfield, *Statistics for Business and Economics: Readings and Cases* (New York: Norton, 1980).

## PROBING DEEPER

1. Does the histogram on page 57 prove that the firm's supplier was buying metal from two firms, each with different standards? What other interpretations of the histogram are possible?

2. According to the firm, this histogram indicated that the hardness of the panels was "surprisingly variable." How would you go about measuring their variability? What criterion would you use to determine whether their hardness really was "surprisingly variable"?

3. Why did Deming regard the histogram on page 58 as "strange"? Is it always true that there are no gaps in histograms?

4. Once the producer of steel rods determined that 105, rather than 40, out of 500 rods were defective, was it obvious from the histogram alone that the large proportion defective was due to improper setting of the machine?

5. The firm producing steel rods reports that the mean diameter of the rods produced this month was 1.003 centimeters, whereas last month it was 1.002 centimeters. If the standard deviation of the diameters was the same last month as this month, do you think that the proportion of rods produced that were defective was higher, lower, or the same this month as last month? Explain.

6. The producer of steel rods reports that the mean diameter of rods produced was 1.003 centimeters in both November and December, and that the standard deviation of the diameters was .0025 centimeters in December and .0015 centimeters in November. Do you think that the proportion defective was higher, lower, or the same in December as in November? Explain.

VERY SMALL OR OBVIOUSLY BIASED SAMPLES. Suppose you receive a report stating that a new device will increase the number of miles an automobile can travel per gallon of gasoline by 5 percent. On the surface this certainly sounds good, but this figure of 5 percent might be based on the effect of the device on the mileage of only three cars. If so, the figure of 5 percent may not apply to your car, because so small a sample may not be at all representative of the entire population of automobiles. Therefore, it is generally wise to ask about the size of the sample on which a particular statistical result is based. This is a good question to ask even in areas like medicine and the natural sciences, since despite the scientific and precise nature of the work, results sometimes are based on very small samples.

Some organizations and people use statistics as a drunk uses a lamppost: for support, not illumination. Thus individuals or firms sometimes carry out studies on one small sample after another until eventually and by chance one of the samples indicates what they want to prove about the product they sell. Then they publicize the results of this small sample, hoping thereby to influence their customers or potential customers. The makers of the new device to increase mileage per gallon may have tested the effect of their invention on one three-car sample after another. (They may also have discarded the results of many earlier small samples that showed the device had no effect!) Finally, when one sample indicated, by chance, a 5 percent increase in mileage per gallon, they may have publicized the results. Clearly, statistics based on such improperly selected samples can be worse than valueless.

IMPROPER CHOICE OF AVERAGE. For a given set of data, a person can use one measure of central tendency to indicate one thing and another measure of central tendency to indicate another. A real estate salesman trying to impress a potential customer with the high income level in a particular suburb quotes the *mean* family income in this suburb as $30,000 per year. His statement is perfectly true, but it is also misleading because the income *distribution* in the suburb happens to be highly skewed to the right. Perhaps 2 percent of the suburb's residents earn about $750,000 per year, and the rest earn between $10,000 and $20,000 per year. Thus, the mean of $30,000 is not a very representative figure.

On the other hand, the mayor of this particular suburb is trying to impress the state government with the *low* income level of the area (and hence, with its need for state aid). In contrast to the real estate salesman, the mayor chooses as a measure of central tendency the *mode* of the frequency distribution of families by income level. Because the income distribution is highly skewed to the right, the mode is likely to be considerably less than the mean and somewhat less than the median. The mode of the income distribution in the community may be only $14,000. Thus, as a measure of central tendency, the real estate salesman quotes $30,000, while the mayor quotes $14,000. Both are right; the difference is due to the fact that different measures of central tendency are being used. The moral, of course, is clear: Whenever someone quotes an "average" or a measure of central tendency, be sure to find out what kind of measure it is and how representative it is likely to be.

NEGLECT OF THE VARIATION ABOUT AN AVERAGE. Even if the proper kind of average is chosen, the average may be surrounded by so much variation that it alone may be misleading. It is sometimes argued that home builders, focusing their attention on the average number of people in the American family, build too many medium-sized homes and too few small and large ones. In other words, if the average family size in the United States is 3.6 persons, home builders may tend to neglect the considerable variation about this average, and build too many homes for families of 3 persons or 4 persons. (According to some observers, this has in fact occurred.)[9]

[9] See Darrell Huff and Irving Geis, *How To Lie with Statistics* (New York: Norton, 1954), part of which is contained in E. Mansfield, *Statistics for Business and Economics: Reading and Cases* (New York: Norton, 1980). Also, see Oskar Morgenstern's article in the latter book.

THE CASE OF THE NEW TOOTH POWDER

**TO ERR IS HUMAN ... BUT COSTLY**

Some years ago, a new brand of tooth powder was introduced in the United States. According to its advertisements, studies showed that it had "considerable success" in improving the health of a person's teeth. Suppose that these studies were designed in the following way. The manufacturer of this tooth powder chose three samples of seven people; the first sample agreed to use the new brand, the second sample used brand X, and the third sample used brand Y. After six months, a dentist examined each person's teeth and scored the health of his or her teeth on a scale of 0 (poorest possible score) to 100 (highest possible score). The results were as follows:

New brand: 65, 71, 53, 55, 34, 82, 77
Brand X   : 58, 60, 63, 90, 95, 89, 62
Brand Y   : 54, 38, 43, 61, 94, 96, 82

The manufacturer of the new tooth powder says that the average score for the new brand is 2 points higher than for brand X and 4 points higher than for brand Y. Would you agree with the manufacturer's advertisements? If not, what mistakes have been made?

SOLUTION: There are at least four errors or problems. (1) The data pertain to the health of a person's teeth, not to the extent of its improvement. (2) The manufacturer seems to be using the median, rather than the mean. One suspects that this choice was prompted by the fact that the median for the new brand (65) is higher than for brand X (63) or brand Y (61). The mean for the new brand (62.4) is less than for brand X (73.9) or brand Y (66.9). Unless some good reason is given for preferring the median over the mean, the results are suspect. (3) The samples are small, and the differences among the averages may be due merely to chance. (Ways of testing whether this is the case will be discussed in Chapter 8.) (4) The respondents who used the new brand "agreed" to do so, which suggests that the sample was not random and that the people who used the new brand may differ systematically in dental health from those who used the other brands. Advertisements based on statistical methods of this sort may be misleading and costly both in terms of consumers' pocketbooks and their health (if the advertised products are in fact less effective than other products).

**Basic Idea #4**    *An average may mean very litle if there is a great deal of variation about it. There is an old story about a man who drowned in a lake which had an average depth of 1 foot.*

Another case of this sort is encountered in ads which announce that a certain toothpaste will result, on the average, in a such-and-such percent reduction in dental cavities. Besides the fact that some of these claims may be subject to some of the other problems discussed above, they may be misleading because there may be such great variation about the average. What is important to a particular consumer is whether the toothpaste will reduce his or her cavities. Because the variation among individuals in the effects of the toothpaste may be so great, the chance that it will have a beneficial effect on one particular person may be scarcely better than 50-50, even though on the average it may offer some protection against tooth decay.

MISINTERPRETATION OF GRAPHS AND CHARTS. Sometimes graphs and charts are presented so as to give a misleading impression. For example, a real estate salesman wants to run an ad showing the increase over time in the average price of houses in the suburb where he works. He decides to use a *pictogram,* a graph in which the size of the object in the picture indicates the relative size of the thing the object represents. The salesman uses the pictogram in Figure 2.5 to show that the average price of a house (represented by the size of the deed) doubled between 1975 and 1986. This chart certainly makes it appear that real estate prices have risen, and indeed, it is misleading in this respect. Why? Because the salesman doubled both the width and the height of

the deed in Figure 2.5, thus *quadrupling* the area of the deed. Thus, based on the *area* of the deed, the chart makes it appear that real estate prices have quadrupled, not doubled.

Another frequently encountered error occurs in the construction of histograms where the class intervals are not of equal width. The frequency distribution of profit rates of the 17 major oil firms that we constructed in Chapter 1 is reproduced in Table 2.6. If we combine the first three class intervals in Table 2.6, we obtain the frequency distribution in Table 2.7. We then construct bars on the line segments of the horizontal axis corresponding to the classes of the frequency distribution, and we make the height of each bar equal to the number of cases in each class interval. The result is shown in panel A of Figure 2.6. Comparing panel A of Figure 2.6 with Figure 1.2, it is clear that panel A gives a distorted picture. The reason is simple: In the class interval from 0.0 to 9.0 percent, there are 5 firms, which means that on the average there are 1⅔ firms for every three percentage points in this class interval. Thus, to be comparable with the other vertical bars, the height of the bar for this class interval should be 1⅔, not 5. Put differ-

| Profit rate (percent) | Number of firms |
|---|---|
| 0.0 and under 3.0 | 1 |
| 3.0 and under 6.0 | 2 |
| 6.0 and under 9.0 | 2 |
| 9.0 and under 12.0 | 5 |
| 12.0 and under 15.0 | 3 |
| 15.0 and under 18.0 | 3 |
| 18.0 and under 21.0 | 1 |
| Total | 17 |

**Table 2.6**
Frequency Distribution of Population of 1984 Profit Rates of 17 Major Petroleum Firms

| Profit rate (percent) | Number of firms |
|---|---|
| 0.0 and under 9.0 | 5 |
| 9.0 and under 12.0 | 5 |
| 12.0 and under 15.0 | 3 |
| 15.0 and under 18.0 | 3 |
| 18.0 and under 21.0 | 1 |
| Total | 17 |

**Table 2.7**
Frequency Distribution of Population of 1984 Profit Rates of 17 Major Petroleum Firms, Unequal Widths of Class Intervals

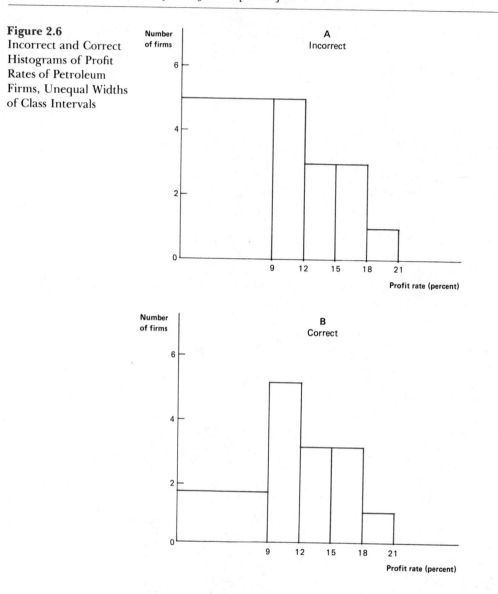

**Figure 2.6**
Incorrect and Correct
Histograms of Profit
Rates of Petroleum
Firms, Unequal Widths
of Class Intervals

ently, the *area* of each bar in a histogram should be proportional to the number of observations in the relevant class interval. Thus, since area equals width times height, the height of the bar in the 0.0–9.0 percent class interval should be 5 divided by 3 because this class interval is three times as wide as the others. The resulting histogram, corrected in this way, is shown in panel B of Figure 2.6. Put bluntly, panel A is wrong, panel B is right.

EXERCISES

**2.21** "There were more civilian than military amputees during the war. During the period of the war, 120,000 civilians suffered amputations, but only 18,000 military personnel."[10] Does this prove that civilians were more likely to suffer amputations? If not, where does the fallacy lie?

[10] W. A. Wallis and H. Roberts, *Statistics* (Glencoe: Free Press, 1956), p. 91.

**2.22** A national magazine once published a story saying that farmers lead other groups in the consumption of alcohol. As evidence, it pointed to the fact that a rehabilitation center in rural Illinois treated more farmers than other occupational groups for alcoholism. Do you regard this evidence as unbiased? Why, or why not?

**2.23** Data have been published which indicate that the more children a couple has, the less likely the couple is to get a divorce. Does this indicate that increases in the number of children are related causally to the likelihood of divorce? Why, or why not?

**2.24** Based on the data in Exercise 2.10, a real estate agent says that the typical size of a lot in the relevant township is .54 acres. What sort of average is the agent using? How many lots are of this "typical" size?

**2.25** In 1973, the total net income per farm was $19,685 in Iowa and $19,456 in Minnesota. Based on these data, a television commentator maintains that Iowa farmers were better off than Minnesota farmers in 1973. Do you consider this statement to be very meaningful? What sort of pitfall is present here? What would be a better way of interpreting the data concerning these two states?

**2.26** "Patents are of little value since the Supreme Court invalidates most of the patents that come before it."[11] Do you agree with this statement? If not, in what way does it represent a misuse of statistics?

**2.27** A newspaper reporter writes that the average snowfall in New York is less than 2 inches and consequently the expense of preparing for snowfalls of more than 10 inches is far in excess of the potential benefits. Do you agree? If not, in what way is this statement a misuse of statistics? What sort of pitfall is present here?

**2.28** The following is a distribution of the prices charged for a particular type of hi-fi speaker by 100 stores in Los Angeles:

| | |
|---|---|
| $80 and under $90 | 41 |
| $90 and under $100 | 9 |
| $100 and under $110 | 6 |
| $110 and under $120 | 44 |
| | 100 |

(a) A newspaper ad says that the median price for such a speaker is about $100. Is this correct?
(b) Is $100 a good measure of central tendency? Why or why not?

## Chapter Review

1. There are several frequently used *measures of central tendency:* the *mean,* the *median,* and the *mode.* The *mean* is the sum of the numbers contained in the body of data divided by how many numbers there are. The *median* is a figure which is chosen so that one-half of the numbers in the body of data are below it and one-half are above it. The *mode* is the number that occurs most often. These three kinds of measures may differ substantially from one an-

[11] Ibid., p. 98.

other. For example, if there are a few extremely high observations (as in most income distributions) the mean will be considerably higher than the median.

2. *Measures of variability or dispersion* tell us how much variation there is among the numbers in a body of data. Perhaps the simplest measure of variability is the *range,* which is defined as the difference between the highest and the lowest number in the body of data. However, the most important measure of dispersion is the *standard deviation,* which is defined as the square root of the mean of the squared deviations of the observations from their mean. The square of the standard deviation is called the *variance.*

3. There are many misuses of descriptive statistics. Figures are sometimes presented in such a way that they seem comparable when in fact they are based on different definitions, concepts, time periods, areas, and so forth. Sometimes figures are presented which are based on very small or obviously biased samples. Individuals and organizations sometimes choose the type of average that supports their case best, even if this information is misleading. An average is sometimes presented which has so much variability about it that the average alone is misleading. In addition, graphs and charts (such as pictograms) are sometimes presented so as to give a misleading impression. Be on your guard against improper statistical procedures of this sort.

## Chapter Review Exercises

**2.29** Do you think that the median amount paid by Americans in income tax in 1986 was less than the mean amount paid? Explain.

**2.30** (a) Given the mean profit rate for nondurable manufacturing industries and the mean profit rate for durable manufacturing industries, describe a simple procedure for obtaining the mean profit rate for all manufacturing industries (if the number of nondurable manufacturing industries is known to equal the number of durable manufacturing industries).

(b) Given the median profit rate for nondurable manufacturing industries and the median profit rate for durable manufacturing industries, can you determine the median profit rate for all manufacturing industries from this information alone?

**2.31** (a) For *any* set of measurements, what is the sum of the deviations of these measurements from their mean?

(b) A set of measurements has a symmetrical frequency distribution, and the median is 3. If there are 1,000 measurements, can you calculate their sum?

**2.32** A salesman made 100 visits to customers. The frequency distribution of the amount of commission he earned per visit is as follows:

| Amount of commission (dollars) | Number of visits |
| --- | --- |
| 0 and under 20 | 60 |
| 20 and under 40 | 30 |
| 40 and under 60 | 10 |

(a) Calculate the mean amount of his commission per visit. Estimate the total commission he earned for all 100 visits.

(b) Calculate the median amount of commission earned by the salesman per visit. Based on information solely concerning this median, can you tell whether the total commission earned for all 100 visits exceeded $2,000? Why, or why not?

**2.33** Based on the data in Exercise 2.9 in this chapter, what is the range of the profit rates among nondurable manufacturing industries in 1974? What is the range among durable manufacturing industries? What is the range among all manufacturing industries? Can the range for nondurable manufacturing industries be greater than that for all manufacturing industries?

**2.34** Based on the data in Exercise 2.32, can you determine the range of the amount of commission earned per visit by the salesman? Can you obtain upper and/or lower bounds for this range? Explain.

**2.35** Use the data in Exercise 2.9 to calculate the standard deviation of the industry profit rates (a) for nondurable manufacturing industries; (b) for durable manufacturing industries; and (c) for all manufacturing industries. (Assume that the entire population is given.)

**2.36** Based on the data in Exercise 2.32, is the distribution of commissions earned by the salesman skewed to the left? To the right?

**2.37** Calculate the standard deviation of the amount of commission per visit in Exercise 2.32. Then express the commissions in cents (not dollars) and calculate the standard deviation. What is the ratio of the latter standard deviation (for commissions expressed in cents) to the former standard deviation (for the commissions expressed in dollars)? Why?

**2.38** (a) Prove that the coefficient of variation of the commissions in Exercise 2.32 is the same whether the commissions are measured in cents or dollars.

(b) From the data in Exercise 2.9 calculate the coefficient of variation of profit rates for nondurable manufacturing industries.

(c) From the data in Exercise 2.9 calculate the coefficient of variation of the profit rates for durable manufacturing industries.

(d) Based on the coefficient of variation, are the profit rates more variable among durable goods industries or nondurable goods industries?

**2.39** In the fall of 1973, 8,442 men and 4,321 women applied for admission to the Graduate Division of the University of California, Berkeley. About 44 percent of the men and 35 percent of the women were admitted.[12]

(a) Was this evidence of a sex bias in admissions?

(b) Admissions were made separately for each major. Upon closer examination, it turned out that, major by major, the percentage of women accepted was about the same as the percentage of men accepted. Is this consistent with the above facts? Does there still appear to be a sex bias in admissions?

**2.40** According to Sheldon and Eleanor Glueck,[13] a large percentage of juvenile delinquents are middle children (not the first or last born).

[12] D. Freedman, R. Pisani, and R. Purves, *Statistics* (New York: Norton, 1978).
[13] S. and E. Glueck, *Unraveling Juvenile Delinquency* (Cambridge: Harvard University Press, 1950).

(a) Does this imply that being a middle child contributes to delinquency?
(b) Studies have shown that there is a strong direct relationship between family size and delinquency. Can this help to explain the Gluecks' results?

**\* 2.41** Economists and demographers have found that there are considerable differences among countries in life expectancy. The life expectancy (in years) in 30 countries is as follows:[14]

| | | | | | |
|---|---|---|---|---|---|
| United States | 73 | Greece | 72 | Spain | 73 |
| Argentina | 65 | India | 47 | Sri Lanka | 64 |
| Australia | 74 | Indonesia | 46 | Sweden | 75 |
| Bangladesh | 46 | Italy | 73 | Thailand | 61 |
| Belgium | 71 | Japan | 76 | Turkey | 57 |
| Brazil | 60 | Mexico | 60 | United Kingdom | 73 |
| Canada | 73 | Nigeria | 41 | Venezuela | 63 |
| Chile | 62 | Pakistan | 48 | West Germany | 72 |
| Egypt | 54 | Poland | 71 | Yugoslavia | 70 |
| France | 73 | Soviet Union | 70 | Zaire | 39 |

Use Minitab (or some other computer package) to
(a) construct a histogram of these data;
(b) calculate the mean of these data;
(c) calculate the standard deviation of these data.
If no computer package is available, do the calculations by hand.

\* Exercise requiring computer package, if available.

# Appendix 2.1

### RULES OF SUMMATION

In this chapter, we encountered $\Sigma$, the mathematical summation sign. Since this sign will be used frequently in later chapters, it is worthwhile to summarize some of the rules of summation. We know from this chapter's discussion that

$$\sum_{i=1}^{n} X_i = X_1 + X_2 + \cdots + X_n.$$

From this fact, we can establish the validity of the following three rules.
The first rule is:

$$\sum_{i=1}^{n} aX_i = a \sum_{i=1}^{n} X_i,$$

[14] *Statistical Abstract of the United States, 1982–83* (Washington, D.C.: Bureau of the Census, 1982), p. 862.

where $a$ is a constant. To prove that this rule is correct, note that

$$\sum_{i=1}^{n} aX_i = aX_1 + aX_2 + \cdots + aX_n$$

$$= a(X_1 + X_2 + \cdots + X_n)$$

$$= a \sum_{i=1}^{n} X_i,$$

which proves the rule.

The second rule is

$$\sum_{i=1}^{n} a = na.$$

To prove that this rule is correct, note that

$$\sum_{i=1}^{n} a = a \sum_{i=1}^{n} 1$$

$$= a\underbrace{(1 + 1 + \cdots + 1)}_{n \text{ terms}}$$

$$= na,$$

which proves the rule.

The third rule is

$$\sum_{i=1}^{n} (X_i + Y_i) = \sum_{i=1}^{n} X_i + \sum_{i=1}^{n} Y_i.$$

To prove that this rule is correct, note that

$$\sum_{i=1}^{n} (X_i + Y_i) = X_1 + Y_1 + X_2 + Y_2 \cdots + X_n + Y_n$$

$$= (X_1 + X_2 + \cdots + X_n) + (Y_1 + Y_2 + \cdots + Y_n)$$

$$= \sum_{i=1}^{n} X_i + \sum_{i=1}^{n} Y_i,$$

which proves the rule.

Finally, let's also consider the concept of double summation. The expression

$$\sum_{i=1}^{n} \sum_{j=1}^{m} X_i Y_j$$

means the sum of the products of $X$ and $Y$ where $X$ takes its first, second, . . . , $n$th values and $Y$ takes its first, second, . . . , $m$th values. For example,

$$\sum_{i=1}^{3} \sum_{j=1}^{2} X_i Y_j = X_1 Y_1 + X_1 Y_2 + X_2 Y_1 + X_2 Y_2 + X_3 Y_1 + X_3 Y_2.$$

The following example illustrates the use of these rules of summation.

EXAMPLE 2.1  If $X_1 = 4$, $X_2 = 6$, and $X_3 = -3$, evaluate the following sums:

(a) $\sum_{i=1}^{3} X_i$

(b) $\sum_{i=1}^{2} X_i^2$

(c) $\sum_{i=1}^{3} 3X_i$

SOLUTION: (a) $\sum_{i=1}^{3} X_i = 4 + 6 - 3 = 7$

(b) $\sum_{i=1}^{2} X_i^2 = 4^2 + 6^2 = 52$

(c) $\sum_{i=1}^{3} 3X_i = 3 \sum_{i=1}^{3} X_i = 3(7) = 21$

# Appendix 2.2

### SHORTCUTS IN CALCULATING THE VARIANCE AND STANDARD DEVIATION

Modern electronic computers are often used to compute the variances and standard deviations of populations and samples. The advent of computers has transferred a great deal of drudgery from human beings to these mechanical aids. However, since some computations of this sort are still done by hand, it is useful to note that modifications of equations (2.8) and (2.9) often simplify the calculations. Specifically, these modifications are

$$s^2 = \frac{\sum_{i=1}^{n} X_i^2 - \frac{1}{n}\left(\sum_{i=1}^{n} X_i\right)^2}{n-1} \qquad (2.11)$$

and

$$s = \sqrt{\frac{\sum_{i=1}^{n} X_i^2 - \frac{1}{n}\left(\sum_{i=1}^{n} X_i\right)^2}{n-1}}. \qquad (2.12)$$

Similarly, if you are calculating the sample standard deviation based on a frequency distribution, it is often quicker and easier to use the following modification of equation (2.10) to carry out the calculations:

$$s = \sqrt{\frac{\sum\limits_{j=1}^{k} f_j X_j'^2 - \frac{1}{n}\left(\sum\limits_{j=1}^{k} f_j X_j'\right)^2}{n-1}} \qquad (2.13)$$

These modifications do not alter the answers given by the formulas in equations (2.8), (2.9), and (2.10). They are merely different ways of obtaining the same result. To verify this, let's apply the formula in equation (2.13) to the frequency distribution of profit rates of chemical firms in Table 2.5. Since

$$\sum\limits_{j=1}^{k} f_j X_j'^2 = 2(4.5^2) + 3(7.5)^2 + 3(10.5)^2 + 4(13.5)^2 + 3(16.5)^2 =$$

2085.75,

and

$$\sum\limits_{j=1}^{k} f_j X_j' = 2(4.5) + 3(7.5) + 3(10.5) + 4(13.5) + 3(16.5) = 166.50.$$

$$s = \sqrt{\frac{2085.75 - \frac{1}{15}(166.5)^2}{14}} = \sqrt{\frac{2085.75 - 1.848.15}{14}} = 4.1.$$

Thus, the answer is the same as for equation (2.10). Except for the effects of rounding errors, these answers must always be identical.

# Appendix 2.3

### INTERPRETING COMPUTER OUTPUT REGARDING DESCRIPTIVE STATISTICS

Computers enable you to obtain a wealth of descriptive statistics in practically no time at all. Consider, for example, the state crime rates (number of offenses known to the police per 1,000 population), which, as explained on pages 21–22, we put in column C1 of the worksheet that Minitab maintains in the computer. To obtain a variety of statistics describing these state crime rates, we simply type the command, "Describe C1," and the computer prints out the following:

```
MTB > describe c1
                N     MEAN  MEDIAN  TRMEAN  STDEV  SEMEAN
C1             50    5.392   5.300   5.382  1.358   0.192

              MIN     MAX      Q1      Q3
C1          2.600   8.600   4.500   6.525
```

This printout says that the number of observations (N) is 50, their mean is 5.392, their median is 5.3, and their trimmed mean (the mean when the smallest 5 percent and largest 5 percent of the values are omitted) is 5.382. Their standard deviation is 1.358, the standard error of the mean (defined and discussed in Chapter 6) is 0.192, the minimum state crime rate is 2.6, and the maximum is 8.6. The first quartile (Q1) is 4.5, and the third quartile (Q3) is 6.525. Clearly, this printout tells us a great deal about the state crime rates in the United States, and the computer provides this information in a matter of seconds.

# Probability, Probability Distributions, and Sampling

# Probability

3

## 3.1 Introduction

Much of statistical theory and practice rests on the concept of probability since, as stressed in Chapter 1, any conclusion concerning a population based only on a sample is subject to a certain amount of uncertainty. The concept of probability is by no means unfamiliar. For example, based on the past examination questions given by a professor, you make rough judgments as to the probability that he or she will include certain kinds of questions in a forthcoming test. Or based on the past performance of certain football teams, you make rough estimates of the probability that one team will defeat another. This chapter will provide an introduction to probability theory; in subsequent chapters, we shall build on this foundation and go deeper into this subject.

## 3.2 Experiments, Sample Spaces, and Events

EXPERIMENTS AND SAMPLE SPACES

Any probability pertains to the results of a situation which we call an experiment. *An experiment is any process by which data are obtained through the observation of uncontrolled events in nature or through controlled procedures in a laboratory.* If you roll a die, this is an experiment which may have one of six outcomes, depending on which number comes up. Or if you flip a coin, this is also an experiment, which has as possible outcomes either a head or a tail. Or if you take a statistics course, this too is an experiment (although you may not prefer to think of it that way), which has as possible outcomes your passing or your not passing.

Any experiment can result in various **outcomes.** For example, one possible outcome of your experiment with the die is that a 4 may come up. And one possible (and hopefully very likely) outcome of your ex-

**Sample Space**

periment with the statistics course is that you will pass it. The **sample space** *is the set of **all** possible outcomes that may occur as a result of a particular experiment. For example, the sample space for the experiment in which you rolled the die is*

$$S = \{1, 2, 3, 4, 5, 6\}. \qquad (3.1)$$

In other words, $S$ is a set composed of the numbers that can come up when a die is thrown; that is, it is a set composed of 1, 2, . . . , 6. The symbol $S$ is conventionally used to designate the sample space, and the outcomes in this set are called *elements* of the sample space.

It is important to recognize that the outcomes in a sample space need not be numbers. In the case of your experiment with the statistics course, there are two possible outcomes: You pass, or you don't pass. In this case, the sample space is

$S = \{$pass, not pass$\}$.

Sometimes it is convenient in cases of this sort to designate the elements of the sample space as 0 and 1, where 0 stands for "pass" and 1 stands for "not pass." Thus, the sample space becomes

$S = \{0, 1\}$.

It frequently is useful to represent the sample space visually. One way to do this is to express the possible outcomes of the experiment as points on a graph. For example, if two dice are rolled simultaneously, then a graph showing the sample space can be drawn where the number coming up on the first die is located along the horizontal axis, and the number coming up on the second die is located along the vertical axis. This is shown in Figure 3.1. Each of the 36 points in the graph represents a possible outcome of the roll of the dice.

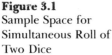

**Figure 3.1**
Sample Space for
Simultaneous Roll of
Two Dice

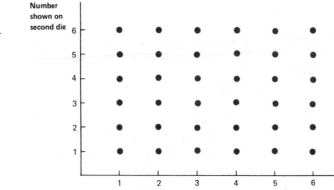

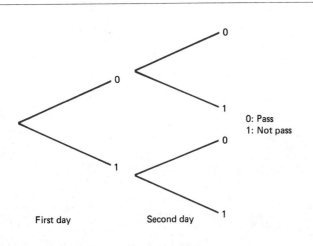

**Figure 3.2**
Sample Space for Two
Days of Examinations

Another way to represent the sample space, particularly when an experiment is carried out in stages, is to use a *tree diagram*. Each fork in such a diagram shows the possible outcomes that may occur at a certain stage of the experiment. For example, if you take examinations two days in a row, on the first day there are two possibilities: You pass, or you don't pass. These possibilities are represented by the first fork in Figure 3.2. On the second day there are again two possibilities: Either you pass or you don't pass. These possibilities are represented by the two forks to the right of the first fork. The upper fork represents the outcomes if you passed on the first day, whereas the bottom fork represents the outcomes if you did not pass on the first day. At the right-hand end of the tree diagram we wind up with four points representing the four elements in the sample space, which are (1) passing on neither the first day nor the second; (2) passing on the first day but not on the second; (3) passing on the second day but not on the first; and (4) passing on both days.

*Tree Diagram*

The following two examples should help illustrate the important concept of sample space.

EXAMPLE 3.1 A market-research firm is interested in the buying habits of families in Topeka, Kansas. The Martin family is among those the firm is studying. The firm is interested in which laundry presoak and detergent boosters the Martins will buy during August of this year. It wants to know whether the Martins buy Axion (a Colgate-Palmolive product), or Biz (a Procter and Gamble product), or both, or neither. Depict graphically the sample space generated by this experiment.

SOLUTION: We can depict this sample space as four points on a graph, shown in the following figure. In this graph, a 0 on the horizontal axis means that the Martins do not buy Axion in August, while a 1 means that they do buy it. On the vertical axis, a 0 means that the Martins do not buy Biz in August, while a 1 means that they do buy it.

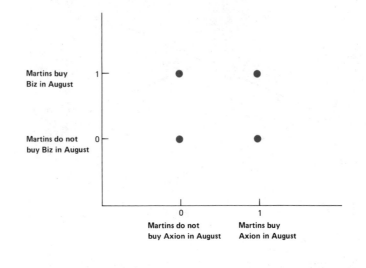

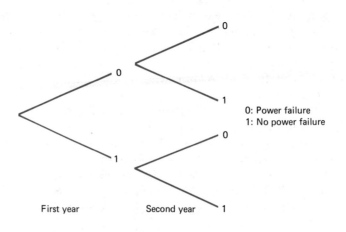

EXAMPLE 3.2 Each year, Americans (and particularly New Yorkers) wonder whether there will be a massive electric power failure in New York City (such as occurred in July 1977). Next year, there may or may not be a failure; and in the year after next, there may or may not be one. (Only God and perhaps Consolidated Edison, the local power company, know for sure.) Use a tree diagram to depict the sample space for the next two years.

SOLUTION: We can depict this sample space by constructing the graph below, which shows a fork for the first year leading to a 0 (if a failure occurs) or a 1 (if a failure does not occur). Whether or not there is a failure in the first year, another such fork is given for the second year. Thus, at the right-hand end of the graph, we arrive at the four points in the sample space.

EVENTS

One of the most important functions of probability theory is to enable us to calculate the probability of an event. *An event is a subset of a sample space.* A *subset* is any part of a set (including the whole set and the empty set, which has no elements at all). Put less formally, an event can be defined as a group of zero, one, two, or more outcomes of an experiment. For example, if the experiment is rolling a particular die, one event is that the number that comes up is odd. Another event is that the number that comes up is a 3 or a 6. Still another event is that the number that comes up is a 5. There are many practical reasons (other than preparing for trips to Las Vegas or Atlantic City) why statisticians need to be able to calculate the probability of an event.

## 3.3 Probabilities

WHAT IS A PROBABILITY?

*The probability of an event is the proportion of times that this event occurs over the long run if the experiment is repeated many times under uniform conditions.* Thus, the probability that a particular die will come up a 1 is the proportion of times this will occur if the die is thrown many, many times; and the probability that the same die will come up a 2 is the proportion of times this will occur if the die is thrown many, many times. And so on.

*Frequency Definition of Probability*

In general, *if an experiment is repeated a very large number of times M, and if event A occurs m times, the probability of A is*

$$P(A) = \frac{m}{M} \tag{3.2}$$

Thus, if a die is "true" (meaning that each of its sides is equally likely to come up when the die is rolled), the probability of its coming up a 1 is 1/6, because if it is rolled many, many times, this will occur one-sixth of the time. Moreover, even if the die is not true, this definition can be applied. Suppose that a local mobster injects some loaded dice into a crap game, and that one of the players (who is suspicious) asks to examine one of them. If he rolls this die, what is the probability that it will come up a 1? To answer the question, we must imagine the die in question being rolled again and again. After many thousands of rolls, if the proportion of times that it has come up a 1 is 0.195, then this is the desired probability.

Based on our definition of a probability, the following three fundamental propositions must be true:

1. *The Probability of an Impossible Event Must be Zero.* This follows from the definition of a probability in equation (3.2), because if an event is impossible, the number of times the event occurs (that is, *m*) must equal zero.

2. *The Probability of an Event that Is Certain Must Equal 1.* This also follows from the definition of a probability in equation (3.2), because if an event is certain, the number of times the event occurs (*m*) must equal the number of times the experiment takes place (that is, *M*). Note that this implies that the probability that some element of the sample space occurs is 1, since the sample space includes all possible outcomes.

3. *The Probability of Any Event Must Be No Less than Zero and No Greater than 1.* This, too, follows from the definition of a probability in equation (3.2). Since the number of times any event occurs (*m*) cannot be negative, its probability cannot be less than zero. Since the number of times any event occurs cannot exceed the number of times the experiment takes place (*M*), its probability cannot exceed 1.

### CALCULATING THE PROBABILITY OF AN EVENT

*The probability that an event will take place is the sum of the probabilities of the outcomes that comprise the event.* For example, if a true die is rolled, what is the probability that it comes up with an odd number? The event in which the die comes up with an odd number can be broken down into three outcomes: (1) the die comes up a 1; (2) the die comes up a 3; and (3) the die comes up a 5. In other words, these three outcomes comprise this event. Since the probability of each of these outcomes is 1/6, the probability of this event is $1/6 + 1/6 + 1/6 = 1/2$. This is true because (1) if any of the outcomes comprising the event occurs, the event itself occurs, and (2) more than one outcome cannot occur simultaneously.

A somewhat more complex illustration involves the buying habits of the Martin family with respect to liquid detergents and scouring powders. If the Martin family can buy up to 3 containers of liquid detergent and up to 3 cans of scouring powder in the next 3 months, the sample space showing the amount of each product purchased during this period is shown in Figure 3.3. For simplicity, let's assume that each

**Figure 3.3**
Sample Space for
Purchases by Martin
Family

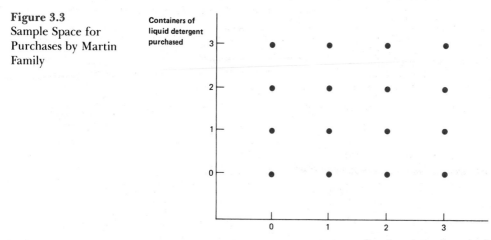

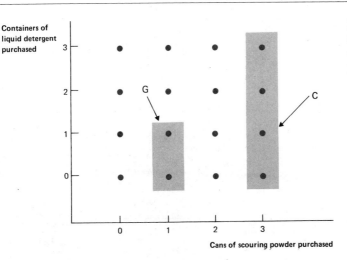

**Figure 3.4**
Subset of Sample Space
Corresponding to Two
Events

of the 16 elements of this sample space is equally likely. This means that the probability of each point in Figure 3.3 is 1/16. Given these probabilities of the outcomes, we can readily calculate the probabilities of various other events.

To illustrate, what is the probability that the Martin family buys 3 cans of scouring powder and up to 3 containers of liquid detergent in the next 3 months? This event includes the 4 elements in the sample space designated as subset *C* in Figure 3.4. Thus, the probability of this event is the sum of the probabilities of these 4 outcomes, or 1/16 + 1/16 + 1/16 + 1/16 = 1/4. What is the probability that the Martin family buys both 1 can of scouring powder and less than 2 containers of liquid detergent? This event includes the 2 elements in the sample space designated as subset *G* in Figure 3.4. Thus, the probability of this event is the sum of the probabilities of these 2 outcomes, or 1/16 + 1/16 = 1/8.

## PROBABILITY OF EITHER EVENT A OR EVENT B OR BOTH

A ***composite event*** is an event that is defined by using combinations of other events. One type of composite event occurs if *either* or *both* of two other events occur. For example, if *A* is the event that you pass your economics course and *B* is the event that you pass your statistics course, your graduation may depend on *either* or *both* of these events occurring. If so, the composite event that you graduate will occur only if *either* event *A* or event *B (or both)* occur. Turning to a nonacademic illustration, consider the composite event that the Martin family buys during the next 3 months *either* (1) less than 2 cans of scouring powder; or (2) less than 2 containers of liquid detergent; or (3) *both* (that is, less than 2 cans of scouring powder *and* less than 2 containers of liquid detergent). Subset *D* of the sample space in Figure 3.5 is the set of outcomes where fewer than 2 cans of scouring powder are bought (event 1). Subset *E* is the set of outcomes where less than 2 containers of liquid detergent are bought (event 2). *The set of outcomes where either or both of these events occurs is composed of all outcomes in ei-*

*Composite Events*

**Figure 3.5**
Subset of Sample Space
Corresponding to
Either of Two Events

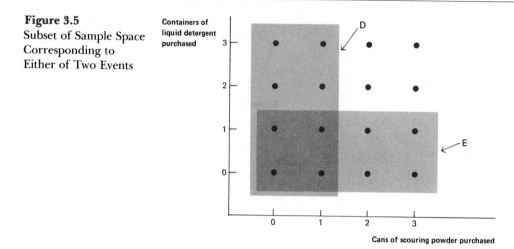

*ther subset* D *or* E. This set consists of the 12 points in the shaded area in Figure 3.5. Since the probability of each of these points is 1/16, the probability of this composite event is 12/16.

PROBABILITY OF BOTH EVENT A AND EVENT B

A second type of composite event occurs *only* if *both* of two other events occur. Thus, if *A* is the event that you pass your economics course and *B* is the event that you pass your statistics course, your receiving a good job offer may be dependent on the occurrence of *both*. If so, the event that you receive a good job offer will occur only if *both* events *A* and *B* occur. Consider the event that the Martin family buys *both* less than 2 cans of scouring powder *and* less than 2 containers of liquid detergent during the next three months. This event occurs only if *both* of the following events occur: (1) the Martins buy less than 2 cans of scouring powder, and (2) the Martins buy less than 2 containers of liquid detergent. Since the set of outcomes comprising the first event is subset *D* of the sample space in Figure 3.6 and the set of outcomes

**Figure 3.6**
Subset of Sample Space
Corresponding to Both
of Two Events

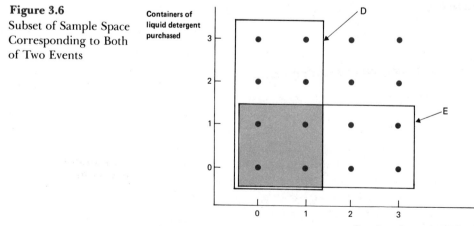

comprising the second event is subset *E* of the sample space in Figure 3.6, *the set of outcomes comprising both of these events is composed of all outcomes in both subsets* D *and* E. In other words, it is the set of 4 points in the shaded area in Figure 3.6. Since the probability of each of these points is 1/16, the probability of this composite event is 4/16.

### EXERCISES

**3.1** List all of the outcomes if a coin is flipped three times. What probability should be attached to each point in the sample space corresponding to this experiment? (Assume it is a fair coin.)

**3.2** A box contains two white balls and one black ball. Two balls are to be taken from this box. After the first ball is selected, it will not be put back in the box. List all possible outcomes of this experiment. What probability would you attach to each outcome?

**3.3** A true die has its 4-spot changed to a 2-spot. When tossed, what is the probability of obtaining (a) 3, (b) 2, (c) 4, (d) 2 or less, (e) 5 or more?

**3.4** An economist says that the odds are 2 to 1 that a recession will not occur next year. He also says that the probability that neither a recession nor a boom will occur next year is 0.70. Are his two statements consistent? Explain.

**3.5** A gambler makes a single roll with a pair of true dice. What is the probability that each of the following numbers comes up?
(a) either a 7 or an 11
(b) either a 2, a 12, or both

## 3.4 Addition Rule

### CASE OF TWO EVENTS

As pointed out in the previous section, statisticians and decision makers frequently must calculate the probability that at least one of two events occurs. For instance, if a fair coin is flipped twice and we want to calculate the probability that it comes up heads at least once, we know that this will occur if (1) the coin comes up heads on the first flip; or (2) the coin comes up heads on the second flip. The probability of the first event is 1/2, and the probability of the second event is 1/2. The probability of both events (that is, that the coin will come up heads *both* times) is 1/4. Can we use these probabilities to determine the likelihood that the coin will come up heads at least once in the two flips?

To solve problems of this sort, statisticians have devised the so-called addition rule, which is as follows.

ADDITION RULE: *If* A *and* B *are two events, and the probability of* A *is denoted by* P(A) *and the probability of* B *is denoted by* P(B), *then the probability of either* A *or* B *(or both), denoted by* P(A *or* B), *equals* P(A) + P(B) − P(A *and* B), *where* P(A *and* B) *is the probability that both* A *and* B *will occur.*

*Formula for P(A or B)*

Let's define $A$ as the event that the coin comes up heads on the first flip, and $B$ as the event that the coin comes up heads on the second flip. Then $P(A)$ is the probability of heads on the first flip, which is $1/2$. And $P(B)$ is the probability of heads on the second flip, which is $1/2$. And $P(A \text{ and } B)$ is the probability of heads on both the first and second flips, which is $1/4$. Thus, the probability of at least one heads is

$$P(A \text{ or } B) = P(A) + P(B) - P(A \text{ and } B) = 1/2 + 1/2 - 1/4 = 3/4.$$

This result accords with Figure 3.7, which shows the relevant sample space. There are four possible outcomes in this sample space. The outcomes correspond to

(1) heads/heads (heads first flip, heads second flip)
(2) heads/tails (heads first flip, tails second flip)
(3) tails/heads (tails first flip, heads second flip)
(4) tails/tails (tails first flip, tails second flip)

**Figure 3.7**
Sample Space for Two
Flips of a Coin

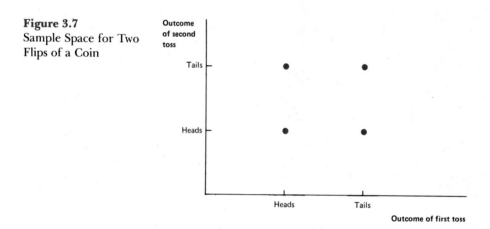

Each outcome has a probability of $1/4$. Since at least one heads comes up in three of these four outcomes, the probability of at least one heads is $3/4$, which accords with the result obtained from the addition rule.

This example indicates why $P(A \text{ and } B)$ must be subtracted from the sum of $P(A)$ and $P(B)$ to obtain $P(A \text{ or } B)$. If $P(A \text{ and } B)$ were not subtracted, the outcome where both events $A$ and $B$ occur (that is heads/heads) would be counted twice, since it is included in the outcomes where $A$ (heads on the first flip) occurs, as well as in the outcomes where $B$ (heads on the second flip) occurs. If it were not subtracted, the answer would be incorrectly given as 1, not $3/4$.

MUTUALLY EXCLUSIVE EVENTS

In some cases, two events are ***mutually exclusive;*** that is, they cannot occur together. For example, if you roll a die once, it is impossible for the die to come up with *both* a 4 *and* a 5, since only one number can appear at a given time. Thus, the event that "the die shows a 4" and the event that "the die shows a 5" are mutually exclusive. *If two events are mutually exclusive, the probability that they both occur is zero; consequently, if the events, A and B, are mutually exclusive P(A and B) = 0*, with the result that we have the following addition rule for mutually exclusive events.

> ADDITION RULE FOR MUTUALLY EXCLUSIVE EVENTS: *If A and B are two mutually exclusive events, the probability of either A or B, denoted by* P(A *or* B), *equals* P(A) + P(B).

Clearly, this addition rule follows directly from the addition rule given earlier in this section since, if $P(A \text{ and } B) = 0$, the earlier addition rule implies that this rule will be true. To illustrate the present addition rule, let's define $A$ as "a die shows a 4" and $B$ as "a die shows a 5." As we have seen, these two events are mutually exclusive (if only one die is thrown only once), so $P(A \text{ or } B) = P(A) + P(B) = 1/6 + 1/6 = 1/3$. Thus, the probability of either a 4 or a 5 is $1/3$.

---

*If the A and B are two events, the probability that at least one of them occurs is **not** in general equal to the sum of their probabilities. Only if they are mutually exclusive is this true.*

*Basic Idea #5*

---

## CASE OF MORE THAN TWO EVENTS

Both the general addition rule and the special addition rule for mutually exclusive events can be extended to cases where more than two events are considered. For present purposes it is not necessary to give the extension of the more general addition rule, but it is important to give the extension of the addition rule for mutually exclusive events, which is as follows.

> ADDITION RULE FOR ANY NUMBER OF MUTUALLY EXCLUSIVE EVENTS: *If* $E_1$, $E_2$ ··· $E_n$ *are* n *events, the probability of* $E_1$ *or* $E_2$ *or* $E_3$ ··· *or* $E_n$, *denoted by* P($E_1$ *or* $E_2$ *or* $E_3$ ··· *or* $E_n$), *equals* P($E_1$) + P($E_2$) + P($E_3$) + ··· + P($E_n$), *if these events are mutually exclusive.*

To illustrate the use of this rule, let's define $E_1$ as "a die shows a 1," $E_2$ as "a die shows a 2," $E_3$ as a "die shows a 3," and $E_4$ as "a die shows a 4." Then since these four events are mutually exclusive (so long as only one die is thrown once), $P(E_1 \text{ or } E_2 \text{ or } E_3 \text{ or } E_4) = P(E_1) + P(E_2) + P(E_3) + P(E_4) = 1/6 + 1/6 + 1/6 + 1/6 = 2/3$. Thus, the probability of a 1, a 2, a 3, or a 4 is $2/3$.

## COMPLEMENT OF AN EVENT

The ***complement*** of an event occurs when the event itself does not occur. Thus, since the event or its complement is sure to occur, and since the event and its complement are mutually exclusive, the addition rule implies that if $P(A)$ is the probability that event $A$ will occur, and if $P$ (not $A$) is the probability that event $A$ will not occur, then

$$P(A) + P \text{ (not } A) = 1,$$

or

$$P(A) = 1 - P \text{ (not } A).$$

This result is useful because, if it is easier to determine $P$ (not $A$) than $P(A)$, then $P(A)$ can be obtained by deducting $P$ (not $A$) from 1. Suppose that we want to determine the probability that if two dice are thrown, the number that comes up *will not equal* 2. The complement of this event is that the number *will equal* 2. Since the probability of this complementary event is 1/36, the probability that the number will not equal 2 must be $1 - 1/36$, or 35/36.

APPLICATIONS

The addition rule has a host of important applications. The following two examples illustrate ways in which it is used.

---

EXAMPLE 3.3 A new chemical product is about to be introduced commercially, and a great many chemical firms are competing to be first to put this product on the market. An industrial analyst believes that the probability is 0.30 that DuPont will be first, 0.15 that Dow will be first, 0.15 that Monsanto will be first, and 0.15 that Union Carbide will be first. Based on the analyst's beliefs, what is the probability that any one of these four firms will be first, assuming that a tie does not occur?

SOLUTION: Let $E_1$ be the event that DuPont is first, $E_2$ be the event that Dow is first, $E_3$ be the event that Monsanto is first, and $E_4$ be the event that Union Carbide is first. If a tie is not allowed, these events are mutually exclusive, since more than one firm cannot be first. Thus, $P(E_1 \text{ or } E_2 \text{ or } E_3 \text{ or } E_4) = P(E_1) + P(E_2) + P(E_3) + P(E_4) = 0.30 + 0.15 + 0.15 + 0.15 = 0.75$. Consequently, the probability that any one of these four firms will be first is 0.75.

---

EXAMPLE 3.4 A bank gives summer jobs to two business school students, Mary Carp and John Minelli. The bank's personnel manager hopes that at least one of these students will decide to go to work for the bank upon graduation. If the probability that each student will decide to work for the bank is 0.3, and the probability that both will decide to work for the bank is 0.1, what is the probability that the personnel manager's hopes will be fulfilled?

SOLUTION: Let $E_1$ be the event that Mary Carp will decide to go to work for the bank, and $E_2$ be the event that John Minelli will decide to do so. Since $P(E_1) = 0.3$, $P(E_2) = 0.3$, and $P(E_1$ and $E_2) = 0.1$,

$$P(E_1 \text{ or } E_2) = 0.3 + 0.3 - 0.1 = 0.5.$$

Thus, the probability that the personnel manager's hopes will be fulfilled equals 0.5.

### EXERCISES

**3.6** Forty percent of the members of Congress favor bill A, and 20 percent favor bill B. Ten percent favor both bills A and B. What is the probability that a member of Congress favors either bill A or bill B (or both)?

**3.7** An industrial psychologist presents a subject with a list of nonsense syllables, the list is removed, and the subject is asked to recall the list. The probability is .12 that the subject will recall the first syllable on the list; the probability is .15 that the subject will recall the second syllable on the list; and the probability is .02 that the subject will recall them both. What is the probability that the subject will recall either syllable (or both)?

**3.8** The Monroe Corporation is located in a high-crime area. A criminologist says that in this area, the probability that an 18-year-old high school dropout will be arrested for theft is 0.05, and that the probability that he or she will be arrested for either theft or homicide (or both) is 0.06.
(a) Is the probability that he or she will be arrested for homicide equal to 0.01? Why or why not?
(b) Is the probability that he or she will be arrested for homicide less than 0.01? Why or why not?

**3.9** Coin A is loaded in such a way that heads are three times as likely to occur as tails. Coin B is loaded in such a way that tails are three times as likely to occur as heads. If both coins are tossed once, what is the probability that *at least* one comes up heads? (Note: The probability that *both* coins come up heads equals .1875.)

## 3.5 Multiplication Rule

The addition rule is very useful in cases where we are interested in the probability that *at least one* of several events will take place. Thus, in the previous section, we used it to determine the probability that *at least one* heads would come up in two flips of a true coin. In many situations, however, the statistician or decision maker is interested in the probability that *all* of several events will occur. We may want to determine the probability that *both* flips of a coin will be heads. Or we may want to determine the probability that *all four* of a firm's salesmen will be sick tomorrow. In cases of this sort, the multiplication rule, not the addition rule, is the one to apply. To understand the multiplication rule, it is essential that you be familiar with the concepts of joint probability, marginal probability, and conditional probability.

JOINT PROBABILITIES

In our discussion of the addition rule, we used $P(A$ and $B)$, the probability that *both* events $A$ and $B$ will occur. This is an example of a joint probability, which is defined as the *probability of the joint occurrence of two or more events*. As a concrete illustration of joint probabilities, consider Table 3.1, which shows the results of a hypothetical survey in which 10,000 individuals were asked whether or not they favored increased defense spending. As indicated in Table 3.1, 6,000 of those asked were Democrats and 4,000 were Republicans. Let $A$ be the event that a person favored increased defense spending, let $C$ be the event that he or she did not favor increased defense spending, let $B$ be the event that the person was a Democrat, and let $D$ be the event that he or she was a Republican.

Table 3.1
10,000 Persons
Classified by Attitude
toward Increased
Defense Spending and
by Political Party

| Political party | Favors increased defense spending (A) | Does not favor increased defense spending (C) | Total |
|---|---|---|---|
| Democrat (B) | 2,500 | 3,500 | 6,000 |
| Republican (D) | 2,500 | 1,500 | 4,000 |
| Total | 5,000 | 5,000 | 10,000 |

Joint probabilities can be illustrated as follows. If a person is chosen at random from this group of 10,000, the joint probability that he or she is a Democrat and favors increased defense spending is

$$P(A \text{ and } B) = \frac{2,500}{10,000} = 0.25.$$

Similarly, the probability that a randomly selected person is a Democrat and does not favor increased defense spending is

$$P(B \text{ and } C) = \frac{3,500}{10,000} = 0.35.$$

The probability that a randomly selected person is a Republican and favors increased defense spending is

$$P(A \text{ and } D) = \frac{2,500}{10,000} = 0.25.$$

The probability that a randomly selected person is a Republican and does not favor increased defense spending is

$$P(C \text{ and } D) = \frac{1,500}{10,000} = 0.15.$$

These joint probabilities can be shown in a *joint probability table* like Table 3.2. The probabilities in Table 3.2 are obtained by dividing each of the numbers in Table 3.1 by the total number of persons in the survey (10,000).

| Political party | Favors increased defense spending (A) | Does not favor increased defense spending (C) | Marginal probabilities |
|---|---|---|---|
| Democrat (B) | 0.25 | 0.35 | 0.60 |
| Republican (D) | 0.25 | 0.15 | 0.40 |
| Marginal probabilities | 0.50 | 0.50 | 1.00 |

**Table 3.2**
Joint Probability Table for 10,000 Persons Classified by Attitude toward Increased Defense Spending and by Political Party

MARGINAL PROBABILITIES

In addition to showing the joint probabilities just mentioned, Table 3.2 shows the probability that a randomly chosen person is a Democrat, is a Republican, is in favor of increased defense spending, or is not in favor of increased defense spending. These probabilities, contained in the margins of the joint probability table, are called *marginal probabilities* or *unconditional probabilities*. (The name *marginal* stems from the position of these probabilities in the margins of the table.) For example, the marginal probability that a randomly chosen person favors increased defense spending is 0.50. The marginal probability that a randomly chosen person in this group is a Democrat is 0.60.

*Unconditional Probability*

Each marginal probability can be derived by summing the appropriate joint probabilities. For example, the marginal probability that a person favors increased defense spending is the sum of (1) the joint probability that a person is a Democrat and favors increased defense spending; and (2) the joint probability that a person is a Republican and favors increased defense spending. That this is true follows from the addition rule.[1] Consequently, if we did not know the marginal probability that a person favors increased defense spending, but knew the joint probabilities, we could find the marginal probability by adding .25 (the joint probability that a person is a Democrat and favors increased defense spending) and .25 (the joint probability that a person is

[1] A person favors increased military spending if at least one of the following events occurs: (1) the person is a Democrat and in favor of increased military spending; (2) the person is a Republican and in favor of increased military spending. Since a person cannot be both a Democrat and a Republican, these two events are mutually exclusive. Thus, according to the addition rule, the probability that a person favors increased military spending equals the sum of the probability of the first of these events and the probability of the second of these events.

a Republican and favors increased defense spending). The result of adding the two is of course .50.

## CONDITIONAL PROBABILITIES

Statisticians frequently are interested in how the probability of one event is influenced by whether or not another event occurs. *The probability that one event will occur, given that another event is certain to occur, is a **conditional probability**.* In Table 3.2 we may be interested in calculating the probability that a person favors increased defense spending, given that he or she is a Democrat. This conditional probability is denoted by $P(A|B)$ and is read "the probability of $A$, given $B$." (Recall that event $A$ is a person's favoring increased defense spending and event $B$ is his or her being a Democrat.) The vertical line in $P(A|B)$ is read "given," and the event following the line ($B$ in this case) is the one that is certain to occur.

Regardless of which events $A$ and $B$ stand for, *the conditional probability of* A, *given that* B *must occur, is*

**Formula for $P(A|B)$**

$$P(A|B) = \frac{P(A \text{ and } B)}{P(B)}.$$ (3.3)

In other words, the conditional probability of $A$, given the occurrence of $B$, is the joint probability of $A$ and $B$ divided by the marginal probability of $B$. (To rule out the possibility of dividing by zero it is assumed, of course, that the marginal probability of $B$ is non-zero.)

To illustrate the use of the above definition, let's return to Table 3.2 and calculate the probability that a person favors increased defense spending, given that he or she is a Democrat. Since $P(A \text{ and } B) = .25$ and $P(B) = .6$,

$$P(A|B) = \frac{P(A \text{ and } B)}{P(B)} = \frac{.25}{.60} = .42.$$

This conditional probability is .42. To verify the correctness of this result we can go back to Table 3.1, which shows that the proportion of Democrats favoring increased defense spending is 2,500/6,000, or .42. Thus, the conditional probability of $A$ given $B$ is simply the proportion of times that $A$ occurs out of the total number of times that $B$ occurs.

## STATEMENT OF RULE

Now that we have the definition of conditional probability in equation (3.3) at hand, the statement of the multiplication rule is quite simple. To obtain this rule we need only multiply both sides of equation (3.3) by $P(B)$, the result being as follows.

MULTIPLICATION RULE: *If* A *and* B *are two events, the joint probability that both* A *and* B *will occur equals the conditional probability of* A, *given* B, *times the probability of* B. *In other words,*

$$P(A \text{ and } B) = P(A \mid B)P(B).$$

Using the multiplication rule, we can determine the probability that a randomly selected person is both a Democrat and not in favor of increased defense spending. If we let $C$ be the event that the person is not in favor of increased defense spending, and if we let $B$ be the event that the person is a Democrat, then what we want to determine is $P(C$ and $B)$. We know that the probability that a person is a Democrat equals .6. In other words, $P(B) = .6$. We also know that the probability that a person is not in favor of increased defense spending, given that he or she is a Democrat, equals .58. That is, $P(C \mid B) = .58$. From these facts alone we can determine $P(C$ and $B)$, because the multiplication rule implies that

$$P(C \text{ and } B) = P(C \mid B)P(B)$$
$$= (.58)(.6) = .35.$$

Thus, the desired probability equals .35.

Going a step further, the multiplication rule can be extended to include more than two events, the result being as follows.

MULTIPLICATION RULE FOR n EVENTS: *If* $E_1, E_2, \ldots, E_n$ *are* n *events, the joint probability that all these events will occur equals the probability of* $E_1$ *times the conditional probability of* $E_2$, *given* $E_1$, *times the conditional probability of* $E_3$, *given* $E_1$ *and* $E_2, \ldots$ *times the conditional probability of* $E_n$, *given* $E_1, E_2, \ldots, E_{n-1}$. *In other words,*

$$P(E_1 \text{ and } E_2 \text{ and } \ldots \text{ and } E_n) = P(E_1) \cdot P(E_2 \mid E_1) \cdot P(E_3 \mid E_1 \text{ and } E_2) \cdot \ldots \cdot P(E_n \mid E_1 \text{ and } E_2 \text{ and } \ldots \text{ and } E_{n-1}).$$

*(3.5)*

Consider the following example: (1) the probability that an undergraduate at the University of Michigan will go to graduate school is $1/2$; (2) the probability that a Michigan undergraduate will get a master's degree, given that he or she goes to graduate school, is $2/3$; and (3) the probability that a Michigan undergraduate will get a Ph.D., given that he or she goes to graduate school and gets a master's degree, is $1/5$. What is the probability that a Michigan undergraduate will go to graduate school, will get a master's degree, and will get a Ph.D.? Using equation (3.5), this probability equals $(1/2)(2/3)(1/5)$, or $1/15$.

STATISTICAL INDEPENDENCE

The probability of the occurrence of an event is sometimes dependent on whether or not another event occurs. In Table 3.2, the probability that a person favors increased defense spending depends on whether the person is a Democrat or a Republican.[2] As we saw in a previous section, the probability that a person favors increased defense spending, given that he or she is a Democrat equals .42. On the other hand, the probability that an individual favors increased defense spending, given that he or she is a Republican, equals .25/.40, or .62. Thus, these two events—a person's attitude toward increased defense spending and his or her political party—are *dependent* in the sense that the probability of one's occurring depends on whether or not the other occurs.

Not all events are dependent. For example, the probability that the Martin family purchases 2 cans of scouring powder, given that they purchase 2 containers of liquid detergent, is equal to the marginal (or unconditional) probability that they purchase 2 cans of scouring powder. (Both probabilities equal 1/4, as is evident from our discussion of Figure 3.3.) In other words, the probability that the Martins will purchase 2 cans of scouring powder is not influenced by whether or not they purchase 2 containers of liquid detergent. Similarly, if a fair coin is flipped a number of times, the probability of its coming up heads does not depend on whether the coin came up heads on the last flip. Why? Because the coin's behavior on one flip is not influenced by its behavior on the last flip (or any other flip, for that matter).

The definition of statistical independence is as follows.

> STATISTICAL INDEPENDENCE. *If events* A *and* B *are statistically independent, the probability of the occurrence of one event is not affected by the occurrence of the other. That is, each of the following equations is true:*

$$P(A \mid B) = P(A) \tag{3.6a}$$

$$P(B \mid A) = P(B). \tag{3.6b}$$

The example below illustrates how this definition can be used.

---

EXAMPLE 3.5 Two machines are drawn at random from a population of 10 machines, 3 of which are defective and 7 of which are not defective. The sampling is *not* carried out with replacement. (In other words, the first machine chosen is *not* put back into the population before the second machine is chosen.) Is whether or not the second machine is defective statistically independent of whether or not the first is defective?

SOLUTION: Given that the first machine selected is defective, what is the probability that the second machine selected will be defective? If the machine selected first is *not* put back into the

---

[2] Note once again that the figures in Table 3.2 are hypothetical. Neither Democrats, Republicans, nor independents should assume that we regard the figures as accurate. They are used only for illustrative purposes.

population (and thus has no chance of being selected again), this probability equals 2/9, since only 2 of the 9 machines left in the population are defective. On the other hand, if the first selection were not defective, this probability would be 3/9. Thus, the probability of the second selection's being defective depends on whether the first selection is defective, which means that the results of each selection are *not* statistically independent. (However, *if the sampling were done with replacement, the results of each selection would be statistically independent, since the probability of getting a defective would be 3/10, regardless of the outcome of earlier selections.*)

## THE MULTIPLICATION RULE WITH INDEPENDENT EVENTS

When two events are statistically independent, the multiplication rule can be simplified in the following way.

MULTIPLICATION RULE FOR TWO INDEPENDENT EVENTS. *If* A *and* B *are statistically independent events, the joint probability that both* A *and* B *will occur equals the unconditional probability of* A *times the unconditional probability of* B. *In other words,*

$$P(A \text{ and } B) = P(A) \cdot P(B) \tag{3.7}$$

*If* **A** *and* **B** *are two events, the probability that both occur is* **not** *in general equal to the product of their probabilities. Only if they are statistically independent is this true.*    *Basic Idea #6*

Moreover, the extension of the multiplication rule to situations where there are more than two events can be simplified in the following way when all of the events are statistically independent.

MULTIPLICATION RULE FOR n INDEPENDENT EVENTS. *If* $E_1$, $E_2$, ..., $E_n$ *are events, the joint probability that all* n *events will occur equals the product of their unconditional probabilities of occurrence if all the events are statistically independent. In other words,*

$$P(E_1 \text{ and } E_2 \text{ and } \ldots \text{ and } E_n) = P(E_1) \cdot P(E_2) \cdot \ldots \cdot P(E_n). \tag{3.8}$$

## APPLICATIONS

The multiplication rule is a powerful aid to the solution of a wide variety of practical problems in business, economics, and other fields.

The following are three examples of how the multiplication rule and the concept of statistical independence can be used. Let's begin with a very simple example.

---

EXAMPLE 3.6 Two cards are chosen at random and without replacement from an ordinary deck of playing cards. What is the probability that both are hearts?

SOLUTION: Let $A$ be the event that the first card is a heart and $B$ be the event that the second card is a heart. Clearly, $P(A) = 13/52$, since 13 cards out of the 52 in the deck are hearts. And $P(B|A) = 12/51$, since the probability that the second card is a heart *(given that the first is a heart)* is $12/51$ because 12 cards out of the remaining 51 in the deck are hearts. Applying the multiplication rule,

$$P(A \text{ and } B) = P(B|A) \cdot P(A)$$

$$= 12/51(13/52) = 1/17.$$

Note that the multiplication rule implies that $P(A \text{ and } B)$ equals $P(B|A) \cdot P(A)$, as well as $P(A|B) \cdot P(B)$, which is the expression in equation (3.4).[3]

[3] Because the definition of a conditional probability in equation (3.3) implies that $P(B|A) = P(A \text{ and } B) \div P(A)$, it follows that $P(A \text{ and } B) = P(B|A)P(A)$.

---

EXAMPLE 3.7 A salesman must call on all four of his customers in a certain area. The probability that he finds each of them in his or her office on a particular day is $1/2$, and whether or not one customer is in is statistically independent of whether any of the others is in. What is the probability that the salesman will find all of the customers in their offices on this day?

SOLUTION: Let $E_1$ be the event that the first customer is in his or her office, $E_2$ be the event that the second customer is in, $E_3$ be the event that the third customer is in, and $E_4$ be the event that the fourth customer is in. Since these events are statistically independent, the multiplication rule implies that

$$P(E_1 \text{ and } E_2 \text{ and } E_3 \text{ and } E_4) = P(E_1) \cdot P(E_2) \cdot P(E_3) \cdot P(E_4)$$

$$= (1/2)(1/2)(1/2)(1/2) = 1/16.$$

THE CASE OF "CHUCK-A-LUCK"

"Chuck-a-luck" is a game that is often played at carnivals and gambling spots. The player pays a dollar to play. Three dice are thrown. If any 6's come up, the player gets back his or her dollar, plus one dollar for each 6 that comes up. Players often argue that this game is advantageous to them. They say that, since the probability of a 6 coming up on each die is 1/6, the probability that the player will win is $1/6 + 1/6 + 1/6 = 1/2$. Thus, the player will win a dollar as often as he or she will lose one, and in addition will get an extra dollar when two 6's come up, and two extra dollars when three 6's come up. Are they right in believing that this game is advantageous to them? If not, what mistakes have they made?

SOLUTION: The probability that at least one 6 will come up is not 1/2, because a 6 on the first die does not prevent a 6 on the second or third dies. Since they are not mutually exclusive, this probability is *not* $1/6 + 1/6 + 1/6 = 1/2$. (Recall the addition theorem.) Instead, the probability of exactly one 6 coming up is 75/216; the probability of exactly two 6's coming up is 15/216; and the probability of exactly three 6's coming up is 1/216. (See footnote 4.) Suppose that a person plays this game repeatedly. In 125/216 of the cases (over the long run), he or she will lose a dollar. In 75/216 of the cases, he or she will win a dollar. In 15/216 of the cases, he or she will win two dollars. In 1/216 of the cases, he or she will win three dollars. Thus, on the average, the person will make

$$\frac{125}{216}(-\$1) + \frac{75}{216}(\$1) + \frac{15}{216}(\$2) + \frac{1}{216}(\$3) = -8 \text{ cents.}$$

In other words if this game is played repeatedly, the player will lose, on the average, 8 cents per play. Clearly, this game is not advantageous to the player.[5]

[4] These probabilities can be derived as follows. There are three mutually exclusive ways in which exactly one 6 can come up. (Specifically, the first, second, or third die can show a 6.) The probability that each of these ways occurs is (1/6)(5/6)(5/6). Thus, the probability of exactly one 6 is (1/6)(5/6)(5/6) + (1/6)(5/6)(5/6) + (1/6)(5/6)(5/6) = 75/216.

There are three mutually exclusive ways in which exactly two 6's can come up. (Specifically, the first, second, or third die can show other than a 6.) The probability that each of these ways occurs is (1/6)(1/6)(5/6). Thus, the probability of exactly two 6's is (1/6)(1/6)(5/6) + (1/6)(1/6)(5/6) + (1/6)(1/6)(5/6) = 15/216.

There is only one way in which exactly three 6's can come up. (Specifically, all three dice must show a 6.) The probability that this occurs is (1/6)(1/6)(1/6) = 1/216.

[5] This game is also discussed in W. A. Wallis and H. Roberts, *Statistics* (Glencoe: Free Press, 1956), pp. 332–33.

EXAMPLE 3.8 An automobile company has 4,000 dealers, who are polled to find out their annual incomes and their ages. The number of dealers in each income and age category is as follows:

| Incomes (dollars) | Dealers under 45 | Dealers 45 and above | Total |
|---|---|---|---|
| Under 25,000 | 1,000 | 1,000 | 2,000 |
| 25,000 and over | 500 | 1,500 | 2,000 |
| Total | 1,500 | 2,500 | 4,000 |

(a) What is the probability that a randomly chosen dealer (1) will have an income of under $25,000; (2) will be under 45; (3) will have an income of under $25,000, given that he or she is under 45?

(b) Are age and income statistically independent?

SOLUTION: (a) The probability that a dealer will have an income of under $25,000 is 2,000/4,000, or 1/2. The probability that a dealer will be under 45 is 1,500/4,000, or 3/8. The probability that a dealer will have an income of under $25,000, given that he or she is under 45, is 1,000/1,500, or 2/3.

(b) Age and income are not statistically independent because the probability that a dealer has an income of under $25,000 is different if he or she is under 45 than if he or she is 45 or over. If a dealer is under 45 this probability is 1,000/1,500, or 2/3. If a dealer is 45 or over this probability is 1,000/2,500, or 2/5.

**GETTING DOWN TO CASES**

THE RELIABILITY OF THE APOLLO SPACE MISSION[6]

The Apollo mission with its objective of landing men on the moon involved severe problems of reliability. For the mission to be successful, all (or at least a great many) components of the Apollo system had to operate properly. A malfunction of any one of many components could have resulted in mission failure. Experts used statistical theory to develop sophisticated methods for estimating the probability that the undertaking would be successful. We are able here to present only simplified versions of what was done. A schematic version of the basic Apollo module is shown in Figure 3.8. We assume that the module works if, and only if, all five components function properly.

(a) Suppose that the probability is 0.99 that each of the five Apollo components functions properly. If the components are independent, what is the probability that a mission will succeed?

[6] This case is based to some extent on a section from G. Lieberman, "Striving for Reliability," in J. Tanur, F. Mosteller, W. Kruskal, R. Link, R. Pieters, and G. Rising, *Statistics: A Guide to the Unknown* (San Francisco: Holden-Day, 1972).

One way of increasing the reliability of the Apollo system was through the use of a parallel configuration of components. For example, the second stage of the Saturn rocket used in the Apollo program had five rocket motors, and if any one of the motors failed, the others could be used for a satisfactory earth orbit. To see how the use of a parallel configuration of components increases reliability, suppose that the main engine component in Figure 3.8 is replaced by two engines, either of which is able to perform the tasks required by the mission. If either of these engines malfunctions, the other can take over.

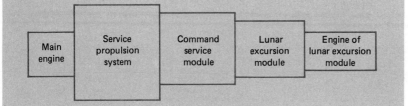

**Figure 3.8**
Simplified Representation of Apollo Module

(b) If the probability is 0.99 that each of the engines functions properly, what is the probability that at least one of the engines functions properly?

(c) If the main engine component in Figure 3.8 is replaced by two such engines, what is the probability that a mission will succeed?

(d) Compare your result in (c) with that in (a) and interpret the difference between the two results.

(e) What are the disadvantages of a parallel configuration of components? Why not add as many redundant components as possible in order to increase reliability?

EXERCISES

**3.10** An article in a New York newspaper by two well-known columnists[7] stated that if the probability of downing an attacking airplane were 0.15 at each of five defense stations, and if a plane had to pass all five stations before arriving at the target, the probability that the plane would be downed before reaching the target was 0.75.
(a) Do you agree with this reasoning?
(b) If not, what is the correct answer?

**3.11** Suppose that $P(A) = 0.6$, $P(B) = 0.3$, and $P(A \text{ and } B) = 0.1$.
(a) Are $A$ and $B$ mutually exclusive events? Why, or why not?
(b) Are $A$ and $B$ statistically independent events? Why, or why not?

**3.12** Grace Jones is selling her house. She believes there is a 0.2 chance that each person who inspects the house will purchase it. What is the probability that more than two people will have to inspect the house before Grace

[7] W. A. Wallis and H. Roberts, *op. cit.*, p. 96.

Jones finds a buyer? (Assume that the decisions of the people inspecting the house are independent.)

**3.13** A true die is rolled twice. What is the probability of getting (a) a total of 6; (b) a total of less than 6; (c) a total of 7 or more?

**3.14** Three cards are to be drawn from a deck with the five of hearts missing.
(a) What is the probability that all three cards will be clubs?
(b) What is the probability that all three cards will be of the same suit?

**3.15** The faces of a true die showing a 5 and 6 are colored green; the other faces are colored red. The faces of another true die are colored the same way. If the two dice are tossed, what is the probability that both show a red face?

**3.16** In a particular industry, 30 percent of the firms lose money and 10 percent have chief executive officers who graduated from the Wharton School. If a firm in this industry is chosen (the probability of choosing each firm being the same for all firms), can we calculate the probability of choosing a firm with a Wharton CEO that loses money by (a) multiplying 0.30 times 0.10; (b) adding 0.30 and 0.10? Why or why not?

**3.17** The probability that a launch of a particular spacecraft will occur on time is 0.4. Whether or not a launch occurs on time is independent of whether previous launches occurred on time. If three such launches take place, compute the probability that
(a) the first two will occur on time, but the third will not.
(b) the number of launches occurring on time exceeds the number not occurring on time.

**3.18** The probability that a lawyer passes the bar exam on his first try is 0.80. On the second try, it is 0.85, and on the third try it is 0.88. Calculate the probability that a lawyer:
(a) does not pass the bar exams after two tries.
(b) takes three tries to pass the bar exam.

**3.19** At a particular textile plant, the number of male employees equals the number of female employees. The probability is $1/2$ that a male employee is a college graduate and $1/4$ that a female employee is one. An employee who is a college graduate is chosen at random (that is, each such employee has the same probability of being chosen). What is the probability that such an employee is female?

**3.20** The Jones family decides to buy a new car and narrows the choice down to a Ford, a Chevrolet, or a Toyota. The probability that they will buy a Ford is 0.3, and the probability that they will buy a Chevrolet is 0.4.
(a) What is the probability that they will purchase either a Ford or a Chevrolet?
(b) What is the probability that they will purchase a Toyota?
(c) If the Joneses were undecided as to whether or not they would purchase a car at all, would this influence your answer to (b)? If so, would you increase or decrease your answer to (b)?
(d) The probability that the Jones family will purchase a Ford is 0.3 and the probability that they will purchase a station wagon is 0.2. If the probability that they will purchase a Ford station wagon is 0.05, what is the probability that they will purchase either a Ford or a station wagon (or both)?

**3.21** The probability of failure of the booster rockets of the U.S. space shuttle has been estimated by some experts as about .03. If there were 4 flights, and if this estimate were true, what would be the probability that the booster rockets would not fail in every case?

# 3.6 Bayes' Theorem

## STATEMENT OF THE THEOREM

The Reverend Thomas Bayes, an eighteenth-century British scholar, put forth a theorem concerned with the calculation of the probability that a particular hypothesis is true, given that particular events occur. Or, put more crudely, the theorem is concerned with the probability of a particular cause, given the observation of a particular effect. For example, suppose that a firm receives goods in lots from three trucking companies. If this firm receives a lot containing numerous defective goods, Bayes' theorem might be used to calculate the probability that each trucking company delivered it. Because it reasons backward from effects to causes, Bayes' theorem has been a controversial part of statistics. Although no one denies that it is formally valid, many great statisticians have felt that it could be applied correctly in very few cases. One reason why it has been viewed with skepticism is that in order to apply this theorem, one must know the probabilities that various hypotheses or causes are true, which often is far from easy. Nonetheless, in recent years more and more statisticians have come to use Bayes' theorem, as we shall see in subsequent chapters.

> BAYES' THEOREM: *Suppose that there are* m *hypotheses—*$H_1$, $H_2$, ... , $H_m$. *Only one of these hypotheses can be true, and one of them must be true. The probability that the* ith *hypothesis is true is* $P(H_i)$. *Then if an event* E *occurs, the conditional probability that the* ith *hypothesis is true equals*

$$P(H_i|E) = \frac{P(E|H_i)P(H_i)}{\sum_{i=1}^{m} P(E|H_i)P(H_i)}.$$

(3.9)    *Formula for* $P(H_i|E)$

The result in equation (3.9) can be derived from our previous findings. From our definition of a conditional probability in equation (3.3), we know that

$$P(H_i|E) = \frac{P(H_i \text{ and } E)}{P(E)}$$

(3.10)

Moreover, since the unconditional probability that $E$ occurs is the sum of the joint probability that $E$ occurs and $H_1$ is true, that $E$ occurs and $H_2$ is true, that $E$ occurs and $H_3$ is true, and so on, it follows that

$$P(E) = \sum_{i=1}^{m} P(H_i \text{ and } E).$$

Substituting the right-hand side of this expression for $P(E)$ in equation (3.10), we have

$$P(H_i|E) = \frac{P(H_i \text{ and } E)}{\sum_{i=1}^{m} P(H_i \text{ and } E)}. \qquad (3.11)$$

Using the multiplication rule in equation (3.4), it is evident that

$$P(H_i \text{ and } E) = P(E|H_i)P(H_i).$$

Thus, substituting the right-hand side of this expression for $P(H_i$ and $E)$ in equation (3.11), we have

$$P(H_i|E) = \frac{P(E|H_i)P(H_i)}{\sum_{i=1}^{m} P(E|H_i)P(H_i)}$$

which is what we set out to derive.

Two things should be noted concerning the conditions underlying Bayes' theorem. First, it is assumed that the $m$ hypotheses are *mutually exclusive* and *exhaustive*. In other words, *only one* of them can be true, but one of them *must* be true. Second, it is assumed that the probability that each of these hypotheses is true is *known*. These probabilities are called ***prior probabilities***. Much more will be said about prior probabilities in Chapters 15 and 16.

*Prior Probabilities*

### AN APPLICATION

To illustrate the application of Bayes' theorem, we return to the case where a firm receives goods in lots from three trucking companies. The probability that each trucking company (Keepon, Superior, or Never-Fail) delivers a lot to this firm is shown in Table 3.3. And the probability that a lot will contain more than 2 percent defectives, given that each trucking company delivers the lot, is also shown in Table 3.3. If this firm receives a lot with more than 2 percent defectives, what is the probability that each trucking company delivered it?

At first glance, the application of Bayes' theorem to this problem may not be obvious. To see how it is applicable, it is important to recognize what the hypotheses are and what the event is. Clearly, there are three hypotheses: (1) Keepon delivered the lot ($H_1$); (2) Superior delivered the lot ($H_2$); and (3) Never-Fail delivered the lot ($H_3$). These hy-

| A. *Probability that lot is delivered by* | |
|---|---|
| Keepon Trucking Company | 0.20 |
| Superior Trucking Company | 0.40 |
| Never-Fail Trucking Company | 0.40 |

B. *Probability that lot contains more than 2 percent defectives if delivered by*

| | |
|---|---|
| Keepon Trucking Company | 0.010 |
| Superior Trucking Company | 0.020 |
| Never-Fail Trucking Company | 0.025 |

**Table 3.3**
Probability that Lot Was Delivered by Each of Three Trucking Companies, and Contains More than 2 Percent Defectives, Given that Each Company Delivered It

potheses are mutually exclusive and exhaustive, as the theorem requires. (Hypotheses of this sort are often called *states of nature*. For example, a particular state of nature—$H_1$—is that Keepon delivered the lot.)

What is the event? It is the fact that a lot with more than 2 percent defectives was delivered. If we call this event $E$, it follows from equation (3.9) that

$$P(H_1|E) = \frac{P(E|H_1)P(H_1)}{P(E|H_1)P(H_1) + P(E|H_2)P(H_2) + P(E|H_3)P(H_3)},$$

$$P(H_2|E) = \frac{P(E|H_2)P(H_2)}{P(E|H_1)P(H_1) + P(E|H_2)P(H_2) + P(E|H_3)P(H_3)},$$

$$P(H_3|E) = \frac{P(E|H_3)P(H_3)}{P(E|H_1)P(H_1) + P(E|H_2)P(H_2) + P(E|H_3)P(H_3)}.$$

And if we insert the numbers in Table 3.3 into these formulas, we have

$$P(H_1|E) = \frac{(.01)(.20)}{(.01)(.20) + (.02)(.40) + (.025)(.40)} = \frac{.002}{.02} = 0.10,$$

$$P(H_2|E) = \frac{(.02)(.40)}{(.01)(.20) + (.02)(.40) + (.025)(.40)} = \frac{.008}{.02} = 0.40,$$

$$P(H_3|E) = \frac{(.025)(.40)}{(.01)(.20) + (.02)(.40) + (.025)(.40)} = \frac{.010}{.02} = 0.50.$$

In other words, the probability that Keepon delivered the lot is 0.10, the probability that Superior delivered the lot is 0.40, and the probability that Never-Fail delivered the lot is 0.50.

## 3.7 Subjective or Personal Probability

The probabilities used in the application of Bayes' theorem often are based on a different definition of probability than that presented in Section 3.3. There the probability of an event was defined as the pro-

*Frequency Definition of Probability*

portion of times that the event will occur in the long run if the relevant experiment is repeated over and over. This is the so-called *frequency definition of probability*. For example, as pointed out in Section 3.3, the probability that a particular die will come up a 1 can be viewed as the proportion of times that this will occur if the die is thrown innumerable times.

Some experiments are not easy to interpret in these terms because they cannot be repeated over and over. A new commercial product may have a different probability of succeeding if it is put on the market this month rather than next month. This is an "experiment" that cannot be performed over and over because market and other conditions vary from month to month. If the new product is *not* introduced this month but next, the experiment will be performed under different conditions and thus will be a different experiment. There are many events of this type in business and economics.

*Subjective Definition of Probability*

In dealing with events and experiments of this sort statisticians and decision makers sometimes use a *subjective* or *personal definition of probability*. According to this definition, *the probability of an event is the degree of confidence or belief on the part of the statistician or decision maker that the event will occur.* For example, if the decision maker believes that event *A* is more likely to occur than event *B*, the probability of *A* is higher than the probability of *B*. If the decision maker believes that the odds are 50-50 that a particular event will occur, the probability attached to this event equals 0.50. The important factor in this concept of probability is what the decision maker believes.

To illustrate subjective probability, suppose that a marketing manager must decide whether to substitute a new product label for an old one. The manager must estimate as accurately as possible the probability that the new label is superior to the old. The relevant experiment cannot be conducted over and over; that is, the new label cannot be substituted for the old one again and again to determine the proportion of cases in which the substitution improves the product's profitability. Instead, the manager must use his or her knowledge, experience, and intuition, together with whatever objective information can be obtained, to estimate the chances that the new label is superior to the old. If the conclusion is that the probability of its being superior is 0.50, then this is the manager's subjective probability of this event.

The same rules of probability apply, regardless of whether probabilities are given a frequency definition or a subjective definition. Therefore, all the principles presented in this chapter apply to both types of probabilities. Since subjective probabilities must conform to the same mathematical rules as probabilities based on the frequency definition, and since they can be manipulated in essentially the same way to solve problems, we shall not distinguish subjective probabilities from frequency-based probabilities in most of the succeeding parts of this book.[8]

[8] For two classic papers concerning alternative definitions of probability, see R. von Mises, "Probability: An Objective View," and L. J. Savage, "Probability: A Subjectivist View," in E. Mansfield, *Statistics for Business and Economics: Readings and Cases* (New York: Norton, 1980).

## 3.8  A Case Study in Marketing[9]

The following is a realistic illustration of how both Bayes' theorem and subjective probabilities are used in business situations.

A firm is concerned about the sales appeal of the label on one of its products. A designer is hired to create some new labels, and when the company's executives see the various new labels, one is consistently regarded as the best. However, the firm's marketing manager is still uncertain as to whether or not the new label will substantially increase the product's sales.

To obtain more information as to whether or not the new label is superior to the old one, the marketing manager sets in motion a survey of consumers. This survey is designed in such a way that if the new label is really superior, the probability that the survey will indicate that it is superior is 0.8, the probability that it will yield ambiguous results is 0.1, and the probability that the survey will indicate that the label is not superior is 0.1. Thus,

$$P(b|B) = 0.8; \; P(a|B) = 0.1; \; P(n|B) = 0.1,$$

where $P(b|B)$ is the probability that the survey will indicate that the label is superior, given that in fact it is superior; $P(a|B)$ is the probability that the survey will yield ambiguous results if the new label is in fact superior; and $P(n|B)$ is the probability that the survey will indicate that the label is not superior, given that it is in fact superior.

The survey is also designed so that if the new label is *not* superior, the probability that the survey will indicate that it is not superior is 0.7, the probability that the survey will be ambiguous is 0.1, and the probability that the survey will indicate that it is superior is 0.2. Thus,

$$P(n|N) = 0.7; \; P(a|N) = 0.1; \; P(b|N) = 0.2,$$

where $P(n|N)$ is the probability that the survey will indicate that the label is not superior given that in fact it is not superior; $P(a|N)$ is the probability that the survey will yield ambiguous results if in fact the new label is not superior; and $P(b|N)$ is the probability that the survey will indicate that the label is superior, given that in fact it is not superior.

Despite the enthusiasm of the firm's executives, the marketing manager believes that the probability that the new label is in fact superior is only 0.5. As we saw in the previous section, this is a subjective probability, based on the manager's experience and judgment. This probability is formulated prior to knowing the results of the survey. When the results of the survey come in, they indicate that the new label is superior. Given these survey results, what is the probability that the

[9] This example is based to some extent on P. Green and R. Frank, "Bayesian Statistics and Marketing Research," *Applied Statistics*, 1966, pp. 173–90. The numbers have been changed.

new label is *in fact* superior? Can you use Bayes' theorem to answer this question?

To apply Bayes' theorem to this problem, one must begin by sorting out the hypotheses and the event. The two hypotheses are (1) the new label is in fact superior (hypothesis B) and (2) the new label is in fact not superior (hypothesis N). The event is the fact that the survey indicates that the new label is superior. Clearly, on the basis of equation (3.9),

$$P(B|b) = \frac{P(b|B)P(B)}{P(b|B)P(B) + P(b|N)P(N)}$$
(3.12)

and

$$P(N|b) = \frac{P(b|N)P(N)}{P(b|B)P(B) + P(b|N)P(N)},$$
(3.13)

where $P(B|b)$ is the probability that the new label is really superior, given that the survey indicates that this is so, and where $P(N|b)$ is the probability that the new label is really not superior, given that the survey indicates that it is superior.

Substituting the figures presented in the several previous paragraphs for the symbols in equations (3.12) and (3.13), we have

$$P(B|b) = \frac{(0.8)(0.5)}{(0.8)(0.5) + (0.2)(0.5)} = 0.8$$

and

$$P(N|b) = \frac{(0.2)(0.5)}{(0.8)(0.5) + (0.2)(0.5)} = 0.2.$$

Thus, given the marketing manager's prior probabilities and the fact that the survey indicates that the new label is superior, the probability that it is in fact superior is 0.8, while the probability that it is not in fact superior is 0.2.

EXERCISES

**3.22** Distinguish between a subjective probability and one based on the frequency definition. Why are subjective probabilties ever used?

**3.23** If a reporter feels that the odds are 2 to 1 that he can get an interview with the governor of his state, what is his personal probability that this will happen?

**3.24** If you believe the odds are 3 to 1 that you will get at least a B in your statistics course, what is the probability (your subjective probability) that this is the case?

**3.25** According to R. A. Fisher, advocates of Bayes' theorem "seem forced to regard mathematical probability, not as an objective quantity measured by observed frequencies, but as measuring merely psychological tendencies, theorems respecting which are useless for scientific purposes."[10] Do you agree? Do you think Bayes' theorem and subjective probability are useless for business decision making? Why, or why not?

**3.26** Five percent of the tires produced by a particular factory are defective. The factory buys a machine to check each tire before shipment. The probability that the machine will fail to recognize a defective tire is 0.10, and the probability that it will consider a nondefective tire to be defective is 0.05. If a tire is checked by this machine, and is considered by it to be nondefective, what is the probability that it really is nondefective?

**3.27** There is a 0.4 probability that the Morgan Company, a paper manufacturer, will send a salesman to call on Mr. Smith. If Morgan does not send a salesman, there is a 0.7 probability that Mr. Smith will buy his paper from Morgan's competitor, the Xerxes Company. But if Morgan does send a salesman, there is only a 0.2 probability that Mr. Smith will buy his paper from Xerxes. If Mr. Smith does buy his paper from Xerxes, what is the probability that Morgan did not send a salesman to call on him?

**3.28** A letter has been put in either file *A* or file *B* or file *C*. The percentage of letters addressed to overseas destinations is 5 percent in file *A*, 10 percent in file *B*, and 15 percent in file *C*. The probability that a letter has been put in file *A* is double the probability that it has been put in file *B*, which in turn is double the probability that it has been put in file *C*. If this letter is addressed to an overseas destination, what is the probability that it has been put in file *A*? In file *B*? In file *C*?

**3.29** A bank gives a test to screen prospective employees. Among those who perform their jobs satisfactorily, 65 percent passed the test. Among those who do not perform satisfactorily (and are fired), 25 percent passed the test. According to the bank's records, 90 percent of its employees perform their jobs satisfactorily. What is the probability that a prospective employee who passed the test will not perform satisfactorily?

# Chapter Review

1. Statisticians use the term *experiment* to describe any process by which data are obtained through the observation of uncontrolled events in nature or through controlled laboratory procedures. The *sample space* is defined as the set of all possible outcomes that may occur as a result of a particular experiment. In each occurrence of the experiment, one and only one outcome takes place. Sample spaces are often depicted as points on a graph or in tree diagrams. *Probabilities* may be regarded as numbers that are assigned to each of the elements of the sample space.

2. The *addition rule* states that, if $P(A)$ is the probability that event *A* occurs and if $P(B)$ is the probability that event *B* occurs, then

---

[10] R. A. Fisher, "Statistical Inference," in E. Mansfield, *Statistics for Business and Economics: Readings and Cases* (New York: Norton, 1980).

$P(A \text{ or } B) = P(A) + P(B) - P(A \text{ and } B)$. Of course, if $A$ and $B$ are mutually exclusive, it follows that $P(A \text{ or } B) = P(A) + P(B)$. If $E_1, E_2, \ldots, E_n$ are mutually exclusive events, then $P(E_1 \text{ or } E_2 \text{ or } \ldots \text{ or } E_n) = P(E_1) + P(E_2) + \ldots + P(E_n)$.

3. The *conditional probability* of event $A$ given the occurrence of event $B$—that is, $P(A \mid B)$—equals $P(A \text{ and } B) \div P(B)$. If $P(A \mid B) = P(A)$, event $A$ is said to be statistically independent of event $B$. The *multiplication rule* states that if $A$ and $B$ are two events, the probability that both occur—$P(A \text{ and } B)$—equals $P(A \mid B)P(B)$. Of course, if $A$ and $B$ are independent, it follows that $P(A \text{ and } B)$ equals $P(A)P(B)$. If $E_1, E_2, \ldots,$ $E_n$ are statistically independent events, $P(E_1 \text{ and } E_2 \text{ and } \ldots \text{ and } E_n) = P(E_1) \cdot P(E_2) \cdot \ldots \cdot P(E_n)$.

4. *Bayes' theorem* is concerned with the calculation of the probabilities that particular hypotheses are true, given that particular events occur. This theorem assumes that there are a number of hypotheses— $H_1, H_2, \ldots, H_m$—that are mutually exclusive and exhaustive. It is assumed that we know the probability that each of these hypotheses is true. These probabilities, called *prior probabilties,* are $P(H_1)$, $P(H_2)$, $\ldots, P(H_m)$. If $E$ is an event and if $P(E|H_i)$ is the probability of the event occurring given that $H_i$ is true, Bayes' theorem states that the probability that $H_i$ is true (given that the event occurs) is

$$P(H_i|E) = \frac{P(E|H_i)P(H_i)}{\sum_{i=1}^{m} P(E|H_i)P(H_i)}.$$

## Chapter Review Exercises

**3.30** If a gambler makes a single roll with a pair of true dice, what is the probability that the numbers on the two dice will differ?

**3.31** The Bona Fide Washing Machine Company has decided to buy a firm which manufactures rivets, but has not decided which one to buy. There are three states in which rivet producers exist: Michigan, Illinois, and New York. In each state, a firm producing rivets may have one of three forms of legal organization; it may be a proprietorship, a partnership, or a corporation.

(a) Draw a diagram similar to Figure 3.1 showing the nine different ways in which the Bona Fide Washing Machine Company can choose a rivet producer to buy. Put the location of the company along the horizontal axis, and its form of legal organization along the vertical axis. On the horizontal axis, let 1 equal Michigan, let 2 equal Illinois, and let 3 equal New York. On the vertical axis, let 1 equal proprietorship, let 2 equal partnership, and let 3 equal corporation.

(b) Using the diagram you drew in (a), designate the subset of the sample space corresponding to each of the following events:
  (i) The Bona Fide Washing Machine Company buys a Michigan rivet producer.
  (ii) The Bona Fide Washing Machine Company buys a corporation.
  (iii) The Bona Fide Washing Machine Company buys a proprietorship in New York.

(iv) The Bona Fide Washing Machine Company buys either a corporation or a proprietorship.

(v) The Bona Fide Washing Machine Company buys either a proprietorship or a New York firm (or both).

(c) Using the results you obtained in (a) and (b), indicate the set of outcomes, one of which must occur if

   (i) The Bona Fide Washing Machine Company buys a Michigan rivet producer;

  (ii) The Bona Fide Washing Machine Company buys a corporation;

 (iii) The Bona Fide Washing Machine Company buys either a corporation or a proprietorship.

(d) Using your results from (a) and (b), indicate the set of events, all of which must occur if (i) the Bona Fide Washing Machine Company buys a proprietorship in New York, (ii) the Bona Fide Washing Machine Company buys a corporation in Michigan.

**3.32** A fair coin is flipped 6 times. What is the probability of 6 heads? What is the probability of 6 tails?

**3.33** The probability that a student will get an A in statistics is 0.20, the probability that he will get an A in English is 0.25, and the probability that he will get an A in both is 0.05. What is the probability that he will get an A in neither subject?

**3.34** According to the *New York Times,* February 11, 1986, "Among the highest risks taken by the general public is in automobiles, where the chance of death for each person is 1 in 4,000 each year." Is this a marginal probability? Is it a conditional probability? How would you determine whether this estimated probability is correct?

**3.35** The Smith family is undecided about whether or not to buy a new car. If the probability is 0.9 that they will buy one, and if the probability is 0.3 that they will buy a Ford, and if the probability is 0.4 that they will purchase a car getting more than 20 miles per gallon, what is the probability

(a) that they will buy either a car getting more than 20 miles per gallon or a Ford, if there are no Ford cars that get more than 20 miles per gallon;

(b) that they will buy either a car getting more than 20 miles per gallon or a Ford (or both), if all Fords get more than 20 miles per gallon?

**3.36** A man has three coins in his pocket. Two are fair coins; the third has tails on both sides. He chooses a coin from his pocket (each coin having the same probability of being chosen), and tosses it twice. What is the probability that two heads will come up?

**3.37** The Black Belt Sewing Machine Company has four plants. The number of motors shipped in 1986 by each supplier to each of Black Belt's plants is as follows:

|  |  |  | *Supplier* |  |  |
|---|---|---|---|---|---|
| *Plant* | *I* | *II* | *III* | *IV* | *V* |
| 1 | 200 | 100 | 300 | 100 | 400 |
| 2 | 300 | 200 | 400 | 200 | 300 |
| 3 | 200 | 300 | 300 | 300 | 100 |
| 4 | 100 | 300 | 200 | 400 | 200 |

(a) If the firm picks a motor from among those received by plant 2 (and if each motor has the same probability of being chosen) what is the probability that this motor came from supplier III?

(b) If the firm picks a motor from among those received at any of its plants, what is the probability that it came from supplier III?

(c) If event $X$ is that a motor comes from supplier III, is this event independent of the plant receiving the motor?

(d) Calculate the probability
  (i) that a motor goes to plant 1, given that it comes from supplier II;
  (ii) that a motor comes from supplier II, given that it goes to plant 1;
  (iii) that a motor goes to plant 1, given that it comes from either supplier II or supplier III;
  (iv) that a motor comes from supplier II, given that it goes to either plant 1 or plant 2.

(e) The Black Belt Sewing Machine Company finds that the probability that a motor both comes from supplier I and is defective is 0.1. What is the probability that a motor purchased from supplier I is defective?

(f) If 20 percent of all the motors received by the Black Belt Sewing Machine Company are defective, what is the probability that a motor comes from supplier I, given that it is found to be defective?

**3.38** The Pound Ridge Company loses a shipment of its goods due to a mistake by one of its executives. The firm has three executives (Tom, Dick, and Jane) who could have made the mistake, and it can't tell with certainty which one was responsible. However, on the basis of past performance, it is known that the probability that Tom would make such a mistake (if he were responsible for a shipment) is .001, the probability that Dick would do so is .002, and the probability that Jane would do so is .001. If Tom is responsible for 50 percent of the shipments, while the rest are split evenly between Dick and Jane, what is the probability that each executive was responsible for the shipment that was lost?

**3.39** The Bona Fide Washing Machine Company is considering John Jones for a key position in marketing. After interviewing Jones and reviewing his credentials, Bona Fide's president feels there is a 75 percent chance that Jones would do well in the job and a 25 percent chance that he would not. The president decides to have a management consulting firm evaluate Jones. According to the experience of other companies, the consulting firm is reasonably accurate in evaluating marketing managers, as indicated by the following table:

| Actual experience | *Conditional probability (given actual experience) that consulting firm predicts:* | |
| --- | --- | --- |
| | *Manager will do well* | *Manager will not do well* |
| Manager does well | 0.9 | 0.1 |
| Manager does not do well | 0.2 | 0.8 |

(a) If the consulting firm predicts Jones will do well, what is the probability that this will be true?

(b) If the consulting firm predicts Jones will not do well, what is the probability that this will be true?

# Appendix 3.1

### COUNTING TECHNIQUES

In many practical situations, it is important to be able to calculate the number of ways in which an event can occur. Such calculations must often be carried out in order to compute the probability of an event.

SELECTION OF ITEMS. Frequently, a selection or a choice must be made in several steps, and at each step there are a number of alternatives. For example, suppose that a student is admitted to three university business schools, Indiana, Michigan, and Northwestern, and that given her choice of business school, she has a choice of specializing in accounting, finance, or marketing. In situations of this sort, one often must be able to calculate the total number of different outcomes that are possible. For example, in how many different ways can the student pick a business school and specialty?

A tree diagram is a useful graphic device for solving such problems because it shows at each step the choices that can be made and the possible outcomes at the end of the selection process. (Recall that at the beginning of this chapter tree diagrams were also used to represent sample spaces.) Figure 3.9 shows a tree diagram for the choices open to the student. As you can see, three branches are open in the first step, each branch corresponding to a business school. Once the young woman has picked one of these initial branches, she goes to another fork where there again are three branches, each corresponding to a different specialty. Once the second step has been completed, there clearly are nine different outcomes corresponding to the points on the right-hand side of the diagram where she can wind up. That is, the young woman can choose accounting at Indiana, finance at Indiana, and so on.

In this case, there are three items of one kind (business schools) and three items of a second kind (specialties), and we find that the number of ways we can pair an item of the first type with an item of the second type is (3)(3), or 9. Let's generalize this result. If there are $m$ kinds of items, and *if there are* $n_1$ *items of the*

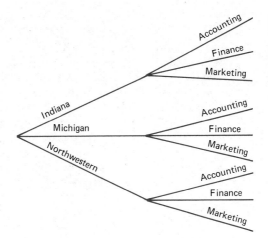

**Figure 3.9**
Tree Diagram of Student's Choices of Business School and Specialty

**Figure 3.10**
Tree Diagram of
Student's Choices with
Five Job Offers

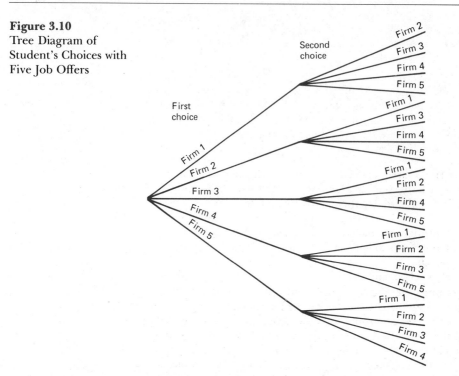

*first kind,* $n_2$ *items of the second kind, . . . , and* $n_m$ *items of the* m*th kind, the number of ways we can select one of each of the* m *kinds of items is* $n_1 \times n_2 \times \ldots \times n_m$. To illustrate, if subsequent to graduation the business student is offered a job by each of five companies, and if she is asked to list her first and second choices, how many different pairs could she list? There are five firms that she could list as her first choice. But once she makes her first choice, there are only four firms left that could be her second choice. Thus, in this case, $m = 2$, $n_1 = 5$, and $n_2 = 4$, so the answer is (5)(4), or 20. Figure 3.10 shows the relevant tree diagram.

The following example provides another illustration of how this result can be used.

---

EXAMPLE 3.9 Thirty firms bid on a particular job, and each firm submits a different bid. The names of the lowest bidder, the second lowest bidder, and the third lowest bidder are published. In how many ways can one select three firms ordered in this way?

SOLUTION: Clearly, any one of the 30 firms may be the lowest bidder. But once one selects any one of them as the lowest bidder, there are only 29 left that can be the second lowest bidder. And once one selects the lowest and second lowest bidders, there are only 28 firms that can be selected as the third lowest bidder. Consequently, one can pick the lowest bidder in 30 different ways, the pair of lowest and second lowest bidders in (30)(29) different ways, and the triad of lowest, second lowest, and third lowest bidders in (30)(29)(28) different ways. Thus, the answer is (30)(29)(28), or 24,360. In this case, $m = 3$, $n_1 = 30$, $n_2 = 29$, and $n_3 = 28$.

---

PERMUTATIONS. At this point, we can define a permutation. *If* x *items are selected (without replacement) from a set of* n *items, any particular sequence of these* x *items is called a* **permutation.** For example, Yale Harvard Brown is a permutation of three of the Ivy League colleges. That is, it is a possible sequence in which three of the eight names of the Ivy League colleges (Harvard, Yale, Princeton, Columbia, Pennsylvania, Dartmouth, Cornell, and Brown) can be listed. In this case, $x = 3$ and $n = 8$, since three items (college names) are being selected from a set of eight items. Turning to another example, *ab* is a permutation of two of the first three letters of the alphabet. In this case, $x = 2$ and $n = 3$ since two items (letters) are being selected from a set of three items. Another such permutation is *bc*. In other words, it is another pair of letters that can be made up out of *a, b,* and *c*. Note that the order of the items is of importance in a permutation. Thus, *ab* and *ba* are different permutations (from one another) even though they are composed of the same letters, *a* and *b*.

In general, *the number of permutations of* x *items that one can select from a set of* n *items is* n(n − 1)(n − 2) . . . (n − x + 1). Based on our previous results, this is not difficult to prove. There are $x$ items to be chosen. Suppose that the first item that is chosen is put into the first slot along the horizontal line in Figure 3.11. Suppose the second item chosen is put into the second such slot, the third into the third slot, and so on. For our first choice, there are $n$ possible items. Once the first selection is made, there are $(n − 1)$ items left for the second slot. Once the first and second selections have been made, there are $(n − 2)$ items left for the third slot and once the $(x − 1)$th selection is made, there are $(n − x + 1)$ items left for the last (that is, the $x$th) slot. So the total number of sequences of $x$ items that can be chosen from $n$ items must be $n(n − 1)(n − 2)$ . . . $(n − x + 1)$.

To check this result, let's see whether it gives the right answer in the case of the business student with the job offers and in Example 3.9. With regard to the student offered a job by the five firms, we want to know the number of permutations of two items (firms) that one can select from a set of five items. According to the previous paragraph, this number should be (5)(4), since $n = 5$ and $x = 2$. This result is just what we got before. In Example 3.9, the case of the 30 firms bidding on the job, we want the number of permutations of three items (firms) that one can select from a set of 30 items. Since $n = 30$ and $x = 3$, the answer according to the previous paragraph is (30)(29)(28), which jibes with our previous result.

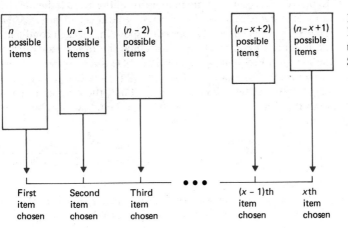

**Figure 3.11**
Number of Permutations of $x$ Items from a Set of $n$ Items

COMBINATIONS. In contrast to a permutation, where the order of the items matters, *a **combination** is a selection of items where the order does not matter.* For example, in the case of the Ivy League colleges, each permutation of three of the colleges shows a possible ranking of the top three schools in football. Clearly, the order here is important, since the first name in the sequence indicates first place and the last name indicates third place. But if instead we were interested in the number of three-way crew races that could occur in the Ivy League, the order wouldn't matter, since Harvard Cornell Brown and Cornell Brown Harvard would mean the same thing. Thus, in the latter case, one would want to calculate the number of combinations of three items that can be selected from a set of eight items.

Using the results we obtained concerning the number of permutations, it is easy to find out how many combinations there are. After all, the crucial point is that, from the point of view of a combination, many permutations are indistinguishable. For example, as we just observed, Harvard Cornell Brown and Cornell Brown Harvard are the same thing. Taking each particular combination, how many permutations correspond to this single combination? Clearly, there are $(3)(2)(1)$, since this is the number of permutations that can be created from three items. (This, of course, is an application of our formula for the number of permutations, both $x$ and $n$ being 3 in this case). To illustrate this, you can see that there are the following six permutations corresponding to the single combination consisting of Harvard, Cornell, and Brown:

Harvard Cornell Brown    Cornell Brown Harvard    Brown Cornell Harvard
Harvard Brown Cornell    Cornell Harvard Brown    Brown Harvard Cornell

Thus, since there are six permutations for each combination and since there are $(8)(7)(6) = 336$ permutations (because $n = 8$ and $x = 3$), it follows that there must be $336 \div 6 = 56$ combinations. In other words, there must be 56 different three-way crew races that can be arranged in the Ivy League.

In general, *the number of combinations of* x *items that one can select from a set of* n *items is*

$$\frac{n(n-1)\ldots(n-x+1)}{x(x-1)\ldots(2)(1)}.$$

$(3.14)$

To prove this, recall that the numerator of this ratio is the number of permutations of $x$ items that can be selected from a set of $n$ items. Since there are $x(x-1)\ldots(2)(1)$ permutations corresponding to each combination of $x$ items, the number of combinations must equal the numerator of $(3.14)$ divided by this amount. Frequently, $x(x-1)\ldots(2)(1)$ is referred to as $x!$, which is read as "$x$ factorial." For example, $4! = (4)(3)(2)(1) = 24$, and $6! = (6)(5)(4)(3)(2)(1) = 720$. (Note that by definition $0! = 1$.) Thus, the number of combinations of $x$ items that can be chosen from a set of $n$ items can also be written as

*Factorials*

$$\frac{n(n-1)\ldots(n-x+1)}{x!} = \frac{n(n-1)\ldots(2)(1)}{(n-x)(n-x-1)\ldots(2)(1)x!} = \frac{n!}{(n-x)!x!}$$

To switch from crew to basketball, if we want to know the number of different basketball teams of five members each that can be formed from a group of eight players, this is the number of combinations of five items that can be selected from a set of eight items. Thus, the answer must be

$$\frac{(8)(7)(6)(5)(4)}{(5)(4)(3)(2)(1)} = \frac{6720}{120} = 56.$$

Expressed in terms of factorials, the answer is

$$\frac{8!}{3!5!} = \frac{40,320}{(6)(120)} = 56.$$

GALILEO ON GAMBLING: A CASE STUDY[11] To illustrate the application of combinations and permutations, let's go back several hundred years to a case involving the famous mathematician and astronomer Galileo. In 1613, Galileo accepted a position as First and Extraordinary Mathematician of the University of Pisa and Mathematician to his Serene Highness Cosimo II of Tuscany. Besides receiving a title that betrayed no false modesty, Galileo received a large salary and no duties (a good combination in any age), except that he was asked by Cosimo to work on certain problems in which his Serene Highness was interested. One problem that Cosimo apparently asked Galileo to solve was the following. Three dice are thrown. Although there are six ways of getting a 9—621 (that is, one die showing a 6, another a 2, and another a 1), 531, 522, 441, 432, and 333—and six ways of getting a 10—631, 622, 541, 532, 442, and 433—the probability of throwing a 9 seems in fact to be less than that of throwing a 10. Why?

Before considering Galileo's solution, it is important to recognize that this was not a purely academic exercise. At that time, there was a popular dice game in which three ordinary six-sided dice were thrown. Cosimo apparently had gambled sufficiently often at this game to find that the probability of getting a 9—that is, the probability that the sum of the numbers on the three dice would be 9—was lower than the probability of getting a 10. He wanted to know why this was the case.

Galileo began his reply by pointing out that if each side of each die is equally likely to come up, there are $(6)(6)(6) = 216$ configurations of the dice that can arise. To see this, note that the first die can have six outcomes (1 to 6). And since the second die can also have six outcomes, the number of pairs of outcomes is $(6)(6)$. Moreover, since the third die can also have six outcomes, the number of triads of outcomes is $(6)(6)(6)$. Further, since each side of each die is equally likely to come up, the probability of each of these triads is $1/216$. This is, of course, a simple illustration of what we discussed in previous sections of this chapter.

Next, Galileo noted that although it was true that both 9 and 10 could be rolled in six ways, the probability of each of these ways was not the same. For example, the probability of 333 was lower than that of 432. To see why, let's consider each of the three dice, and let's denote the outcome of a throw by three numbers in brackets, the first being the number on the first die, the second being the number on the second die, and the third being the number on the third die. Clearly, 432 can occur on the basis of six of these triads—[432], [423], [342], [324], [243], [234]—whereas 333 can occur on the basis of only one triad, [333]. Why six triads in the case of 432? Because this is the number

[11] The discussion in this section is based on F. N. David, *Games, Gods, and Gambling* (New York: Hafner, 1962).

**Table 3.4**
**Number of Triads**
**Resulting in Dice**
**Showing a 9**

| Way of rolling the number | Total number of triads corresponding to this way of rolling the number | (Individual triads) | | |
|---|---|---|---|---|
| | | First die | Second die | Third die |
| 621 | 6 | 6 | 2 | 1 |
| | | 6 | 1 | 2 |
| | | 2 | 6 | 1 |
| | | 2 | 1 | 6 |
| | | 1 | 6 | 2 |
| | | 1 | 2 | 6 |
| 531 | 6 | 5 | 3 | 1 |
| | | 5 | 1 | 3 |
| | | 3 | 5 | 1 |
| | | 3 | 1 | 5 |
| | | 1 | 5 | 3 |
| | | 1 | 3 | 5 |
| 522 | 3 | 5 | 2 | 2 |
| | | 2 | 5 | 2 |
| | | 2 | 2 | 5 |
| 441 | 3 | 4 | 4 | 1 |
| | | 4 | 1 | 4 |
| | | 1 | 4 | 4 |
| 432 | 6 | 4 | 3 | 2 |
| | | 4 | 2 | 3 |
| | | 3 | 4 | 2 |
| | | 3 | 2 | 4 |
| | | 2 | 4 | 3 |
| | | 2 | 3 | 4 |
| 333 | 1 | 3 | 3 | 3 |
| Total | 25 | | | |

of permutations of three items selected from a set of three items. Why 1 triad in the case of 333? Because there is only one triad in which all dice show a 3.

Galileo constructed a table showing the number of triads resulting in each of the six ways that 9 could be formed. As indicated in Table 3.4, there are 25 such triads. Similarly, he showed that there are 27 triads resulting in the six ways that 10 can be formed (as shown in Table 3.5). Since each triad has a probability of 1/216, it follows that the probability of rolling a 9 is 25/216, whereas the problem of rolling a 10 is 27/216. This was how Galileo solved the problem.

**EXERCISES**

**3.40** There are four roads (A, B, C, and D) from Avon to Burgundy and three roads (a, b, and c) from Burgundy to Coventry. The Gamma Machine Company sends a shipment from Avon to Coventry via Burgundy. This shipment gets lost after reaching Coventry, and the traffic agent is asked what route it took from Avon to Coventry. He doesn't know, and says that each possible route was equally likely. Eventually, he is forced to guess, and he guesses that the shipment went via road A to Burgundy and via road b to Coventry. What is the probability that he is correct?

| Way of rolling the number | Total number of triads corresponding to this way of rolling the number | (Individual triads) | | |
|:---:|:---:|:---:|:---:|:---:|
| | | First die | Second die | Third die |
| 631 | 6 | 6 | 3 | 1 |
| | | 6 | 1 | 3 |
| | | 3 | 6 | 1 |
| | | 3 | 1 | 6 |
| | | 1 | 6 | 3 |
| | | 1 | 3 | 6 |
| 622 | 3 | 6 | 2 | 2 |
| | | 2 | 6 | 2 |
| | | 2 | 2 | 6 |
| 541 | 6 | 5 | 4 | 1 |
| | | 5 | 1 | 4 |
| | | 4 | 5 | 1 |
| | | 4 | 1 | 5 |
| | | 1 | 5 | 4 |
| | | 1 | 4 | 5 |
| 532 | 6 | 5 | 3 | 2 |
| | | 5 | 2 | 3 |
| | | 3 | 5 | 2 |
| | | 3 | 2 | 5 |
| | | 2 | 5 | 3 |
| | | 2 | 3 | 5 |
| 442 | 3 | 4 | 4 | 2 |
| | | 4 | 2 | 4 |
| | | 2 | 4 | 4 |
| 433 | 3 | 4 | 3 | 3 |
| | | 3 | 4 | 3 |
| | | 3 | 3 | 4 |
| Total | 27 | | | |

Table 3.5
Number of Triads
Resulting in Dice
Showing a 10

**3.41** The number of combinations of $x$ items that one can select from a set of $n$ items is often represented as $\binom{n}{x}$. Prove that $\binom{n}{x} = \binom{n}{n-x}$.

**3.42** How many different poker hands can be drawn in which there is a straight (or straight flush)?

**3.43** A panel of experts is asked to rank eight possible designs for a new type of shopping mall. How many different possible rankings could the panel give (assuming no ties)?

**3.44** A shipment of goods contains 50 items, 5 of which are defective. To determine whether the shipment should be accepted, the buyer picks one item at random and then (without replacing the first item) picks another at random. If neither is defective, the shipment will be considered acceptable. What is the probability that the buyer will accept the shipment?

**3.45** A publishing firm has six books in physics. Each year, it mails an advertisement to physics professors focusing on three of its physics books. The firm does not want to focus on the same three books more than once; it does not mind repeating one or two of them, but it does not want all three to be the same as in a previous mailing. How long can the firm go on before it will no longer be able to abide by this rule?

**3.46** In the opinion of the chairman of the board, there are seven persons on the board of directors of the Bona Fide Washing Machine Corporation who are competent to serve on a committee to investigate personnel problems. In how many ways can the chairman choose a committee of three members of the board, all of whom he regards as being competent to serve on it?

**3.47** If there are five men and two women on the board of directors who are competent to serve on a committee, in how many ways can the chairman of the board choose a committee of three if he wants two men and one woman on the committee (and all three are competent)?

**3.48** A firm wants to locate six new stores somewhere in the United States (including Alaska and Hawaii). It does not want more than one of the new stores to be located in the same state. In each state, the best location is determined. The firm's president wants to make sure that each possible set of locations is evaluated. If each such evaluation takes 2 hours of an executive's time, how many manhours (of executives' time) would it take to carry out the president's wishes?

**3.49** A factory must carry out seven different manufacturing operations (*A*, *B*, *C*, *D*, *E*, *F*, and *G*). Operations *F* and *G* must be the last two operations, *G* following *F*. The other operations can be performed in any order at all. The factory's manager wants to evaluate each possible sequence of operations to see which is least costly. If it takes one hour of an engineer's time to evaluate each sequence, how many manhours (of engineers' time) would it take to do what the manager wants?

**3.50** The Black Belt Sewing Machine Company relies on five suppliers (I, II, III, IV, and V) to provide it with motors. Each of these suppliers can ship motors to each of Black Belt's four plants (1, 2, 3, and 4). Black Belt stamps a letter on each motor to indicate which supplier it came from and which plant used it. For example, it stamps an *A* on a motor from supplier I that is put into a sewing machine coming out of plant 1, a *B* on a motor from supplier II that is put into a sewing machine coming out of plant 2, a *C* on a motor from supplier II that is put into a sewing machine coming out of plant 1, and so on.
  (a) How many letters will it take to represent all possible combinations of motors and plants?
  (b) If the Black Belt Sewing Machine Company bought the same number of motors in 1987 from each supplier, and if each supplier shipped the same number of motors to each of Black Belt's plants, what proportion of all Black Belt's motors had an *A* stamped on them in 1987?
  (c) Every week the Black Belt Sewing Machine Company tests one motor shipped by each of its five suppliers. The motors tested are ranked according to performance, from highest to lowest. The firm then sends this ranking to its plants. After repeating these tests hundreds of times, all possible orderings of the five suppliers have occurred. How many different rankings has the firm sent out to its plants?

**3.51** John Monroe takes a test consisting of eight true-false questions. In how many different ways can he answer the exam? If he has no idea of the answer to any of the questions, what is the probability of his getting a perfect score?

# Probability Distributions, Expected Values, and the Binomial Distribution

## 4.1 Introduction

In the previous chapter, we discussed basic concepts of probability. To understand statistical techniques, it is necessary to go more deeply into probability theory. In this chapter we will discuss the nature and characteristics of random variables and probability distributions. Then we will describe expected values and take up the most important discrete probability distribution, the binomial distribution. We shall also indicate some of the ways in which the binomial distribution is applied.

## 4.2 Random Variables

One of the basic concepts in probability theory is that of the random variable, which is defined as follows.

> RANDOM VARIABLE: A **random variable** is a numerical quantity the value of which is determined by an experiment. In other words, its value is determined by chance. More formally, it is a numerically valued function defined on a sample space.

In succeeding paragraphs we shall go into considerable detail to clarify what is meant by a random variable, but it is sufficient here to note two points concerning this definition. First, the definition assumes that some sort of an experiment is conducted. (As stressed in Chapter 3, the term *experiment* should be interpreted very broadly and covers more than just laboratory procedures.) The outcomes of this experiment constitute a sample space, as pointed out in Chapter 3. Second, our definition implies that a random variable is one whose value is defined for each element of the sample space. This means that its value is defined for each outcome of the experiment, which is equivalent to saying that its value is determined by chance (because the outcome of the experiment is determined by chance).

**Figure 4.1**
Sample Space for Roll
of Two Dice

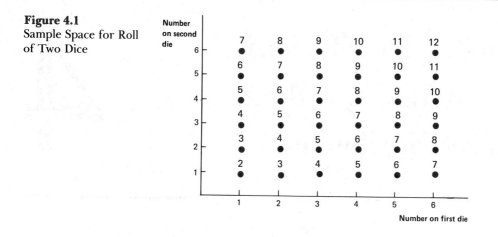

To illustrate, if we take the sum of the numbers showing on two dice, this sum is clearly a random variable since its value is determined by the "experiment" of throwing the pair of dice. As pointed out in the previous chapter, the sample space generated by this experiment can be represented by the 36 points in Figure 3.1, which are reproduced in Figure 4.1. The value of this sum corresponding to each element of the sample space is shown above the point representing this element in Figure 4.1. Obviously, this sum fulfills the condition of the definition that a random variable must assume a value for each element of the sample space.

As another illustration, suppose that a gambler proposes the following bet on the outcome of the throw of a single die. He will give you $10 if you throw a 4, a 5, or a 6, and you will give *him* $10 if you throw a 1, a 2, or a 3. The amount you win (or lose)—+$10 (or −$10)—is a random variable since its value is determined by the outcome of an experiment. (That is, its value is determined by chance.) This experiment can result in six possible outcomes, corresponding to the die's coming up 1, 2, 3, 4, 5, or 6. For each of these outcomes (or points in the sample space), the amount you win is defined, which agrees with our definition of a random variable.

At this point, the various characteristics of a random variable should be coming into focus. First, *a random variable must take on numerical values.* In the experiment with the statistics course discussed in the previous chapter, the outcomes "pass" and "not pass" are not the values of a random variable because they are not in numerical form. However, if we arbitrarily let "pass" equal 0 and "not pass" equal 1, the result is a random variable. Second, *the value of a random variable must be defined for all possible outcomes of the experiment in question (that is, for all elements of the sample space).* For example, in the $10 bet based on the throw of the die, suppose that you win $10 if you throw a 4, a 5, or a 6, and that you lose $10 if you throw a 1 or a 2. But suppose also that *the amount you win or lose is undefined if you throw a 3.* Under these new rules the amount you win (or lose) is no longer a random variable. Why? Because how much you win (or lose) is no longer defined for *all* possible outcomes of the experiment. Specifically, if the outcome of the experiment is a 3, the value of the amount you win (or lose) is unde-fined.

It is important to recognize that the value of a random variable is unknown *before* the experiment in question is carried out. *After* the experiment is carried out, the value of the random variable is always known. For example, if you roll a pair of dice, the sum of the numbers on them is unknown before the roll; but after the roll, the value of the sum is known. Similarly, if you accept the gamble involving the roll of a single die, the amount you win or lose is unknown before you roll the die; after the roll, this amount is known.

## DISCRETE AND CONTINUOUS RANDOM VARIABLES

Statisticians distinguish between two types of random variables: discrete and continuous. A *discrete random variable* can assume *only a finite or countable[1] number of distinct values*. For example, the sum of the numbers showing on two dice is a discrete random variable since the sum can only assume one of 11 possible values: 2, 3, 4, . . . , 11, or 12. The amount at stake in the gamble with the single die is also a discrete random variable since it can assume only one of two possible values: +\$10 and −\$10. Many—but by no means all—important random variables are discrete. Some random variables assume *any numerical value on a continuous scale*. Such random variables are called *continuous random variables*.

To illustrate the concept of a continuous random variable, let's return to the case study we took up in Chapter 1 (page 26), which dealt with a manufacturing plant that produces pieces of metal in which two holes are stamped. The specifications require that the distance between the hole centers should be 3.000 ± .004 inches. In fact, how-ever, the distance between the hole centers varies from piece to piece, because of differences in machines, dies, workers, and other factors. The productive process of turning out each piece of metal can be regarded as an experiment, and the outcome of this experiment is the distance between the hole centers. This distance is a random variable; but because it can vary continuously it is not a discrete random variable. This means that the distance between the hole diameters is not confined to certain rounded values like 3.001 inches, 3.002 inches, and so on, but can also assume *any* value in between them. Of course, in practice we might only measure the distance to the nearest .001 inch, in which case the rounded distances are discrete. But the true distance is continuous.

In this chapter we shall be concerned entirely with discrete random variables, and in the next we shall take up continuous random variables. Meanwhile, the following examples should be useful in illus-trating the meaning and characteristics of a random variable.

---

[1] Some discrete random variables, like the Poisson variable described in Chapter 5, assume a countably infinite number of values.

**EXAMPLE 4.1** A fair coin is flipped three times. If it comes up heads three times you receive $100; if it comes up heads twice, you receive $50; if it comes up heads once, you lose $75; and if it never comes up heads, you lose $100.

(a) Is the number of heads that comes up a random variable? If so, is it discrete or continuous?

(b) Is the amount you win (or lose) a random variable? If so, is it discrete or continuous?

SOLUTION: (a) The number of heads that comes up is a random variable, since there are eight possible outcomes (listed below) of this experiment and to each outcome there corresponds a certain number of heads:

heads/heads/heads    tails/tails/heads
heads/heads/tails     tails/heads/tails
heads/tails/heads     heads/tails/tails
tails/heads/heads     tails/tails/tails

This random variable is discrete because it can assume only four possible values: 0, 1, 2, and 3.

(b) The amount you win is a random variable, because for each of the outcomes of this experiment you win (or lose) a corresponding amount. The amount you win (or lose) is a discrete random variable because it can assume only four possible values: +$100, +$50, −$75, and −$100.

**EXAMPLE 4.2** A shipment contains 20 machines, 4 of which are defective. The firm receiving the shipment chooses a random sample of 3 machines (without replacement). If any of the machines in the sample is defective, the firm rejects the shipment.

(a) Is the number of defective machines in the sample a random variable? If so, is it discrete or continuous?

(b) Is whether or not the shipment is rejected a random variable? If so, is it discrete or continuous?

SOLUTION: (a) The number of defective machines in the sample is a random variable since there are eight possible outcomes (listed below) of this experiment:

N N N     N N D     N D N     D N N
N D D     D N D     D D N     D D D

The first letter stands for the first machine in the sample and is D if it is defective or N if it is not defective. The second letter stands for the second machine in the sample and is D or N. The third letter, which is also D or N, stands for the third machine. To each of these outcomes there corresponds a number of defective machines in the sample. This random variable is discrete since it can assume only four possible values: 0, 1, 2, and 3.

(b) Whether or not the shipment is rejected is not a random variable because it is not in numerical form. However, it can be turned into a random variable by letting zero stand for rejection of the shipment and 1 stand for acceptance of it. (Obviously, any other pair of arbitrarily chosen numbers could also be used for this purpose.)

# 4.3 Probability Distributions

Based on the previous section, we know that the value of a random variable is determined by the outcome of a corresponding experiment. For example, the value of one particular sum of the numbers shown on a pair of dice is determined by one particular roll of the dice. And the amount you win or lose in the gamble involving the single die is determined by each roll of the die. Thus, since there is a certain probability of each outcome of the experiment, there must also be a certain probability of each value of the random variable. For example, since there is a 1/36 chance that each of the points in Figure 4.1 will occur (if the dice are true), it must be possible to deduce the probability that the sum of the numbers showing on the dice will equal 2, 3, 4, and so on. Similarly, since there is a 1/6 chance that a true die will show a 1, 2, 3, 4, 5, or 6, it must be possible to deduce the probability that the amount you will win in your gamble involving the single die will be +\$10 or −\$10. Such probabilities are provided by the probability distribution of the random variable, which is defined as follows.

> PROBABILITY DISTRIBUTION: *The probability distribution of a random variable* X *provides the probability of each possible value of the random variable. If* P(x) *is the probability that* x *is the value of the random variable, we can be sure that* $\Sigma P(x) = 1$, *where the summation is over all values that* X *takes on. This is because these values of* X *are mutually exclusive and one of them must occur.*

To illustrate a probability distribution, consider once again the sum of the numbers showing on two true dice. What is the probability distribution of this sum? Clearly, this sum can assume only the following values: 2, 3, 4, . . . , 11, and 12. *What is the probability of a 2?* This value of the sum can arise only if each die shows a 1, and the probability of this is 1/36. *What is the probability of a 3?* This value of the sum can arise if the first die shows a 1 and the second shows a 2, or if the first shows a 2 and the second shows a 1. Since the probability of each is 1/36, the probability of either one is 2/36. (Remember the addition rule.) *What is the probability of a 4?* This value of the sum can arise in three ways (a 1 on the first die and a 3 on the second, a 2 on the first die and a 2 on the second, or a 3 on the first die and a 1 on the second), each of which has a probability of 1/36. Thus, the probability of a 4 is 3/36. Based on similar reasoning, it is easy to figure out the probability of a 5, 6, . . . , 12. (One way to do this is to count the number of points

**Table 4.1**
Probability Distribution
of Sum of Numbers
Showing on Two True
Dice

| Sum of numbers | Probability |
|:---:|:---:|
| 2 | 1/36 |
| 3 | 2/36 |
| 4 | 3/36 |
| 5 | 4/36 |
| 6 | 5/36 |
| 7 | 6/36 |
| 8 | 5/36 |
| 9 | 4/36 |
| 10 | 3/36 |
| 11 | 2/36 |
| 12 | 1/36 |
| Total | 1 |

in Figure 4.1 that result in a particular sum and multiply this number by 1/36, which is the probability of each point.) The results are shown in Table 4.1. As you can see, the sum of the probabilities in the probability distribution equals 1, which agrees with the definition above.[2]

Although it sometimes is convenient to represent a probability distribution by a *table* like Table 4.1, it is sometimes even more convenient to use a *graph*. We can present the probability distribution of the sum of two dice in a *line chart,* as shown in Figure 4.2. Along the horizontal axis are ranged the various possible values of the random variable. At each point on the horizontal axis corresponding to such a value a vertical line is erected, the length of which is equal to the probability of this value's occurrence. Thus, in Figure 4.2 the vertical line at

*Line Chart*

**Figure 4.2**
Probability Distribution
of Sum of Numbers
Showing on Two True
Dice

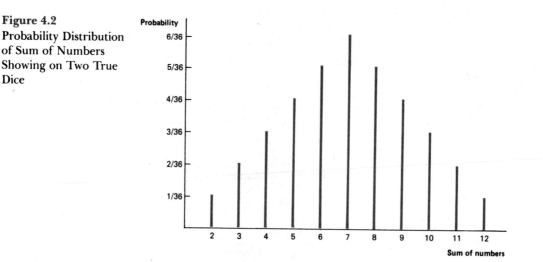

[2] Why must the sum of $P(x)$ equal 1? Since the various values of $x$ are mutually exclusive, the probability that *some* value of the random variable occurs is the sum of $P(x)$, as we know from the addition rule. Since some value of the random variable *must* occur, the probability that some value of the random variable occurs must equal 1. Thus, it follows that the sum of $P(x)$ must equal 1.

3 is twice as long as that at 2 since the probability of a 3 is twice that of a 2. As you can see from Figure 4.2, the highest vertical line is at 7, and the line chart is symmetrical about this line in this case. (That is, the probability of a 6 equals the probability of an 8, the probability of a 5 equals the probability of a 9, and so on.)

Still another way to represent a probability distribution is by a mathematical function. As pointed out in our definition of a probability distribution, the probability that a random variable equals $x$ is denoted by $P(x)$. In some cases we can express $P(x)$ as a relatively simple mathematical function of $x$. For example, if you flip a fair coin twice, the number of heads that comes up is a random variable whose probability distribution can be represented by the following equation:

$$P(x) = \frac{1/2}{x!(2-x)!}, \text{ for } x = 0, 1, 2, \qquad (4.1)$$

where $x!$ is defined as $x(x-1)(x-2) \ldots (2)(1)$. Thus, $2! = (2)(1) = 2$, and $1! = 1$. (Also, note that $0! = 1$.)[3] The derivation of equation (4.1) is discussed in detail in Section 4.7. For now, it is essential only that we make sure that this equation really does represent the probability distribution of the number of heads. To see that this is true, note that the equation says that $P(0) = 1/2 \div [0!2!] = 1/4$; $P(1) = 1/2 \div [1!1!] = 1/2$; $P(2) = 1/2 \div [2!0!] = 1/4$. Clearly, each of these probabilities is right. Note, too, that $P(0) + P(1) + P(2) = 1$, as it should according to our definition of a probability distribution.

## PROBABILITY DISTRIBUTIONS AND RELATIVE FREQUENCY DISTRIBUTIONS

It is important to note the similarity between a probability distribution and a frequency distribution (discussed in Chapter 1). To see how similar they are, consider the probability distribution of the number coming up on a single die. If the die is true, this probability distribution is

$$P(x) = 1/6, \text{ for } x = 1, 2, 3, 4, 5, 6. \qquad (4.2)$$

Suppose that this die is thrown 1,000 times and that the number of times each number comes up is as shown in Table 4.2. If we divide the number of times each number come up by the total number of throws (1,000), we get the proportion of times each number comes up (as shown in the third column of Table 4.2). Each of these proportions is the empirical counterpart of the corresponding probability. That is, the proportion of cases in which a 3 comes up is the empirical counter-

---

[3] Readers of Appendix 3.1 will already have encountered $x!$, which is read as "$x$ factorial."

**Table 4.2**
Number of Times
Each Number Shows
on a Die Cast 1,000
Times

| Number on die | Number of times number comes up | Proportion of times number comes up |
|---|---|---|
| 1 | 170 | .170 |
| 2 | 159 | .159 |
| 3 | 172 | .172 |
| 4 | 158 | .158 |
| 5 | 160 | .160 |
| 6 | 181 | .181 |
| Total | 1,000 | 1.000 |

part of the probability of a 3. If the frequency concept of probability applies, each of these proportions will tend to get closer and closer to the corresponding probabilities as the number of cases included in the frequency distribution becomes larger and larger.

*Relative Frequency Distribution*

Another way of stating this is to say that a probability distribution of a certain type is often regarded as the *relative frequency distribution* of the population. *A relative frequency distribution shows the proportion (not the number) of cases falling within each class interval.* For example, if we were to roll a true die again and again until finally data were collected concerning millions of rolls, what would be the relative frequency distribution of this population? After dividing the number of throws resulting in a 1, 2, 3, 4, 5, or a 6 by the total number of throws, we would obtain the proportion of throws coming up a 1, 2, 3, 4, 5, or a 6. This is the relative frequency distribution in this situation. In essence, this relative frequency distribution would be like the probability distribution in equation (4.2). Thus, *probability distributions are often used to represent or approximate population relative frequency distributions,* as we shall see in later chapters.

Since the concept of a probability distribution is so important in statistics, it is essential that it be well understood. The following two examples should help to clarify and illustrate its meaning.

EXAMPLE 4.3 As in Example 4.1, a fair coin is flipped three times. If it comes up heads three times you receive $100; if it comes up heads twice you receive $50; if it come up heads once you lose $75; if it never comes up heads you lose $100.

(a) Construct a line chart of the probability distribution of the number of heads.

(b) Construct a line chart of the probability distribution of the amount you win.

SOLUTION: (a) There are the following eight possible outcomes of this experiment, each with a probability of 1/8:

| | |
|---|---|
| heads/heads/heads | tails/tails/heads |
| heads/heads/tails | tails/heads/tails |
| heads/tails/heads | heads/tails/tails |
| tails/heads/heads | tails/tails/tails |

Thus, the probability of no heads is 1/8; the probability of heads once is 3/8; the probability of heads twice is 3/8; and the probability of heads three times is 1/8. The line chart is as follows.

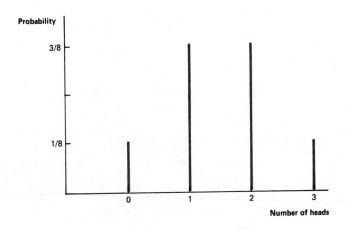

(b) Since the probability of no heads is 1/8, the probability that you will lose $100 is 1/8. Since the probability of heads once is 3/8, the probability that you will lose $75 is 3/8. Since the probability of heads twice is 3/8, the probability that you will win $50 is 3/8. Since the probability of heads three times is 1/8, the probability that you will win $100 is 1/8. Thus, the line chart is as follows.

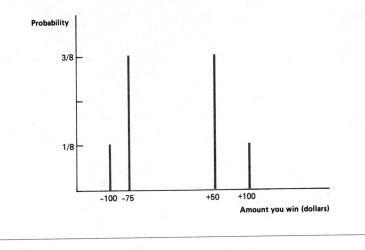

EXAMPLE 4.4 John Gillicuddy buys two tickets at $1 each on a car that is being raffled off. The organizers of the raffle sell a total of 5,000 tickets, and one ticket is picked at random to determine the winner. The car is worth $5,000.

(a) Construct a table showing the probability distribution of the outcome of the raffle, if 0 represents Gillicuddy's not winning the car and 1 represents his winning it.

(b) Construct a table showing the probability distribution of the amount that Gillicuddy wins or loses in the raffle.

SOLUTION: (a) Since John Gillicuddy has two tickets out of a total of 5,000, his probability of winning the car is $2/5,000 = 1/2,500$ and his probability of not winning the car is $2,499/2,500$. Thus, if 0 stands for his not winning and 1 stands for his winning, the probability of each value of this random variable is

| Value of random variable | Probability |
|---|---|
| 0 | 2,499/2,500 |
| 1 | 1/2,500 |

(b) If Gillicuddy wins the car, he makes $4,998 (the value of the car less the amount spent on the tickets); if he does not win the car, he loses $2 (the amount spent on the tickets). Thus, the amount he wins or loses can assume two possible values, with the following probabilities:

| Amount he wins or loses | Probability |
|---|---|
| −$2 | $\dfrac{2,499}{2,500}$ |
| +$4,998 | $\dfrac{1}{2,500}$ |

## EXERCISES

**4.1** Three cards—the three, four, and five of spades—are in a box. One card is drawn at random. Then a second card is drawn at random. (The first card is *not* returned to the box before the second is drawn.) Let $X$ represent the sum of the numbers on the two cards drawn from the box.
(a) What is the probability distribution of $X$?
(b) What is the probability that $X$ is less than 8? Greater than 7?

**4.2** Indicate whether the following random variables are continuous or discrete:
(a) the weight of a randomly chosen heavyweight boxer.
(b) the number of heavyweight boxers knocked out in a seven-day period.
(c) the height of a randomly chosen heavyweight boxer.

**4.3** Which of the following cannot be a probability distribution?
(a) $P(x) = x/4$      $x = 1,2,3$
(b) $P(x) = x^2/8$      $x = 1,2,3$
(c) $P(x) = x/3$      $x = -1,+1,+3$

**4.4** A box contains twice as many red marbles as green marbles. One marble is drawn at random from the box and replaced; then a second marble is drawn at random from the box. If both marbles are green, you win $5; if both are red, you lose $1; and if they are of different colors, you win or lose nothing.
(a) What is the probability distribution of the amount you win or lose?
(b) What is the probability that you at least break even?

**4.5** The Alpha Corporation sells bicycles. Based on past experience, it feels that in the summer months it is equally likely that it will sell 0, 1, 2, 3, or 4 bicycles in a day. (The firm has never sold more than 4 bicycles per day.)

(a) Is the number of bicycles sold in a day a random variable?

(b) If so, what values can this random variable assume?

(c) If the number of bicycles sold in a day is a random variable, construct a table showing its probability distribution.

(d) Plot the probability distribution of the number of bicycles sold in a day in a line chart. What mathematical function can represent this probability distribution?

(e) What is the probability that the number of bicycles sold on a given day will be less than 1? Less than 3? Less than 4? Less than 6?

(f) The Alpha Corporation has only one salesman, whose income depends on the number of bicycles he sells per day. Specifically, he receives no commission on the first bicycle sold per day, a $20 commission on the second bicycle sold in a day, a $30 commission on the third, and a $40 commission on the fourth. (Thus, if he sells 3 bicycles in a given day, his commissions for this day total $50.) His income consists entirely of these commissions. Is his income on a particular day a random variable? If so, what values can this random variable assume? If his income is a random variable, construct a table showing its probability distribution.

(g) Plot the probability distribution of the salesman's income on a particular day in a line chart.

(h) What is the probability that his income in a particular day will exceed $20? $30? $40?

(i) Suppose that the number of bicycles sold one day is independent of the number sold the next day. Construct a table showing the probability distribution of the total number of bicycles sold in a two-day period. What is the probability that this number will exceed 2? 4? 6? 8?

(j) Unfortunately, the bicycle salesman has a weakness for gambling and is in debt to the Mob for $100. The Mob's local enforcer tells the salesman that he has two days to pay up, or he'll need a hospital bed. If the number of bicycles sold one day is independent of the number sold the next day, what is the probability that the salesman will earn enough during this two-day period to stay out of the hospital? (Assume that commissions are net of taxes and other deductions.)

(k) At the end of the two-day period described in (j) the salesman has earned only $70. The Mob's enforcer says that he will give the salesman a chance to win the extra $30. He will allow the salesman to flip a coin twice. If it comes up heads both times, he will give the salesman the $30 he needs; if it doesn't come up heads both times, he will take the salesman's $70 and begin the hospital-filling mayhem. Plot in a line chart the probability distribution for the salesman's monetary gains or losses in this game.

## 4.4 Expected Value of a Random Variable

As pointed out in the previous section, the probability distribution of a random variable is analogous to a frequency distribution. It is therefore not surprising that just as we found it useful to calculate the mean of a frequency distribution, we now find it useful to calculate the mean of a probability distribution. Another term often used to denote the mean of a probability distribution is the *expected value of the random variable,* which is defined as follows.

> EXPECTED VALUE OF A RANDOM VARIABLE: *The expected value of a discrete random variable* X, *denoted by* E(X), *is the weighted mean of the possible values that the random variable can assume, where the weight attached to each value is the probability that the random variable will assume this value. In other words,*

*Formula for E(X)*

$$E(X) = \sum_{i=1}^{m} x_i P(x_i),$$
(4.3)

*where the random variable* X *can assume* m *possible values,* $x_1$, $x_2$, . . . , $x_m$, *and the probability of its equaling* $x_i$ *is* $P(x_i)$.

Clearly, the expected value of a random variable is analogous to the mean of a frequency distribution. If we compare the definition in equation (4.3) with our definition of the mean of a set of data calculated from a frequency distribution in equation (2.3), we see that the definitions are essentially the same.[4]

For the reason just stated, the expected value of any random variable *X* is often called the *mean* of *X.* If, as described in the previous section, the probability distribution in question is viewed as the population relative frequency distribution (which shows the proportion of cases with each value), the expected value of the random variable is really the population mean. To illustrate: We roll a true die a million times, and we regard the resulting relative frequency distribution of outcomes as the population. Clearly, the expected value of the number showing on a true die is

$$E(X) = \sum_{i=1}^{6} x_i P(x_i) = 1(1/6) + 2(1/6) + 3(1/6) + 4(1/6) +$$

$$5(1/6) + 6(1/6) = 3\frac{1}{2}.$$

---

*Basic Idea #7*    **If the experiment determining the value of a random variable were performed over and over, the random variable's expected value would be its mean value over the long run.**

---

Why? Because the random variable in question (the number coming up) can assume 6 possible values (thus, *m* = 6). These possible values

---

[4] To see how similar these definitions are, suppose that in each class interval the observations are all equal. (In other words, assume that every observation equals its class mark.) Then, since $f_j \div n$ in equation (2.3) is the relative frequency with which $x_j$ occurs in the set of data, it can be regarded as the probability that $x_j$ occurs in this set of data. And if $P(x_j)$ is substituted for $f_j \div n$ in equation (2.3), the result is equivalent to the right side of equation (4.3).

are 1 to 6: thus $x_1 = 1, x_2 = 2, \ldots, x_6 = 6$. And the probability of each value is $1/6$: thus, $P(x_1) = P(x_2) = \ldots = P(x_6) = 1/6$. The mean of the population of 1 million outcomes of rolling a true die is 3 1/2, since the number of rolls in this population is so large that the proportion of cases when each number comes up is equal, for all practical purposes, to the probability of the number's coming up.

## ROLE OF EXPECTED VALUES IN DECISION MAKING

The concept of a random variable's expected value (or mean, or *mathematical expectation,* as it is sometimes called) is extremely important in statistics. For example, consider the following gamble, which has already been described in Examples 4.1 and 4.3. A fair coin is flipped three times. If it comes up heads three times, you receive $100; if it comes up heads twice, you receive $50; if it comes up heads once, you lose $75; if it never comes up heads, you lose $100. Should you accept this gamble? Or put differently, if you were to repeat this gamble again and again, would you tend to come out ahead or would you tend to lose? The answer depends on the expected value of the amount you win, an amount which (as we saw in Example 4.1) is a random variable. Why does it depend on this expected value? Because this expected value is the mean amount you will win (or lose) if you accept the gamble again and again for an indefinite number of times. In other words, the expected value is the mean of the infinite population of amounts you would win (or lose) if you accepted this gamble an infinite number of times.

What is the expected value of the amount you would win (or lose) from this gamble? Once again, let $X$ stand for the relevant random variable, which in this case is the amount you would win (or lose). This random variable can assume only four values, $x_1 = 100$, $x_2 = 50$, $x_3 = -75$, and $x_4 = -100$, with the result that

$$E(X) = \sum_{i=1}^{4} x_i P(x_i) = 100(1/8) + 50(3/8) - 75(3/8) - 100(1/8)$$

$$= -9\frac{3}{8},$$

since (as we know from Example 4.3) $P(x_1) = P(x_4) = 1/8$ and $P(x_2) = P(x_3) = 3/8$. Because the expected value of the amount you would win from this gamble is negative, it is clear that you would lose money if you were to repeat the gamble over and over. More specifically, *if you averaged the results of repeating this gamble over and over, you would lose $9.375 per gamble.* Thus, the gamble is not a "fair bet."

Whether the expected value of one's winnings from a particular wager is positive or negative is obviously of importance in determining whether or not one should accept the wager. For example, if you have the choice of accepting a gamble where the expected value of your winnings is +$10 (as opposed to -$9.375 for the gamble described above), you may decide to accept the former rather than the latter. As you would expect, a decision maker often chooses the action or gamble that has the largest expected value of winnings or profits. But in general, although the expected value of one's winnings or profits is of importance, this alone is not the sole indicator of what action a decision maker should take. Much more will be said on this subject in Chapter 15.

The two examples below illustrate the calculation and interpretation of expected values.

---

EXAMPLE 4.5 At a certain gambling casino the following game is played: The dealer, who works for the house, allows to pick two cards from a full deck. If both are hearts, you win $15; otherwise, you lose $1. What is the expected value of the amount you win (or lose) each time you play this game?

SOLUTION: The amount you would win is a random variable that can assume two possible values, +$15 or −$1. The probability that it equals $15 is the probability of your getting two hearts, which is $12/51(13/52) = 1/17$, or .059. (Recall Example 3.6.) The probability that the amount will be −$1 must therefore be $1 − .059$, or .941. Thus, the expected value of this random variable is

$$(\$15)(.059) + (-\$1)(.941) = -\$.056.$$

---

EXAMPLE 4.6 A marketing executive must decide whether or not to use a new label on a product. The firm will gain $800,000 if he adopts the new label and it turns out to be superior to the old label. The firm will lose $500,000 if the executive adopts the new label and it proves to be not superior to the old one. The firm will neither gain nor lose money if the executive sticks with the old label. The executive feels there is a 50-50 chance that the new label is superior to the old and a 50-50 chance that it is not. If he wants to take the action with the higher expected gain to the firm, should he decide to use the new label or not?

SOLUTION: If the executive decides to adopt the new label, the expected value of the firm's gain is

$$(\$800,000)(1/2) + (-\$500,000)(1/2) = \$150,000,$$

because the firm's gain is a random variable that can assume two possible values, +$800,000 and −$500,000, and the probability of each value is $1/2$. If the executive decides not to adopt the new label, the expected value of the firm's gain is zero, since zero is the only possible value that it can assume. Since $150,000 exceeds zero, the expected gain if he adopts the new label is higher than if he does not. Thus, if he wants to take the action with the higher expected gain, he should adopt the new label.

## 4.5 Variance and Standard Deviation of a Random Variable

Just as the expected value of a random variable is analogous to the mean of a frequency distribution, the variance of a random variable is analogous to the variance of a frequency distribution. In particular, the variance of a random variable is defined as follows.

VARIANCE OF A RANDOM VARIABLE: *The variance of a random variable* X, *denoted by* $\sigma^2(X)$, *is the expected value of the squared deviations of the random variable from its expected value. In other words,*

$$\sigma^2(X) = E([X - E(X)]^2) = \sum_{i=1}^{m} [(x_i - E(X)]^2 P(x_i),$$

(4.4)    *Formula for* $\sigma^2(X)$

*where the random variable can assume* m *possible values,* $x_1, x_2, \ldots,$ $x_m$, *and* $P(x_i)$ *is the probability of its equaling* $x_i$.

To see the similarity between the variance of a random variable and the variance of a body of data, recall from Chapter 2 that the latter was defined as the mean of the squared deviations of the observations from their mean. Substitute "expected value" for "mean" in the previous sentence, and you have the definition of the variance of a random variable.

Similarly, the standard deviation of a random variable is analogous to the standard deviation of a body of data. Its definition is as follows.

STANDARD DEVIATION OF A RANDOM VARIABLE: *The standard deviation of a random variable is the positive square root of the random variable's variance. In other words, the standard deviation of a random variable* X, *denoted by* $\sigma(X)$, *is*

$$\sigma(X) = \sqrt{E([X - E(X)]^2)} = \sqrt{\sum_{i=1}^{m} [x_i - E(X)]^2 P(x_i)}.$$

(4.5)    *Formula for* $\sigma(X)$

Recall from Chapter 2 that the standard deviation of a set of data is the square root of the variance of the data. Thus, the definition here is analogous to the one given there.

Just as the standard deviation of a set of data indicates the extent of the dispersion or variability among the individual measurements within the set of data, *the standard deviation of a random variable indicates the extent of the dispersion or variability among the values that the random variable may assume*. For example, if it is *certain* that a random variable $X$ will equal a certain number, then the difference between $X$

and its expected value—that is, $X - E(X)$—will always be zero. Thus, the expected value of $[X - E(X)]^2$ will also be zero. Consequently, the variance and standard deviation of $X$ will be zero. On the other hand, if it is likely that $X$ will assume values *far removed* from its expected value, then the difference between $X$ and its expected value—that is $X - E(X)$—will have a large probability of being big. Therefore, the expected value of $[X - E(X)]^2$ will also be big; and consequently, the variance and standard deviation of $X$ will both be big.

To illustrate how one can calculate the variance and standard deviation of a random variable, let's calculate the variance and standard deviation of the number coming up on a true die. If this random variable is designated as $X$, it willl be recalled from the previous section of this chapter that the expected value of this random variable, $E(X)$, equals $3\frac{1}{2}$. Thus, the variance of this random variable is

$$\sigma^2(X) = \sum_{i=1}^{6} \left[ x_i - 3\frac{1}{2} \right]^2 P(x_i)$$

$$= \left[1 - 3\frac{1}{2}\right]^2 \left(\frac{1}{6}\right) + \left[2 - 3\frac{1}{2}\right]^2 \left(\frac{1}{6}\right) + \left[3 - 3\frac{1}{2}\right]^2 \left(\frac{1}{6}\right)$$

$$+ \left[4 - 3\frac{1}{2}\right]^2 \left(\frac{1}{6}\right) + \left[5 - 3\frac{1}{2}\right]^2 \left(\frac{1}{6}\right)$$

$$+ \left[6 - 3\frac{1}{2}\right]^2 \left(\frac{1}{6}\right)$$

$$= \left(6\frac{1}{4}\right)\left(\frac{1}{6}\right) + \left(2\frac{1}{4}\right)\left(\frac{1}{6}\right) + \left(\frac{1}{4}\right)\left(\frac{1}{6}\right) + \left(\frac{1}{4}\right)\left(\frac{1}{6}\right)$$

$$+ \left(2\frac{1}{4}\right)\left(\frac{1}{6}\right) + \left(6\frac{1}{4}\right)\left(\frac{1}{6}\right)$$

$$= \frac{35}{12}.$$

And since the standard deviation of a random variable is the square roots of its variance,

$$\sigma(X) = \sqrt{35/12}.$$

The following example provides additional practice in calculating the variance and standard deviation of a random variable.

---

**EXAMPLE 4.7** A fair coin is flipped three times. What is the variance and standard deviation of the number of times it comes up heads?

SOLUTION: If $X$ denotes the number of heads that come up, we know from Example 4.3 that the probability that $X = 0$ is $1/8$, the probability that $X = 1$ is $3/8$, the probability that $X = 2$ is

3/8, and the probability that $X = 3$ is 1/8. The expected value of $X$ is

$$E(X) = (0)(1/8) + (1)(3/8) + (2)(3/8) + (3)(1/8)$$

$$= 3/2.$$

Thus, the variance of $X$ is

$$\sigma^2(X) = [0 - 3/2]^2(1/8) + [1 - 3/2]^2(3/8) + [2 - 3/2]^2(3/8)$$

$$+ [3 - 3/2]^2(1/8)$$

$$= (9/4)(1/8) + (1/4)(3/8) + (1/4)(3/8) + (9/4)(1/8)$$

$$= 3/4.$$

And the standard deviation of $X$ is

$$\sigma(X) = \sqrt{3/4}.$$

## 4.6 Chebyshev's Inequality

Once we know the standard deviation of a random variable, we can make some interesting statements about the extent of the dispersion or variability among the values that the random variable can assume. In particular, we can apply the following theorem developed by the nineteenth-century Russian mathematician P. Chebyshev.

> CHEBYSHEV'S INEQUALITY: *For any random variable, the probability that the random variable will assume a value within* k *standard deviations of the random variable's expected value is at least* $1 - 1/k^2$.

Thus, the probability that a random variable will assume a value within *two* standard deviations of its expected value (that is, its mean) is *at least* $1 - 1/2^2$, or 3/4. And the probability that a random variable will assume a value within *three* standard deviations of its expected value is *at least* $1 - 1/3^2$, or 8/9.

In other words, this theorem tells us that the probability that a random variable will assume a value *more than* k standard deviations from the random variable's expected value is *less than* $1/k^2$. Consider the following example. You are given the expected value and standard deviation of the profits to be made from a particular business venture. The expected value is $400,000 and the standard deviation is $100,000. What is the probability that the profits from this venture will be below zero or above $800,000? If you know the probability distribution of the profits you can figure out this probability exactly, but suppose that *you are not given this probability distribution.* Using Chebyshev's inequality, you can still determine the *maximum* amount this probability can possibly be: $1/4^2$, or $1/16$. Since the probability that the profits will be below zero or above \$800,000 is the same as the probability that the profits will assume a value more than four standard deviations from the profits' expected value, the maximum amount this probability can be is $1/4^2$.

A key point to remember is that Chebyshev's inequality is true for *all* probability distributions.[5] The following is another illustration of its usefulness.

---

EXAMPLE 4.8 A manufacturer (cited in the case study we took up in Chapter 1) produces pieces of metal in which two holes are stamped. The specifications call for the distance between the hole centers to be $3.000 \pm .004$ inches. It is believed that the plant is turning out parts such that the expected value of this distance is 3.000 inches and the standard deviation of this distance is .0005. What is the maximum probability that a piece of metal produced by this plant will fail to meet the specifications?

SOLUTION: If the distance between hole centers in a piece of metal is less than 2.996 inches or greater than 3.004 inches, the piece does not meet specifications. Since the expected value of this distance is 3.000 inches and the standard deviation is .0005 inches, a piece will fail to meet the specifications if the distance between its hole centers assumes a value greater than eight standard deviations from the expected value. (Why? Because 2.996 inches is eight standard deviations below the expected value and 3.004 is eight standard deviations above the expected value.) According to Chebyshev's inequality, the probability of this occurring is less than $1/8^2$, or $1/64$.

---

EXERCISES

**4.6** A pair of true dice are thrown. If the total number rolled is less than 6, you win \$1; if it is greater than 8, you lose \$1; otherwise you neither gain nor lose any amount. What is the expected value of the amount gained or lost?

**4.7** If $Y = 2X$, what is the expected value of $Y$ if the expected value of $X$ is (a) 14; (b) $-3$? (For further discussion of problems of this kind, see Appendix 4.1 of this chapter.)

**4.8** A firm manufactures metal rods which must be rejected if they are not between 8.250 and 8.500 inches in diameter. The expected value of the diameters of the rods is 8.375 inches, and the standard deviation of these diameters is $1/40$ inch. What is the maximum probability that a diameter will be rejected?

---

[5] It is also worth emphasizing that Chebyshev's inequality can be applied to any body of data. Thus, if you know the mean and standard deviation of a particular set of data Chebyshev's inequality says that the proportion of observations (in the set of data) within $k$ standard deviations of the mean must be at least $1 - 1/k^2$.

**4.9** Suppose that the probability distribution of $X$ is as shown below:

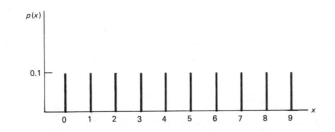

(a) Is $X$ a continuous or discrete random variable?
(b) What real-life phenomena might the random variable $X$ represent? (Hint: any number must end in 0, 1, . . . , 9.)
(c) Calculate the expected value, variance, and standard deviation of $X$.

**4.10** There is a 0.97 probability that no accident will occur at a particular race track during each day. The probability of one accident is 0.02, and the probability of two accidents is 0.01.
(a) What is the expected number of accidents in a day?
(b) What is the expected number of accidents in 10 days?
(c) What is the variance of the number of accidents in a day?
(d) What is the standard deviation of the number of accidents in a day?

**4.11** The Alpha Corporation is equally likely to sell 0, 1, 2, 3, or 4 bicycles in a day. (a) What is the expected value of the number of bicycles sold by the Alpha Corporation in a particular day? (b) Is this a value that the random variable in question can assume? (c) What is the variance of the number of bicycles sold in a day? (d) What is the standard deviation of the number of bicycles sold in a day?

**4.12** (a) Under the circumstances described in Exercise 4.5, what is the expected value of the income of the Alpha Corporation's salesman in a particular day? (b) Is this a value that the random variable in question can assume? (c) What is the standard deviation of the salesman's income in a day? (d) What is the expected value of the income of the Alpha Corporation's salesman in a particular two-day period? (e) Is this answer twice the answer to (a)? Why, or why not?

**4.13** A bag contains 10 packages worth $20 each, 5 packages worth $50 each, and 1 package worth $100.
(a) If one package is chosen at random, what is the expected value of its worth?
(b) What is the standard deviation of its worth?
(c) In this case, is it possible for the random variable (the worth of the chosen package) to equal the expected value of the random variable?
(d) If you want to take the action that maximizes expected gain, should you pay $40 for the opportunity to pick a package at random from this bag?

## 4.7 The Binomial Distribution

Earlier in this chapter we discussed the nature and characteristics of random variables and probability distributions in general. We turn now to the description and analysis of the most important discrete probability distribution in statistics: the binomial distribution. Specifically, we indicate the conditions that generate the binomial distribution, the formula for the distribution (as well as its mean and standard deviation), and some of the ways in which this distribution has been used to help solve important problems in business and economics.

### BERNOULLI TRIALS

To understand the circumstances under which the binomial distribution arises, it is convenient to begin by describing a *Bernoulli process* or, what is the same thing, a series of *Bernoulli trials.* (James Bernoulli was a seventeenth-century Swiss mathematician who performed some of the early work on the binomial distribution.) Each Bernoulli trial takes place under the following circumstances. First, *each trial results in one of two possible outcomes, which is termed either "success" or "failure."* Second, *the probability of a success remains the same from one trial to the next.* Third, *the outcomes of the trials are independent of one another.*

An example of a Bernoulli trial is a game in which a true die is thrown, and if a 4, a 5, or a 6 comes up, you win. If a 1, a 2, or a 3 comes up, you lose. To make sure that this is a Bernoulli trial, let's see whether it meets the three conditions specified in the previous paragraph. Certainly, the first condition is met, since there are only two possible outcomes—win (success) or lose (failure). And the second condition is met, since the probability of winning (that is, of a success) remains constant (at 1/2 from one trial to the next). Moreover, there is no reason to believe that your chances of winning are influenced in any way by the outcome of previous trials, since the die has no memory. Thus, all three conditions are met.

### THE BINOMIAL DISTRIBUTION

Suppose that $n$ Bernoulli trials occur and that the probability of a success on each trial equals $\Pi$. Under these circumstances[6] the number of successes occurring in these $n$ trials has a binomial probability distribution, which means

*Formula for Binomial Distribution*

$$P(x) = \frac{n!}{x!(n-x)!}\Pi^x(1-\Pi)^{n-x}, \text{ for } x = 0, 1, 2, \ldots, n \qquad (4.6)$$

where $P(x)$ is the probability that the number of successes equals $x$. The number of successes, $X$, is a random variable, whereas $n$ and $\Pi$ are constants. It is customary to refer to $X$ as a binomial random variable.

---

[6] Of course, the probability of failure on each trial equals $1 - \Pi$, since success and failure on each trial are complementary events (as defined in Chapter 3).

*If there are* **n** *independent trials, and if the probability is* $\Pi$ *that each trial is a success, the number of successes in the* **n** *trials conforms to the binomial distribution.*

Basic Idea #8

To illustrate the calculation of binomial probabilities, consider a case where a fair coin is flipped four times. What is the probability distribution of the number of heads? Since there are two mutually exclusive outcomes each time the coin is flipped, since the probability of heads is 1/2 each time, and since the outcomes of the tosses are statistically independent, it follows that this is a situation where there are four Bernoulli trials and where the probability of a success (heads) equals 1/2. Substituting 4 for $n$ and 1/2 for $\Pi$ in equation (4.6), we can calculate the probability distribution of the number of heads, with the following result:

$$P(0) = \frac{4!}{0!4!} \left(\frac{1}{2}\right)^0 \left(\frac{1}{2}\right)^4 = \frac{1}{16},$$

$$P(1) = \frac{4!}{1!3!} \left(\frac{1}{2}\right)^1 \left(\frac{1}{2}\right)^3 = \frac{1}{4},$$

$$P(2) = \frac{4!}{2!2!} \left(\frac{1}{2}\right)^2 \left(\frac{1}{2}\right)^2 = \frac{3}{8},$$

$$P(3) = \frac{4!}{3!1!} \left(\frac{1}{2}\right)^3 \left(\frac{1}{2}\right)^1 = \frac{1}{4},$$

$$P(4) = \frac{4!}{4!0!} \left(\frac{1}{2}\right)^4 \left(\frac{1}{2}\right)^0 = \frac{1}{16}.$$

Thus, the probability of no heads is 1/16, of heads once is 1/4, of heads twice is 3/8, of heads three times is 1/4, and of heads four times is 1/16. This probability distribution is plotted in Figure 4.3.

It is important to recognize that many random variables of considerable practical importance in business and economics have a binomial distribution. For example, later in this chapter we show how this probability distribution has played a central role in quality control in the manufacture of railway-car side frames. Much more will be said on this score in Chapter 8.

TESTING THE FORMULA FOR $P(x)$

Before undertaking some practical applications of the binomial distribution, we want to verify the correctness of the formula for the binomial probability distribution in equation (4.6). To do this, let's show that the probabilities of 0, 1, or 2 heads (in 4 tosses) based on equation (4.6) are correct.

**Figure 4.3**
Binomial Probability
Distribution, $n = 4$ and
$\Pi = 1/2$

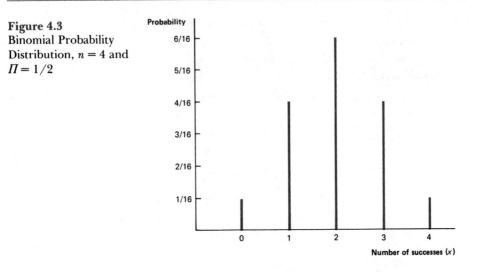

PROBABILITY OF NO HEADS. Clearly, there is only one way for this outcome to occur: The coin must come up tails *four times in a row*. Recalling the multiplication rule, the probability of tails four successive times must be equal to $(1/2)^4$ or $1/16$, since the probability of tails each time is $1/2$ and the outcomes of various tosses are statistically independent of one another. This result agrees with our calculation of $P(0)$, as computed from equation (4.6).

PROBABILITY OF HEADS ONCE. There are four mutually exclusive ways that one heads can occur in four tosses; namely, heads can come up on the first, second, third, or fourth toss. The probability of each of these ways equals $1/16$. Why? Because the probability both of heads on a *particular* toss and tails on the *other three* tosses equals $(1/2)^4$, or $1/16$, since the probability of heads and the probability of tails equal $1/2$, and the outcomes of various tosses are statistically independent. (Again, recall the multiplication rule.) Since the probability of the occurrence of *each* of these ways is $1/16$ (and since they are mutually exclusive), the addition rule dictates that the probability that *any one* of these ways occurs must be $4(1/16)$, or $1/4$. This result is in accord with our calculation of $P(1)$ as computed from equation (4.6).

PROBABILITY OF HEADS TWICE. If we designate heads by $H$ and tails by $T$, there are the following six ways that two successes can be distributed among the four tosses of the coin: (1) *HHTT;* (2) *HTHT;* (3) *HTTH;* (4) *THHT;* (5) *THTH;* and (6) *TTHH.* The first way occurs when the first and second tosses come up heads and the third and fourth tosses come up tails; the second way occurs when the first and third tosses come up heads and the second and fourth tosses come up tails; and so on. Each of these six ways has a probability of occurrence of $(1/2)^4$ or $1/16$, because of the multiplication rule. Thus, because of the addition rule, the probability that any one of these (mutually exclusive) ways occurs must be $6(1/16)$, or $3/8$. This result is in accord with our calculation of $P(2)$, as computed from equation (4.6).

## DERIVATION[7] OF THE FORMULA FOR $P(x)$

Although the foregoing discussion demonstrates that the formula for $P(x)$ results in the correct probabilities of 0, 1, and 2 heads out of four tosses of a coin, it does not demonstrate that this formula is always valid. To show that this is the case, it is necessary to derive the formula in equation (4.6). In the following four paragraphs, we provide such a derivation. Since the derivation is somewhat more technical than other parts of this chapter, some readers may want to skip these paragraphs. An understanding of subsequent sections does not depend on reading them.

To derive this formula, the first step is to note that there are a variety of ways that $n$ Bernoulli trials can give rise to exactly $x$ successes and $n - x$ failures. One way is that successes occur on each of the first $x$ trials, after which all the failures occur. If $S$ is a success and $F$ is a failure, this sequence can be represented by

$$\overbrace{SSS \ldots S}^{x} \quad \overbrace{FFF \ldots F}^{n-x}$$

Another possible sequence is the following, where the first $x - 1$ trials result in successes, the next $n - x$ trials result in failures, and the last trial results in a success:

$$\overbrace{SS \ldots S}^{x-1} \quad \overbrace{FFF \ldots FS}^{n-x}$$

Because the trials are independent, the probability that the first of these two sequences will occur is

$$\overbrace{\Pi\Pi\Pi \ldots \Pi}^{x} \quad \overbrace{(1 - \Pi)(1 - \Pi)(1 - \Pi) \ldots (1 - \Pi)}^{n-x} = \Pi^x(1 - \Pi)^{n-x}.$$

And the probability of occurrence of the second sequence is:

$$\overbrace{\Pi\Pi \ldots \Pi}^{x-1} \quad \overbrace{(1 - \Pi)(1 - \Pi)(1 - \Pi) \ldots (1 - \Pi)\Pi}^{n-x} = \Pi^x(1 - \Pi)^{n-x}.$$

which equals the probability of the first sequence. Clearly, the probability of obtaining *any* particular sequence of $x$ successes and $(n - x)$ failures must be the same as the probability of each of these two sequences.

How many ways can $n$ Bernoulli trials give rise to exactly $x$ successes and $n - x$ failures? In other words, how many different sequences of this type can occur? There are $n$ trials, which we can

---

[7] Some instructors may prefer to omit the next five paragraphs, which can be skipped without loss of continuity.

represent by $n$ different cards (with a 1 on the first card, a 2 on the second card, . . . , and an $n$ on the $n$th card). The problem is to determine how many different ways $x$ of these cards can be chosen, since any set of $x$ cards can be used to represent a sequence where successes occur on the trials corresponding to these cards (and where failures occur on the rest of the trials). Expressed in the language of Appendix 3.1, this problem amounts to asking how many combinations of $x$ cards can be drawn from a set of $n$ cards, since the order in which the cards are drawn does not matter. From Appendix 3.1 we know that the answer is $n! \div [(n - x)!x!]$. Consequently, this is the number of sequences in which exactly $x$ successes occur.

Because each of these sequences constitutes one of the mutually exclusive ways in which $x$ successes can occur in $n$ trials, and because each such sequence has a probability of occurrence of $\Pi^x(1 - \Pi)^{n-x}$, the probability of $x$ successes is obtained by adding $\Pi^x(1 - \Pi)^{n-x}$ as many times as there are sequences. Since there are $n! \div [(n - x)!x!]$ such sequences, it follows that this sum equals $\Pi^x(1 - \Pi)^{n-x}$ multiplied by $n! \div [(n - x)!x!]$. Thus, the probability that $x$ successes will occur in $n$ trials must equal

$$\frac{n!}{(n - x)!x!} \Pi^x(1 - \Pi)^{n-x},$$

which is the expression in equation (4.6). We have therefore proved that this formula is correct.

## THE BINOMIAL DISTRIBUTION: A FAMILY OF DISTRIBUTIONS

It is important to recognize that the binomial probability distribution is not a single distribution, but a family of distributions. Depending on the values of $n$ and $\Pi$, one can obtain a wide variety of probability distributions, all of which are binomial. For example, Figure 4.4 shows the binomial probability distribution when $n = 6$ and $\Pi = 0.2, 0.3, 0.7,$ and $0.8$. As you can see, it is skewed to the right when $\Pi$ is less than $1/2$ and skewed to the left when $\Pi$ is greater than $1/2$. Figure 4.5 shows the binomial probability distribution when $\Pi = 0.4$ and $n = 5, 10,$ and $20$. As you can see, the binomial probability distribution becomes increasingly bell-shaped as $n$ increases in value. More will be said about this in the next chapter.

If $n$ is fairly large, it is laborious to carry out the calculations involved in using equation (4.6) to compute $P(x)$. Fortunately, there are tables which give the value of $P(x)$ corresponding to various values of $n$ and $\Pi$. Appendix Table 1 supplies the values of $P(x)$ for values of $n$ from 1 to 20, and for values of $\Pi$ from 0.05 to 0.50. For large values of $n$, approximations of the binomial distribution can be found, as we shall see in the following chapter.

If the value of $\Pi$ exceeds 0.5, Appendix Table 1 can still be used. We need only switch the definitions of success and failure. For example, if the probability of a success is 0.8, what is the probability of four successes in 10 trials? To use Appendix Table 1, let's reverse the defini-

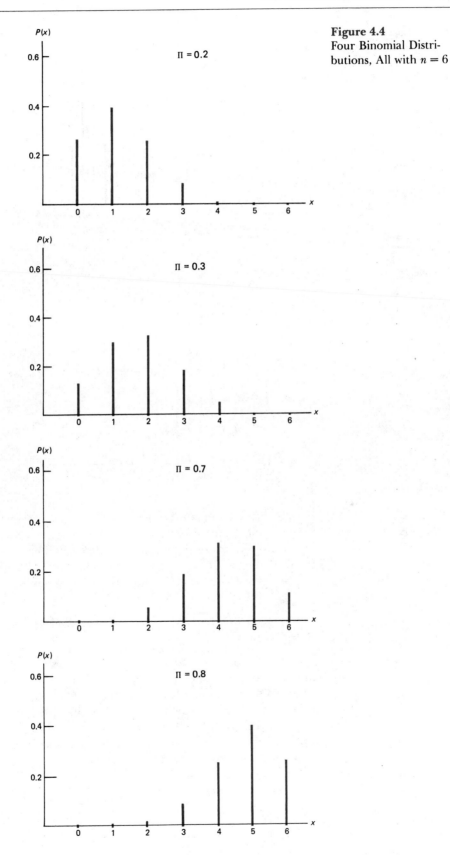

**Figure 4.4**
Four Binomial Distri-
butions, All with $n = 6$

**Figure 4.5**
Three Binomial
Probability Distribu-
tions, All with $\Pi = 0.4$

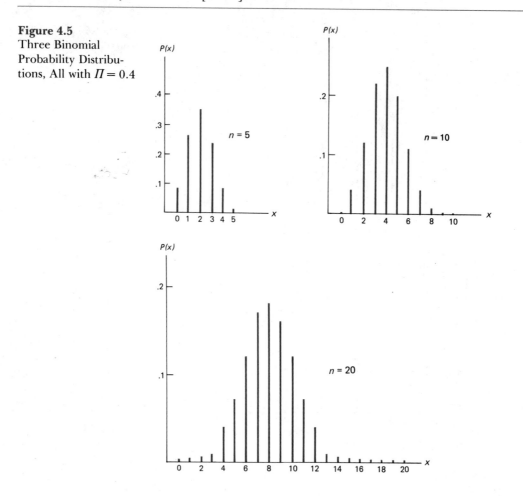

tions of success and failure; that is, let's call a success what we formerly
called a failure and vice versa. If we do this, what we want is the probability of six successes (formerly regarded as failures) in 10 trials. Since
the probability of a success (formerly a failure) is 0.2, Appendix Table
1 shows that this probability equals .0055. (Look it up and see.)

## MEAN AND STANDARD DEVIATION OF A BINOMIAL RANDOM VARIABLE

In Sections 4.4 and 4.5, we defined the expected value and standard deviation of a random variable. What are the expected value and
standard deviation of a binomial random variable? First, let's consider
the expected value, which (you will recall) is defined as the weighted
mean of the possible values that a random variable can assume, each
value being weighted by its probability of occurrence. Thus, the mean
of a binomial random variable must equal

$$E(X) = \sum_{x=0}^{n} xP(x),$$

(4.7)

since the possible values of $X$ are from 0 to $n$.

To illustrate the calculation of the expected value of a binomial random variable, suppose once again that $n = 4$ and $\Pi = 1/2$. Then it follows from equation (4.7) that

$$E(X) = (0)P(0) + (1)P(1) + (2)P(2) + (3)P(3) + (4)P(4).$$

Substituting into this formula the values of $P(0), \ldots, P(4)$ obtained on page 137, we find that

$$E(X) = 0(1/16) + 1(1/4) + 2(3/8) + 3(1/4) + 4(1/16) = 2.$$

Thus, the expected value (or mean) of $X$, if $n = 4$ and $\Pi = 1/2$, is 2; or put somewhat differently, the mean number of successes under these conditions is 2.

Calculating the expected value of a binomial random variable in this way shows clearly what the basic computations are. In practical work, however, this is a highly inefficient procedure because it can be shown that for a binomial random variable

$$E(X) = n\Pi. \qquad (4.8)$$

*Expected Value of a Binomial Random Variable*

*Thus, all one has to do is multiply the number of trials by the probability of success on each trial.* Regardless of the values of $n$ or $\Pi$, this simple computation will give the expected value of a binomial random variable. To make sure that this shortcut is correct, let's apply it to the case where $n = 4$ and $\Pi = 1/2$. Using equation (4.8), the expected value is $4(1/2)$, or 2, which is precisely the result we obtained by the more laborious procedure carried out in the previous paragraph. Of course, equation (4.8) certainly appeals to common sense. If $n$ trials are carried out, and if the probability of a success in each trial is $\Pi$, it stands to reason that the mean number of successes will be $n\Pi$.

Next we turn to the standard deviation of a binomial random variable. From Section 4.5 we know that the standard deviation of any random variable is defined as

$$\sqrt{\Sigma(x_i - E(X))^2 P(x_i)},$$

where the summation is over all possible values of $X$. Since the possible values of a binomial random variable are $0, 1, \ldots, n$, its standard deviation must therefore equal

$$\sigma(X) = \sqrt{\sum_{x=0}^{n} (x - E(X))^2 P(x)}. \qquad (4.9)$$

To illustrate the calculation of the standard deviation of a binomial random variable, suppose once again that $n = 4$ and $\Pi = 1/2$. Then it follows from equation (4.9) that

$$\sigma(X) = [(0-2)^2(1/16) + (1-2)^2(1/4) + (2-2)^2(3/8)$$

$$+ (3-2)^2(1/4) + (4-2)^2(1/16)]^{1/2}$$

$$= \sqrt{1} = 1.$$

As in the case of the expected value, this rather laborious computation is useful because it shows clearly what the definition means. But in practical work there is no need to carry out these extensive calculations, since it can be shown that for a binomial random variable

*Standard Deviation of a Binomial Random Variable*

$$\sigma(X) = \sqrt{n\Pi(1-\Pi)}. \qquad (4.10)$$

Thus, *all one has to do is multiply the number of trials by the product of the probability of success and probability of failure, and find the square root of the result.* Regardless of the values of $n$ and $\Pi$, this will yield the standard deviation of a binomial random variable. For example, if $n = 4$ and $\Pi = 1/2$, $\sigma(X) = \sqrt{4(1/2)(1/2)} = 1$, according to equation (4.10), which is precisely the answer we got by the more laborious process earlier in this paragraph.

The following two examples are designed to illustrate how binomial probabilities are calculated and interpreted, as well as to show how the expected value and standard deviation of a binomial random variable are computed.

**GETTING DOWN TO CASES**

QUALITY CONTROL IN THE MANUFACTURE OF RAILWAY-CAR SIDE FRAMES[8]

A producer of railway-car side frames wanted to establish a system to control the proportion of defective frames produced. Based on past data, the firm knew that 20 percent of the frames it produced were defective. It wanted to establish a procedure which would signal when the fraction of defectives jumped above 20 percent. After considerable discussion the firm decided to sample 10 frames from each day's output and find out the number of defectives. If this number exceeded 2 out of 10, the firm would stop its productive process to attempt to find out why such a relatively high percent was defective.

[8] The first half of this case is based in part on a section from A. J. Duncan, *Quality Control and Industrial Statistics* (Homewood, Ill.: Irwin, 1959). The numbers have been changed for pedagogical reasons. The reader may be surprised that so large a percentage (20 percent) of the frames was defective, but this was actually the case.

(a) Assuming that the 10 frames sampled were a very small proportion of the day's output, what was the probability that the firm's productive process would be stopped when in fact 20 percent of the day's output was defective?

(b) If defectives were to increase to 40 percent of a particular day's output, what is the probability that this inspection procedure would result in a stoppage of the firm's productive process?

(c) Based on your results in (a) and (b), write a one-paragraph report concerning the adequacy of the firm's sampling plan.

The firm sends a customer a shipment of five frames chosen at random from the day's output under the conditions described in (b). According to the terms of the firm's agreement with the customer, the firm must pay the customer $100 for every defective frame shipped as compensation for expenses incurred in receiving and handling defective materials.

(d) Graph the probability distribution of the amount the firm will have to pay the customer with regard to this shipment of five frames.

(e) What is the expected value of the amount the firm will have to pay the customer with regard to this shipment?

(f) What is the standard deviation of the amount the firm will have to pay the customer with regard to this shipment?

---

EXAMPLE 4.9 An oil exploration firm plans to drill six holes. It is believed that the probability that each hole will yield oil is 0.1. Since the holes are in quite different locations, the outcome of drilling one hole is statistically independent of that of drilling any of the other holes.

(a) If the firm will be able to stay in business only if two or more holes produce oil, what is the probability of its staying in business?

(b) Give the expected value and the standard deviation of the number of holes that result in oil.

SOLUTION: (a) If the firm can stay in business only if two or more holes produce oil, it follows that the probability that it will stay in business equals 1 minus the probability that the number of holes resulting in oil is 0 or 1. Each hole drilled can be viewed as a Bernoulli trial where the probability of success is 0.1. Thus, the probability that the number of successes is 0 or 1 equals:

$$P(0 \text{ or } 1) = P(0) + P(1) = \frac{6!}{6!0!}(.9^6) + \frac{6!}{5!1!}(.1)(.9^5)$$

$$= .531 + .354 = .885$$

Consequently, the probability that the firm will be able to stay in business is $1 - .885 = .115$.

(b) The expected value of the number of holes yielding oil is 6(.1), or 0.6, since $n = 6$ and $\Pi = .1$. The standard deviation of the number of holes yielding oil is $\sqrt{6(.1)(.9)} = \sqrt{.54}$, or .73, since $n = 6$, $\Pi = .1$, and $(1 - \Pi) = .9$.

EXAMPLE 4.10 A drug firm administers a new drug to 20 people with a certain disease. If the probability is 0.15 that the drug will cure each person of the disease, and if the result for one person is independent of that for another person, what is the probability that 3 or more of the 20 people will be cured? Use Appendix Table 1 to find the answer.

SOLUTION: In Appendix Table 1, look at the section where $n = 20$, and find the column corresponding to $\Pi = .15$. Let $X$ be the number of persons cured by the drug. The probability that $X$ is greater than or equal to 3 equals 1 minus the probability that $X = 0, 1,$ or 2. According to the relevant column of Appendix Table 1, $P(0) = .0388$, $P(1) = .1368$, and $P(2) = .2293$. Thus, the probability that $X$ is greater than or equal to 3 is $1 - (.0388 + .1368 + .2293) = 1 - .4049$, or .5951. In other words, the probability that 3 or more people will be cured is approximately .60.

**TO ERR IS HUMAN ... BUT COSTLY**

THE CASE OF RAINED-OUT ROCK CONCERTS

The producers of an outdoor rock concert must make provisions for some performances being rained out. A rock concert will be held in Baltimore on September 4–6. There will be performances at 3 P.M. and 8 P.M. on each day. The afternoon performances take place in a stadium which is partly covered; the evening performances take place in a park with no protection against rain. Since there are six performances (three days times two performances per day) and since the producers of the concert feel that the probability that each performance will be rained out is 0.1, they calculate that the probability that more than one of these performances will be rained out is only .1143. To obtain this result, they use the binomial distribution where $n = 6$ and $\Pi = 0.1$. (See Appendix Table 1.) Do you agree with their result? If not, why not?

SOLUTION: The binomial distribution does not seem to be appropriate in this situation for at least two reasons. First, the probability that a performance will be rained out does not seem to be the same from one performance to another. A given amount of rain seems less likely to rain out the afternoon performances than the evening performances because of the difference in the amount of cover. Whereas there is no cover for the evening performances, there is a partial covering for the afternoon performances. Second, whether or not one performance is rained out does not seem to be independent of whether another performance is rained out. For example, if the afternoon performance on September 4 is rained out, the probability that the evening performance on that day will also be rained out seems higher than if the afternoon performance was not rained out (because rainy afternoons often are followed by rainy evenings). The moral is that one must be sure that the conditions underlying the binomial distribution are met; if they are not met, the binomial distribution should not be used.

## EXERCISES

**4.14** The ratio of the variance of a binomial random variable to its expected value is $1/2$. Can you determine $n$? Can you determine $\Pi$?

**4.15** In how many ways can four successes be distributed among five trials? In how many ways can one success be distributed among five trials? Are your answers to these two questions the same?

**4.16** (a) In Example 4.10, what is the expected value of the number of people in the sample cured by the drug? What is the standard deviation of this number?

(b) In Example 4.10, what is the expected value of the number of people *not* cured by the drug? What is the standard deviation of this number?

**4.17** The Monroe Corporation allows a 2 percent discount on any invoice it sends out that is paid within 30 days. In the past, 20 percent of its invoices have been paid within 30 days. If it sends out 12 invoices in the second week in November, and if the payment of these invoices is independent, calculate the probability that

(a) all receive a discount.

(b) at least half receive a discount.

(c) none receive a discount.

(d) less than 2 receive a discount.

(e) the number receiving a discount is less than the expected number receiving a discount.

**4.18** The Maroni Corporation bids on 10 jobs, believing that its chances of getting each one is 0.1. (a) What is the expected value of the number of jobs it will get? (b) What is the standard deviation of the number of jobs it will get? (Assume that the outcomes are independent.) (c) Calculate the probability that the number of jobs it gets is

(i) less than the expected value.

(ii) above the expected value, but less than one standard deviation above it.

(iii) above the expected value, but less than two standard deviations above it.

**4.19** The football team at the local state university has a 0.4 probability of winning each of the nine games that remain to be played this season. (The outcomes of the games are independent.) If the team wins at least eight of these games, the coach will receive a pay raise of $5,000; otherwise, he will receive a raise of $2,000.

(a) Is his pay raise a random variable? If so, does it have a binomial distribution?

(b) What is the expected value of his pay raise?

(c) What is the standard deviation of his pay raise?

**4.20** John Martin, a plumber, installs ten hot water heaters in a particular housing development. The chance that each heater will last more than 10 years is 0.3, and their lives are independent.

(a) Construct a line chart showing the probability distribution of the number of hot water heaters lasting more than ten years.

(b) Is this probability distribution skewed to the right or the left?

(c) Construct a line chart showing the probability distribution of the number of hot water heaters not lasting more than ten years.

(d) Is the probability distribution in (c) skewed to the right or the left?

**4.21** A firm receives a shipment of 500 hi-fi speakers. It chooses a sample of 9 speakers, and rejects the shipment if 2 or more of the speakers are defective. What is the probability that this firm will accept the shipment if the proportion defective is (a) 0.05; (b) 0.10; (c) 0.15?

**\*4.22** Use Minitab (or some other computer package) to calculate the binomial distribution for $n = 30$ and $\Pi = 0.5$.

**\*4.23** Use Minitab (or some other computer package) to calculate the binomial distribution for $n = 21$ and $\Pi = 0.4$.

## Chapter Review

1. A *random variable* is a quantity the value of which is determined by an experiment; in other words, its value is determined by chance. A random variable must assume numerical values, and its value must be defined for all possible outcomes of the experiment in question; that is, for all elements of the sample space. Random variables are of two types: discrete and continuous. *Discrete random variables* can assume only a finite or countable number of numerical values, whereas *continuous random variables* can assume any numerical value on a continuous scale.

2. The *probability distribution of a discrete random variable* provides the probability of each possible value of the random variable. If $P(x)$ is the probability that $x$ is the value of the random variable, the sum of $P(x)$ for all values of $x$ must be 1. Probability distributions are represented by tables, graphs, or equations. A probability distribution can often be viewed as the relative frequency distribution (which shows the proportion of cases with each value) of a population.

3. If $x_1, x_2, \ldots, x_m$ are the possible values of a random variable $X$ and if $P(x_1)$ is the probability that $x_1$ occurs, $P(x_2)$ is the probability that $x_2$ occurs, and so on, then the expected value of this random variable $E(X)$ is $\Sigma x_i P(x_i)$. In other words, the *expected value* is the weighted mean of the values that the random variable can assume, each value being weighted by the probability that it occurs. To shed light on decision problems, statisticians frequently calculate and compare the expected value of monetary gain if various courses of action are followed. However, such a comparison alone often cannot provide a complete solution to these problems.

4. The *variance of a random variable X*, denoted by $\sigma^2(X)$, is the expected value of the squared deviation of the random variable from its expected value. The *standard deviation of a random variable* is the positive square root of the

random variable's variance. The standard deviation of a random variable is the most frequently used measure of the extent of dispersion or variability of the values assumed by the random variable. If it is certain that a random variable will equal a certain value, its standard deviation (and variance) is zero.

5. According to *Chebyshev's inequality,* the probability that any random variable will assume a value within $k$ standard deviations of the random variable's expected value is at least $1 - 1/k^2$. Stated differently, the probability that any random variable will assume a value more than $k$ standard deviations from the random variable's expected value is less than $1/k^2$.

6. The most important discrete probability distribution is the *binomial distribution,* which represents the number of successes in $n$ Bernoulli trials. That is, if each trial can result in a "success" or a "failure," and if the probability of a success equals $\Pi$ on each trial (and if the outcomes of the trials are independent), the number of successes, $x$, that occur in $n$ trials has a binomial distribution. Specifically,

$$P(x) = \frac{n!}{x!(n-x)!} \Pi^x (1 - \Pi)^{n-x}, \text{ for } x = 0, 1, 2, \ldots n.$$

The expected value of $x$ is $n\Pi$, and its standard deviation is $\sqrt{n\Pi(1 - \Pi)}$.

## Chapter Review Exercises

**4.24** The probability distribution of $X$ is

$$P(x) = \frac{1}{10} \qquad x = 1, 2, \ldots, 10.$$

Draw a line chart representing this probability distribution.

**4.25** Which of the following are random variables?
(a) the birthday of President Reagan;
(b) the number of years in a century;
(c) the number of presidents in the United States in the nineteenth century.

**4.26** An insurance company offers a 50-year-old woman a $1,000 one-year term insurance policy for an annual premium of $18. If the annual number of deaths per 1,000 is six for women in this age group, what is the expected gain for the insurance company from a policy of this kind?

**4.27** An insurance company offers a 52-year-old man a $1,000 one-year term insurance policy for an annual premium of $25. If the annual number of deaths per 1,000 is seven for men in this age group, what is the standard deviation of the gain for the insurance company from a policy of this kind?

**4.28** The mean foot length of men who buy shoes at the Aurora Shoe Store is 10.2 inches. The standard deviation of their foot lengths is 0.9 inches. The owner of the shoe store says that the probability that a man who buys his shoes at this store will have a foot length exceeding 12.0 inches or less than 8.4 inches is less than 0.25. Is he correct?

**4.29** A firm carried out 1,000 R and D projects in 1987. The mean and standard deviation of the amount spent on each project in 1987 was $51,000 and $11,000, respectively. Use Chebyshev's inequality to determine a maximum value of the proportion of these projects on which the firm spent less than $18,000 or more than $84,000.

**4.30** (a) Considering only the monetary gains and losses of the gamble in part (k) of Exercise 4.5, is this a fair gamble? (b) What is the expected value of the salesman's winnings (or losses) in this game? (c) Omitting the non-monetary considerations (namely, the threat of injury), would the salesman agree to this gamble if he is interested in maximizing the expected value of his monetary gains?

**4.31** The Grover Corporation is considering the purchase of a large number of chairs being sold by another firm which is going out of business. If these chairs are free of any major defects the Grover Corporation would make $5,000 by buying and reselling them. On the other hand, if these chairs contain major defects the Grover Corporation would lose $10,000 by having to repair the defects before selling them. Grover's president believes the probability is 0.9 that these chairs are free of major defects (and hence the probability is 0.1 that they do contain major defects). If he is interested in maximizing the expected value of his firm's gains, should he purchase the chairs? Why, or why not?

**4.32** The Uphill Manufacturing Company, a maker of bicycle pedals, has seven suppliers that provide it with materials. The materials producers are faced with work stoppages because of labor problems, and Uphill's management feels that in the next six months there is a 10 percent probability that each supplier will be unable to provide Uphill with materials. Because the prospective labor problems are quite different from one supplier to another and the employees of the various suppliers do not act together, Uphill's management also thinks that whether any one supplier is unable to provide materials is independent of whether any other supplier can do so.
(a) What is the probability that none of the suppliers will be able to provide Uphill with materials?
(b) What is the probability that more than half of the suppliers will be unable to provide Uphill with materials?
(c) What is the expected number of suppliers that will be unable to provide Uphill with materials?
(d) What is the standard deviation of the number of suppliers that will be unable to provide Uphill with materials?
(e) If the employees of the various suppliers band together to negotiate with all their employers, what effect do you think this will have on your answers to (a), (b), (c), and (d)?

**4.33** The Wallingford Company wants to know whether its own employees are satisfied with their working conditions. Since it would be impossible for Wallingford's management to talk in depth with each of the company's 20,000 employees, the president decides to pick 20 employees and see how well each is satisfied with existing conditions. Suppose that the probability that each employee in this sample will express dissatisfaction is 0.20, and that whether or not one employee expresses dissatisfaction is statistically independent of whether or not another employee does so.
(a) What is the probability that a majority of the sample will express dissatisfaction with existing working conditions?
(b) What is the probability that 2 or less of the sample will express dissatisfaction with existing conditions?
(c) Wallingford's personnel director proposes the following gamble to the company president: if the number of employees expressing dissatisfaction is more than 20 percent of the sample, he will pay the president $100. If this is not the case, the president will pay him $100. Is this a fair bet? Why or why not?

(d) What is the expected number of employees who will express dissatisfaction? What is the standard deviation?

(e) Using Chebyshev's inequality, determine an upper bound for the probability that the number of employees will be more than three standard deviations from the expected number you determined in (d). By how much does this upper bound exceed the true probability?

**4.34** It is important that the Durham Manufacturing Company maintain careful control over the quality of the bolts it uses. Each shipment (which contains 10,000 bolts) is subjected to the following acceptance-sampling procedure: A sample of 15 bolts is taken from the shipment and each is tested for defects. If more than one of the bolts are found to be defective, the shipment is rejected.

(a) If a shipment contains 20 percent defectives, what is the probability that it will be accepted?

(b) If a shipment contains 10 percent defectives, what is the probability that it will be accepted?

(c) If a shipment contains 5 percent defectives, what is the probability that it will be accepted?

(d) Durham's suppliers protest that this inspection procedure is inappropriate because they guarantee only that no more than 10 percent of a shipment is defective. Do you agree that it is inappropriate? Why, or why not?

# Appendix 4.1

## EXPECTED VALUE OF A LINEAR FUNCTION OF A RANDOM VARIABLE

Frequently, one is interested in the expected value of a linear function of a random variable. (If $X$ is a random variable and $Y = a + bX$, where $a$ and $b$ are constants, then $Y$ is a *linear function* of $X$.) For example, suppose that a firm's total annual costs are a linear function of the number of units of output it sells per year, and that the price per unit of output is constant. Then the firm's annual profit is a linear function of the number of units of output it sells per year, as shown in the illustration below. If the number of units of output sold per year is a random variable with a known expected value, what is the expected value of the firm's annual profit? To solve this sort of problem, we use the following result:

EXPECTED VALUE OF A LINEAR FUNCTION OF A RANDOM VARIABLE. *If* Y *is a linear function of a random variable* X—*that is, if* Y $= a + bX$—*the expected value of* Y, E(Y), *equals* a $+ b$E(X), *where* E(X) *is the expected value of* X.

This result is not difficult to prove. Given the definition of an expected value in equation (4.3), it follows that

$$E(Y) = \sum_{i=1}^{m} y_i P(y_i),$$ (4.11)

where the summation is over all possible values of $Y$, and $P(y_i)$ is the probability that $Y$ equals $y_i$. Since $y_i$ equals $a + bx_i$, the probability that $Y = y_i$ must equal the probability that $X = x_i$. (Why? Because $Y$ can equal $y_i$ if and only if $X$ equals $x_i$.) Thus, $P(y_i)$ must equal $P(x_i)$. Substituting $P(x_i)$ for $P(y_i)$ and $(a + bx_i)$ for $y_i$ in equation (4.11), we have

$$E(Y) = \sum_{i=1}^{m} (a + bx_i)P(x_i),$$

and since

$$(a + bx_i)P(x_i) = aP(x_i) + bx_iP(x_i),$$

$$E(Y) = \sum_{i=1}^{m} [aP(x_i) + bx_i P(x_i)] = \sum_{i=1}^{m} aP(x_i) + \sum_{i=1}^{m} bx_iP(x_i).$$

Since $a$ and $b$ are constants, they can be moved in front of the summation signs, so

$$E(Y) = a \sum_{i=1}^{m} P(x_i) + b \sum_{i=1}^{m} x_iP(x_i).$$

Finally, since

$$\sum_{i=1}^{m} P(x_i) = 1 \text{ and } \sum_{i=1}^{m} x_iP(x_i) = E(X),$$

$$E(Y) = a + bE(X),$$

which is the result we set out to prove.

To illustrate the application of this result, consider the following. A certain plant manufacturing TV sets has a fixed cost of $1 million per year. The gross profit from each TV set sold—that is, the price less the unit variable cost—is $20. The number of sets the plant sells per year is a random variable with an expected value of 100,000. What is the expected value of this plant's annual profit? Let $Y$ equal the plant's annual profit. Since this profit equals its gross profit less its fixed costs,

$$Y = -1,000,000 + 20X,$$

where $X$ is the number of TV sets sold per year. Thus,

$$E(Y) = -1,000,000 + 20E(X) = -1,000,000 + 20(100,000)$$

$$= 1,000,000.$$

Consequently, the expected value of the plant's annual profit is $1 million.

In some cases, one also is interested in the standard deviation of a linear function of a random variable. To solve such a problem, we use the following result.

STANDARD DEVIATION OF A LINEAR FUNCTION OF A RANDOM VARIABLE. *If* Y *is a linear function of a random variable* X—*that is, if* Y $= a + bX$—*the standard deviation of* Y, $\sigma(Y)$, *equals* $b\sigma(X)$, *where* $\sigma(X)$ *is the standard deviation of* X.

To illustrate the above, in the case of the TV manufacturer, what is the standard deviation of the annual profit if the standard deviation of the number of TV sets sold per year is 10,000? Since $Y = -1,000,000 + 20X$, the standard deviation of $Y$ equals $20\sigma(X)$, or \$200,000, since $\sigma(X)$ equals 10,000.

# Appendix 4.2

## THE HYPERGEOMETRIC DISTRIBUTION[9]

The hypergeometric distribution is an important probability distribution. To understand the conditions under which this distribution arises, suppose that a finite population consists of two types of items which we call successes and failures. If there are $A$ successes and $(N - A)$ failures in the population, and if we draw a random sample (without replacement) of $n$ items from the population, what is the probability distribution of the number of successes in the sample? Letting $x$ equal the number of successes in the sample, the answer is

$$P(x) = \frac{\dfrac{A!}{x!(A - x)!}\,\dfrac{(N - A)!}{(n - x)!(N - A - n + x)!}}{\dfrac{N!}{(N - n)!n!}}, \text{ for } x = 0, 1, 2, \ldots, n. \qquad (4.12)$$

This is called the hypergeometric distribution.[10]

It is not too difficult to derive the result in equation (4.12). How many different combinations of $x$ successes can be made from the $A$ successes in the population? As we know from Appendix 3.1, the answer is $A! \div [x!(A - x)!]$. How many different combinations of $(n - x)$ failures can be made from the $(N - A)$ failures in the population? As we know from Appendix 3.1, the answer is $(N - A)! \div [(n - x)!(N - A - n + x)!]$. Thus, the total number of different ways that we can combine the $x$ successes with $(n - x)$ failures is

$$\left(\frac{A!}{x!(A - x)!}\right)\left(\frac{(N - A)!}{(n - x)!(N - A - n + x)!}\right)$$

As we also know from Appendix 3.1, the total number of different combinations of $n$ items that can be made from the $N$ items in the population is $N! \div [(N - n)!n!]$. Thus, since the probability of $x$ successes in a sample of $n$ is the number of equally likely samples with $x$ successes and $(n - x)$ failures divided by the total number of equally likely samples of size $n$, this probability is

$$\frac{\dfrac{A!}{x!(A - x)!}\,\dfrac{(N - A)!}{(n - x)!(N - A - n + x)!}}{\dfrac{N!}{(N - n)!n!}},$$

which is what we set out to prove.

[9] We assume here that the reader has read Appendix 3.1.

[10] Note that $x$ cannot exceed $A$, and that $n - x$ cannot exceed $N - A$.

To illustrate the use of the hypergeometric distribution, suppose that a shipment of motors contains 10 motors, 2 of which are defective and 8 of which are not. What is the probability that a sample of three motors chosen at random from this shipment *without replacement* will contain one defective? Since $A = 2$, $N = 10$, $n = 3$, and $x = 1$, equation (4.12) implies that this probability equals

$$\frac{\left(\dfrac{2!}{1!1!}\right)\left(\dfrac{8!}{2!6!}\right)}{\dfrac{10!}{7!3!}} = \frac{2\left[\dfrac{8(7)}{2}\right]}{\dfrac{10(9)(8)}{3(2)(1)}} = \frac{56}{120} = \frac{7}{15}.$$

*If one is sampling without replacement from a finite population, the hypergeometric distribution, not the binomial distribution, is the correct one to use. (If the sampling is with replacement, the binomial is the correct one.) However, if the sample is a small percentage of the population, the binomial distribution provides a very good approximation to the hypergeometric distribution.* For example, suppose that a shipment of goods contains 1,000 items, 20 of which are defective and 980 of which are not defective. A random sample of 20 items is drawn from the shipment without replacement, and we want to know the probability of $x$ defectives in this sample. Even though the sample is without replacement, the sample is so small a proportion of the shipment that for an adequate approximation we should use the binomial probability distribution based on $n = 20$ and $\Pi = .02$ (that is, $20/1000$). In general, if the sample is 5 percent or less of the population, this approximation is adequate.

If sampling is without replacement, it can be shown that the mean and standard deviation of the number of successes is

$$E(X) = n\Pi$$

$$\sigma(X) = \sqrt{n\Pi(1 - \Pi)\left(\frac{N - n}{N - 1}\right)},$$

where $\Pi = A/N$, the proportion of successes in the population. In other words, these are the mean and standard deviation of the hypergeometric distribution. If the sample is a small percent of the population, $(N - n)/(N - 1)$ will be approximately equal to 1, since $n/N$ is very small. Thus, under these conditions, the mean and standard deviation of the number of successes will be approximately equal to those given for the binomial distribution in equations (4.8) and (4.10).

## Appendix 4.3

### JOINT PROBABILITY DISTRIBUTIONS AND SUMS OF RANDOM VARIABLES

JOINT PROBABILITY DISTRIBUTIONS. In Chapter 3, we discussed experiments involving two events, and characterized the probability of each outcome in terms of a joint probability table. (Recall the joint probability table for persons classified by attitude toward increased defense spending and by political party, shown in Table 3.2.) Let's now consider cases where each event is represented by a numerical value. Experiments of this sort involve two random vari-

**Table 4.3**
Joint Probability
Distribution for *X*
and *Y*

| Value of Y | Value of X | | |
|---|---|---|---|
| | *0*<br>*(no tails)* | *1*<br>*(one tail)* | *Total*<br>*(marginal*<br>*probabilities)* |
| 0 (no tails) | 1/4 | 1/4 | 1/2 |
| 1 (one tail) | 1/4 | 1/4 | 1/2 |
| Total (marginal probabilities) | 1/2 | 1/2 | 1 |

ables. For example, suppose that two coins (coin A and coin B) are each flipped once. Let *X* be the random variable corresponding to the number of times that coin A comes up tails, and let *Y* be the random variable corresponding to the number of times that coin B comes up tails. Table 4.3 shows the *joint probability distribution* for *X* and *Y*. The joint probabilities in Table 4.3 show the probabilities that *X* and *Y* assume particular values. For example, the probability that $X = 1$ and $Y = 0$, denoted by $P(X = 1$ and $Y = 0)$, equals 1/4.

A joint probability distribution can be represented graphically, as illustrated in Figure 4.6. The value of *X* is measured along one axis, the value of *Y* is measured along another axis, and the joint probability is measured along the third (vertical) axis. Each possible outcome is represented by a spike from the floor of the graph (at the point corresponding to this outcome's value of *X* and *Y*); the height of the spike measures the probability of this outcome. (Of course, the sum of these heights for all possible outcomes equals 1.) In the particular case in Figure 4.6, the height of each of the four spikes equals 1/4, in accord with Table 4.3.

In Chapter 3 we referred to the row and column totals in a joint probability table as marginal probabilities. The column totals in Table 4.3 correspond to $P(x)$ and constitute the *marginal probability distribution of* X. The row

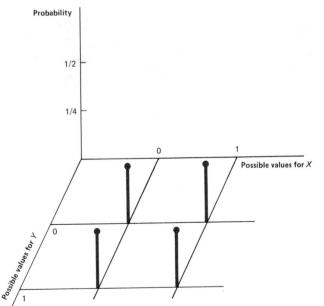

**Figure 4.6**
Joint Probability
Distribution for *X*
and *Y*

**Table 4.4**
Marginal Probability
Distributions of *X*
and *Y*

| Marginal probability distribution of X | | Marginal probability distribution of Y | |
| --- | --- | --- | --- |
| *Value of X* | *Probability* | *Value of Y* | *Probability* |
| 0 | 1/2 | 0 | 1/2 |
| 1 | 1/2 | 1 | 1/2 |
| Total | 1 | Total | 1 |

totals in Table 4.3 correspond to $P(y)$ and constitute the *marginal probability distribution of* Y. In other words, the marginal probability distributions are as shown in Table 4.4.

It is important to distinguish between a marginal probability distribution and a conditional probability distribution. To obtain the conditional probability that *Y* assumes a particular value (say, 1) given that *X* assumes a particular value (say, 0), we divide the joint probability that $Y = 1$ and $X = 0$ by the marginal probability that $X = 0$. In other words,

$$P(Y = 1 \mid X = 0) = \frac{P(Y = 1 \text{ and } X = 0)}{P(X = 0)} = \frac{.25}{.50} = .50.$$

To obtain the conditional probability that $Y = 0$ given that $X = 0$, we divide the joint probability that $Y = 0$ and $X = 0$ by the marginal probability that $X = 0$:

$$P(Y = 0 \mid X = 0) = \frac{P(Y = 0 \text{ and } X = 0)}{P(X = 0)} = \frac{.25}{.50} = .50.$$

Thus, the *conditional probability distribution of* Y *given that* X $= 0$ is as shown in Table 4.5. (Prove, as an exercise, that in this case the conditional probability distribution of *Y*, given that $X = 1$, is the same as the conditional probability distribution of *Y*, given that $X = 0$.)

**Table 4.5**
Conditional Probability
Distributions of *Y*

| Conditional probability distribution of Y, given that X = 0 | | Conditional probability distribution of Y, given that X = 1 | |
| --- | --- | --- | --- |
| *Value of Y* | *Probability* | *Value of Y* | *Probability* |
| 0 | 1/2 | 0 | 1/2 |
| 1 | 1/2 | 1 | 1/2 |
| Total | 1 | Total | 1 |

The conditional probability distribution of *X* can be derived in a similar way. To obtain the conditional probability that *X* assumes a particular value (say, 1), given that *Y* assumes a particular value (say, 1), we divide the joint probability that $X = 1$ and $Y = 1$ by the marginal probability that $Y = 1$. In other words,

$$P(X = 1 \mid Y = 1) = \frac{P(X = 1 \text{ and } Y = 1)}{P(Y = 1)} = \frac{.25}{.50} = .50.$$

To obtain the conditional probability that $X = 0$, given that $Y = 1$, we divide the joint probability that $X = 0$ and $Y = 1$ by the marginal probability that $Y = 1$:

$$P(X = 0 \mid Y = 1) = \frac{P(X = 0 \text{ and } Y = 1)}{P(Y = 1)} = \frac{.25}{.50} = .50.$$

Thus the *conditional probability distribution of* X *given that* $Y = 1$ is as shown in Table 4.6. (Prove, as an exercise, that in this case the conditional probability distribution of $X$, given that $Y = 0$, is the same as the conditional probability distribution of $X$, given that $Y = 1$.)

In Chapter 3 we took up the definition of statistical independence of events. Now we define statistical independence of random variables.

STATISTICAL INDEPENDENCE OF RANDOM VARIABLES: *Two random variables* X *and* Y *are statistically independent if the conditional probability distribution of* X, *given any value of* Y, *is identical to the marginal probability distribution of* X, *and if the conditional probability distribution of* Y, *given any value of* X, *is identical to the marginal probability distribution of* Y.

In the case of the two coins, a comparison of Table 4.4 with Tables 4.5 and 4.6 shows that $X$ and $Y$ are independent. Why? Because both conditional probability distributions of $Y$ in Table 4.5 are identical to the marginal probability distribution of $Y$ in Table 4.4, and both conditional probability distributions of $X$ in Table 4.6 are identical to the marginal probability distribution of $X$ in Table 4.4.

**Table 4.6** Conditional Probability Distributions of $X$

| *Conditional probability distribution of* X, *given that* Y = 0 | | *Conditional probability distribution of* X, *given that* Y = 1 | |
|---|---|---|---|
| *Value of* X | *Probability* | *Value of* X | *Probability* |
| 0 | 1/2 | 0 | 1/2 |
| 1 | 1/2 | 1 | 1/2 |
| Total | 1 | Total | 1 |

EXAMPLE 4.11 Let $X$ equal 0 if a particular flight from Seattle to Chicago is on time, and 1 if it is not. Let $Y$ equal 1 if the flight encounters severe turbulence, and 0 if it does not. The joint probability distribution for $X$ and $Y$ is as follows:

| | *Value of* Y | |
|---|---|---|
| *Value of* X | *0 (No turbulence)* | *1 (Turbulence)* |
| 0 (on time) | .75 | .05 |
| 1 (not on time) | .15 | .05 |

What is the marginal probability distribution of $X$? What is the conditional probability distribution of $X$, given that $Y = 0$? Are $X$ and $Y$ statistically independent?

SOLUTION: The marginal probability distribution of $X$ is given by the horizontal row totals; that is, the marginal probability that $X = 0$ is

.75 + .05 = .80, and the marginal probability that $X = 1$ is .15 + .05 = .20. The conditional probability that $X = 0$, given that $Y = 0$, is

$$P(X = 0 \mid Y = 0) = \frac{P(X = 0 \text{ and } Y = 0)}{P(Y = 0)} = \frac{.75}{.90} = .83,$$

and the conditional probability that $X = 1$, given that $Y = 0$, is

$$P(X = 1 \mid Y = 0) = \frac{P(X = 1 \text{ and } Y = 0)}{P(Y = 0)} = \frac{.15}{.90} = .17.$$

Since the marginal probability distribution of $X$ is not identical to the conditional probability distribution of $X$, given that $Y = 0$, $X$ and $Y$ are not statistically independent.

SUMS OF RANDOM VARIABLES. To solve problems in business and economics, statisticians often find it necessary to determine the expected value and variance of the sum of a number of random variables. The following propositions are very helpful in solving such problems.

EXPECTED VALUE OF A SUM OF RANDOM VARIABLES: *If* $X_1, X_2, \ldots, X_m$ *are* m *random variables,*

$$E(X_1 + X_2 + \ldots + X_m) = E(X_1) + E(X_2) + \ldots + E(X_m).$$

*That is, the expected value of the sum of random variables is equal to the sum of the expected values of the random variables. This proposition is true whether or not the random variables are statistically independent.*

VARIANCE OF A SUM OF RANDOM VARIABLES: *If* $X_1, X_2, \ldots, X_m$ *are* m *statistically independent random variables,*

$$\sigma^2(X_1 + X_2 + \ldots + X_m) = \sigma^2(X_1) + \sigma^2(X_2) + \ldots + \sigma^2(X_m).$$

*That is, the variance of the sum of statistically independent random variables is equal to the sum of the variances of the random variables.*

The example below illustrates how these propositions can be applied.

EXAMPLE 4.12 In order to commercialize a new product, the Ozone Chemical Company must carry out research, do pilot-plant work, and build a production facility. The cost of each of these three steps is a random variable. The expected value and standard deviation of each of these random variables are as follows:

| Random variable | Expected value (dollars) | Standard deviation (dollars) |
|---|---|---|
| Cost of carrying out research | 25,000 | 20,000 |
| Cost of pilot-plant work | 50,000 | 20,000 |
| Cost of production facility | 200,000 | 30,000 |

What is the expected value of the total cost incurred in these three steps? If the costs incurred in various steps are statistically independent, what is the standard deviation of the total cost incurred in these three steps?

SOLUTION: Let $X_1$ be the cost of carrying out research, $X_2$ be the cost of pilot-plant work, and $X_3$ be the cost of building a production facility. Based on the above proposition,

$$E(X_1 + X_2 + X_3) = E(X_1) + E(X_2) + E(X_3).$$

Since $E(X_1) = \$25,000$, $E(X_2) = \$50,000$, and $E(X_3) = \$200,000$,

$$E(X_1 + X_2 + X_3) = \$25,000 + \$50,000 + \$200,000 = \$275,000.$$

That is, the expected value of the total cost incurred in these three steps is \$275,000. If $X_1$, $X_2$, and $X_3$ are statistically independent,

$$\sigma^2(X_1 + X_2 + X_3) = \sigma^2(X_1) + \sigma^2(X_2) + \sigma^2(X_3).$$

Since $\sigma^2(X_1) = (\$20,000)^2$, $\sigma^2(X_2) = (\$20,000)^2$, and $\sigma^2(X_3) = (\$30,000)^2$,

$$\sigma^2(X_1 + X_2 + X_3) = (\$20,000)^2 + (\$20,000)^2 + (\$30,000)^2,$$

and

$$\sigma(X_1 + X_2 + X_3) = \sqrt{(\$20,000)^2 + (\$20,000)^2 + (\$30,000)^2}$$
$$= \$41,231.$$

That is, the standard deviation of the total cost incurred in these three steps is \$41,231.

# 5

# The Normal and
# Poisson Distributions

## 5.1 Introduction

In the previous chapter we studied the nature and characteristics of the binomial distribution, the most important discrete probability distribution. We now turn to the normal distribution, a continuous probability distribution that plays a central role in statistics. We will take up the nature and characteristics of the normal distribution and some of the ways in which this distribution is used in business and industry. We will also describe the Poisson distribution, another important probability distribution. In the appendix to this chapter we show how the Poisson distribution and the exponential distribution are used to analyze queuing problems in business and government.

## 5.2 Continuous Distributions

First we will describe how the probability distribution of a continuous random variable can be characterized graphically. To see how this is done, let's return to the actual case of a manufacturing plant that produces pieces of metal in which two holes are stamped. (see p. 26). The distance between the hole centers varies from one piece of metal to another and is a continuous random variable since it can assume any numerical value on a continuous scale. If we were to examine 1,000 pieces of metal produced by this plant, we could find the proportion of these pieces where the distance between the hole centers is 2.994 inches and under 2.995 inches, the proportion where the distance is 2.995 inches and under 2.996 inches, and so on. Then, as shown in panel A of Figure 5.1, we could construct a histogram in which the area of the bar representing each class interval is equal to the proportion of all pieces within this class interval. (In other words, the height of the bar equals the proportion of all pieces within this class interval divided by the width of the class interval.) The area under this histogram must

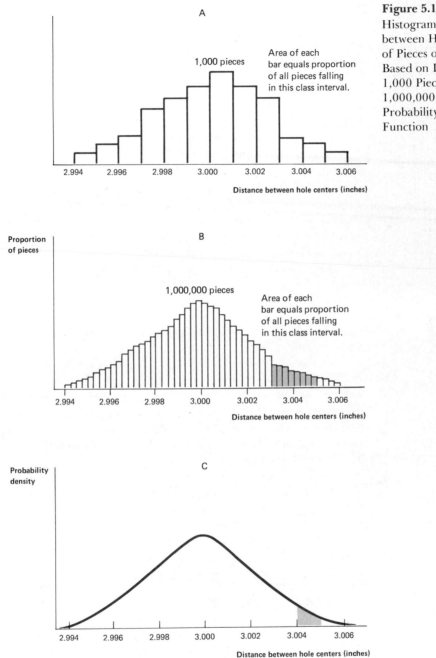

**Figure 5.1**

Histograms of Distances between Hole Centers of Pieces of Metal, Based on Data for 1,000 Pieces and 1,000,000 Pieces; and Probability Density Function

equal 1 because the sum of the proportions in all class intervals must equal 1. Such a histogram is said to use a *density* scale.

*Density Scale*

To obtain more complete information about the relative frequency distribution of these distances, we could examine 1,000,000 rather than 1,000 pieces of metal and construct another histogram like the one in panel A of Figure 5.1. Because of the greatly increased number of observations, we would now be able to divide the class intervals *more finely*. As a result, we would have *more* bars, each of which is *narrower* than in the histogram in panel A. This second histogram is shown in panel B of Figure 5.1. The proportion of pieces of metal in which the distance between hole centers lies in a particular range can be read from this histogram by measuring the area of the bars lying within this range. For example, the proportion of pieces of metal where the distance between the hole centers is 3.003 inches and under 3.005 inches is measured by the shaded area of the histogram in panel B of Figure 5.1.

Finally, in order to obtain even more complete information, we could examine countless numbers of these pieces of metal, with the result that the class intervals could (and would) be made ever finer and more numerous. In the limit, the histogram would become a *smooth curve,* as shown in panel C of Figure 5.1. As in panels A and B, the total area under this smooth curve would equal 1, since the proportion of observations in all class intervals must total 1. Also, in panel C, as in the other panels of Figure 5.1, the proportion of pieces of metal in which the distance between the hole centers lies in a particular range can be found by measuring the area under the smooth curve in this range. Thus, if we wanted to know the proportion of pieces of metal in which the distance between the hole centers is between 3.004 inches and 3.005 inches, we would measure the shaded area under the smooth curve in panel C.

The smooth curve in panel C of Figure 5.1 is important because we can use it to determine the probability that the distance between hole centers lies within a particular range, such as between 3.004 inches and 3.005 inches. As we have just seen, the proportion of cases in the long run where the distance lies in this range is equal in value to the area under the smooth curve in this range. Thus, *the probability that the distance between hole centers lies within a particular range is equal in value to the area under the smooth curve in this range.* In Figure 5.1 the probability that the distance between hole centers is between 3.004 inches and 3.005 inches equals the shaded area under the smooth curve in panel C.

## 5.3 Probability Density Function of a Continuous Random Variable

The smooth curve in panel C of Figure 5.1 is called a probability density function. We turn now from the specific case in Figure 5.1 to some generalizations concerning the probability density function of a continuous random variable.

PROBABILITY DENSITY FUNCTION OF A CONTINUOUS RANDOM VARIABLE. *In the limit, as more and more observations are gathered concerning a continuous variable, and as class intervals become narrower and more numerous, the histogram (using a density scale) of the variable becomes a smooth curve (as in Figure 5.1) called a* probability density function. *The total area under any probability density function must equal 1. The probability that a random variable will assume a value between any two points,* a *and* b, *equals the area under the random variable's probability density function over the interval from* a *to* b.[1]

To illustrate the use and interpretation of a random variable's probability density function, consider the diameters of tires produced by a particular manufacturer. The diameter of a randomly chosen tire can be viewed as a random variable. Suppose the tire's probability density function is as shown in Figure 5.2. This curve is of basic importance because it enables us to calculate the probability that a tire's diameter will be in any particular range in which we are interested. If we want to know the probability that a tire's diameter will be between $C$ and $D$ in Figure 5.2, all we have to do is determine the area between $C$ and $D$ under the probability density function. This is the shaded area in Figure 5.2.

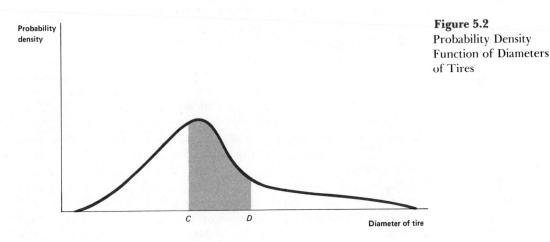

**Figure 5.2**
Probability Density
Function of Diameters
of Tires

## 5.4 The Normal Distribution

The most important continuous probability distribution is the normal distribution. The formula for the probability density function of a normal random variable (that is, a random variable with a normal distribution) is

$$f(x) = \frac{1}{\sqrt{2\pi}\,\sigma}\, e^{-\frac{1}{2}[(x-\mu)/\sigma]^2},$$

(5.1)    ***The Normal Curve***

where $\mu$ is the random variable's expected value (or mean), $\sigma$ is its standard deviation, $e$ is approximately 2.718 and is the base of the natural logarithms, and $\pi$ is approximately 3.1416. Like the binomial distribution, the normal distribution is really a family of distributions.

---

[1] The probability that a continuous random variable is precisely equal to a particular value is zero since the area under the probability density function at this particular value is a line of zero width.

**Figure 5.3**
Three Normal Curves, with $\mu = 15$ and $\sigma = 2.5$; $\mu = 40$ and $\sigma = 5$; and $\mu = 60$ and $\sigma = 1$

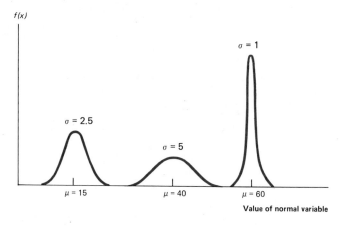

Depending on its mean and standard deviation, the location and shape of the normal probability density function—or ***normal curve,*** as we shall call it for short—can vary considerably.

To show how much the normal curve can vary, Figure 5.3 presents three normal curves, one with a mean of 15 and a standard deviation of 2.50, one with a mean of 40 and a standard deviation of 5, and one with a mean of 60 and a standard deviation of 1. As you can see, all three are bell-shaped and symmetrical, but the curves are located at quite different points along the horizontal axis because they have different means, and they exhibit quite different amounts of spread or dispersion because they have different standard deviations. Because of the differences in their means and standard deviations, some normal curves (like the middle one in Figure 5.3) are short and squat, whereas others (like the one on the right in Figure 5.3) are tall and skinny. But in accord with the definition of a probability density function in the previous section, the total area under any normal curve must equal 1.

Although normal curves vary in shape because of differences in mean and standard deviation, all normal curves have the following characteristics in common:

1. SYMMETRICAL AND BELL-SHAPED. All normal curves are symmetrical about the mean. In other words, the height of the normal curve at a value that is a certain amount *below* the mean is equal to the height of the normal curve at a value that is the same amount *above* the mean. Because of this symmetry, the mean of a normal random variable equals both its median and its mode. (Recall the discussion of the relative position of the mean, median, and mode in Chapter 2.) Besides being symmetrical, the normal curve is bell-shaped, as in Figure 5.3. And a normal random variable can assume values ranging from $-\infty$ to $+\infty$.

2. PROBABILITY THAT A VALUE WILL LIE WITHIN $k$ STANDARD DEVIATIONS OF THE MEAN. Regardless of its mean or standard deviation, the probability that the value of a normal random variable will lie within *one* standard deviation of its mean is 68.3 percent, the probability that it will lie within *two* standard deviations of its mean is 95.4 percent, and

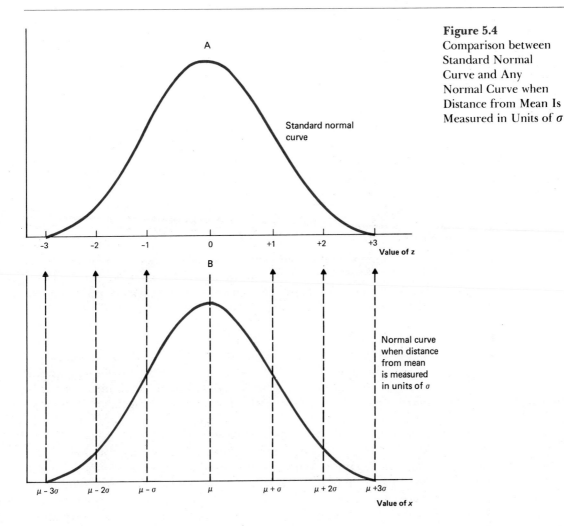

**Figure 5.4**
Comparison between
Standard Normal
Curve and Any
Normal Curve when
Distance from Mean Is
Measured in Units of $\sigma$

the probability that it will lie within *three* standard deviations of its mean is 99.7 percent. Panel B of Figure 5.4 shows the distance from the mean, $\mu$, in units of the standard deviation, $\sigma$. Clearly, almost all the area under a normal curve lies within 3 standard deviations of the mean.

3. LOCATION AND SHAPE DETERMINED ENTIRELY BY $\mu$ AND $\sigma$. The location of a normal curve along the horizontal axis is determined *entirely* by its mean $\mu$. For example, if the mean of a normal curve equals 4, it is centered at 4; if its mean equals 400, it is centered at 400. The amount of spread in a normal curve is determined *entirely* by its standard deviation $\sigma$. If $\sigma$ increases, the curve's spread widens; if $\sigma$ decreases, the curve's spread narrows.

Why is the normal distribution so important in statistics? Basically, for three reasons. First, *the normal distribution is a reasonably good approximation to many populations.* Experience has shown that many (but by no means all) population histograms (using a density scale) are approximated quite well by a normal curve. For example, the histogram (using a density scale) of heights, weights, or IQs is likely to be reasonably close to a normal curve. Second, it can be shown that under circumstances described in the following chapter, *the probability distribution of the sample mean should be close to the normal distribution.* This is one of the most fundamental results in statistics. We must postpone discussing it until the next chapter, but it is an important reason for the key role played by the normal distribution in statistics.[2] Third, and related to the previous point, *the normal distribution can be used in many instances to approximate the binomial distribution.* In Section 5.8, we shall show how this approximation can be employed.

## 5.5 The Standard Normal Curve

As stressed in the previous section, normal curves vary greatly in shape because of differences in the mean $\mu$, and in the standard deviation $\sigma$. However, if one expresses any normal random variable as a deviation from its mean, and measures these deviations in units of its standard deviation, the resulting random variable, called a ***standard normal variable,*** has the probability distribution shown in panel A of Figure 5.4. This probability distribution is called the ***standard normal curve.***

*Standard Units*     If the weights of adult males are normally distributed, with a mean of 170 pounds and a standard deviation of 20 pounds, it is possible to express the weight of each adult male in *standard units* by finding the deviation of his weight from the mean and expressing this deviation in units of the standard deviation. For example, if William Morris's weight is 190 pounds, it is $+1.0$ in standard units. Why? Because his weight is 20 pounds above the mean, and since the standard deviation is 20 pounds, this amounts to a positive $(+)$ deviation from the mean of 1 standard deviation. On the other hand, if John Jarvis's weight is 160 pounds, it is $-0.5$ in standard units because his weight is 10 pounds below the mean, which amounts to a negative $(-)$ deviation from the mean of 0.5 standard deviations.

The important point to note is that if any normal variable is expressed in standard units, its probability distribution is given by the standard normal curve. Thus, if the weights of adult males are normally distributed and if we express them in standard units, their probability distribution is given by the standard normal curve. Put more formally, if $X$ is a normally distributed random variable, then

*The Standard*
*Normal Variable*

$$Z = \frac{X - \mu}{\sigma}$$

(5.2)

has the standard normal distribution regardless of the values of $\mu$ and $\sigma$. Thus, if $X$ is the weight of an adult male, $(X - 170) \div 20$ has the standard normal distribution.

[2] Under some circumstances, statistics other than the sample mean also have a normal distribution. We single out the sample mean only because of its great importance in statistical applications.

As a further illustration, given the fact that the heights of adult females are normally distributed with a mean of 66 inches and a standard deviation of 2 inches, if $X$ is the height of an adult female, what is $Z$? $Z$ is $X$ expressed in standard units. That is,

$$Z = \frac{X - 66}{2}.$$

What is the value of $Z$ corresponding to a height of 67 inches? It is $(67 - 66) \div 2$, or 0.5. What is the height corresponding to a $Z$ value of $-2.0$? Since $(X - 66) \div 2 = -2.0$, $X$ must equal 62 inches.

---

*If any normal random variable is expressed in standard units, its probability distribution is the standard normal curve.*    **Basic Idea #9**

---

Figure 5.4 shows what happens when we express a normal variable, $X$, in standard units. Panel B shows the probability distribution of $X$. Note that in this panel the value of $X$ is measured in units of the standard deviation ($\sigma$) from the mean ($\mu$). When we express $X$ in standard units, the value of $Z$ corresponding to each value of $X$ is shown by the arrows. Thus, if $X$ equals $\mu - 3\sigma$, the corresponding $Z$ value is $-3$; if $X$ equals $\mu - 2\sigma$, the correponding $Z$ value is $-2$; and so on. Clearly, *the mean of the standard normal distribution is zero,* since zero in panel A corresponds to $\mu$ in panel B. Also, *the standard deviation of the standard normal distribution is 1,* since a distance of $\sigma$ along the horizontal axis in panel B corresponds to a distance of 1 in panel A.[3]

## 5.6 Calculating Normal Probabilities

It frequently is necessary to calculate the probability that the value of a normal random variable lies between two points. To calculate this probability, two steps must be carried out:

1. FIND THE POINTS ON THE STANDARD NORMAL DISTRIBUTION CORRESPONDING TO THESE TWO POINTS. For example, if the heights of adult women are normally distributed with a mean of 66 inches and a stan-

---

[3] Using the results of Appendix 4.1, it is easy to prove that the mean of the standard normal variable is zero. Since $Z = (X - \mu) \div \sigma$, it follows that

$$Z = -\frac{\mu}{\sigma} + \frac{1}{\sigma} X.$$

Thus, in accord with Appendix 4.1,

$$E(Z) = -\frac{\mu}{\sigma} + \frac{1}{\sigma} E(X) = -\frac{\mu}{\sigma} + \frac{\mu}{\sigma} = 0.$$

Also, the results of Appendix 4.1 imply that the standard deviation of $Z$ must equal the standard deviation of $X$ (that is, $\sigma$) multiplied by $1/\sigma$, or 1.

dard deviation of 2 inches and if we want to know the probability that the height of an adult woman lies between 65 and 68 inches, our first step is to find the points on the standard normal distribution corresponding to 65 and 68 inches. Since $\mu = 66$ and $\sigma = 2$, these points are $(65 - 66) \div 2$ and $(68 - 66) \div 2$, respectively. Simplifying terms, they are $-0.5$ and $+1.0$.

2. DETERMINE THE AREA UNDER THE STANDARD NORMAL CURVE BE-TWEEN THE TWO POINTS WE HAVE FOUND. If we want to know the probability that the height of an adult woman lies between 65 and 68 inches, we determine the area under the standard normal curve between the points on the curve corresponding to 65 and 68 inches. Since these points are $-0.5$ and $1.0$ (as we know from the preceding paragraph), we must determine the area under the standard normal curve between $-0.5$ and $1.0$.

Why does this procedure give the correct answer? *Because the area under any normal curve between two points is equal to the area under the standard normal curve between the corresponding two points.* A comparison between panels A and B in Figure 5.5 demonstrates that this is true.

**Figure 5.5**
Normal Distribution of Female Heights and Standard Normal Curve

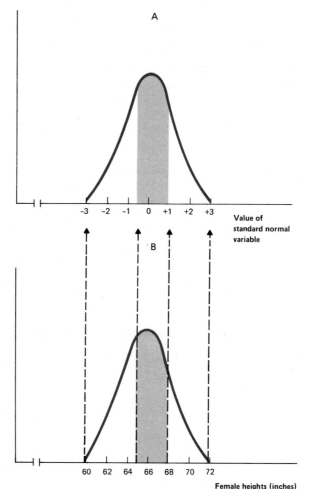

A

Value of standard normal variable

B

Female heights (inches)

In panel B the normal curve shows the distribution of adult female heights. The probability that the height of an adult woman lies between 65 and 68 inches equals the area under the curve between 65 and 68 inches (that is, the shaded area in panel B). In panel A we show the standard normal distribution. The points on the standard normal distribution corresponding to 65 and 68 inches are −0.5 and 1.0, as we know from previous paragraphs. The shaded area under the standard normal curve equals the probability that the standard normal variable lies between −0.5 and 1.0. As you can see, the two shaded areas are equal. Thus, the probability that a height lies between 65 and 68 inches equals the area under the standard normal curve between −0.5 and 1.0.

## 5.7 Using the Table of the Standard Normal Distribution

The area under the standard normal distribution between various points is tabled. To carry out the second step in the procedure described above, one must be able to use this table, which is contained in Appendix Table 2. In the following paragraphs, we indicate how this table is used in various situations.

1. AREA BETWEEN ZERO AND SOME POSITIVE VALUE. Each number in the body of Appendix Table 2 shows the area between zero (the mean of the standard normal distribution) and the positive number $(z)$ given in the left-hand column (and top) of the table. For example, to determine the area between zero and 1.10, look at the row labeled 1.1 and the column labeled .00; the area is .3643. This is the shaded area in panel A of Figure 5.6. Similarly, to determine the area between zero and 1.63, look at the row labeled 1.6 and the column labeled .03; the area is .4484.

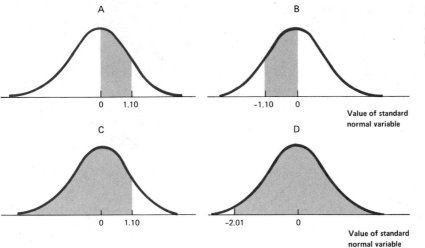

**Figure 5.6**
Areas under the Standard Normal Curve

2. Area between zero and some negative value. Because the standard normal curve is symmetrical, the area between zero and any negative value is equal to the area between zero and the same positive value. Hence Appendix Table 2 can readily be used to evaluate the desired area. For example, the area between zero and −1.10 equals that between zero and +1.10; thus, this area (shaded in panel B of Figure 5.6) must be .3643. Similarly, the area between zero and −1.63 equals that between zero and +1.63, which we know to be .4484.

3. Area to the left of some positive value. Suppose that we want to determine the area to the left of 1.10. This area (shaded in panel C of Figure 5.6) is composed of two parts: the area to the left of zero, and the area between zero and 1.10. The area to the left of zero is .5 because the standard normal curve is symmetrical about zero and because the area under the entire curve equals 1. The area between zero and 1.10 can be determined from Appendix Table 2, as we already know. Since it is .3643, the area we want is .5000 + .3643 = .8643.

4. Area to the right of some negative number. What is the area to the right of −2.01? This area (shaded in panel D of Figure 5.6) is composed of two parts: the area to the right of zero, and the area between zero and −2.01. The area to the right of zero is .5 because the standard normal curve is symmetrical about zero and because the area under the entire curve equals 1. The area between zero and −2.01 can be determined from Appendix Table 2, since it equals the area between zero and 2.01, which is .4778. Thus, the area we want equals .5000 + .4778 = .9778.

5. Area to the right of some positive value. What is the area to the right of 1.65? This area (shaded in panel A of Figure 5.7) plus the area between zero and 1.65 must equal .5000, because the total area to the right of zero equals .5000. Thus, the area we want equals .5000

**Figure 5.7**
Areas under the
Standard Normal
Curve

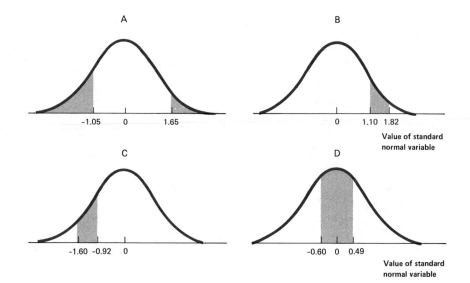

minus the area between zero and 1.65. Since Appendix Table 2 shows that the area between zero and 1.65 is .4505, the area we want equals .5000 − .4505 = .0495.

6. AREA TO THE LEFT OF SOME NEGATIVE VALUE. What is the area to the left of −1.05? This area (shaded in panel A of Figure 5.7) plus the area between zero and −1.05 must equal .5000, because the total area to the left of zero is .5000. Thus, the area we want equals .5000 minus the area between zero and −1.05. The area between zero and −1.05 equals the area between zero and 1.05, which is .3531, according to Appendix Table 2. Hence the area we want equals .5000 − .3531 = .1469.

7. AREA BETWEEN TWO POSITIVE VALUES. What is the area between 1.10 and 1.82? This area (shaded in panel B of Figure 5.7) equals the difference between (a) the area between zero and 1.82, and (b) the area between zero and 1.10. Appendix Table 2 shows that the former area is .4656 and that the latter area is .3643. Thus, the area we want is .4656 − .3643 = .1013.

8. AREA BETWEEN TWO NEGATIVE VALUES. What is the area between −1.60 and −0.92? This area (shaded in panel C of Figure 5.7) equals the difference between (a) the area between zero and −1.60, and (b) the area between zero and −0.92. The area between zero and −1.60 equals the area between zero and 1.60, which is .4452, according to Appendix Table 2. The area between zero and −0.92 equals the area between zero and 0.92, which is .3212, according to Appendix Table 2. Thus, the area we want is .4452 − .3212 = .1240.

9. AREA BETWEEN A NEGATIVE AND A POSITIVE VALUE. Finally, suppose we want to determine the area between −0.60 and 0.49. This area (shaded in panel D of Figure 5.7) equals the sum of (a) the area between zero and −0.60, and (b) the area between zero and 0.49. The area between zero and −0.60, equals the area between zero and 0.60, which is .2257, according to Appendix Table 2. The area between zero and 0.49 is .1879, according to Appendix Table 2. Thus, the area we want is .2257 + .1879 = .4136.

Because of the central importance of the normal distribution in statistics, it is essential that you be able to calculate the probability that a normal random variable lies in a given range. The following three examples are designed to illustrate how this is done.

---

EXAMPLE 5.1 Find the probability that the value of the standard normal variable will lie between −1.23 and +1.14.

SOLUTION: Appendix Table 2 shows that the area under the standard normal curve between 0 and 1.23 is .3907, so the area between 0 and −1.23 must also be .3907. Appendix Table 2 shows that the area between 0 and 1.14 is .3729. Thus, the area between −1.23 and +1.14 equals .3907 + .3729 = .7636, which means that the probability we want equals .7636.

---

EXAMPLE 5.2 The diameters of the tires produced by a tire manufacturer are normally distributed, with a mean of 36 inches and a standard deviation of .001 inches. What is the probability that a tire produced by this firm will have a diameter that is (a) between 35.9990 and 36.0005 inches; (b) less than 35.9985 inches; (c) greater than 36.0004 inches?

SOLUTION: (a) The first step is to find the points on the standard normal distribution corresponding to 35.9990 and 36.0005. These points are $(35.9990 - 36) \div .001$ and $(36.0005 - 36) \div .001$, or $-1.0$ and $+0.5$, respectively. According to Appendix Table 2, the area under the standard normal curve between zero and 1.0 is .3413, which means that the area between zero and $-1.0$ also is .3413. According to Appendix Table 2, the area between zero and 0.5 is .1915. Thus, the area between $-1.0$ and 0.5 is $.3413 + .1915 = .5328$. This is the probability that a tire's diameter is between 35.9990 and 36.0005 inches.

(b) The point on the standard normal distribution corresponding to 35.9985 inches is $(35.9985 - 36) \div .001$, or $-1.5$. According to Appendix Table 2, the area under the standard normal curve between 0 and 1.5 is .4332, so the area between 0 and $-1.5$ also is .4332. Thus, the area to the left of $-1.5$ equals $.5000 - .4332 = .0668$. This is the probability that a tire's diameter is less than 35.9985 inches.

(c) The point on the standard normal distribution corresponding to 36.0004 inches is $(36.0004 - 36) \div .001$, or $+0.4$. According to Appendix Table 2, the area under the standard normal curve between zero and 0.4 is .1554, so the area to the right of 0.4 must equal $.5000 - .1554 = .3446$. This is the probability that a tire's diameter is more than 36.0004 inches.

EXAMPLE 5.3 The president of the tire manufacturing firm in Example 5.2 makes a statement to reporters that 90 percent of the tires produced by the firm have diameters of 36.0020 inches or less, while 10 percent of the tires have diameters of more than 36.0020 inches. He is incorrect. What figure should be substituted for 36.0020?

SOLUTION: The first step is to find the number which the value of the standard normal variable will exceed with a probability of .10. Since the probability must be $.50 - .10$ (or .40) that the value of the standard normal variable lies between zero and this number, we must look in Appendix Table 2 for the value of $Z$ corresponding to a probability of .40. This value is 1.28. The next step is to find the value of the tire diameter that corresponds to this value of the standard normal variable. In other words, we must find the value of $X$ in equation (5.2) that corresponds to $Z = 1.28$. Clearly, this value is $\mu + 1.28\sigma$, which equals $36 + 1.28(.001)$, or 36.00128 inches. Thus, the probability is 0.10 that a tire's diameter will exceed 36.00128 inches. The figure of 36.00128 inches, not the president's, is correct.

Vendors of Demon Rum

In 1984, mean sales of the top 20 liquor brands was 2.930 million cases. The standard deviation of these brands' sales was 1.512 million cases.[4] Based on these data, a market researcher wanted to estimate the number of brands that sold over 3 million cases in 1984. Assuming that the sales of these brands were normally distributed, he calculated the proportion of brands with sales exceeding 3 million. Since the point on the standard normal distribution corresponding to 3 million is (3 million − 2.930 million) ÷ 1.512 million, or 0.05, he found the area under the standard normal curve to the right of 0.05, which is .5000 − .0199, or .4801. (See Appendix Table 2.) Then he multiplied .4801 times 20 (the number of brands), the result being about 10. Thus, he estimated that 10 brands sold over 3 million cases in 1984. Do you agree with this result? If not, what mistakes did he make?

SOLUTION: The crucial mistake is his assumption that the sales of these 20 brands were normally distributed. In fact, their distribution was far from normal, as shown below. (The area of

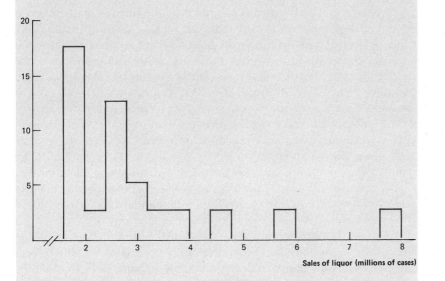

Sales of liquor (millions of cases)

each bar equals the number of brands falling in each class interval.) Consequently, his calculation (based on the assumption that the distribution was normal) is very much in error. Rather than 10 brands with sales exceeding 3 million cases, there really were only 6 (Bacardi rum, Smirnoff vodka, Seagram's 7 Crown, Canadian Mist, Jim Beam, and Jack Daniel's). The moral here is that one cannot assume without any good reason that a variable is normally distributed. A great many variables are *not* normally distributed, as we shall see in this and subsequent chapters.

[4] "Liquor Industry Scoreboard," *Business Week*, May 13, 1985.

## 5.8 The Normal Distribution as an Approximation to the Binomial Distribution

As pointed out in an earlier section of this chapter, one reason why the normal distribution is so important is that it can be used as an approximation to the binomial distribution under certain circumstances. These are described below.

> NORMAL APPROXIMATION TO THE BINOMIAL DISTRIBUTION. *If* n *(the number of trials) is large and $\Pi$ (the probability of success) is not too close to 0 or 1, the probability distribution of the number of successes occurring in* n *Bernoulli trials can be approximated by a normal distribution. Experience indicates that the approximation is fairly accurate as long as* $n\Pi > 5$ *when* $\Pi \leqslant 1/2$ *and* $n(1 - \Pi) > 5$ *when* $\Pi > 1/2$.

The fact that the normal distribution can approximate the binomial distribution under the circumstances described above is useful because, as noted in Chapter 4, it is tedious to calculate the binomial probabilities when $n$ is large.

The following illustration shows how the normal distribution is used to estimate binomial probabilities. If a true coin is flipped 1,600 times, what is the probability distribution of the number of times that the coin comes up heads? Since there are 1,600 Bernoulli trials, the number of times heads comes up is clearly a binomial random variable. Moreover, since $n = 1,600$ and $\Pi = 1/2$, its mean is 800 $(=n\Pi)$ and its standard deviation is 20 $(= \sqrt{n\Pi(1 - \Pi)})$. The probability distribution of the number of times heads comes up is shown (as a histogram) in panel A of Figure 5.8. Since it is difficult to evaluate each of the binomial probabilities (in equation 4.6) when $n$ is as large as 1,600, we would like to approximate this probability distribution with another that is easier to calculate. Fortunately, as noted above, the normal distribution—with the same mean (800) and standard deviation (20) as the binomial distribution—is a good approximation.

A visual comparison of the normal distribution in panel B of Figure 5.8 with the binomial distribution in panel A (each of which has a mean of 800 and a standard deviation of 20) certainly indicates that the former is shaped much like the latter. But to make sure that this is a good approximation, we must investigate in greater detail. Let's look carefully at the segment of both probability distributions between $x = 788$ and $x = 790$ (where $x$ is the number of heads that comes up). Figure 5.9 shows a "blow up" of each of the probability distributions over the relevant range. If the approximation is accurate, the area under the normal curve between 787.50 and 790.50 must be approximately equal to the area under the binomial distribution between 787.50 and 790.50. In other words, the shaded area under the continuous curve should be approximately equal to the sum of the areas of

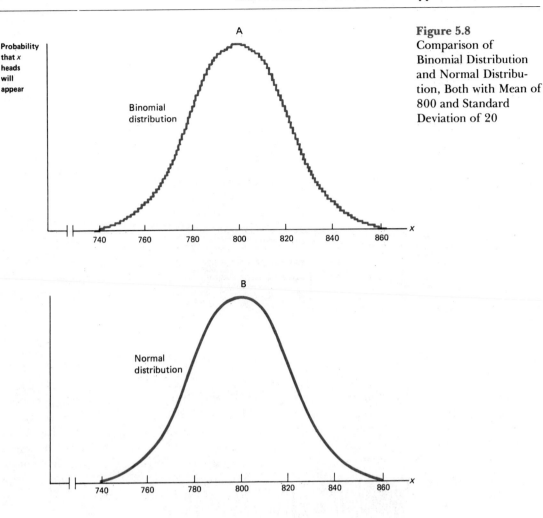

**Figure 5.8**
Comparison of
Binomial Distribution
and Normal Distribu-
tion, Both with Mean of
800 and Standard
Deviation of 20

the three rectangles (*A, B,* and *C*) shown in Figure 5.9. The sum of the areas of these three rectangles equals the true probability that the number of heads is 788, 789, or 790, whereas the shaded area is the approximation to this probability. Based on Figure 5.9, it certainly appears that the approximation is good.

To find the probability that the number of heads is 788, 789, or 790, we do *not* find the area under the normal curve between 788 and 790; instead, we find the area *between 787.50 and 790.50.* As shown in Figure 5.9, in order to approximate the three rectangles (*A, B,* and *C*) corresponding to the probabilities of 788, 789, and 790 heads, we must include the area under the continuous curve from 787.50 to 790.50. This is often called a *continuity correction,* a correction due to the fact that a discrete probability distribution is being approximated by a continuous one. In general, to find the probability that a binomial variable equals at least *c* but no more than *d* (where *c* < *d* ), we find the probability that a normal variable (with mean *nΠ and standard deviation* $\sqrt{n\Pi(1-\Pi)}$) lies between $(c - 1/2)$ and $(d + 1/2)$.

*Continuity*
*Correction*

**Figure 5.9**
Normal Approximation
to the Binomial
Distribution in the
Range between 787.50
and 790.50

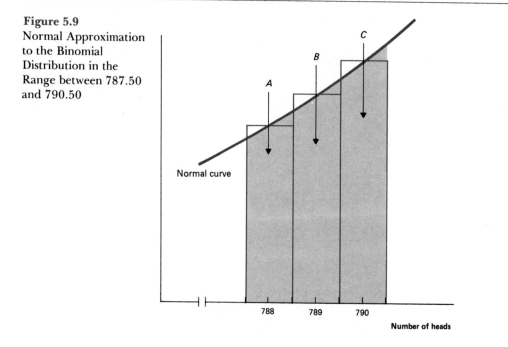

Based on the normal approximation, what is the probability that the number of heads is 788, 789, or 790? As we have seen, this probability is approximately equal to the probability that the value of a normal random variable with mean equal to 800 and standard deviation equal to 20 lies between 787.50 and 790.50. The value of the standard normal variable corresponding to 787.50 is −.625, and that corresponding to 790.50 is −.475. Using Appendix Table 2, we find that the area under the standard normal curve between zero and 0.625 is approximately .234, which means that the area between zero and −0.625 also is approximately .234. Similarly, the area between zero and 0.475 is approximately .183, which means that the area between zero and −0.475 also is approximately .183. Thus, the probability that the number of heads is 788, 789, or 790 equals (approximately) .234 − .183 = .051.

The following is another illustration of how the normal distribution can be used to approximate the binomial distribution.

EXAMPLE 5.4 The probability that a machine will be down for repairs next week is 1/2. A firm has 100 such machines, and whether one is down is statistically independent of whether another is down. What is the probability that at least 60 machines will be down?

SOLUTION: The number of machines down for repair has a binomial distribution with mean equal to 100(1/2), or 50, and standard deviation equal to $\sqrt{100(1/2)(1/2)}$, or 5. Because of the continuity correction, the probability that the number down for repairs is 60 or more can be approximated by the probability that the value of a normal variable with mean equal to 50 and standard deviation equal to 5 exceeds 59.50. The value of the standard normal variable corresponding to 59.50 is $(59.50 - 50) \div 5$, or 1.9. Appendix Table 2 shows that the area under the standard normal curve between zero and 1.9 is .4713, so the area to the right of 1.9 must equal $.5000 - .4713 = .0287$. This is the (approximate) probability that at least 60 machines will be down for repair.

**GETTING DOWN TO CASES**

### A Trunking Problem in the Telephone Industry

Bell Telephone has been a pioneer in using probability theory to solve many kinds of engineering problems. The following case (with some simplifications) is derived from actual practice.

A telephone exchange at $A$ was to serve 2,000 telephones in a nearby exchange at $B$. Since it would have been too expensive to install 2,000 trunklines from $A$ to $B$, it was decided to install enough trunklines so that only 1 out of every 100 calls would fail to find an unutilized trunkline immediately at its disposal.

During the busiest hour of the day, each of these 2,000 telephone subscribers requires a trunkline to $B$ for an average of two minutes. Thus, at a fixed moment during the busiest hour, there are 2,000 telephone subscribers each of which has a probability of 1/30 that it will require a trunkline to $B$. Under normal conditions, whether or not one subscriber requires a trunkline to $B$ is independent of whether another subscriber does so. (Under abnormal conditions, such as a flood, or an earthquake, this assumption of independence is unlikely to hold, since many people are likely to want to make calls; however, the telephone company was interested in solving the problem under typical conditions.)

As stated above, the telephone company wanted to determine how many trunklines it should install so that when 1 out of the 2,000 subscribers puts through a call requiring a trunkline to $B$ during the busiest hour of the day, he or she would find an unutilized trunkline to $B$ immediately at his or her disposal in 99 out of 100 cases.

(a) Solve this problem using the normal distribution.

(b) Write a one-paragraph report to the telephone company describing the results you have obtained.

## 5.9 Wind Changes for Hurricanes: A Case Study[5]

Since hurricanes annually cause property damage totaling billions of dollars, government agencies and private organizations have a strong interest in studying hurricanes in order to reduce their impact. In 1970, statisticians at the Stanford Research Institute (now SRI International) began a study for the U.S. Department of Commerce of the behavior and impact of hurricanes. The normal distribution was used to compute the probability that wind changes of various magnitudes would occur for an unseeded hurricane during a 12-hour period before landfall. Based on various meteorological studies, it appears that *the percentage change in maximum sustained wind speed during this period is normally distributed with a mean of zero and a standard deviation of 15.6 percent.* In other words, the distribution of the percentage change in an unseeded hurricane's wind speed during such a period is as shown in Figure 5.10.

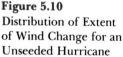
**Figure 5.10**
Distribution of Extent
of Wind Change for an
Unseeded Hurricane

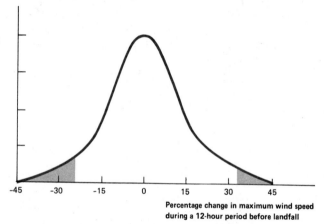

-45    -30    -15    0    15    30    45

Percentage change in maximum wind speed
during a 12-hour period before landfall

    The amount of property damage caused by a hurricane is related to the extent of the change in its maximum sustained wind speed. Thus, if a change of +32 percent or more occurs, the Stanford statisticians estimate that property damage of well over $300 million might be expected. On the other hand, if a change of −34 percent or less occurs, the estimated property damage is less than $20 million. From this we see that it is important to know the probability that the change in max-

---

[5] This case study is based on R. Howard, J. Matheson, and D. North, "The Decision to Seed Hurricanes," *Science*, June 1972. An abridged version of this article is contained in E. Mansfield, *Statistics for Business and Economics: Readings and Cases* (New York: Norton, 1980). The present discussion is simplified, and some numbers have been changed. For a much more complete and accurate (but also more technical) account, see the above article in *Science*.

    It is assumed that *if no change occurs in maximum sustained wind speed,* a hurricane of the sort considered here will cause $100 million in property damage.

imum sustained wind speed will lie in various ranges. What is the probability that the change in wind speed will exceed +32 percent, thus resulting in well over $300 million in property damage? Obviously, this is an important question for many firms and government agencies.

If $X$ is the percentage change in maximum sustained wind speed, we want to evaluate the probability that $X > 32$. This probability is denoted by

$Pr\ \{X > 32\}.$

Using the procedures described in earlier sections of this chapter, it is relatively simple to evaluate this probability. To do so, we must find the value of the standard normal variable corresponding to 32 percent. This value is $(32 - 0) \div 15.6$, or 2.05. (Why? Because the mean of $X$ is zero and its standard deviation is 15.6, as pointed out above.) Appendix Table 2 shows that the area under the standard normal curve between zero and 2.05 is .4798, so the area to the right of 2.05 equals $.5000 - .4798 = .0202$. Thus, the probability is about .02 that an increase of more than 32 percent in wind speed will occur. (The shaded area to the right of zero in Figure 5.10 is equal to this probability.)

It is also useful to be able to answer a somewhat different type of question: What is the percentage change in wind speed that will be exceeded with a specified probability? For example, what is the percentage change in wind speed that will be exceeded with a probability of .95? To answer this question, we must find the value of $x_0$ such that

$Pr\ \{X > x_0\} = .95.$

As a first step, note that Appendix Table 2 shows that the area under the standard normal curve between zero and 1.64 is .45. Thus, the area between zero and $-1.64$ also is .45, which means that the area to the right of $-1.64$ is .95. Having determined that the probability is .95 that a standard normal variable will exceed $-1.64$, we must now find the value of the normal variable corresponding to this value of the standard normal variable. In other words, we must find the value of $X$ in equation (5.2) when $Z = -1.64$. Clearly, the desired value of the normal variable is $\mu - 1.64\sigma$, which here equals $0 - 1.64(15.6) = -25.6$ percentage points. Thus, a $-25.6$ percentage point change in wind speed is the value that will be exceeded with a probability of .95. (The shaded area to the left of zero in Figure 5.10 equals .05.)

## EXERCISES

**5.1** Find the probability that the standard normal variable lies (a) above 2.3; (b) below $-3.0$; (c) above 0.7; (d) between 1 and 2; (e) between $-1$ and 2.

**5.2** If $X$ is a normal variable with $\mu = 2$ and $\sigma = 3$, show how it can be converted into the standard normal variable.

**5.3** "If you know that the probability that a normal variable exceeds a certain number, $Q$, is .10, you can be sure that the probability that this variable is less than $-Q$ is also .10." Do you agree? Why, or why not?

**5.4** Find the area under the standard normal curve which lies: (a) between 0 and 1.82; (b) between $-1.32$ and 0; (c) between $-1.08$ and 1.08; (d) between 1.32 and 1.46; (e) between $-1.08$ and $-0.23$; (f) between $-0.48$ and 2.01.

**5.5** Find the value of $z$ if
(a) the area under the standard normal curve between 0 and $z$ is 0.1985.
(b) the area under the standard normal curve to the right of $z$ is 0.2776.
(c) the area under the standard normal curve between $-z$ and 0 is 0.0910.
(d) the area under the standard normal curve to the left of $z$ is 0.8051.
(e) the area under the standard normal curve between $-z$ and $z$ is 0.1820.

**5.6** A random variable is normally distributed with mean equal to 300 and standard deviation equal to 60. Calculate the probability that the value of this random variable
(a) is less than 280.
(b) exceeds 350.
(c) lies between 185 and 265.
(d) lies between 305 and 375.

**5.7** The IQs of students at a particular college are normally distributed with mean equal to 125 and standard deviation equal to 10. Determine the percentage of the students with IQs
(a) below 115.
(b) above 140.
(c) between 120 and 130.
(d) between 100 and 115.

**5.8** Suppose $X$ is a normal random variable with mean $\mu$ and standard deviation $\sigma$.
(a) Under what circumstances is $X/\sigma$ the standard normal variable?
(b) Under what circumstances is $(X - \mu)$ the standard normal variable?

**5.9** If the weights of adult males are normally distributed with mean equal to 170 pounds and standard deviation equal to 20 pounds, the probability that a certain weight will be exceeded is .05. What is this weight?

**5.10** In the previous exercise, the probability that an adult male's weight will be less than a certain amount is .10. What is this amount?

**5.11** The probability that a marksman will hit the target is $1/3$. If he takes 50 shots, what is the probability that he will hit the target less than 10 times? (Use the normal distribution as an approximation to the binomial distribution.)

**5.12** The Martin Company announces that it will give $500 bonuses to its sales people who are among the top 10 percent in sales in 1988. It believes that its salespeople will be normally distributed with respect to their 1988 sales, the mean being $400,000 and the standard deviation being $100,000. If this is correct, how big must a salesperson's sales be to get him or her the bonus?

**5.13.** A manufacturing process turns out 20 percent defective items. A sample of 50 items is taken from the 3,000 produced on a particular day. Determine the probability that
(a) 7 items or less in the sample are defective.
(b) 21 items or more in the sample are defective.
(c) more than 10 and less than 19 items in the sample are defective.
(d) more than 8 and less than 12 items in the sample are defective.

**5.14** The XYZ Oil Company drills 200 wells in Canada. The probability that each will be a dry hole is 0.9. Calculate the probability that
(a) 8 or more will be dry holes.
(b) 20 or more will not be dry holes.
(c) more than 7 but less than 20 will not be dry holes.
(d) the number of dry holes will differ by one or less from its expected value.

**5.15** There is a 1/3 probability that a bicycle pedal produced by the Uphill Manufacturing Company will not survive more than 10 years of normal wear. The firm sells these pedals in cartons of 300. What is the probability that in any such carton less than 85 or more than 115 pedals will not survive more than 10 years of normal wear? (Use the normal approximation to the binomial distribution.)

# 5.10 The Poisson Distribution[6]

Another important probability distribution is the Poisson distribution, which is named after a nineteenth-century Swiss mathematician. The *Poisson distribution* is a discrete probability distribution which has the following formula:

$$P(x) = \frac{\mu^x e^{-\mu}}{x!}, \text{ for } x = 0, 1, 2, \ldots \qquad (5.3)$$

*Formula for Poisson Distribution*

where $P(x)$ is the probability that a variable with a Poisson distribution equals $x$, $\mu$ is the mean or expected value of the Poisson distribution, and $e$ is approximately 2.718 and is the base of the natural logarithms.[7] Like the binomial and normal distributions, the Poisson distribution is really a family of distributions. Depending on the value of $\mu$, the shape of the probability distribution will vary considerably.

---

[6] This material and the remainder of this chapter (including the Appendix) can be omitted without loss of continuity.

[7] A Poisson random variable, like a binomial random variable, is discrete. Unlike the binomial, it does not assume a finite number of possible values; instead, it assumes a countably infinite number of possible values.

One reason why the Poisson distribution is so important in statistics is that it can be used as an approximation to the binomial distribution under the circumstances described below.

POISSON APPROXIMATION TO THE BINOMIAL DISTRIBUTION. *If* n *(the number of trials) is large and* $\Pi$ *(the probability of success) is small, the probability of* x *successes occurring in* n *Bernoulli trials can be approximated by the Poisson distribution where* $n\Pi = \mu$*. Experience indicates that this approximation is adequate for most practical purposes if* n *is at least 20 and* $\Pi$ *is no greater than .05.*

Whereas the normal distribution approximates the binomial distribution when $\Pi$ is *not* very small, the Poisson distribution approximates it when $\Pi$ *is* very small; thus, the two approximations complement one another.

To illustrate how the Poisson distribution can be used in this way, let's consider the following situation. You drive to work 15,000 times in a 30-year period, and the probability of your having an accident each time you drive to work is .0001. In this case, each trip can be considered a "trial" and each accident can be considered a "success" (although only for your garage mechanic and/or mortician). Thus, $n = 15{,}000$ and $\Pi = .0001$. Since $n$ is very large and $\Pi$ is very small, the Poisson distribution should be a good approximation to the binomial distribution. Since $\mu = n\Pi = 15{,}000(.0001)$, or 1.5, equation (5.3) can be used to obtain the probability of 0, 1, 2, . . . accidents during this 30-year period, the results being:

$$P(0) = \frac{1.5^0 e^{-1.5}}{0!} = e^{-1.5} = 0.22,$$

$$P(1) = \frac{1.5^1 e^{-1.5}}{1!} = 1.5e^{-1.5} = 0.33,$$

$$P(2) = \frac{(1.5)^2 e^{-1.5}}{2!} = \frac{2.25e^{-1.5}}{2} = 0.25,$$

$$P(3) = \frac{(1.5)^3 e^{-1.5}}{3!} = \frac{3.375e^{-1.5}}{6} = 0.13,$$

$$P(4) = \frac{(1.5)^4 e^{-1.5}}{4!} = \frac{5.062e^{-1.5}}{24} = 0.05,$$

$$P(5) = \frac{(1.5)^5 e^{-1.5}}{5!} = \frac{7.594e^{-1.5}}{120} = 0.01.$$

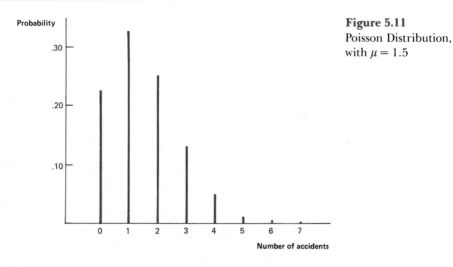

**Figure 5.11**
Poisson Distribution,
with $\mu = 1.5$

We could, of course, compute the probability of 6, 7, 8, . . . accidents, but these probabilities are less than .005. Figure 5.11 shows this Poisson distribution graphically. As you can see, the distribution is not symmetrical, but skewed to the right. (Recall our discussion of skewness in Chapter 2.)

As pointed out earlier in this section, *the expected value of any Poisson random variable equals $\mu$*. To demonstrate this, in the present illustration the expected number of accidents equals

$$E(X) = (0)(0.22) + (1)(0.33) + (2)(0.25) + (3)(0.13) + (4)(.05)$$
$$+ (5)(.01) + (6)(.004) + (7)(.001) + . . .$$
$$= 1.50.$$

Also, *the standard deviation of any Poisson random variable equals $\sqrt{\mu}$*. (As an exercise, prove that this is true in this illustration; that is, prove that the standard deviation of the number of accidents equals $\sqrt{1.50}$. If you have difficulty, consult the footnote on this page).[8]

The computations involved in evaluating $P(x)$ can be onerous. To reduce the computational burden, Appendix Table 3 can be used. This table shows $P(x)$ for selected values of $\mu$. The following example will provide additional practice in using the Poisson distribution.

---

[8] To determine the standard deviation, first determine the variance by inserting the values of $P(x)$ into the following expression:

$$\Sigma(x - \mu)^2 P(x).$$

The result is

$$\sigma^2 = (0 - 1.5)^2(0.22) + (1 - 1.5)^2(0.33) + (2 - 1.5)^2(0.25) + (3 - 1.5)^2(0.13)$$
$$+ (4 - 1.5)^2(.05) + (5 - 1.5)^2(.01) + (6 - 1.5)^2(.004)$$
$$+ (7 - 1.5)^2(.001) + . . .$$

(continued)

EXAMPLE 5.5. A machine turns out engine parts, 2 percent of which are defective. These parts are packaged in boxes of 100 and are shipped to the plant where they are used.

   (a) What is the probability of 0, 1, 2, or 3 defectives in a box?

   (b) What is the expected number of defectives in a box?

   (c) What is the standard deviation of the number of defectives in a box?

   SOLUTION: (a) Since $\mu = n\Pi = 100(.02)$ or 2.0, equation (5.3) yields the following probabilities:

$$P(0) = 2^0 e^{-2} \div 0! = .1353$$

$$P(1) = 2^1 e^{-2} \div 1! = .2707$$

$$P(2) = 2^2 e^{-2} \div 2! = .2707$$

$$P(3) = 2^3 e^{-2} \div 3! = .1804$$

These probabilities can be found in the column of Appendix Table 3 where $\mu = 2.0$.

   (b) The expected number of defectives in a box equals $\mu = n\Pi = 100(.02)$ or 2.

   (c) The standard deviation of the number of defectives in a box equals $\sqrt{\mu} = \sqrt{n\Pi} = \sqrt{2}$.

## 5.11 Additional Uses for the Poisson Distribution

Besides being a useful approximation to the binomial distribution, the Poisson distribution is very important in its own right. Assume that events of a particular kind occur at random during a particular time span. To make things more concrete, suppose that the events in question are demands by a firm's customers for a particular type of spare part. If the following four conditions are met, the probability distribution of the number of such events (that is, the number of demands for this type of spare part) in a fixed period of time will be a Poisson distribution:

---

Footnote 8 (continued)

Thus,

$$\sigma^2 = 2.25(0.22) + 0.25(0.33) + 0.25(0.25) + 2.25(0.13) +$$

$$6.25(.05) + 12.25(.01) + 20.25(.004) + 30.25(.001) + \ldots$$

$$= 0.495 + .082 + .063 + .292 + .313 + .122 + .081 + .030 + \ldots$$

$$= 1.50.$$

The standard deviation is the square root of the variance, or $\sqrt{1.50}$.

1. THE PROBABILITY THAT EACH EVENT OCCURS IN A VERY SHORT TIME INTERVAL MUST BE PROPORTIONAL TO THE LENGTH OF THIS TIME INTERVAL. Thus, the probability that a spare part of this type is demanded in a two-minute interval must be double the probability that a spare part of this type is demanded in a one-minute interval.

2. THE PROBABILITY THAT TWO OR MORE EVENTS OF THE RELEVANT KIND OCCUR IN A VERY SHORT TIME INTERVAL MUST BE SO SMALL THAT IT CAN BE REGARDED AS ZERO. Thus, the probability that more than one order will occur for this type of spare part in a one-second time interval must be essentially zero. This assumption seems reasonable in this case. If the time interval is only one second long, it would be difficult indeed for two different orders for this type of spare part to be received by the company.

3. THE PROBABILITY THAT A PARTICULAR NUMBER OF THESE EVENTS OCCURS IN A PARTICULAR TIME INTERVAL MUST NOT DEPEND ON WHEN THIS TIME INTERVAL BEGINS. Thus, the probability that an order is received by the company in a one-minute time interval beginning at noon tomorrow must be the same as the probability that an order will be received in a one-minute time interval beginning at 2 P.M. today. This is because it is assumed that this probability depends only on the length of the time interval, not on when the time interval begins. The fact that the one time interval begins at noon tomorrow and the other time interval begins at 2 P.M. today must not influence this probability at all.

4. THE PROBABILITY THAT A PARTICULAR NUMBER OF THESE EVENTS (DEMANDS) OCCURS IN A PARTICULAR TIME INTERVAL MUST NOT DEPEND ON THE NUMBER OF THESE EVENTS THAT OCCURRED PRIOR TO THE BEGINNING OF THIS TIME INTERVAL (OR IN SOME SHORTER TIME INTERVAL PRIOR TO THE BEGINNING OF THIS TIME INTERVAL). For example, suppose that five orders for this type of spare part were received prior to 2 P.M. today. The fact that five orders (rather than four, six, or some other number) were received prior to that time should not influence the probability of receiving an order in the one-minute time interval beginning at 2 P.M. today. Of course, this assumption may be violated if there is a tendency for these events to bunch together in time. For example, if orders tend to bunch together, the probability of receiving an order in the one-minute time interval beginning at 2 P.M. today may be dependent on whether an order was received just before 2 P.M. If so, the Poisson distribution is not appropriate.

If these four conditions are met, it can be shown that *the probability that* x *such events will occur in a time interval of length* $\Delta$ *(delta) is*

$$P(x) = \frac{(\lambda\Delta)^x e^{-\lambda\Delta}}{x!},$$    (5.4)

*where λ (lambda) is the mean number of such events per unit of time.* This, of course, is the same probability distribution as in equation (5.3), the only difference being that $\lambda\Delta$ is used here in place of $\mu$. However, since $\lambda\Delta$ is the expected value of $x$, it is the same as what we formerly called $\mu$.[9]

## 5.12 Replacement of Parts on Polaris Submarines: A Case Study[10]

The U.S. Department of Defense faced a problem in the operation of Polaris submarines that illustrates the practical utility of the Poisson distribution. Each submarine goes out on a mission of relatively fixed length (about 60 days), after which it is resupplied by a tender. During each mission a submarine must rely upon its own supply of spare parts. At the end of the mission, the tender replenishes the items that have been taken out of the submarine's supplies. How many spare parts should each tender carry in order to replace the spare parts that are used up in the preceding mission? Analysts have made substantial use of the Poisson distribution in solving this problem.

Obviously, the number of spare parts of a particular type that a tender must replenish for a particular submarine equals the number of such parts that failed during the preceding mission. Thus, the answer to the Defense Department's problem depends in considerable part (but not wholly)[11] on the probability distribution of the number of parts of a particular type which will fail during a mission. To estimate this probability distribution, analysts have found it useful to assume that failures meet the four conditions that were discussed in the previous section. In other words, the probability that a particular kind of part will fail during a short time interval is proportional to the length of time interval and is independent of when the interval occurs; it is also independent of how many such parts have failed prior to that time interval. Also, the probability that more than one part of a particular type will fail in a very short interval is so small that it can be regarded as zero.

Given that failures of a particular type of part can be represented in this way, we know (from the previous section) that the probability that $x$ failures of a particular type of part will occur during a mission is

$$P(x) = \frac{(\lambda\Delta)^x e^{-\lambda\Delta}}{x!} \, ,$$

---

[9] Sometimes the Poisson distribution is used to characterize events distributed at random in space rather than in time. For example, the Poisson might be used to find the probability of a submarine's being located in a particular area.

[10] This section is based in part on S. Haber and R. Sitgreaves, "An Optimal Inventory Model for the Intermediate Echelon when Repair Is Possible," *Management Science,* February 1975.

[11] Ibid.

where $\Delta$ is the length of the mission and $\lambda$ is the average number of failures per unit of time for the particular type of part. According to studies based on the first 61 patrols of Polaris submarines, the value of $\lambda\Delta$ varies from very close to zero to as high as 5.0, depending on the particular part.

If the Defense Department is specifically interested in a part where the value of $\lambda\Delta$ is estimated to be 1.0, what is the probability distribution of the number of such parts that will fail during a mission? Based on the column of Appendix Table 3 where $\mu = 1.0$, the answer is

$$P(0) = \frac{1^0 e^{-1}}{0!} = .3679,$$

$$P(1) = \frac{1^1 e^{-1}}{1!} = .3679,$$

$$P(2) = \frac{1^2 e^{-1}}{2!} = .1839,$$

$$P(3) = \frac{1^3 e^{-1}}{3!} = .0613,$$

$$P(4) = \frac{1^4 e^{-1}}{4!} = .0153.$$

We can ignore $P(5)$, $P(6)$, ... since each of these probabilities is less than .005.

Obviously, decision makers in the Defense Department have found it very helpful to know that the chances are about 37 out of 100 that *no* spare parts of this type will need to be replaced after a mission, that the chances are about 37 out of 100 that *one* will have to be replaced, that the chances are about 18 out of 100 that *two* will have to be replaced, and so on. Information of this sort, if properly applied, can promote a much more effective and economical inventory policy for submarine tenders.

## EXERCISES

**5.16** If $\mu = 2$, what is the probability that a Poisson random variable, $X$, equals (a) 1; (b) 2; (c) 3?

**5.17** If $\Pi = 1/3$ and $n = 100$, should you use the normal distribution or the Poisson distribution as an approximation to the binomial distribution?

**5.18** Given $\Pi = .01$ and $n = 300$, should you use the normal distribution or the Poisson distribution as an approximation to the binomial distribution?

**5.19** If a Poisson random variable has an expected value of 3.0, what is its variance?

**5.20** If a Poisson random variable's coefficient of variation—that is, its standard deviation divided by its mean—equals 2, what is its mean?

**5.21** Given $n = 20$ and $\Pi = .05$, what is the probability that $X = 0$, based on (a) the binomial distribution; and (b) the Poisson approximation to the binomial distribution?

**5.22** R. D. Clarke reported the following data concerning the number of hits by buzz bombs during World War II in south London. (Each area covers 1/4 square kilometer.)

| Number of hits | Number of areas |
|---|---|
| 0 | 229 |
| 1 | 211 |
| 2 | 93 |
| 3 | 35 |
| 4 | 7 |
| 5 or more | 1 |
| Total | 576 |

Are the results what might be expected on the basis of the Poisson distribution, where the mean number of hits per area was 1.0?

**5.23** The number of accidents occurring in a given month at one of the Morris Company's plants is known to conform to the Poisson distribution. The standard deviation of this distribution is 1.732 accidents per month. What is the probability that no accidents will occur in this plant during this month?

**5.24** One hundred packages are mailed to customers in New York City. Each is mailed on a different day, and the probability that each will arrive within 48 hours of mailing is 0.05. How long it takes for one package to arrive is independent of how long it takes for another package to arrive. Determine the probability that
(a) four of the packages arrive within 48 hours of mailing.
(b) five of the packages arrive within 48 hours of mailing.
(c) six of the packages arrive within 48 hours of mailing.

**5.25** The box office of a Broadway play has on the average five incoming calls per minute. Using the Poisson distribution, find the probability that there will be
(a) exactly two incoming calls during any given minute.
(b) exactly three incoming calls during any given minute.
(c) no incoming calls during any given minute.

**5.26** According to the records of the relevant airline, the number of people who buy tickets but fail to show up for the airline's early morning flight between San Francisco and Denver is a random variable with a Poisson distribution, its mean being 4. Determine the probability that the number of no shows
(a) equals 5.
(b) is less than 3.
(c) is greater than 6.

**5.27** The number of automobile accidents on a particular turnpike between 1 P.M. and 2 P.M. has a mean of 3, and is distributed according to the Poisson distribution. What is the probability that it will exceed or fall short of its mean by more than one standard deviation?

**5.28** The mean number of defects in 10 yards of cloth produced by the ABC Textile Company is two. The number of defects is distributed according to the Poisson distribution. If more than four defects are present in 10 yards of cloth, it is substandard, and the company makes no profit on it. What is the probability that 10 yards of cloth are substandard?

**5.29** A stock broker receives an average of six telephone calls while he is away during his lunch hour (from noon to 1 P.M.). The number of calls he receives then is distributed according to the Poisson distribution. During his lunch hour, another broker agrees to answer this broker's telephone twice, but not more. No one else is available to answer his phone.

(a) What is the probability that this broker will receive at least one call during his lunch hour that is not answered?

(b) What is the probability that this broker will receive exactly one call during his lunch hour that is not answered?

# Chapter Review

1. In the limit, as more and more observations are gathered concerning a continuous random variable, and as class intervals become narrower and more numerous, the histogram (using a density scale) of the variable becomes a smooth curve called a *probability density function.* The total area under any probability density function must equal 1. The probability that a random variable will assume a value between any two points is equal in value to the area under the random variable's probability density function between these two points.

2. The most important continuous probability distribution is the *normal distribution,* whose probability density function is called the *normal curve.* The location and spread of a normal curve depend on its mean and standard deviation, but all normal curves are symmetrical and bell-shaped. If one expresses any normal variable as a deviation from its mean and measures these deviations in units of its standard deviation, the result is called the *standard normal variable.* The standard normal variable is a normal variable with a mean of zero and a standard deviation of 1. The areas under the standard normal curve are tabled.

3. To calculate the probability that the value of any normal variable (with mean $\mu$ and standard deviation $\sigma$) lies between two points, $a$ and $b$, find the points on the standard normal distribution corresponding to $a$ and $b$. These points are $(a - \mu) \div \sigma$ and $(b - \mu) \div \sigma$. Then use Appendix Table 2 to determine the area under the standard normal curve between $(a - \mu) \div \sigma$ and $(b - \mu) \div \sigma$. This area equals the probability that the value of the normal variable is between $a$ and $b$.

4. If $n$ (the number of trials) is large and $\Pi$ (the probability of success) is not too close to zero or 1, the probability distribution of the number of successes in $n$ Bernoulli trials can be approximated by a normal distribution. Specifically, to approximate the probability that the number of successes is from $c$ to $d$, find the probability that the value of a normal variable (with mean $n\Pi$ and standard deviation $\sqrt{n\Pi(1 - \Pi)}$) lies between $(c - 1/2)$ and $(d + 1/2)$. (Of course, $c$ is presumed to be less than $d$.)

5. An important discrete probability distribution is the Poisson distribution, which is $P(x) = \mu^x e^{-\mu} \div x!$. The mean of the Poisson distribution is $\mu$, and its standard deviation is $\sqrt{\mu}$. If $n$ (the number of trials) is large and $\Pi$ (the probability of success) is small, the probability distribution of the number of successes in $n$ trials can be approximated by the Poisson distribution where $n\Pi = \mu$. Experience indicates that this approximation is adequate for most practical purposes when $n$ is at least 20 and $\Pi$ is no greater than .05.

6. Besides being a useful approximation to the binomial distribution, the Poisson distribution is of importance in its own right. For example, the Poisson distribution is the probability distribution of the number of events that occur in a time interval under the following circumstances: (a) The probability that an event occurs in a very short time interval is proportional to the length of the time interval, and does not depend on when the interval occurs or on how many events occurred before the beginning of the interval. (b) The probability of more than one event occurring in a very short time interval is negligible. For intance, the Poisson distribution has been used to represent the probability distribution of the number of parts of a particular type which fail during one mission of a Polaris submarine.

## Chapter Review Exercises

**5.30** The probability that any customer who enters a particular supermarket will purchase a can of orange juice is 0.10. If 1,000 customers enter the supermarket on a particular day, what is the minimum number of cans of orange juice that the supermarket must have in stock if the probability is to be at most 5 percent that it will run out of orange juice? (Assume that the customers act independently.)

**5.31** If an automobile stops at Joe's Gas Station, there is a .25 probability that its driver will buy more than $10 worth of gasoline. Between 9 A.M. and 10 A.M. yesterday, 20 cars stopped at Joe's Gas Station. What was the probability that more than 3 but less than 8 of these cars' drivers bought more than $10 worth of gasoline? (Assume that the drivers act independently.)

(a) Use the normal approximation to the binomial distribution to answer this question.

(b) Use the binomial distribution to obtain an exact answer to this question.

(c) How close is the answer based on the normal approximation to the exact answer?

**5.32** The diameter of screws produced by a particular machine is normally distributed. Based on past experience, it is known that 30 percent of the screws are less than 1.30 inches in diameter and that 40 percent of the screws are greater than 1.71 inches in diameter. What is the mean diameter of screws produced by this machine? What is the standard deviation of the diameters of the screws produced by this machine?

**5.33** From past experience, the Uphill Manufacturing Corporation knows that the deviations of the width of its bicycle pedals from their mean width are normally distributed with a standard deviation of .02 inches.

(a) What is the probability that a pedal's width is more than .03 inches above the mean?

(b) What is the probability that a pedal's width is more than .05 inches below the mean?

(c) What is the probability that a pedal's width differs (either positively or negatively) from the mean by less than .015 inches?

(d) During 1986, the mean width of all bicycle pedals produced by Uphill was exactly equal to what the design called for. However, during 1987, the mean width of all bicycle pedals produced by the firm was .01 inches greater than the design called for. In both years, the standard deviation of the pedal widths was .02 inches, and the pedal widths were normally distributed. In 1986, what was the probability that a pedal would be wider than called for by the design?

(e) In 1987, what was the probability that a pedal would be wider than called for by the design?

(f) In 1986, what was the probability that a pedal would be more than .04 inches wider than called for by the design?

(g) In 1987, what was the probability that a pedal would be more than .04 inches wider than called for by the design?

**5.34** A factory contains 20 machine tools. The probability that one of them is malfunctioning on a particular day is 0.05. Whether one of them is malfunctioning is independent of whether another is malfunctioning.

(a) Use the binomial distribution to determine the exact probability that two of the machines malfunction on a particular day.

(b) Use the Poisson approximation to the binomial distribution to approximate this probability.

(c) Based on your answers to (a) and (b), how close is the Poisson approximation?

**5.35** The Howe Company wants no more than 1/10 of 1 percent of the motors it produces to be defective. Production quality is checked by examining a certain number of motors chosen at random from each day's output, and the manufacturing process is stopped if any motor is defective. If the firm wants the probability to be about .05 that the process will be stopped when it is producing 1/10 of 1 percent defectives, how many motors must be examined from each day's output?

**5.36** The Uphill Manufacturing Company's switchboard receives an average of four incoming calls per minute.

(a) What conditions must be satisfied if the number of incoming calls in any given minute is to be represented by a Poisson distribution?

(b) If these conditions are met, what is the probability that there will be exactly five incoming calls in a given minute? Exactly six incoming calls in a given minute? Exactly seven calls in a given minute?

(c) Describe how the probability distribution of the number of incoming calls in a given minute might be useful in deciding how much capacity the firm should have to handle calls.

(d) What is the standard deviation of the number of incoming calls in a given minute?

(e) Using Chebyshev's inequality, determine an upper bound for the probability that the number of incoming calls in a given minute is more than 2 standard deviations above or below is mean. By how much does this upper bound exceed the true probability?

**\*5.37** Use Minitab (or some other computer package) to calculate the Poisson distribution when $\mu = 1.45$.

**\*5.38** Use Minitab (or some other computer package) to calculate the Poisson distribution when $\mu = 2.82$.

**5.39** In a 1983 report for the U.S. Air Force,[12] R. K. Weatherwax estimated that the probability of a catastrophic accident involving the space shuttle's solid-fuel booster rockets was about 1/35. (On the other hand, the

---

\* Exercise requiring computer package, if available.

[12] "Study of Rockets by Air Force Said Risks Were 1 in 35," *New York Times*, February 11, 1986.

National Aeronautics and Space Administration estimated in 1985 that the probability of booster failure was 1 in 60,000 flights.) On January 28, 1986, the shuttle exploded on its 25th flight. If Weatherwax's estimate was correct, what was the probability of at least one catastrophic accident of this sort in 25 flights?

# Appendix 5.1

### WAITING LINES AND THE EXPONENTIAL DISTRIBUTION

Many business situations are concerned with waiting lines or *queues*. People wait in line at railroad stations to buy tickets and at airports to have themselves and their baggage inspected for weapons. Cars must wait in line at toll booths, and airplanes must wait in line to take off and land. Statisticians and management scientists are interested in such waiting lines because they want to be sure that the capacity to service the people, cars, or airplanes is in the proper relation to the rate at which they arrive. If this service capacity is too small, the queue will be disproportionately large and the people, cars, or airplanes will waste an inordinate amount of time waiting for service. On the other hand, if the service capacity is too large, the waiting time will tend to be very short; but much of the service capacity will be underutilized or perhaps not utilized at all.

In this appendix, we shall be concerned with queuing theory, a branch of statistics that analyzes the factors determining how long a waiting line will be, and how much time a person, car, or airplane must spend in order to get through the line. We shall present some simple analytical results that apply under special but not unusual circumstances.

POISSON ARRIVALS. To begin with, let's use the term **customers** to designate the persons, cars, airplanes, or whatever else that is arriving. Further, let's regard the facility that services and eventually releases the customers as a set of **service counters.** Thus, the general situation that we are analyzing is as shown in Figure 5.12. Clearly, *the average length of time that a customer waits in line depends, among other things, on how frequently customers arrive at the service counters and on how rapidly the service counters can perform the service that is required.* This section will focus on the arrivals; the service times will be taken up in a later section.

For many types of situations, it is realistic to assume that customers arrive *at random.* Put somewhat crudely, this means that the probability that a customer will arrive in one small interval of time is no different from the probability that a customer will arrive in any other small interval of time of equal duration. This means also that the probability that a certain number of cus-

**Figure 5.12**
**General Queuing**
**Situation**

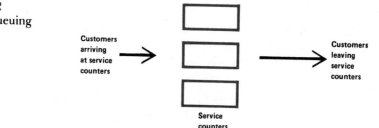

tomers will arrive in a certain time interval does not depend on the number of customers that arrived prior to the beginning of this time interval. Thus, the probability distribution of arrivals conforms to the four assumptions described in Section 5.11, so that the probability that $x$ customers will arrive in a time interval of length $\Delta$ is

$$P(x) = \frac{(\lambda\Delta)^x e^{-\lambda\Delta}}{x!},$$

(5.5)

where $\lambda$ is the mean number of arrivals per unit of time. In other words, *the number of arrivals in a given time interval has a Poisson distribution.*

THE EXPONENTIAL DISTRIBUTION. If the number of customers arriving within a given time interval has a Poisson distribution, it can be shown that *the interval of time between consecutive arrivals* has an exponential probability distribution, which is defined below.

EXPONENTIAL DISTRIBUTION. *If* Y *is the time interval between consecutive arrivals,* Y *is a continuous random variable with the following probability density function:*

$$P(y) = \lambda e^{-\lambda y},$$

(5.6)

*where $\lambda$ is the mean number of arrivals per unit of time. The probability that* Y *lies between any two numbers,* c *and* d, *equals*

$$e^{-\lambda c} - e^{-\lambda d}.$$

(5.7)

Figure 5.13 shows the exponential probability density function, assuming that $\lambda = 1.0$. The shaded area under the curve equals the probability that the value of the exponential random variable will lie between $c$ and $d$; as indicated in

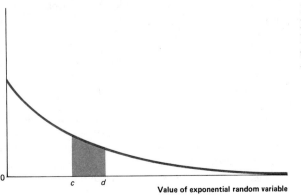

**Figure 5.13**
Probability Density Function for Exponential Random Variable, with $\lambda = 1$

0

*c*    *d*

**Value of exponential random variable**

equation (5.7), this probability equals $e^{-\lambda c} - e^{-\lambda d}$. The mean of an exponential random variable equals $1/\lambda$.

To illustrate the use of the exponential distribution, suppose that customers arrive randomly at a shoe store, and that the mean number of arrivals per hour is 6. What is the probability that the time interval between two consecutive arrivals is between $1/4$ and $1/2$ hour? Since the mean number of arrivals per hour is 6, $\lambda$ equals 6. Consequently, using equation (5.7) the probability that the time interval between two consecutive arrivals is between $1/4$ and $1/2$ hour is

$$e^{-6(1/4)} - e^{-6(1/2)} = e^{-1.5} - e^{-3.0} = .223 - .050 = .173,$$

since $\lambda = 6$, $c = 1/4$, and $d = 1/2$. (Appendix Table 4 shows the value of $e^{-x}$ for various values of $x$.) Thus, this probability equals .173.

EXPONENTIAL SERVICE TIMES. Now let's turn to the probability distribution of service time. Once service begins, a certain length of time elapses before a customer has received service and leaves the service counter. This length of time, known as the *service time,* has a probability distribution *which is often assumed to be exponential.* In other words, the probability that the service time is between $u$ and $v$ is assumed to equal

$$e^{-mu} - e^{-mv}, \tag{5.8}$$

where $m$ is the mean number of customers that can be serviced per unit of time. This assumption is adopted frequently because it is analytically convenient and because it often conforms quite well to reality.

To keep things simple, we also assume that there is only one service counter, and that the *line discipline*—that is, the rules governing the behavior of the customers in line—is that a single line is formed, that customers are served on a first come, first served basis, and that no customer leaves the line before being served. Obviously, the line discipline as well as the probability distribution of arrivals, the probability distribution of service times, and the number of service counters, influences the probability that a customer will have to wait as well as the expected length of the waiting line. The rules that are adopted are clearly of importance, and the results presented below must be modified if the line discipline departs significantly from these assumptions.

Based on these assumptions, a number of interesting and important conclusions can be derived concerning the waiting lines that will occur under various circumstances. First, it can be shown that the *probability that a newly arrived customer will have to wait is*

$$P_w = \lambda/m. \tag{5.9}$$

Thus, if the mean number of customers arriving per hour is 3, and if the mean number of customers that can be serviced per hour is 4, the probability is $3/4$, or 0.75, that a newly arrived customer will have to wait. The ratio, $\lambda/m$, is often called the *utilization factor.* We assume that $\lambda$ is less than $m$; if not, the waiting line will grow beyond any finite bounds.

Second, *the mean time that a customer spends waiting in line is*

$$\frac{\lambda/m}{m-\lambda} = \frac{P_w}{m-\lambda}. \qquad (5.10)$$

Thus, under the circumstances described in the previous paragraph, the mean time that a customer spends waiting is $0.75 \div (4-3) = 0.75$ hours. *The mean time that a customer spends waiting in line and being served is*

$$\frac{1}{m-\lambda} \qquad (5.11)$$

So, under the circumstances described in the previous paragraph, the mean time that a customer spends waiting and being served is $1 \div (4-3) = 1$ hour.

Third, *the mean length of the waiting line is*

$$E_L = \frac{(\lambda/m)^2}{1-\lambda/m} = \frac{P_w^2}{1-P_w}. \qquad (5.12)$$

Thus, under the circumstances described above, the mean length of the waiting line is $0.75^2 \div (1-.75) = 2.25$ customers. Note that *the total time lost (due to waiting) by all customers during a unit of time is also given by* $E_L$, since the total time spent waiting by all customers is equal to the average number of customers waiting at any point in time multiplied by the unit of time used in the analysis. So, under the circumstances described above, 2.25 customer hours are lost (spent on line) every hour.

DOCKING SHIPS: AN ILLUSTRATION. Queuing theory can be applied to a wide variety of practical situations, as in the case of a shipping company that must dock and service its ships at a facility it owns. The ships arrive randomly at this facility; five ships, on the average, arrive per month. The probability distribution of the length of time that a ship must stay at the facility for docking and servicing is known to be an exponential distribution. The mean number of ships that the facility can dock and service per month is 10. Given these circumstances, it follows from equation (5.9) that the probability that a ship will have to wait before being docked and serviced is $5/10$. It also follows from equation (5.10) that the mean time a ship will spend waiting before being docked and serviced is $5/10 \div (10-5) = 1/10$ months, or about three days. From equation (5.12) it follows that the mean length of the waiting line is $(5/10)^2 \div (1-5/10) = 1/2$ ships and that about $1/2$ ship-months are lost (in waiting) each month.

Faced with this situation, the firm must decide whether or not to increase the capacity of its docking and servicing facility. In particular, if it is willing to spend an extra $10,000 per month, it can increase the mean number of ships that the facility can dock and service per month from 10 to 12. What effect will this have on the amount of time lost? Using equation (5.12) again, it is clear that the number of ship-months which will be lost each month if this extra expenditure is made equals $(5/12)^2 \div (1-5/12) = 25/84$. In other words, if the extra expenditure is made, there will be a monthly reduction of lost time of $(1/2 - 25/84) = 17/84$ ship-months (about six ship-days). Assume that \$3,000 is the cost of losing each ship-day since this is the amount the firm could earn if a ship were working for a day rather than waiting before being docked and serviced. If so, the firm should spend the extra \$10,000 on the facility, since this will reduce the expected monthly costs due to lost time by about \$18,000. (Reducing lost time by about six days is worth about \$18,000.)

EXERCISES

**5.40** A firm owns and operates a large number of machines. During each eight-hour shift, machines needing repair are brought to the firm's service facility. The mean arrival rate at the facility is five machines per hour, and the mean service rate is six machines per hour. The cost of a machine's being at the service facility and not in production is \$20 per hour. If the firm's management can increase the mean service rate to seven machines per hour by adding personnel costing \$40 per hour, would it be worthwhile to hire the additional personnel?

**5.41** The mean number of customers who arrive at the checkout counter of a supermarket during a one-minute interval is three. The number of arrivals is a random variable with the Poisson distribution. Find the probability that the interval between the arrival of two customers will be at least (a) one minute; (b) two minutes.

**5.42** The Alpha Corporation, which sells bicycles, has only one salesman, who waits on customers on a first come, first served basis. Customers arrive at the firm's sales office at random, and the mean number arriving in an hour is three.
(a) Alpha's salesman (who, as we know from Exercise 4.5, has a weakness for gambling) is considering leaving the sales office untended for an hour to visit a local gambling parlor. If a customer arrives while the salesman is out, there is a good chance that he will be fired. What is the probability that a customer will arrive in his absence?
(b) A customer arrives at 11 A.M. What is the probability that the next customer will arrive between 11:30 A.M. and noon?
(c) The length of time the Alpha Corporation's salesman spends with a customer conforms to the exponential distribution, and the mean number of customers which can be serviced per hour is six. What is the probability that the amount of time he spends with a particular customer exceeds 20 minutes?
(d) What is the probability that the amount of time he spends with a particular customer is between 10 and 20 minutes?
(e) What is the probability that a newly arrived customer will have to wait before the salesman can see him or her?
(f) What is the mean time a customer spends waiting? What is the mean time a customer spends waiting and being served?
(g) What is the mean length of the waiting line? What is the expected number of customer-hours lost (spent waiting) each hour?
(h) The president of the Alpha Corporation, after receiving complaints about the poor service at its sales office, reprimands the salesman, who admits that he has been slow in servicing customers. The salesman promises to double the mean number of customers that can be serviced per hour (that is, increase it from 6 to 12). If the salesman keeps his promise and if the service times continue to conform to the

exponential distribution, how much of a reduction will occur in the probability that a newly arrived customer will have to wait?

(i) If he keeps his promise and if the service times continue to conform to the exponential distribution, how much of a reduction will occur in the mean length of the waiting line?

**5.43** People arrive at a receptionist's desk where they are serviced on a "first come, first served" basis. Arrivals are distributed according to the Poisson distribution, the mean arrival rate being one per minute. Service time is distributed according to the exponential distribution, the mean service rate being two per minute.

(a) What is the mean time that a person spends waiting in line?

(b) What is the probability that a newly arrived person will have to wait?

(c) What is the mean time that a person spends waiting in line and being served?

(d) What is the mean length of the waiting line?

# Sample Designs and Sampling Distributions

## 6.1 Introduction

As emphasized throughout the previous chapters, the field of statistics is concerned with the nature and effectiveness of sampling techniques. To comprehend how statistical methods are used in business and economics, it is essential that you be familiar with the major kinds of sample designs and that you understand the concept of a sampling distribution. In this chapter, we begin by describing the various kinds of commonly used sample designs. Then we discuss the concept of a sampling distribution and present some fundamental results concerning the sampling distributions of the sample mean and proportion. Our treatment of these topics makes extensive use of the probability theory contained in Chapters 3–5.

## 6.2 Probability Samples and Judgment Samples

At the outset of any sampling investigation, one must determine whether the sample is to be a probability sample or a judgment sample. These are the two broad classes of sample designs that can be used, and each is defined as follows.

> A **probability sample** is one where the probability that each element (that is, each member) of the population is included in the sample is known. In a **judgment sample,** personal judgment plays a major role in determining which elements of the population are selected, and this probability is not known.

For example, in constructing a sample of steel firms in the United States, if we pick randomly 5 of the 10 biggest steel firms and if we pick randomly 10 of the other steel firms, the resulting sample of 15 steel firms is a probability sample because we know the probability that each steel firm in the population is included. On the other hand, if we go down the list of steel firms and choose 15 firms that we consider "typical" or "average," this is a judgment sample because it is not based on random methods of selection and we do not know the probability that each steel plant in the population will be included.

---

**The most important advantage of a probability sample is that one can estimate how large the sampling error is likely to be.**     *Basic Idea #11*

---

*The most important disadvantage of a judgment sample is that there is no way to tell how "far off" a sample result is likely to be.* That is, one cannot estimate the difference between the sample result and the population parameter one is trying to measure. In contrast, *if a probability sample is taken, we can estimate how large the sampling error is likely to be,* as we shall see in subsequent sections of this chapter. Nonetheless, there are situations where judgment samples are used. In some cases, a probability sample is too expensive or impractical. For example, if our sample of steel firms must be confined to a single city because of budget limitations, we might well decide to choose the city on the basis of expert judgment rather than on the basis of chance. In a case of this sort, the geographical coverage of the sample will be so narrow that what can be inferred concerning the population as a whole will be largely a matter of judgment in any event, regardless of whether a probability or a judgment sample is used.

One type of judgment sample encountered frequently is a quota sample. *In a quota sample, the population as a whole is split into a number of groups or strata, and whoever is charged with drawing the sample is instructed to include a certain number of members of each group.* For example, if it is known that 20 percent of the individuals in a particular town are black men, 25 percent are black women, 30 percent are white men, and 25 percent are white women, a quota sample of 100 people in the town might specify that the interviewers pick 20 black men, 25 black women, 30 white men, and 25 white women. In this way an attempt is made to construct a sample that seems representative of the population as a whole. *The reason why quota sampling is a form of judgment sampling is that the choice of members from each group or stratum to be included in the sample is not determined by random selection, but by the interviewers.* Since interviewers tend to choose members of each group that can be contacted most readily, a variety of unknown biases may result. And like any form of judgment sample, there is no way to determine how large the sampling errors are likely to be.[1] Nonetheless, quota sampling is often used because it is convenient and inexpensive.

*Quota Sampling*

---

[1] For further comparison of probability and judgment samples, see M. Hansen and W. Hurwitz, "Dependable Samples for Market Surveys," in E. Mansfield, *Statistics for Business and Economics: Readings and Cases* (New York: Norton, 1980).

## 6.3  Types of Probability Samples

Four types of probability samples are frequently used: simple random samples, systematic samples, stratified random samples, and cluster samples. Following are brief descriptions of each type.

### SIMPLE RANDOM SAMPLE

Simple random sampling is the method that serves as the best introduction to probability sampling.

> *If the population contains* N *elements, a simple random sample of* n *elements is a sample chosen so that every combination of* n *elements has an equal chance of selection. Assuming that the sampling is without replacement, this means that each element in the population has a probability of* $1/N$ *of being the first chosen, that each of the* $(N-1)$ *elements not chosen on the first draw has a probability of* $1/(N-1)$ *of being the second chosen, . . . , and that each of the* $(N-n+1)$ *elements not chosen on the* $(n-1)$*th draw has a probability of* $1/(N-n+1)$ *of being the last chosen.*

An alternative and equally useful definition is as follows.

> *A **simple random sample** is a sample chosen so that the probability of selecting each element in the population is the same for each and every element, and the chance of selecting one element is independent of whether some other element is chosen.*

To illustrate what we mean by a simple random sample, suppose that a firm has four manufacturing plants (*A, B, C,* and *D*) and that it wants to select a simple random sample of two of these plants. There are six different samples of size 2 that can be drawn from this population of four plants.[2] (As shown in Appendix 3.1, the number of different samples of size *n* that can be drawn without replacement from a population of *N* elements equals $N! \div [(N-n)!n!]$.) Specifically, these six samples are (*AB*); (*AC*); (*AD*); (*BC*); (*BD*); and (*CD*). For this sample to be a simple random sample, each of these six samples must have a probability of 1/6 of being selected. If sampling is without replacement, this can be achieved by choosing the first plant in such a way that each of the four plants has a probability of 1/4 of being chosen, and then by choosing the second plant in such a way that each of the remaining three plants has a probability of 1/3 of being chosen.

If the population is infinite, the number of possible samples of size *n* is also infinite. Thus, the alternative definition (given above) of a

---

[2] More accurately, in keeping with definitions in Chapter 1, the population consists of *measurements* concerning these four plants, and the sample consists of *measurements* concerning two of them. Also, we assume here that sampling is without replacement.

simple random sample must be used: A simple random sample is one where the probability of selecting each element in the population is the same, and the chance of selecting one element is independent of whether some other element is chosen. For example, if a coin were tossed repeatedly for an indefinite period of time, an infinite population would result and each element contained in this population would be heads or tails. If the coin were tossed six times, this would be a simple random sample of size 6 if the probability of heads (or tails) remained the same from one toss to the next and if the result of one toss was independent of the result of another toss.

How do statisticians actually pick a simple random sample? In cases where the number of elements in the population is small, we could (1) number each of the elements in the population; (2) record each number on a slip of paper; (3) place all the slips of paper in a hat or bowl, where they are mixed well; and (4) draw $n$ slips of paper from the hat or bowl. The numbers on the slips of paper would indicate which elements in the population are included in the sample. Although this procedure is straightforward, we would run into problems if the number of elements in the population is large or if the slips of paper are not well mixed. To avoid such problems statisticians frequently use a table of random numbers in choosing a simple random sample. We shall describe how such a table is used in Section 6.5.

### SYSTEMATIC SAMPLE

To illustrate how a systematic sample is chosen, suppose that a statistician has a list of 1,000 tool and die firms from which to pick a sample of 50. Since there are 1,000 firms, this can be accomplished by taking every 20th firm on the list. Once a choice has been made among the first 20 firms on the list, the entire sample has been chosen since there are only 20 samples that can be drawn. The first possible sample consists of the 1st, 21st, 41st, . . . firms on the list. The second possible sample consists of the 2nd, 22nd, 42nd, . . . firms on the list. And so on. To determine which of these possible samples will be drawn, the statistician draws at random a number between 1 and 20. The chosen number identifies which of the first 20 firms he or she will begin with, which in turn identifies which sample will be drawn. In general, a systematic sample is defined as follows.

> A **systematic sample** is obtained by taking every k*th element on a list of all elements in the population. To determine which of the first* k *elements is chosen, a number from* 1 *to* k *is chosen at random.*

A systematic sample is often viewed as being essentially the same as a simple random sample. It is important to recognize that this is true only *if the elements of the population are in random order on the list.* One cannot always be sure that the phenomenon being measured does not have a periodicity or other type of pattern on the list. For example, W. A. Wallis and H. Roberts have pointed out that on census record sheets "the first names on the sheets tend to be predominantly male, gainfully employed, and above average in income. The reason is that the enumerators are instructed to start in a certain block at the corner house (which tends to have a higher rental value than houses in the middle of the block), and in the household to start with the head (usually male and the breadwinner)."[3] If there is a periodicity of this sort, a systematic sample may be far from random and may be less precise than a simple random sample.

On the other hand, circumstances exist where because the elements of the population are not listed in random order, a systematic sample may be *more* precise than a simple random sample. Basically, the reason is that a systematic sample insures that the elements in the sample are distributed evenly throughout the list. If there is a strong tendency for the characteristic being measured to increase or decrease steadily as one progresses from the beginning to the end of the list, this "even distribution" increases the precision of the sample estimate.

### STRATIFIED RANDOM SAMPLE

In designing a sample, it frequently is useful to recognize that the population can be divided into various groups or *strata*. For example, if a sample of students is to be drawn in order to estimate the mean height of freshmen at your college or university, it would probably be wise to stratify the population (in this case, the freshman class) into two groups: males and females.

> *In general, **stratified random sampling** is sampling in which the population is divided into strata and a random sample is taken from the elements in each stratum.*

Why is stratified random sampling employed? Because a more precise estimate can often be obtained from a sample of a given size if stratified random sampling is used rather than simple random sampling.

To illustrate the application of stratified random sampling, consider once again the problem of sampling the tool and die firms. A sample survey is being made in order to estimate the proportion of all tool and die firms which have introduced numerically controlled machine tools. If the statistician believes that the proportion of tool and die firms that have introduced numerically controlled machine tools is higher among larger firms than smaller ones, he or she can divide the population into two *strata*, one containing firms with more than 20 employees, and one containing firms with 20 or fewer employees. Then, if 50 percent of all tool and die companies have more than 20 employees, and 50 percent have 20 employees or fewer, the proportion of all firms in the population with numerically controlled machine tools is

$$\Pi = 0.5\Pi_1 + 0.5\Pi_2$$

where $\Pi_1$ is the proportion of *all* firms with more than 20 employees having introduced numerically controlled machine tools, and $\Pi_2$ is the

---

[3] W. Allen Wallis and Harry V. Roberts, *op. cit.*, pp. 488–489.

proportion of *all* firms with 20 or fewer employees having introduced numerically controlled machine tools. Consequently, *if a simple random sample is taken from each stratum,* an estimate of the proportion of all tool and die firms that have introduced numerically controlled machine tools can be obtained by computing

$$p = 0.5\, p_1 + 0.5\, p_2, \qquad (6.1)$$

where $p_1$ is the proportion of the *sample* of firms taken from the stratum with more than 20 employees having introduced numerically controlled machine tools, and $p_2$ is the proportion of the *sample* of firms taken from the stratum with 20 or fewer employees having introduced such machine tools.

*The basic idea in formulating strata is to subdivide the population so that these subdivisions differ greatly with regard to the characteristic being measured, and so that there is as little variation as possible within each stratum (or subdivision) with regard to the characteristic under measurement.* For example, in the problem discussed above, the statistician should stratify firms so that the differences *among* the strata in the proportion using numerically controlled machine tools is *large,* while the variation *within* each stratum in this regard is *small.*

**How to Form Strata**

If properly constructed, a stratified sample can generally result in more precise results than those obtained by using a simple random sample. This is why stratified sampling is used, as we stressed above.[4]

Once the strata have been defined, there remains the problem of determining how the total sample is to be divided among the strata. In other words, how many elements are to be chosen from each stratum? Two possible answers are proportional allocation and optimum allocation.

**Allocating the Sample**

PROPORTIONAL ALLOCATION. This method of allocating the sample *makes the sample size in each stratum proportional to the total number of elements in the stratum.* For example, in the case of the tool and die firms, this would mean that $1/2$ of the sample would be chosen from

---

[4] Stratified sampling results in greater precision because the error in the estimate for the population as a whole is due only to errors in the estimates for each stratum, whereas in simple random sampling there are also errors due to weighting the strata incorrectly. For example, in the case described above, the proportion of firms that have introduced numerically controlled machine tools *in a simple random sample* would be

$$\frac{n_1}{n} p_1 + \frac{n_2}{n} p_2,$$

where $n_1$ is the number of firms in the *sample* with more than 20 employees, $p_1$ is the proportion of these larger firms that have introduced numerically controlled machine tools, $n_2$ is the number of firms in the *sample* with 20 or less employees, $p_2$ is the proportion of these smaller firms having introduced numerically controlled machine tools, and $n = n_1 + n_2$. This expression is the same as equation (6.1) except that $n_1/n$ will differ (due to chance variation) from the true proportion of firms with more than 20 employees, whereas the true proportion, 0.5, is used in equation (6.1). Similarly, $n_2/n$ in this expression will differ (due to chance variation) from the true proportion of firms with 20 or less employees, whereas the true proportion, 0.5, is used in equation (6.1).

firms with more than 20 employees while 1/2 would be chosen from firms with 20 or fewer employees. Why? Because 1/2 of *all* tool and die companies have more than 20 employees, and 1/2 have 20 employees or fewer. At first glance, it certainly seems reasonable to sample the same proportion of elements in each stratum, and this allocation method is frequently used in business and economic surveys. But if the statistician has some knowledge of the population standard deviation in each stratum, it may be preferable to use optimum allocation, described below.

OPTIMUM ALLOCATION. This method prescribes that *the sample size in each stratum be proportional to the product of the number of elements (in the population) in the stratum and the standard deviation of the characteristic being measured in the stratum.* For example, suppose that we want to estimate the average assets of banks in a given state, and that we subdivide banks into two strata; national banks and state banks. If there are 300 national banks and 500 state banks, and if the standard deviation of the assets of the national banks is $100 million and that of the assets of the state banks is $20 million, then the number of national banks in the sample should be proportional to 300 × $100 million, while the number of state banks in the sample should be proportional to 500 × $20 million. In other words, the number of national banks in the sample should be proportional to $30 billion, while the number of state banks in the sample should be proportional to $10 billion, which means that 3/4 of the banks in the sample should be national banks and 1/4 should be state banks. The reason why optimum allocation is "optimum" is that it minimizes the expected sampling errors in the estimate of the population mean (or proportion). Despite its advantages, it may be impractical to use optimum allocation if little or nothing is known about the population standard deviation in each stratum.

### CLUSTER SAMPLE

Still another important sampling technique is cluster sampling.

*In a **cluster sample**, one divides the elements in the population into a number of clusters or groups. One then begins by choosing at random a sample of these clusters, after which a simple random sample of the elements in each chosen cluster is selected.*

The problem of sampling tool and die firms can be used to illustrate the application of a cluster sample. In this case, clusters can be formed by geographical location. All tool and die firms can be classified by the cities or areas in which they are located, and a simple random sample of these cities or areas can be chosen. Then, within each of the chosen cities or areas a simple random sample of the firms themselves can be picked.

The major advantage of cluster sampling is that *it is cheaper to sample elements that are physically or geographically close to one another.* Thus, it is cheaper to sample 50 tool and die firms, all of which are concentrated in five cities, than to sample 50 tool and die firms that are scattered all over the country. In general, the results of a cluster sample are less precise than those of a simple random sample (assuming that sample size is constant) because the elements in a particular cluster tend to be relatively similar. However, *per dollar spent on the survey* a cluster sample may be more effective than simple random sampling. Why? Because, for the same total cost, cluster sampling provides a much larger sample than simple random sampling.

## 6.4 Inventory Valuation: A Case Study

To illustrate how the techniques described in previous sections have been applied successfully to actual business problems, consider the case of a major manufacturing firm that had a large inventory of materials. The firm wanted to sample the items in this inventory (rather than attempting to include them all) in order to estimate the change during the previous year in the total monetary value of the inventory. Information of this sort was important in estimating the firm's earnings and its taxes. After studying the problem, the firm's statisticians and consultants recommended a sampling plan that combined stratified sampling (with optimum allocation) and systematic sampling.

First, this sampling plan stipulated that all items be divided into four strata: (1) Items worth $10,000 or more; (2) items worth $1,000–$9,999; (3) items worth $100–$999; and (4) items worth worth $99 or less. The reason for this stratification was that the increase in monetary value during the year of a high-priced item (in current prices) was liable to be greater than that of a low-priced item. Given the four strata, the sampling plan called for the selection of all 395 items in the first stratum, 1,350 of the 3,800 items in the second stratum, 260 of the 7,280 items in the third, and 110 of the 7,700 items in the fourth.[5] In all, 2,115 out of the approximately 19,000 items in the inventory were to be included in the sample.

How did the firm's statisticians determine this allocation of the sample among the strata? Basically, by using the principles of optimum allocation discussed previously in this chapter. To see this, consider the three strata from which samples were taken. (In the first stratum, all items were selected.) Table 6.1 shows the number of items in each of these three strata, as well as the estimated standard deviation of the change in the monetary values of items in each. In the last column of Table 6.1 the product of the number of items and the standard deviation is shown for each stratum. Optimum allocation requires that the sample size in each stratum be proportional to the figures in the last column of Table 6.1. The firm's statisticians picked the sample size in each stratum to conform to this requirement. That is, since $1,900,000 \div \$364,000 = 5.2$, the sample size in the second stratum is

---

[5] The numbers have been changed slightly to simplify the exposition. This case is from W. Edwards Deming, *Sample Design in Business Research* (New York: Wiley, 1960). It has been simplified in various respects.

**Table 6.1**
Selected Characteristics
of Three Strata, Case
Study of Inventory
Valuation

| Stratum (dollar value of item) | (1) Number of items in stratum | (2) Standard deviation of change in monetary value of items in stratum (dollars) | (1) × (2) (dollars) |
|---|---|---|---|
| 1,000 to 9,999 | 3,800 | 500 | 1,900,000 |
| 100 to 999 | 7,280 | 50 | 364,000 |
| 99 or less | 7,700 | 20 | 154,000 |

about 5.2 times as large as the sample size in the third stratum. And since $1,900,000 ÷ $154,000 = 12.3, the sample size in the second stratum is about 12.3 times as large as the sample size in the fourth stratum.

To carry out the sampling in each stratum, all items were listed and numbered. Then a systematic sample was chosen in each stratum. For example, in the third stratum a number between 1 and 28 was chosen at random. If this number turned out to be 11, then the 11th, 39th, 67th, . . . items in this stratum were chosen, the result being a sample consisting of 260 items. Since there was every reason to believe that the items were listed at random, the result could be considered a simple random sample.

Once the sample in each stratum had been gathered,[6] the firm could compute the mean change in monetary value of the sample items in each stratum. Then, to estimate the total change in the monetary value of the inventory, it could compute

$$395\bar{x}_1 + 3800\bar{x}_2 + 7280\bar{x}_3 + 7700\bar{x}_4,$$

where $\bar{x}_1$ is the mean change in monetary value of the sample items in the first stratum, $\bar{x}_2$ is the mean change in monetary value of the sample items in the second stratum, and so on. This, of course, was the estimate that the firm wanted, the "pay dirt" of the analysis. Why was this an estimate of the total change in monetary value of the inventory? Because it equaled the sum of the number of items in each stratum multiplied by the mean change in monetary value of the sample items in each stratum. Clearly, it was the estimate the firm wanted.

## 6.5 Using a Table of Random Numbers

Here and in subsequent chapters we shall concentrate on simple random sampling because it provides the simplest and best introduction to probability sampling. In order to select a simple random sample, statisticians generally use a table of random numbers.

A *table of random numbers* is a table of numbers generated by a

---

[6] Actually, of course, the sample in the first stratum includes all the items, not just some of them.

random process. For example, suppose that we want to construct a table of five-digit random numbers. To do so, we could write 0 on one slip of paper, a 1 on a second slip, a 2 on a third, ... , and a 9 on a tenth slip. We could then put the 10 slips of paper into a hat and draw out one at random. The number on this slip of paper would be the first digit of a random number. After replacing the slip of paper in the hat, we could make another random draw. The number on the new slip of paper would be the second digit of the random number. This procedure could then be repeated three more times, yielding the third, fourth, and fifth digits of the random number. The result is the first five-digit random number. Given enough time and persistence, we could formulate as many such random numbers as needed. Fortunately, it is not necessary to go through such a time-consuming process because tables of random numbers have already been formulated. Although these tables generally have been calculated on a computer rather by drawing slips of paper from a hat, the principles of their construction are essentially the same as if slips of paper had been used.

A table of random numbers is given in Appendix Table 5. As you can see, it contains column after column of five-digit random numbers. (The left-hand column shows the line number and should not be confused with the random numbers.) To illustrate how the table can be used to pick a simple random sample, suppose that we want to draw a simple random sample of 50 tool and die firms from the population of 1,000 tool and die firms. Our first step would be to number each of the tool and die firms from 0 to 999. We would then turn to any page of the table and proceed in any systematic way to pick our sample. For example, we might begin at the top, left-hand column of page A14 and work down. Since we need three-digit, not five-digit, random numbers, we would pay attention to only the first three digits of each five-digit random number. The topmost five-digit number in the left-hand column of page A14 is 53535; its first three digits are 535. Thus, the first firm to be picked is number 535. Reading down, the next five-digit number in the left-hand column of page A14 is 41292; its first three digits are 412. Thus, the second firm to be picked is number 412. This procedure should be repeated until the entire sample of 50 has been selected.

Several points should be noted concerning this procedure. (1) If any previously chosen number comes up again, it should be ignored. For example, if 412 were to come up again before the 50 firms were chosen, it should be ignored. (2) If a number comes up which is not admissible because it does not correspond to any item in the population, it also should be ignored. For example, if there were 500, not 1,000, tool and die firms, and if 680 came up, it should be ignored. (3) It makes no difference whether you read the table down, up, to the left, to the right, or diagonally. As long as you proceed from number to number in a systematic fashion the table will work properly.

Finally, it is worth noting that tables of random numbers are used in statistics in so-called Monte Carlo studies, as well as in the selection of random samples. (In the appendix to this chapter, we present a brief discussion of Monte Carlo studies.)

**TO ERR IS
HUMAN ... BUT
COSTLY**

THE 1970 DRAFT LOTTERY

In the late 1960s, there were charges of unfairness in the procedures by which some local draft boards decided who should be
drafted into the U.S. armed forces. Thus, for the 1970 draft, it
was decided that the eligible candidates would be randomly ordered for induction. Capsules representing each of the 366 days
of the year[7] were put in a cage and were randomly withdrawn by
an individual. Since September 14 was chosen first, men born on
that date were the first ones inducted in 1970. Since April 24 was
chosen second, men born on that date were the second ones inducted; and so on.

Each day of the year was given a number from 1 to 366. For
example, September 14 received the number 1 and April 24 received the number 2, as we have just seen. The average numbers
for the days in each month are shown in the figure below. Is there
any indication that the numbers were really not chosen at random? What sort of departure from randomness, if any, seems to
exist?

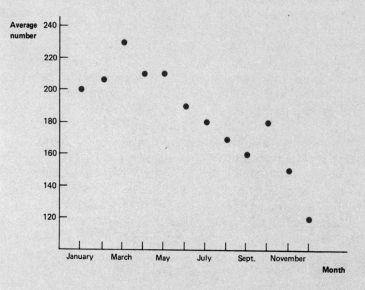

SOLUTION: Later months (like November and December)
tend to have lower numbers than earlier months (like January
and February). Thus, men with birthdays in later months tended
to be drafted before men with birthdays in earlier months. Had
the drawing been completely random, one would have expected
the averages to be about the same, on the average, in the later
months as in the early ones. Some statisticians believe that the
capsules were put into the cage in monthly order and that they

[7] This includes February 29, which occurs every four years.

were not thoroughly mixed, the result being that the capsules for the later months tended to be withdrawn before those for the earlier months. The moral is that, even if a sample is drawn by picking capsules out of a cage or by picking numbers out of a hat, unsuspected biases may occur. Thus, it is wise to use a table of random numbers.[8]

[8] For further discussion, see S. Fienberg, "Randomization and Social Affairs: The 1970 Draft Lottery," *Science*, January 22, 1971; and B. Williams, *A Sampler on Sampling* (New York: Wiley, 1978), p. 7.

## EXERCISES

**6.1** (a) If a finite population consists of the elements *A, B, C, D*, and *E*, list all possible samples of size 2 that can be drawn (without replacement) from this population.

(b) List all possible samples of size 3 that can be drawn (without replacement) from this population.

(c) If each of these possible samples of size 3 has the same probability of being selected, what is the probability that *A* will be included in the sample that is selected? What is the probability that both *A* and *B* will be included in this sample?

**6.2** How many different samples of size 4 can be selected (without replacement) from a finite population consisting of the five boroughs of New York City (Manhattan; the Bronx; Brooklyn; Queens; Staten Island)? How many different samples of size 3 can be selected (without replacement) from this population?

**6.3** Use a table of random numbers to select a sample of two drugstores from among those listed in the yellow pages of your telephone directory.

**6.4** The tickets issued by a movie theater are numbered serially. On October 28, 1987, the theater sold tickets with numbers beginning at 20860 and ending at 23102. Use a table of random numbers to select a sample of 10 of these tickets.

**6.5** Hotels often leave questionnaires concerning the quality of their service in their rooms, and ask their guests to fill them out. Can the results be regarded as a random sample? Why or why not?

**6.6** The Educational Testing Service, in one of its studies, wanted to obtain a representative sample of U.S. college students. To do so, they stratified all colleges and universities into a number of groups (e.g., public universities with 25,000 or more students, and private four-year colleges with 1,000 or fewer students). Then they picked one college or university in each stratum that they regarded as representative. Each of the chosen colleges or universities was asked to select a sample of its students. Write a one-paragraph evaluation of this procedure.[9]

**6.7** In a television appearance, a congressman asks voters in his district to send him a postcard indicating whether he should vote for or against a

[9] For further discussion of this case, see D. Freedman, R. Pisani, and R. Purves, *Statistics* (New York: Norton, 1978).

particular proposed piece of legislation. He receives 4,100 replies. Since voters from the western part of his district tend to be Democrats, while those from the eastern part tend to be Republicans, he divides the replies on the basis of their geographical origin. Based on the results, he finds that 68 percent of the replies from the western part of his district are against the legislation, while 36 percent of the replies from the eastern part are against it.

(a) Did the congressman carry out a stratified sample?

(b) Did he carry out a stratified random sample?

(c) What criticisms would you make concerning his sample design?

(d) What improvements would you suggest?

6.8 A firm wants to determine the mean income of the families in a particular suburb of Washington, D.C. The firm divides the suburb into seven parts, each of which contains the same number of families. After picking two of these parts at random, it chooses a random sample of 50 families in each of these two parts.

(a) What are the advantages of this sample design over a simple random sample of 100 families drawn from the entire suburb?

(b) Is the result of this sample design likely to be as precise as the result of a random sample of 100 families drawn from the entire suburb?

(c) Regardless of the sample design, can you identify some problems in obtaining the desired information?

6.9 As pointed out in Chapter 1, Philadelphia's school teachers went on strike for a long period in 1981. Suppose that Philadelphia's school board had carried out a sample of public opinion concerning its behavior regarding the strike by picking a random sample of Philadelphia school children and sending a questionnaire to their parents.

(a) Would such a sample be a random sample of the opinions of Philadelphia residents?

(b) From what population would such a sample be drawn?

(c) What biases might result from such a sample?

6.10 A lawyer must estimate the mean value of the pieces of furniture contained in a 15-room house after the house's owner dies. Such information is needed to fill out the estate tax return. In all, there are 135 pieces of furniture, including sofas, chairs, tables, and so on. The lawyer decides to estimate the mean value on the basis of a sample.

(a) Would you advise him to stratify the relevant population?

(b) If so, what strata would you advise him to use?

6.11 A population consists of two strata, A and B. There are 100 items in A, and the standard deviation is 20. There are 50 items in B, and the standard deviation is 100.

(a) If you want to carry out a stratified random sampling with optimum allocation, what proportion of the sample will you take from stratum A? From stratum B?

(b) If you want to use proportional allocation, what proportion of the sample will you take from stratum A? From stratum B?

(c) Which of the allocation schemes in (a) and (b) will result in smaller expected sampling errors?

6.12 In 1948, the Gallup Poll predicted that Harry Truman would get 44 percent of the vote and that Thomas Dewey, his Republican opponent for president, would get 50 percent of the vote. These results were based on a quota sample of 50,000 potential voters.

(a) How close were these results?

(b) In the presidential elections of 1936, 1940, and 1944, the Gallup Poll always overestimated the Republican percentage of the vote. Can you give some reasons for this?

(c) Since 1948, the Gallup Poll has used probability samples in its estimates of the outcome of presidential elections. Its estimate of the percent of votes going to the winner has been in error, on the average, by about 2 percentage points since 1948. By the 1970s, its sample size was under 4,000. How can you explain its better performance in recent years than in 1948, even though its sample size has decreased greatly?

**6.13** A polling organization asks a sample of voters which of two presidential candidates they favor. Do you think that the order in which the names of the candidates are presented will affect the results? If so, do you think that the candidate named first would be at an advantage or disadvantage?

## 6.6 Concept of a Sampling Distribution

Sampling distributions are of central importance in statistics. The idea of a sampling distribution stems directly from the fact that there generally are a large number of possible samples that can be selected from a population. Thus, the value of any statistic computed from a sample will vary from sample to sample. For example, suppose that a population consists of the eight numbers in Table 6.2, that we are going to take a simple random sample of two of these numbers, and that the sample statistic in which we are interested is the sample mean. Since the sample is randomly chosen, the probability that each possible sample will be selected is known and it is possible to deduce the probability distribution of the value of the sample statistic. This is an example of a sampling distribution, which is defined as follows.

*A **sampling distribution** is the probability distribution of the value of a statistic.*

What is the sampling distribution of the mean of a simple random sample of size 2 from the population in Table 6.2? There are 28 different samples of size 2 that can be drawn without replacement from this population (since there are only eight numbers in the population). Table 6.3 lists each of these samples, as well as its mean. Since the probability of selecting each of these samples is 1/28, the probability that the sample mean assumes each of the 28 values (not all of which are different from one another) is 1/28. Thus, the probability distribution of the value of the sample mean in this case is as shown in Figure 6.1. This probability distribution is a sampling distribution.

Sampling distributions can be constructed for any sample statistic, not just the sample mean. For example, Table 6.3 also shows the sample range for each of the 28 samples of size 2 from the population in Table 6.2. Since the probability of each of the 28 values (not all of which are different from one another) is 1/28, the probability distri-

**Table 6.2**
Population of Eight
Numbers

| Measurements included in the population |
|:---:|
| 2 |
| 4 |
| 5 |
| 7 |
| 8 |
| 0 |
| 1 |
| 3 |

**Table 6.3**
Mean and Range of 28
Possible Samples of
Size 2 from the
Population in Table 6.2

| Sample | Numbers included in sample | Sample mean | Sample range |
|:---:|:---:|:---:|:---:|
| 1 | 2,4 | 3.0 | 2 |
| 2 | 2,5 | 3.5 | 3 |
| 3 | 2,7 | 4.5 | 5 |
| 4 | 2,8 | 5.0 | 6 |
| 5 | 2,0 | 1.0 | 2 |
| 6 | 2,1 | 1.5 | 1 |
| 7 | 2,3 | 2.5 | 1 |
| 8 | 4,5 | 4.5 | 1 |
| 9 | 4,7 | 5.5 | 3 |
| 10 | 4,8 | 6.0 | 4 |
| 11 | 4,0 | 2.0 | 4 |
| 12 | 4,1 | 2.5 | 3 |
| 13 | 4,3 | 3.5 | 1 |
| 14 | 5,7 | 6.0 | 2 |
| 15 | 5,8 | 6.5 | 3 |
| 16 | 5,0 | 2.5 | 5 |
| 17 | 5,1 | 3.0 | 4 |
| 18 | 5,3 | 4.0 | 2 |
| 19 | 7,8 | 7.5 | 1 |
| 20 | 7,0 | 3.5 | 7 |
| 21 | 7,1 | 4.0 | 6 |
| 22 | 7,3 | 5.0 | 4 |
| 23 | 8,0 | 4.0 | 8 |
| 24 | 8,1 | 4.5 | 7 |
| 25 | 8,3 | 5.5 | 5 |
| 26 | 0,1 | 0.5 | 1 |
| 27 | 0,3 | 1.5 | 3 |
| 28 | 1,3 | 2.0 | 2 |

bution of the value of the sample range in this case is as shown in Figure 6.2 This is another example of a sampling distribution.

Any sampling distribution is based on the assumption that the sample size and the sample design remain fixed. Thus, the sampling distributions in Figures 6.1 and 6.2 are based on the assumption that samples of size 2 are taken, and that the samples are simple random

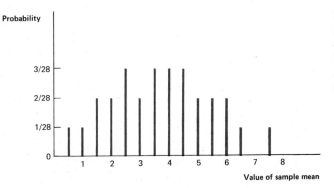

**Figure 6.1**
Probability Distribution of the Mean of a Simple Random Sample of Size 2 from Population in Table 6.2

samples. In general, it is not possible to derive sampling distributions in the way employed in Table 6.3 because most populations contain so many elements that it is essentially impossible to list all possible samples. Even if there are only 1,000 elements in a population, there are 1,000!/(998! 2!) or 499,500 different samples of size 2 that can be drawn from this population. Obviously, the listing of all such samples would be a formidable task; and fortunately (as we shall see in later sections) there are other ways of finding the sampling distribution of a sample statistic. The reason why we presented this laborious procedure is not that it is used to obtain sampling distributions in actual cases, but that it is useful in explaining the concept of a sampling distribution.

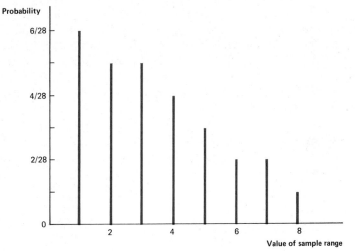

**Figure 6.2**
Probability Distribution of the Range of a Simple Random Sample of Size 2 from Population in Table 6.2

## 6.7 The Sampling Distribution of the Sample Proportion

Many statistical investigations are aimed at estimating the proportion of the members of some population who have a specified characteristic. For example, we may be interested in the proportion of St. Louis women who prefer Brand A beer to Brand B, or in the proportion of automobiles of a particular make that have a certain defect. In investigations of this sort, the sample proportion is generally used as an estimate of the population proportion. If the sample is a small percentage of the whole population (say, 5 percent or less), we can determine the sampling distribution of the sample proportion without much difficulty, based on our results in Chapter 4 concerning the binomial distribution.

If the sample is a simple random sample, and if it is a small percentage of the population, each observation has a probability $\Pi$ of having the specified characteristic, where $\Pi$ is the proportion of the population with this characteristic. If the sample contains $n$ observations, each observation selected can be viewed as a Bernoulli trial where there is a probability $\Pi$ of success, where success is defined as the observation's having the specified characteristic. Thus, as we know from Chapter 4, the number of observations in the sample that have the specified characteristic must be a binomial random variable. In other words,

$$P(x) = \frac{n!}{(n-x)!x!} \, \Pi^x (1-\Pi)^{n-x}, \text{ for } x = 0,1,2, \dots n$$

where $P(x)$ is the probability that $x$ observations in the sample will have the specified characteristic.

The sample statistic whose probability distribution we want to derive is the sample proportion. It is important to note that the value of the sample proportion is determined entirely by the number of observations in the sample with the specified characteristic. For example, if the sample size is five the sample proportion can be $1/5$ if and only if exactly one observation in the sample has the desired characteristic; it can be $2/5$ if and only if exactly two observations in the sample have the desired characteristic; and so on. Thus, *the probability that the value of the sample proportion is $x/n$ must equal the probability that the number of observations in the sample with the specified characteristic is $x$.* Consequently, the sampling distribution of the sample proportion must be as follows:

***Formula for Sample Proportion's Probability Distribution***

$$Pr\left(p = \frac{x}{n}\right) = \frac{n!}{(n-x)!x!} \, \Pi^x (1-\Pi)^{n-x}, \tag{6.2}$$

where $Pr(p = x/n)$ denotes the probability that the sample proportion $p$ equals $x/n$.

To illustrate how this formula can be used, suppose that a market-research firm is about to ask a sample of beer drinkers whether they prefer Budweiser over Miller (presenting both in unmarked cans). If the firm selects a simple random sample of 20 beer drinkers, and if the proportion of the population preferring Budweiser over Miller is 0.50, what is the sampling distribution of the sample proportion? The formula in equation (6.2) tells us that the probability that the sample

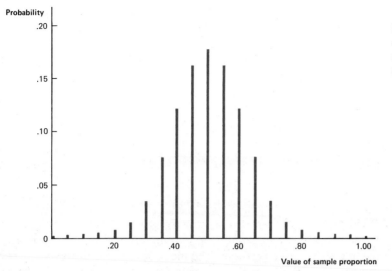

**Figure 6.3**
Sampling Distribution of the Sample Proportion, $n = 20$ and $\Pi = 0.50$

proportion equals zero can be determined by finding (in Appendix Table 1) the probability that a binomial variable (with $n = 20$ and $\Pi = 0.5$) equals zero, that the probability that the sample proportion equals $1/20$ can be determined by finding the probability that this binomial variable equals 1, that the probability that the sample proportion equals $2/20$ can be determined by finding the probability that this binomial variable equals 2, and so on. The resulting sampling distribution is shown in Figure 6.3

The sampling distribution of a sample statistic is useful in indicating the amount of error the statistic is likely to contain. The following example shows how the sampling distribution of the sample proportion can be used in this way.

EXAMPLE 6.1 A shipment of 1,000 engines arrives at a firm, which draws a random sample of 20 engines and calculates the proportion of defectives in the sample. If the proportion of defective engines in the entire shipment equals .15, what is the probability that the sample proportion will be in error by more than .10? (That is, what is the probability that the sample proportion will differ by more than .10 from the population proportion?)

SOLUTION: Since the population proportion is .15, the sample proportion will *not* differ by more than .10 from the population proportion if the sample proportion equals .05, .10, .15, .20, or .25. Using Appendix Table 1, we can obtain the probability of the sample proportion's equaling each of these values. The probability that it equals .05 is .1368, since this is the probability of one success (in this case, one defective) in 20 trials, given that $\Pi = .15$. The probability that it equals .10 is .2293, since this is the probability of two successes in 20 trials, given that $\Pi = .15$. The probability that it equals .15 is .2428, since this is the probability of three successes in 20 trials, given that $\Pi = .15$. The probability that it equals .20 is .1821, since this is the probability of four successes in 20 trials, given that $\Pi = .15$. The probability that it equals .25 is .1028, since this is the probability of five successes in 20 trials, given that $\Pi = .15$. Thus, the probability that the sample proportion differs from the population proportion by more than .10 is $1 - (.1368 + .2293 + .2428 + .1821 + .1028)$, or .1062.

## 6.8 The Sampling Distribution of the Sample Mean

Perhaps the most frequent objective of statistical investigations is to estimate the mean of some population. For example, a government agency may want to estimate the mean profit rate of firms in a certain industry, or a firm may want to estimate the mean longevity of a particular type of material or piece of equipment. In investigations of this sort the sample mean is generally used as an estimate of the population mean. To measure the extent of the sampling errors that may be present in the sample mean, it is essential to know the sampling distribution of the sample mean, which is the subject of this section. We will begin with the case where the population is known to be normal, then take up the case where it is not normal, and conclude with the case where the population contains relatively few observations.

WHERE THE POPULATION IS NORMAL

Suppose that simple random samples of size 10 are selected repeatedly from a normal population and that the mean of each such sample is calculated. If a very large number of such sample means are calculated, what does the probability distribution of their values look like? As shown in Figure 6.4, this probability distribution has the same mean as the population, namely $\mu$. Thus, if a large number of sample means were calculated, *on the average* the sample mean would equal the population mean. Also, the probability distribution of the sample mean is bell-shaped and symmetrical about the population mean, which means that it is equally likely that a sample mean will fall below or above the population mean. Further, because of the averaging, the dispersion of the sampling distribution of the sample mean is less than the dispersion of the population. In other words, the standard deviation of the sampling distribution of the sample mean, denoted by $\sigma_{\bar{x}}$, is less than the standard deviation of the population, denoted by $\sigma$.

Now suppose that simple random samples of size 40 are selected repeatedly from the same normal population, and that the mean of each such sample is calculated. If a very large number of such sample means are calculated, what does the probability distribution of their values look like? As shown in Figure 6.4, this probability distribution, like the one for samples of size 10, has the same mean as the population and is bell-shaped and symmetrical. The most obvious difference be-

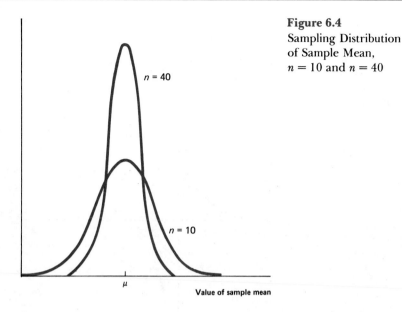

**Figure 6.4**
Sampling Distribution
of Sample Mean,
$n = 10$ and $n = 40$

tween this sampling distribution and the one for samples of size 10 is this distribution's much smaller standard deviation. Because the sample size is larger, it is likely that the mean of a sample of size 40 will be closer to the population mean than the mean of a sample of size 10. This, of course, accounts for the smaller dispersion in the sampling distribution of the sample mean for samples of size 40.

A close inspection of both sampling distributions in Figure 6.4 suggests that they are normal distributions. At least this is how they look, and mathematicians have proved that this is indeed the case. A remarkable feature of a normal population is that the sampling distribution of means of simple random samples from such a population is also normal. This is true for a sample of any size. Given this result as well as the others noted above, it is possible to provide the following complete description of the sampling distribution of the sample mean from a normal population.

> SAMPLING DISTRIBUTION OF THE SAMPLE MEAN (NORMAL POPULA-
> TION): *The sample mean is normally distributed, the mean of its sam-
> pling distribution equals the mean of the population ($\mu$), and the
> standard deviation of its sampling distribution ($\sigma_{\bar{x}}$) equals the stan-
> dard deviation of the population divided by the square root of the sam-
> ple size. That is,*

$$\sigma_{\bar{x}} = \frac{\sigma}{\sqrt{n}},$$  (6.3)  *Formula for $\sigma_{\bar{x}}$*

*where $\sigma$ is the population standard deviation and* n *is the sample size.*

Equation (6.3) is a very important result. *The standard deviation of the sampling distribution of the sample mean, $\sigma_{\bar{x}}$, is a measure of "how far*

**Figure 6.5**
Sampling Distribution
of the Sample Mean,
when $\sigma_{\bar{x}}$ Is Relatively
Large and Relatively
Small

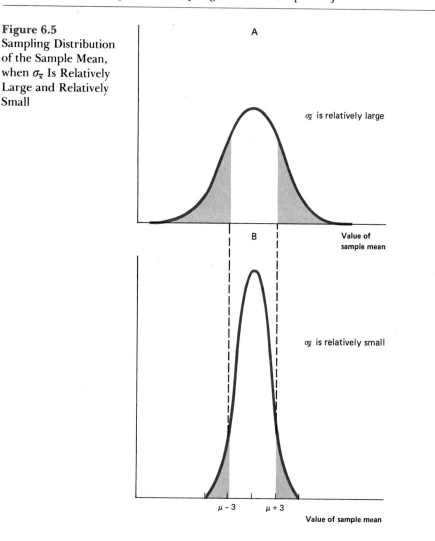

A

$\sigma_{\bar{x}}$ is relatively large

B

Value of
sample mean

$\sigma_{\bar{x}}$ is relatively small

$\mu - 3$     $\mu + 3$

Value of sample mean

*off" the sample mean is likely to be from the population mean. If $\sigma_{\bar{x}}$ is large,*
the distribution of the sample mean contains *a great deal of dispersion,*
which means there is a relatively *high* probability that the sample mean
will depart considerably from the population mean. If $\sigma_{\bar{x}}$ is *small,* the
distribution of the sample mean contains *relatively little dispersion,*
which means there is a relatively *low* probability that the sample mean
will depart considerably from the population mean. For example,
panel A of Figure 6.5 shows the distribution of the sample mean in a

*Standard Error of*
*the Sample Mean*

case where its standard deviation—often called the **standard error of**
**the sample mean**—is relatively large. Panel B shows the distribution of
the sample mean in a case where its standard deviation is relatively
small. As shown in Figure 6.5, the probability that the sample mean
differs from the population mean by more than an arbitrary amount
(say 3, which is shown in Figure 6.5) is much higher in the case shown
in panel A than in that shown in panel B. (In each case this probability is
equal in value to the shaded area in Figure 6.5.)

Given that $\sigma_{\bar{x}}$ is a measure of how "far off" the sample mean is likely to be, it is obviously important to specify what determines $\sigma_{\bar{x}}$. According to equation (6.3), $\sigma_{\bar{x}}$ is determined by two things: the standard deviation of the population, $\sigma$, and the sample size, $n$. *Holding sample size constant, $\sigma_{\bar{x}}$ is proportional to the standard deviation of the population.* This is reasonable because one would expect that sample means drawn from more variable populations would themselves tend to be more variable. *Holding constant the standard deviation of the population, $\sigma_{\bar{x}}$ is inversely proportional to the square root of the sample size.* This means that, in a sense, diminishing returns set in as the sample size is increased. If the sample size is quadrupled, $\sigma_{\bar{x}}$ is cut only in half; if the sample size is multiplied by 25, $\sigma_{\bar{x}}$ is cut only to one-fifth of its previous amount.

To prevent confusion, it is extremely important that you understand the distinction between $\sigma_{\bar{x}}$ (the standard deviation of the sampling distribution of the sample mean) and $\sigma$ (the standard deviation of the population). The former is the standard deviation of the distribution of the sample mean. (Such distributions have been presented in Figure 6.4 and Figure 6.5) The latter is the standard deviation of the population from which the samples are drawn. Since the distribution of the sample mean is obviously quite different from the population being sampled, the standard deviations of the two are quite different, although not unrelated. (As pointed out in the previous paragraph, $\sigma_{\bar{x}}$ is proportional to $\sigma$, if sample size is held constant.)

Our results concerning the sampling distribution of the sample mean have important applications, some of which are illustrated by the following example.

---

**EXAMPLE 6.2** The diameters of tires produced by a firm are known to be normally distributed with a standard deviation of .01 inches. A simple random sample of 100 tires is selected from the plant's output, and the mean of the diameters is computed. What is the probability that this sample mean will exceed the population mean by more than .0015 inches?

SOLUTION: The distribution of the sample mean is normal with mean equal to the population mean $\mu$ and standard deviation equal to $\sigma/\sqrt{n}$, which in this case is $.01 \div \sqrt{100}$, or .001 inches. Figure 6.6 shows the distribution of the sample mean. The probability we want equals the area under this curve to the right of $\mu +$

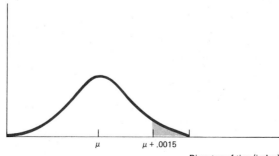

μ     μ + .0015

Diameters of tires (inches)

**Figure 6.6**
Probability that the Sample Mean Exceeds the Population Mean by More than .0015 Inches

.0015. The value of the standard normal variable corresponding to $\mu + .0015$ is $[(\mu + .0015) - \mu] \div \sigma_{\bar{x}} = .0015 \div .001 = 1.5$. Thus, we must evaluate the area under the standard normal curve to the right of 1.5. Appendix Table 2 shows that the area between zero and 1.5 is .4332, so the area to the right of 1.5 must equal $.5000 - .4332 = .0668$. This is the probability that the sample mean will exceed the population mean by more than .0015 inches.

### WHERE THE POPULATION IS NOT NORMAL

As we know from the previous paragraphs, if the population is normal the sampling distribution of the sample mean is normal. But what if, as frequently occurs in business and economics, the population is *not* normal? A remarkable mathematical theorem, known as the central limit theorem, has shown that if the sample size is moderately large, the distribution of the sample mean can be approximated by the normal distribution, even if the population is not normal. This very important theorem can be stated as follows.

*Central Limit Theorem*

CENTRAL LIMIT THEOREM. *As the sample size* n *becomes large, the sampling distribution of the sample mean can be approximated by a normal distribution with a mean of μ and a standard deviation of* $\sigma/\sqrt{n}$, *where μ is the mean of the population and σ is its standard deviation.*

What is particularly impressive about this result is that the normal approximation seems to be quite good so long as the sample size is larger than about 30, regardless of the nature of the population. And if the population is reasonably close to normal, sample sizes of much less than 25 are likely to result in the normal approximation's being serviceable.[10]

---

**Basic Idea #12**  *If the sample size is moderately large, the distribution of the sample mean is approximately normal, with a mean of μ and a standard deviation of* $\sigma/\sqrt{n}$, *where μ is the mean of the population and σ is its standard deviation.*

---

[10] Using the results of Appendix 4.3 and 4.1, we can prove that the expected value of $\bar{x}$ equals $\mu$. From Appendix 4.3 we know that $E(\Sigma x_i) = E(x_1) + E(x_2) + \ldots + E(x_n)$. Since the expected value of each observation in a random sample equals $\mu$, it follows that $E(\Sigma x_i) = n\mu$. And using the results of Appendix 4.1, $E(\Sigma x_i/n) = \mu$. Note that this proof does not depend on an assumption that the population is normal.

Based on Appendix 4.3 and 4.1, it is also possible to prove that $\sigma_{\bar{x}} = \sigma/\sqrt{n}$. Since the observations in a random sample are statistically independent, we know from Appendix 4.3 that $\sigma^2(\Sigma x_i) = \sigma^2(x_1) + \sigma^2(x_2) + \ldots + \sigma^2(x_n)$. Since the variance of each observation equals $\sigma^2$, it follows that $\sigma^2(\Sigma x_i) = n\sigma^2$. And using the results in Appendix 4.1, $\sigma^2(\Sigma x_i/n) = n\sigma^2/n^2 = \sigma^2/n$. This proof does not assume that the population is normal.

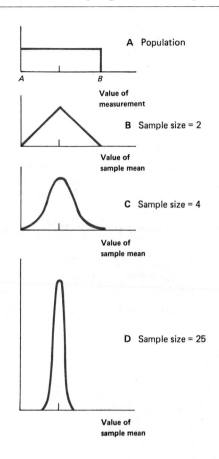

**Figure 6.7**
Sampling Distribution
of Means of Samples of
Sizes 2, 4, and 25 from
a Population with a
Uniform Distribution

To illustrate the working of the central limit theorem, suppose that the population frequency distribution is as shown in panel A of Figure 6.7. This population is such that it is equally likely that an observation in the population falls anywhere between the lower limit $A$ and the upper limit $B$. (A population of this sort is said to have a *uniform distribution.*) Clearly, this population is far from normally distributed. Panels B, C, and D of Figure 6.7 show the sampling distribution of the sample mean for samples of size 2, 4, and 25 from this population. Even for samples of size 4, this sampling distribution is bell-shaped and close to normal. For samples of size 25, it is very close to normal.

The central limit theorem applies to discrete populations as well as to continuous ones, a fact which has important implications for the sampling distribution of the sample proportion. *The sample proportion is really a sample mean from a population where a success is denoted by a 1 and a failure is denoted by a 0.* The mean of such a population is $\Pi$, the proportion of successes in the population; and the mean of the sample is $p$, the proportion of successes in the sample. Since the sample proportion is really a sample mean from a population consisting of 0's and 1's, its sampling distribution must tend toward normality as the sample size increases, according to the central limit theorem. This fact will be used frequently in subsequent chapters.

The following example illustrates some of the practical implications of the central limit theorem. In subsequent chapters this theorem will have repeated application.

EXAMPLE 6.3 A manufacturer of light bulbs tests a random sample of 64 of the light bulbs produced today to determine the longevity of each. From past experience, the firm knows that the standard deviation of the longevity of its light bulbs is 160 hours. What is the probability that the sample mean will differ from the population mean by more than 40 hours?

SOLUTION: Because the sample size is well above 30, we can be reasonably certain that the sampling distribution of the sample mean is very close to normal, regardless of the nature of the population. More specifically, the distribution of the sample mean should be essentially normal with a mean equal to the population mean $\mu$ and a standard deviation equal to $\sigma/\sqrt{n}$, where $\sigma$ is the population standard deviation. Since $\sigma = 160$ and $n = 64$, $\sigma_{\bar{x}} = 160 \div \sqrt{64} = 20$ hours. What we want is the probability that the sample mean lies above $\mu + 40$ or below $\mu - 40$. Figure 6.8 shows

**Figure 6.8**
Probability that the Sample Mean Differs from the Population Mean by More than 40 Hours

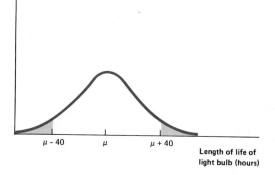

$\mu - 40$    $\mu$    $\mu + 40$

**Length of life of light bulb (hours)**

the distribution of the sample mean. The probability we want equals the sum of the values of the two shaded areas. The value of the standard normal variable corresponding to $\mu + 40$ is $[(\mu + 40) - \mu] \div \sigma_{\bar{x}} = 40 \div 20 = 2$; the value corresponding to $\mu - 40$ is $[(\mu - 40) - \mu] \div \sigma_{\bar{x}} = -40 \div 20 = -2$. Thus, we must evaluate the areas under the standard normal curve to the right of 2 and to the left of $-2$. Since Appendix Table 2 shows that the area between zero and 2 is .4772, the area to the right of 2 must be .5000 − .4772 = .0228. Because of the symmetry of the standard normal curve, the area to the left of $-2$ must also be .0228. Thus, the total area to the right of 2 and to the left of $-2$ equals .0228 + .0228 = .0456. This is the probability that the sample mean will differ from the population mean by more than 40 hours.

## WHERE THE POPULATION IS SMALL

In this section we have assumed that the population is infinite, while in fact many populations in business and economics are not. If the population is large relative to the size of the sample, our previous results can be applied without modification even though the population is finite. However, if the population is less than 20 times the sample size, the following simple correction should be applied to the standard deviation of the sample mean.

STANDARD DEVIATION OF THE SAMPLE MEAN (FINITE POPULATIONS). *Whereas $\sigma_{\bar{x}}$ equals $\sigma/\sqrt{n}$ for infinite populations, in the case of finite populations the following formula applies:*

$$\sigma_{\bar{x}} = \frac{\sigma}{\sqrt{n}} \sqrt{\frac{N-n}{N-1}},$$

(6.4)     *Formula for $\sigma_{\bar{x}}$ if the Population is Small*

*where $\sigma$ is the standard deviation of the population, N is the number of observations in the population, and n is the sample size.*

In other words, the value of $\sigma_{\bar{x}}$ for an infinite population must be multiplied by $\sqrt{(N-n)/(N-1)}$ to make it appropriate for a finite population. The multiplier $\sqrt{(N-n)/(N-1)}$ is often called the *finite population correction factor*. When $n$ is small relative to $N$, this correction factor is so close to 1 that it can be ignored.[11]

*Finite Population Correction Factor*

If sampling is with replacement (that is, if each element that is included in the sample is put back in the population so that it can be chosen more than once), there is no need to use the finite population correction factor, even if the number of elements in the population is small relative to the sample size. In other words, equation (6.3) is valid under these circumstances. However, in most practical situations in business and economics, sampling is not carried out with replacement.

The following illustrates the application of equation (6.4). A taxi company has 17 cabs and chooses a simple random sample of 13 of them. On the basis of the sample, the firm computes the mean number of miles that a taxi has been driven since its tires were inspected. If the population standard deviation is 5,000 miles, the standard deviation of the sample mean equals

$$\sigma_{\bar{x}} = \frac{\sigma}{\sqrt{n}} \sqrt{\frac{N-n}{N-1}} = \frac{5,000}{\sqrt{13}} \sqrt{\frac{17-13}{17-1}}$$

$$= \left(\frac{5,000}{3.606}\right)\left(\frac{1}{2}\right) = 693.$$

---

[11] In Appendix 4.2 it was stated that the standard deviation of the hypergeometric distribution equals the standard deviation of the binomial distribution times the finite population correction factor.

Thus, the standard deviation of the sample mean is 693 miles, which is one-half of what it would have been had the population been infinite.

Since the standard deviation of the sample mean is a measure of how far off the sample mean is likely to be, an important implication of equation (6.4) is that *if the sample is small relative to the population, the accuracy of the sample mean depends entirely on the sample size and not on the fraction of the population included in the sample.* To test this, we can draw a simple random sample of size 100 from two populations, both of which have a standard deviation of 1,000. One population contains 10,000 observations, so the sample is 1 percent of the population; the other contains 100,000 observations, so the sample is 1/10 of 1 percent of the population. In the first population, the standard deviation of the sample mean is

$$\sigma_{\bar{x}} = \frac{1,000}{\sqrt{100}} \sqrt{\frac{10,000 - 100}{10,000 - 1}} = 100 \sqrt{\frac{9,900}{9,999}} \doteq 100.$$

In the second population, it is

$$\sigma_{\bar{x}} = \frac{1,000}{\sqrt{100}} \sqrt{\frac{100,000 - 100}{100,000 - 1}} = 100 \sqrt{\frac{99,900}{99,999}} \doteq 100.$$

Thus, the standard deviation of the sample mean is essentially the same, although the fraction of the population included in the sample is 10 times bigger in the first population than in the second. This is an important point, and one some people find hard to believe. To repeat, the sample size, not the fraction of the population included in the sample, determines the accuracy of the sample mean if the sample is small relative to the population.

Once the modification of $\sigma_{\bar{x}}$ in equation (6.4) has been made, the results for infinite populations can be applied to finite populations where the sample size is large relative to the population. The following example shows how this modification can be made and applied.

EXAMPLE 6.4 A shipment of 100 chairs is received by a furniture store. A simple random sample of 36 chairs is chosen, and each is tested for its closeness to specifications. Based on this test, each chair is given a rating from 0 to 10 points. If the store knows from past experience that the standard deviation of the ratings in a shipment is 2 points, what is the probability that the mean rating in the sample will be more than 1 point below the mean rating for all chairs in the shipment?

SOLUTION. Since the sample is considerably more than 5 percent of the population, equation (6.4) must be used to calculate $\sigma_{\bar{x}}$, which is $(2/\sqrt{36}) \sqrt{(100 - 36)/99} = (2/6) \sqrt{64/99} = .27$ points. Because the sample size is well above 25, the sampling distribution of the sample mean should be close to normal. The probability that the sample mean is more than 1 point below the population mean equals the probability that the sample mean is more than $1/.27 = 3.70$ standard deviations below the population mean. To evaluate this probability, we find the area under the standard normal curve to the left of $-3.70$. Appendix Table 2 shows that the area between zero and 3.70 exceeds .499, so the area to the right of 3.70 must be less than .001. Because of the symmetry of the standard normal curve, the area to the left of $-3.70$ must also be less than .001. Thus, the probability that the mean rating for the sample will be more than 1 point below the mean rating for all chairs in the shipment is less than .001.

## 6.9 Accounts Receivable in a Department Store: An Experiment

To illustrate the fact that the central limit theorem (which played an important role in the previous section) can be counted on to work, consider the following experiment that was carried out by Robert Trueblood, a partner in the accounting firm of Touche, Ross and Company, and Richard Cyert, now president of Carnegie-Mellon University. Trueblood and Cyert obtained data from a department store concerning the total dollar-amounts of 59 accounts-receivable ledgers. (Each ledger contained the accounts of 200–300 individual customers of the department store.) The amount for each ledger is shown in Table 6.4. Based on these data, the mean dollar-amount for a ledger is $10,263, and the standard deviation is $1,670.

Trueblood and Cyert were interested in what would happen if the department store, rather than going through all the ledgers, drew a random sample of 10 and used the sample mean to estimate the population mean.[12] To find out, 100 random samples (each containing 10 ledgers) were drawn, and the sample mean for each was computed. From these 100 samples was constructed the relative frequency distribution of the sample mean, which showed the proportion of these samples where the sample mean fell into various class intervals. The results are plotted in the histogram (using a density scale) in Figure 6.9.

According to our discussion in the previous section, the sample mean should be distributed approximately normally, the mean of its sampling distribution should equal the population mean, and the standard deviation of its sampling distribution should equal the population standard deviation divided by the square root of the sample size (because the sample was drawn with replacement). Thus, since we know

[12] Actually, they were more concerned with the use of the sample in estimating the population *total*. But since the population total is simply 59 times the population mean, the population total can be estimated by multiplying the sample mean by 59. Thus, the reliability of the estimate of the population total is directly related to the reliability of the sample mean. Note, too, that the samples were drawn *with replacement*. Thus, the finite population correction factor should not be used. See R. Trueblood and R. Cyert, *Sampling Techniques in Accounting* (New York: Prentice-Hall, 1957).

**Table 6.4**
Total Dollar-Amounts
of 59 Accounts-
Receivable Ledgers

| Ledger number | Amount (dollars) | Ledger number | Amount (dollars) |
|---|---|---|---|
| 1 | 10,811 | 31 | 9,534 |
| 2 | 13,977 | 32 | 11,453 |
| 3 | 10,167 | 33 | 10,432 |
| 4 | 10,956 | 34 | 11,211 |
| 5 | 9,983 | 35 | 9,474 |
| 6 | 16,026 | 36 | 8,774 |
| 7 | 9,468 | 37 | 11,338 |
| 8 | 11,362 | 38 | 10,644 |
| 9 | 10,034 | 39 | 11,779 |
| 10 | 9,281 | 40 | 10,438 |
| 11 | 9,395 | 41 | 10,853 |
| 12 | 10,221 | 42 | 8,211 |
| 13 | 10,873 | 43 | 11,244 |
| 14 | 8,611 | 44 | 13,698 |
| 15 | 12,795 | 45 | 10,841 |
| 16 | 11,436 | 46 | 8,753 |
| 17 | 8,342 | 47 | 10,067 |
| 18 | 8,882 | 48 | 12,325 |
| 19 | 10,731 | 49 | 12,039 |
| 20 | 8,192 | 50 | 12,147 |
| 21 | 12,541 | 51 | 9,013 |
| 22 | 8,844 | 52 | 9,441 |
| 23 | 12,420 | 53 | 10,479 |
| 24 | 11,098 | 54 | 8,476 |
| 25 | 8,050 | 55 | 9,029 |
| 26 | 8,100 | 56 | 8,871 |
| 27 | 8,754 | 57 | 7,125 |
| 28 | 9,922 | 58 | 9,561 |
| 29 | 9,877 | 59 | 9,541 |
| 30 | 7,593 | | |
| Total dollar amount of accounts receivable | | | 605,533 |

that the mean of the population is \$10,263 and that the population standard deviation is \$1,670, it follows that the mean of the sampling distribution of the sample mean in this case should be \$10,263 and that its standard deviation should be $\$1,670 \div \sqrt{10} = \$528$.

The point of Trueblood and Cyert's experiment was to demonstrate that the distribution of the sample means in Figure 6.9 would be quite close to theoretical expectations. To test this, compare the actual distribution of sample means (shown in the histogram in Figure 6.9) with the normal curve with mean of \$10,263 and standard deviation of \$528 (also shown in Figure 6.9). The two are quite close, as you can see. They are not exact replicas of one another, but no statistician would expect exact agreement, since the distribution in Figure 6.9 is based only on 100 random samples. If one had the patience to draw 1,000 such samples, the approximation would be closer than in Figure 6.9; and if one were to draw 10,000 such samples, the approximation would be even better.

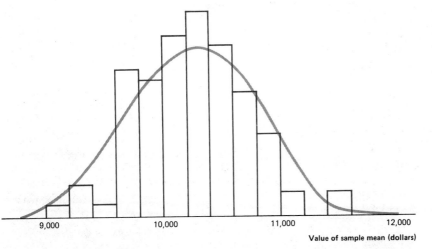

**Figure 6.9**
Histogram of 100
Sample Means, and
Normal Approxima-
tion*

Value of sample mean (dollars)

* The height of each bar equals the proportion of sample means falling within the relevant class interval divided by the width of the class interval ($200). The area of each bar equals the proportion of sample means falling within this class interval.

As pointed out in the previous section, statisticians frequently use the sampling distribution of the sample mean to compute the probability that a sample mean will differ from the population mean by at least a certain amount. The results in Figure 6.9 provide some idea of the validity of these computations. For example, as we shall see in greater detail in the next chapter, statisticians are fond of stating that the probability that the sample mean will differ from the population mean by more than $1.96\sigma_{\bar{x}}$ is approximately .05. This follows from the fact that the probability is approximately .95 that the value of any normal variable will lie within 1.96 standard deviations of its mean.

Using the results in Figure 6.9, we can see how close this statement is to being true for the 100 samples drawn by Trueblood and Cyert. Since $\sigma_{\bar{x}} = \$528$, and since $(1.96)(\$528) = \$1,035$, this statement amounts to saying that the probability is .05 that the sample mean will differ by more than $1,035 from the population mean. And since the population mean is $10,263, this amounts to saying that the probability is about .05 that the sample mean will fall below $9,228 (that is, $10,263 − $1,035) or will exceed $11,298 (that is, $10,263 + $1,035). In fact, the sample mean did fall below $9,228 or exceed $11,298 in 5 of the 100 samples (or 5 percent of the time). Of course, the fact that this experimental result is so close to what theory predicts is to some extent a coincidence, but it does illustrate the usefulness of our propositions in the previous section concerning the sampling distribution of the sample mean.

# Playing the Slots at the Sands*

The Sands Hotel and Casino is one of the nine casinos operating in Atlantic City, New Jersey. The casino floor of the Sands consists of areas devoted to slot machines and areas for table games (like black-jack, craps, and roulette). Table games are ordinarily located in the center of the casino floor, while the slot machines are put around the perimeter. The slot machines are of five denominations—$0.05, $0.10, $0.25, $0.50, and $1.00. The Sands Hotel and Casino receives about $4 million per month in net revenues from slot machines. As shown in Table 1, over half of these revenues come from the $0.25 machines.

Table 1
Percentage Distribution of Slot Machines and Percentage Distribution of Total Net Revenue, by Denomination of Slot Machine, Sands Hotel and Casino, February 1983.

| *Demonination of machine* | *Percent of all machines* | *Percent of total net revenue* |
|---|---|---|
| $0.05 machines | 5.4 | 3.9 |
| $0.25 machines | 64.3 | 58.2 |
| $1.00 machines | 4.9 | 8.1 |
| Other | 25.4 | 29.8 |
| Total | 100.0 | 100.0 |

*Source:* Bayus, Banker, Gupta, and Stone, *op. cit.*

During the mid-1980s, the management of the Sands Hotel and Casino set in motion a study of the preferences and behavior of consumers with respect to slot machines. One question of interest to management was: How do customers that prefer $0.05 or $1.00 machines differ from those that prefer $0.25 machines with regard to the frequency of visits to the casinos, and how much are they willing to spend? Based on a sample of about 200 slot-machine players in the Sands, those preferring $0.05 or $1.00 machines visit the casinos somewhat more frequently than those preferring $0.25 machines (Table 2). On the average, players preferring $1.00 machines are willing to spend about one-third more than those preferring $0.25 machines, and almost double as much as those preferring $0.05 machines.

Another question of interest to management was: To what extent do players tend to be "loyal" to the first machine played? (That is, to what extent do they play it over and over again?) Based on the sample described above, the proportion of time spent by players on the same machine is about 60 percent (Table 3). Thus, there is considerable loyalty of this sort. At the same time, players who begin with a machine of

* For further discussion of this case, see B. Bayus, R. Banker, S. Gupta, and B. Stone, "Evaluating Slot Machine Placement on the Casino Floor," *Interfaces*, March–April 1985.

**Table 2**

Average Number of
Visits to Casinos and
Average Amount
Players Are Willing to
Spend, by Player's
Preferences with
Regard to
Denomination of Slot
Machine

| Preferred denomination of slot machine | Average number of visits | Average amount player is willing to spend (dollars) |
|---|---|---|
| $0.05 machines | 6.0 | 70 |
| $0.25 machines | 5.3 | 98 |
| $1.00 machines | 5.4 | 131 |

*Source:* Bayus, Banker, Gupta, and Stone, *op. cit.*

**Table 3**

Percentage of Time
Spent by Players on
Machine Played First
and on Other Machines
of Various
Denominations, Sands
Hotel and Casino

| | Denomination of first machine played | | | | |
| | $0.05 | $0.10 | $0.25 | $0.50 | $1.00 |
| | (percentage of time spent) | | | | |
|---|---|---|---|---|---|
| First machine played | 61 | 80 | 59 | 61 | 44 |
| Other machines: | | | | | |
| $0.05 machines | 15 | 0 | 8 | 0 | 0 |
| $0.10 machines | 7 | 6 | 4 | 4 | 2 |
| $0.25 machines | 5 | 14 | 21 | 9 | 16 |
| $0.50 machines | 11 | 0 | 2 | 21 | 5 |
| $1.00 machines | 1 | 0 | 6 | 5 | 33 |
| Total | 100 | 100 | 100 | 100 | 100 |

*Source:* Bayus, Banker, Gupta, and Stone, *op. cit.*

one denomination do sometimes switch to other denominations. For example, players beginning with a $1.00 machine spend about 16 percent of their time on $0.25 machines.

The Sands Hotel and Casino was interested in these questions because they are relevant to decisions concerning how the casino's limited floor space should be utilized. New Jersey state law limits the slot area on the casino floor to 30 percent of the total. The management of the Sands must determine how many slot machines of each denomination should occupy this space and how they should be arranged. According to the top management of the Sands, this sample survey was useful in guiding such decisions.

### PROBING DEEPER

1. If the Sands wants to maximize its profit from slot machines, and if the amount of floor space that it can devote to slot machines is fixed, what factors will determine how many slot machines of each type it should operate?

2. If 60 percent of all slot-machine players prefer $0.25 machines, and if the Sands has 300 such machines, what is the probability, if there are 400 people playing the slot machines there (simultaneously), that some of those who prefer such machines will not find them available?

3. Does it appear that the amount of net revenue per machine is the same from each type of machine? If not, which type of machine has

the highest net revenue per machine? How big is the sampling error in this regard?

4. Suppose that the sample of slot-machine players on which Tables 2 and 3 are based was chosen in the following way: A random sample of 200 people registered at the Sands Hotel during the second week of June was selected. All answered the relevant questions. Does this sample contain possible biases? If so, what are they?

5. If the sample is a properly selected random sample, and if 100 of the people in the sample prefer $0.25 machines, can you determine the probability that the mean amount ($98) each is willing to spend will differ by more than $20 from the mean amount that all such players are willing to spend?

6. If the standard deviation of the amount that all slot-machine players are willing to spend is $35, does this allow you to answer the previous question?

EXERCISES

**6.14** If we are sampling from an infinite population, how much reduction occurs in the standard deviation of the sample mean if (a) the sample size is increased from 2 to 4; (b) the sample size is increased from 4 to 6; (c) the sample size is increased from 100 to 102?

**6.15** The Energetic Corporation wants to estimate the mean life of the light bulbs it produced on a particular day. It is known that the standard deviation of the longevity of the light bulbs produced on any given day is about 100 hours. A random sample of 100 bulbs produced on the day in question is drawn and the mean length of life of this sample is determined. What is the probability that this sample mean will differ from the population mean by more than 15 hours?

**6.16** A simple random sample of size 5 is taken (without replacement) from a population of 10 items. If the standard deviation of the population is 10, what is the standard error of the sample mean?

**6.17** (a) Show that for a finite population the standard error of the sample mean equals approximately

$$\frac{\sigma}{\sqrt{n}} \sqrt{1 - \frac{n}{N}}$$

if $N$ is of reasonable size. The quantity, $n/N$, is often called the *sampling fraction*.
(b) Using the above formula, find the standard error of the sample mean if $\sigma = 2$, $n = 16$, and (i) $N = 100$; (ii) $N = 10,000$.

**6.18** A furniture factory produces tables, the mean length of its tables being 28.0 inches, the standard deviation being 0.02 inches.
(a) The factory's manager says that 99.7 percent of the tables produced are between 27.94 and 28.06 inches long. Do you agree? Why or why not?
(b) The factory's foreman says that, if simple random samples of 100 tables are drawn from the factory's output, the mean length of the sample will be between 27.994 and 28.006 inches in about 99.7 percent of the samples. Do you agree? Why or why not? (Assume that each such sample is a very small percentage of the factory's output.)

**6.19** There are 3,000 gas stations in a particular region of the United States. Their mean annual sales in 1987 equals $950,000, and the standard deviation of their 1987 sales is $310,000. Random samples of 16 gas stations are taken repeatedly, and the mean 1987 sales of the gas stations in each sample is computed.
(a) What is the mean of the sampling distribution of the sample mean?
(b) What is the standard deviation of the sampling distribution of the sample mean?
(c) Can we be sure that the sampling distribution of the sample mean is normal?

**6.20** (a) If the variance of an infinite population is 100, what is the probability of drawing a simple random sample of size 64 with a mean of 40 or more from a population with a mean of 38?
(b) If the variance of the population is 10,000 rather than 100, what is the above probability?
(c) Why does the answer to (b) exceed that of (a)?

**6.21** (a) A random sample of size 100 is drawn from an infinite population with mean equal to 80 and standard deviation equal to 9. What is the probability that the sample mean is less than 78?

(b) If the sample size is 49 rather than 100, what is the above probability?

(c) Why does the answer to (b) exceed that of (a)?

**6.22** The Polymer Company has 50 spark plugs. The average life of these spark plugs is 40,000 miles, the standard deviation being 3,000 miles. Eight of these spark plugs will be chosen at random and installed on one of the firm's cars.

(a) What is the standard deviation of the sampling distribution of the average life of these eight spark plugs?

(b) If the lives of the 50 spark plugs are (approximately) normally distributed, what is the probability that the average life of these eight spark plugs is less than 38,000 miles?

**6.23** A class contains 85 students. The mean and standard deviation of their IQ's are 115 and 10, respectively. A random sample of 50 students is drawn from this class, and the mean IQ of the students in the sample is calculated.

(a) What is the probability that this mean IQ exceeds 120?

(b) If the class had contained twice as many students, how would the answer to (a) have changed? Why?

**6.24** A shipment of 2,000 rugs is delivered to a large store, which selects a random sample of 20 of the rugs to inspect. The shipment contains 200 defective rugs. What is the probability that the proportion defective in the sample will depart from the true proportion defective by more than 0.05?

**6.25** A ship with 3,000 refugees arrives in Miami, Florida. In fact, 1,000 of these refugees have relatives in the United States. A random sample of 100 refugees is questioned to determine whether they have relatives in the United States. What is the probability that the proportion of the sample having relatives in the U.S. differs from the population proportion by more than 0.02?

**6.26** The Martin Corporation draws a simple random sample of its bills to customers to determine what proportion of bills contain numerical errors. In the population as a whole, 10 percent contain such errors.

(a) If there are 10,000 such bills, and if the sample contains 100 bills, what is the probability that the proportion of bills in the sample containing numerical errors exceeds 16 percent?

(b) How large must the sample be if the standard deviation of the sample proporton equals .06?

## Chapter Review

1. Sample designs can be divided into two broad classes: probability samples and judgment samples. A *probability sample* is one where the probability that each element of the population will be chosen in the sample is known. A *judgment sample* is chosen in such a way that this probability is not known. One of the most important disadvantages of a judgment sample is that one cannot calculate the sampling distribution

of a sample statistic and thus there is no way of knowing how big the sampling errors in the sample results are likely to be.

2. If a population contains $N$ elements, a *simple random sample* is one chosen so that each of the different samples of size $n$ that could be chosen has an equal chance of selection. Alternatively, a simple random sample can be defined as one in which the probability of selecting each element in the population is the same and in which the chance that any one element is chosen is independent of the choice of any other element. A table of random numbers is used in choosing a simple random sample. A *systematic sample* is obtained by taking every $k$th element of the population on a list. If the elements of the population are in random order, this is tantamount to a simple random sample.

3. One of the most important ways that expert judgment can be used to design a sample is in the construction of strata, or subdivisions of the population. In *stratified random sampling* (where the population is stratified first, and a simple random sample is chosen from each stratum), the resulting estimate is often more precise for a sample of given size than if simple random sampling is used. In formulating strata, one should subdivide the population so that the strata differ greatly with regard to the characteristic being measured, and so that there is as little variation as possible *within* each stratum with regard to this characteristic. To maximize the precision of the sample results, the sample size in each stratum should be proportional to the product of the number of elements in the stratum and the standard deviation of the characteristic being measured in the stratum. In *cluster sampling,* the population is divided into a number of clusters and a sample of these clusters is chosen, after which a simple random sample of the elements in each chosen cluster is selected. The major advantage of cluster sampling is that it is relatively cheap to sample elements that are physically or geographically close together.

4. A *sampling distribution* is the probability distribution of the value of a particular sample statistic. If the sample is a small proportion of the population (or if the sampling is with replacement), the binomial distribution can be used to derive the sampling distribution of the sample proportion. If the population is normal, the sampling distribution of the sample mean is normal with a mean equal to the population mean and with a standard deviation equal to the population standard deviation divided by the square root of the sample size. Even if the population is not normal, the sampling distribution of the sample mean will have approximately these properties as long as the sample size is larger than about 30. (If the population is at all close to normal, a sample size of much less than 30 often will suffice.)

5. The *standard deviation of the sampling distribution of the sample mean* $\sigma_{\bar{x}}$ is a measure of the extent of the sampling error that is likely to be contained by the sample mean. This is often called the *standard error of the sample mean.* If the population is finite, it equals the value of $\sigma_{\bar{x}}$ for an infinite population multiplied by $\sqrt{(N-n)/(N-1)}$; this multiplier is called the *finite population correction factor.* If the sample is a small percentage of the population, this multiplier is so close to 1 that it can be ignored. Thus, if the sample is small relative to the population,

the accuracy of the sample mean depends entirely on the sample size, not on the fraction of the population included in the sample.

## Chapter Review Exercises

**6.27** An agricultural research firm wants to estimate the total acres of corn planted in a certain county of 3,000 farms where corn is the major crop.

(a) If the firm can obtain an alphabetical list of all the names of farm owners in the county, describe how the firm can draw a random sample of 100 farms. (Assume each farm has a single owner.)

(b) The research firm decides to pick a number from 1 to 30 at random, and the number turns out to be 16. The firm then picks the 16th, 46th, 76th, 106th, . . . farms on the list. Is this a random sample? If the number of acres of a farm planted with corn is not related in any way to the name of its owner, will a sample of this sort be essentially equivalent to a simple random sample?

(c) A statistician working for a local university criticizes the agricultural research firm's survey design. The statistician says that the firm could obtain more precise results by stratifying the sample according to last year's farm size. Do you agree? Why, or why not?

(d) Available data indicate that the number of farms, classified by last year's size, is as shown in the middle column below:

| Farm size last year (acres) | Number of farms | Standard deviation of number of acres of corn |
|---|---|---|
| 0–50 | 1,000 | 10 |
| 51–100 | 500 | 10 |
| 101–150 | 500 | 10 |
| 151–200 | 400 | 15 |
| 201–250 | 400 | 15 |
| Over 250 | 200 | 50 |
| Total | 3,000 | |

If proportional allocation is used, and if the total sample size is 100, what will be the number of farms chosen at random from each of the six strata?

(e) The right-hand column in the table in (d) shows the estimated standard deviation of the number of acres of corn among the farms in each stratum. (For example, among farms with a total acreage last year of 50 or less, the standard deviation of the number of acres planted with corn this year is estimated to be 10.) If optimum allocation is used, and if the total sample size is 100, what will be the number of farms chosen at random in each of the six strata?

(f) Still another statistician takes issue with the stratified sample proposed by the statistician in (c). The second statistician suggests that the list of farms in alphabetical order by owner's names be used to construct a cluster sample. All farms owned by people with names beginning with *A* would make up one cluster, all farms owned by people with names beginning with *B* would constitute a second cluster, and so on. The statistician suggests that a random sample of five clusters be chosen and that 20 farms be chosen at random in each cluster. If the principal cost of the survey is the expense of driving from one farm to another to obtain information concerning the number of acres planted with corn, is this suggestion likely to reduce the survey's costs? Why, or why not?

**6.28** The Uphill Manufacturing Company is interested in determining how consumers in a particular town rate the performance of various makes of bicycles. It hires a polling organization to call every hundredth number listed in the telephone directory and ask whoever answers to rate the comparative performance of various makes.
(a) Is this a random sample? Why, or why not?
(b) What pitfalls and disadvantages do you see in the design of this sample?
(c) The polling organization carries out this survey. The report of its findings indicates that the telephone calls were made between 10 A.M. and 2 P.M. on Mondays, Tuesdays, and Wednesdays; if there was no answer at a particular telephone number, that number was dropped from the sample. Give several reasons why this survey might yield distorted results.

**6.29** If the population is roughly normal, the *standard deviation of the sample median* equals $\sqrt{\pi/2} \cdot \sigma/\sqrt{n}$. If a sample median is to have a standard deviation equal to that of a sample mean (from the same population), how much bigger must the sample size be?

**6.30** A statistician calculates the standard error of a sample mean (from an infinite population) on the basis of incorrect information concerning the sample size. His result is double what it should be. He originally thought that the sample size was 100. What was the true sample size?

**6.31** The Energetic Corporation is a producer of electric light bulbs. The company's production process does not result in precisely similar bulbs. Instead, bulb length is normally distributed, with a mean of 3.00 inches and a standard deviation of .10 inches.
(a) What is the sampling distribution of the mean length of a simple random sample of four bulbs? That is, what is the mean of this sampling distribution? What is its standard deviation? What is its shape?
(b) If the population is normal, the distribution of the sample mean is normal, regardless of how small the sample may be. Using this fact, determine the probability that the sample mean in (a) will exceed 3.01 inches.
(c) In (a), how large a sample must the firm take to make the standard deviation of the sample mean equal .01 inches?

**6.32** The Bel Air Corporation wants to estimate the mean amount spent on paper clips in 1987 by the nation's 100 largest firms. A random sample of 25 of these firms is drawn (without replacement).
(a) If the actual mean expenditure by these 100 firms was $50,000 and the standard deviation was $5,000, what is the expected value of the sample mean? What is the standard deviation of the sample mean?
(b) Using Chebyshev's inequality, determine an upper-bound for the probability that the sample mean in (a) will *not* be between $48,000 and $52,000. If the population in (a) is reasonably close to normal, what is a good approximation to this probability?

# Appendix 6.1

### MONTE CARLO METHODS: ANOTHER APPLICATION OF A TABLE OF RANDOM NUMBERS[13]

Here and in previous chapters we have shown cases where mathematical analysis has been used to determine probability distributions. (For example, the central limit theorem uses mathematical analysis to determine the sampling distribution of the sample mean.) What techniques can be applied when the situation is too complicated for mathematical analysis? In such a case the statistician can use Monte Carlo methods, which are based on experimentation or simulation of the relevant situation. Although Monte Carlo methods do not result in an *analytical* solution to a *general* type of problem, they can provide an *approximate numerical answer* to a *particular concrete problem*—and in many practical cases this is all that is needed.

To illustrate how Monte Carlo methods work, suppose that a firm receives a lot of 100 items, 30 of which fail within a year. The firm would like to determine the probability that 30 or more out of 100 items will fail, given that the supplier is correct in its claim that the probability is 1/6 that each item will fail within a year. One way of solving this problem is to take 100 dice and throw them simultaneously. The probability that 30 or more dice will come up with a 1 is the same as the probability the firm requires, since the probability that each (true) die will come up with a 1 is 1/6. If the firm throws the 100 dice 1000 times and records the number of throws on which 30 or more dice come up with a 1, then this number divided by 1000 should be a fairly accurate approximation of the desired probability.

The essence of Monte Carlo methods is that *an analogue of the relevant situation is created, and one simulates the relevant process to generate a body of data from which the desired information can be estimated.* Let's take the following situation as an example. The shipping firm discussed in Appendix 5.1 is contemplating a change in its facilities such that it will take exactly 0.1 months to dock and service each ship. Given that ships arrive at random and that the mean number of arrivals per month is 5, the firm wants to know the probability that a ship will wait more than 0.2 months under this proposed setup. How can Monte Carlo methods be used to answer this question?

To simulate this situation, the firm's statisticians must generate a hypothetical "history" (record) of arrivals which conforms to the fact that ships arrive at random and that the mean number of arrivals per month is 5. Under these circumstances, as we know from Appendix 5.1, the probability that the interval between arrivals is between $c$ and $d$ months equals

$$e^{-5c} - e^{-5d}.$$

Letting $c = 0$, we obtain the probability that this time interval, $Y$, is less than $d$ months:

$$Pr\{Y < d\} = 1 - e^{-5d}. \tag{6.5}$$

---

[13] Some of the material in this Appendix assumes that the reader has covered the latter part of Chapter 5 and Appendix 5.1. Readers who have not covered these topics will probably want to skip this appendix as well.

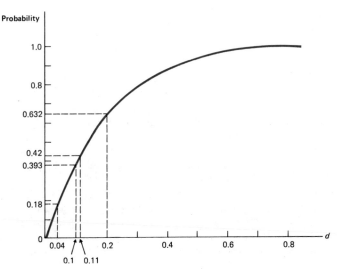

**Figure 6.10**
Probability that Time Interval between Successive Arrivals Is Less than *d* Months[a]

[a] To derive the function graphed here, we insert $d = 0, 0.1, 0.2, \ldots$ into $1 - e^{-5d}$, the result being

| d | $1 - e^{-5d}$ | d | $1 - e^{-5d}$ |
|---|---|---|---|
| 0 | .000 | 0.5 | .918 |
| 0.1 | .393 | 0.6 | .950 |
| 0.2 | .632 | 0.7 | .970 |
| 0.3 | .777 | 0.8 | .982 |
| 0.4 | .865 | 0.9 | .989 |

Figure 6.10 shows this probability for various values of *d*.

In generating a history of arrivals, we want the probability distribution of the values we assign to the intervals between successive arrivals to agree with equation (6.5). This can be done by picking a point between 0 and 1 at random on the vertical axis of the graph in Figure 6.10, and finding the horizontal coordinate of the point on the curve corresponding to this randomly chosen point. The resulting interval will have the desired probability distribution.[14] To pick a point at random along the vertical axis of Figure 6.10. we draw a number from a table of two-digit random numbers and divide the number by 100. Thus, to generate a history of arrivals, we must draw a large number of two-digit random numbers. Suppose that they are as follows:

18, 42, 00, 02, 24, 43, 64, 33, 42, 65, 11, 11, 32, 40,

62, 28, 35, 20, 86, 51, 49, 62, 01, and so forth.

[14] To see that an interval chosen in this way has the desired probability distribution, let's begin by considering two probabilities: (1) the probability that an interval chosen in this way will be less than .1 months; and (2) the probability that an interval chosen in this way will be less than .2 months. The probability that the time interval between successive arrivals will be less than 0.1 months should be .393, according to equation (6.5). What is the probability that an interval chosen in this way will be less than 0.1 months? It equals .393, because if we pick a random number between zero and 1, there is a .393 chance that it will be less than .393. If it is less than .393, Figure 6.10 shows that this procedure will result in a interval that is less than .1 months. The probability that the time interval between successive intervals is less than 0.2 months should be .632, according to equation (6.5). What is the probability that an interval chosen in this way will be less than 0.2 months? It equals .632, because if we pick a random number between zero

Dividing each of these numbers by 100, we can convert them into probabilities; and, as we have indicated, each such probability can be used to generate a time interval between ship arrivals. The first random number is 18, so the first probability is 0.18. Corresponding to 0.18 on the vertical axis in Figure 6.10, the time interval on the horizontal axis is 0.04 months.[15] Thus, assuming that the first ship arrives at time zero, the next arrives at 0.04 months. Since the next random number is 42, the next probability is 0.42. Figure 6.10 shows that the time interval on the horizontal axis corresponding to 0.42 on the vertical axis is 0.11 months, so the third ship arrives 0.11 months after the second, or at time 0.15 months. Repeating this procedure, we can generate as long a history of arrivals as we like. Table 6.5 shows the first 23 such results.

Once the firm's statisticians have established this history of arrivals, they can determine how long each ship must wait before it can be docked and serviced. As shown in Table 6.5, the second ship to arrive appears at time 0.15 and has to wait until time 0.20, when the first ship has been serviced. Then the third ship arrives at approximately the same time as the second ship, and must wait until time 0.30 when the second ship has been serviced. Table 6.5 shows

**Table 6.5**

Monte Carlo Study of Waiting Line in Shipping Example

| Random number | Time between arrivals | Time of arrival | Time wait ends | Time of servicing | Waiting time |
|---|---|---|---|---|---|
| 18 | .04 | .04 | .10 | .20 | .06 |
| 42 | .11 | .15 | .20 | .30 | .05 |
| 00 | .00 | .15 | .30 | .40 | 15 |
| 02 | .00 | .15 | .40 | .50 | .25 |
| 24 | .05 | .20 | .50 | .60 | .30 |
| 43 | .11 | .31 | .60 | .70 | .29 |
| 64 | .20 | .51 | .70 | .80 | .19 |
| 33 | .08 | .59 | .80 | .90 | .21 |
| 42 | .11 | .70 | .90 | 1.00 | .20 |
| 65 | .21 | .91 | 1.00 | 1.10 | .09 |
| 11 | .02 | .93 | 1.10 | 1.20 | .17 |
| 11 | .02 | .95 | 1.20 | 1.30 | .25 |
| 32 | .08 | 1.03 | 1.30 | 1.40 | .27 |
| 40 | .10 | 1.13 | 1.40 | 1.50 | .27 |
| 62 | .19 | 1.32 | 1.50 | 1.60 | .18 |
| 28 | .07 | 1.39 | 1.60 | 1.70 | .21 |
| 35 | .09 | 1.48 | 1.70 | 1.80 | .22 |
| 20 | .04 | 1.52 | 1.80 | 1.90 | .28 |
| 86 | .39 | 1.91 | 1.91 | 2.01 | .00 |
| 51 | .15 | 2.06 | 2.06 | 2.16 | .00 |
| 49 | .14 | 2.20 | 2.20 | 2.30 | .00 |
| 62 | .19 | 2.39 | 2.39 | 2.49 | .00 |
| 1 | .00 | 2.39 | 2.49 | 2.59 | .10 |

and 1, there is a .632 chance that it will be less than .632. If it is less than .632, Figure 6.10 shows that this procedure will result in an interval that is less than .2 months. It should now be clear that the probability of an interval chosen in this way being less than *any* length of time equals what is dictated by equation (6.5). This means that an interval chosen in this way has the desired probability distribution.

[15] To keep things as simple as possible, we use only two decimal places here for the time intervals.

the history of arrivals and waiting times for the first 23 ships (after the first ship, which is assumed to appear at time 0.00). The firm's statisticians can determine the percentage of ships that have to wait more than 0.2 months to be serviced, and this proportion can be used to estimate the desired probability. In a real Monte Carlo study, a history of this sort might be run for several hundred or even several thousand ships, depending on how precise the results needed to be.

It is important to emphasize that the results in Table 6.5 are intended only to demonstrate the methods used, not to provide the solution to a real problem. As stated above, many more than 23 ships would have to be included to obtain the solution (and a computer almost certainly would be used to cut down on the drudgery). Monte Carlo methods are used in many aspects of business and economics, not just in the analysis of waiting lines, and modern computers make it possible to carry out Monte Carlo studies in situations where they would otherwise be too complex or too expensive.

## EXERCISES

**6.33** Use one-digit random numbers to simulate 100 flips of a fair coin.

**6.34** Use one-digit random numbers (excluding 7, 8, 9, and 0) to simulate 100 rolls of a single true die.

**6.35** The Jones Used Car Agency, based on its records, has the following probabilities of selling various numbers of used cars in a day.

| Number of cars sold | Probability |
| --- | --- |
| 0 | 0.10 |
| 1 | 0.15 |
| 2 | 0.30 |
| 3 | 0.20 |
| 4 | 0.15 |
| 5 | 0.10 |

Devise a Monte Carlo sampling scheme to simulate the daily number of used cars sold by this firm during a 100-day period.

**6.36** The Marietta Ambulance Service has the following probability distribution of emergency calls in a day:

| Number of emergency calls | Probability |
| --- | --- |
| 0 | 0.45 |
| 1 | 0.25 |
| 2 | 0.20 |
| 3 | 0.10 |

Devise a Monte Carlo sampling scheme to simulate the daily number of emergency calls received during a 200-day period.

**\*6.37** The weights of metal pieces produced at the Marshall Company are normally distributed with mean equal to 15 ounces and standard deviation equal to 0.2 ounces. Use Minitab (or some other computer package) to

* Exercise requiring computer package, if available.

simulate the experiment of sampling (randomly) the weights of 30 metal pieces. Construct a histogram of the sample observations. Find the mean, median, standard deviation, minimum, and maximum of the observations. Have the computer print out the observations.

# PART III

# Estimation and Hypothesis Testing

# Statistical Estimation

7

## 7.1 Introduction

Having covered the necessary aspects of probability theory and sampling techniques in previous chapters, we can now go on to statistical inference, the branch of statistics that shows how rational decisions can be made on the basis of sample information. Statistical inference deals with two types of problems: *estimation* and *hypothesis testing*. This chapter covers estimation, and the next two chapters deal with hypothesis testing.

Since statistical estimation is of enormous importance, we will provide a rather detailed discussion of various estimation techniques and how they can be used.

## 7.2 Point Estimates and Interval Estimates

Many statistical investigations are carried out in order to estimate a parameter of some population. For example, at the end of Chapter 6 we considered a case where the object was to estimate the mean dollar-amount of a number of accounts-receivable ledgers; earlier in that chapter we examined a case where a firm wanted to estimate the total change in the value of its inventory. Decision makers are interested in estimating a particular parameter because their proper course of action depends on the value of this parameter. For example, a breakfast food firm may want to estimate the proportion of consumers who prefer brand C to brand D because this information will influence whether or not the firm tries to develop a competing breakfast food that imitates brand C.

In estimating a particular parameter, the decision maker uses a statistic calculated from a sample. For example, in the investigation of the accounts-receivable ledgers, the sample mean was used as an estimate of the population mean; and in the study to estimate the proportion of people preferring brand C breakfast food to brand D the sample proportion might well be the statistic used to estimate the population proportion. A statistic which is used to estimate a parameter is an **estimator.** An **estimate** is the *numerical value* of the estimator that is used. It is important to distinguish between an *estimator* and an *estimate*. For example, if a sample mean is used to estimate a population mean, and if the sample mean equals 10, the estimator used is the sample mean, whereas the estimate is 10.

Statisticians differentiate between two broad classes of estimates: point estimates and interval estimates. A *point estimate is a single number*. For example, if a firm estimates that 9 percent of the items in an incoming shipment are defective, 9 percent is a point estimate. Or if the proportion of individuals preferring brand C breakfast food to brand D is estimated to be .81 this, too, is a point estimate. An **interval estimate,** on the other hand, *is a range of values within which the parameter is thought to lie*. Thus, if a firm estimates that between 8 and 10 percent of the items in an incoming shipment are defective, 8 to 10 percent is an interval estimate. If the proportion of people preferring brand C breakfast food to brand D is estimated at between .75 and .90 this also is an interval estimate.

In some circumstances, only a point estimate is required. The estimate may be used in a complex series of computations, and the users may want only a single number. More often, however, an interval estimate is preferable to a point estimate because the former indicates how much error is likely to be in the estimate. For example, consider the point estimate that 9 percent of a certain shipment of items are defective. Such an estimate provides no idea of how much error it is likely to contain. That is, it is impossible to tell whether the data indicate that the proportion defective is likely to be very close to 9 percent, or whether it may well depart considerably from 9 percent. The advantage of interval estimates is that they provide such information. The decision maker can construct an interval estimate so that he or she has a specified amount of confidence that it will include the desired parameter. For example, in constructing its interval estimate of the percent of defective items, the firm in question can establish a .95 probability that such an interval will include the population percent defective.

At the outset of an attempt to estimate a parameter of any given population, the statistician must make basic decisions about the nature of the sample design and the type of estimator to be used. (For reasons pointed out in Chapter 6, we assume that simple random sampling is used here.) Once the sample design has been chosen, the statistician must determine *how large a sample to take*. The sample size will depend on the costliness of errors in the estimate and on the costliness of sampling. If substantial errors in the estimate will result in large penalties, the optimal sample size will tend to be large because the cost of increased sample size is likely to be outweighed by the resulting reduction in the sampling errors contained in the estimate. Also, if it is relatively inexpensive to increase the sample size, the optimal sample size will tend to be larger than if it is relatively expensive to do so.

The choice of *which estimator to use* will depend on the nature of the sampling distributions of various estimators. As stressed in the previous chapter, an estimator's sampling distribution shows the extent to which the estimator will vary from sample to sample and how far we can expect it to depart from the parameter it is intended to measure. An estimator with a sampling distribution showing a low probability of its departing greatly from this parameter is preferable to one whose sampling distribution indicates a high probability of its departing greatly. Why? Because for the same sample size and cost one can obtain greater *accuracy* with the former rather than the latter estimator. More will be said about this in the following section.

## 7.3 Point Estimation

Generally, there are a variety of estimators that can be used to make a point estimate of a particular parameter of a given population. For example, if we want to estimate the mean of a certain population, we can use either the mean or the median of a sample drawn from this population. Which should we use? Obviously, we want to choose the one that will be closer to the population mean, but there is no way of knowing which is closer because we do not know the value of the population mean. All we can do is compare the sampling distribution of the sample mean with that of the sample median. Such a comparison will show which of these estimators is more likely to depart considerably from the population mean. In choosing among estimators, statisticians consider the following three criteria: unbiasedness, efficiency, and consistency. Each of these is described below.

### UNBIASEDNESS

As indicated in previous chapters, the expected value of a statistic's sampling distribution is a measure of central tendency which shows the statistic's long-term mean value. If a statistic's sampling distribution is as shown in Figure 7.1, and if this statistic is used as an estimator of $\theta$, this statistic clearly is not a very reliable estimator. Why? Because the expected value of this estimator is $3\theta$, which means that if such an estimator were used repeatedly, the average estimate would be about three times the parameter we wish to estimate. To avoid estimators of this sort, statisticians use unbiasedness as one criterion.

> UNBIASEDNESS. *An unbiased estimator is a statistic the expected value of which equals the parameter being estimated.*

To illustrate how this definition can be applied, the sample mean is an unbiased estimator of the population mean because, as we saw in the previous chapter, the mean of its sampling distribution equals the population mean.

**Figure 7.1**
Sampling Distribution
of a Biased Estimator
of $\theta$

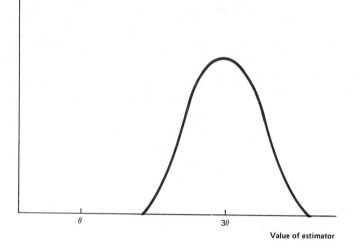

The concept of bias used here is somewhat different from that discussed in Chapter 1 where bias was described as a systematic, persistent sort of error due to faulty selection of a sample. Even if a sample is a properly chosen random sample, a bias can result if the estimator does not, on the average, equal the parameter being estimated. For example, had we defined the sample variance as

$$\sum_{i=1}^{n} \frac{(x_i - \bar{x})^2}{n},$$

it would have been a biased estimator of the population variance, $\sigma^2$. As shown in panel A of Figure 7.2, its expected value would have been $[(n-1)/n]\sigma^2$, not $\sigma^2$. It is for this reason (as we pointed out in Chapter 2) that we define the sample variance with $(n-1)$, not $n$, in the denominator. If $(n-1)$ is the denominator, the sample variance is an unbiased estimate of $\sigma^2$, as shown in panel B of Figure 7.2.

EFFICIENCY

The unbiasedness of an estimator does not necessarily mean that the estimator is likely to be close to the parameter we want to estimate. For example, the estimator whose sampling distribution is shown in panel A of Figure 7.3 has a high probability of differing considerably from its mean, which (since it is unbiased) is the parameter we want to estimate. This is because its sampling distribution contains a great deal of dispersion or variability. On the other hand, the estimator whose sampling distribution is shown in panel B of Figure 7.3 has a low probability of differing considerably from its mean because its sampling distribution exhibits little dispersion or variability. Statisticians would say that the estimator in panel B is "more efficient" than the one in panel A, since its sampling distribution is concentrated more tightly about the parameter we want to estimate. Efficiency, defined as follows, is an important criterion used by statisticians to choose among estimators.

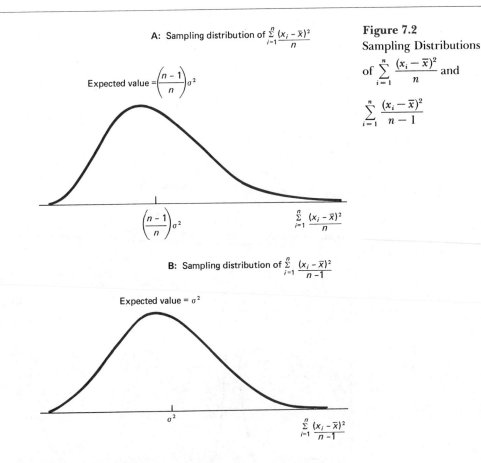

**Figure 7.2**
Sampling Distributions
of $\sum_{i=1}^{n} \frac{(x_i - \bar{x})^2}{n}$ and

$$\sum_{i=1}^{n} \frac{(x_i - \bar{x})^2}{n-1}$$

A: Sampling distribution of $\sum_{i=1}^{n} \frac{(x_i - \bar{x})^2}{n}$

Expected value $= \left(\frac{n-1}{n}\right)\sigma^2$

$\left(\frac{n-1}{n}\right)\sigma^2$      $\sum_{i=1}^{n} \frac{(x_i - \bar{x})^2}{n}$

B: Sampling distribution of $\sum_{i=1}^{n} \frac{(x_i - \bar{x})^2}{n-1}$

Expected value $= \sigma^2$

$\sigma^2$      $\sum_{i=1}^{n} \frac{(x_i - \bar{x})^2}{n-1}$

EFFICIENCY. *If two estimators are unbiased, one is more efficient than the other if its variance is less than the variance of the other.*[1]

To illustrate, let's go back to the question of whether the sample mean or the sample median should be used to estimate the population mean. As we know from Chapter 6, the variance of the sample mean is $\sigma^2 \div n$; and it can be shown that if the population is normal, the variance of the sample median is approximately $1.57\sigma^2 \div n$ if the sample is large. (See Exercise 6.29.) Thus, the sample mean is *more efficient* than the sample median because the sample median's variance is 57 percent greater than that of the sample mean. This implies that the sample median is more likely than the sample mean to differ considerably from the population mean. (In other words, holding constant the sample size, the sample mean is more likely than the sample median to be close to the population mean.) Because it is more efficient, the sample mean is generally preferred over the sample median as an estimator of the population mean.

[1] For biased estimators, one estimator is more efficient than another if its mean-square error is less than that of the other. If $X$ is an estimator of $\theta$, its mean-square error is the expected value of its squared deviation from $\theta$. That is, it equals

$E[(X - \theta)^2]$.

**Figure 7.3**
Sampling Distributions
of Two Unbiased
Estimators of $\theta$

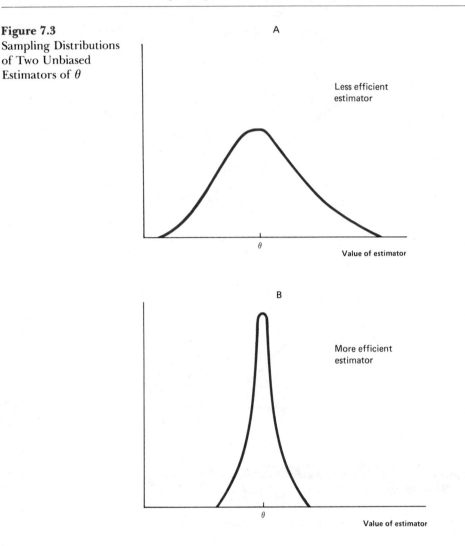

A

Less efficient
estimator

$\theta$

Value of estimator

B

More efficient
estimator

$\theta$

Value of estimator

## CONSISTENCY

Still another criterion used for choosing among estimators is consistency. Some estimators are consistent, while others are not. The statistical definition of consistency is as follows.

> CONSISTENCY. *A statistic is a consistent estimator of a parameter if the probability that the statistic's value is very near the parameter's value increasingly approaches* 1 *as the sample size increases.*

In other words, if a statistic is a consistent estimator of a particular parameter, the statistic's probability distribution becomes increasingly concentrated on this parameter as the sample size increases.

It is desirable for an estimator to be consistent because this means that the estimator becomes more reliable as the sample size increases. For example, consider the sample mean, which is a consistent estimator of the population mean. Since the standard deviation of the sample mean's sampling distribution equals $\sigma/\sqrt{n}$, it clearly tends to zero as $n$ increases in value. This means that the probability distribution of the sample mean becomes concentrated ever more tightly about the population mean. If the population is normal, the sample median is also a consistent estimator; thus, both the sample mean and the sample median satisfy this criterion.

# 7.4 Point Estimates for $\mu$, $\sigma$, and $\Pi$

We have just described the criteria that statisticians frequently use to choose among estimators. Based on these criteria, certain statistics are generally preferred over others as estimators of the population mean ($\mu$), population variance ($\sigma^2$), population standard deviation ($\sigma$), or population proportion ($\Pi$). This does not mean that other estimators are not preferred under special circumstances or that it is incorrect to use other estimators. It does mean that based on the criteria given in the previous section, the following estimators are the standard ones used by statisticians to estimate these parameters.

SAMPLE MEAN. This is the most common estimator of the population mean. As we know, it is unbiased and consistent. Moreover, it can be shown that if the population is normal the sample mean is the most efficient unbiased estimator available. For these reasons the sample mean is generally the preferred estimator of the population mean.

SAMPLE VARIANCE AND STANDARD DEVIATION. The sample variance is an unbiased and consistent estimator of the population variance. It is relatively efficient as compared with other estimators. Its square root, the sample standard deviation, is generally used as an estimator of the population standard deviation even though it is not unbiased. The sample standard deviation is also relatively efficient.

SAMPLE PROPORTION. This is an unbiased, consistent, and relatively efficient estimator of the population proportion. For these reasons, it is generally the preferred estimator of the population proportion.

## EXERCISES

**7.1** Does each of the following statements pertain to an estimate or an estimator? If the statement pertains to an estimate, is it an interval estimate or a point estimate?

(a) The head of a publishing company says that the firm's physics textbook will sell about 9,000 copies in 1988.

(b) The U.S. Bureau of the Census prefers the median to the mean as a measure of central tendency for the income distribution of families in a standard metropolitan area.

(c) A stock market analyst who appears on the television program, *Wall Street Week,* estimates that the price of a particular common stock will be between 50 and 60 at the end of six months.

**7.2** "The sample standard deviation is biased; thus, it should not be used, except under exceptional circumstances." Comment and evaluate.

**7.3** The Bona Fide Washing Machine Company chooses a random sample of 25 motors from those it receives from one of its suppliers (supplier I). It determines the length of life of each of the motors. The results (expressed in thousands of hours) are as follows:

| | | | | |
|---|---|---|---|---|
| 4.1 | 4.6 | 4.6 | 4.6 | 5.1 |
| 4.3 | 4.7 | 4.6 | 4.8 | 4.8 |
| 4.5 | 4.2 | 5.0 | 4.4 | 4.7 |
| 4.7 | 4.1 | 3.8 | 4.2 | 4.6 |
| 3.9 | 4.0 | 4.4 | 4.0 | 4.5 |

The firm's management is interested in estimating the mean length of life of the motors received from supplier I. Provide a point estimate of this population parameter.

**7.4** Is the sample proportion a consistent estimator of the population proportion? Why, or why not?

**7.5** A public-interest law firm picks a random sample of 60 hi-fi stores in a particular area, and asks each of them to repair a hi-fi set. In each case the law firm determines whether the store makes unnecessary repairs in order to inflate its bill. The law firm finds that eight of the stores are guilty of this practice.
(a) Provide a point estimate of the proportion of all such stores in the area that inflate bills in this way.
(b) After the law firm has presented the results of its study, an attorney representing the hi-fi repair stores objects that a sample of this size is quite unreliable. The attorney also maintains that the sample percentage of hi-fi repair stores engaging in such shady practices is a biased estimate of the percentage of all such stores engaging in such practices. Evaluate the attorney's objections.

**7.6** A statistic's mean-square error can be used as a measure of its reliability as an estimator. If $X$ is a statistic that is used as an estimator of $\theta$, $X$'s mean-square error is

$$E[(X - \theta)^2].$$

(a) If $X$ is an unbiased estimator of $\theta$, what is another name for its mean-square error?
(b) It can be shown that an estimator's mean-square error equals

$$(\mu - \theta)^2 + \sigma^2,$$

where $\mu$ is its mean, $\theta$ is the parameter to be estimated, and $\sigma$ is its standard deviation. Explain why $(\mu - \theta)$ is often called its bias.
(c) Based on the formula in (b), explain why a biased estimator may be preferred to an unbiased one if the former has a much smaller variance than the latter.

## 7.5 Confidence Intervals for the Population Mean

Interval estimates are generally preferred over point estimates because
the latter provide no information concerning how much error they are
likely to contain. Interval estimates, on the other hand, do provide
such information. To illustrate an interval estimate and how it is constructed, in this section we show how such an estimate is made of the
population mean. We begin with the case where the population standard deviation is known and the sample size is large ($n > 30$). Then we
take up the more realistic case where the standard deviation is unknown, both when the sample size is large and when it is small.

### WHERE $\sigma$ IS KNOWN: LARGE SAMPLE

In Chapter 5 we defined $Pr\,\{X > 32\}$ as the probability that $X$ is
greater than 32. Now we define $Pr\,\{a < X < b\}$ as the probability that $X$
lies between $a$ and $b$. Thus, the probability that the value of the sample
mean lies between $\mu - 1.96\sigma/\sqrt{n}$ and $\mu + 1.96\sigma/\sqrt{n}$ is denoted by
$Pr\,\{\mu - 1.96\sigma/\sqrt{n} < \overline{X} < \mu + 1.96\sigma/\sqrt{n}\}$. To construct an interval estimate of the population mean, we begin by noting that our results from
the previous chapter concerning the sampling distribution of the sample mean imply that

$$Pr\left\{\mu - 1.96\,\frac{\sigma}{\sqrt{n}} < \overline{X} < \mu + 1.96\,\frac{\sigma}{\sqrt{n}}\right\} = 0.95, \qquad (7.1)$$

where $\mu$ is the population mean, $\sigma$ is the population standard deviation,
$n$ is the sample size, and $\overline{X}$ is the sample mean. What equation (7.1) says
is that the probability that the sample mean will lie within 1.96 standard errors of the population mean equals 0.95. (Recall that the standard error of the sample mean equals $\sigma/\sqrt{n}$.) Since we know from the
previous chapter that if $n > 30$ (and if the population is large relative to
the sample size), the sample mean is normally distributed with a mean
of $\mu$ and a standard deviation of $\sigma/\sqrt{n}$, and since we know (from Appendix Table 2) that the probability that any normal random variable
will lie within 1.96 standard deviations of its mean is 0.95, it follows
that equation (7.1) is true.[2]

To construct an interval estimate for the population mean, we rearrange the terms inside the brackets on the left side of equation (7.1).
If we subtract $\mu$ from $\mu - 1.96\sigma/\sqrt{n}$, $\overline{X}$, and $\mu + 1.96\sigma/\sqrt{n}$, we get

---

[2] How do we know (from Appendix Table 2) that the probability that any normal
variable will be within 1.96 standard deviations of its mean is 0.95? The reasoning is as
follows. We want to determine the probability that a normal variable lies between $\mu -
1.96\sigma$ and $\mu + 1.96\sigma$. The points on the standard normal curve corresponding to these
points are $[(\mu - 1.96\sigma) - \mu] \div \sigma$ and $[(\mu + 1.96\sigma) - \mu] \div \sigma$. That is, they are $-1.96$ and
$+1.96$. Appendix Table 2 shows that the area under the standard normal curve between
zero and 1.96 equals .4750. Thus, the area between $-1.96$ and $+1.96$ equals $2(.4750) =$
.95.

$$Pr\left\{-1.96\frac{\sigma}{\sqrt{n}} < \overline{X} - \mu < 1.96\frac{\sigma}{\sqrt{n}}\right\} = .95.$$

And if we subtract $\overline{X}$ from all three terms of the inequality in brackets and multiply each of the resulting terms by $-1$, we get

$$Pr\left\{\overline{X} - 1.96\frac{\sigma}{\sqrt{n}} < \mu < \overline{X} + 1.96\frac{\sigma}{\sqrt{n}}\right\} = .95. \qquad (7.2)$$

(See footnote 3 if you have difficulty deriving equation [7.2.][3])

Since equation (7.2) is of basic importance in the construction of an interval estimate for a population mean, it is essential that you know exactly what it means. Does it mean that the probability is .95 that the population mean lies between $\overline{X} - 1.96\sigma/\sqrt{n}$ and $\overline{X} + 1.96\sigma/\sqrt{n}$? In one sense yes, in another no. Before the sample has been drawn and while the value of the sample mean is still unknown, there is a .95 probability that the interval between $\overline{X} - 1.96\sigma/\sqrt{n}$ and $\overline{X} + 1.96\sigma/\sqrt{n}$ will include the population mean. Thus, if we were to draw one sample after another, in 95 percent of the samples (over the long run) this interval—which varies, of course, from sample to sample because $\overline{X}$ varies—would include the population mean.

To see that this is true, look at Figure 7.4, which shows what would occur if we were to construct a large number of interval estimates for a certain population mean $\mu$. Panel A of Figure 7.4 shows the sampling distribution of the sample mean, $\overline{X}$. If the sample mean lies between $\mu - 1.96\sigma/\sqrt{n}$ and $\mu + 1.96\sigma/\sqrt{n}$, the interval estimate will contain $\mu$. Since the area under the sampling distribution between $\mu - 1.96\sigma/\sqrt{n}$ and $\mu + 1.96\sigma/\sqrt{n}$ equals .95, the probability that $\overline{X}$ will lie in this range equals .95. Panel B of Figure 7.4 shows the interval estimates that would result from repeated sampling. Over the long run, 95 percent of these intervals would contain $\mu$.

Note, however, that the previous discussion has pertained only to the situation before the sample has been drawn. Once a particular sam-

---

[3] If we subtract $\overline{X}$ from all three terms, we get

$$Pr\left\{-\overline{X} - 1.96\frac{\sigma}{\sqrt{n}} < -\mu < -\overline{X} + 1.96\frac{\sigma}{\sqrt{n}}\right\} = .95.$$

The important fact to note at this point is that if all three terms inside the brackets are multiplied by $-1$ the inequalities must be reversed. (To see that this is true, consider the simple inequality $y \geq 1$. If this is true, it follows that $-y \leq -1$. As we indicated, the inequality is reversed after multiplication by $-1$.) Applying this fact, it follows that

$$Pr\left\{\overline{X} - 1.96\frac{\sigma}{\sqrt{n}} < \mu < \overline{X} + 1.96\frac{\sigma}{\sqrt{n}}\right\} = .95,$$

which is what we set out to derive.

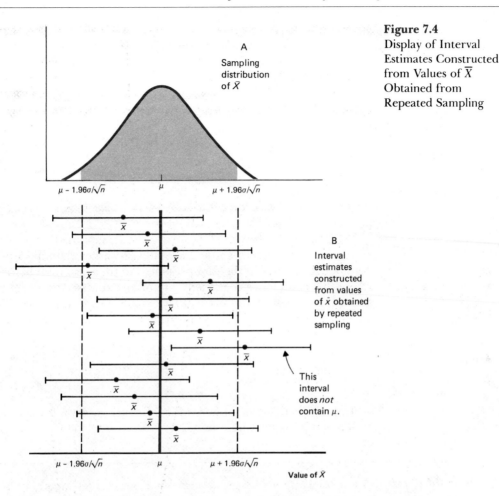

**Figure 7.4**
Display of Interval
Estimates Constructed
from Values of $\overline{X}$
Obtained from
Repeated Sampling

ple has been drawn and the sample mean has been calculated, it no longer is correct to say that the probability is 0.95 that the interval between $\overline{x} - 1.96\sigma/\sqrt{n}$ and $\overline{x} + 1.96\sigma/\sqrt{n}$ includes the population mean. To see why, suppose that $\sigma = 10$, $n = 100$, and $\overline{x} = 2.00$. Under these circumstances, equation (7.2) may be incorrectly interpreted to say that

$$Pr\left\{2.00 - 1.96\,\frac{10}{\sqrt{100}} < \mu < 2.00 + 1.96\,\frac{10}{\sqrt{100}}\right\} = .95,$$

or

$$Pr\left\{.04 < \mu < 3.96\right\} = .95.$$

But since the population mean $\mu$ is a constant, it makes no sense to say that the probability that it is between .04 and 3.96 is .95. Either its value lies in this interval or it doesn't. *All that one can say is that if intervals of this sort are calculated repeatedly, they will include the population mean in about 95 percent of the cases.*

The interval within the brackets on the left side of equation (7.2) is called a **confidence interval.** As you can see, this is an interval which has a certain probability of including the population mean, this probability being called the **confidence coefficient.** Thus, the confidence coefficient in equation (7.2) is .95. Although .95 is frequently used as a confidence coefficient, there is no reason why other confidence coefficients should not be chosen. In general, *if the confidence coefficient is set equal to* $(1 - \alpha)$, *the confidence interval for the population mean is*

**Confidence Interval for $\mu$ ($\sigma$ Known)**

$$\bar{x} \pm z_{\alpha/2} \frac{\sigma}{\sqrt{n}}, \qquad (7.3)$$

*where $z_{\alpha/2}$ is the value of the standard normal variable that is exceeded with a probability of $\alpha/2$.*

To illustrate the use of expression (7.3), suppose that we want to construct a 90 percent confidence interval rather than a 95 percent confidence interval in the case presented in the paragraph before last. (Note that confidence coefficients are often expressed as percentages.) Since $1 - \alpha = .90$, it follows that $\alpha/2 = .05$. From Appendix Table 2, we find that $z_{.05} = 1.64$. Thus, inserting $z_{.05}, \bar{x}, \sigma,$ and $n$ into expression (7.3), we get

$$2.00 \pm 1.64 \frac{10}{\sqrt{100}},$$

which implies that the 90 percent confidence interval for the population mean is .36 to 3.64.

**Basic Idea #13** | *If you calculate a 95 percent confidence interval for the population mean, it will include the population mean about 95 percent of the time.*

Two important points should be noted here. First, *holding constant the size of the sample, the width of the confidence interval tends to increase as the confidence coefficient increases.* For example, if the confidence coefficient is 90 percent, the width of the confidence interval is $2(1.64)\sigma/\sqrt{n}$, whereas if the confidence interval is 95 percent, the width[4] of the confidence interval is $2(1.96)\sigma/\sqrt{n}$. This makes sense. If you want to be more and more confident that the interval estimate includes the population mean, you must widen the interval if the sample size is fixed. Second, *for a fixed confidence coefficient (and a fixed population standard deviation), the only way to reduce the width of the confidence interval is to increase the sample size.* For example, if the confidence coefficient is 95 percent, the width of the confidence interval is $2(1.96)\sigma/\sqrt{n}$,

[4] The 95 percent confidence interval is $\bar{x} - 1.96\sigma/\sqrt{n}$ to $\bar{x} + 1.96\sigma/\sqrt{n}$; thus, the difference between its upper and lower limits (that is, its width) is $2[1.96\sigma/\sqrt{n}]$. Similarly, since the 90 percent confidence interval is $\bar{x} - 1.64\sigma/\sqrt{n}$ to $\bar{x} + 1.64\sigma/\sqrt{n}$, the difference between its upper and lower limits (that is, its width) is $2[1.64\sigma/\sqrt{n}]$.

as we saw above. If $\sigma$ is fixed, the only way to reduce this width is to increase $n$.

## WHERE $\sigma$ IS UNKNOWN: LARGE SAMPLE

In most actual cases, the standard deviation of the population is unknown. If the sample size exceeds 30 it is a relatively simple matter to adapt the results presented previously to the situation where $\sigma$ is unknown. As indicated earlier in this chapter, the sample standard deviation $s$ is generally used as an estimator of the population standard deviation. Mathematicians have shown that if the sample is large, we can simply substitute the sample standard deviation for the population standard deviation in the results obtained in the previous part of this section. Thus, if we want to construct a 95 percent confidence interval—that is, a confidence interval with a confidence coefficient of 95 percent—we can substitute $s$ for $\sigma$ in equation (7.2), the result being

$$ Pr\left\{ \overline{X} - 1.96\,\frac{s}{\sqrt{n}} < \mu < \overline{X} + 1.96\,\frac{s}{\sqrt{n}} \right\} = .95. $$

Consequently, the interval estimate is from $\overline{x} - 1.96s/\sqrt{n}$ to $\overline{x} + 1.96s/\sqrt{n}$.

In general, *if the confidence coefficient is set equal to $(1 - \alpha)$, the confidence interval for the population mean is*

$$ \overline{x} \pm z_{\alpha/2}\,\frac{s}{\sqrt{n}}, \qquad (7.4) $$

*Confidence Interval for $\mu$ ($\sigma$ Unknown and $n > 30$)*

*where s is the sample standard deviation, and $z_{\alpha/2}$ is the value of the standard normal variable that is exceeded with a probability of $\alpha/2$.* Like expression (7.3), expression (7.4) is applicable only if the population is large relative to the sample or if sampling is with replacement.[5] As we saw in the previous chapter, if these two conditions are both not true the finite population correction factor must be included. Thus, *if sampling is without replacement, and if the population is not large,* the confidence interval for the population mean is

$$ \overline{x} \pm z_{\alpha/2}\,\frac{s}{\sqrt{n}}\,\sqrt{\frac{N-n}{N-1}}, \qquad (7.5) $$

*Confidence Interval for $\mu$ ($\sigma$ Unknown and $n > 30$) When Population Is Small and Sampling Is without Replacement*

where $N$ is the number of items in the population.

These results concerning the confidence interval for a population mean are of enormous importance. The following example based on a

---

[5] If the population is not large relative to the sample and if sampling is without replacement, $\sigma/\sqrt{n}$ in (7.3) should be multiplied by the finite population correction factor $\sqrt{(N-n)/(N-1)}$.

real case[6] (although the numbers have been changed) should help illustrate how these results are used and interpreted.

---

EXAMPLE 7.1 A chemist wants to estimate the mean strength of a new synthetic fiber. To measure this fiber's strength, she determines the number of pounds that can be supported by one strand before breaking. A random sample of 36 strands of the fiber is taken, with the following results:

| Strand | Breaking load (pounds) | Strand | Breaking load (pounds) | Strand | Breaking load (pounds) |
|--------|------------------------|--------|------------------------|--------|------------------------|
| 1  | 2.2 | 13 | 2.2 | 25 | 2.3 |
| 2  | 2.2 | 14 | 2.3 | 26 | 2.4 |
| 3  | 2.2 | 15 | 2.3 | 27 | 2.3 |
| 4  | 2.3 | 16 | 2.3 | 28 | 2.4 |
| 5  | 2.3 | 17 | 2.2 | 29 | 2.4 |
| 6  | 2.3 | 18 | 2.2 | 30 | 2.3 |
| 7  | 2.3 | 19 | 2.2 | 31 | 2.3 |
| 8  | 2.3 | 20 | 2.2 | 32 | 2.3 |
| 9  | 2.4 | 21 | 2.2 | 33 | 2.4 |
| 10 | 2.4 | 22 | 2.4 | 34 | 2.3 |
| 11 | 2.4 | 23 | 2.3 | 35 | 2.4 |
| 12 | 2.2 | 24 | 2.4 | 36 | 2.3 |

Construct a 95 percent confidence interval for the mean breaking load of a strand of this new fiber.

SOLUTION: If $x_i$ is the breaking load (in pounds) of the $i$th strand in the sample, we find that

$$\sum_{i=1}^{36} x_i = 82.8$$

$$\bar{x} = 82.8/36 = 2.3$$

$$\sum_{i=1}^{36} (x_i - \bar{x})^2 = 0.20$$

$$\sum_{i=1}^{36} (x_i - \bar{x})^2/(n-1) = 0.20/35 = .00571$$

$$s = \sqrt{.00571} = .0756.$$

Since the population is very large (because a very large number of such strands of fiber can be produced), expression (7.4) is appro-

---

[6] Owen Davies, *The Design and Analysis of Industrial Experiments* (London: Oliver and Boyd, 1956), p. 72. The actual analysis was more extensive and complicated than our discussion in this section, which is simplified to emphasize the basic points considered here.

priate. Because a 95 percent confidence interval is wanted, $z_{\alpha/2} = z_{.025} = 1.96$. Thus, the desired confidence interval is

$$2.30 \pm 1.96 \left( \frac{.0756}{\sqrt{36}} \right).$$

Simplifying terms, this confidence interval is 2.275 to 2.325 pounds.

As stressed before, the proper interpretation of this result is *not* that the probability is 95 percent that the population mean lies between 2.275 and 2.325 pounds. Instead, this result means that if confidence intervals of this sort were constructed in a great number of cases they would include the population mean 95 percent of the time.

---

THE EFFECT OF A NEW ENZYME ON A PHARMACEUTICAL MANUFACTURING PROCESS[7]

GETTING DOWN TO CASES

A drug firm was attempting to estimate the extent to which a new enzyme altered the yield of a certain manufacturing process. For a given batch of product, the yield was measured by the ratio of the actual output to the theoretical output as calculated from formulas based on past experience. Thus a yield of 1.03 meant that 3 percent more output was gotten from the batch than the formula indicated; 0.98 meant that 2 percent less output was gotten from the batch than the formula indicated.

The drug firm tested the new enzyme on 36 batches and obtained the following yields:

| 1.28 | 1.31 | 1.48 | 1.10 | 0.99 | 1.22 |
| 1.65 | 1.40 | 0.95 | 1.25 | 1.32 | 1.23 |
| 1.43 | 1.24 | 1.73 | 1.35 | 1.31 | 0.92 |
| 1.10 | 1.05 | 1.39 | 1.16 | 1.19 | 1.41 |
| 0.98 | 0.82 | 1.22 | 0.91 | 1.26 | 1.32 |
| 1.71 | 1.29 | 1.17 | 1.74 | 1.51 | 1.25 |

(a) Calculate a 90 percent confidence interval for the true mean yield gotten from the enzyme.

(b) Calculate a 95 percent confidence interval for the true mean yield gotten from the enzyme.

(c) Calculate a 99 percent confidence interval for the true mean yield gotten from the enzyme.

(d) Specify the assumptions underlying your results in (a), (b), and (c).

(e) Provide an unbiased point estimate of the true mean yield from the enzyme.

[7] This case is based on a section from W. A. Wallis and H. V. Roberts, *op cit.* Some of the numbers have been changed for expository purposes.

(f) If the yields are normally distributed is the estimator you used in (e) at least as efficient as any other unbiased estimator?

(g) If the assumptions specified in (d) are correct, would you conclude that the true mean yield gotten from the enzyme almost certainly exceeds 1.00? Explain.

(h) What is the mistake (if any) in the following argument, put forth by one member of the firm's research department: "There is no good theoretical reason for believing that the mean yield should be higher than 1.00 with the enzyme. Moreover, in the 36 batches studied, the standard deviation of individual yields was .228. In my opinion, .228 represents a large fraction of the difference between [the sample mean] 1.268 and 1.00. There is no real evidence that the enzyme increases yield. . . ."[8]

[8] *Ibid.* Some of the numbers have been changed.

## WHERE $\sigma$ IS UNKNOWN: SMALL SAMPLE

In many cases, a confidence interval must be constructed on the basis of a sample where $n \leqslant 30$. In such cases, the expressions given in the previous part of this section are not appropriate. No longer can we simply substitute the sample standard deviation for the population standard deviation as we did in expression (7.4). However, if the population is normal it is possible to construct a confidence interval for the population mean even if the sample size is 30 or less. Such a confidence interval is based on the $t$ distribution described below.

THE $t$ DISTRIBUTION. *If the population sampled is normally distributed, $(\overline{X} - \mu) \div s/\sqrt{n}$ has the t distribution. The t distribution is symmetrical, bell-shaped and has zero as its mean.*

The $t$ distribution is a sampling distribution: Specifically, it is the distribution of the statistic $(\overline{X} - \mu) \div s/\sqrt{n}$. Suppose that simple random samples of size $n$ are taken repeatedly from a normal population with expected value of $\mu$ and standard deviation of $\sigma$. If we calculate the value of $(\overline{X} - \mu) \div s/\sqrt{n}$ for each sample, we can construct a sampling distribution for this statistic. If a sufficient number of samples are chosen this sampling distribution will conform to the $t$ distribution. The $t$ distribution is really a family of distributions, each of which corresponds to a particular *number of degrees of freedom*. In this context the number of degrees of freedom equals $(n - 1)$; in other contexts it will equal other amounts, as we shall see in later chapters.

*Degrees of Freedom*

It is not easy to give an adequate intuitive, nonmathematical interpretation of the number of degrees of freedom. From a mathematical point of view, the number of degrees of freedom is simply a parameter in the formula for the $t$ distribution. However, one way to interpret the number of degrees of freedom (that is, [$n - 1$]) in the

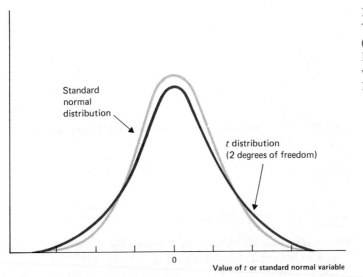

**Figure 7.5**
The *t* Distribution
(with 2 Degrees of
Freedom) Compared
with the Standard
Normal Distribution

Standard
normal
distribution

*t* distribution
(2 degrees of freedom)

0

Value of *t* or standard normal variable

present context is to say that it equals the number of independent deviations from the sample mean (that is, $[x_i - \bar{x}]$) in the computation of the sample standard deviation $s$. Since the sum of these deviations—that is, $\Sigma(x_i - \bar{x})$—equals zero, it follows that if we know $(n - 1)$ of these deviations we can determine the value of the remaining deviation by using the fact that their sum equals zero. Thus, only $(n - 1)$ of the deviations are independent. In other words, as the statistician might put it, there are $(n - 1)$ degrees of freedom.

The shape of the *t* distribution is rather like that of the standard normal distribution. Figure 7.5 compares the *t* distribution (with 2 degrees of freedom) to the standard normal distribution. As you can see, both are symmetrical, bell-shaped, and have a mean of zero. The *t* distribution is somewhat flatter at the mean and somewhat higher in the tails than the standard normal distribution. As the number of degrees of freedom becomes larger and larger, the *t* distribution tends to become exactly the same as the standard normal distribution. The *t* distribution is often called Student's *t* distribution because the statistician W. S. Gosset, who first derived this distribution, published his findings under the pseudonym Student.[9]

To find the probability that the value of *t* exceeds a certain number, we can use Appendix Table 6. As you can see, each row of this table corresponds to a particular number of degrees of freedom. The numbers in each row are the numbers that are exceeded with the indicated probability by a *t* variable. For example, the first row indicates that if a *t* variable has one degree of freedom, there is a .40 probability that its value will exceed .325, a .25 probability that its value will exceed 1.000, a .05 probability that its value will exceed 6.314, a .01 probability that its value will exceed 31.821, and so on. Since the *t* dis-

---

[9] Gosset used this pseudonym because his employer, Guinness Brewery, forbade publication of scientific research of this sort by employees under their own names.

tribution is symmetrical, it follows that if a *t* variable has one degree of freedom there is a .40 probability that its value will lie below $-$.325, a .25 probability that its value will lie below $-$1.000, and so on.

If the sample size is 30 or less and the population is normal (and large relative to the sample), a confidence interval for the population mean can be constructed by using the *t* distribution in place of the standard normal distribution in expression (7.4). In other words, *if the confidence coefficient is set equal to $(1 - \alpha)$, the confidence interval for the population mean is:*

**Confidence Interval for μ (σ Unknown and n ≤ 30)**

$$\bar{x} \pm t_{\alpha/2}\frac{s}{\sqrt{n}}, \tag{7.6}$$

*where* $t_{\alpha/2}$ *is the value of a* t *variable (with* n $-$ 1 *degrees of freedom) that is exceeded with a probability of* $\alpha/2$. Thus, if a sample of 16 observations is chosen, and if the sample mean is 20 and the sample standard deviation is 4, the 95 percent confidence interval for the population mean is

$$20 \pm 2.131\left(\frac{4}{\sqrt{16}}\right),$$

since Appendix Table 6 shows that for 15 degrees of freedom the value of *t* that will be exceeded with a probability of .025 is 2.131. Simplifying terms, it follows that the confidence interval in this case is 17.869 to 22.131. Statisticians frequently construct confidence intervals in this way. The following example illustrates how it is done.

EXAMPLE 7.2 A manufacturer of light bulbs wants to estimate the mean length of life of a new type of bulb which is designed to be extremely durable. The firm's engineers test nine of these bulbs and find that the length of life (in hours) of each is as follows:

| | | |
|---|---|---|
| 5,000 | 5,100 | 5,400 |
| 5,200 | 5,400 | 5,000 |
| 5,300 | 5,200 | 5,200 |

Previous experience indicates that the lengths of life of individual bulbs of a particular type are normally distributed. Construct a 90 percent confidence interval for the mean length of life of all bulbs of this new type.

SOLUTION: If $x_i$ is the length of life of the *i*th light bulb in the sample, we find that

$$\sum_{i=1}^{9} x_i = 46,800$$

$$\bar{x} = 5,200$$

$$\sum_{i=1}^{9} (x_i - \bar{x})^2 = 180,000$$

$$\sum_{i=1}^{9} (x_i - \bar{x})^2/(n-1) = 22{,}500$$

$$s = \sqrt{22{,}500} = 150.$$

Since $n = 9$, expression (7.6) is appropriate. Because a 90 percent confidence interval is wanted, $t_{\alpha/2} = t_{.05}$; and the number of degrees of freedom $(n - 1)$ is 8. Appendix Table 6 shows that if there are 8 degrees of freedom, $t_{.05} = 1.86$. Thus, the desired confidence interval is

$$5200 \pm 1.86 \left(\frac{150}{\sqrt{9}}\right).$$

Simplifying terms, the confidence interval is 5107 to 5293 hours.

---

OBTAINING A CONFIDENCE INTERVAL

**WHAT COMPUTERS CAN DO**

As pointed out in previous chapters, computer packages like Minitab are available to carry out many statistical calculations. To illustrate how Minitab can be used to calculate a confidence interval, suppose that one wants to compute a 95 percent confidence interval for a mean (the standard deviation of the population being unknown). The sample is 12, 18, 15, 16, 20, 21.

To calculate such a confidence interval, all that one has to do is to type the following four lines:

```
MTB > set into c4
DATA> 12 18 15 16 20 21
DATA> end
MTB > tinterval 95 percent c4
```

The first line says that the numbers should be put in column c4 of the worksheet that Minitab maintains in the computer. The next line contains the numbers themselves. The fourth line commands the computer to calculate the desired confidence interval.

After the above four lines are typed, the computer automatically prints out the following:

|    | N | MEAN | STDEV | SE MEAN | 95.0 PERCENT C.I. | |
|----|---|------|-------|---------|---------|---------|
| C4 | 6 | 17.00 | 3.35 | 1.37 | ( 13.49, | 20.51) |

This line contains the number of observations and the sample mean and standard deviation, as well as the standard error of the mean and the desired confidence interval. Thus, the 95 percent confidence interval for the mean is 13.49 to 20.51. This is the answer we want.

EXERCISES

**7.7** What is the probability that the value of a random variable with the $t$ distribution with 4 degrees of freedom will lie
(a) above 3.747;
(b) below −4.604;
(c) between 2.132 and 2.776?

**7.8** Compare the $t$ distribution with an infinite number of degrees of freedom in Appendix Table 6 to the standard normal distribution in Appendix Table 2. In particular, show that the probability is the same that each will exceed (a) 1.645; (b) 1.960; (c) 0.674. Do you find this surprising? Why, or why not?

**7.9** A hospital's records indicate that, in a simple random sample of 90 cases, the mean fee of the physician for a particular kind of operation was $810, the sample standard deviation being $85. These 90 cases are a very small percentage of the operations of this type carried out in the hospital.
(a) Construct a 90 percent confidence interval for the mean fee of the physician for this type of operation in all such cases in this hospital.
(b) Construct a 95 percent confidence interval of this sort.
(c) Construct a 99 percent confidence interval of this sort.

**7.10** It is known that the difference between a person's true weight and his or her weight as indicated by a particular scale is normally distributed with a standard deviation of 0.8 ounces. Thirty-one people weigh themselves on this scale. If we subtract each person's true weight from his or her weight according to this scale, the results are 0.6, 0.4, −0.1, 0.2, 0.5, 0.7, 0.4, −0.2, −0.1, 0.6, 0.8, 0.9,1.7, 1.2, 1.5, 2.0, 1.5, −1.4, −1.3, −1.2, −1.1, −1.8, −1.2, 1.4, 0.9, −0.7, 0.8, 0.5, 1.0, 1.2, and 0.2 ounces.
(a) Construct a 95 percent confidence interval for the mean difference between a person's true weight and his or her weight according to this scale.
(b) Construct a 90 percent confidence interval of this sort.

**7.11** A social worker wants to estimate the mean number of years of school completed by the residents of a particular neighborhood. A simple random sample of 90 residents is taken, the mean years of school completed being 8.4 and the sample standard deviation being 1.8. The neighborhood contains about 2,500 residents.
(a) Calculate a 90 percent confidence interval for the mean years of school completed by the residents of the neighborhood.
(b) Calculate a 98 percent confidence interval of this sort.

**7.12** The Mesa Company draws a random sample of ten employees from its labor force of 10,000 people. The number of years that each of these employees has been with the company is 8.2, 5.6, 4.7, 9.6, 7.8, 9.1, 6.4, 4.2, 9.1, and 5.6. The number of years an employee has been with this firm is normally distributed.
(a) Calculate an 80 percent confidence interval for the mean number of years that all the company's employees have been with the company.
(b) Calculate a 90 percent confidence interval for this mean.

**7.13** An oil company wants to determine the mean weight of a can of its oil. It takes a random sample of 80 such cans (from several thousand cans in its warehouse), and finds the mean weight is 31.15 ounces and the standard deviation is 0.08 ounces.

(a) Compute a 95 percent confidence interval for the mean weight of the cans in the firm's warehouse.

(b) Is your answer to (a) based on the assumption that the weights of the cans of oil in the warehouse are normally distributed? Why, or why not?

**7.14** A firm with 50 overseas plants chooses (without replacement) a random sample of 40 plants. For each plant in this sample the firm determines the number of days the plant was shut down in 1987 by labor disputes. The sample mean turns out to be 9.8 days.

(a) If the standard deviation of the number of days the firm's overseas plants were shut down by labor disputes in 1987 was 2, calculate a 90 percent confidence interval for the mean number of days that all the firm's overseas plants were shut down for this reason in 1987.

(b) If the firm had 100 rather than 50 overseas plants, calculate a 90 percent confidence interval for the mean number of days that all the firm's overseas plants were shut down by labor disputes in 1987. Explain the difference between your answer here and in (a).

**7.15** (a) The Bona Fide Washing Machine Company knows that the standard deviation of the lengths of life of motors received from supplier I is 400 hours. Calculate the 95 percent confidence interval for the mean length of life of the motors received from this supplier, based on a sample of 40 motors where the mean length of life is 4,500 hours.

(b) The Bona Fide Washing Machine Company's statistician says that the 90 percent confidence interval for the mean length of life of motors received from supplier II is 4,500 to 4,800 hours, based on a sample of 36 motors. The statistician also says that the standard deviation of the lengths of life of motors received from supplier II is 500 hours. Is there any contradiction between these statements? If so, what is the contradiction?

(c) The statistician in (b) says that if the standard deviation of the lengths of life of motors received from supplier II is known, the width of the 95 percent confidence interval for the mean is always about 20 percent greater than the width of the 90 percent confidence interval (if the sample size is held constant). Is this true? Why, or why not?

**7.16** The Mercer Company does not know the standard deviation of the lengths of life of a particular component. It therefore chooses a random sample of 36 of these components and obtains the following lengths of life (in thousands of hours):

| | | | | | |
|---|---|---|---|---|---|
| 4.2 | 5.0 | 4.6 | 4.9 | 5.0 | 5.1 |
| 4.3 | 4.9 | 4.5 | 4.8 | 4.9 | 4.6 |
| 4.4 | 5.1 | 4.7 | 4.4 | 4.8 | 4.6 |
| 4.8 | 4.7 | 4.4 | 4.5 | 4.8 | 4.8 |
| 4.9 | 4.8 | 4.3 | 4.6 | 4.7 | 4.5 |
| 5.1 | 4.8 | 4.6 | 4.6 | 4.7 | 5.0 |

(a) Compute a 90 percent confidence interval for the mean length of life of these components.

(b) Compute a 95 percent confidence interval for the mean length of life of these components.

(c) Since Mercer also does not know the standard deviation of the lengths of life of motors received from a particular supplier, it chooses a random sample of 9 motors from this supplier and determines the life of each. The results (in thousands of hours) are as follows:

| | | |
|-----|-----|-----|
| 4.3 | 4.6 | 3.8 |
| 4.2 | 4.3 | 3.9 |
| 4.1 | 3.9 | 4.0 |

Compute a 90 percent confidence interval for the mean length of life of motors received from this supplier.

(d) Compute a 95 percent confidence interval for the mean length of life of motors received from this supplier.

(e) What major assumption underlies your calculations in (c) and (d)?

**7.17** For sufficiently large samples it can be shown that the sample standard deviation $s$ is approximately normally distributed, with a mean equal to the population standard deviation $\sigma$, and with a standard deviation equal to $\sigma \div \sqrt{2n}$. Use these results to show that if $n$ is sufficiently large *a confidence interval for $\sigma$ is*

$$\frac{s}{1 + \frac{z_{\alpha/2}}{\sqrt{2n}}} < \sigma < \frac{s}{1 - \frac{z_{\alpha/2}}{\sqrt{2n}}}$$

where the confidence coefficient equals $(1 - \alpha)$. (In Chapter 9 we take up the $\chi^2$ distribution, which can be used to construct a confidence interval for $\sigma$ when $n$ is small.)

## 7.6 Confidence Intervals for the Population Proportion

Statisticians find it important to estimate the population proportion as well as the population mean. We have seen that the purpose of a statistical investigation may be to estimate the proportion of people in a particular city who prefer brand C breakfast food to brand D. Or a political pollster may want to estimate the percentage of voters who say they will vote for a particular candidate in the next election. In this section we describe how confidence intervals are constructed for population proportions. We will begin with the case where the normal distribution can be used and then take up the case where special graphs are required.

### USE OF THE NORMAL DISTRIBUTION

The sample proportion, $p$, equals $X/n$, where $X$ is the number of successes in the sample and $n$ is the sample size. A success occurs when a person prefers brand C breakfast food to brand D, or when a voter says he or she will vote for the candidate in the next election. As we know from Chapter 4, the expected value of $X$, the number of successes, is $n\Pi$, where $\Pi$ is the population proportion. Since the sample proportion is $X$ divided by $n$, its expected value must equal the expected value of $X$ divided by $n$. (For a proof, see Appendix 4.1.)[10] Consequently,

$$E(p) = \Pi.$$

In other words, the expected value of the sample proportion equals the population proportion. For example, if 70 percent of the population prefers brand C breakfast food to brand D, the expected value of the sample proportion equals 0.70.

The standard deviation of the sampling distribution of the sample proportion equals

$$\sigma(p) = \sqrt{\frac{\Pi(1 - \Pi)}{n}}.$$

To see why, recall from Chapter 4 that the standard deviation of $X$ equals $\sqrt{n\Pi(1 - \Pi)}$. Since $p$ equals $X/n$, the standard deviation of $p$ equals the standard deviation of $X$ divided by $n$—that is, $\sqrt{n\Pi(1 - \Pi)}$ divided by $n$ or $\sqrt{\Pi(1 - \Pi)}/n$. (This follows from Appendix 4.1). For example, if a random sample of 10 people is taken (with replacement) and if $\Pi = .70$, the standard deviation of the sample proportion saying that they will vote for the candidate in the next election equals $\sqrt{.7(.3)/10} = .145$.

Thus, the sampling distribution of the sample proportion has a mean equal to $\Pi$ and a standard deviation equal to $\sqrt{\Pi(1 - \Pi)/n}$. We know from previous chapters that if the sample size is sufficiently large (and if $\Pi$ is not very close to zero or 1) the sampling distribution can be approximated by the normal distribution. Thus, under these conditions, the sample proportion is approximately normally distributed with a mean of $\Pi$ and a standard deviation of $\sqrt{\Pi(1 - \Pi)/n}$, which means that

$$Pr\left\{\Pi - z_{\alpha/2}\sqrt{\frac{\Pi(1 - \Pi)}{n}} < p < \Pi + z_{\alpha/2}\sqrt{\frac{\Pi(1 - \Pi)}{n}}\right\} = 1 - \alpha,$$

where $z_{\alpha/2}$ is the value of the standard normal variable that is exceeded with the probability of $\alpha/2$. If we rearrange terms within the brackets on the left side of this equation, it follows that[11]

$$Pr\left\{p - z_{\alpha/2}\sqrt{\frac{\Pi(1 - \Pi)}{n}} < \Pi < p + z_{\alpha/2}\sqrt{\frac{\Pi(1 - \Pi)}{n}}\right\} = 1 - \alpha.$$

---

[10] In Appendix 4.1 we showed that if a random variable is a linear function of another random variable, its expected value is a linear function of the expected value of the other random variable. The sample proportion is a linear function of $X$; specifically, it equals $(1/n)X$. Thus, its expected value equals $1/nE(X) = 1/n(n\Pi) = \Pi$.

[11] The reasoning here is like that leading up to equation (7.2). Specifically, if we subtract $\Pi$ from all three terms inside the brackets,

$$Pr\left\{-z_{\alpha/2}\sqrt{\frac{\Pi(1 - \Pi)}{n}} < p - \Pi < z_{\alpha/2}\sqrt{\frac{\Pi(1 - \Pi)}{n}}\right\} = 1 - \alpha.$$

As it stands, this equation cannot be used to construct a confidence interval for $\Pi$ because without a knowledge of $\Pi$ we cannot compute $\sqrt{\Pi(1 - \Pi)/n}$. However, if the sample is sufficiently large it is permissible to substitute the sample proportion for $\Pi$ in this expression, the result being

$$Pr\left\{ p - z_{\alpha/2} \sqrt{\frac{p(1 - p)}{n}} < \Pi < p + z_{\alpha/2} \sqrt{\frac{p(1 - p)}{n}} \right\} = 1 - \alpha \quad (7.7)$$

Based on equation (7.7), it is clear that *if the sample is sufficiently large and if the confidence coefficient is set equal to* $(1 - \alpha)$ *the confidence interval for the population proportion is*

*Confidence Interval for $\Pi$ (If n is Large)*

$$p \pm z_{\alpha/2} \sqrt{\frac{p(1 - p)}{n}} \qquad\qquad (7.8)$$

Of course, this expression assumes that the population is large relative to the sample or that sampling is carried out with replacement; otherwise $\sqrt{p(1 - p)/n}$ must be multiplied by $\sqrt{(N - n)/(N - 1)}$. This result is of widespread usefulness in business and economic statistics. The following example illustrates how it is applied.

---

EXAMPLE 7.3 A polling organization selects a random sample of 400 metalworking firms in Texas and asks the president of each firm whether it is using robots. Sixty percent of the firms in the sample are using robots. Calculate a 95 percent confidence interval for the percentage of Texas metalworking firms using robots.

SOLUTION: Since $(1 - \alpha) = .95$, $z_{\alpha/2} = z_{.025} = 1.96$. Inserting .60 for $p$ and 400 for $n$ in expression (7.8), we have

$$.60 \pm 1.96 \sqrt{\frac{.60(.40)}{400}}.$$

Subtracting $p$ from all three terms inside the brackets, we get

$$Pr\left\{ -p - z_{\alpha/2} \sqrt{\frac{\Pi(1 - \Pi)}{n}} < -\Pi < -p + z_{\alpha/2} \sqrt{\frac{\Pi(1 - \Pi)}{n}} \right\} = 1 - \alpha.$$

Multiplying all three terms inside the brackets by $-1$, and reversing the inequalities, we get

$$Pr\left\{ p - z_{\alpha/2} \sqrt{\frac{\Pi(1 - \Pi)}{n}} < \Pi < p + z_{\alpha/2} \sqrt{\frac{\Pi(1 - \Pi)}{n}} \right\} = 1 - \alpha,$$

which is what we set out to derive.

Simplifying terms, this confidence interval is

.60 ± .048,

which means that our interval estimate of the percentage of Texas metalworking firms using robots is 55.2 percent to 64.8 percent.

## USE OF SPECIAL GRAPHS

Expression (7.8) provides a confidence interval for the population proportion when the sample size is sufficiently large. But how large is "sufficiently large"? According to William Cochran,[12] the minimum sample size needed to insure the validity of this expression is about 30 if the population proportion is about .50; about 50 if the population proportion is about .40 or .60; about 80 if it is about .30 or .70; about 200 if it is about .20 or .80; and about 600 if it is about .10 or .90. If the sample size does not meet these standards one must then use special graphs that have been computed for this purpose. Appendix Tables 7a and 7b respectively contain 95 percent confidence intervals and 99 percent confidence intervals for the population proportion. These graphs are used in the following way.

*If the sample proportion is less than 50 percent,* one finds the point on the bottom horizontal axis that equals the sample proportion. For example, if the sample proportion is 45 percent, the appropriate point on the horizontal axis is .45. Then, using the two curves that pertain to the relevant sample size, one finds the two points on the vertical axis that correspond to this point on the horizontal axis. For example, if $n = 100$, the two curves are as shown in Figure 7.6 for a 95 percent confidence interval. If the sample proportion is 0.45, the two points on the vertical axis corresponding to 0.45 on the horizontal axis are .35 and .55, as shown in Figure 7.6. Thus, the 95 percent confidence interval is 35 percent to 55 percent.

*If the sample proportion is greater than 50 percent,* one uses the top (rather than the bottom) horizontal scale and the right-hand (rather than the left-hand) vertical scale in Appendix Tables 7a and 7b. For example, suppose that the sample proportion is 0.65 and $n = 100$. As shown in Figure 7.7, we begin by finding .65 on the *top* horizontal scale, after which we find the two points on the *right-hand* vertical scale corresponding to this location on the two curves. Since these two points are 55 and 74 percent, the 95 percent confidence interval is 55 percent to 74 percent.

These graphs are of considerable use to statisticians. The following is a further illustration of how they are employed.

---

[12] See W. G. Cochran, *Sampling Techniques* (New York: Wiley, 1953), p. 41. As Cochran points out, smaller sample sizes than those given above may be acceptable if the statistician is willing to accept somewhat larger risks of error than is Cochran. As pointed out in Chapter 5, many statisticians regard the normal approximation as acceptable so long as $n\Pi > 5$ *when* $\Pi \leq 1/2$ *and* $n(1 - \Pi) > 5$ *when* $\Pi > 1/2$.

**Figure 7.6**
Derivation of a
Confidence Interval
for the Population
Proportion, when the
Sample Proportion Is
0.45

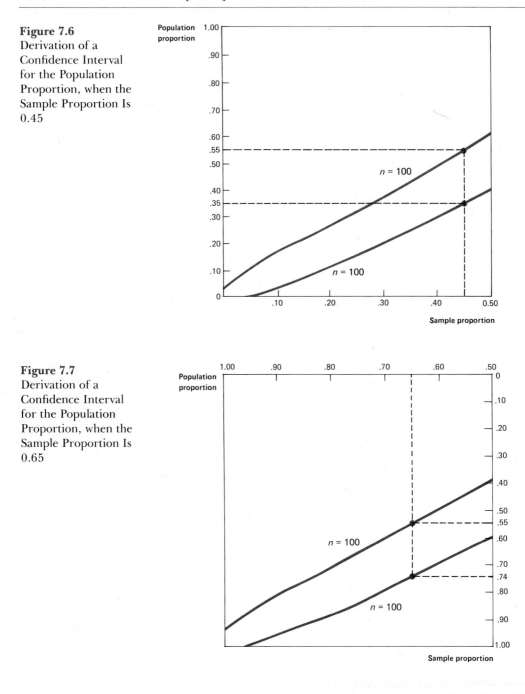

**Figure 7.7**
Derivation of a
Confidence Interval
for the Population
Proportion, when the
Sample Proportion Is
0.65

EXAMPLE 7.4 An advertising agency shows two television commercials
to a sample of 60 people. Seventy percent of these individuals
prefer the first commercial over the second. Obtain a 95 percent
confidence interval for the proportion of all members of the relevant population who feel this way.

SOLUTION: Since the sample proportion exceeds 50 percent,
we must use the top horizontal and right-hand vertical scales in

Appendix Table 7a. First, we find .70 (the sample proportion) on the top horizontal scale. Then we must find the points on the two curves (for $n = 60$) that correspond to this horizontal location. Since these two points correspond to 57 percent and 81 percent on the right-hand vertical axis, the desired confidence interval is 57 percent to 81 percent.

## 7.7 Confidence Intervals for the Difference Between Two Means or Two Proportions[13]

Frequently, the purpose of a statistical investigation is to estimate the difference between two population means or two population proportions. For example, suppose that a firm buys thread from two suppliers and wants to estimate the difference between the mean strength of the thread of one supplier and that of the other. In this section we will show how a confidence interval can be constructed for the difference between two population means. We will also show how a confidence interval can be constructed for the difference between two population proportions.

### DIFFERENCE BETWEEN TWO MEANS: INDEPENDENT SAMPLES

Let's take two populations, one with a mean of $\mu_1$, and a standard deviation of $\sigma_1$, the other with a mean of $\mu_2$ and a standard deviation of $\sigma_2$. A simple random sample of $n_1$ observations is chosen from the first population and a simple random sample of $n_2$ observations is chosen from the second. These two random samples are entirely *independent;* in particular, the observations in one sample are not paired in any way with those in the other sample.[14] *If both samples are large, and if the confidence coefficient is set equal to* $(1 - \alpha)$, *the confidence interval for the difference between the population means is*

$$\bar{x}_1 - \bar{x}_2 \pm z_{\alpha/2} \sqrt{\frac{s_1^2}{n_1} + \frac{s_2^2}{n_2}},$$

(7.9) *Confidence Interval for* $\mu_1 - \mu_2$

*where* $s_1^2$ *is the variance of the sample taken from the first population and* $s_2^2$ *is the variance of the sample taken from the second population.*

[13] Some instructors may want to skip this section. If so, this can be done without loss of continuity. Note that the methods discussed in this selection are by no means the only ones that can be applied to obtain a confidence interval for the difference between two means or two proportions.

[14] If they are paired, the results of this paragraph are not applicable. For a description of how a confidence interval can be obtained from paired comparisons, see Section 7.9. Also, it is assumed here and in expression (7.10) that the population is large relative to the sample or that sampling is with replacement.

The following example shows how this result can be used under practical circumstances.

EXAMPLE 7.5 A pea-canning company has two plants and wants to estimate the difference between the mean drained weight of the contents of the cans filled at the two plants. It is suspected that the mean drained weight is higher at plant 1 than at plant 2 because of the gradual deterioration of some of the equipment at plant 1. A random sample of 100 cans is drawn at each plant, and the sample mean is found to be 23.02 ounces at plant 1 and 22.83 ounces at plant 2. The sample variance is .64 (ounces)$^2$ at plant 1 and .36 (ounces)$^2$ at plant 2. Construct a 90 percent confidence interval for the difference between the mean drained weight of cans filled at plant 1 and the mean drained weight of cans filled at plant 2.

SOLUTION: Since $(1 - \alpha) = .90$, $z_{\alpha/2} = z_{.05} = 1.64$. Inserting 23.02 for $\bar{x}_1$, 22.83 for $\bar{x}_2$, .64 for $s_1^2$, .36 for $s_2^2$, and 100 for $n_1$ and $n_2$, expression (7.9) becomes the following:

$$.19 \pm 1.64(.1).$$

Thus, the desired confidence interval for the difference between the mean drained weight of cans filled at plant 1 and the mean drained weight of cans filled at plant 2 is .026 to .354 ounces.

### DIFFERENCE BETWEEN TWO PROPORTIONS: INDEPENDENT SAMPLES

Suppose that there are two populations, one where the proportion having a certain characteristic is $\Pi_1$, the other where the proportion with this characteristic is $\Pi_2$. A simple random sample of $n_1$ observations is chosen from the first population and a simple random sample of $n_2$ observations is chosen from the second. These two random samples are entirely *independent;* in particular, the observations in one sample are not paired in any way with those in the other. *If both samples are sufficiently large, and if the confidence coefficient is set at $(1 - \alpha)$, the confidence interval for the difference between the population proportions is*

***Confidence Interval for $\Pi_1 - \Pi_2$***

$$p_1 - p_2 \pm z_{\alpha/2}s_{p_1 - p_2}, \qquad (7.10)$$

*where $p_1$ is the sample proportion in the first population, $p_2$ is the sample proportion in the second population, and* $s_{p_1 - p_2} =$
$$\sqrt{\frac{p_1(1 - p_1)}{n_1} + \frac{p_2(1 - p_2)}{n_2}}.$$
The following example shows how this result can be used.

EXAMPLE 7.6 A polling organization wants to estimate the difference between the proportion of Republicans favoring the abolition of double taxation of dividends and the proportion of Democrats favoring this measure. A simple random sample of 400 Republicans is drawn, and it is found that 80 percent favor the measure. A random sample of 400 Democrats is also drawn, 40 percent of whom favor this measure. Construct a 95 percent confidence interval for the difference between the proportion of Republicans and the proportion of Democrats favoring the tax-abolition proposal.

SOLUTION: Since $(1 - \alpha) = .95$, $z_{\alpha/2} = z_{.025} = 1.96$. If we substitute .80 for $p_1$, .40 for $p_2$, and 400 for $n_1$ and $n_2$ in expression (7.10), we get

$$.40 \pm \frac{1.96}{20} \sqrt{.40}.$$

Thus, the desired confidence interval for the difference between the percentage of Republicans and percentage of Democrats favoring the tax-abolition measure is 33.8 to 46.2 percentage points.

## EXERCISES

**7.18** A trucking firm has a large inventory of spare parts. In order to estimate the proportion of these parts that have deteriorated to the point of being no longer usable, the firm draws a random sample of 24 of the parts and finds that 25 percent are no longer usable. Construct a 95 percent confidence interval for the nonusable proportion of the entire inventory.

**7.19** A market-research firm plans to estimate the difference between the mean rating of two television programs. It asks 100 viewers to rate the first program on a scale of 0 to 10; the sample mean is 5.3 and the sample variance is 1.6. Another sample of 400 viewers are asked to rate the second program on the same scale; the sample mean is 5.8 and the sample variance is 1.8. Calculate a 90 percent confidence interval for the difference between the mean ratings of the two programs in the population as a whole.

**7.20** An airline wants to determine the proportion of passengers on its New York-Chicago flights who carry only hand luggage. The airline picks a random sample of 40 passengers traveling on these flights and finds that 14 percent carry only hand luggage. Calculate a 95 percent confidence interval for the proportion of hand-luggage passengers on these flights.

**7.21** A U.S. steel firm wants to determine the proportion of its customers who bought Japanese steel during the past year. It selects a random sample of 100 of its 3,500 customers, and finds that 36 percent of them bought Japanese steel then.
(a) Calculate a 90 percent confidence interval for this proportion.
(b) Calculate a 95 percent confidence interval for this proportion.

**7.22** A Philadephia bank wants to estimate the proportion of its depositors who have safe deposit boxes at other banks. It picks a random sample of

250 of its depositors, and finds that 19 percent of them have such boxes at other banks.
(a) Calculate a 98 percent confidence interval for this proportion.
(b) Calculate a 90 percent confidence interval for this proportion.

**7.23** An Atlanta insurance agency selects a random sample of 40 people from those to whom it has sold life insurance. It finds out that 16 of them also have bought auto insurance from this agency.
(a) Determine a 95 percent confidence interval for the proportion of people to whom the agency has sold life insurance who have also bought auto insurance from it.
(b) Determine a 99 percent confidence interval of this sort.

**7.24** A San Francisco department store picks a random sample of 12 people from those who owe the store more than $50. It finds that 5 of these people are 65 years old or more.
(a) Determine a 95 percent confidence interval for the proportion of people who owe the store more than $50 who are 65 years old or more.
(b) Determine a 99 percent confidence interval of this sort.

**7.25** A New York market research firm wants to compare the average price of a pair of shoes (of a particular type) in Chicago with that in New York. It picks a random sample of 50 shoe stores in Chicago, and finds that the mean price is $56.35, the standard deviation being $3.42. It picks a random sample of 50 shoe stores in New York, and finds that the mean price is $58.15, the standard deviation being $4.13.
(a) Compute a 95 percent confidence interval for the difference between the mean price of such a pair of shoes in New York and the mean price of such a pair of shoes in Chicago.
(b) Is your answer to (a) based on the assumption that the prices in each city are normally distributed? Why or why not?

**7.26** A bank wants to determine the proportion of its depositors who also have deposits in any of the local savings and loan associations. A random sample is constructed of 100 depositors, and it is determined that 46 percent have deposits in a local savings and loan association.
(a) Use the normal approximation to determine a 95 percent confidence interval for this proportion.
(b) Use Appendix Table 7a to determine this confidence interval, and compare the result with your finding in (a).
(c) The bank finds that the sample standard deviation of the size of its depositors' deposits in local savings and loan associations is $8,200. Calculate a 95 percent confidence interval for the population standard deviation. (Use the information provided in Exercise 7.17.)

## 7.8 Determining the Size of the Sample

Earlier in this chapter we learned that in setting out to estimate a particular parameter a statistician must choose the kind of estimator to be used and the sample size. We have already discussed the various kinds of estimators, and now we take up the determination of the size of the sample. First, we will indicate how this decision can be made in estimating a population mean; then we will indicate how to pick a sample size when estimating a population proportion.

## WHERE THE POPULATION MEAN IS ESTIMATED

In many cases, the decision maker authorizing the experiment or survey wants the resulting estimate to have a specified degree of precision.[15] For example, consider the manufacturer of light bulbs (Example 7.2) which needed to estimate the mean longevity of a new type of bulb. This firm might want the probability to be .90 that the sample mean will differ from the population mean by no more than 30 hours. Given this specified degree of precision, we can determine how large the sample must be if we know the standard deviation of the life of bulbs of the new type.

To see how the sample size can be determined, recall once again that for reasonably large samples the sample mean $\overline{X}$ is distributed normally with mean equal to the population mean $\mu$ and standard deviation equal to $\sigma/\sqrt{n}$ (where $\sigma$ is the population standard deviation and $n$ is the sample size). Thus,

$$Pr\left\{\mu - 1.64\frac{\sigma}{\sqrt{n}} < \overline{X} < \mu + 1.64\frac{\sigma}{\sqrt{n}}\right\} = .90, \qquad (7.11)$$

which says that the probability that the sample mean will lie within 1.64 standard deviations of the population mean equals .90. As we know from previous discussions, this is true of any normal variable. Another way of stating the same thing is

$$Pr\left\{-1.64\frac{\sigma}{\sqrt{n}} < \overline{X} - \mu < 1.64\frac{\sigma}{\sqrt{n}}\right\} = .90.$$

In other words, the probability that the sample mean will differ from the population mean by less than $1.64\sigma/\sqrt{n}$ is .90.

If the desired precision is to be obtained, the probability that the sample mean will differ from the population mean by less than 30 hours must equal .90. This means that

$$1.64\sigma/\sqrt{n} = 30,$$

since we know from the previous paragraph that the probability that the sample mean will differ from the population mean by less than $1.64\sigma/\sqrt{n}$ equals .90. Suppose (contrary to Example 7.2) that the firm knows that $\sigma$ equals 160 hours. Then

$$\frac{1.64(160)}{\sqrt{n}} = 30$$

[15] Ideally, the degree of precision specified should reflect the basic factors described in Section 7.2: the costliness of errors in the estimate and the costliness of sampling. However, it frequently is difficult to quantify the cost of errors of a particular size; and in practice decision makers often specify degrees of precision based on intuitive and informal judgments concerning these more basic factors.

Basic Idea #14    *If you have some idea of the standard deviation of the population, you can estimate how big a sample is required so that the population mean can be estimated with a particular degree of accuracy.*

or

$$n = \left[\frac{1.64(160)}{30}\right]^2,$$

which means that $n$ must equal 77.

In general, *if it is desired that the probability be* $(1 - \alpha)$ *that the sample mean differ from the population mean by no more than some number* $\delta$ *(delta), the sample size must equal*

**Formula for n in Estimating μ**

$$n = \left(\frac{z_{\alpha/2}\sigma}{\delta}\right)^2, \qquad (7.12)$$

where $\sigma$ is the population standard deviation and $z_{\alpha/2}$ is the value of the standard normal variable which has a probability $\alpha/2$ of being exceeded.[16]

Equation (7.12) is of great practical importance. Even if you do not know the value of $\sigma$, rough estimates of its value can be inserted into (7.12) to get some idea of how large the sample must be. The following example shows how one can determine the sample size using this formula.

EXAMPLE 7.7 A bank wants to estimate the mean balance in the checking accounts of its depositors 65 years old or over. There are a very large number of such accounts, and the bank manager believes that the standard deviation of the balances held by such individuals is about $160. If the bank wants the probability to be .95 that the sample mean will differ from the population mean by no more than $20, how big a sample must be taken?

SOLUTION: Since $(1 - \alpha) = .95$, $z_{\alpha/2} = z_{.025} = 1.96$. Substituting 160 for $\sigma$ and 20 for $\delta$ in (7.12), we have

$$n = \left[\frac{(1.96)(160)}{20}\right]^2 = 246.$$

Thus, the sample size should be 246.

[16] This assumes that the sample is large and that the population is large relative to the sample (or that sampling is with replacement).

### WHERE THE POPULATION PROPORTION IS ESTIMATED

In investigations aimed at estimating a population proportion, a degree of precision is generally specified. For example, consider a polling organization that needs to estimate the proportion of Dallas residents who favor stronger federal measures to stem illegal immigration into the United States. The sponsors of this survey may decide that for the results to be useful to them the probability must be .90 that the sample percentage differs from the population percentage by no more than five percentage points. To see how this statement of desired precision can be used to determine the sample size, recall that for sufficiently large samples the sample proportion $p$ is approximately normally distributed with mean equal to the population proportion $\Pi$ and standard deviation equal to $\sqrt{\Pi(1 - \Pi)/n}$. Thus,

$$Pr\left\{- 1.64 \sqrt{\frac{\Pi(1 - \Pi)}{n}} < p - \Pi < 1.64 \sqrt{\frac{\Pi(1 - \Pi)}{n}}\right\} = .90,$$

which says that the probability that the sample proportion will differ from the population proportion by no more than 1.64 standard deviations equals .90. This, of course, is true of all normal variables.

If the desired precision is to be obtained, the probability that the sample proportion will differ from the population proportion by .05 (that is, five percentage points) must be .90. This means that

$$1.64 \sqrt{\frac{\Pi(1 - \Pi)}{n}} = .05,$$

since we know from the previous paragraph that the probability that the sample proportion will differ from the population proportion by less than $1.64\sqrt{\Pi(1 - \Pi)/n}$ equals .90. Although the polling organization does not know the value of $\Pi$, it is likely to have some idea of its approximate value. For example, suppose that $\Pi$ is believed to be in the neighborhood of 0.5. Then

$$1.64 \sqrt{\frac{(.5)(.5)}{n}} = .05,$$

or

$$n = \frac{(1.64)^2}{(.05)^2} (.5)(.5),$$

which means that $n$ must equal about 269.

In general, *if it is desired that the probability be* $(1 - \alpha)$ *that the sample proportion differs from the population proportion by no more than some number* $\delta$, *and if the population proportion is believed to be approximately equal to* $\hat{\Pi}$, *the sample size must equal*

$$n = \left(\frac{z_{\alpha/2}}{\delta}\right)^2 \hat{\Pi}(1 - \hat{\Pi}).$$

(7.13)

*Formula for n in Estimating $\Pi$*

This result assumes that the sample is large enough for the normal approximation to be used and that the population is large relative to the sample (or that sampling is with replacement). It is worth noting that if one wants a conservative estimate of $n$ (that is, an estimate that tends to err on the high side), it is best to shade one's estimate of $\Pi$ in the direction of 0.5. Why? Because the right side of equation (7.13) gets larger as $\Pi$ approaches 0.5. Thus, if $\Pi$ is shaded toward 0.5, the estimate of $n$ will tend to err on the high side.

Equation (7.13), like equation (7.12), is of great practical importance. The following example illustrates how this formula is used.

**TO ERR IS HUMAN ... BUT COSTLY**

SAMPLING TECHNIQUES IN RETAILING

The Merton Company uses sampling techniques to check on the amount of a particular product that it has on hand. The company has 1,000 stores located in various parts of the nation. At the beginning of each year, the company's controller asks that a random sample of 100 of these stores be asked to determine the number of boxes of this product that each store has on hand. To estimate the total number of boxes of this product that the company has on hand, the controller computes the mean number of boxes of the product per store in the sample and multiplies the result by 1,000. He wants the probability to be 0.90 that his estimate (of the total number of boxes of the product on hand in all the firm's stores) will be in error by no more than 500 boxes. From past experience, he knows that the standard deviation of the number of boxes of this product on hand in a store is about 2. The cost of determining how many boxes of this product are on hand in a store is about $25. When the company statistician hears of what the controller is doing, she says that the controller is wasting money. Is this true? If so, how?

SOLUTION: The controller wants the probability to be 0.90 that the mean number of boxes of this product per store in the sample will differ from the true mean by 500/1000, or 0.5. For this to be true, the sample size must be

$$\left(\frac{1.64(2)}{0.5}\right)^2 = 43,$$

according to equation (7.12). Thus, the sample size is much too large. By reducing the sample size from 100 to 43 stores, the controller could save 57 times $25, or $1,425, without having the precision fall below the desired level.

Going a step further, it is possible that the Merton Company really does not need the probability to be 0.90 that this estimate be in error by no more than 500 boxes. Perhaps a lower probability (such as 0.75 or 0.60) will do. There are costs as well as benefits in obtaining greater accuracy. If a lower probability will do, the sample size can be reduced further.

---

EXAMPLE 7.8 A government agency plans to estimate the percentage of welfare recipients in a particular area who are over 60 years of age. A reasonable estimate is that this percentage is about 30. The agency wants the probability to be .99 that the sample percentage differs from the population percentage by less than 5 percentage points. How large a sample should the agency take?

SOLUTION: Since $(1 - \alpha) = .99$, $z_{\alpha/2} = z_{.005} = 2.58$. Substituting .30 for $\Pi$ and .05 for $\delta$ in equation (7.13), we have

$$n = \left(\frac{2.58}{.05}\right)^2 (.30)(.70)$$

$$= 2662.56(.21)$$

$$= 559.$$

Thus, the sample size should be about 559.

---

## 7.9 Statistical Estimation in the Chemical Industry: A Case Study

In Chapter 1 we pointed out that the huge British chemical firm Imperial Chemical Industries (ICI) carried out the following experiment to estimate the effect of a chlorinating agent on the abrasion resistance of a certain type of rubber.[17] Ten pieces of this type of rubber were cut in half, and one half-piece was treated with the chlorinating agent, while the other half-piece was untreated. Then the abrasion resistance of each half was evaluated on a machine, and the difference between the abrasion resistance of the treated half-piece and the untreated half-piece was computed. Table 7.1 shows the 10 differences (1 corresponding to each of the pieces of rubber in the sample). Based on this experiment, ICI was interested in estimating the mean difference between the abrasion resistance of a treated and untreated half-piece of this type of rubber. In other words, if this experiment were performed again and again, an infinite population of such differences would result. ICI was interested in estimating the mean of this population, since the mean is a good measure of the effect of the chlorinating agent on this type of rubber's abrasion resistance.

[17] O. Davies, *op cit.*, p. 13.

**Table 7.1**
Differences in Abrasion
Resistance (Treated
Material Minus
Untreated Material),
10 Pieces of Rubber

| Piece | Difference |
|---|---|
| 1 | 2.6 |
| 2 | 3.1 |
| 3 | −0.2 |
| 4 | 1.7 |
| 5 | 0.6 |
| 6 | 1.2 |
| 7 | 2.2 |
| 8 | 1.1 |
| 9 | −0.2 |
| 10 | 0.6 |

$$\sum_{i=1}^{10} x_i = 12.7$$

$$\bar{x} = 1.27$$

$$s = 1.1265$$

*Source:* Owen L. Davies, *The Design and Analysis of Industrial Experiments* (London: Oliver and Boyd, 1956), p. 13.

If you were a statistical consultant for ICI, how would you analyze these data? Recalling the material in Section 7.4, you would recognize that a good point estimate of the mean of this population is the sample mean, which Table 7.1 shows to be 1.27. Thus, your first step would be to advise ICI that if they want a single number as an estimate, 1.27 is a good number to use. Next, mindful of one of the central points of this chapter, you would point out that such a point estimate contains no indication of how much error it may contain, whereas a confidence interval does contain such information. Since the population standard deviation is unknown and the sample is small, expression (7.6) should be used in this case to calculate a confidence interval. Assuming that the firm wants a confidence coefficient of 95 percent, the confidence interval is 0.464 to 2.076, because $t_{.025} = 2.262$, $s = 1.1265$, and $n = 10$. The chances are 95 out of 100 that such a confidence interval would include the population mean.[18]

The above analysis is, in fact, exactly how ICI's statisticians proceeded. Despite the fact that the sample consisted of only 10 observations, the evidence was very strong that the chlorinating agent had a positive effect on abrasion resistance. After all, the 95 percent confidence interval was that the mean difference between abrasion resistance of rubber with and without treatment was an increase of between 0.464 and 2.076. (For that matter, the statisticians found that the 98 percent confidence interval was that the mean difference was an increase of between 0.265 and 2.275.) The best estimate was that the

[18] Note that this analysis assumes that the population is normally or approximately normally distributed.

chlorinating agent resulted in an increase of about 1.27 in abrasion resistance.

In conclusion, note that it would have been incorrect to have viewed the abrasion resistance of the 10 treated half-pieces as one sample and the abrasion resistance of the 10 untreated half-pieces as another, and to have used expression (7.9) to obtain a confidence interval for the difference between the mean abrasion resistance of treated half-pieces and the mean abrasion resistance of untreated half-pieces. This would have been incorrect because the two samples are *paired* or *matched;* that is, each half-piece in one sample has a mate (the other half of the piece it comes from) in the other sample. Thus, the observations in one sample are not independent of those in the other sample, because if one half-piece is relatively resistant to abrasion due to chance variation in its production or other factors, its mate in the other sample is likely to be resistant as well. Because the samples are not independent, expression (7.9) is not appropriate, since it assumes independence.[19] Instead, the proper technique is to consider the difference between each observation in one sample and its mate in the other sample as a single observation, as we did in Table 7.1.

*Matched Pairs*

## EXERCISES

**7.27** A magazine for business executives plans to determine the mean annual income of its subscribers. For the results to be useful, it is believed that the probability that the sample mean will differ by less than $5,000 from the population mean should equal .95. A rough estimate of the standard deviation of the annual income of the magazine's subscribers is $15,000.
(a) How large a sample should be drawn?
(b) What sorts of biases may be present in such a survey?
(c) How can bias be present when the sample mean is an unbiased estimator?

**7.28** A local government plans to estimate the percentage of vacant buildings in a particular area of several square miles. A reasonable guess is that about 20 percent of the buildings in this area are vacant. The government wants the probability to be .90 that the sample percentage differs from the population percentage by no more than 2 percentage points. How large should the sample of buildings be?

**7.29** The president of the Bona Fide Washing Machine Company tells the firm's statistician that it is important he be able to estimate the mean longevity of motors received from supplier I with a .95 probability of an error of no more than 20 hours. Based on the information in Exercise 7.15, how large a sample must the firm take of the motors received from supplier I to be certain this is true?

**7.30** A spokesman for a hi-fi repair shop claims that in about 40 percent of the cases where a hi-fi set is repaired the customer is undercharged due to clerical errors. A public-interest law firm decides to estimate the proportion of cases where undercharging occurs. The law firm plans to con-

---

[19] Also, the formula in expression (7.9) is not appropriate because it assumes that the sample sizes are large.

struct a random sample so that the probability of the sample proportion's being in error by more than .01 is .05. How large should the sample be?

**7.31** Suppose that in the experiment described in Section 7.9 the chlorinated half-pieces of rubber were evaluated on a different machine than the untreated half-pieces. Would this make the results of the experiment more difficult to interpret? If so, how?

**7.32** A clothing store wants to estimate the percentage of its customers that have seen a certain magazine ad. It intends to take a simple random sample of its customers to estimate this percentage. It wants the probability to be 0.95 that this estimate differs by no more than 1 percentage point from the true percentage. The store thinks that about 30 percent of its customers may have seen the ad. How big a sample should be taken by the store?

**7.33** An airline wants to estimate the percentage of people traveling on its flights between Boston and Miami whose tickets are bought more than a week before their flight. It believes that this percentage is about 40. It is about to select a random sample of people traveling on its flights between Boston and Miami. If it wants the probability to be 0.01 that the sample percentage will be in error by more than 2 percentage points, how big a sample should it select?

## Chapter Review

1. Estimates are of two types: *point estimates* and *interval estimates.* A statistic used to estimate a population parameter is called an *estimator.* In choosing among estimators, statisticians consider the following three criteria: lack of bias, efficiency, and consistency. An *unbiased estimator* is one whose expected value equals the parameter being estimated. If two estimators are unbiased, one is more efficient than the other if its variance is less than the variance of the other. A statistic is a consistent estimator of a parameter if the probability that the statistic's value is very near that of the parameter approaches 1 as the sample size increases. Based on these criteria, the sample mean, sample proportion, and sample standard deviation are judged to be very good estimators of the population mean, population proportion, and population standard deviation, respectively.

2. If the population standard deviation $\sigma$ is known, the large-sample confidence interval for the population mean $\mu$ is

$$\bar{x} \pm z_{\alpha/2} \frac{\sigma}{\sqrt{n}},$$

where $z_{\alpha/2}$ is the value of the standard normal variable that is exceeded with a probability of $\alpha/2$. Before the sample is drawn, there is a probability of $(1 - \alpha)$ that this interval will include $\mu$. Thus, the confidence coefficient is said to be $(1 - \alpha)$. If $\sigma$ is unknown, the sample standard deviation $s$ can be substituted for $\sigma$ in this expression if the sample size $n$ is at least 30.

3. When the population standard deviation is unknown and the sample size is less than 30, the $t$ distribution can be used to formulate a confidence interval for the population mean, providing the population is at least approximately normal. The *t distribution* is a family of distributions, each of which corresponds to a certain number of degrees of freedom. As the number of degrees of freedom increases, the $t$ distribution moves increasingly closer to the standard normal distribution. In this context, the number of degrees of freedom equals $(n-1)$. If the confidence coefficient is set equal to $(1-\alpha)$, the confidence interval for the population mean is

$$\bar{x} \pm t_{\alpha/2} \frac{s}{\sqrt{n}},$$

where $t_{\alpha/2}$ is the value of a $t$ variable (with $[n-1]$ degrees of freedom) that is exceeded with a probability of $\alpha/2$.

4. If the sample size is sufficiently large, and if the confidence coefficient is set equal to $(1-\alpha)$, the confidence interval for the population proportion $\Pi$ is

$$p \pm z_{\alpha/2} \sqrt{\frac{p(1-p)}{n}},$$

where $p$ is the sample proportion. Otherwise, one must use special graphs (in Appendix Tables 7a and 7b) to obtain a confidence interval for the population proportion.

5. If independent samples of sizes $n_1$ and $n_2$ are chosen from two populations, we can calculate a confidence interval for the difference between the population means $\mu_1 - \mu_2$. The following formula is applicable if both $n_1$ and $n_2$ are large:

$$\bar{x}_1 - \bar{x}_2 \pm z_{\alpha/2} \sqrt{\frac{s_1^2}{n_1} + \frac{s_2^2}{n_2}},$$

where the confidence coefficient equals $(1-\alpha)$. Also, we can calculate a confidence interval for the difference between the population proportions, $\Pi_1 - \Pi_2$. The following formula is applicable if both $n_1$ and $n_2$ are sufficiently large:

$$p_1 - p_2 \pm z_{\alpha/2} s_{p_1 - p_2},$$

where $s_{p_1 - p_2} = \sqrt{\dfrac{p_1(1-p_1)}{n_1} + \dfrac{p_2(1-p_2)}{n_2}}$.

Neither of these formulas is correct if the two samples are matched, since this violates the assumption that the samples are independent.

6. If it is desired that the probability be $(1-\alpha)$ that the sample mean differs from the population mean by no more than a given

number $\delta$, the sample size must equal

$$n = \left(\frac{z_{\alpha/2}\sigma}{\delta}\right)^2.$$

If it is desired that the probability be $(1 - \alpha)$ that the sample proportion differs from the population proportion by no more than some number $\delta$, the sample size must equal

$$n = \left(\frac{z_{\alpha/2}}{\delta}\right)^2 \hat{\Pi}(1 - \hat{\Pi}),$$

where $\hat{\Pi}$ is an estimate of $\Pi$. These results assume that simple random sampling is used, that the sample is large enough so that the normal distribution can be used, and that the population is large relative to the sample (or that sampling is with replacement). The last assumption is made throughout this Chapter Review.

## Chapter Review Exercises

**7.34** A telephone company wants to determine the mean height of its telephone poles. It takes a random sample of 12 poles, and finds that the heights (in feet) are: 35.9, 35.8, 36.2, 36.4, 36.6, 36.2, 35.8, 35.9, 36.0, 36.1, 35.9, 36.0.
   (a) Compute a 90 percent confidence interval for the mean height of the firm's telephone poles.
   (b) Is your answer to (a) based on the assumption that the heights of the company's telephone poles are normally distributed? Why, or why not?

**7.35** There are 60 apartments in a San Diego apartment building. The owner of the building wants to estimate the mean number of people living in an apartment. He draws a random sample (without replacement) of 36 apartments in the building. The number of people living in each apartment is as follows:

| 1 | 2 | 1 | 1 | 2 | 1 | 2 | 3 | 1 |
|---|---|---|---|---|---|---|---|---|
| 2 | 2 | 1 | 2 | 2 | 1 | 1 | 2 | 3 |
| 2 | 3 | 1 | 2 | 3 | 2 | 2 | 3 | 2 |
| 3 | 2 | 2 | 3 | 1 | 1 | 2 | 2 | 1 |

   (a) Compute the desired 95 percent confidence interval.
   (b) Is your answer to (a) based on the assumption that the numbers of people in the apartments are normally distributed? Why, or why not?

**7.36** A school board is responsible for two elementary schools. It wants to determine how the mean IQ of students at school A compares with the mean IQ of those at school B. It chooses a random sample of 90 students from each school. At school A, the sample mean IQ is 109 and the sample standard deviation is 11. At school B, the sample mean IQ is 98 and the standard deviation is 9.

(a) Compute a 90 percent confidence interval for the difference between the mean IQ at school A and mean IQ at school B.

(b) Compute a 99 percent confidence interval of this sort.

**7.37** A U.S. senator from Michigan believes that the percentage of voters favoring a particular proposal is higher in Detroit than in other parts of the state. He picks a random sample of 200 Detroit voters and finds that 59 percent favor the proposal. He picks a random sample of 200 Michigan voters from outside Detroit and finds that 52 percent favor the proposal.

(a) Construct a 90 percent confidence interval for the difference between the proportion in Detroit favoring the proposal and the proportion outside Detroit favoring it.

(b) Construct a 99 percent confidence interval of this sort.

**7.38** There are two methods of reroofing a house, method A and method B. A research organization chooses a random sample of 200 houses re-roofed according to method A and finds that 18 percent experienced leaks within four years. Another random sample of 200 houses re-roofed according to method B is chosen. It is found that 29 percent of these houses experienced leaks within four years. Compute a 95 percent confidence interval for the difference between method B and method A in the percentage of houses experiencing leaks within four years.

**7.39** A public-interest law firm picks a random sample of 100 hi-fi repair stores and asks them to fix a hi-fi set. In 34 of the cases the customer is undercharged. (That is, the bill is lower than the law firm considers appropriate.)

(a) Use the normal approximation to obtain a 90 percent confidence interval for the proportion of cases where such undercharging occurs.

(b) If a sample is large, the sample standard deviation is approximately normally distributed, with a mean equal to the population standard deviation $\sigma$ and with a standard deviation equal to $\sigma \div \sqrt{2n}$. In the circumstances described above, the sample standard deviation of the amounts charged by the 100 hi-fi repair stores to fix the hi-fi set was $8.70. Provide a 90 percent confidence interval for the population standard deviation. (See Exercise 7.17.)

**7.40** There are 10,000 people in a particular neighborhood in Brooklyn. A cable television station wants to estimate the average number of hours on a particular day that a person in this neighborhood spent watching its programs. Its executives think that the standard deviation of the number of hours a person spent watching its programs was about 3.2 hours. The station's president proposes that a random sample be taken of the people in the neighborhood. If it is desired that the probability be 0.98 that the sample mean differs by no more than 0.5 hours from the true mean, how big must such a sample be?

**7.41** A state university wants to estimate the average amount that its students paid in state income taxes in 1987. The standard deviation of state income tax paid by the students is estimated to be about $60. The university wants the probability to be 0.95 that the sample mean will differ by no more than $10 from the true mean. If a random sample of the students is selected, how big should it be?

**\*7.42** Minitab was used to calculate a 99 percent confidence interval for the mean difference in abrasion resistance between treated and untreated half-pieces of rubber. The data are in Table 7.1. Interpret the computer printout, which is as follows:

```
MTB > set into c5
DATA> 2.6 3.1 -.2 1.7 .6 1.2 2.2 1.1 -.2 .6
DATA> end
MTB > tinterval 99 percent c5
         N    MEAN    STDEV   SE MEAN    99.0 PERCENT C.I.
C5      10   1.270   1.126    0.356    (   0.112.   2.428)
```

**\*7.43** An insurance company wants to determine the average age of women having legal abortions. A random sample of ten women who had legal abortons in 1986 indicates that their ages (in years at the time of the abortion) were 21, 18, 28, 16, 22, 26, 31, 21, 19, 30. Minitab is used to calculate a 95 percent confidence interval for the mean age of women having legal abortions in 1986. Interpret the computer printout, which is as follows:

```
MTB > set into c6
DATA> 21 18 28 16 22 26 31 21 19 30
DATA> end
MTB > tinterval 95 percent c6
         N    MEAN    STDEV   SE MEAN    95.0 PERCENT C.I.
C6      10   23.20    5.22     1.65    (  19.46.    26.94)
```

# Appendix 7.1

### BAYESIAN ESTIMATION

The basic difference between *Bayesian estimation* and the estimation procedures discussed in previous sections of this chapter is that Bayesian techniques view the parameter to be estimated as a random variable, not a constant. In the case of the new fiber in Example 7.1, suppose that the firm's statisticians feel strongly (based on previous experience with the new fiber and on *a priori* reasoning) that the mean breaking load of a strand of this fiber should be about 2.5 pounds. Specifically, suppose that the statisticians' subjective probability distribution of the population mean is as shown in Figure 7.8. In other words, their prior probabilities regarding the mean can be represented by this distribution. (Recall our discussion of prior probabilities in Chapter 3). If this is the case, then Bayesian estimation techniques allow the firm's statisticians to take these prior probabilities into account, whereas the techniques discussed previously in this chapter make no allowance for such prior probabilities.

---

\* Exercise requiring some familiarity with Minitab.

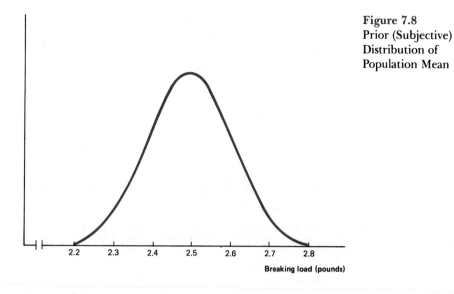

**Figure 7.8**
Prior (Subjective)
Distribution of
Population Mean

Breaking load (pounds)

If the prior distribution of the population mean is normal, as in Figure 7.8, then the Bayesian estimate of the population mean is

$$\tilde{\mu} = \frac{n\sigma_p^2 \bar{x} + \sigma^2 \mu_p}{n\sigma_p^2 + \sigma^2},$$
(7.14)

where $\mu_p$ is mean of the prior (subjective) probability distribution of the population mean, $\sigma_p$ is the standard deviation of the prior (subjective) probability distribution of the population mean, $\sigma$ is the population standard deviation, $\bar{x}$ is the sample mean, and $n$ is the sample size.

At first glance, the reasonableness of this estimate is by no means obvious. But another way of writing this estimate is

$$\tilde{\mu} = \left(\frac{n\sigma_p^2}{n\sigma_p^2 + \sigma^2}\right)\bar{x} + \left(\frac{\sigma^2}{n\sigma_p^2 + \sigma^2}\right)\mu_p.$$

Written this way, it is clear that this estimate of the mean is a weighted average of (1) the sample mean and (2) the mean of the prior distribution of the population mean. In other words, this estimate is a compromise between $\mu_p$ (based on the decision maker's prior feelings) and $\bar{x}$ (what the sample actually turns up). The larger the sample size and the greater the decision maker's uncertainty concerning the population mean (that is, the greater $n$ and $\sigma_p^2$), the more weight is given to $\bar{x}$ and the less weight is given to $\mu_p$. This, of course, is sensible.

To show how Bayesian estimation works, let's return once more to the case of the new synthetic fiber. If the firm's research department's prior distribution concerning the mean breaking load of a strand of this fiber is as shown in Figure 7.8, then $\mu_p = 2.5$ and $\sigma_p = 0.1$. (Why? Because the mean and standard deviation of the distribution in

Figure 7.8 are 2.5 and 0.1, respectively.) Suppose that the firm's statisticians know that the standard deviation of the population equals .075. Then if the firm draws a sample of 36 strands and if the sample mean is 2.3, the Bayesian estimate is

$$\bar{\mu} = \left\{ \frac{36(.01)}{36(.01) + .0056} \right\} (2.3) + \left\{ \frac{.0056}{36(.01) + .0056} \right\} (2.5)$$

$$= (.98)(2.3) + (.02)(2.5) = 2.30.$$

Thus, in this case the Bayesian estimate is very close to $\bar{x}$. This is because $n\sigma_p^2$ is large relative to $\sigma^2$.

# Hypothesis Testing

## 8.1 Introduction

As pointed out in the previous chapter, statistical inference, the part of statistics that indicates how probability theory can be used to help make decisions based on sample data, deals with two types of problems: estimation and hypothesis testing. Having discussed estimation problems in the previous chapter, we turn now to hypothesis testing. Since statisticians are continually engaged in testing hypotheses, and since the methods they use and the results they obtain play a major role in business and economics, it is important to understand the concepts involved in hypothesis testing and how these concepts can be applied to practical problems.

## 8.2 Hypothesis Testing: An Illustration

Hypothesis testing deals with decision making—that is, with the rules for choosing among alternatives. Since statistical decisions must be made under conditions of uncertainty, there is a non-zero probability of error; and the object of the statistician's theory of hypothesis testing is to develop decision rules that will control and minimize the probability of error.

To introduce some of the essential features of the statistical theory of hypothesis testing, let's look at a problem that confronted a firm manufacturing aluminum rods. Based on previous experience, the firm's statisticians knew that, on the average, the rods should be 2 centimeters in diameter. A test was needed for detecting any change in this mean diameter so that whatever factors were responsible for such a change could be corrected.

The company's statisticians picked a random sample from each day's output. Based on this sample, they wanted to test the hypothesis that a change had *not* occurred in the mean diameter of all the rods produced in a day. In other words, the statisticians wanted to test the hypothesis that the mean diameter still was 2 centimeters. Let's denote this hypothesis by $H_0$, and call it the null hypothesis. In statistics the *null hypothesis* is the basic hypothesis that is being tested for possible rejection. The firm had to choose between this hypothesis and the *alternative hypothesis*, denoted by $H_1$, which maintained that a change *had* occurred in the mean diameter of the rods produced in a day. In other words, the alternative hypothesis maintained that the mean diameter no longer was 2 centimeters. (Much more will be said in the following section about the ways in which the null and alternative hypotheses are defined.)

To understand the nature of the decision that faced this firm, it is essential to recognize that, on any given day, four possible situations could arise.

1. No change occurs in the mean diameter of the rods produced ($H_0$ is true), and the firm concludes that this is the case. Thus, the firm does not reject $H_0$, which is the correct decision under these circumstances.

2. No change occurs in the mean diameter of the rods produced ($H_0$ is true), but the firm concludes that such a change has occurred. Thus, the firm rejects $H_0$, which is the wrong decision under the circumstances. (One cost of this wrong decision is the output that is lost while the firm's productive process is shut down needlessly so that the nonexistent deterioration in quality can be corrected.)

3. A change occurs in the mean diameter of the rods produced ($H_0$ is not true), and the firm concludes that this is the case. Thus, the firm rejects $H_0$, which is the correct decision under the circumstances.

4. A change occurs in the mean diameter of the rods produced ($H_0$ is not true), but the firm concludes that no change has occurred. Thus, the firm does not reject $H_0$, which is the wrong decision under the circumstances. (One cost of this wrong decision is that the deterioration in quality is allowed to go uncorrected, with the result that an unusually high percentage of the firm's output must be scrapped because it fails to meet specifications.)

Based on the listing of the possibilities above, it is clear that two kinds of error can be committed. First, one can reject the null hypothesis when it is true—which is possibility 2 above. Second, one can fail to reject the null hypothesis when it is false—which is possibility 4 above. These two kinds of error are called Type I and Type II errors and are defined as follows:

A **Type I error** *occurs if the null hypothesis is rejected when it is true.*
A **Type II error** *occurs if the null hypothesis is not rejected when it is false.*

---

*Basic Idea #15*    ***Any test procedure can result in two types of error. The null hypothesis may be rejected when it is true (Type I error), or the null hypothesis may not be rejected when it is false (Type II error).***

| Alternative courses of action | State of nature | |
|---|---|---|
| | $H_0$ *is true* | $H_0$ *is not true** |
| Do not reject $H_0$ | Correct decision | Type II error |
| Reject $H_0$ | Type I error | Correct decision |

**Table 8.1**
Possible Outcomes of Firm's Decision Problem

*In other words, $H_1$ is true.

Thus, if this firm concludes that a change occurs in the mean diameter of the rods produced when in fact no such change takes place, it commits a Type I error. If it fails to recognize such a change, it commits a Type II error.

Table 8.1 provides a condensed description of the situation facing the firm. The two possible *states of nature* shown in the table are that hypothesis $H_0$ is true or that it is not true. It is not known which of these states of nature—or states of the world—is valid, but one of the two alternative *courses of action* (rejection of the null hypothesis or failure to reject it) must be chosen. The *outcome*—that is, the result of each course of action when each state of nature is true—is shown in Table 8.1. As you can see, a correct decision is the result of the two cases corresponding to possibilities 1 and 3 above; a Type I error is the result of the case corresponding to possibility 2 above; and a Type II error is the result of the case corresponding to possibility 4 above.

## 8.3 Basic Concepts of Hypothesis Testing

To carry out the desired test, this firm needs some criterion or decision rule for choosing between the null hypothesis (that the mean diameter is still 2 centimeters) and the alternative hypothesis (that the mean diameter is no longer 2 centimeters). The classical statistical theory of hypothesis testing presented in this chapter provides such decision rules based on the results of a random sample. These rules are constructed so that the probability of a Type I error and that of a Type II error can each be measured (if possible) and, to some extent at least, be reduced. Before describing these rules, we must discuss how one specifies the null hypothesis and the alternative hypothesis, and we must distinguish between one-tailed and two-tailed tests.

### NULL AND ALTERNATIVE HYPOTHESES

As mentioned above, the null hypothesis is the basic hypothesis that is being tested for possible rejection. Typically, the null hypothesis corresponds to the *absence* of the effect that is being investigated. For example, the firm in the previous section wants to detect a change in the mean diameter of the aluminum rods produced. If such a change is *absent*, the mean diameter is still 2 centimeters. Thus, the null hypothesis is that the mean diameter is still 2 centimeters. In testing any null hypothesis, there is an alternative hypothesis which, as its name indicates, is the alternative to the null hypothesis. For example, the alternative hypothesis in the previous section is that the mean diameter differs from 2 centimeters. There are two quite different kinds of alternative hypotheses, which correspond to one-tailed tests and two-tailed tests.

TWO-TAILED TESTS

In many statistical investigations, the purpose is to see whether a certain population parameter has changed or whether it differs from a particular value. This is true in the case of this firm, which wants to detect changes in *either direction* in the mean diameter of the rods produced. Because the proportion of its output not meeting specifications will increase if the mean diameter is *either* too large *or* too small, the firm wants to detect changes in either direction. Thus, the null hypothesis and alternative hypothesis in this case are

$$H_0: \mu = 2 \qquad H_1: \mu \neq 2$$

where $\mu$ is the population mean.

How can the firm carry out such a test? The firm can calculate the sample mean of the heights, $\bar{x}$. If the null hypothesis is true (that is, if the population mean is 2 centimeters), the sampling distribution of the sample mean is as shown in Figure 8.1. Specifically, as we know from Chapter 6, the sample mean is approximately normally distributed and has a mean of 2 centimeters and a standard deviation of $\sigma/\sqrt{n}$ (where $\sigma$ is the population standard deviation and $n$ is the sample size). Thus, if $\sigma$ is known, the firm can establish certain values of the sample mean that are very unlikely to occur, given that the null hypothesis is true. In particular, as pointed out in Figure 8.1, the probability is only .05 that the

**Figure 8.1**
Sampling Distribution of the Mean Diameter of Rods in Sample (If Null Hypothesis Is True)

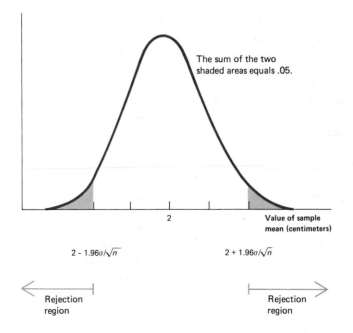

The sum of the two shaded areas equals .05.

2

Value of sample mean (centimeters)

$2 - 1.96\sigma/\sqrt{n}$        $2 + 1.96\sigma/\sqrt{n}$

Rejection region                    Rejection region

value of the sample mean will differ from 2 centimeters by more than $1.96\sigma/\sqrt{n}$ under these circumstances. Consequently, if the value of the sample mean does in fact differ by that much from 2 centimeters, it is evidence that the null hypothesis is not true.

Several things should be noted about this hypothesis-testing procedure. First, *the decision maker or statistician bases his or her decision on the value of a **test statistic**, which is a statistic computed from the sample.* In this case, the test statistic is the sample mean since this is the statistic used by the firm to test the null hypothesis. Second, *the testing procedure designates certain values of the test statistic as the **rejection region.** If the test statistic's value falls in this region, the null hypothesis is rejected; otherwise, it is not rejected.* For example, in this case the rejection region consists of values of the sample mean that differ from 2 centimeters by more than $1.96\sigma/\sqrt{n}$.

*Test Statistic*

*Rejection Region*

Why is this hypothesis-testing procedure called a two-tailed test? The answer is indicated in Figure 8.1, which shows that the rejection region contains values of the test statistic (that is, the sample mean) that lie under *both* tails of its (that is, the sample mean's) sampling distribution. Basically, the reason why the rejection region is of this sort is that the firm wants to reject the null hypothesis (that is, $\mu = 2$ centimeters) *either if $\mu$ is less than 2 centimeters or if $\mu$ is greater than 2 centimeters.* Regardless of whether the mean diameter is too small or too large, the firm wants to detect departures from the specified value of 2 centimeters.

## ONE-TAILED TESTS

In many statistical investigations, the decision maker or statistician is concerned about detecting departures from the null hypothesis *in only one direction.* For example, suppose that an automobile firm wants to test whether the proportion of defective tires in a particular shipment is .06, as claimed by the manufacturer. In this case, the firm does not want to reject the null hypothesis (that the proportion defective is .06) if the proportion defective is in fact less than .06 since, regardless of whether this proportion is .06 or less than .06, the firm wants to keep the shipment. *Only if the proportion defective exceeds .06 does the firm want to reject the shipment.* Thus, in this case the null and alternative hypotheses are

$$H_0: \Pi = .06 \qquad H_1: \Pi > .06,$$

where $\Pi$ is the proportion of the shipment that is defective.[1]

Whereas the automobile firm wants to reject the null hypothesis only if the parameter in question is too high, other situations call for a rejection of the null hypothesis only if the parameter in question is too

---

[1] In this case, it is possible to define the null hypothesis as $\Pi \leq .06$ rather than $\Pi = .06$. (And in the case in the next paragraph, one can define the null hypothesis as $\mu \leq 2,000$ rather than $\mu = 2,000$.) However, it simplifies the exposition to define the null hypothesis in the manner shown above. More is said about this in footnotes 2 and 6.

low. For example, a company wants to test whether the mean length of life of the light bulbs in a particular incoming shipment is less than 2,000 hours. In this case, the firm does not want to reject the null hypothesis (that the mean length of life is 2,000 hours) if the population mean is in fact greater than 2,000 hours, since, regardless of whether the mean is 2,000 hours or more than 2,000 hours, the firm wants to keep the shipment. *Only if the population mean is less than 2,000 hours does the firm want to reject the shipment.* Thus, in this case the null and alternative hypotheses are

$$H_0: \mu = 2{,}000 \qquad H_1: \mu < 2{,}000,$$

where $\mu$ is the population mean.

When the decision maker or statistician is interested only in detecting departures from the null hypothesis in one direction, the hypothesis-testing procedure is a one-tailed test. To see why, consider the case of the company that wants to test whether the mean length of life of the light bulbs in the shipment equals 2,000 hours. If the population mean equals 2,000 hours, the sampling distribution of the sample mean is as shown in Figure 8.2. That is, if a random sample of $n$ light bulbs is selected from the shipment, the mean length of life of the bulbs in the sample will be distributed approximately normally with a mean of 2,000 hours and a standard deviation of $\sigma/\sqrt{n}$ (where $\sigma$ is the population standard deviation). Thus, if the null hypothesis is true (that is, if $\mu = 2{,}000$), the probability is only .05 that the value of the sample mean will fall below 2,000 hours by an amount exceeding $1.64\sigma/\sqrt{n}$. Consequently, if the value of the sample mean does in fact fall below 2,000 hours by an amount exceeding $1.64\sigma/\sqrt{n}$, this is evidence that the null hypothesis is not true. In other words, the sample mean is the test sta-

**Figure 8.2**
Sampling Distribution of Mean Length of Life of Bulbs in Sample (if Null Hypothesis Is True)

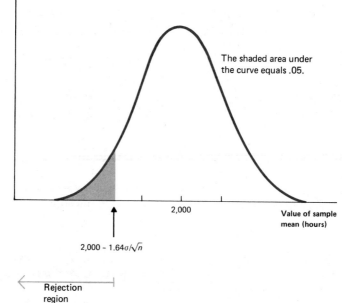

The shaded area under the curve equals .05.

2,000

Value of sample mean (hours)

2,000 − 1.64$\sigma/\sqrt{n}$

Rejection region

tistic and the rejection region contains all values of the sample mean falling below 2,000 hours by an amount exceeding $1.64\sigma/\sqrt{n}$. This rejection region (shown in Figure 8.2) is under only *one* tail of the test statistic's (that is, the sample mean's) sampling distribution, which accounts for the term *one-tailed test*.

## DECISION RULES AND THE NULL HYPOTHESIS

It is important to recognize that *a test procedure (whether a one-tailed or two-tailed test) specifies a **decision rule** indicating whether the null hypothesis should be rejected or not.* For example, in the case of the firm producing aluminum rods, the decision rule is (1) reject the null hypothesis if the sample mean differs by more than $1.96\sigma/\sqrt{n}$ from 2 centimeters; (2) otherwise, do not reject the null hypothesis. In the case of the firm receiving the shipment of light bulbs, the decision rule is (1) reject the null hypothesis if the sample mean falls below 2,000 hours by more than $1.64\sigma/\sqrt{n}$; otherwise, do not reject the null hypothesis.

If the null hypothesis is not rejected, this doesn't mean that we are sure that it is true. All it means is that, on the basis of the available evidence, we have no good reason to reject the null hypothesis. Each decision rule can be expected to result in a wrong decision for a certain percentage of the time. In other words, each decision rule contains a certain probability of Type I error and a certain probability of a Type II error. In later sections, we shall indicate how these probabilities can be calculated.

It is also important to note that *the null hypothesis always states that a population parameter is equal to some value, not that it is unequal to some value.*[2] For example, a beer manufacturer needs to determine whether beer drinkers prefer a new brand of its beer over the old brand. To find out, a sample of beer drinkers is given some of the new brand, and these beer drinkers are asked to rate the new brand on a scale from 1 (very poor) to 90 (perfect). Then the mean rating in the sample is compared with 60, which is known to be the population mean rating for the old brand. In a situation of this sort the preferred procedure, as noted earlier, is to regard the existence of *no difference* between the mean rating of the new brand and that of the old brand as the null hypothesis. One reason why statisticians prefer this procedure is that, if we regard the null hypothesis as being that the population mean for the new brand *equals* 60, we can calculate the sampling distribution of the sample mean. Thus, we can control the probability of a Type I error (as will be indicated in greater detail in the next section). But if we regard the null hypothesis as being that the mean rating for the new brand *exceeds* that of the old brand, we cannot calculate the probability of a Type I error unless we know how large the difference between the

---

[2] In footnote 1 we pointed out that in the one-tailed tests discussed there, the null hypothesis may be defined as $\Pi \leq .06$ or $\mu \geq 2,000$. If such a definition is used, the point in the text still stands: The null hypothesis cannot be that a population parameter is unequal to some value. For example, neither the hypothesis that $\Pi \leq .06$ nor that $\mu \geq 2,000$ is a statement that a population parameter is *unequal* to some value; they are both statements that a population parameter is *at most* or *at least* some value.

means is. Since statisticians find it desirable to specify and control the probability of a Type I error, they have a preference for making the null hypothesis specific in this way.[3]

EXERCISES

**8.1** The Crooked Arrow National Bank is interested in testing the hypothesis that a particular applicant is qualified for a teller's position.
(a) If this is the null hypothesis, what is the alternative hypothesis?
(b) What is the consequence of a Type I error and of a Type II error?
(c) What sorts of evidence can the bank use to test this hypothesis?

**8.2** A firm's engineers test the hypothesis that 2 percent of the items coming off its assembly line are defective. They pick a random sample of 5 items each hour. If any of the items is defective, they reject the null hypothesis.
(a) What is the test statistic?
(b) What is the rejection region?
(c) Is the test a one-tailed test or a two-tailed test?

**8.3** "The statistician should set the probability of Type I error as low as possible."
(a) Do you agree with this statement? Why or why not?
(b) In Figure 8.2, is the decision rule in accord with this statement?

**8.4** Which of the following are two-tailed tests? Which are one-tailed tests?
(a) Determine whether $\mu = 3,000$, the alternative hypothesis being that $\mu \neq 3,000$.
(b) Determine whether $\mu = 10$, the alternative hypothesis being that $\mu < 10$.
(c) Determine whether $\mu = 0$, the alternative hypothesis being that $\mu > 0$.

## 8.4 One-Sample Test of a Mean: Large Samples

Having described some of the basic concepts in the statistical theory of hypothesis testing, we are ready now for a detailed consideration of the most important statistical tests. In this and the following sections, we are concerned with the case where data are available concerning a single sample. This section covers the test of a mean; the following section discusses the test of a proportion. In both sections,[4] we assume that the sample is large ($n > 30$).

It is important to recognize that the following steps must be carried out in testing any hypothesis:

---

[3] For a classic discussion of these and related matters, see R. A. Fisher, "Statistical Inference," reprinted in E. Mansfield, *Statistics for Business and Economics: Readings and Cases* (New York: Norton, 1980). As pointed out in footnote 2, the null hypothesis may sometimes be defined as a population parameter's being at most or at least some value. More is said about this in footnote 6.

[4] In the next section, it is sometimes assumed that $n$ is considerably in excess of 30. See footnote 10.

1. Formulate the null and alternative hypotheses.
2. Specify the significance level of the test.
3. Choose a test statistic.
4. State the decision rule.
5. Collect the data and perform the calculations.
6. Make the statistical and administrative decisions.

In this section, we describe each of these steps in testing a mean.

1. *Formulate the null and alternative hypotheses.* The first step is to formulate the null hypothesis (the hypothesis that is being tested) and the alternative hypothesis. Exactly what parameter is relevant for the problem at hand, and what value of this parameter constitutes the null hypothesis? Is the alternative hypothesis a *one-sided alternative,* which means that the decision maker cares about departures from the null hypothesis in one direction? Or is the alternative hypothesis a *two-sided alternative,* which means that the decision maker wants to detect departures from the null hypothesis in both directions?[5]

To be specific, suppose that we take the case of the firm that receives the shipment of light bulbs. In this case, as we know from the previous section, the null hypothesis is that the mean length of life of the bulbs in the shipment is 2,000 hours; and the alternative hypothesis is one-sided (namely, that the mean length of life is less than 2,000 hours). Suppose that the firm knows from past experience that the standard deviation of the length of life of light bulbs in a shipment of this sort is 200 hours, and that it tests a random sample of 100 bulbs. What sort of test procedure or decision rule should the firm use to determine whether, once it obtains the results of the sample, it should reject the null hypothesis (and reject the shipment)? Clearly, the firm should be more inclined to reject the null hypothesis if the sample mean is relatively low than if it is relatively high, but how low must the sample mean be to justify rejecting the null hypothesis?

To answer this question, the decision maker or statistician must first recognize that any such test procedure or decision rule can make either a Type I error or a Type II error. In this case, a Type I error will arise if the firm rejects a shipment in which the mean length of life of the bulbs is 2,000 hours; a Type II error will arise if the firm does not reject a shipment where the mean length of life of the bulbs is less than 2,000 hours. Table 8.2 shows the possibilities in this case, much as Table 8.1 did in the case of the firm manufacturing aluminum rods. As stressed in an earlier section, any test procedure or decision rule is characterized by a certain probability of a Type I error and a certain probability of a Type II error, denoted as follows:

> The **probability of a Type I error** is designated as $\alpha$ (alpha), and the    $\alpha$ and $\beta$
> **probability of a Type II error** is designated as $\beta$ (beta).

---

[5] Of course, one-sided alternative hypotheses lead to one-tailed tests, and two-sided alternative hypotheses lead to two-tailed tests.

**Table 8.2**
Possible Outcomes of
the Decision Problem
concerning the
Shipment of Light
Bulbs

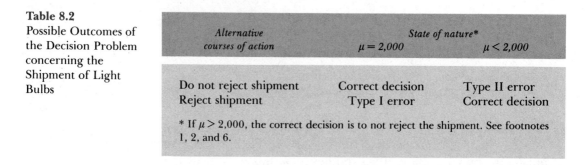

| Alternative courses of action | State of nature* $\mu = 2{,}000$ | State of nature* $\mu < 2{,}000$ |
|---|---|---|
| Do not reject shipment | Correct decision | Type II error |
| Reject shipment | Type I error | Correct decision |

\* If $\mu > 2{,}000$, the correct decision is to not reject the shipment. See footnotes 1, 2, and 6.

**Basic Idea #16**    *To tell whether a particular test procedure is appropriate, it is important to see whether $\alpha$ (the probability of Type I error) and $\beta$ (the probability of Type II error) are set at the right levels.*

Whether a particular test procedure or decision rule is appropriate is dependent on whether $\alpha$ and $\beta$ are set at the right levels from the standpoint of the decision maker. In this case, whether a particular test procedure or decision rule is appropriate depends on whether $\alpha$ and $\beta$ are set at levels that are satisfactory to the firm receiving the shipment of light bulbs.[6]

2. *Specify the significance level of the test.* The value assigned to $\alpha$ is called the **significance level** of the test. Typically, the significance level is set at .05 or .01, which means that the probability of a Type I error is .05 or .01. The decision maker can set the significance level at any amount that he or she chooses. However, in setting this level it generally is wise to keep in mind the relative costs of committing various types of errors. If a Type I error is very costly, a very low level for $\alpha$ should be chosen; on the other hand, if a Type I error is not very costly, a higher value for $\alpha$ is acceptable. At the same time it should be recognized that for a fixed sample size the probability of a Type I error cannot be reduced without increasing the probability of a Type II error. (For example, if the firm establishes a decision rule whereby the shipment is rejected if the sample mean is less than 1,600 hours, this rule will have a lower value of $\alpha$ than if it rejects only those where the sample mean is less than 1,800 hours; but the value of $\beta$ will be higher for the former rule.) Thus, in choosing $\alpha$, the decision maker should keep $\beta$ in mind as well, picking a value of $\alpha$ such that $\alpha$ and $\beta$ bear a sensible relation to the relative costs of Type I and Type II errors. (In a subsequent part of this section, we shall indicate how the value of $\beta$ can be calculated.)

3. *Choose a test statistic.* As pointed out in a previous section, the sample mean is used to test whether the population mean equals a particular value. In particular, if the population standard deviation is known, the test statistic that is generally used is $(\overline{X} - \mu_0) \div \sigma / \sqrt{n}$, where

[6] In footnotes 1 and 2 we noted that the null hypothesis can be defined as $\mu \geq 2{,}000$ rather than $\mu = 2{,}000$. If this definition is used, $\alpha$ is the *maximum* probability of a Type I error. That is, if the population mean exceeds 2,000 hours, the probability of a Type I error will be less than $\alpha$.

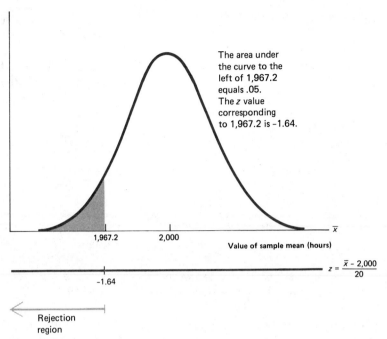

**Figure 8.3**
Sampling Distribution
of Mean Length of
Life of Bulbs in Sample
(If Null Hypothesis Is
True)

The area under
the curve to the
left of 1,967.2
equals .05.
The $z$ value
corresponding
to 1,967.2 is −1.64.

Value of sample mean (hours)

$$z = \frac{\bar{x} - 2,000}{20}$$

Rejection
region

$\mu_0$ is the population mean if the null hypothesis is true. Thus, in the light bulb case, the test statistic is $(\overline{X} - 2,000) \div 20$.

4. *State the decision rule.* Once the test statistic and the value of $\alpha$ have been specified, one can determine the decision rule that should be applied to test whether a population mean equals a specified value. For example, in the case of the firm that is testing whether the mean length of life of the light bulbs is 2,000 hours, suppose that this firm wants $\alpha$ to equal .05. If the null hypothesis is true, the sampling distribution of the sample mean is as shown in Figure 8.3. Since the sample size is large, the central limit theorem assures us that the sample mean will be approximately normally distributed. And since $\sigma = 200$ and $n = 100$, $\sigma / \sqrt{n} = 20$. Thus, if the null hypothesis is true (that is, if the population mean is 2,000), the sample mean must be (approximately) normally distributed, with a mean of 2,000 and a standard deviation of 20, as shown in Figure 8.3.

Since (for reasons given in the previous section) this is a one-tailed test, the rejection region is entirely in the lower tail of the sampling distribution in Figure 8.3. In other words, our test consists of determining whether the value of the sample mean is below 1,967.2 hours, as shown in Figure 8.3. If the sample mean is below this number, the null hypothesis is rejected; if it is not below this number, the null hypothesis is not rejected.

Why 1,967.2 hours? Because there is a .05 probability that the sample mean's value will fall below 1,967.2 hours when the null hypothesis is true. Thus, $\alpha$ equals .05, which (as pointed out in the paragraph before last) is what the firm wants. To verify that $\alpha = .05$, it is convenient to express the sample mean in standard units (that is, in $z$ values). The $z$ value corresponding to 1,967.2 hours is (1,967.2 −

$2,000) \div 20$, or $-1.64$. Using Appendix Table 2, you can confirm that the probability that the value of a standard normal variable will fall below $-1.64$ is .05.

In general, suppose that we want to test the null hypothesis that a population mean equals a certain amount $\mu_0$ against the alternative hypothesis that it is less than $\mu_0$. Since the expected value of the sample mean is $\mu_0$ (if the null hypothesis is true) and its standard deviation is $\sigma/\sqrt{n}$, its observed value in standard units is

$$z = \frac{\bar{x} - \mu_0}{\sigma/\sqrt{n}}$$

Letting $z_\alpha$ be the value of the standard normal variable that is exceeded with a probability of $\alpha$, the decision rule is as follows.

*Test that $\mu = \mu_0$, the Alternative Hypothesis Being $\mu < \mu_0$.*

**DECISION RULE WHEN ALTERNATIVE HYPOTHESIS IS $\mu < \mu_0$:** *Reject the null hypothesis if $z < -z_\alpha$; otherwise do not reject it.*

Thus, in the light bulb example, the null hypothesis should be rejected if the observed value of $z$ is less than $-1.64$ (which equals $-z_{.05}$), as shown in Figure 8.3.

In addition, we should present the decision rule for testing the null hypothesis that a population mean equals a certain amount $\mu_0$ against the alternative hypothesis that it is *more than* $\mu_0$.

*Test that $\mu = \mu_0$, the Alternative Hypothesis Being $\mu > \mu_0$.*

**DECISION RULE WHEN ALTERNATIVE HYPOTHESIS IS $\mu > \mu_0$:** *Reject the null hypothesis if $z > z_\alpha$; otherwise do not reject it.*

The reasoning leading up to this decision rule is, of course, precisely the same as that leading up to the decision rule in the case where the alternative hypothesis is $\mu < \mu_0$. The only difference is that the rejection region is in the upper tail, not the lower tail, of the sampling distribution of the sample mean.

Finally, if the test is two-tailed—that is, if the alternative hypothesis is that the population mean differs from $\mu_0$—the decision rule is as follows.

*Test that $\mu = \mu_0$, the Alternative Hypothesis Being $\mu \neq \mu_0$.*

**DECISION RULE WHEN ALTERNATIVE HYPOTHESIS IS $\mu \neq \mu_0$:** *Reject the null hypothesis if $z < -z_{\alpha/2}$ or if $z > z_{\alpha/2}$; otherwise do not reject it.*

Since the probability of a Type I error must equal $\alpha$, the probability that the sample mean will fall in the rejection region in *each* tail of its sampling distribution, given that the null hypothesis is true, is set equal to $\alpha/2$.[7]

---

[7] Note that there is a close connection between the test procedure described above and the confidence interval described in Chapter 7. *If the null hypothesis is that the population mean equals $\mu_0$, this hypothesis will be rejected if and only if a confidence interval (with confidence coefficient equal to $1 - \alpha$) for $\mu$ does not include $\mu_0$.* For example, if one computes a 95 percent confidence interval for the population mean, and it turns out to

5. *Collect the data and perform the calculations.* Suppose that the firm receiving the shipment of light bulbs tests a random sample of 100 of the bulbs, and that the mean length of life of the bulbs in the sample is 1,972 hours. Then

$$z = \frac{\bar{x} - \mu_0}{\sigma/\sqrt{n}} = \frac{1972 - 2,000}{20} = -1.40.$$

6. *Make the statistical and administrative decisions.* Since the observed value of $z$ (which is $-1.40$) is not less than $-1.64$, the firm should not reject the null hypothesis that the mean length of life of all the light bulbs in the shipment is 2,000 hours. Thus, based on this test, the firm should not reject the shipment.

## THE OPERATING-CHARACTERISTIC CURVE OF THE TEST

Having described each step in testing whether the population mean equals $\mu_0$, we must consider the operating-characteristic curve of this test. The decision rules presented in this section have been constructed so that the probability of a Type I error—that is, $\alpha$—is fixed at a predetermined level. But how large is $\beta$, the probability of a Type II error? For example, if the firm receiving the shipment of light bulbs sets $\alpha$ equal to .05, how large is the probability that the firm will not reject the shipment, given that the mean length of life of the bulbs is less than 2,000 hours? Clearly, the answer depends upon the extent to which the mean length of life of the bulbs falls below 2,000 hours. To see this, recall from Figure 8.3 that if $\alpha = .05$, the test procedure is the following: Reject the null hypothesis (and thus the shipment) if $\bar{x}$ is less than 1.967.2 hours; otherwise, do not reject the null hypothesis. Thus, the probability that the firm will not reject the shipment, given that the mean length of life of the light bulbs is less than 2,000 hours equals the probability that the sample mean will be 1,967.2 hours or more under these circumstances. It is obvious that this probability will decrease as the population mean falls increasingly farther below 2,000 hours.

Figure 8.4 shows the value of $\beta$ at each value of the population mean. In accord with our argument in the previous paragraph, the value of $\beta$ decreases as the population mean falls farther and farther below 2,000 hours. To show how the value of $\beta$ is calculated, suppose that the mean length of life of the bulbs in the shipment is 1,960 hours. The probability that the shipment will not be rejected equals the probability that the sample mean will be 1,967.2 hours or more. (See panel B of Figure 8.5.) Since $\bar{X} \geqslant 1,967.2$ if and only if $(\bar{X} - 1,960) \div 20 \geqslant (1,967.2 - 1,960) \div 20$, the probability that $\bar{X} \geqslant 1,967.2$ equals the probability that $(\bar{X} - 1,960) \div 20 \geqslant 0.36$ (because $(1,967.2 - 1,960) \div 20 = 0.36$). Since $\bar{X}$ is distributed approximately normally with a mean of 1,960 hours and a standard deviation of 20 hours, $(X - 1,960) \div 20$

---

be 15 to 20, then it follows that a two-tailed test (with $\alpha = .05$) will reject the null hypothesis that the population mean is any value below 15 or above 20. This intimate connection between hypothesis testing and confidence intervals is present for other tests too.

**Figure 8.4**
Operating-Characteristic Curve of Test that Mean Length of Life of Bulbs Equals 2,000 Hours

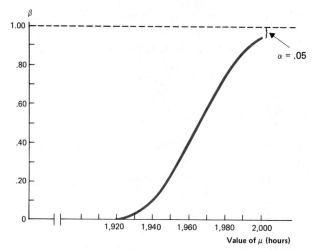

has the standard normal distribution. Thus, the probability that $(\overline{X} - 1,960) \div 20 \geq 0.36$ equals the area under the standard normal curve to the right of .36, which equals .36. (See Appendix Table 2.)

What if the population mean is 1,940 hours, not 1,960 hours? Under these conditions, what is $\beta$? The probability that the shipment will not be rejected equals the probability that the sample mean will be 1,967.2 hours or more. (See panel c of Figure 8.5.) Since $\overline{X} \geq 1,967.2$ if and only if $(\overline{X} - 1,940) \div 20 \geq (1,967.2 - 1,940) \div 20$, the probability that $\overline{X} \geq 1,967.2$ equals the probability that $(\overline{X} - 1,940) \div 20 \geq 1.36$ (because $(1,967.2 - 1,940) \div 20 = 1.36$). Since $\overline{X}$ is distributed approximately normally with a mean of 1,940 hours and a standard deviation of 20 hours, $(\overline{X} - 1,940) \div 20$ has the standard normal distribution. Thus, the probability that $(\overline{X} - 1,940) \div 20 \geq 1.36$ equals the area under the standard normal curve to the right of 1.36, which equals .09. (See Appendix Table 2.)

Using the methods described in the previous two paragraphs, we can calculate the value of $\beta$ at a large number of values of the population mean. (For example, panel A of Figure 8.5 corresponds to the case where the population mean is 2,000 hours.) If we plot each value of $\beta$ against the corresponding value of the population mean, we get the curve in Figure 8.4, which is called the operating-characteristic curve of this test. The definition of an operating-characteristic curve (or *OC curve*) is as follows.

> OPERATING-CHARACTERISTIC CURVE: *A test's operating-characteristic curve shows the value of $\beta$ (that is, the probability of a Type II error) if the population parameter assumes various values other than that specified by the null hypothesis.*

Sometimes statisticians use the *power curve* rather than the operating-characteristic curve. The power curve shows the value of $(1 - \beta)$—whereas the operating-characteristic curve shows the value of $\beta$—at each value of the parameter under test. Clearly, one can easily deduce the power curve from the operating-characteristic curve, and vice versa, since they provide the same information.

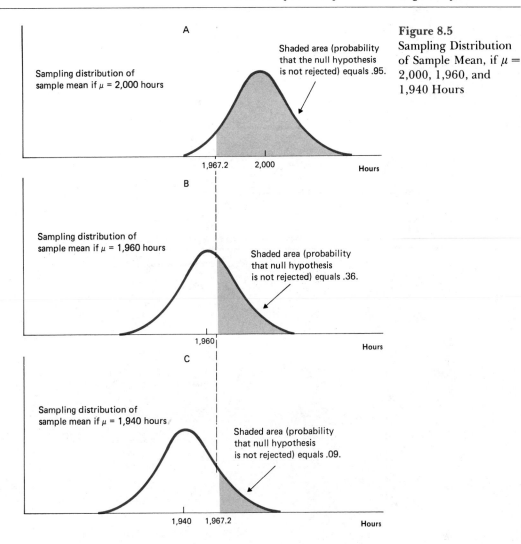

**Figure 8.5**
Sampling Distribution of Sample Mean, if $\mu =$ 2,000, 1,960, and 1,940 Hours

A test's operating-characteristic curve can be used to determine the value of $\alpha$ as well as the value of $\beta$. At the point where the parameter equals the value specified by the null hypothesis, the OC curve shows the probability that the null hypothesis will not be rejected. Thus 1 minus this probability must equal the probability that the null hypothesis will be rejected, given that it is true (which is $\alpha$, the probability of a Type I error). Consequently, at the point where the parameter equals the value specified by the null hypothesis, the distance between the OC curve and 1 equals $\alpha$. For example, Figure 8.4 shows that at the point where the mean length of life of the bulbs is 2,000 hours, the distance between the OC curve and 1 equals .05, which we know equals $\alpha$ in this case.[8]

[8] Recall that we pointed out in an earlier part of this section that if the sample size is fixed, a reduction in the probability of a Type I error can be achieved only at the expense of an increase in the probability of a Type II error. The reason for this can be explained more fully now. Basically, the reason is that, if $\alpha$ is reduced, the OC curve must be shifted upwards, which means an increase in $\beta$.

**Figure 8.6**
Shape of Operating-
Characteristic Curve If
Alternative Hypothesis
Is $\mu > \mu_0$ or $\mu \neq \mu_0$

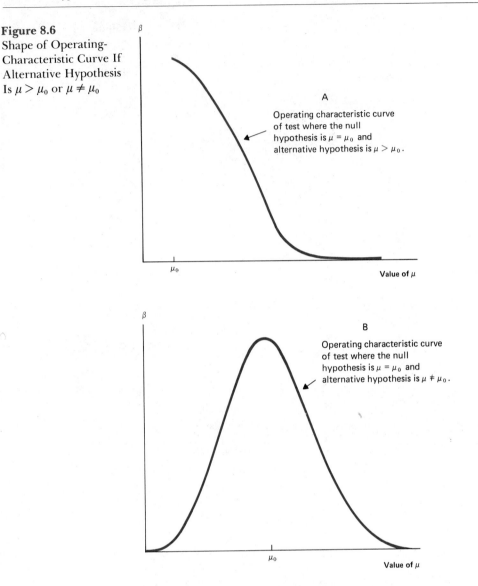

A

Operating characteristic curve
of test where the null
hypothesis is $\mu = \mu_0$ and
alternative hypothesis is $\mu > \mu_0$.

$\mu_0$

**Value of $\mu$**

B

Operating characteristic curve
of test where the null
hypothesis is $\mu = \mu_0$ and
alternative hypothesis is $\mu \neq \mu_0$.

$\mu_0$

**Value of $\mu$**

Not all tests have operating-characteristic curves shaped like that
in Figure 8.4. If a test is designed to detect positive rather than negative departures of the population mean from the value specified by the
null hypothesis, its operating-characteristic curve will be shaped like
that in panel A of Figure 8.6. If a test is designed to detect departures of
the population mean from the value specified by the null hypothesis in
either direction (both positive and negative) the operating-characteristic curve will be shaped like that in panel B of Figure 8.6.

Once a test's OC curve is available, one is in a much better position to determine whether the test is appropriate. For example, if the
executives of the firm receiving the light bulbs are informed of the OC
curve for this test, they may feel that it results in too high a probability
of both Type I error and Type II error. Figure 8.7 shows the OC curve
for another test, one whereby the firm would choose a sample of 400

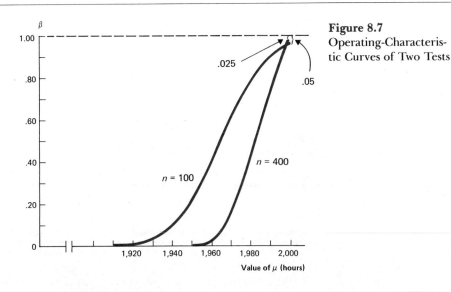

**Figure 8.7**
Operating-Characteristic Curves of Two Tests

bulbs and reject the shipment if the sample mean were less than 1,980.4 hours. As you can see, this test results in a much smaller probability of Type I error (.025 rather than .05) and of Type II error (for practically all values of the population mean) than the earlier test. Why does this test have smaller probabilities of both types of error? Because it is based on a much larger sample. Whether or not the reduction in risk is worth the extra cost of inspecting the larger sample must be determined by the firm's management. After weighing the relevant costs, the firm can specify the sort of OC curve it wants. Then, by a proper choice of sample size and significance level a test can be chosen that is best suited to its requirements.

### EXTENSIONS OF THE TEST PROCEDURES

At this point we must indicate how the test procedures described in this section can be modified to suit circumstances somewhat different from those specified above. In particular, we will describe how a test of this sort can be constructed when the standard deviation is unknown and when the population is not large relative to the sample.

WHERE THE POPULATION STANDARD DEVIATION IS UNKNOWN. If $\sigma$ is unknown and if the sample size is large ($n > 30$), the sample standard deviation $s$ can be substituted for the population standard deviation $\sigma$ in the decision rules given above. In other words, to obtain $z$, one can calculate $(\bar{x} - \mu_0) \div s/\sqrt{n}$, rather than $(\bar{x} - \mu_0) \div \sigma/\sqrt{n}$. In Section 8.8 we shall indicate how a test can be constructed if $\sigma$ is unknown and the sample size is small.

WHERE THE POPULATION IS SMALL. If the size of the population is less than 20 times the size of the sample (and if sampling is without replacement), the decision rules given above must be modified. In particular, $\sigma/\sqrt{n}$ (or $s/\sqrt{n}$ if $\sigma$ is unknown) must be multiplied by the finite population correction factor $\sqrt{(N - n)/(N - 1)}$. In other words, to obtain $z$, one must calculate $(\bar{x} - \mu_0) \div (\sigma/\sqrt{n}) \sqrt{(N - n)/(N - 1)}$.

## INTERPRETATION OF TEST RESULTS

When we test the hypothesis that the mean of a particular population equals a specified value $\mu_0$, we base our test on the value of the sample mean. Because of sampling variation, we cannot expect the mean of the sample to be exactly equal to the population mean. Thus, if the sample mean is different from $\mu_0$ we cannot conclude that the population mean does not equal $\mu_0$. What matters is the probability that the size of the difference between the sample mean and $\mu_0$ could have arisen by chance. If this probability is so great that the null hypothesis is not rejected, we say that the difference between the sample mean and $\mu_0$ (the value of the population mean specified by the null hypothesis) is *not statistically significant.* That is, this difference could be attributable to chance. On the other hand, if this probability is so low that the null hypothesis is rejected this difference between the sample mean and $\mu_0$ is termed *statistically significant.*

*Statistical vs. Practical Significance*

The concept of *statistical* significance should not be confused with *practical* significance. Even if there is a statistically significant difference between the sample mean and the postulated value of the population mean, this difference may not really matter. For example, a market researcher wants to determine whether the mean income of families in a certain town is \$24,000. Based on a very large sample, the researcher finds that the sample mean is \$24,152, which differs significantly (in a statistical sense) from the postulated \$24,000. Whether this difference is of any practical significance depends on whether the market researcher (or his client) really cares about a difference of \$152. For many purposes, such a difference may not be important. For instance, the researcher may only want to determine whether the mean falls between \$23,500 and \$24,500, or whether it falls outside this range. If so, the observed difference of \$152 between the sample mean and the postulated value of the population mean is of no practical importance even though it is statistically significant.

It is also worth noting that the fact that the null hypothesis is rejected by any of the decision rules given above does not *prove* that the null hypothesis is false. Moreover, the fact that the null hypothesis is not rejected by any one of these decision rules does not *prove* that the null hypothesis is true. These test procedures are designed to detect cases where it is unlikely that the sample evidence could have occurred if the null hypothesis was true: In such cases the null hypothesis is rejected. But as long as there is *some chance* that the sample evidence could have occurred if the null hypothesis was true, there is no way to *guarantee* that the results of the test will be absolutely correct. As stressed throughout this section, statistical test procedures are generally designed so that there is a non-zero probability of a Type I or a Type II error. To eliminate such errors entirely would generally be impossible or foolishly expensive.

THE EFFECTS OF A CHANGE IN PRODUCTION LAYOUT

To measure the effects on productivity of a particular change in production layout, the director of engineering of a large firm carries out an experiment in 50 of his company's manufacturing facilities. In each of these plants, he asks his subordinates to obtain data concerning the levels of productivity with the old and the new production layout. In two of these plants, he finds that there is a statistically significant difference (at the 0.05 probability level based on a one-tailed test) between the mean levels of productivity with the new and old layouts. The results for these two plants (at Albany, New York, and Austin, Texas) are as follows:

| | Mean level of productivity | |
| Plant | Old layout | New layout |
| --- | --- | --- |
| Albany | 121.03 | 121.13 |
| Austin | 121.32 | 121.41 |

In the other plants, the mean level of productivity with the new layout is not significantly (in a statistical sense) higher than with the old layout. The director of engineering issues a report saying that this new layout has a positive effect on productivity, as indicated by the above table. Since these differences are statistically significant, he says that they are not due to chance, and that they are important. When his report is circulated, it is criticized severely by other executives of the firm, who imply (in a not-so-polite way) that he is a statistical ignoramus. Where did he go wrong?

SOLUTION: If tests at 50 plants are carried out, and if there is a 5 percent chance that each of them will reject the null hypothesis (that the new layout results in no higher productivity than the old layout) when it is true, one would expect that about 50 times 0.05, or 2.5 plants will show such results even if the new layout results in no higher productivity than the old layout. In other words, when one looks at this battery of tests as a whole, the probability of getting two "statistically significant" differences by chance is much greater than if only these two plants had been included in the study. After all, if one carries out enough tests, eventually one or two "statistically significant" results will arise by chance. This is an important point, since some people publish worthless results of this kind and conveniently forget to mention that they carried out many other such tests where the results were not significant.[9]

---

[9] Also, the differences in the mean level of productivity in the table above are less than 1/10 of 1 percent. Even if such differences could not be attributed to chance, they may be of little practical importance. As pointed out above, statistical significance should not be confused with practical significance.

TO ERR IS HUMAN ... BUT COSTLY

EXERCISES

**8.5** The number of cough drops included in a box varies from box to box. A manufacturer of cough drops wants to test whether the mean number in a box of its cough drops equals 12. It picks a random sample of 60 boxes, and finds that the mean number in a box equals 11.95, the sample standard deviation being 0.09. It sets $\alpha$ equal to 0.05.
(a) If the firm is only interested in detecting whether the true mean is under 12, what is the relevant decision rule?
(b) If the true mean equals 11.9, and if the decision rule in (a) is applied, what is the value of $\beta$?
(c) What should the firm's decision be?

**8.6** A firm produces metal wheels. The mean diameter of the wheels should be 4 inches. Because of chance variation and other factors, the diameters of the wheels vary, the standard deviation being 0.05 inches. To test whether the mean is really 4 inches, the firm selects a random sample of 50 wheels and finds that the sample mean diameter equals 3.97 inches. (The sample is less than 1 percent of the population.)
(a) If the firm is interested in detecting whether the true mean is above or below 4 inches, and if $\alpha$ is set equal to 0.01, what is the relevant decision rule?
(b) If the true mean equals 3.99 inches, and if the decision rule in (a) is applied, what is the value of $\beta$?
(c) What should the firm's decision be?

**8.7** A class contains 60 students. The teacher wants to test whether the mean IQ in the class equals 120. He chooses a random sample of 36 students (without replacement), and finds that the mean and standard deviation of the IQs in the sample are 122.8 and 10.9, respectively.
(a) If the teacher is only interested in detecting whether the true mean exceeds 120, and if $\alpha$ is set at 0.02, what is the relevant decision rule?
(b) If the true mean equals 121, and if the decision rule in (a) is applied, what is the value of $\beta$?
(c) What should the teacher's decision be?

**8.8** A statistician is testing whether $\mu = 100$ on the basis of a random sample of 49 observations from a very large population. She knows that $\sigma = 5$, and the test is two-tailed.
(a) If she uses the test procedure described in the text, and if $\alpha$ is set equal to 0.05, what is the probability of a Type II error if $\mu = 96, 98, 100, 102$, and 104?
(b) Draw five points on the operating-characteristic curve of the test.
(c) Draw five points on the power curve of the test.

**8.9** A statistician is testing whether the mean age of employees at a large factory equals 50 years on the basis of a random sample of 64 employees. He knows that the standard deviation of the ages of the employees at the plant equals 10 years. The alternative hypothesis is that $\mu < 50$ years.
(a) If he uses the test procedure described in the text, and if $\alpha$ is set at 0.01, what is the probability of a Type II error if $\mu = 47, 48, 49$, and 50 years?
(b) Draw four points on the operating-characteristic curve of the test.
(c) Draw four points on the power curve of the test.

# 8.5 One-Sample Test of a Proportion: Large Samples

Many statistical investigations are aimed at testing whether a population proportion equals a specified value. For example, firms frequently use statistical techniques to test whether the proportion of defective items in a shipment equals a specified value. (A detailed discussion of this topic is provided on pages 331–335 below.) In this section we will describe the statistical test procedures that can be used in such investigations. As in the previous section, we describe each of the six steps that must be carried out.

1. *Formulate the null and alternative hypotheses.* As emphasized in the previous section, the first step in any test procedure (whether of a mean, a proportion, or whatever) is to formulate the hypothesis that is being tested. This is a crucial aspect of the work, since, if the null hypothesis and alternative hypothesis are formulated incorrectly, the test procedure will be designed to answer the wrong question. Suppose that an auto company receives a report from a Washington lobbyist stating that 70 percent of American voters favor limitations on Japanese auto imports into the United States. The auto company is skeptical of this report since it believes that the percentage favoring such limitations is lower than this figure. The company's statisticians draw a simple random sample of 400 voters and ask these voters' preferences in this regard. What is the null hypothesis in this case? It is that the proportion $\Pi$ of all voters favoring such limitations equals .70. What is the alternative hypothesis in this case? It is that $\Pi$ is less than .70 since the company is interested only in determining whether $\Pi$ is below .70. It makes no difference to the auto company whether $\Pi$ equals .70 or is greater than .70.

2. *Specify the significance level of the test.* Having formulated the null and alternative hypotheses, the next step is to determine the proper values of $\alpha$ (the probability of a Type I error) and of $\beta$ (the probability of a Type II error). As stressed in the previous section, these probabilities should be chosen on the basis of the relative costs of a Type I and a Type II error. In the present case, a Type I error occurs if the auto company rejects the hypothesis that $\Pi = .70$ when in fact this hypothesis is true. A Type II error occurs if the auto company does not reject the hypothesis that $\Pi = .70$ when in fact $\Pi$ is less than .70. If a Type I error is much more costly than a Type II error, the value of $\alpha$ should be low relative to the value of $\beta$. If a Type II error is much more costly than a Type I error, the value of $\beta$ should be low relative to the value of $\alpha$. For each possible value of $\alpha$ one can calculate the test's operating-characteristic curve which shows $\beta$. (The method for calculating the operating-characteristic curve is demonstrated later in this section.) The value of $\alpha$ should be chosen so that the OC curve is in line with the relative costs in the case at hand. For the sake of concreteness, suppose that in this case $\alpha$ is set equal to .05.

3. *Choose a test statistic.* As one might guess, the test statistic is

based on the sample proportion. In particular, if we are testing whether the population proportion is $\Pi_0$, the test statistic is $(p - \Pi_0) \div \sqrt{\Pi_0(1 - \Pi_0)/n}$. If the null hypothesis is true, the standard deviation of the sample proportion equals $\sqrt{\Pi_0(1 - \Pi_0)/n}$. Thus, this test statistic, like that used to test a population mean, is of the form:

$$\frac{\text{sample statistic} - \text{value of parameter (if null hypothesis is true)}}{\text{standard error of sample statistic}}$$

As we shall see, the test statistics used in other tests frequently are of the same form.

4. *State the decision rule.* Once the test statistic and the value of $\alpha$ have been specified, we can determine the decision rule that should be applied to test whether a population proportion equals the specified value $\Pi_0$. If the null hypothesis is true (that is, if $\Pi = \Pi_0$), the sample proportion is approximately normally distributed with a mean of $\Pi_0$ and a standard deviation of $\sqrt{\Pi_0(1 - \Pi_0)/n}$, if the sample size $n$ is sufficiently large. (Recall Chapter 7.) Thus, if the null hypothesis is true, the sampling distribution of the sample proportion is as shown in Figure 8.8. In the case of the auto company the rejection region is entirely in the lower tail of this sampling distribution. (It is a one-tailed test.) Our test consists of determining whether the value of the sample proportion is below 0.662, as shown in Figure 8.8.

Why 0.662? Because there is a .05 probability that the sample proportion's value will fall below 0.662 when the null hypothesis is true. Thus, $\alpha$ equals .05, as it must to meet the auto company's specifications for the test. To see that this is true, it is convenient to express the sample proportion in standard units (that is, in $z$ values). The $z$ value corresponding to 0.662 is $(0.662 - 0.70) \div \sqrt{(.70)(.30)/400}$, since, if the null hypothesis is true, the expected value of the sample proportion equals 0.70 and its standard deviation equals $\sqrt{(.70)(.30)/400}$. (Recall once more that the standard deviation of the sample proportion equals $\sqrt{\Pi(1 - \Pi)/n}$.) Thus, this $z$ value equals $-1.64$, and based on Appendix Table 2, you can confirm that the probability that the value of a standard normal variable will fall below $-1.64$ is .05.

In general, suppose that we want to test the null hypothesis that a population proportion equals a certain amount $\Pi_0$ against the alternative hypothesis that it is less than $\Pi_0$. Since the expected value of the sample proportion is $\Pi_0$ (if the null hypothesis is true) and its standard deviation is $\sqrt{\Pi_0(1 - \Pi_0)/n}$, its observed value in standard units is

$$z = \frac{p - \Pi_0}{\sqrt{\Pi_0(1 - \Pi_0)/n}}.$$

Under these circumstances, when $n$ is sufficiently large,[10] the decision rule is as follows.

[10] In Chapters 5 and 7, we discussed how large the sample size must be for the normal approximation to be useful.

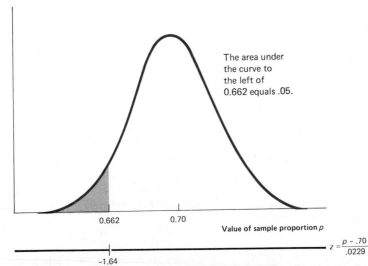

**Figure 8.8**
Sampling Distribution of Proportion of Voters in Sample Favoring Import Limitations, if Null Hypothesis is True

The area under the curve to the left of 0.662 equals .05.

$$z = \frac{p - .70}{.0229}$$

DECISION RULE WHEN ALTERNATIVE HYPOTHESIS IS $\Pi < \Pi_0$: *Reject the null hypothesis if z < −z_α; otherwise do not reject it.*

*Test that $\Pi = \Pi_0$, the Alternative Hypothesis Being $\Pi < \Pi_0$.*

Thus, in the Japanese auto example, the null hypothesis should be rejected if the observed value of z is less than −1.64 (which equals −$z_{.05}$), as shown in Figure 8.8.

The decision rule for testing the null hypothesis that a population proportion equals a certain amount $\Pi_0$ against the alternative hypothesis that it is *greater* than $\Pi_0$ is as follows.

DECISION RULE WHEN ALTERNATIVE HYPOTHESIS IS $\Pi > \Pi_0$: *Reject the null hypothesis if z > z_α; otherwise do not reject it.*

*Test that $\Pi = \Pi_0$, the Alternative Hypothesis Being $\Pi > \Pi_0$.*

The reasoning behind this decision rule is the same as that for the decision rule in the previous paragraph. The only difference is that the rejection region is in the upper tail, not the lower tail, of the sampling distribution of the sample proportion.

Finally, if the test is two-tailed—that is, if the alternative hypothesis is that the population proportion *differs* from $\Pi_0$—the decision rule is the following:

DECISION RULE WHEN ALTERNATIVE HYPOTHESIS IS $\Pi \neq \Pi_0$: *Reject the null hypothesis if z > z_{α/2} or if z < −z_{α/2}; otherwise do not reject it.*

*Test that $\Pi = \Pi_0$, the Alternative Hypothesis Being $\Pi \neq \Pi_0$.*

Since the probability of a Type I error must equal $\alpha$, the probability that the sample proportion will fall in the rejection region in *each* tail of its sampling distribution, given that the null hypothesis is true, is set equal to $\alpha/2$.[11]

---

[11] Of course, in calculating z, if the population is small relative to the sample, $\sqrt{\Pi_0(1 - \Pi_0)/n}$ should be multiplied by $\sqrt{(N - n)/(n - 1)}$.

5. *Collect the data and perform the calculations.* Suppose that the auto company's statisticians draw a random sample of 400 voters and that 223 of them favor limitations on Japanese auto imports into the United States. Since $p = 223/400$, or .5575,

$$z = \frac{p - \Pi_0}{\sqrt{\Pi_0(1 - \Pi_0)/n}} = \frac{.5575 - .70}{.0229} = -6.22.$$

6. *Make the statistical and administrative decisions.* Since the observed value of $z$ (which is $-6.22$) is less than $-1.64$, the auto company should reject the null hypothesis that 70 percent of American voters favor limitations on Japanese auto imports. Apparently, the report from the Washington lobbyist (which claimed this to be true) was wrong.

### THE OPERATING-CHARACTERISTIC CURVE OF THE TEST

Having described each step in testing whether the population proportion equals $\Pi_0$, we must consider the operating-characteristic curve of this test. The decision rules presented in the previous part of this section are constructed so that the probability of a Type I error equals $\alpha$. For example, in the case of the auto company $\alpha$ was set equal to .05. But how large is $\beta$, the probability of a Type II error? The answer is given by the test's operating-characteristic curve. Figure 8.9 shows the operating-characteristic curve of the auto company's test. This curve indicates that the probability that the company will not reject the hypothesis that $\Pi = .70$ when in fact $\Pi = 2/3$ is .58, and the probability that the company will not reject the hypothesis that $\Pi = .70$ when in fact $\Pi = .60$ is less than .01. As we pointed out at the beginning of this section, the decision maker should set the value of $\alpha$ (and $n$) so that the operating-characteristic curve of the test is suitable for the problem at hand.

To illustrate how the OC curve in Figure 8.9 was calculated, consider the point on this curve corresponding to $\Pi = .64$. To determine the value on the OC curve that corresponds to $\Pi = .64$, recall that the null hypothesis ($\Pi = .70$) will not be rejected by the auto company if the sample proportion $p$ is .662 or greater. Thus, the probability that the auto company will not reject the null hypothesis equals the probability that the sample proportion will be .662 or greater. If $\Pi = .64$, the sample proportion will be approximately normally distributed with mean equal to .64 and standard deviation equal to $\sqrt{(.64)(.36)/400}$, or .024. Consequently, the probability that the sample proportion will be .662 or more is equal to the probability that a normal variable with a mean of .64 and a standard deviation of .024 will be .662 or more. Applying the techniques we learned in Chapter 5, this probability is .18.[12] Thus, .18 is the value of the OC curve when $\Pi = .64$.

---

[12] The probability that the value of a normal variable with a mean of .64 and a standard deviation of .024 will be at least .662 equals the probability that the value of the standard normal variable will be at least $(.662 - .64) \div .024$, or .92. Using Appendix Table 2, we find that this probability is .18.

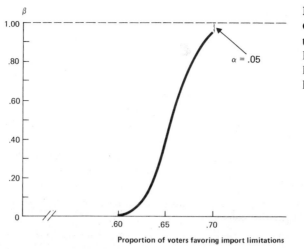

**Figure 8.9**
Operating-Characteristic Curve of Test that Proportion of Voters Favoring Import Limitations Is 0.70

The following example further illustrates the application of these test procedures.

EXAMPLE 8.1 A ketchup producer samples a large shipment of tomatoes received from a supplier. A random sample of 100 tomatoes is inspected, and each tomato is rated either "superior" or "less-than-superior." The ketchup producer wants to test the hypothesis that 40 percent of the tomatoes in the shipment are "superior." Whether the population proportion is greater or less than 40 percent, the firm wants the test to detect this fact. If 34 of the tomatoes in the sample are rated "superior," should the ketchup producer reject the hypothesis that 40 percent of the tomatoes in the shipment fall in this category? (Assume that $\alpha = .05$.)

SOLUTION: Let $\Pi$ be the proportion of "superior" tomatoes in the shipment. Then $H_0$: $\Pi = .40$ and $H_1$: $\Pi \neq .40$. (That is, the null hypothesis is that $\Pi = .40$ and the alternative hypothesis is that $\Pi \neq .40$.) Since $z_{.025} = 1.96$, the null hypothesis should be rejected if $z$ is greater than 1.96 or less than $-1.96$. In fact,

$$z = \frac{p - \Pi_0}{\sqrt{\Pi_0(1 - \Pi_0)/n}}$$

$$= \frac{.34 - .40}{\sqrt{(.40)(.60)/100}} = \frac{-.06}{.049} = -1.22.$$

Thus, the null hypothesis should not be rejected.

## REPORTING THE PROBABILITY ($p$) VALUE OF A TEST

Sometimes it is not possible to specify the value of $\alpha$ before taking the sample, because the statistical analyst does not know what value of

$\alpha$ the decision maker wants to use. Frequently, a statistical report is sent to many clients and readers, and each may want to use a different value of $\alpha$. In such cases, it is customary for the statistician to report the *p-value* of the test.

> The **p-value** *is the probability of getting a value of the test statistic as extreme or more extreme than that actually observed, given that the null hypothesis is true. It is the lowest significance level at which the null hypothesis can be rejected.*

The *p*-value for a given test depends on what test statistic is used, on whether the alternative hypothesis is one-sided or two-sided, and on the computed value of the test statistic. In Example 8.1, suppose that the alternative hypothesis was that $\Pi < .40$ (rather than $\Pi \neq .40$). If so, the *p*-value of this test would be the probability that $z$ (the test statistic) is less than $-1.22$ (the observed value of the test statistic), given that the null hypothesis is true. Since the probability that a standard normal variable is less than $-1.22$ equals .11 (see Appendix Table 2), the *p*-value for this test would equal .11. Thus, users of these test results should reject the null hypothesis if they want to use a significance level of .11 or more; otherwise they should not reject it.

If the alternative hypothesis is two-sided, as in Example 8.1, we double the *p*-value that would be reported if the alternative hypothesis were one-sided.[13] Thus, the *p*-value is .22, rather than .11 (the *p*-value given in the previous paragraph). Doubling the one-sided *p*-value reflects the fact that either a sufficiently large or a sufficiently small value of the test statistic causes rejection of the null hypothesis. Consequently, users of the results in Example 8.1 should reject the null hypothesis if they want to use a significance level of .22 or more; otherwise they should not reject it.

### EXERCISES

**8.10** Suppose that the Quarryville National Bank is testing the hypothesis that the proportion of deposit slips filled out incorrectly is 1 percent.
   (a) In this case, what is $H_0$?
   (b) Under what circumstances will the bank incur a Type I error?
   (c) Under what circumstances will the bank incur a Type II error?
   (d) In this case, what considerations will determine the proper value of $\alpha$ and $\beta$?

**8.11** The president of Toonerville's Second National Bank makes a speech to the bank employees in which he hints that the existing system of sampling transactions is too likely to overlook tellers with a high incidence of errors. The bank's statistician, in response to this speech, suggests that a random sample of 20 of the transactions performed by each teller should be inspected each day and that the teller should be reprimanded if any transaction contains an error.
   (a) What is the decision rule if this suggestion is accepted?

---

[13] This assumes that the distribution of the test statistic is symmetrical, which is true of the distributions taken up in this chapter.

(b) What is the value of $\alpha$ if this suggestion is accepted (and if the hypothesis being tested is that 1 percent of the transactions contain errors)?

(c) If this suggestion is adopted, what is the value of $\beta$ if 5 percent of the transactions contain errors?

**8.12** The president of the Crooked Arrow National Bank wants to test whether 60 percent of the bank's loans are made to persons who reside in the city where the bank is located. The bank's statistician chooses a random sample of 200 of the people to whom the bank has made loans and finds that 52 percent reside in this city.

(a) If a 5 percent significance level is used, should this hypothesis be rejected? (Use a two-tailed test).

(b) If a 1 percent significance level is used, should this hypothesis be rejected? (Use a two-tailed test.)

(c) What is the $p$-value of this test?

**8.13** A survey is to be carried out to determine whether the proportion of high school students in Chester, Pennsylvania, who pass a particular achievement test is different from the past. In previous years, 40 percent passed this test. It is desired to set a 0.01 probability of concluding that a change has occurred when in fact this proportion has not changed. The survey consists of a simple random sample of 120 high school students (out of the 2,500 high school students in Chester).

(a) Set up the decision rule for this test.

(b) What is the probability of failing to detect a change, if the true proportion who pass the test is now 42 percent?

(c) The survey shows that 43 percent of the sample now pass the test. What should one conclude from this result?

(d) What is the $p$-value of this test?

**8.14** William Malone is responsible for maintaining the quality of parts produced by the factory for which he works. Since inspection destroys the part, he must employ sampling techniques. If the proportion of parts produced that are defective exceeds 8 percent, it is important that he be aware of it. He draws a random sample of 100 parts (from 20,000 parts produced) and finds that 16 are defective.

(a) Is this persuasive evidence that the defective rate exceeds 8 percent? Why, or why not?

(b) If he uses the test described in the text, what is the probability of not rejecting the hypothesis that the defective rate is 8 percent when in fact it is 12 percent?

**8.15** The Internal Revenue Service wants to determine whether the percentage of personal income tax returns filed on time by taxpayers is less than last year's percentage, 81 percent. It selects a random sample of 400 of this year's tax returns and finds that 317 of them were filed on time.

(a) Can the Internal Revenue Service be reasonably sure that the percentage filed on time has declined?

(b) If the significance level (that is, $\alpha$) is set at 0.05, is the apparent decline statistically significant?

(c) If the test described in the text is used, what is the probability of failing to reject the hypothesis of no decline in the percentage, if in fact it now equals 79 percent?

**8.16** George Moriarty (the professor's grandson) is testing whether $\Pi = 0.65$ on the basis of a random sample of 100 observations from a very large population. The alternative hypothesis is that $\Pi < 0.65$.

(a) If he uses the test procedure described in the text, and if $\alpha$ is set at 0.025, what is the probability of a Type II error if $\Pi = 0.60, 0.62,$ and 0.65?

(b) Draw three points on the operating-characteristic curve of the test.

(c) Draw three points on the power curve of the test.

**8.17** A public-interest law firm picks a random sample of 100 hi-fi repair stores and asks them to repair a hi-fi set. In 34 of the cases the customer is undercharged.

(a) Use these results to test (at the 10 percent significance level) the hypothesis that in one-half of all such cases customers are undercharged. (Use a two-tailed test.)

(b) Use a one-tailed test, where the alternative hypothesis is that $\Pi < .50$.

(c) What is the *p*-value of this test?

**8.18** To test whether its tellers are performing adequately, the Crooked Arrow National Bank takes a random sample of 10 of the transactions performed by each teller each day. If 1 or more of a certain teller's transactions contains an error, the teller is reprimanded.

(a) What is the decision rule in this case?

(b) If the hypothesis being tested is that 1 percent of the transactions contain errors, what is $\alpha$ in this case? (*Hint:* The normal distribution should *not* be used here.)

(c) What is the probability of a Type II error if in fact 2 percent of the transactions contain errors?

(d) What is the probability if 5 percent of the transactions contain errors?

(e) Draw the operating-characteristic curve of the test used by the Crooked Arrow National Bank.

## 8.6 Two-Sample Test of Means: Large Samples

In previous sections, we have described how a statistical procedure can be formulated to test whether a population mean or a population proportion equals a specified value. We turn now to a case where a random sample is drawn from each of two populations. In this section we describe how one can test the hypothesis that the means of two populations are equal. In the next section, we will describe how one can test the hypothesis that two population proportions are equal. In both sections,[14] we assume that the sample size in each population is large (that is, $n > 30$). Although each of the six steps described in earlier sections must be carried out, it is unnecessary to repeat all of them in all cases, since the general procedure should be clear at this point.

SETTING UP THE TEST

Suppose that a market research firm wants to determine whether the mean rating consumers give to their favorite beer differs from that

---

[14] In the next section, $n$ may have to be substantially greater than 30 for the normal approximation to apply. As pointed out in Chapter 7, this depends on the value of $\Pi$. In the rest of this chapter, we assume that the population is large relative to the sample (or that sampling is with replacement).

which is given to brand X, when both beers are unidentified. Specifically, the firm picks a random sample of $n_1$ people, gives each person his or her favorite beer in an unmarked can, and asks the individual to rate it on a scale from 1 (very poor) to 90 (perfect). The firm then picks another random sample of $n_2$ people, gives each person brand X (also in an unmarked can) and asks the person to rate it on the same scale. The market research firm would like to test whether the mean rating of the favorite beer differs from the mean rating of brand X.

In contrast to the previous two sections, we are concerned here with two populations, not one. The first population is the population of ratings that beer drinkers will give to their favorite beer if it is served to them in an unmarked can. The mean of this population is $\mu_1$ and the standard deviation is $\sigma_1$. The second population is the population of ratings that beer drinkers will give to brand X when it is given to them in an unmarked can. The mean of this population is $\mu_2$ and the standard deviation is $\sigma_2$. The market-research firm draws a random sample of $n_1$ ratings from the first population. In other words, $n_1$ individuals are asked to rate their favorite beer in an unmarked can. Also, the market-research firm draws a random sample of $n_2$ ratings from the second population. This means that $n_2$ people are asked to rate brand X beer in an unmarked can. There is no relation between the individuals chosen in the two samples, which are completely independent. How can the market research firm use these two samples to test whether $\mu_1$ equals $\mu_2$?

As we have stressed before, the first step in constructing a test is to formulate the null hypothesis and the alternative hypothesis. The null hypothesis here is that $\mu_1 = \mu_2$—or that $\mu_1 - \mu_2 = 0$. The alternative hypothesis is that $\mu_1 - \mu_2 \neq 0$, since the firm is interested in rejecting the null hypothesis both when $\mu_1 - \mu_2 < 0$ and when $\mu_1 - \mu_2 > 0$. In other words, the firm wants to reject the null hypothesis both when brand X's mean rating is higher than the mean rating of the favorite beer and when the reverse is true. Stated differently, what the firm wants is a two-tailed test.

The next step is to specify the value of $\alpha$, the significance level of the test. As we have emphasized in previous sections, the value of $\alpha$ should be determined on the basis of the relative costs of both Type I and Type II errors. An effort should be made to strike a proper balance between $\alpha$ and $\beta$, even if (as is sometimes the case) one has only a vague idea of the relative costs. One important point is that $\alpha$ should always be specified *before the data are examined*. (This is true for any type of test, not just those described in this section.) Otherwise, it would be possible for the investigator to choose a significance level small enough so that the null hypothesis is not rejected (if the investigator does not want to reject it) or a significance level large enough so that the null hypothesis is rejected (if the investigator wants to reject it). Such a procedure would be a mockery of correct statistical practice. For the sake of concreteness, suppose that $\alpha$ in this case is set equal to .01.

DECISION RULES

Once the value of $\alpha$ has been specified, we can determine the decision rule that should be applied to test whether the difference between the two population means (that is, $\mu_1 - \mu_2$) is zero. If both $n_1$ and $n_2$ are large, it can be shown that the sampling distribution of the difference between the sample means (that is, $\bar{X}_1 - \bar{X}_2$) is approximately normal, with a mean of $(\mu_1 - \mu_2)$ and a standard deviation of $\sqrt{\sigma_1^2/n_1 + \sigma_2^2/n_2}$. Thus, if the null hypothesis is true (which means that $\mu_1 - \mu_2 = 0$), the difference between the sample means is approximately normally distributed, with a mean of zero and a standard deviation of $\sqrt{\sigma_1^2/n_1 + \sigma_2^2/n_2}$. Consequently, if the market-research firm knows that $\sigma_1$ (the standard deviation of the ratings of the favorite beer) is 12 and $\sigma_2$ (the standard deviation of the ratings of brand X) is 10, and if $n_1 = 100$ and $n_2 = 200$, the difference between the sample means is approximately normally distributed, with a mean of 0 and a standard deviation of $\sqrt{12^2/100 + 10^2/200}$ or 1.39. In other words, if the null hypothesis is true, the sampling distribution of $(\bar{X}_1 - \bar{X}_2)$ is as shown in Figure 8.10.

Since the market-research firm wants to reject the null hypothesis either when $\mu_1 - \mu_2 < 0$ or when $\mu_1 - \mu_2 > 0$, it must set up a rejection region under each tail of the sampling distribution of $(\bar{X}_1 - \bar{X}_2)$, as shown in Figure 8.10. The rejection region in *each* tail must be constructed so that the probability is $\alpha/2$ that the difference between the sample means will fall in this region if the null hypothesis is true. (Since there are two tails, the total probability of a Type I error will be $\alpha$.) If the rejection region in the lower tail is $(\bar{x}_1 - \bar{x}_2) < -3.59$, and if the rejection region in the upper tail is $(\bar{x}_1 - \bar{x}_2) > 3.59$, this condition is met. That is, the probability is .005 that $(\bar{X}_1 - \bar{X}_2)$ will be less than $-3.59$ if

**Figure 8.10**
Sampling Distribution of Difference between Sample Means, If Null Hypothesis Is True

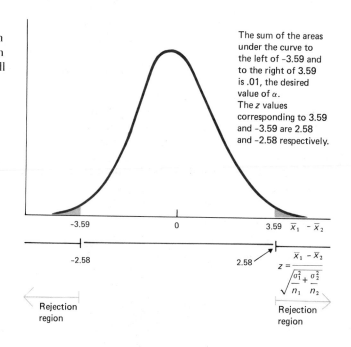

The sum of the areas under the curve to the left of –3.59 and to the right of 3.59 is .01, the desired value of $\alpha$.
The $z$ values corresponding to 3.59 and –3.59 are 2.58 and –2.58 respectively.

the two population means are equal, and the probability is .005 that $(\overline{X}_1 - \overline{X}_2)$ will be greater than 3.59 if the two population means are equal. Thus, this test has the desired probability of a Type I error, namely .01.

Like any random variable, the difference between the sample means can be expressed in standard units (that is, in $z$ values). Since its mean is 0 (if the null hypothesis is true) and its standard deviation is $\sqrt{\sigma_1^2/n_1 + \sigma_2^2/n_2}$, its observed value in standard units is

$$z = (\overline{x}_1 - \overline{x}_2) \div \sqrt{\frac{\sigma_1^2}{n_1} + \frac{\sigma_2^2}{n_2}}.$$

Stated in terms of $z$, the rejection region in the lower tail is

$$z < -z_{\alpha/2} = -2.58$$

and the rejection region in the upper tail is

$$z > z_{\alpha/2} = 2.58.$$

(Why? Because if $(\overline{x}_1 - \overline{x}_2) < -3.59$, it follows that $z < -3.59 \div 1.39$, or $-2.58$. Similarly, if $(\overline{x}_1 - \overline{x}_2) > 3.59$, it follows that $z > 3.59 \div 1.39$, or 2.58. This is because $z$ equals $(\overline{x}_1 - \overline{x}_2) \div 1.39$, since $\sqrt{\sigma_1^2/n_1 + \sigma_2^2/n_2} = 1.39$.) Thus, the market-research firm should reject the null hypothesis if $z < -2.58$ or if $z > 2.58$.

In general, the decision rules for testing the null hypothesis that $\mu_1 - \mu_2 = 0$ are as follows.

> **DECISION RULES:** [15] *When the alternative hypothesis is $\mu_1 - \mu_2 \neq 0$, reject the null hypothesis if z exceeds $z_{\alpha/2}$ or is less than $-z_{\alpha/2}$. When the alternative hypothesis is $\mu_1 - \mu_2 < 0$, reject the null hypothesis if z $< -z_\alpha$. When the alternative hypothesis is $\mu_1 - \mu_2 > 0$, reject the null hypothesis if z $> z_\alpha$.*

**Test that $\mu_1 - \mu_2 = 0$**

The following example is provided to further illustrate the application of these tests.

---

EXAMPLE 8.2 A firm has two plants, each of which produces screws. It is necessary to test whether the mean diameter of the screws produced at one plant equals the mean diameter of those produced at the other plant. If the mean diameter of screws at either plant is larger than that at the other plant, the hypothesis that the mean diameters at the two plants are equal will have to be rejected. At both plants the standard deviation of the diameters of the screws

---

[15] If the population standard deviations are unknown, the sample standard deviations can be substituted in these decision rules if both samples are large.

is .1 inches. A random sample of 100 screws is taken from each plant's output, and the sample mean is found to be .41 inches at the one plant and .45 inches at the other. Should the firm reject the hypothesis that the means are equal? Since the report of the statistical findings is to be sent to the firm's top managers who will have varying opinions as to what significance level to use, the firm's statistician wants to calculate a $p$-value.

SOLUTION: Let $\mu_1$ be the mean of the diameters of the screws produced at the one plant and $\mu_2$ be the mean of the diameters of the screws produced at the other plant. The null hypothesis is that $\mu_1 - \mu_2 = 0$ and the alternative hypothesis is that $\mu_1 - \mu_2 \neq 0$. The test statistic is $z$, which equals $(.41 - .45) \div \sqrt{.1^2/100 + .1^2/100} = -.04/.014 = -2.86$. If the null hypothesis is true, the probability that $z < -2.86$ is .0021. Since this is a two-tailed test, the $p$-value is twice this amount, or .0042. Thus, the null hypotheses should be rejected if the significance level is .0042 or more.

## 8.7 Two-Sample Test of Proportions: Large Samples

In this section we will describe a statistical procedure to test the hypothesis that two population proportions are equal. This test procedure is based on two independent samples, one from each population. It is assumed that each of these samples is large.

### SETTING UP THE TEST

Suppose that a chair manufacturer needs to find out whether a new technique should be substituted for an old one. If the proportion of chairs meeting specifications is greater using the new technique, it is economical for the firm to make the switch; otherwise, it is not economical to do so. The firm uses the new technique to produce 200 chairs and the old technique to produce 400 chairs. It then determines the proportion of chairs in each sample which meets specifications. How can the firm test whether the proportion of chairs meeting specifications is the same for both techniques?

As we have stressed, the first step in setting up a test is to formulate the null hypothesis and the alternative hypothesis. Clearly, the null hypothesis is that $\Pi_1$, the proportion of chairs meeting specifications when the new technique is used, equals $\Pi_2$, the proportion meeting specifications when the old technique is used. The alternative hypothesis is that $\Pi_1 > \Pi_2$. Since the firm wants to switch to the new technique only if $\Pi_1 > \Pi_2$, this is the kind of departure from the null hypothesis that it wants to detect. It is of no concern whether $\Pi_1 < \Pi_2$ since, if this is the case, the implications are no different than if the null hypothesis is true: In either case the firm will not switch to the new technique.

The next step, as we know, is to set a significance level $\alpha$. The considerations influencing this choice are no different here than in earlier sections. Suppose that the chair producer decides to set $\alpha$ equal to .05.

## DECISION RULES

Given that the value of $\alpha$ has been chosen, we can determine the decision rule that should be applied to determine whether the difference between the two population proportions (that is, $\Pi_1 - \Pi_2$) is zero. If the null hypothesis is true and if both $n_1$ and $n_2$ are large, the difference between the sample proportions ($p_1 - p_2$) is approximately normally distributed with a mean of zero and a standard deviation of $\sqrt{\Pi(1 - \Pi)(1/n_1 + 1/n_2)}$, where $\Pi$ is the value of the population proportion in both populations. (In other words, if the null hypothesis is true, the sampling distribution of the difference between the sample proportions is as shown in Figure 8.11.)

If the null hypothesis is true, a good estimate of $\Pi$ is

$$p = \frac{n_1 p_1 + n_2 p_2}{n_1 + n_2}$$

Clearly, $p$ is the weighted mean of the sample proportions in the two samples, each being weighted by the sample size. If $n_1$ and $n_2$ are large we can substitute the above estimate for $\Pi$ in the expression in the previous paragraph. Thus, if the null hypothesis is true, the difference between the sample proportions is approximately normally distributed, with a mean of 0 and a standard deviation of $\sqrt{p(1 - p)(1/n_1 + 1/n_2)}$.

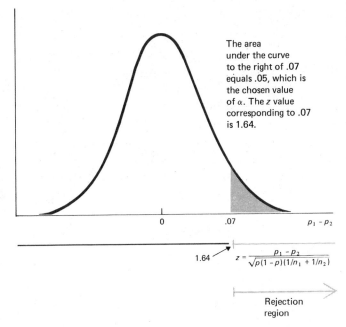

The area under the curve to the right of .07 equals .05, which is the chosen value of $\alpha$. The $z$ value corresponding to .07 is 1.64.

**Figure 8.11**
Sampling Distribution of Difference between Sample Proportions, If Null Hypothesis Is True

Since the chair manufacturer wants to reject the null hypothesis only if $\Pi_1 > \Pi_2$, the rejection region must be set up under the upper tail of the sampling distribution of $(p_1 - p_2)$, as shown in Figure 8.11. This rejection region must be constructed so that the probability is $\alpha$ that the difference between the sample proportions will fall in this region, if the null hypothesis is true. If this rejection region is

$$p_1 - p_2 > z_\alpha \sqrt{p(1-p)\left(\frac{1}{n_1} + \frac{1}{n_2}\right)},$$

this condition is met. Thus, since $\alpha = .05$, the chair company should reject the null hypothesis if $p_1 - p_2 > 1.64 \sqrt{p(1-p)(3/400)}$.

Like any random variable, the difference between the sample proportions can be expressed in standard units. Since its mean is 0 (if the null hypothesis is true) and its standard deviation is $\sqrt{p(1-p)(1/n_1 + 1/n_2)}$, its observed value in standard units is

$$z = (p_1 - p_2) \div \sqrt{p(1-p)(1/n_1 + 1/n_2)}.$$

Stated in terms of $z$, the chair manufacturer should reject the null hypothesis if $z > 1.64$. (Why? Because the $z$ value exceeds 1.64 if $p_1 - p_2 > 1.64 \sqrt{p(1-p)(1/n_1 + 1/n_2)}$ which, according to the previous paragraph, is the condition for rejecting the null hypothesis.)

To illustrate the application of this decision rule, suppose that the chair company finds that the proportion of chairs meeting specifications in the sample where the new technique is used is .60 and that the proportion in the sample where the old technique is used is .50. Then,

$$p = \frac{(200)(.60) + (400)(.50)}{600} = \frac{320}{600} = .53,$$

and

$$z = (.60 - .50) \div \sqrt{.53(.47)(3/400)}$$

$$= .10 \div .043 = 2.33.$$

Consequently, the chair manufacturer should reject the null hypothesis (that $\Pi_1 = \Pi_2$) because $z > 1.64$.

In general, when $n_1$ and $n_2$ are sufficiently large, the decision rules for testing the null hypothesis that $\Pi_1 - \Pi_2 = 0$ are as follows.

*Test that $\Pi_1 - \Pi_2$*
*= 0*

DECISION RULES: *When the alternative hypothesis is $\Pi_1 - \Pi_2 \neq 0$, reject the null hypothesis if $z$ exceeds $z_{\alpha/2}$ or is less than $-z_{\alpha/2}$. When the alternative hypothesis is $\Pi_1 - \Pi_2 < 0$, reject the null hypothesis if $z < -z_\alpha$. When the alternative hypothesis is $\Pi_1 - \Pi_2 > 0$, reject the null hypothesis if $z > z_\alpha$.*

A further illustration of the application of these tests follows.

EXAMPLE 8.3 A polling organization is interested in determining whether candidate Jones will run better in urban or in rural areas. A random sample is drawn of 400 urban voters and 400 rural voters, and it is determined that 55 percent of the urban voters and 49 percent of the rural voters prefer candidate Jones over the others. Should the polling organization conclude that the popularity of this candidate differs between urban and rural areas? (Assume that $\alpha = .05$.)

SOLUTION: Let $\Pi_1$ be the proportion of urban voters who prefer candidate Jones, and let $\Pi_2$ be the proportion of rural voters who prefer Jones. Since the polling organization wants to reject the null hypothesis (that $\Pi_1 = \Pi_2$) when either $\Pi_1 > \Pi_2$ or $\Pi_2 > \Pi_1$, the correct decision rule is to reject the null hypothesis if $z$ is greater than 1.96 or less than $-1.96$ (because $z_{\alpha/2} = 1.96$). Since $n_1 = n_2 = 400$,

$$p = \frac{(400)(.55) + (400)(.49)}{800} = .52,$$

and

$$z = (.55 - .49) \div \sqrt{.52(.48)(.005)}$$

$$= .06 \div .035 = 1.71.$$

Thus, the polling organization should not reject the null hypothesis (that $\Pi_1 = \Pi_2$), since $z$ is not greater than 1.96 or less than $-1.96$.

EXERCISES

**8.19** (a) A personnel agency decides to test whether the mean aptitude test score of engineering graduates differs from that of business graduates. The agency knows that the standard deviation of the scores among engineering graduates is 10 points and that the same is true for business graduates. A random sample is taken of 100 engineering graduates and of 100 business graduates. The mean score of the engineering graduates is 80 and the mean score of the business graduates is 78.
   (i) What is the null hypothesis? What is the alternative hypothesis?
   (ii) What is the appropriate decision rule?
   (iii) If $\alpha = .10$, should the agency reject the null hypothesis?
   (iv) If $\alpha = .05$, should the agency reject the null hypothesis?
   (v) If $\alpha = .01$, should the agency reject the null hypothesis?
   (b) Suppose that the personnel agency wants to reject the hypothesis that the mean scores of engineering and business graduates are the same only if the mean engineering score exceeds the mean business score.
   (i) What is the null hypothesis? What is the alternative hypothesis?
   (ii) What is the appropriate decision rule?
   (iii) If $\alpha = .10$, should the agency reject the null hypothesis?

      (iv) If $\alpha = .05$, should the agency reject the null hypothesis?

      (v) If $\alpha = .01$, should the agency reject the null hypothesis?

  (c) Suppose that the personnel agency wants to reject the hypothesis that the mean scores of engineering and business graduates are the same only if the mean business score exceeds the mean engineering score.

      (i) What is the null hypothesis? What is the alternative hypothesis?

      (ii) What is the appropriate decision rule?

      (iii) If $\alpha = .10$, should the agency reject the null hypothesis?

      (iv) If $\alpha = .05$, should the agency reject the null hypothesis?

      (v) If $\alpha = .01$, should the agency reject the null hypothesis?

**8.20** (a) An economist wants to determine whether the proportion of tool and die firms now using numerically controlled machine tools is different in Canada than in the United States. The economist draws a random sample of 81 tool and die firms in Canada and 100 tool and die firms in the United States, and finds that 20 of the Canadian firms and 30 of the American firms have introduced numerically controlled machine tools.

      (i) What is the null hypothesis? What is the alternative hypothesis?

      (ii) What is the appropriate decision rule?

      (iii) If $\alpha = .10$, should the null hypothesis be rejected?

      (iv) If $\alpha = .05$, should the null hypothesis be rejected?

      (v) If $\alpha = .01$, should the null hypothesis be rejected?

  (b) Suppose that the economist wants to reject the hypothesis that the proportion of tool and die firms now using numerically controlled machine tools is the same in Canada as in the United States only if the proportion is higher in Canada.

      (i) What is the null hypothesis? What is the alternative hypothesis?

      (ii) What is the appropriate decision rule?

      (iii) If $\alpha = .10$, should the null hypothesis be rejected?

      (iv) If $\alpha = .05$, should the null hypothesis be rejected?

      (v) If $\alpha = .01$, should the null hypothesis be rejected?

  (c) Suppose that the economist wants to reject the hypothesis that the proportion of tool and die firms now using numerically controlled machine tools is the same in Canada as in the United States only if the proportion of users is higher in the United States.

      (i) What is the null hypothesis? What is the alternative hypothesis?

      (ii) What is the appropriate decision rule?

      (iii) If $\alpha = .10$, should the null hypothesis be rejected?

      (iv) If $\alpha = .05$, should the null hypothesis be rejected?

      (v) If $\alpha = .01$, should the null hypothesis be rejected?

**8.21** It is claimed that men are better than women at a certain clerical task. To see whether this is the case, a firm chooses a random sample of 100 men and a random sample of 100 women. Each person is given this task to do, and his or her performance is graded from 0 to 100. For the men, the sample mean is 60.8 and the sample standard deviation is 9.9. For the women, the sample mean is 58.4 and the sample standard deviation is 8.7.

  (a) If the significance level is set at 0.05, does the evidence indicate that men are better than women at this task?

  (b) Does this test depend on the assumption that the standard deviation of the grades is the same among men as among women?

**8.22** A chemical firm has two very large research laboratories, one in the United States and one in Europe. It wants to determine whether the chemists at its U.S. laboratory tend to be older or younger than those at its European laboratory. It selects a random sample of 80 chemists from each laboratory, the results being as follows:

|  | Mean age (years) | Standard deviation (years) |
|---|---|---|
| United States | 50.6 | 8.7 |
| Europe | 47.3 | 7.6 |

Each laboratory hires 1,700 chemists.
(a) If the significance level is set at 0.01, does the evidence indicate that the mean age is the same at the two laboratories?
(b) If the significance level is set at 0.05 instead, what does the evidence indicate?
(c) What is the $p$-value of the test? Explain its meaning.

8.23 A firm is about to market a new type of diaper. Before doing so, it wants to determine whether the proportion of mothers preferring this new product to existing products is different in the East than in the West. Based on a random sample of 100 Eastern mothers and a random sample of 100 Western mothers, the results are as follows:

|  | East | West |
|---|---|---|
| Number preferring new product | 71 | 56 |
| Number not preferring new product | 29 | 44 |
| Total | 100 | 100 |

(a) If the significance level is set at 0.02, does the evidence indicate that the proportion of mothers preferring the new product is the same in the East as in the West?
(b) If the significance level is set at 0.05 instead, what does the evidence indicate?
(c) What is the $p$-value of the test? Explain its meaning.

# 8.8 One-Sample Test of a Mean: Small Samples

In previous sections of this chapter we have assumed that the sample was large. We now turn to the case where the sample is small ($n \leq 30$). In this section, we indicate how one can test the hypothesis that the population mean equals a specified value. In the next section, we will indicate how one can test the hypothesis that two population means are equal. In both sections we assume that the populations are normal or approximately normal. In neither section do we assume that any population standard deviation is known.

SETTING UP THE TEST

Suppose that the beer manufacturer (discussed briefly in Section 8.3) wants to test whether the mean rating for a new brand of its beer

**Table 8.3**

Ratings of a New Brand of Beer by 16 Randomly Selected Consumers

| Person | Rating | Person | Rating | Person | Rating |
|--------|--------|--------|--------|--------|--------|
| 1 | 46 | 7 | 71 | 12 | 54 |
| 2 | 64 | 8 | 77 | 13 | 64 |
| 3 | 62 | 9 | 69 | 14 | 63 |
| 4 | 58 | 10 | 67 | 15 | 57 |
| 5 | 54 | 11 | 59 | 16 | 68 |
| 6 | 65 | | | | |

equals 60 (which is known to be the population mean rating for its old brand). If the mean rating for the new brand is higher than 60, the beer manufacturer will substitute the new brand for the old. If the mean rating for the new brand is less than 60, the implications are the same as if it equals 60. That is, in either of these two latter cases the new brand will be dropped. Suppose that the beer manufacturer samples 16 beer drinkers and obtains ratings on a scale of 1 to 90 for the new brand of beer, the results being those in Table 8.3. Should the manufacturer reject the hypothesis that the mean rating for the new brand is 60?

As usual, we begin by formulating the null hypothesis and the alternative hypothesis. In this case, the null hypothesis is that the population mean rating for the new brand $\mu$ equals 60. The alternative hypothesis is that this mean exceeds 60. (The beer manufacturer will not reject the null hypothesis if $\mu < 60$, since if $\mu < 60$ the implications are the same as if $\mu$ equals 60.) To sum up, $H_0$ (the null hypothesis) is that $\mu = 60$, and $H_1$ (the alternative hypothesis) is that $\mu > 60$. We also need to know the significance level: Assume that the beer manufacturer wants $\alpha$ to equal .05.

DECISION RULES

In contrast to our earlier discussion (in Section 8.4) of tests that $\mu$ equals a specified value, we do not know the population standard deviation, and the sample size is small. The decision rules in Section 8.4 are therefore not appropriate here. Nonetheless, if the population is normal we can obtain the decision rule we want by using the $t$ distribution. As pointed out in Chapter 7, $(\overline{X} - \mu) \div s/\sqrt{n}$ has the $t$ distribution (with $n - 1$ degrees of freedom) if the population is normal. (Of course, $s$ is the sample standard deviation, and $n$ is the sample size.) Thus, if $\mu = \mu_0$, the value specified by the null hypothesis, $(\overline{X} - \mu_0) \div s/\sqrt{n}$ has the $t$ distribution. To test the null hypothesis, we compute

$$t = (\overline{x} - \mu_0) \div s/\sqrt{n},$$

which is used as follows.

DECISION RULES: *When the alternative hypothesis is $\mu \neq \mu_0$, reject the null hypothesis if t exceeds $t_{\alpha/2}$ or is less than $-t_{\alpha/2}$. When the alternative hypothesis is $\mu < \mu_0$, reject the null hypothesis if t $< -t_\alpha$. When the alternative hypothesis is $\mu > \mu_0$, reject the null hypothesis if t $> t_\alpha$.*

*Test that $\mu = \mu_0$, where $n \leqslant 30$*

Since the alternative hypothesis in the beer manufacturer's case is that $\mu > 60$, the third decision rule is the applicable one. Using the data in Table 8.3, we find that the sample standard deviation $s$ is 7.60. Thus, the decision rule is: Reject the null hypothesis if $t$ is greater than 1.753, since, as shown in Appendix Table 6, $t_{.05} = 1.753$. (Note that since $n = 16$, there are $16 - 1$, or 15, degrees of freedom.) Because $\bar{x} = 62.38$, $t = (62.38 - 60) \div 7.6/\sqrt{16}$, or 1.25. Since $t$ is not greater than 1.753, there is no reason to reject the null hypothesis that the mean rating of the new brand of beer is 60.

The example below provides a further application of these tests.

EXAMPLE 8.4 A construction firm samples nine pieces of pipe in a shipment and finds that their diameters (in inches) are as follows:

| | | |
|---|---|---|
| 7.7 | 8.0 | 7.4 |
| 7.8 | 8.4 | 7.2 |
| 7.9 | 8.2 | 7.6 |

The firm would like to test (at the .05 significance level) the hypothesis that the mean diameter in the shipment is 8.0 inches. The firm wants to reject this hypothesis if the mean is less than 8.0 inches, but not if it is greater than 8.0 inches. The diameters are believed to be normally distributed. Should the firm reject or not reject this hypothesis?

SOLUTION: The null hypothesis is that the mean diameter $\mu$ equals 8.0 inches and the alternative hypothesis is that $\mu < 8.0$ inches. The sample standard deviation $s$ is .38 inches. The null hypothesis should be rejected if $t < -1.86$, since Appendix Table 6 shows that $t_{.05}$ equals 1.86 when there are 8 degrees of freedom. Because $\mu_0 = 8.0$ and $\bar{x} = 7.8$,

$$t = (7.8 - 8.0) \div .38/\sqrt{9}$$

$$= -1.57.$$

Since the observed value of $t$ is not less than $-1.86$, the null hypothesis should not be rejected. That is, the firm should not reject the hypothesis that the mean pipe diameter is 8.0 inches.[16]

---

[16] For a case study in which an economic theory was subjected to this sort of test, see L. Fouraker and S. Siegel, "Tests of Hypotheses Concerning Bilateral Monopoly," reprinted in E. Mansfield, *op. cit.*

## 8.9 Two-Sample Test of Means: Small Samples

We will now describe how one can test the hypothesis that the means of two populations are equal, if the samples are small and the population standard deviations are unknown. (The tests assume that the population standard deviations in the two populations are equal.)

### SETTING UP THE TEST

Suppose that the market-research firm in Section 8.6 wants to test whether the mean rating given by consumers to their favorite beer differs from that given to brand Y. The firm samples nine people and asks them to rate a beer (in reality their favorite brand) in an unmarked can. Suppose that the ratings are as shown in Table 8.4. Also, the firm samples another nine people and asks them to rate a beer (in reality brand Y) in an unmarked can. These ratings are also shown in Table 8.4. What should the firm conclude from these results?

Again, the first step is to specify the null and alternative hypotheses. The null hypothesis is that $\mu_1$, the population mean of the ratings for the favorite beer, equals $\mu_2$, the population mean of the ratings for brand Y. The alternative hypothesis is $\mu_1 - \mu_2 \neq 0$. The next step is to set the significance level, $\alpha$. Suppose that the firm sets $\alpha$ at .05.

**Table 8.4**
Results of Beer
Manufacturer's Study

| Person | Rating of favorite brand in unmarked can | Person | Rating of brand Y in unmarked can |
|---|---|---|---|
| 1 | 58 | 10 | 62 |
| 2 | 60 | 11 | 60 |
| 3 | 62 | 12 | 60 |
| 4 | 60 | 13 | 62 |
| 5 | 60 | 14 | 63 |
| 6 | 62 | 15 | 59 |
| 7 | 58 | 16 | 58 |
| 8 | 59 | 17 | 64 |
| 9 | 61 | 18 | 61 |

$\bar{x}_1 = 60$

$s_1 = \sqrt{\dfrac{18}{8}} = 1.5$

$\bar{x}_2 = 61$

$s_2 = \sqrt{\dfrac{30}{8}} = 1.94$

HYPOTHESIS TESTING AND $p$-VALUES

Computer packages like Minitab can be used to test hypotheses. Suppose, for example, that one wants to test the hypothesis that a population mean equals 19. The sample is 12, 18, 15, 16, 20, 21, and the population standard deviation is unknown. After putting these numbers in column C7 of the worksheet that Minitab maintains in the computer (recall page 22), all that one has to do is to type the following command.

`MTB > ttest mu=19,c7`

After this command is typed, the computer automatically prints out the following:

```
TEST OF MU = 19.00 VS MU N.E. 19.00
```

| | N | MEAN | STDEV | SE MEAN | T | P VALUE |
|---|---|------|-------|---------|---|---------|
| C7 | 6 | 17.00 | 3.35 | 1.37 | −1.46 | 0.20 |

The first line says that this is a test that $\mu = 19$, where the alternative hypothesis is that $\mu$ is not equal to 19. The second line contains the number of observations (6), the mean of the sample (17.00), the standard deviation of the sample (3.35), and the standard error of the mean (1.37). Also, the value of $t$ is printed out (−1.46).

The last number shown on the second line of the printout (.20) is the $p$-value of the test. As pointed out above, *the $p$-value is the lowest significance level at which the test would result in the rejection of the null hypothesis.* Thus, in this case the observed value of $t$ (that is, −1.46) is such that the null hypothesis should be rejected if $\alpha$ is 0.20 or more. Of course, the advantage of reporting the $p$-value is that the reader of the printout can use whatever significance level he or she wants. If the computer were to report only that the null hypothesis was rejected at a particular significance level, the reader could not tell whether the same conclusion would result if some other significance level (which the reader believes to be more appropriate) were used instead.

## DECISION RULES

In contrast to our earlier discussion in Section 8.6 of tests of whether two population means are equal, we do not know the standard deviation of each population, and the sample size is small. Nonetheless, if both populations are normal, and if their standard deviations are equal, we can obtain the decision rules we need. First, an estimate of the variance in each population is

$$s^2 = \frac{(n_1 - 1)s_1^2 + (n_2 - 1)s_2^2}{n_1 + n_2 - 2}.$$

If the null hypothesis is true, it can be shown that $(\overline{X}_1 - \overline{X}_2) \div \sqrt{s^2(1/n_1 + 1/n_2)}$ has the $t$ distribution with $n_1 + n_2 - 2$ degrees of freedom. Thus, to test the null hypothesis, we compute

$$t = (\overline{x}_1 - \overline{x}_2) \div \sqrt{s^2(1/n_1 + 1/n_2)},$$

which is used as follows.

*Test that $\mu_1 - \mu_2 = 0$,*
*where $n \leqslant 30$*

DECISION RULES: *When the alternative hypothesis is $\mu_1 - \mu_2 \neq 0$, reject the null hypothesis if* t *exceeds* $t_{\alpha/2}$ *or is less than* $-t_{\alpha/2}$. *When the alternative hypothesis is $\mu_1 - \mu_2 < 0$, reject the null hypothesis if* t $< -t_\alpha$. *When the alternative hypothesis is $\mu_1 - \mu_2 > 0$, reject the null hypothesis if* t $> t_\alpha$.

Because the alternative hypothesis is $\mu_1 - \mu_2 \neq 0$, the first decision rule is the appropriate one for the market-research firm. Using the data in Table 8.4, we must calculate $s^2$, which is

$$\frac{(n_1 - 1)s_1^2 + (n_2 - 1)s_2^2}{n_1 + n_2 - 2} = \frac{8(1.5)^2 + 8(1.94)^2}{16} = 3.$$

Thus,

$$t = (60 - 61) \div \sqrt{3(1/9 + 1/9)}$$

$$= -1 \div .82 = -1.22.$$

Since $t_{.025} = 2.12$ when there are 16 degrees of freedom (see Appendix Table 6), the observed value of $t$ does not exceed $t_{.025}$ or fall below $-t_{.025}$. Consequently, the null hypothesis should not be rejected.

The above illustration is based on an actual study carried out by the director of research of Carling Brewing Company.[17] The actual study was more complicated and involved larger samples than those considered here, but the principles are essentially the same. Also, it is interesting to note that the results were much the same as the hypothetical results analyzed here. In particular, the actual study concluded that "participants, in general, did not appear to be able to discern the taste differences among the various beer brands."[18]

The example below again illustrates the application of these tests.

---

EXAMPLE 8.5 The Department of Defense tests two types of aircraft, type A and type B, to determine whether their mean speed under a particular set of conditions is the same. In four tests of type A aircraft the mean speed under the prescribed conditions was 590 miles per hour, and the sample standard deviation was 100 miles per hour. In four tests of type B aircraft the mean speed under these conditions was 750 miles per hour, and the standard deviation was 80 miles per hour. The Defense Department wants to detect differences in mean speed, regardless of which type of aircraft may be faster. There is good reason to believe that the

---

[17] R. Allison and K. Uhl, "Influence of Beer Brand Identification on Taste Perception," *Journal of Marketing Research,* August 1964, pp. 36–39.
[18] *Ibid.,* p. 39.

speeds of each type of aircraft are normally distributed with the same standard deviation. If $\alpha = .05$, should the Department of Defense reject the hypothesis that the mean speeds are equal?

SOLUTION: Let $\mu_1$ be the population mean speed of type A aircraft, and let $\mu_2$ be the population mean speed of type B aircraft. Since the alternative hypothesis is $\mu_1 - \mu_2 \neq 0$, the first decision rule given above is the appropriate one. To test the null hypothesis that $\mu_1 = \mu_2$, we compute the following:

$$s^2 = \frac{(n_1 - 1)s_1^2 + (n_2 - 1)s_2^2}{n_1 + n_2 - 2} = \frac{3(100^2) + 3(80^2)}{6} = 8{,}200,$$

and

$$t = (\bar{x}_1 - \bar{x}_2) \div \sqrt{s^2\left(\frac{1}{n_1} + \frac{1}{n_2}\right)} = (590 - 750) \div \sqrt{8200\left(\frac{1}{4} + \frac{1}{4}\right)}$$

$$= -160 \div 64.03 = -2.50.$$

As shown in Appendix Table 6, $t_{.025} = 2.447$ (because there are 4 + 4 − 2, or 6, degrees of freedom). Thus, the Department of Defense should reject the null hypothesis if the observed value of $t$ exceeds 2.447 or is less than −2.447. Hence, since $t = -2.50$, it follows that the null hypothesis (that $\mu_1 = \mu_2$) should be rejected.

## 8.10 Limitations of Classical Hypothesis Testing

Our purpose in this chapter has been to present the basic principles of classical testing procedures and to describe and apply some of the most commonly used tests. Although these test procedures are of great importance, they are subject to a number of limitations described below. Many of these limitations will be remedied in later chapters.

DEPENDENCE UPON SAMPLE INFORMATION. The test procedures described in this chapter are totally dependent upon sample information. While sample information is often available, this is not always the case. For example, such information is not available concerning the effects of a merger of two firms because two companies cannot merge with one another a number of times in order to see what happens. Other kinds of statistical procedures, which will be covered in Chapters 15 and 16, allow us to come to decisions based on evidence other than purely sample information.

INCOMPLETE ATTENTION TO COSTS OF ERROR AND OF SAMPLING. As we have seen, the test procedures described here are based on a choice by the decision maker of $\alpha$ and $\beta$, the probabilities of Type I and Type II error. However, the classical techniques do not indicate in any detail how $\alpha$ and $\beta$ should be chosen. If one can specify the costs of various types and degrees of errors, it is possible to extend the analysis so that this choice of $\alpha$ and $\beta$ is in effect subsumed within a more inclusive analysis. This kind of extension of classical theory falls under the heading of decision theory and is discussed in the last two chapters of this book.

DECIDING WHETHER OR NOT TO SAMPLE. The test procedures described in this chapter are designed to indicate whether, on the basis of a given sample, a certain hypothesis should be accepted or rejected. These procedures do not tell us whether a sample should be taken in the first place. In some cases, the cost of collecting and analyzing a sample outweighs the value of the information the sample is likely to provide. If this is the case, the decision maker is better off not sampling at all. This kind of decision, which is not covered by the test procedures discussed here, is analyzed in detail in Chapter 16.

INCLUSION OF PRIOR PROBABILITIES. The test procedures described here make no use of whatever personal or subjective judgments the decision maker may be willing to make. If reasonably dependable judgments of this sort can be made, it is obviously useful to be able to include them in one's analysis. In our discussion of decision theory in Chapters 15 and 16, we shall see how the analysis can be extended to include such judgments.

## 8.11 Statistical Process Control at the American Stove Company: A Case Study

One of the most important industrial applications of the statistical theory of hypothesis testing is in *process control*. In many kinds of manufacturing plants, the production process is repetitive. A bottle manufacturing plant turns out batch after batch of bottles, and a light bulb manufacturing plant turns out batch after batch of light bulbs. Obviously, it is extremely important that the quality of these bottles or light bulbs be maintained, but because of gradual changes in the underlying process, the shape or thickness of the bottles or light bulbs may change somewhat. The manufacturer wants to maintain a close surveillance over product quality so that whenever there is evidence that the plant is turning out an inferior product corrections can be made, thus avoiding the financial loss associated with the production of substandard items.

In carrying out this objective firms commonly use statistical process control. To illustrate this technique let's consider the actual case of the American Stove Company, which some time ago was producing a metal piece whose height, according to the specifications, was supposed to be about .83 inches. The firm studied the performance of its production process in order to establish a statistical process control procedure. The firm's statisticians calculated the mean value of the heights of the metal pieces produced during a long period when the production process was known to be in control. During this period the mean

# Acceptance Sampling at the Pentagon and Dow Chemical

Acceptance sampling is an important field of quality control. As an illustration of acceptance sampling, suppose that a firm purchases a large number of metal fixtures from a particular supplier. To make sure that only a reasonably small percentage of the fixtures in any particular shipment is defective, *the firm takes a sample of 30 fixtures from the shipment and tests each to determine whether it is defective. If none is defective, the firm accepts the shipment; otherwise, it rejects it.* This is a very simple type of acceptance-sampling scheme, but it is a convenient starting point.

If the sample is chosen at random, the probability that each fixture in the sample will be defective will equal the proportion of defective fixtures in the shipment. (This assumes that the sample is only a small proportion of the shipment, which is generally true, or that sampling is with replacement.[19]) For example, if 10 percent of the fixtures in a particular shipment are defective, the probability that a randomly chosen fixture in the shipment will be defective is .10. Thus, each choice of a fixture by the inspector can be regarded as a Bernoulli trial, where a success occurs if the fixture is found to be defective. On each such trial, the probability of a success equals $\Pi$, the proportion of defective fixtures in the shipment. Consequently, if the inspector chooses a sample of 30 fixtures from a shipment, the number of defective fixtures in the sample will be a binomial random variable. Specifically, the probability that $x$ of these fixtures will be defective is

$$P(x) = \frac{30!}{(30 - x)!x!} \, \Pi^x (1 - \Pi)^{30 - x},$$

and thus the probability that none is defective is

$$P(0) = (1 - \Pi)^{30}.$$

This result is important because it enables the firm to calculate the probability that it will accept a shipment with various percentages of fixtures that are defective. For example, if 20 percent of the fixtures in a shipment are defective, what is the probability that the firm will accept such a shipment? Substituting .20 for $\Pi$, we find that this probability equals $(1 - .20)^{30}$, or .001. (Why? Because the firm will accept the shipment only if none are defective.) On the other hand, suppose that 10 percent of the fixtures in a shipment are defective. Under these circumstances, what is the probability that the firm will accept the ship-

---

[19] If the sampling is without replacement and if the sample is not a small proportion of the shipment, the hypergeometric distribution, not the binomial distribution, should be used. For a discussion of the hypergeometric distribution, see Appendix 4.2.

ment? Substituting .10 for $\Pi$, we find that this probability equals $(1 - .10)^{30}$, or .04. Finally, if 5 percent of the fixtures in a shipment are defective, what is the probability that the firm will accept the shipment? Substituting .05 for $\Pi$, we find that this probability equals $(1 - .05)^{30}$, or .21.

A very important set of sampling plans has been established by the Department of Defense, which requires that they be used where applicable by the Army, Navy, and Air Force. Since the Department of Defense purchases goods and services so extensively, these sampling plans have had a very wide and significant influence throughout American industry. The purpose of these plans is to determine whether the Department of Defense or some component thereof should accept a shipment or lot of goods received from a supplier. Each plan is characterized by (1) a sample size; (2) an acceptance number; and (3) a rejection number. If the number of defective items in the sample is equal to or less than the *acceptance number,* the lot or batch is accepted. If the number of defective items in the sample is equal to or greater than the *rejection number,* the lot or batch is rejected. Of course, the acceptance and rejection numbers and the sample size are under the control of the statistician. It is not necessary to set the acceptance number equal to zero, as was done in the previous paragraph.

Suppose that the Department of Defense receives a shipment of 1,000 motors. According to its sampling plans, it would draw a random sample of 80 motors, and the acceptance and rejection numbers would depend on the *acceptance quality level* (AQL), which is defined as the maximum percentage of items in a shipment that can be defective and yet have the shipment acceptable. For example, if the acceptable quality level is 1 percent, the acceptance number is 2 and the rejection number is 3. On the other hand, if the acceptable quality level is 4 percent, the acceptance number is 7 and the rejection number is 8. Table 1 shows the acceptance and rejection numbers corresponding to various acceptable quality levels for the shipment of motors.

Once the sample size and the acceptance and rejection numbers have been chosen, the binomial distribution can be used to determine the probability that a shipment of goods will be accepted (given that it

**Table 1**
Acceptance and
Rejection Numbers for
Various Acceptable
Quality Levels*

| Acceptable quality level (percentage of defective items in shipment) | Acceptance number | Rejection number |
|---|---|---|
| 0.1 | 0 | 1 |
| 1.0 | 2 | 3 |
| 2.5 | 5 | 6 |
| 4.0 | 7 | 8 |
| 6.5 | 10 | 11 |
| 10.0 | 14 | 15 |

*Source:* Department of Defense, *Military Standard 105D.*

* This table assumes that there are 501–1,200 items in the shipment and that the sample size is 80.

| Percentage of defective items in shipment | Probability of accepting shipment |
|---|---|
| 5 | .95 |
| 6 | .90 |
| 7.5 | .75 |
| 11.9 | .25 |
| 14.2 | .10 |
| 15.8 | .05 |

**Table 2**
Probability of Accepting Shipment of Items (Based on the Defense Department's Acceptance-Sampling Plan when the AQL = 4 Percent), Given that the Shipment Contains Various Percentages of Defectives*

*Source:* See Table 1.

* See Table 1.

contains a certain percentage of defectives). The Department of Defense statisticians have calculated these probabilities and have put them into tabular form for use in characterizing and quantifying the risks associated with the department's sampling plan. For example, if the acceptable quality level is 4 percent (and thus the acceptance number is 7 and the rejection number is 8) the probability of accepting the shipment of motors is as shown in Table 2, given that the shipment contains various percentages of defectives. Thus, if the shipment contains 5 percent defectives, this probability equals .95; if the shipment contains 14.2 percent defectives, this probability equals .10. These and other acceptance-sampling plans of the Department of Defense are a major application of statistics.[20]

To illustrate the civilian applications of acceptance sampling, consider the Dow Chemical Company, which, like any large company, must audit invoices to determine whether they contain clerical errors. After some study, the company found that with regard to invoices covering purchases of bulk mailing and duplicating items, about 325 invoices are filled out each week, and each invoice contains 10 places where an error can be made. Thus, there are about 3,250 opportunities for invoice errors per week. Dow decided to sample these error opportunities in order to determine the number of cases where errors actually existed. The results were then used to determine whether or not the invoices for a given week should be accepted.[21]

More specifically, Dow used the Department of Defense's standard procedures for acceptance sampling. Dow's statisticians wanted to test whether the proportion of error opportunities where there in fact was an error was equal to 0.25 percent; and they further wanted the probability of rejecting a week's batch of invoices with a .25 percentage of error to equal 5 percent. Also, they wanted the probability to be 90 percent that a batch of invoices would be rejected if 2.5 percent of its error opportunities contained errors.

[20] For further discussion of the Defense Department's acceptance-sampling plans, see E. Mansfield, "Acceptance Sampling by the Defense Department," in E. Mansfield, *Statistics for Business and Economics: Readings and Cases* (New York: Norton, 1980).

[21] This case is based on E. Yehle, "Accuracy in Clerical Work," which appeared in *Systems and Procedures*.

**Figure 1**
Operating-Characteristic Curve for Specified Sampling Plan

*Source:* U.S. Department of Defense.

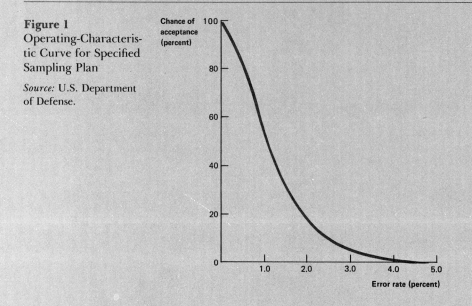

Put in the language of hypothesis testing, the null hypothesis was that the proportion of error opportunities where there was an error was 0.25 percent, and the alternative hypothesis was that this proportion exceeded 0.25 percent. Dow's statisticians wanted $\alpha$ to equal .05, and $\beta$ to equal .10 when the proportion of error opportunities where there is an error equals 2.5 percent.

The Defense Department's standard acceptance-sampling tables indicated that what Dow wanted could be provided by taking a random sample of 225 error opportunities each week and rejecting the batch of

**Figure 2**
Error Rate in Invoices Sampled

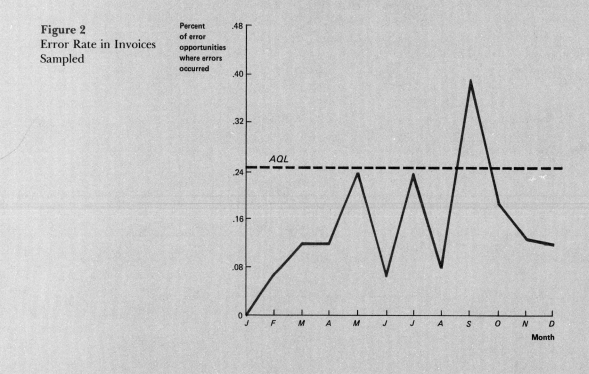

invoices if more than two contained errors. Figure 1 contains the operating-characteristic curve for this test. For a year after adopting this plan, Dow continued to carry out a 100 percent audit of all invoices in order to check the accuracy of this plan. The results are shown in Figure 2.

## PROBING DEEPER

1. If the acceptable quality level is 4 percent, and if the Defense Department's acceptance sampling plan in Tables 1 and 2 is used, what is the null hypothesis? What is the alternative hypothesis? Is this sampling plan a one-tailed or a two-tailed test?

2. How large is $\beta$, if the percentage defective is really 7.5 percent?

3. Does the decision rule implicit in Table 1 correspond with that on page 309 for a test that a proportion equals a specified amount? If not, what are the differences?

4. With regard to Dow's acceptance sampling plan, how can one determine the cost of a Type I or Type II error?

5. In its sampling plan, did Dow give different treatment to invoices where errors can be very costly than to invoices where errors cannot be very costly? If so, how might this be done?

6. The graph on page 334 shows that the actual error rate for September exceeded 0.25 percent; yet Dow's test procedure resulted in acceptance of all weekly invoice batches in September. Does this indicate that this statistical test doesn't work? Why, or why not?

height of the metal pieces being produced was found to be .8312 inches. The standard deviation of the heights of the metal pieces produced during this period also was measured and was found to be .006169 inches.

The firm wanted a test procedure for detecting changes in the mean height of the metal pieces produced. For example, if (because of machine failure or mistakes by workers) the mean height increased to .841 inches or decreased to .828 inches, the firm wanted to be able to ascertain this change. Thus, what was really required was a test procedure for determining whether the mean height equaled .8312 inches against the two-sided alternative that the mean was greater than or less than .8312. Moreover, the firm wanted $\alpha$ (the probability of a Type I error) to be very low—about .003. The reason for setting such a low value of $\alpha$ was that the cost of a Type I error was very high. The cost was high because a Type I error in this context would have necessitated halting the production process (because the test indicates that the mean differs from .8312 inches) when in fact there was no reason to stop production. Such false alarms can be very costly.

The firm decided to pick a simple random sample of five metal pieces from each day's output and compute the mean height of the metal pieces in the sample. The following decision rule was chosen: Reject the null hypothesis (that the mean equals .8312 inches) if the sample mean $\bar{x}$ lies below $.8312 - 3(.006169/\sqrt{5})$ or above $.8312 + 3(.006169/\sqrt{5})$—in other words, if the sample mean lies below .8229 inches or above .8395 inches. This is a straightforward application of the results in Section 8.4, which indicate that if the alternative hypothesis is $\mu \neq \mu_0$, the null hypothesis should be rejected if $z < -z_{\alpha/2}$ or if $z > z_{\alpha/2}$. Since $z = (\bar{x} - \mu_0) \div \sigma/\sqrt{n}$, and since $\mu_0 = .8312$, $\sigma = .006169$, $n = 5$, and $z_{\alpha/2} \doteq 3$, this amounts to precisely the same thing as the firm's decision rule. (Since $z_{.0015} \doteq 3$, 3 is the value of $z_{\alpha/2}$ that is customarily used in quality control.)[22]

*Control Chart*

Quality control procedures of this kind are generally embodied in a **control chart**. For example, the American Stove Company's inspectors plotted each day's sample mean on the control chart in Figure 8.12. This chart contains one horizontal line which shows the mean height when the production process is in control (.8312 inches) and two other horizontal lines, one at $\mu_0 + 3\sigma/\sqrt{n}$ (that is, .8395 inches) and one at $\mu_0 - 3\sigma/\sqrt{n}$ (that is, .8229 inches). These latter two lines are called the upper and lower control limits. When a sample mean falls above the **upper control limit** (.8395 inches) or below the **lower control limit** (.8229 inches), the decision rule in the previous paragraph says we should reject the null hypothesis.

To illustrate the actual performance of this control chart, let's

---

[22] In Section 8.4, the relevant decision rule was stated in terms of $z \left( = \dfrac{\bar{x} - \mu_0}{\sigma/\sqrt{n}} \right)$. If $z > z_{\alpha/2}$ or if $z < -z_{\alpha/2}$, it follows that $\bar{x} > \mu_0 + z_{\alpha/2}\,\sigma/\sqrt{n}$ or $\bar{x} < \mu_0 - z_{\alpha/2}\,\sigma/\sqrt{n}$. Thus, the decision rule in the paragraph above is the same as that in Section 8.4.

Since the sample size is only 5, the sampling distribution of the sample mean may not be exactly normal. (See Chapter 6.) But even if the approximation is rough, there is bound to be a very small probability of a Type I error, given the manner in which the control limits are constructed.

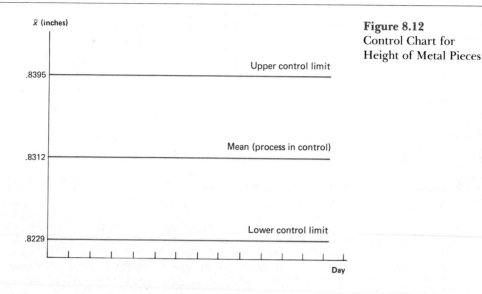

**Figure 8.12**
Control Chart for
Height of Metal Pieces

look at the first nine days of its use. During the first day the sample height was .8312 inches; thus, .8312 inches was plotted for day 1 on the chart. During the second day the sample mean was .8304; thus, .8304 inches was plotted for day 2 on the chart. A similar procedure was carried out during the next seven days, as shown in Figure 8.13. On none of these days did the sample mean fall outside the control limits. Thus, there were no grounds for rejecting the null hypothesis (that $\mu = .8312$ inches) on any of these days. Of course, this brief period only illustrates the practical application of the control chart, which in fact was used over a much longer period of time.

When a sample mean falls outside the control limits, the firm must look for the reasons why the population mean seems to have shifted. In some cases the cause may be machine wear or a breakdown of equipment. (Such a degradation in machine performance can occur gradually without being apparent to the operatives.) In other instances

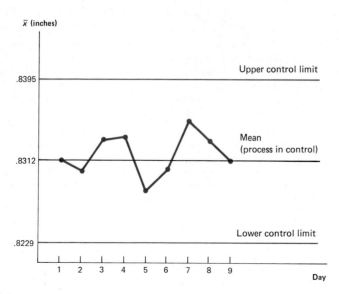

**Figure 8.13**
First Nine Days of
Operation of Control
Chart

the reason for a shift in the mean may be poor workmanship by the labor force. The control chart by itself cannot indicate the reason for the shift in the mean. It can only signal that such a shift seems to have occurred, and it is then up to the firm to find and remedy the cause.[23]

Besides establishing a control chart for the sample mean (a so-called $\bar{x}$ *chart*), firms often establish a control chart to signal changes in the variability of the quality of the production output. For example, the American Stove Company set up a control chart to detect whether, for each day's output, the variation among the heights of the metal pieces differed from what it was when the process was within control. Such charts can be of considerable importance, since an increase in the amount of variation among the items produced can be a serious problem. For example, if the heights of the metal pieces become more variable, a larger proportion of the output will fall outside the tolerances. Generally, the sample range rather than the sample standard deviation is used in these charts because it is easier to compute and because in small samples it is approximately as reliable as the sample standard deviation. Control charts based on the range are often called *R charts*.

### EXERCISES

**8.24** The Internal Revenue Service wants to determine the mean amount of income that waiters fail to report on their income tax. A random sample of 12 waiters is selected, and very detailed and accurate records are kept of their actual incomes by IRS agents. In this sample, the mean amount of unreported income is $4,896, and the standard deviation is $1,592. The Service would like to test whether the true mean is $6,000. If the significance level is set at 0.05, and a two-tailed test is used, are these data consistent with the Service's hypothesis?

**8.25** A pharmaceutical firm wants to determine whether a weight-reducing drug has a different effect on adults over 40 years old than on adults that are no more than 40 years old. Twelve people over 40 are given the drug; their mean weight loss is 8.9 pounds, and the standard deviation is 4.1 pounds. Twelve people no more than 40 are given the drug; their mean weight loss is 11.3 pounds, and the standard deviation is 3.8 pounds.

(a) Based on a two-tailed test with $\alpha$ equal to 0.05, are these data consistent with the hypothesis that the average effect of this drug is the same in both age groups?

(b) Would your answer to (a) change if the significance level were 0.01 rather than 0.05?

(c) What assumptions are you making in (a) and (b)?

---

[23] In some cases a shift in the mean may be favorable, not undesirable. For example, consider a manufacturer of metal parts that establishes a control chart based on the average amount of waste per part. Suppose that the control chart signals a downward shift in the mean. In other words, the mean amount of waste per part seems to be less than was formerly the case. In instances of this sort, the firm will stop the production process in an attempt to find the reasons for this shift in the mean so that this improvement in performance can be continued. Thus, even if a shift in the mean is favorable, the firm wants to detect it.

For further discussion and case studies of statistical quality control, see W. Edwards Deming, "Making Things Right," reprinted in E. Mansfield, *Statistics for Business and Economics: Readings and Cases* (New York: Norton, 1980).

**8.26** A firm that cans tomatoes wants the mean weight of tomatoes in a can to equal 15 ounces. When the firm's plant is operating properly, the mean weight equals 15 ounces, and the standard deviation equals 0.02 ounces. The firm picks five cans at random from each hour's output, and measures the weight of the tomatoes in each can.

(a) If $\alpha$ is set equal to 0.003, plot the control chart the firm should use.

(b) The mean weights of tomatoes in the samples drawn in six successive hours are 15.01, 15.02, 14.99, 14.98, 14.96, 15.00. Should the firm stop the production process? If so, when and why?

**8.27** A manufacturer of ball bearings draws a random sample of ball bearings from each hour's output, and measures their diameters. The firm wants the mean diameter to be 0.2 inches. When the firm's plant is operating properly, the standard deviation of the diameters is 0.0015 inches. The firm uses a control chart where the upper limit exceeds the lower limit by 0.0030 inches.

(a) If $\alpha$ is set equal to 0.003, how big a sample does the firm select from each hour's output?

(b) Suppose that 20 consecutive sample means lie between 0.2000 and 0.2014 inches. Will the control chart signal that the mean diameter has shifted?

(c) Under the circumstances described in (b), does the behavior of the sample means seem to reflect only random variation?

**8.28** The Bona Fide Washing Machine Company chooses a random sample of 25 motors received from supplier I. It determines the length of life of each of the motors, the results (in thousands of hours) being as follows:

| | | | | |
|---|---|---|---|---|
| 4.1 | 4.6 | 4.6 | 4.6 | 5.1 |
| 4.3 | 4.7 | 4.6 | 4.8 | 4.8 |
| 4.5 | 4.2 | 5.0 | 4.4 | 4.7 |
| 4.7 | 4.1 | 3.8 | 4.2 | 4.6 |
| 3.9 | 4.0 | 4.4 | 4.0 | 4.5 |

Suppose that the firm wants to test the hypothesis that the mean length of life of motors received from supplier I equals 4,900 hours. If the Bona Fide Washing Machine Company does not know the standard deviation of the length of life of motors received from supplier I, should it reject this hypothesis (if it sets $\alpha$ equal to .05)? (Use a one-tailed test, the alternative hypothesis being that the mean is less than 4,900 hours.)

**8.29** The Crooked Arrow National Bank wants to test whether the mean income of the individuals holding deposits at its southern branch equals $15,000, the alternative hypothesis being that the mean income does not equal $15,000. The bank takes a random sample of 36 of the depositors at this branch and finds that their incomes in 1986 (in dollars) were as follows:

| | | | | | |
|---|---|---|---|---|---|
| 20,100 | 8,200 | 13,200 | 5,200 | 11,100 | 51,100 |
| 19,400 | 8,900 | 14,300 | 10,100 | 10,800 | 9,600 |
| 10,100 | 10,300 | 15,800 | 12,300 | 9,100 | 10,900 |
| 23,000 | 26,000 | 16,100 | 14,000 | 7,200 | 12,000 |
| 24,200 | 11,400 | 17,200 | 15,100 | 4,300 | 13,200 |
| 25,100 | 12,900 | 18,900 | 16,000 | 38,000 | 15,100 |

(a) If the bank sets a 5 percent significance level, should it reject this hypothesis? (Use a two-tailed test.)

(b) If it sets a 1 percent significance level, should it reject this hypothesis? (Use a two-tailed test.)

**8.30** The manager of the northern branch of the Crooked Arrow National Bank believes that the mean income of the depositors at his branch is $20,000. He wants to test this hypothesis against the alternative that this mean is less than $20,000. A random sample of nine of the depositors at his branch is chosen, and their incomes (in dollars) turn out to be

| | | |
|---|---|---|
| 24,000 | 13,400 | 18,400 |
| 22,900 | 13,800 | 8,200 |
| 11,100 | 9,300 | 14,600 |

(a) If the bank sets a 5 percent significance level, should it reject the manager's hypothesis?

(b) If the bank sets a 1 percent significance level, should it reject the manager's hypothesis?

**8.31** The president of the Crooked Arrow National Bank asks the firm's statistician to test whether the mean income of the depositors at the southern branch is equal to the mean income of the depositors at the northern branch.

(a) If the significance level is set at 5 percent, should the statistician reject this hypothesis, based on the data in Exercises 8.29 and 8.30? (Use a two-tailed test.)

(b) If the significance level is set at 1 percent should the statistician reject this hypothesis, based on the data in Exercises 8.29 and 8.30? (Use a two-tailed test.)

**8.32** Recall from Exercise 7.16 that the Mercer Company has no idea of the standard deviation of the life of the motors it receives from a particular supplier, and that it chose a random sample of 9 motors received from this supplier and determined the length of life of each. The results (in thousands of hours) were as follows:

| | | |
|---|---|---|
| 4.3 | 4.6 | 3.8 |
| 4.2 | 4.3 | 3.9 |
| 4.1 | 3.9 | 4.0 |

(a) Use these results to test the hypothesis that the mean life of the motors received from this supplier equals 4,200 hours (against the alternative that it does not equal 4,200 hours) at the 5 percent level of significance. Do the same at the 10 percent level of significance. (Use two-tailed tests.)

(b) Compare your results in (a) with your results in Exercise 7.16. Can you see how the results you obtained in Exercise 7.16 could have been used to provide the answer to (a)?

**8.33** A manufacturing plant is interested in establishing a control chart for a particular metal part that it produces. The part averages .1020 inches in length when the plant is operating normally, its standard deviation being .0004 inches. The plant decides to pick a random sample of four parts from each day's output and determine their mean length.

(a) Plot the control chart the firm should use under these circumstances. (Let $\alpha = .003$.)

(b) Suppose that the firm adopts your control chart and that the mean length of parts (in inches) on eight successive days is as follows: .10203; .10204; .10198; .10186; .10188; .10181; .10180; and .10139. Plot these results on your control chart, and indicate whether (and if so, when) the firm should reject the null hypothesis.

## Chapter Review

1. The basic hypothesis that is being tested for possible rejection is called the *null hypothesis, $H_0$*. In testing any null hypothesis, there is an *alternative hypothesis, $H_1$*, that is the alternative to the null hypothesis. A Type I error occurs when the null hypothesis is rejected when it is true. A Type II error occurs when the null hypothesis is not rejected when it is false. A test procedure specifies a *decision rule* indicating the conditions under which the null hypothesis should be rejected. This decision rule is based on the value of a test statistic, which is a statistic computed from a sample. If the test statistic's value falls within the rejection region, the null hypothesis is rejected.

2. The probability of a Type I error is designated as $\alpha$, and the probability of a Type II error is designated as $\beta$. Whether or not a particular test procedure is appropriate depends on whether $\alpha$ and $\beta$ are set at the right levels from the point of view of the decision maker. The value of $\alpha$ that is chosen is called the *significance level* of the test. A test's *operating-characteristic curve* shows the value of $\beta$ if the population parameter assumes various values other than that specified by the null hypothesis. The shape of a test's operating-characteristic curve is influenced by the sample size and by the chosen level of $\alpha$. With increases in sample size, it is possible to reduce the probability of both types of error (that is, both $\alpha$ and $\beta$).

3. If the alternative hypothesis is that the value of the parameter differs in either direction from the value specified by the null hypothesis, the test is a *two-tailed test*. If the alternative hypothesis is that the value of the parameter differs in only one direction from the value specified by the null hypothesis, the test is a *one-tailed test*.

4. For a two-tailed test of the null hypothesis that a population mean equals a specified value $\mu_0$, the decision rule is the following: Reject the null hypothesis if $z < -z_{\alpha/2}$ or if $z > z_{\alpha/2}$. For purposes of this test, $z = (\bar{x} - \mu_0) \div \sigma/\sqrt{n}$. If the population standard deviation $\sigma$ is unknown, it can be replaced with $s$, the sample standard deviation. This decision rule applies to large samples (that is, when $n > 30$). For small samples, the $t$ distribution (with $n - 1$ degrees of freedom) must be used in place of the standard normal distribution. For small samples the decision rule is: Reject the null hypothesis if $t < -t_{\alpha/2}$ or if $t > t_{\alpha/2}$. For purposes of this test, $t = (\bar{x} - \mu_0) \div s/\sqrt{n}$. This small-sample test assumes that the population is approximately normal.

5. For a two-tailed test of the null hypothesis that a population proportion equals a specified value $\Pi_0$, the decision rule is: Reject the null hypothesis if $z < -z_{\alpha/2}$ or if $z > z_{\alpha/2}$. For purposes of this test, $z = (p - \Pi_0) \div \sqrt{\Pi_0(1 - \Pi_0)/n}$. For a two-tailed test of the null hypothesis that two population proportions are equal, the decision rule is: Reject the null hypothesis if $z < -z_{\alpha/2}$ or if $z > z_{\alpha/2}$. For purposes of this test, $z = (p_1 - p_2) \div \sqrt{p(1 - p)(1/n_1 + 1/n_2)}$. Both tests assume that the samples are large.

6. For a two-tailed test of the null hypothesis that the mean of one population $\mu_1$ equals the mean of another population $\mu_2$, the decision

rule is: Reject the null hypothesis if $z < -z_{\alpha/2}$ or if $z > z_{\alpha/2}$. For purposes of this test $z = (\bar{x}_1 - \bar{x}_2) \div \sqrt{\sigma_1^2/n_1 + \sigma_2^2/n_2}$. This rule assumes that the samples are large and that the standard deviation of each population is known. If neither of these assumptions is true, the decision rule is: Reject the null hypothesis if $t < -t_{\alpha/2}$ or if $t > t_{\alpha/2}$. For purposes of this test, $t = (\bar{x}_1 - \bar{x}_2) \div \sqrt{s^2(1/n_1 + 1/n_2)}$, where

$$s^2 = \frac{(n_1 - 1)s_1^2 + (n_2 - 1)s_2^2}{n_1 + n_2 - 2}.$$

The latter test assumes that both populations are normal and that their standard deviations are equal. (The number of degrees of freedom of the $t$ distribution is $n_1 + n_2 - 2$.)

7. An important industrial application of statistical testing procedures is *statistical process control*. A control chart is established with a central horizontal line at $\mu_0$, the value of the mean when the process is in control, and with two other lines at the upper and lower control limits. The upper control limit is at $\mu_0 + 3\sigma/\sqrt{n}$; the lower control limit is at $\mu_0 - 3\sigma/\sqrt{n}$. Sample means are plotted on the chart, and the process is stopped and examined when a mean falls outside these limits. There is approximately a .003 probability of the process being stopped if the population mean remains equal to $\mu_0$. Besides control charts (so-called $\bar{x}$ charts) for the sample mean, firms often use so-called R charts, which plot the sample range and which are used to detect changes in the variability of the output.

## Chapter Review Exercises

**8.34** The Connelly Company is engaged in the direct mail of advertisements. Because of the importance of its mailing costs, the firm continually scrutinizes the average weight of the letters it mails. A simple random sample of 12 letters mailed by the Connelly Company shows a mean weight of 2.7 ounces and a standard deviation of 1.1 ounces.
  (a) Using a two-tailed test with $\alpha$ equal to 0.02, are these data consistent with the company's belief that the mean weight of all letters it mails is 2.0 ounces?
  (b) Suppose that the sample size had been 120 rather than 12. Would you have reached the same conclusion as in (a)?

**8.35** In an investigation of the legal costs of a particular kind of antitrust suit, a random sample of 16 such antitrusts suits is selected. The mean legal cost is $2.3 million, and the standard deviation is $1.4 million.
  (a) Based on a one-tailed test with $\alpha$ equal to 0.05, are these data consistent with the statement that the mean legal cost of such antitrust cases is $2 million? (The alternative hypothesis is that it exceeds $2 million.)
  (b) If the sample size had been 160 rather than 16, would you have reached a different conclusion in (a)? (Assume that the total number of such antitrust suits is 4,000.)

**8.36** In October 1981, banks began to offer the All Savers Certificate, a new type of one-year certificate with interest that was tax-free (up to $1,000). A Virginia bank believes that about 70 percent of the people who bought these certificates obtained the money by withdrawing it from their savings accounts. It selects a random sample of 150 people who bought such certificates (which is a very small proportion of all such people).

(a) The bank says that it will reject this belief if more than 114 or less than 96 of the people in the sample obtained the money in this way. What value of $\alpha$ is the bank establishing?

(b) It turns out that 110 of the people in the sample obtained the money in this way. If the bank had set $\alpha$ at 0.05, would the difference between the sample proportion and 70 percent be statistically significant?

(c) What is the $p$-value of this test? Explain its meaning.

**8.37** An Atlantic City gambling casino wants to make sure that the probability of winning on a certain slot machine is 0.4 (no more, no less). The machine is operated 200 times, and the operator wins 89 times.

(a) If $\alpha$ is set at 0.01, is the difference between the sample proportion and 0.4 statistically significant?

(b) If the test described in the text is used, what is the probability of not rejecting the hypothesis that $\Pi = 0.4$ when in reality $\Pi = 0.38$, 0.40, and 0.42?

(c) Draw three points on the operating-characteristic curve of the test.

**8.38** In a poll of television viewers in Los Angeles, 130 out of 200 Democrats disliked a particular news program, while 158 out of 300 Republicans disliked it. Is there a real difference of opinion along party lines on this matter? What is the $p$-value of this test?

**8.39** In an agricultural experiment to determine the effects of a particular insecticide, a field was planted with corn. Half of the plants were sprayed with the insecticide, and half were not sprayed with it. Several weeks later, a random sample of 200 sprayed plants was selected, and a random sample of 200 unsprayed plants was selected. The number of healthy plants in each sample was as follows:

|             | *Sprayed* | *Unsprayed* |
|-------------|-----------|-------------|
| Healthy     | 121       | 109         |
| Not healthy | 79        | 91          |
| Total       | 200       | 200         |

(a) If the significance level is set at 0.05, does the evidence indicate that a higher proportion of sprayed than of unsprayed plants was healthy?

(b) If the significance level is set at 0.02 instead, what does the evidence indicate?

**8.40** (a) John Jerome finds that the difference between the mean IQ of a sample of students at the local high school and 100 is not statistically significant. Explain what this means.

(b) Is the finding in (a) independent of the level at which $\alpha$ is set?

(c) John Jerome also finds that the difference between the mean IQ of the sample of students and 140 is statistically significant. Explain what this means.

(d) Is the finding in (c) independent of the level at which $\alpha$ is set?

**8.41** It is necessary for an automobile producer to test the hypothesis that the mean number of miles per gallon achieved by its cars is 28 against the alternative hypothesis that it is not 28.

(a) If the standard deviation of the number of miles per gallon achieved by the company's cars is 6, and if the company decides to base its test on a random sample of 100 of its cars, provide a suitable test procedure if $\alpha$ is set equal to .05.

(b) Suppose that the mean number of miles per gallon for the sample of 100 cars is 26.2. On the basis of this result, should the company reject the hypothesis that the population mean is 28? Why, or why not?

(c) Suppose that the automobile producer is interested in rejecting the null hypothesis only if the mean number of miles per gallon achieved by its cars is less than 28. Provide the firm with a suitable test procedure for this set of circumstances.

(d) If the automobile producer is interested in rejecting the null hypothesis only if the mean number of miles per gallon achieved by its cars is less than 28, should it reject the null hypothesis? Why, or why not?

**8.42** The U.S. Army receives a shipment of 1,000 tires. If more than 10 of the tires are defective, the Army wants to reject the shipment.
(a) What is the acceptable quality level?
(b) According to the Defense Department's sampling plans, what should be the acceptance number and rejection number, assuming that the sample size is 80?

**8.43** The U.S. Navy receives a shipment of 1,000 shells. If more than 40 of the shells are defective, the Navy wants to reject the shipment.
(a) What is the acceptable quality level?
(b) According to the Defense Department's sampling plans, what should be the acceptance number and the rejection number, given that the sample size is 80?
(c) If the sampling plan described in your answer to (b) is used, what is the probability of accepting the shipment if it contains (i) 119 defective shells (ii) 142 defective shells; (iii) 158 defective shells?

**8.44** In the *New York Times* in late 1985, Congressman William Green attributed the rejection of the Westway project, a road proposed to be constructed in New York, in part to a draft statement by the Corps of Engineers which stated that Westway would have a "significant adverse effect" on striped bass in the Hudson River. Afterward, the Corps said that the effect would be small. The Court of Appeals responded that this was "Orwellian-like 'doublespeak.' "[24] Do you agree? Explain.

***8.45** An insurance company wants to test whether the mean age of women having legal abortions is 20 years. A random sample of ten women who had legal abortions in 1986 indicated that their ages (in years at the time of the abortion) were 21, 18, 28, 16, 22, 26, 31, 21, 19, 30. Minitab is used to test the hypothesis that the mean age of women having legal abortions in 1986 was 20 years. Interpret the computer printout, which is as follows:

```
MTB > ttest mu=20,c6

TEST OF MU = 20.00 VS MU N.E. 20.00
```

|     | N  | MEAN  | STDEV | SE MEAN | T    | P VALUE |
|-----|----|-------|-------|---------|------|---------|
| C6  | 10 | 23.20 | 5.22  | 1.65    | 1.94 | 0.085   |

[24] Bill Green, "When a Significant Difference Becomes Critical," *New York Times*, November 9, 1985.

* Exercise requiring some familiarity with Minitab.

# PART IV

# Chi-Square Tests, Nonparametric Techniques, and the Analysis of Variance

# Chi-Square Tests and Nonparametric Techniques*

## 9.1 Introduction

The procedures described in the previous chapter are only a small (albeit very important) sample of the many tests available for coping with a wide variety of problems that arise in business and economics. In this chapter we will present some additional tests which have a wide application. These tests are based on the principles described in Chapter 8, but they deal with different hypotheses and assumptions than those encountered there.

The $\chi^2$ (chi-square) tests taken up in this chapter are based on a very important sampling distribution that we have not encountered before—the $\chi^2$ or chi-square distribution. Besides describing these tests, we will also discuss the $\chi^2$ distribution itself. Then we will take up nonparametric techniques, which make fewer assumptions about the nature of the population being sampled than the tests described in the previous chapter. Like the $\chi^2$ tests, nonparametric techniques are very widely used in practical work.

## 9.2 Target Practice and Pearson's Test Statistic

To begin our discussion of $\chi^2$ tests, let's consider the following sporting situation. Suppose that a marksman fires 200 shots at a target, and that each shot can result in three outcomes: (1) a bull's-eye; (2) a hit on the target, but not a bull's-eye; and (3) a miss of the target. If the probability that a shot results in each of these outcomes is $\Pi_1$, $\Pi_2$, and $\Pi_3$, respectively, and if the outcomes of successive shots are statistically independent, what is the expected number of bull's-eyes? Clearly, that

---

* Some instructors may want to skip the latter part of this chapter, which deals with nonparametric techniques. Others may want to skip this chapter entirely. Subsequent chapters have been written so that this may be done without loss of continuity.

answer is $200\varPi_1$. What is the expected number of hits (but not bull's-eyes)? Clearly, $200\varPi_2$. What is the expected number of misses? Clearly, $200\varPi_3$.

Given the outcomes of the 200 shots, we want to test the null hypothesis that $\varPi_1 = .2$, $\varPi_2 = .3$, and $\varPi_3 = .5$. To test this hypothesis, it seems reasonable to compare the actual number of shots having each outcome with the number that would be expected, given that these specified values of $\varPi_1$, $\varPi_2$, and $\varPi_3$ are true. For example, if $\varPi_1$ equals .2, we would expect about .2(200), or 40, shots to result in bull's-eyes. If the actual number of bull's-eyes is only 14, there is a seemingly large difference (14–40) between the actual and expected number of bull's-eyes. To tell whether we should view this difference as evidence that the hypothesis that $\varPi_1 = .2$ is untrue, we must know the probability of a difference of this size, given that $\varPi_1 = .2$.

In 1900 Karl Pearson, an English statistician who was one of the fathers of modern statistics, proposed that in a situation of this sort the following procedure be used to test the null hypothesis that $\varPi_1$, $\varPi_2$, and $\varPi_3$ equal the specified values (.2, .3, and .5 in this case). First, we determine the expected number of shots that would be bull's-eyes if $\varPi_1 = .2$. This expected number (which equals .2(200), or 40) is denoted by $e_1$. Second, we determine the expected number of shots that would be hits (but not bull's-eyes) if $\varPi_2 = .3$. This expected number (which equals .3(200), or 60) is denoted by $e_2$. Third, we determine the expected number of shots that would be misses if $\varPi_3 = .5$. This expected number (which equals .5(200), or 100) is denoted by $e_3$. Then we calculate the following test statistic:

**Chi-Square Test Statistic**

$$\sum_{i=1}^{3} \frac{(f_i - e_i)^2}{e_i},$$

(9.1)

where $f_1$ is the actual number of shots that are bull's-eyes, $f_2$ is the actual number that are hits (but not bull's-eyes), and $f_3$ is the actual number of misses. Pearson showed that *if the null hypothesis is true (that is, if $\varPi_1 = .2$, $\varPi_2 = .3$, and $\varPi_3 = .5$), and if the sample size is large enough so that the smallest value of $e_i$ is at least 5, this test statistic will have a sampling distribution which can be approximated adequately by the $\chi^2$ distribution,* a very important probability distribution to which we now turn.

## 9.3 The Chi-Square Distribution

The chi-square ($\chi^2$) distribution, which is the probability distribution of a $\chi^2$ random variable, is defined as follows.

$\chi^2$ DISTRIBUTION: *The $\chi^2$ distribution, with $\nu$ (nu) degrees of freedom, is the probability distribution of the sum of squares of $\nu$ (nu) independent standard normal variables.*

To clarify this definition, let's begin by taking a single standard normal variable. (Recall from Chapter 5 that the standard normal variable is $(X - \mu)/\sigma$, where $X$ is a normal variable with mean equal to $\mu$ and standard deviation equal to $\sigma$.) Instead of considering the probability distribution of its value, let's consider the probability distribution of the *square* of its value. If one knows that this variable has a standard normal distribution, it should be possible to figure out the distribution of the square of its value. This distribution, derived many years ago, is a $\chi^2$ distribution with 1 degree of freedom. Next, suppose that we consider two independent standard normal variables. Let's square the value of each and add the squares together. The distribution of this sum is a $\chi^2$ distribution with 2 degrees of freedom. Finally, suppose that we consider four independent standard normal variables. Let's square the value of each and add the squares together. The distribution of this sum, shown in panel A of Figure 9.1, is a $\chi^2$ distribution (with 4 degrees of freedom).

Like the $t$ distribution, the $\chi^2$ distribution is a family of distributions, each of which is characterized by a certain number of *degrees of freedom*.[1] The number of degrees of freedom is the number of squares of standard normal variables that are summed up. Thus, in Figure 9.1, panel A shows a $\chi^2$ distribution with 4 degrees of freedom; panel B shows a $\chi^2$ distribution with 10 degrees of freedom; and panel C shows a $\chi^2$ distribution with 20 degrees of freedom.

As is evident from Figure 9.1, the $\chi^2$ distribution generally is skewed to the right. Since the $\chi^2$ random variable is a sum of squares, the probability that it is negative is zero. If there are $v$ degrees of freedom, it can be shown that

$$E(\chi^2) = v, \qquad (9.2)$$
$$\sigma(\chi^2) = \sqrt{2v}. \qquad (9.3)$$

*Mean and Standard Deviation of a $\chi^2$ Random Variable*

In other words, the mean of a $\chi^2$ random variable equals its number of degrees of freedom, and its standard deviation equals the square root of twice its number of degrees of freedom.

In applying the test procedures described in the following sections, it is essential that we be able to find the value of $\chi^2$ that is exceeded with a probability of .05, .10, or some other such amount. That is, we need to be able to calculate $\chi_\alpha^2$, which is the value of $\chi^2$ that is exceeded with a probability of $\alpha$. Appendix Table 8 shows the value of $\chi_\alpha^2$ for various numbers of degrees of freedom and for various values of $\alpha$. Each row in this table corresponds to a certain number of degrees of freedom. For example, the tenth row shows that, if there are 10 de-

---

[1] There is a close relationship between the $t$ and $\chi^2$ distributions. It can be shown that $t$ equals the ratio of a standard normal variable to the square root of a $\chi^2$ variable divided by its number of degrees of freedom. For further explanation one should consult a more advanced statistics text.

**Figure 9.1**
$\chi^2$ Probability Density Functions with 4, 10, and 20 Degrees of Freedom

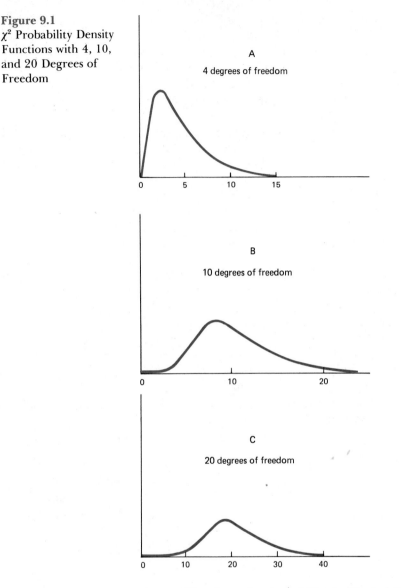

A
4 degrees of freedom

B
10 degrees of freedom

C
20 degrees of freedom

grees of freedom, the probability is .10 that the value of $\chi^2$ will exceed 15.987, that the probability is .05 that it will exceed 18.307, and that the probability is .01 that it will exceed 23.2093. As the number of degrees of freedom becomes very large, the $\chi^2$ distribution can be approximated by the normal distribution.

## 9.4 Test of Differences among Proportions

One very important application of the $\chi^2$ distribution is to problems where the decision maker wants to determine whether a number of proportions are equal. In the previous chapter, we showed how one can test whether two population proportions are equal. Now we show how one can test whether any number of population proportions are

| Day | Number defective | Number accepted | Sample size |
|---|---|---|---|
| April 27 | 4 | 46 | 50 |
| 28 | 9 | 41 | 50 |
| 29 | 10 | 40 | 50 |
| 30 | 11 | 39 | 50 |
| May 1 | 13 | 37 | 50 |
| 2 | 30 | 20 | 50 |
| 3 | 26 | 24 | 50 |
| 4 | 13 | 37 | 50 |
| 5 | 8 | 42 | 50 |
| 6 | 23 | 27 | 50 |
| 7 | 34 | 16 | 50 |
| 8 | 25 | 25 | 50 |
| 9 | 18 | 32 | 50 |
| 10 | 12 | 38 | 50 |
| 11 | 4 | 46 | 50 |
| 12 | 3 | 47 | 50 |
| 14 | 11 | 39 | 50 |
| 15 | 8 | 42 | 50 |
| 16 | 14 | 36 | 50 |
| 17 | 21 | 29 | 50 |
| 18 | 25 | 25 | 50 |
| 19 | 18 | 32 | 50 |
| 21 | 10 | 40 | 50 |
| 22 | 8 | 42 | 50 |
| 23 | 18 | 32 | 50 |
| 24 | 19 | 31 | 50 |
| 25 | 4 | 46 | 50 |
| 26 | 8 | 42 | 50 |

$$\sum \frac{(f-e)^2}{e} \doteq 182$$

**Table 9.1**
Number of Defective
Railway-Car Side
Frames, 28 Days

equal. To illustrate the sort of situation to which this test is applicable, we will describe an actual case study involving the production of side frames of railway cars.[2] Each day, a foundry producing these frames sampled 50 at random and, after inspecting them, determined how many were defective. Table 9.1 shows the number determined defective for each day during a 28-day period. The question that the executives of the foundry wanted answered was: Is the proportion defective constant from day to day, or does it vary? If the proportion was found to vary from day to day, this meant the production process was not in control; and the firm would then try to determine the reasons for the variation.

In a problem of this sort, *the null hypothesis is that the population proportions are all equal.* Thus, in this case study, the null hypothesis is that $\Pi_1 = \Pi_2 = \Pi_3 = \ldots \Pi_{28}$, and *the alternative hypothesis is that these proportions are not all equal.* (Of course, $\Pi_1$ is the population proportion defective on the first day, $\Pi_2$ is the population proportion defec-

[2] This case is reported in A. Duncan, *op. cit.*

tive on the second day, and so on.) If the null hypothesis is true, the proportion defective on all days can be estimated by pooling the data for all days. Clearly, the common proportion defective equals

$$\frac{\sum_{i=1}^{28} x_i}{\sum_{i=1}^{28} n_i},$$

where $x_i$ is the number of defectives on the $i$th day, and $n_i$ is the size of the total sample in the $i$th day. Thus, in this case the expected proportion defective on any day, if the null hypothesis is correct, is

$$\frac{4 + \ 9 + 10 + 11 + 13 + \ldots + 19 + \ 4 + \ 8}{50 + 50 + 50 + 50 + 50 + \ldots + 50 + 50 + 50} = .29.$$

To test whether the null hypothesis is correct, we calculate the expected number of defectives and the expected number of nondefectives on each day. Since the common proportion defective is .29, we would expect that $50(.29) = 14.5$ frames would be defective each day, and that $50(.71) = 35.5$ frames would not be defective each day. To test the null hypothesis, we compare these theoretical, or expected, frequencies with the actual ones. Clearly, the greater the difference between the theoretical frequencies and the actual ones, the less likely it is that the null hypothesis is true.

Specifically, the test procedure is as follows. Having calculated each expected frequency (shown in Table 9.2), we must compute the following test statistic,

$$\sum \frac{(f-e)^2}{e}, \qquad \qquad (9.4)$$

where $f$ is the actual frequency, $e$ is the corresponding expected frequency, and the summation is over all items in Table 9.2.[3] If the null hypothesis is true, this test statistic has a sampling distribution that can be approximated by the $\chi^2$ distribution with degrees of freedom equal to $(r - 1)$, where $r$ is the number of population proportions that are being compared.

Why is $(r - 1)$ the appropriate number of degrees of freedom? The general rule is that *the appropriate number of degrees of freedom equals the number of comparisons between actual and expected frequencies, less the number of independent linear restrictions placed upon the frequencies.* Because the number of such restrictions varies from one case to another in this chapter, the number of degrees of freedom will be given by different formulas from one case to another. Here the appro-

---

[3] Note that this test statistic is the same as that in expression (9.1).

| Day | Number defective Actual ($f$) | Expected ($e$) | $\frac{(f-e)^2}{e}$ | Number not defective Actual ($f$) | Expected ($e$) | $\frac{(f-e)^2}{e}$ |
|---|---|---|---|---|---|---|
| April 27 | 4 | 14.5 | $(4-14.5)^2/14.5$ | 46 | 35.5 | $(46-35.5)^2/35.5$ |
| 28 | 9 | 14.5 | $(9-14.5)^2/14.5$ | 41 | 35.5 | $(41-35.5)^2/35.5$ |
| 29 | 10 | 14.5 | $(10-14.5)^2/14.5$ | 40 | 35.5 | $(40-35.5)^2/35.5$ |
| 30 | 11 | 14.5 | $(11-14.5)^2/14.5$ | 39 | 35.5 | $(39-35.5)^2/35.5$ |
| May 1 | 13 | 14.5 | $(13-14.5)^2/14.5$ | 37 | 35.5 | $(37-35.5)^2/35.5$ |
| 2 | 30 | 14.5 | $(30-14.5)^2/14.5$ | 20 | 35.5 | $(20-35.5)^2/35.5$ |
| 3 | 26 | 14.5 | $(26-14.5)^2/14.5$ | 24 | 35.5 | $(24-35.5)^2/35.5$ |
| 4 | 13 | 14.5 | $(13-14.5)^2/14.5$ | 37 | 35.5 | $(37-35.5)^2/35.5$ |
| 5 | 8 | 14.5 | $(8-14.5)^2/14.5$ | 42 | 35.5 | $(42-35.5)^2/35.5$ |
| 6 | 23 | 14.5 | $(23-14.5)^2/14.5$ | 27 | 35.5 | $(27-35.5)^2/35.5$ |
| 7 | 34 | 14.5 | $(34-14.5)^2/14.5$ | 16 | 35.5 | $(16-35.5)^2/35.5$ |
| 8 | 25 | 14.5 | $(25-14.5)^2/14.5$ | 25 | 35.5 | $(25-35.5)^2/35.5$ |
| 9 | 18 | 14.5 | $(18-14.5)^2/14.5$ | 32 | 35.5 | $(32-35.5)^2/35.5$ |
| 10 | 12 | 14.5 | $(12-14.5)^2/14.5$ | 38 | 35.5 | $(38-35.5)^2/35.5$ |
| 11 | 4 | 14.5 | $(4-14.5)^2/14.5$ | 46 | 35.5 | $(46-35.5)^2/35.5$ |
| 12 | 3 | 14.5 | $(3-14.5)^2/14.5$ | 47 | 35.5 | $(47-35.5)^2/35.5$ |
| 14 | 11 | 14.5 | $(11-14.5)^2/14.5$ | 39 | 35.5 | $(39-35.5)^2/35.5$ |
| 15 | 8 | 14.5 | $(8-14.5)^2/14.5$ | 42 | 35.5 | $(42-35.5)^2/35.5$ |
| 16 | 14 | 14.5 | $(14-14.5)^2/14.5$ | 36 | 35.5 | $(36-35.5)^2/35.5$ |
| 17 | 21 | 14.5 | $(21-14.5)^2/14.5$ | 29 | 35.5 | $(29-35.5)^2/35.5$ |
| 18 | 25 | 14.5 | $(25-14.5)^2/14.5$ | 25 | 35.5 | $(25-35.5)^2/35.5$ |
| 19 | 18 | 14.5 | $(18-14.5)^2/14.5$ | 32 | 35.5 | $(32-35.5)^2/35.5$ |
| 21 | 10 | 14.5 | $(10-14.5)^2/14.5$ | 40 | 35.5 | $(40-35.5)^2/35.5$ |
| 22 | 8 | 14.5 | $(8-14.5)^2/14.5$ | 42 | 35.5 | $(42-35.5)^2/35.5$ |
| 23 | 18 | 14.5 | $(18-14.5)^2/14.5$ | 32 | 35.5 | $(32-35.5)^2/35.5$ |
| 24 | 19 | 14.5 | $(19-14.5)^2/14.5$ | 31 | 35.5 | $(31-35.5)^2/35.5$ |
| 25 | 4 | 14.5 | $(4-14.5)^2/14.5$ | 46 | 35.5 | $(46-35.5)^2/35.5$ |
| 26 | 8 | 14.5 | $(8-14.5)^2/14.5$ | 42 | 35.5 | $(42-35.5)^2/35.5$ |

$$\sum \frac{(f-e)^2}{e} \doteq 182$$

priate formula is $(r-1)$, since the sum of the frequencies (50) in each day's population is given and the total number of defectives in the sample is also given. Thus, there are $(r+1)$ restrictions on the frequencies.[4] Subtracting $(r+1)$ from $2r$ (the number of comparisons between actual and expected frequencies), we get $(r-1)$, the number of degrees of freedom.

As we just noted, if the null hypothesis is true, the distribution of $\Sigma(f-e)^2/e$ can be approximated by the $\chi^2$ distribution with $(r-1)$ degrees of freedom, which is shown in Figure 9.2. The rejection region is under the upper tail of this distribution, since large values of this test statistic indicate large discrepancies between the actual and expected

[4] What are the $(r+1)$ restrictions? For each of the $r$ populations (in this case, 28 days), the number of defectives plus the number of non-defectives must equal 50. Besides these $r$ restrictions, the total number of defectives is taken as given. (That is, $\sum_{i=1}^{n} x_i = 407$.) Thus, there are $(r+1)$ restrictions in all.

**Figure 9.2**
Sampling Distribution
of $\sum \frac{(f-e)^2}{e}$, If Null
Hypothesis Is True

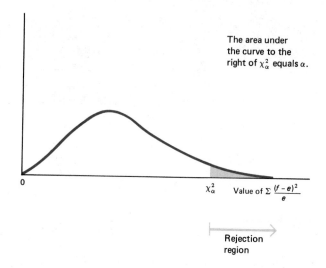

The area under
the curve to the
right of $\chi_\alpha^2$ equals $\alpha$.

0

$\chi_\alpha^2$    Value of $\sum \frac{(f-e)^2}{e}$

Rejection
region

frequencies. In general, to test whether $r$ population proportions are
equal, the appropriate decision rule is as follows.

*Test that* $\Pi_1 = \Pi_2 =$
$\dots \Pi_r$

DECISION RULE: *Reject the null hypothesis that* $\Pi_1 = \Pi_2 = \dots = \Pi_r$ *if*
$\Sigma(f-e)^2/e > \chi_\alpha^2$ *where* $\alpha$ *is the desired significance level of the test*
*(and the number of degrees of freedom is* r − 1*).*

Table 9.2 shows that the value of $\Sigma(f-e)^2/e$ is approximately
182. Appendix Table 8 shows that $\chi_{.05}^2 = 40.1$ if the number of degrees
of freedom equals 27. (Why are there 27 degrees of freedom? Because
the number of degrees of freedom equals the number of population
proportions to be compared—28 in this case—minus 1.) Since 182 exceeds 40.1, it is clear that the null hypothesis should be rejected. Even
if the decision maker sets the significance level at .01, the null hypothesis should be rejected since $\chi_{.01}^2 = 46.96$ if the number of degrees of
freedom equals 27. Thus, there is every indication that the proportion
defective varies from day to day, which in turn indicates that the foundry's performance was not in control. Based on this evidence, the firm
was led to study the reasons for the observed variation in performance
since it clearly was more than merely chance variation.

The test of differences among proportions that we have just described is commonly employed in business and economics. The following is another illustration of how it is used.

EXAMPLE 9.1 In 1981, Andrew Cherlin and Pamela Walters published a
study estimating the proportion of men in 1972, 1975, and 1978
who approved of a married woman earning money (if she has a
husband capable of supporting her).[5] The number of men in their
sample that approved and disapproved each year was as follows:

[5] A. Cherlin and P. Walters, "Trends in United States Men's and Women's
Sex-Role Attitudes: 1972 to 1978," *American Sociological Review*, August 1981.

|  | *1972* | *1975* | *1978* |
|---|---|---|---|
| Approve | 410 | 412 | 409 |
| Disapprove | 252 | 176 | 151 |

On the basis of these data, should one reject the hypothesis that the proportion of men who approved has remained constant during these three years? (Assume that the significance level equals .05.)

SOLUTION: Summing up the data for all three years, 1,231 of the 1,810 men approved. Thus, the overall proportion is 1,231/1,810 = 0.68. The expected number approving and disapproving, if the null hypothesis is true, is

|  | *1972* | *1975* | *1978* |
|---|---|---|---|
| Approve | 450 | 400 | 381 |
| Disapprove | 212 | 188 | 179 |

Thus, $\sum \dfrac{(f-e)^2}{e} = \dfrac{(410-450)^2}{450} + \dfrac{(252-212)^2}{212} + \dfrac{(412-400)^2}{400}$

$+ \dfrac{(176-188)^2}{188} + \dfrac{(409-381)^2}{381} + \dfrac{(151-179)^2}{179} = 18.68.$

Since there are three population proportions being compared, the number of degrees of freedom is $3 - 1$, or 2. According to Appendix Table 8, $\chi^2_{.05} = 5.991$ when there are 2 degrees of freedom. Thus, since the observed value of $\Sigma(f-e)^2/e$ exceeds 5.991, the null hypothesis should be rejected. In other words, we should reject the hypothesis that the proportion of men who approved of a married woman earning money (if her husband can support her) was the same in 1972, 1975, and 1978.

## EXERCISES

**9.1** Find the value of $\chi^2_{.05}$ when there are (a) 5 degrees of freedom; (b) 10 degrees of freedom; (c) 20 degrees of freedom.

**9.2** Find the value of $\chi^2_{.01}$ when there are (a) 8 degrees of freedom; (b) 14 degrees of freedom; (c) 26 degrees of freedom.

**9.3** If the coefficient of variation of a $\chi^2$ random variable equals 1, how many degrees of freedom does it have?

**9.4** If $\chi^2_{.01} = 29.1413$, what is the number of degrees of freedom? If $\chi^2_{.95} = 10.8508$, what is the number of degrees of freedom?

**9.5** We add up the squared values of 15 independent standard normal variables. (a) What sort of probability distribution does this sum have? (b) What is the expected value of this sum? (c) What is the variance of this sum?

**9.6** The Crooked Arrow National Bank obtains data concerning a random sample of 100 deposit slips each day. The number of these slips containing errors is as follows for a period of 10 consecutive days: 9; 10; 8; 6; 12; 15; 12; 9; 8; 6. Test whether the proportion of deposit slips containing an error is the same each day. (Let $\alpha = .05$.)

**9.7** The Quarryville National Bank obtains data concerning a sample of 200 loan applications at each of its four branches in 1987. The number of these loan applications containing falsified information at each branch is 6; 8; 9; and 12. Test whether the proportion of loan applications containing falsified information is the same at each branch. (Let $\alpha = .05$.)

**9.8** The Alpha Corporation, a bicycle maker, obtains tires from two suppliers, firm A and firm B. During 1987, it received the following numbers of acceptable and defective tires from each supplier:

| Supplier | Number of acceptable tires | Number of defective tires | Total number of tires received |
|---|---|---|---|
| A | 940 | 89 | 1,029 |
| B | 780 | 32 | 812 |

Is the probability that the Alpha Corporation would receive a defective tire the same, regardless of the supplier? Test this hypothesis, based on a significance level of .05.

**9.9** A firm has four manufacturing plants. To determine whether the proportion of items produced that are too small (relative to specifications) is the same at each plant, the firm has 80 items chosen at random off each plant's assembly line, the results being as follows:

| | Plant I | Plant II | Plant III | Plant IV |
|---|---|---|---|---|
| Too small | 9 | 15 | 20 | 21 |
| Not too small | 71 | 65 | 60 | 59 |
| Total | 80 | 80 | 80 | 80 |

(a) Can the observed differences among plants be attributed to chance variation? (Set $\alpha = .05$.)

(b) If $\alpha$ is set at .01, can the observed differences be attributed to chance variation?

## 9.5 Contingency Tables

Decision makers are often concerned with problems involving contingency tables. A *contingency table* indicates whether two characteristics or variables are dependent on one another. In other words, a contingency table contains two variables of classification, and the point is to determine whether these two variables are related. For example, Table 9.3 shows a simple contingency table where the vertical columns represent Republicans and Democrats and the horizontal rows show the number of persons earning more or no more than $50,000 per year. Based on a random sample of 300 individuals, the results are as shown in Table 9.3. This is a $2 \times 2$ contingency table since there are two rows

| Income | Republicans | Democrats | Total |
|---|---|---|---|
| More than $50,000 | 30(20) | 30(40) | 60 |
| No more than $50,000 | 70(80) | 170(160) | 240 |
| Total | 100 | 200 | 300 |

**Table 9.3**
A 2 × 2 Contingency
Table

and two columns. *A table of this sort can be used to test whether a person's income is independent of his or her political affiliation.*

The null hypothesis here is that the probability that a Republican will have an income above $50,000 is the same as the probability that a Democrat will have an income above $50,000. That is, the null hypothesis is that income is *independent* of political party. To test this hypothesis, we compute the expected frequencies in Table 9.3, assuming that the hypothesis is true. Since the probability of a person making over $50,000 is 60 ÷ 300 or .20 in the sample as a whole, we would expect that 20 percent of the Republicans—that is, .20(100), or 20 Republicans—would earn over $50,000 per year, and that 20 percent of the Democrats—that is, .20(200), or 40 Democrats—would earn over $50,000 per year, if the null hypothesis is true. By the same token, we would expect that 80 percent of the Republicans—that is, .80(100), or 80 Republicans—would make no more than $50,000 per year, and that 80 percent of the Democrats—that is, .80(200), or 160 Democrats—would make no more than $50,000 per year, if the null hypothesis is true.

Given these expected frequencies, which are shown in parentheses in Table 9.3, we test the null hypothesis by computing $\Sigma(f - e)^2/e$. In other words $\Sigma(f - e)^2/e$ is the test statistic. If $\alpha$ is the desired significance level, the decision rule is as follows.

DECISION RULE: *Reject the null hypothesis (of independence) if $\Sigma(f - e)^2/e > \chi^2_\alpha$, where there are $(r - 1)(c - 1)$ degrees of freedom (r being the number of rows and c being the number of columns).*

*Test of Independence in a Contingency Table*

Why is the appropriate number of degrees of freedom equal to $(r - 1)(c - 1)$? As noted in the previous section, the appropriate number of degrees of freedom equals the number of comparisons between actual and expected frequencies less the number of restrictions placed upon these frequencies. Since the number of entries in a contingency table equals *cr,* there are *cr* actual frequencies to be compared with the corresponding expected frequencies. Because the sum of the frequencies in each row and each column is a given quantity, there are $r + c - 1$ such restrictions.[6] Thus, the appropriate number of degrees of freedom is $cr - (r + c - 1)$, or $(r - 1)(c - 1)$.

[6] There is a restriction for each row and column, since the sum of the frequencies in each row or column is given. However, one of these restrictions must hold if all the others are met, so the number of independent restrictions is $r + c - 1$.

In this case the value of the test statistic is

$$\sum \frac{(f-e)^2}{e} = \frac{(30-20)^2}{20} + \frac{(30-40)^2}{40} + \frac{(70-80)^2}{80} + \frac{(170-160)^2}{160}$$

$$= \frac{100}{20} + \frac{100}{40} + \frac{100}{80} + \frac{100}{160} = 9.375.$$

Suppose that the significance level is set at .05. If so, the null hypothesis should be rejected, since $\Sigma(f-e)^2/e$ exceeds $\chi^2_{.05}$, which is 3.84. (Note that there is only 1 degree of freedom here, since $(r-1) \cdot (c-1) = 1$ in a $2 \times 2$ contingency table.)[7] In other words based on this evidence, the probability that a person's income is above \$50,000 does not seem to be independent of his or her political affiliation.

## 9.6 Contingency Tables and Metal Castings: A Case Study

Contingency tables are of great practical importance, and in this section we will describe their application to an actual industrial problem.[8] Table 9.4 provides actual data for a three-week period concerning the causes of rejects of metal castings in a certain manufacturing plant. The managers of this plant wanted to know whether the probability

**Table 9.4**
Causes of Rejects of Metal Castings, Three Weeks

| Cause of rejection | First | Week Second | Third | Total |
|---|---|---|---|---|
| Sand | 97 (93.9) | 120 (111.1) | 82 (93.9) | 299 |
| Misrun | 8 (8.5) | 15 (10.0) | 4 (8.5) | 27 |
| Shift | 18 (9.4) | 12 (11.2) | 0 (9.4) | 30 |
| Drop | 8 (10.4) | 13 (12.3) | 12 (10.4) | 33 |
| Corebreak | 23 (25.8) | 21 (30.5) | 38 (25.7) | 82 |
| Broken | 21 (19.8) | 17 (23.4) | 25 (19.8) | 63 |
| Other | 5 (12.3) | 15 (14.5) | 19 (12.3) | 39 |
| Total | 180 | 213 | 180 | 573 |

[7] When there is only 1 degree of freedom, one should use a slightly different formula:

$$\sum \frac{(|f-e| - 1/2)^2}{e}.$$

In this case, the result would be

$$\frac{(9.5)^2}{20} + \frac{(9.5)^2}{40} + \frac{(9.5)^2}{80} + \frac{(9.5)^2}{160} = 90.25\left(\frac{3}{32}\right) = 8.46.$$

Thus, the results are essentially the same as in the text. Note that this so-called *continuity correction factor* also applies to the $\chi^2$ tests in Sections 9.4 and 9.7, as well as to contingency tables, when there is only 1 degree of freedom.

[8] This case is from A. Duncan, *op. cit.*

distribution of causes of rejects was the same from one week to the next. Clearly, the question is of importance, since increases in the probability of a particular cause are likely to indicate that the production process is becoming increasingly deficient in certain ways. A $\chi^2$ test was used to answer this question. The null hypothesis is that the probability distribution of causes is the same for each week. Based on this assumption, the firm's statisticians calculated the expected frequencies for each item in the table, these frequencies being shown in parentheses to the right of the corresponding actual frequencies in Table 9.4.

To obtain each of the expected frequencies, the (horizontal) row total for the item is multiplied by the (vertical) column total for the item, and the result is then divided by the overall total. For example, to get the expected frequency in the first row in the first column, we multiply the total for (horizontal) row 1 (299) by the total for (vertical) column 1 (180) and divide by the overall total (573). To see why this is correct, note that the total for row 1 divided by the overall total is an estimate of the probability of being in row 1, and the total for column 1 divided by the overall total is an estimate of the probability of being in column 1. Thus, to estimate the probability of being in *both* row 1 and column 1, assuming that these events are statistically independent, we obtain the product of these two estimated probabilities. This product times the overall total equals the expected frequency in the first row in the first column if the null hypothesis is true. Clearly, the same result can be obtained by multiplying the total for row 1 by the total for column 1 and dividing by the overall total.

Once the firm had obtained the expected frequencies in Table 9.4, it was a simple matter to test the null hypothesis. All that the firm had to do was compute $\Sigma(f-e)^2/e$, which equals

$$\frac{(97-93.9)^2}{93.9}+\frac{(8-8.5)^2}{8.5}+\frac{(18-9.4)^2}{9.4}+\frac{(8-10.4)^2}{10.4}+\frac{(23-25.8)^2}{25.8}+$$

$$\frac{(21-19.8)^2}{19.8}+\frac{(5-12.3)^2}{12.3}+\frac{(120-111.1)^2}{111.1}+\frac{(15-10.0)^2}{10.0}+$$

$$\frac{(12-11.2)^2}{11.2}+\frac{(13-12.3)^2}{12.3}+\frac{(21-30.5)^2}{30.5}+\frac{(17-23.4)^2}{23.4}+$$

$$\frac{(15-14.5)^2}{14.5}+\frac{(82-93.9)^2}{93.9}+\frac{(4-8.5)^2}{8.5}+\frac{(0-9.4)^2}{9.4}+\frac{(12-10.4)^2}{10.4}+$$

$$\frac{(38-25.7)^2}{25.7}+\frac{(25-19.8)^2}{19.8}+\frac{(19-12.3)^2}{12.3}=45.2.$$

Since $(r-1)(c-1)=(7-1)(3-1)=12$, the null hypothesis should be rejected at the .05 significance level if $\Sigma(f-e)^2/e$ exceeds $\chi^2_{.05}$, where $\chi^2$ has 12 degrees of freedom. Since Appendix Table 8 shows that $\chi^2_{.05}$ is 21.03 when there are 12 degrees of freedom, it follows that the firm should reject the null hypothesis—which is what it did. Thus, contrary to the null hypothesis, it appears that the probability distribution of causes of rejects did vary significantly from week to week.

## 9.7 Tests of Goodness of Fit

Still another important application of the $\chi^2$ distribution is where the decision maker wants to determine whether an observed frequency distribution conforms to a theoretical distribution. For example, suppose that a stock broker believes that the probability is .50 that one of his employees will sell over \$100,000 of securities in a day. Since the broker hires four such employees, he can use the binomial distribution to calculate the theoretical probabilities that 0, 1, 2, 3, or 4 of his employees will sell over \$100,000 of securities in a given day. (Recall Chapter 4.) These probabilities are shown in the first column of Table 9.5.

Let's assume that the broker, after collecting data concerning the actual number of employees selling over \$100,000 on each of 160 days, wants to determine whether the binomial distribution is an accurate representation of the actual distribution. The theoretical frequency distribution is provided in the third column of Table 9.5. To obtain this distribution, all that one has to do is multiply each of the numbers in the second column of Table 9.5 by 160. The actual frequency distribution is provided in the fourth column of Table 9.5. Of course, the actual and theoretical distributions do not coincide exactly, but this does not prove that the theoretical distribution is inappropriate, since some discrepancy between the two distributions would be expected due to chance. The question facing the stock broker is: Are the discrepancies between the actual and theoretical distributions so large that they cannot reasonably be attributed to chance?

To answer this kind of question, statisticians use a procedure quite similar to those described in the previous two sections. *The null hypothesis is that the actual distribution can in fact be represented by the theoretical distribution, and that the discrepancies between them are due to chance.* To test this hypothesis, we calculate $\Sigma(f-e)^2/e$, where $f$ is the observed frequency in a particular class interval of the frequency distribution and $e$ is the theoretical, or expected, frequency in the same class interval of the frequency distribution. If the null hypothesis is true, it can be shown that $\Sigma(f-e)^2/e$ has approximately a $\chi^2$ distribution, the number of degrees of freedom being 1 less than the number of values of $(f-e)^2/e$ that are summed up. (If some of the parameters of the theoretical distribution are estimated from the sample, the

**Table 9.5**
Goodness-of-Fit Test

| Number of employees selling over \$100,000 in securities in a day | Theoretical probability | Theoretical frequency | Actual frequency |
|---|---|---|---|
| 0 | 1/16 | 10 | 12 |
| 1 | 1/4 | 40 | 35 |
| 2 | 3/8 | 60 | 60 |
| 3 | 1/4 | 40 | 45 |
| 4 | 1/16 | 10 | 8 |

number of degrees of freedom is less than this amount by the number of parameters that are estimated.[9])

Once this test statistic has been calculated, the decision rule is as follows.

DECISION RULE: *Reject the null hypothesis that the discrepancy between the actual and theoretical frequency distributions is due to chance if $\Sigma(f - e)^2/e > \chi_\alpha^2$ where the number of degrees of freedom equals the number of class intervals of the frequency distribution minus* (h + 1), *where* h *is the number of parameters estimated from the sample.*

*Test that Discrepancies between Actual and Theoretical Distributions Are Due to Chance*

Given the actual and theoretical frequency distributions in Table 9.5, the value of the test statistic is

$$\sum \frac{(f-e)^2}{e} = \frac{(12-10)^2}{10} + \frac{(35-40)^2}{40} + \frac{(60-60)^2}{60}$$

$$+ \frac{(45-40)^2}{40} + \frac{(8-10)^2}{10}$$

$$= 0.4 + 0.625 + 0 + 0.625 + 0.4 = 2.05.$$

Since no parameter of the theoretical distribution is estimated from the data, the number of degrees of freedom is 1 less than the number of values of $(f-e)^2/e$ that are summed up; that is, $5 - 1 = 4$. Thus, if the .05 significance level is chosen, the null hypothesis should be rejected if $\Sigma(f-e)^2/e$ exceeds 9.488, since this is the value of $\chi_{.05}^2$ when there are 4 degrees of freedom. (See Appendix Table 8.) Since $\Sigma(f-e)^2/e$ does not exceed 9.488, it follows that there is no reason to reject the null hypothesis. In other words, the probability is greater than .05 that the observed discrepancies between the actual distribution and the binomial distribution could be due to chance.

In carrying out goodness-of-fit tests, statisticians sometimes force the mean of the theoretical distribution to equal the mean of the actual distribution. Frequently, the mean of the theoretical distribution is not stipulated on *a priori* grounds, and therefore this seems the sensible thing to do. For example, if the stock broker wanted to test whether the actual distribution in Table 9.5 was binomial, but was not willing to specify that the mean of the binomial distribution was 2.0, then the mean of the theoretical distribution would be set equal to the mean of the actual distribution. Since this would enable us to specify the entire theoretical distribution, we could calculate $\Sigma(f-e)^2/e$ and see whether it exceeds $\chi_{.05}^2$. But it is important to note that $\chi^2$ would have one less degree of freedom, because we estimated an additional param-

---

[9] In this case there are two parameters ($n$ and $\Pi$) of the theoretical probability distribution, and their values (4 and 0.5) are given. However, often we want to test whether the data conform to a binomial distribution without specifying the value of $\Pi$, since we have no *a priori* information concerning its value. More will be said about this below.

eter of the theoretical distribution (namely, the mean) from the actual distribution.[10]

Finally, in carrying out all such goodness-of-fit tests, one should define the class intervals of the frequency distribution so that the *theoretical, or expected, frequency in each and every class interval is at least 5*. The reason is that the $\chi^2$ distribution is not a good approximation of the distribution of $\Sigma(f-e)^2/e$ when the null hypothesis is correct, if the theoretical frequencies in particular class intervals are very small. To make sure that this rule of thumb is met, all that one has to do is to combine adjacent class intervals when any of them has a theoretical frequency of less than 5. (Example 9.2 illustrates this procedure.)

This rule of thumb should also be observed in the other applications of the $\chi^2$ distribution described in the preceding sections. In the case of tests of differences among proportions, each expected or theoretical frequency must equal 5 or more; in contingency tables, the theoretical frequency in each cell must equal 5 or more. Recall that, when we discussed Karl Pearson's findings in Section 9.2, we pointed out that his results assumed that the smallest expected or theoretical frequency (that is, the smallest value of $e_i$) is at least 5. This rule of thumb is important and should be observed.

[10] In testing whether a frequency distribution conforms to a normal population, statisticians frequently force the population mean to equal the sample mean and the population standard deviation to equal the sample standard deviation. Under these circumstances, the number of degrees of freedom is the number of class intervals in the frequency distribution minus 3 since two parameters are estimated.

## TO ERR IS HUMAN . . . BUT COSTLY

### SKULLDUGGERY IN THE PEA PATCH

Gregor Mendel, one of the great figures of science, published in 1865 a seminal article on heredity, which put forward basic theories concerning the ways in which various characteristics are transmitted from one generation to the next. His experiments were all conducted on garden peas. Mendel cross-fertilized two pure strains of pea plants, one producing only wrinkled green seeds, the other producing only round yellow seeds. Then he mated these hybrids with one another. According to his theories (in which he postulated the existence of what are now called *genes*), such a mating should result in plants with four types of seeds: round yellow, round green, wrinkled yellow, wrinkled green. Moreover, these types should occur in definite proportions: for every plant with wrinkled green seeds, there should be three plants with round green seeds, three plants with wrinkled yellow seeds, and nine plants with round yellow seeds.

Mendel carried out many experiments and found that his experimental results were very close to what his theories would predict. For example, among 556 plants resulting from the mating of the above hybrids, he obtained the following results:

| Type of pea | Observed number of plants | Expected number of plants |
|---|---|---|
| Round yellow | 315 | 313 |
| Wrinkled yellow | 101 | 104 |
| Smooth green | 108 | 104 |
| Wrinkled green | 32 | 35 |

The expected number of plants of each type, if his theories are correct, is shown in the second column above. If we compute $\Sigma(f-e)^2/e$, the result is $(315-313)^2/313 + (101-104)^2/104 + (108-104)^2/104 + (32-35)^2/35$, or about 0.5.

Decades later, the great British statistician R. A. Fisher said that something was wrong with Mendel's findings. What was it?

SOLUTION: Fisher pointed out that the data fit the theoretical distribution *too well*. Based on chance, one would expect *larger* deviations from the theoretical distribution than Mendel reported. For example, the above value of $\chi^2$ is unusually low. (According to Appendix Table 8, the probability that a $\chi^2$ random variable with 3 degrees of freedom would equal 0.5 or less is about 0.08.) Moreover, Fisher found that the agreement between the observed and theoretical distributions was unusually close in all of Mendel's experiments. The implication clearly was that Mendel's data were fudged, although Fisher delicately suggested that it was due to Mendel's assistant, who knew what Mendel expected and altered the data accordingly.[11]

[11] See *Experiments in Plant Hybridisation* (Edinburgh: Oliver and Boyd, 1965), as well as Freedman, Pisani, and Purves, *op. cit.*

Goodness-of-fit tests are of considerable importance in business and economics. The following example illustrates how they can be used by government agencies also.

EXAMPLE 9.2 Let's return to the replacement of parts on Polaris submarines, an actual case study presented in Chapter 5. Suppose that the Department of Defense believes that the probability distribution of the number of submarine parts of a certain type that will fail during a mission is as follows:[12]

[12] Readers who have covered the latter part of Chapter 5 should note that this probability distribution is a Poisson distribution in which the mean number of failures per mission is 1.0.

| Number of failures per mission | Theoretical probability |
|:---:|:---:|
| 0 | .368 |
| 1 | .368 |
| 2 | .184 |
| 3 | .061 |
| 4 or more | .019 |

Data for 500 missions indicate that the number of these missions in which each number of failures occurred was as follows:

| Number of failures per mission | Number of missions |
|:---:|:---:|
| 0 | 190 |
| 1 | 180 |
| 2 | 90 |
| 3 | 30 |
| 4 or more | 10 |

Test the hypothesis that the discrepancies between the actual and theoretical frequency distributions are due to chance. (Let the significance level equal .05.)

SOLUTION: The theoretical frequency distribution can be obtained by multiplying the theoretical probabilities by 500, the results being as follows:

| Number of failures per mission | Expected number of missions |
|:---:|:---:|
| 0 | .368(500) = 184.0 |
| 1 | .368(500) = 184.0 |
| 2 | .184(500) = 92.0 |
| 3 | .061(500) = 30.5 |
| 4 or more | .019(500) = 9.5 |

Note that we have lumped together all numbers of failures exceeding 3 in one class interval to hold the theoretical frequency in the class interval to a level of five or more.

The value of the test statistic is

$$\sum \frac{(f-e)^2}{e} = \frac{(190-184)^2}{184} + \frac{(180-184)^2}{184} + \frac{(90-92)^2}{92} +$$

$$\frac{(30-30.5)^2}{30.5} + \frac{(10-9.5)^2}{9.5}$$

$$= .196 + .087 + .043 + .008 + .026 = .360.$$

Since no parameters of the theoretical frequency distribution are estimated from the sample, the number of degrees of freedom is

5 − 1, or 4. Appendix Table 8 shows that if there are 4 degrees of freedom, $\chi^2_{.05} = 9.488$. Since the observed value of $\Sigma(f - e)^2/e$ is less than 9.488, the null hypothesis should not be rejected. The probability is much higher than .05 that the observed value of the test statistic could have arisen, given that the theoretical distribution was valid. In other words, the Defense Department should not reject the hypothesis that the data conform to the theoretical distribution.

TESTING FOR NORMALITY AT THE AMERICAN STOVE COMPANY

GETTING DOWN TO CASES

A number of years ago, the American Stove Company collected data concerning the heights of a metal piece that the firm manufactured. The firm measured the heights of 145 of these metal pieces, and the results are shown in the following frequency distribution:[13]

| Height of metal piece (inches) | Number of metal pieces |
|---|---|
| Less than .8215 | 9 |
| .8215 and under .8245 | 5 |
| .8245 and under .8275 | 14 |
| .8275 and under .8305 | 21 |
| .8305 and under .8335 | 55 |
| .8335 and under .8365 | 23 |
| .8365 and under .8395 | 7 |
| .8395 and under .8425 | 6 |
| .8425 and over | 5 |
| Total | 145 |

For many purposes it is important to test whether a certain random variable is normally distributed. For example, as we saw in Chapter 8, some statistical tests assume normality. Suppose that the American Stove Company had given you the task of using the above data to test whether the heights of the metal pieces are normally distributed.

(a) Express the upper limit of each class interval in the frequency distribution as a deviation from the sample mean (.8314 inches), and divide this deviation by the sample standard deviation (.0059 inches).

(b) Use the results in (a)—which correspond to points on the standard normal distribution—to determine the expected number of metal pieces that should fall into each class interval if the heights are normally distributed.

(c) Calculate the difference between the actual and expected

frequency in each class, square this difference, divide it by the
expected frequency, and sum the results for all classes.

(d) Use the $\chi^2$ distribution to test the hypothesis that the
heights are normally distributed. (Let $\alpha = .05$.)

(e) Explain your choice of the number of degrees of freedom.

(f) Write a three-sentence memorandum summarizing your
results.

[13] See A. J. Duncan, *op. cit.*

## 9.8 Tests and Confidence Intervals Concerning the Variance

Before turning to nonparametric techniques, one final application of
the $\chi^2$ distribution should be discussed. If the population is normal, the
$\chi^2$ distribution can be used to test hypotheses concerning the variance
or to construct confidence intervals for the variance. For example, a
firm that produces coats wants to test the hypothesis that the variance
of the prices charged by retailers for its coats equals 100 (dollars
squared). A random sample of 11 prices (charged by retailers for the
firm's coats) is drawn, and it is found that the sample variance $s^2$ equals
300 (dollars squared). If the significance level is set at .05, should the
firm reject the hypothesis that the population variance equals 100
(dollars squared)?

The first step is to formulate the null hypothesis and the alternative hypothesis. In this case, the null hypothesis is that the population
variance equals 100 (dollars squared), and the alternative hypothesis is
that the population variance is either less than or greater than 100
(dollars squared). If the population is normal, and if its variance equals
$\sigma_0^2$, it can be shown that $(n-1)s^2 \div \sigma_0^2$ has the $\chi^2$ distribution with
$(n-1)$ degrees of freedom, where $n$ is the sample size. Thus, to test the
null hypothesis that the variance equals $\sigma_0^2$, we compute

$$\frac{(n-1)s^2}{\sigma_0^2},$$

which is used as follows:

*Test that $\sigma^2 = \sigma_0^2$*

DECISION RULE: *When the alternative hypothesis is $\sigma^2 \neq \sigma_0^2$, reject the
null hypothesis if $(n-1)s^2/\sigma_0^2$ exceeds $\chi^2_{\alpha/2}$ or is less than $\chi^2_{1-\alpha/2}$.
When the alternative hypothesis is $\sigma^2 > \sigma_0^2$, reject the null hypothesis if
$(n-1)s^2/\sigma_0^2$ exceeds $\chi^2_{\alpha}$. When the alternative hypothesis is $\sigma^2 < \sigma_0^2$,
reject the null hypothesis if $(n-1)s^2/\sigma_0^2$ is less than $\chi^2_{1-\alpha}$.*

The alternative hypothesis in this case is that $\sigma^2 \neq \sigma_0^2$. Appendix Table 8 shows that $\chi_{.025}^2 = 20.48$ and $\chi_{.975}^2 = 3.25$ because the number of degrees of freedom equals $(n - 1) = 11 - 1 = 10$. Thus, the decision rule is: Reject the null hypothesis if $(n - 1)s^2/\sigma_0^2$ is greater than 20.48 or less than 3.25. In fact,

$$\frac{(n - 1)s^2}{\sigma_0^2} = \frac{10(300)}{100} = 30,$$

because $n = 11$, $s^2 = 300$, and $\sigma_0^2 = 100$. Thus, the firm should reject the null hypothesis that the variance of the prices equals 100 (dollars squared).

---

*The $\chi^2$ distribution is used to test whether a number of proportions are equal, whether two characteristics are dependent on one another (in a contingency table), whether an observed frequency distribution conforms to a theoretical distribution, and whether the variance of a normal population equals a particular amount.*    *Basic Idea #17*

---

In many circumstances, statisticians must estimate the variance of a normal population based on the results of a random sample of $n$ observations. *If the confidence coefficient is $(1 - \alpha)$, the confidence interval for the population variance is*

*Confidence Interval for $\sigma^2$*

$$\frac{(n - 1)s^2}{\chi_{\alpha/2}^2} < \sigma^2 < \frac{(n - 1)s^2}{\chi_{1-\alpha/2}^2}, \tag{9.5}$$

*where $\chi_{\alpha/2}^2$ is the value of a $\chi^2$ variable (with n − 1 degrees of freedom) that is exceeded with a probability of $\alpha/2$, and $\chi_{1-\alpha/2}^2$ is the value that is exceeded with a probability of $1 - \alpha/2$.* Thus, if a sample of 20 observations is chosen, and the sample variance equals 5, the 95 percent confidence interval for the population variance is

$$\frac{19(5)}{32.85} < \sigma^2 < \frac{19(5)}{8.91},$$

since Appendix Table 8 shows that for 19 degrees of freedom, $\chi_{.025}^2 = 32.85$ and $\chi_{.975}^2 = 8.91$. Simplifying terms, it follows that the confidence interval in this case is 2.89 to 10.66.

## EXERCISES

**9.10** A soap manufacturer is trying to determine whether or not to market a new type of soap. It chooses a random sample of 300 people in the United

States, and samples of equivalent size in England and France. It asks each of the people in each sample to try the new soap, and see whether he or she likes it better than other soaps. The results are as follows:

|  | United States | England | France |
|---|---|---|---|
| Prefer new soap | 81 | 43 | 26 |
| Do not prefer it | 219 | 257 | 274 |
| Total | 300 | 300 | 300 |

Does it appear that there are international differences in the proportion of people who prefer the new soap?
(a) Use the .01 level of significance.
(b) Use the .05 level of significance.

**9.11** The mayor of a large city wants to determine the proportion of voters that support an increase in schoolteachers' salaries. He asks that a random sample of 100 voters be chosen from each of the city's four districts, and that each person in these samples be asked whether he or she supports such an increase. The results are as follows:

|  | District I | District II | District III | District IV |
|---|---|---|---|---|
| Supports an increase | 38 | 43 | 78 | 53 |
| Does not support an increase | 62 | 57 | 22 | 47 |
| Total | 100 | 100 | 100 | 100 |

Can the differences among districts in the proportion supporting such an increase be attributed to chance?
(a) Use the .10 level of significance.
(b) Use the .05 level of significance.
(c) Use the .01 level of significance.

**9.12** According to a theory put forth by plant geneticists, the number of peas of each of four types should be as shown in the table below. The actual numbers are also given.

| Type of pea | Expected | Actual |
|---|---|---|
| Smooth green | 321 | 303 |
| Wrinkled green | 105 | 109 |
| Smooth yellow | 105 | 98 |
| Wrinkled yellow | 36 | 57 |
| Total | 567 | 567 |

Are the data consistent with the theory?
(a) Use the .01 level of significance.
(b) Use the .05 level of significance.

**9.13** A gambler wants to determine whether a die is true. He throws the die 100 times, with the following results:

| Number on face of die | Number of outcomes |
|---|---|
| 1 | 15 |
| 2 | 18 |
| 3 | 20 |
| 4 | 17 |
| 5 | 13 |
| 6 | 17 |
| Total | 100 |

Is there any evidence that the die is not true? (Let $\alpha = .05$.)

**9.14** A detective wants to determine whether a pair of dice is "loaded." He rolls the dice 200 times, with the following results:

| Number on face of die | Number of outcomes |
|---|---|
| 2 | 0 |
| 3 | 2 |
| 4 | 2 |
| 5 | 16 |
| 6 | 30 |
| 7 | 100 |
| 8 | 30 |
| 9 | 14 |
| 10 | 2 |
| 11 | 4 |
| 12 | 0 |
| Total | 200 |

Is there any evidence that these dice are not true? (Let $\alpha = .01$.)

**9.15** A teacher claims that over the long run the grades that she gives are normally distributed and that she gives 8 percent As, 25 percent Bs, 34 percent Cs, 25 percent Ds, and 8 percent Fs. Last year, she gave the 110 students in her class the following grades:

| Grade | A | B | C | D | F |
|---|---|---|---|---|---|
| Number of students | 5 | 22 | 50 | 27 | 6 |

Are these data consistent with her claim?
(a) Use the .05 significance level.
(b) Use the .01 significance level.

**9.16** The following table shows the number of daughters in 50 families, each of which contains five children.

| Number of daughters | Number of families |
|---|---|
| 0 | 7 |
| 1 | 12 |
| 2 | 13 |
| 3 | 8 |
| 4 | 6 |
| 5 | 4 |
| Total | 50 |

(a) Use the binomial distribution with $n = 5$ and $\Pi = 0.5$ to compute the expected number of families with each number of daughters.

(b) Based on a $\chi^2$ test, does it appear that these data conform to the binomial distribution? (Let $\alpha = .05$.)

**9.17** A random sample of 30 observations is taken from a normal population. The sample variance equals 10.

(a) Test the hypothesis that the population variance equals 20, the alternative hypothesis being that $\sigma^2 \neq 20$. Use the .05 significance level.

(b) Calculate a 98 percent confidence interval for the population variance.

**9.18** If a population is normal, and if the sample standard deviation (based on a sample of 18 observations) equals 3.2, test the hypothesis that $\sigma = 4$.

(a) Use a two-tailed test with $\alpha = .05$.

(b) Use a two-tailed test with $\alpha = .02$.

**9.19** A firm draws a random sample of 18 ball bearings from today's output. The sample standard deviation of their diameters is 0.003 inches.

(a) Construct a 95 percent confidence interval for the population standard deviation.

(b) Construct a 90 percent confidence interval for the population standard deviation.

(c) What assumptions underlie your answers to (a) and (b)?

# 9.9 Nonparametric Techniques

Many of the tests discussed here and in the previous chapter are based on the assumption that the population or populations under consideration are normal. Although this assumption frequently is close enough to the truth so that these tests are good approximations, cases often arise where statisticians prefer instead to use various kinds of nonparametric, or distribution-free, tests. The hallmark of these tests is that they avoid the assumption of normality. For practically every test discussed in this and the previous chapter, there is a nonparametric analogue. Besides having the advantage that normality is not assumed, nonparametric tests often are easier to carry out because the computations are simpler. In addition, some experiments yield responses that can only be ranked, not measured along a cardinal scale. Nonparametric techniques are designed to analyze data of this sort, which is another important reason for their use.

In the following sections we shall describe three of the most commonly used nonparametric techniques: the sign test, the Mann-Whitney test, and the runs test. Each of these has important applications in business and economics.

---

*Basic Idea #18*    **Nonparametric techniques do not assume that populations are normally distributed, and they are relatively easy to apply.**

## 9.10 The Sign Test

The sign test is used to test the null hypothesis that the population median equals a certain amount. For example, suppose that the Federal Trade Commission has 36 tires from a manufacturer tested to determine their length of life. The purpose is to find out whether the manufacturer's advertising claim that the tires average 25,000 miles of use is correct. If the median length of life is 25,000 miles, the probability that a tire chosen at random will last for less than 25,000 miles is 1/2, and the probability that it will last for more than 25,000 miles is also 1/2. If the results for the 36 tires are as shown in Table 9.6, we can place

| | | | |
|---|---|---|---|
| 27,100(+) | 24,500(−) | 24,800(−) | 25,100(+) |
| 24,200(−) | 25,600(+) | 24,600(−) | 24,800(−) |
| 23,300(−) | 24,600(−) | 25,500(+) | 24,700(−) |
| 23,800(−) | 24,800(−) | 26,100(+) | 25,900(+) |
| 22,600(−) | 24,100(−) | 23,900(−) | 23,900(−) |
| 24,800(−) | 23,900(−) | 24,100(−) | 24,800(−) |
| 24,100(−) | 24,900(−) | 24,300(−) | 24,700(−) |
| 24,800(−) | 24,600(−) | 24,700(−) | 24,900(−) |
| 25,500(+) | 26,100(+) | 24,900(−) | 25,100(+) |

**Table 9.6**
Length of Life of
Sample of 36 Tires

a minus sign to the right of the figure for each tire that lasts less than 25,000 miles. By the same token, we place a plus sign to the right of the figure for each tire that lasts more than 25,000 miles. If the null hypothesis holds (that is, if the median life is 25,000 miles), the number of pluses will have a binomial distribution:

$$P(x) = \frac{36!}{x!(36-x)!}\left(\frac{1}{2}\right)^x\left(\frac{1}{2}\right)^{36-x},$$

where $x$ is the number of pluses. (Why? Because there is a probability of 1/2 that each tire in the sample will last more than 25,000 miles and thus will have a plus beside it.) Since the sample size is relatively large ($n = 36$), we know from Chapter 5 that this binomial distribution can be approximated by a normal distribution with a mean of 18 and a standard deviation of 3. (Because $n = 36$ and $\Pi = 1/2$, $n\Pi = 36(1/2)$ or 18, and $\sqrt{n\Pi(1-\Pi)} = \sqrt{36(1/2)(1/2)}$, or 3.) Thus, the sampling distribution of the number of pluses is as shown in Figure 9.3.

The null hypothesis, as noted above, is that the median length of life is 25,000 miles. The alternative hypothesis is that the median length of life is less than 25,000 miles. (Why is the alternative hypothesis one-sided? Because the FTC is interested in taking action against the advertising claims only if the median is less than 25,000 miles.) Thus, the FTC's statisticians will reject the null hypothesis only if the number of pluses—that is, $x$—is so small that it is unlikely that the sampling distribution in Figure 9.3 is valid. If the significance level is set at .05, it is clear that the null hypothesis should be rejected if the

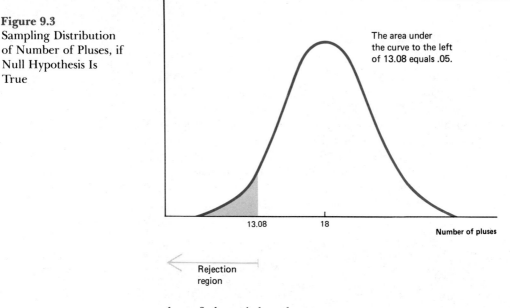

The area under
the curve to the left
of 13.08 equals .05.

13.08          18

**Number of pluses**

Rejection
region

number of pluses is less than

$$18 - 1.64(3) = 13.08,$$

since there is a .05 probability that the number of pluses will be less than 13.08, given that the null hypothesis is true.[14] (See Figure 9.3.) Based on the data in Table 9.6, it follows that the FTC should reject the hypothesis that the median length of life is 25,000 miles, because the number of pluses is 9, which is considerably less than 13.08.

The sign test is based on the fact that if the population median ($M$) is equal to the value specified by the null hypothesis (say, $M_0$), $x$, the number of observations in the sample above $M_0$, has a binomial distribution with $\Pi$ equal to $1/2$ and $n$ equal to the sample size. Thus, if the sample size is large, the following decision rules can be used.

*Test that M = $M_0$*

DECISION RULE IF THE ALTERNATIVE HYPOTHESIS IS THAT $M < M_0$: *Reject the null hypothesis if* $x < n/2 - z_\alpha\sqrt{n/4}$.

DECISION RULE IF THE ALTERNATIVE HYPOTHESIS IS THAT $M > M_0$: *Reject the null hypothesis if* $x > n/2 + z_\alpha\sqrt{n/4}$.

DECISION RULE IF THE ALTERNATIVE HYPOTHESIS IS THAT $M \neq M_0$: *Reject the null hypothesis if* $x < n/2 - z_{\alpha/2}\sqrt{n/4}$ *or if* $x > n/2 + z_{\alpha/2}\sqrt{n/4}$.

If the sample size is small, the binomial distribution, not the normal distribution, must be used; thus, these decision rules must be altered.

Note that the sign test does not depend on any assumptions about the nature or shape of the population. It rests only on the fact that the probability that a randomly chosen observation falls above the median

[14] We have ignored the continuity correction discussed in Chapter 5 because it has no material effect on our results.

is 1/2; and this is true for any population, normal or otherwise. (Basically, this is why the sign test is nonparametric.) Also, note that the computations involved in this test are comparatively simple. As pointed out previously, nonparametric tests are often used for this reason.

The sign test is useful in a wide variety of circumstances. The following example illustrates how it is employed in a sample consisting of matched pairs.

EXAMPLE 9.3 A paint company conducts an experiment in which one half of a wood plank is painted with its paint and the other half is painted with its competitor's paint. The sample consists of 100 planks treated in this way. Each plank is placed outdoors in order to determine the difference between (1) the time it takes for the half covered with the company's paint to crack and (2) the time it takes for the half covered with the competitor's paint to crack. The paint firm wants to test the null hypothesis that the median difference equals zero. For 62 boards tested, it takes longer for the half covered with the company's paint to crack than for the half covered with the competitor's paint: for 38 boards the reverse is true. Use the sign test to test the null hypothesis. (Let the significance level equal .05.)

SOLUTION: If the null hypothesis is true (that is, if the median difference is zero), the probability that the half of each board covered with the company's paint will last longer than the other half is 1/2. Thus, if we put a plus next to each board where the half covered by the company's paint lasts longer than the other half, the probability distribution of the number of pluses should be binomial with $n = 100$ and $\pi = 1/2$, if the null hypothesis is true. Since $n$ is large, the binomial distribution can be approximated by the normal distribution, and the null hypothesis should be rejected if the number of pluses $x$ is less than $n/2 - z_{.025} \sqrt{n/4}$ or greater than $n/2 + z_{.025} \sqrt{n/4}$. Since $n = 100$, the null hypothesis should be rejected if $x < 50 - 1.96(5)$, or 40.2, or if $x > 50 + 1.95(5)$, or 59.8. Since $x = 62$, the null hypothesis should be rejected. In other words, the evidence seems to indicate that the company's paint lasts longer than its competitor's.

Note that if the cracking times for both halves of any boards were exactly equal, they should be omitted from the analysis.

## 9.11 The Mann-Whitney Test

Another important nonparametric technique is the Mann-Whitney test, also known as the Wilcoxon test or the $U$ test. This technique is used to find out whether two populations are identical. For example, the Federal Trade Commission might need to determine whether there is a difference between the population of lifetimes of firm C's tires and the population of lifetimes of firm D's tires. A random sample of 14 of firm C's tires and a random sample of 14 of firm D's tires are taken, the results being shown in Table 9.7. In this case, *the null hypothesis is that there is no difference between the two populations. The alternative hypothesis is that they are different.* Note that in contrast to Example 9.3, the two samples are entirely independent; *they are not matched pairs.*

The first step in applying the Mann-Whitney test is to rank all observations in the two combined samples. Table 9.7 shows the 28 observations ranked from lowest to highest, with the letter under each observation (a *C* or a *D*) indicating the firm from which it came. The members of firm C's sample are ranked 1, 2, 3, 4, 5, 7, ···, and the members of firm D's sample are ranked 6, 8, 9, 11, 14, 16, 17 ··· .[15] If the null hypothesis is true, we would expect that the average rank for the items from one sample would be approximately equal to the average rank for the items from the other sample. On the other hand, if the alternative hypothesis is true, we would expect a marked difference between the average ranks in the two samples.

**Table 9.7**
Lengths of Life of Sample of Tires from Two Firms

| Firm C | | Firm D | |
|---|---|---|---|
| 19,800 | 19,700 | 19,850 | 20,850 |
| 19,900 | 19,600 | 31,250 | 20,650 |
| 20,100 | 19,500 | 20,380 | 21,050 |
| 20,200 | 20,400 | 19,950 | 19,820 |
| 31,300 | 20,300 | 20,260 | 20,350 |
| 18,700 | 21,000 | 19,750 | 31,150 |
| 18,900 | 21,100 | 20,450 | 20,750 |

*Observations Ranked from Lowest to Highest*

| Observation | 18,700, | 18,900, | 19,500, | 19,600, | 19,700, | 19,750, | 19,800, | 19,820, |
|---|---|---|---|---|---|---|---|---|
| Firm | C | C | C | C | C | D | C | D |
| Rank | 1 | 2 | 3 | 4 | 5 | 6 | 7 | 8 |

| Observation | 19,850, | 19,900, | 19,950, | 20,100, | 20,200, | 20,260, | 20,300, | 20,350, |
|---|---|---|---|---|---|---|---|---|
| Firm | D | C | D | C | C | D | C | D |
| Rank | 9 | 10 | 11 | 12 | 13 | 14 | 15 | 16 |

| Observation | 20,380, | 20,400, | 20,450, | 20,650, | 20,750, | 20,850, | 21,000, | 21,050, |
|---|---|---|---|---|---|---|---|---|
| Firm | D | C | D | D | D | D | C | D |
| Rank | 17 | 18 | 19 | 20 | 21 | 22 | 23 | 24 |

| Observation | 21,100, | 31,150, | 31,250, | 31,300 |
|---|---|---|---|---|
| Firm | C | D | D | C |
| Rank | 25 | 26 | 27 | 28 |

[15] If there are ties, we must assign the tied items the mean of the ranks they have in common. For example, if two items are tied for third and fourth place, each would receive the rank of 3.5.

To carry out the Mann-Whitney test, it is more convenient to use the *rank-sum* instead of the average rank in a sample. Let $R_1$ be the sum of the ranks for one of the samples. (It doesn't matter which sample one chooses.) Then we compute

$$U = n_1 n_2 + \frac{n_1(n_1 + 1)}{2} - R_1, \qquad (9.6)$$

where $n_1$ is the number of observations in the sample on which $R_1$ is based and $n_2$ is the number of observations in the other sample. If the null hypothesis is true, the sampling distribution of $U$ has a mean equal to

$$E_u = \frac{n_1 n_2}{2} \qquad (9.7)$$

and a standard deviation equal to

$$\sigma_U = \sqrt{\frac{n_1 n_2 (n_1 + n_2 + 1)}{12}}. \qquad (9.8)$$

And if $n_1$ and $n_2$ are at least 10, the sampling distribution of $U$ can be approximated quite well by the normal distribution.

Based on these facts concerning the sampling distribution of $U$ (if the null hypothesis is true), it is easy to formulate an appropriate decision rule.

**DECISION RULE**: *Reject the null hypothesis if* $(U - E_u) \div \sigma_u < -z_{\alpha/2}$ *or* ***Mann-Whitney Test***
*if* $(U - E_u) \div \sigma_u > z_{\alpha/2}$.

In applying this decision rule, note that

$$\frac{U - E_u}{\sigma_u} = \frac{n_1 n_2 + \frac{n_1(n_1+1)}{2} - R_1 - \frac{n_1 n_2}{2}}{\sqrt{n_1 n_2 (n_1 + n_2 + 1)/12}}. \qquad (9.9)$$

Also, it should be recognized that this decision rule applies only to a two-tailed test. (The extension to a one-tailed test is straightforward.) Further, this decision rule assumes that both $n_1$ and $n_2$ are at least 10; if not, the test must be based on special tables, and the procedure described here is not appropriate.

To illustrate the use of the Mann-Whitney test, let's return to the data in Table 9.7. To compute $R_1$, let's use the sample of firm C's tires. Since the observations in this sample have ranks of 1, 2, 3, 4, 5, 7, 10,

12, 13, 15, 18, 23, 25, and 28, it is clear that $R_1 = 166$. Thus, since $n_1 = n_2 = 14$,

$$U = 14(14) + \frac{14(15)}{2} - 166 = 135,$$

and

$$E_u = \frac{14(14)}{2} = 98$$

$$\sigma_u = \sqrt{\frac{14(14)(29)}{12}} = 21.8.$$

Consequently,

$$\frac{U - E_u}{\sigma_u} = \frac{135 - 98}{21.8} = 1.70.$$

If the null hypothesis is true, the probability that $(U - E_u) \div \sigma_u > 1.70$ equals .045. (See Appendix Table 2.) Thus, since this is a two-tailed test, the p-value is 2 x .045, or .09. Unless the desired significance level is .09 or more, the Federal Trade Commission should conclude that the available evidence does not indicate that there is a difference between the population of lifetimes of firm C's tires and the population of lifetimes of firm D's tires.

Finally, it is worth noting that this test does not assume that the populations in question are normal. That is, there is no need to assume that the population of lifetimes of firm C's tires is normal, or that the population of lifetimes of firm D's tires is normal. Thus, this test avoids the assumption of normality underlying the $t$ test, which was used in the previous chapter to test a similar kind of hypothesis. Moreover, this test requires far less computation than the $t$ test, an important advantage under some circumstances.

The Mann-Whitney test is useful as a possible substitute for the $t$ test. The following example further illustrates how it is used.

---

EXAMPLE 9.4 A personnel agency gives an aptitude test to 12 men and 12 women, the scores being as follows:

Men    80, 79, 92, 65, 83, 84, 95, 78, 81, 85, 73, 52
Women    82, 87, 89, 91, 93, 76, 74, 70, 88, 99, 61, 94

Use these data to test the null hypothesis that the distribution of scores on this test is the same for men as for women. (Let the significance level equal .05.)

SOLUTION: Rank the scores from lowest to highest, and put an M under each score if it is a man's and a W under each score if it is a woman's:

```
52 61 65 70 73 74 76 78 79 80 81 82 83 84 85 87 88
M  W  M  W  M  W  W  M  M  M  M  W  M  M  M  W  W

89 91 92 93 94 95 99
W  W  M  W  W  M  W
```

The sum of the ranks for men is $1 + 3 + 5 + 8 + 9 + 10 + 11 + 13 + 14 + 15 + 20 + 23 = 132$. Thus $U = 12(12) + [12(13)/2] - 132 = 90$. Also, $E_u = 12(12)/2 = 72$, and $\sigma_u = \sqrt{12(12)(25)/12} = \sqrt{300} = 17.3$. Thus, since $z_{\alpha/2} = 1.96$, the null hypothesis should be rejected if $(U - E_u) \div \sigma_u$ is less than $-1.96$ or greater than $1.96$. In fact, $(U - E_u) \div \sigma_u = (90 - 72) \div 17.3 = 1.0$. Thus, the null hypothesis should not be rejected. The probability is greater than .05 that these results could have occurred, given that the distribution of scores on this test was the same for men as for women.

## 9.12 The Runs Test

Another important nonparametric technique is the runs test, which is designed to test the hypothesis that a sequence of numbers, symbols, or objects is in random order. A tennis fan interested in the outcomes of the matches between John McEnroe and Ivan Lendl finds that these outcomes are as follows (where M stands for a McEnroe victory and L stands for a Lendl victory):

L M L M L M L M L M L M L M.

Does this sequence appear to be in a random order? No, since the winner seems to alternate back and forth. (If this were the case, one might suspect that tennis players, like some phony wrestlers, decide in advance who will win and alternate the winner.) On the other hand, suppose that the sequence is

M M M M M M M M L L L L L L L.

Does this sequence appear to be in a random order? No, since the winner of all the early matches is McEnroe, and the winner of all the later matches is Lendl. (If this were the case, we might suspect that Lendl had greatly improved—or that McEnroe's play had deteriorated—at about the middle of the sequence of matches.)

To understand the runs tests, you must know the statistician's definition of a run, which is the following:

A RUN *is a sequence of identical numbers, symbols, objects, or events preceded and followed by different numbers, symbols, objects, or events (or by nothing at all).*

In the first version of the McEnroe-Lendl matches given above, there are 14 runs; in the second version there are only 2 runs. Suppose that a third version is as follows:

M̲ L̲ M̲ M̲ M̲ M̲ L̲ L̲ L̲ M̲ L̲ M̲ L̲ L̲.

How many runs are there? The answer is 8. (The lines under the letters designate each run.)

The runs test is based on the idea that if the probability of the occurrence of one number, symbol, object, or event is constant throughout the sequence, then there should be neither a very large number of runs (as in our first "history" of the McEnroe-Lendl matches) nor a very small number of runs (as in our second such "history"). If there are two possible outcomes (such as L and M), and if the probability of each outcome remains constant throughout the sequence (and if successive outcomes are independent), it can be shown that the expected number of runs is:

$$E_r = \frac{2n_1 n_2}{n_1 + n_2} + 1, \qquad (9.10)$$

and the standard deviation of the number of runs is

$$\sigma_r = \sqrt{\frac{2n_1 n_2 (2n_1 n_2 - n_1 - n_2)}{(n_1 + n_2)^2 (n_1 + n_2 - 1)}}, \qquad (9.11)$$

where $n_1$ is the number of outcomes of one type (such as L) and $n_2$ is the number of outcomes of the other type (such as M). If either $n_1$ or $n_2$ is larger than 20, it can be shown that the number of runs $r$ is approximately normally distributed. When both $n_1$ and $n_2$ exceed 10, this approximation is good.

The runs test is designed to test the null hypothesis that a given sequence of outcomes is in random order—that is, that the probability of each outcome remains constant throughout the sequence and that successive outcomes are independent. If the significance level is set equal to $\alpha$, it follows from the previous paragraph that the following decision rule can be applied.

*Runs Test*

DECISION RULE: *Reject the null hypothesis (of randomness) if* $(r-E_r)/\sigma_r < -z_{\alpha/2}$ *or if* $(r - E_r)/\sigma_r > z_{\alpha/2}$; *otherwise do not reject it.*

This assumes that $n_1$ and $n_2$ meet the requirements in the previous paragraph.

To illustrate the application of this test, suppose that McEnroe and Lendl play 37 times, with the following results:

MLMMLMMLMMLLLMLLMMMLMLMMLMMLLMMMLLMMML.

If $n_1$ is the number of McEnroe victories and $n_2$ is the number of Lendl victories, $n_1 = 21$ and $n_2 = 16$. Thus,

$$E_r = \frac{2(21)(16)}{37} + 1 = 19.16$$

and

$$\sigma_r = \sqrt{\frac{2(21)(16)[2(21)(16) - 21 - 16]}{37^2(36)}} = \sqrt{\frac{672(635)}{49,284}} = \sqrt{8.7} = 2.94.$$

Thus, since the actual number of runs $r$ equals 22,

$$(r - E_r)/\sigma_r = \frac{22 - 19.16}{2.94} = 0.97.$$

If $\alpha$ is set equal to .05, $z_{\sigma/2}$ equals 1.96. Thus, the null hypothesis should not be rejected since $(r - E_r)/\sigma_r$ is not greater than 1.96 or less than $-1.96$.

The runs test has many applications in business and economics. The example below shows how it can be used with quantitative rather than qualitative measurements.

EXAMPLE 9.5 A winery measures daily the acidity of the wine it produces. During a 32-day period, it is found that the mean acidity equals 8. Each day during this period is classified as A (above average in acidity) or B (below average in acidity), the results being as follows:

B A A A A B B A B B A A A A B B B B A B A A A A A A A B A A A A.

Use the runs test to determine whether there is a departure from randomness in the sequence. (Let the significance level equal .05.)

SOLUTION: Since $n_1 = 21$ and $n_2 = 11$,

$$E_r = \frac{2(21)(11)}{32} + 1, \text{ or } 15.4,$$

and

$$\sigma_r = \sqrt{\frac{2(21)(11)[2(21)(11) - 21 - 11]}{32^2(31)}}$$

$$= \sqrt{\frac{198,660}{31,744}} = \sqrt{6.26}, \text{ or } 2.50.$$

Since the number of runs $r$ equals 12,

$$\frac{r - E_r}{\sigma_r} = \frac{12 - 15.4}{2.50} = -1.36,$$

which is neither greater than $z_{.025}$ nor less than $-z_{.025}$. (As shown in Appendix Table 2, $z_{.025} = 1.96$.) Thus, the winery should not reject the null hypothesis. In other words, it should not reject the hypothesis that the sequence is random.

## 9.13 Nonparametric Techniques: Pros and Cons

Based on the discussion in the previous sections, nonparametric techniques are clearly a very important statistical tool. Frequently, as we have seen, the statistician has the choice of using a nonparametric technique or a parametric procedure of the sort discussed in the previous chapter. For example, the sign test is to some extent a substitute for the test (in the previous chapter) of a specified value of the mean, and the Mann-Whitney test is to some extent a substitute for the parametric test (also in the previous chapter) of equality between two population means. The following factors should guide the choice between nonparametric and parametric techniques in situations where both can be used.

SIMPLICITY OF CALCULATIONS. Under some circumstances, it is very important that the computations in a statistical test be simple and rapid. For example, it may be necessary for a person with a relatively limited mathematical background to carry out the computations; or there may be a premium on speed. In such cases, nonparametric techniques like the sign test may be preferred over parametric tests.

REALISM OF ASSUMPTIONS. Many parametric tests assume that the relevant populations are normal. Two such tests are the small-sample test that a population mean equals a specified value and the small-sample test for equality between two population means. Although these procedures are dependable in the face of moderate departures from normality, they can be quite misleading in the face of gross departures of this sort. Thus, if the populations are very far from normality, statisticians often prefer nonparametric techniques like the sign test and Mann-Whitney test.

POWER OF THE TEST. If the assumptions underlying the parametric tests in the previous chapter are reasonably close to reality, these tests generally are more powerful than nonparametric techniques. In other words, the parametric techniques are less likely to produce a Type II error, given that the probability of a Type I error is the same for both tests. This is an important advantage of parametric tests, as long as their assumptions are reasonably valid. Nonparametric techniques, as we have seen, often use rankings or orderings as distinct from the actual values of the observations, thus losing a certain amount of the information in the sample.

## EXERCISES

**9.20** A firm manufactures a component of a television set. This component is supposed to weigh 3.5 ounces. The firm selects a random sample of 100 components, and finds that 58 percent of them weigh more than 3.5 ounces.
  (a) Test whether the median weight of these components is 3.5 ounces. Let $\alpha = .05$.
  (b) Do your results change if $\alpha$ is set equal to .01 rather than .05?

**9.21** The Department of Defense wants to determine whether alloy A withstands heat better or worse than alloy B. This information is important in deciding which alloy should be used in a new weapons system. Thirty pieces of alloy A are matched against thirty pieces of alloy B. In each of these 30 cases, one piece of each type of alloy is subjected to the same intense heat, and engineers determine which type of alloy fares better. In 19 of the 30 cases, alloy B fares better than alloy A.
  (a) Use these results to test whether there is any difference between alloy A and alloy B in how well they withstand heat. Let $\alpha = .05$.
  (b) Do your results change if $\alpha$ is set equal to .01 rather than .05?

**9.22** A company is interested in comparing the accuracy of two methods of measuring weight. An object is weighed 10 times based on the first method and 10 times based on the second method. The size of the error is recorded in each of the 20 cases. When these 20 errors are ranked (from smallest to largest), the results are as follows:

  Method 1: 2, 3, 5, 6, 9, 11, 13, 14, 15, 16
  Method 2: 1, 4, 7, 8, 10, 12, 17, 18, 19, 20

  (a) Test whether the accuracy of the two methods is the same. (Let $\alpha = .05$.)
  (b) Does the assumption of normality of the errors underlie your results in (a)?
  (c) Calculate a *p*-value for this test. Explain its meaning.

**9.23** A psychologist claims that the length of time a person takes to do a particular job is not influenced by whether the person is male or female. To test this claim a statistician asks 11 men and 11 women to do this job, and measures each one's time. The 22 times are ranked (from lowest to highest), with the following results:

  Men:   1, 2, 4, 6, 8, 9, 10, 11, 14, 15, 16
  Women: 3, 5, 7, 12, 13, 17, 18, 19, 20, 21, 22

  (a) Are the data consistent with the psychologist's claim, if $\alpha = .05$?
  (b) What are the results if $\alpha = .01$ rather than .05?

**9.24** Having installed a new procedure to reduce clerical errors, the Crooked Arrow National Bank wants to determine whether this new procedure has resulted in a difference in the distribution of the number of clerical errors per day in one of the bank's departments. The bank chooses a random sample of 11 days prior to the installation of the new procedure and finds that the number of errors per day was as follows: 7, 8, 10, 6, 5, 8, 9, 11, 10, 7, 9. The bank also chooses a random sample of 11 days following the installation of the new procedure and finds that the number of errors per day was as follows: 6, 5, 8, 7, 9, 6, 5, 4, 9, 7, 6.

(a) Use the Mann-Whitney test to determine whether the population distribution of the number of errors per day was different after vs. before the installation of the new procedure. (Let $\alpha = .05$.)
(b) Use a $t$ test to test this hypothesis, based on the data given above.
(c) Are the results consistent?
(d) Are the two tests aimed at testing the same hypothesis?
(e) Are the two tests based on the same assumptions? If not, what are the differences in the assumptions?

**9.25** The Crooked Arrow National Bank decides that a better way to test error reduction under its new procedure is to choose 10 employees randomly and determine how many errors each made in a week before the introduction of the new procedure and how many each made in a week after its introduction. The results are as follows:

| Type of procedure | Person |   |   |   |   |   |   |   |   |    |
|---|---|---|---|---|---|---|---|---|---|---|
|  | 1 | 2 | 3 | 4 | 5 | 6 | 7 | 8 | 9 | 10 |
| Old | 40 | 38 | 37 | 34 | 39 | 37 | 42 | 33 | 39 | 40 |
| New | 37 | 37 | 32 | 30 | 37 | 35 | 39 | 32 | 37 | 38 |

(a) Use a nonparametric technique to test the hypothesis that there is no difference between the old and new procedure in the incidence of errors committed. (Let $\alpha = .05$)
(b) Calculate a $p$-value for this test. Explain its meaning.

**9.26** There is a sequence containing 3 $Ss$ and 17 $Ts$. (a) What is the minimum number of runs that can occur in this sequence? (b) What is the maximum number?

**9.27** A table of one-digit random numbers begins as follows:

```
0 1 3 4 6 8 9 7 6 5 4 1 0 0 9 8 7 6 5 4 3 2 1
1 4 3 8 7 6 6 7 5 4 5 0 2 5 8 7 6 9 4 3 2 1 0
9 7 6 5 4 3 2 2 1 1 4 5 6 7 7 8 9 2 4 4 5
```

(a) If an even number (or zero) is denoted by $E$ and an odd number is denoted by $O$, how many runs of $Es$ or $Os$ are there in this sequence?
(b) Based on the number of runs of even (including zero) and odd numbers in the sequence, is there any evidence that the sequence of numbers given there is not random? (Let $\alpha = .05$.)
(c) Let any number from zero to 4 be denoted by $A$, and any number from 5 to 9 be denoted by $B$. How many runs of $As$ or $Bs$ are there in the sequence?
(d) Based on the number of runs of $As$ and $Bs$ in the sequence, is there any evidence that the sequence of numbers given there is not random? (Let $\alpha = .01$.)

**9.28** The price of a share of DuPont common stock behaves in the following way over a 40-day period. (A "+" indicates that the price rose; a "−" indicates that it did not rise.)

$$+++--+-+++--+-+-+-+++--+-+-$$
$$+-+--++--+-+-$$

(a) If $\alpha = .05$, is there evidence of a departure from randomness in this sequence?
(b) Does your answer to (a) change if $\alpha$ equals .01 rather than .05?
(c) Calculate a $p$-value for this test. Explain its meaning.

**9.29** The following sequence indicates whether or not a robbery occurred in a particular residential neighborhood in a 24-day period. (A "1" means that a robbery occurred on a particular day; a "0" means that no robbery occurred on this day.)

$$1\ 0\ 1\ 1\ 0\ 1\ 1\ 0\ 1\ 1\ 1\ 0\ 0\ 0\ 0\ 1\ 0\ 1\ 0\ 1\ 0\ 0\ 1\ 1$$

(a) If $\alpha = .02$, is there evidence of a departure from randomness in this sequence?
(b) Does your answer to (a) change if $\alpha$ equals .01 rather than .02?
(c) Calculate a $p$-value for this test. Explain its meaning.

**9.30** The following sequence indicates whether output per hour of labor in the Chidester Company is above average or below average on a particular day. (An $A$ means that it is above average; a $B$ means that it is below average.) The data pertain to 25 consecutive days.

$$A\ B\ B\ A\ A\ B\ B\ A\ B\ A\ A\ A\ B\ B\ A\ A\ B\ A\ A\ B\ A\ B\ A$$

(a) If $\alpha = .01$, is there any evidence of a departure from randomness in this sequence?
(b) Why should the managers of the Chidester Company care whether a sequence of this sort is random?

# Chapter Review

1. The $\chi^2$ *distribution* (with $v$ degrees of freedom) is the probability distribution of the sum of squares of $v$ independent standard normal variables. The $\chi^2$ distribution is a continuous probability distribution and is generally skewed to the right; but as the number of degrees of freedom increases, it approaches normality. There is a zero probability that a $\chi^2$ random variable will be negative.

2. To test whether a number of population proportions are equal (against the alternative hypothesis that they are unequal), the sample proportion (in the samples from all populations combined) should be computed. This proportion should then be multiplied by the sample size from each population to obtain the expected frequency of "successes" in this sample. The test statistic is $\Sigma(f-e)^2/e$, where $f$ is the actual frequency of "successes" or "failures" in each sample and $e$ is the corresponding expected frequency. If the test statistic exceeds $\chi^2_\alpha$ (where the number of degrees of freedom is one less than the number

of populations being compared), the null hypothesis should be rejected.

3. A *contingency table* contains a certain number of rows and columns, and the decision maker wants to know whether the probability distribution in one column differs from that in another column. In other words, is the variable represented by the rows independent of that represented by the columns? Assuming such independence, we can compute the expected frequencies in the table. The test statistic is $\Sigma(f-e)^2/e$, where $f$ is an actual frequency and $e$ is the corresponding expected frequency. If the test statistic exceeds $\chi_\alpha^2$ (where the number of degrees of freedom equals $[r-1][c-1]$, $r$ being the number of rows and $c$ being the number of columns), the null hypothesis (of independence) should be rejected.

4. To test whether an *observed frequency distribution* conforms to a *theoretical frequency distribution*, we calculate the test statistic $\Sigma(f-e)^2/e$, where $f$ is the observed frequency in a particular class interval and $e$ is the theoretical, or expected, frequency in the same class interval. If the test statistic exceeds $\chi_\alpha^2$—where the number of degrees of freedom equals the number of class intervals minus $(1+h)$, and $h$ is the number of parameters estimated from the sample—the null hypothesis (that the observed frequency distribution conforms to the theoretical one) should be rejected.

5. To test whether the variance of a normal population equals $\sigma_0^2$, we compute the test statistic: $(n-1)s^2 \div \sigma_0^2$, where $n$ is the sample size and $s^2$ is the sample variance. If the test statistic exceeds $\chi_{\alpha/2}^2$ or is less than $\chi_{1-\alpha/2}^2$ (where $n-1$ is the number of degrees of freedom), the null hypothesis should be rejected if the alternative hypothesis is two-sided.

6. The *sign test* is used to test the null hypothesis that the population median equals a certain amount. A plus sign is placed next to a sample observation exceeding this amount; a minus sign is placed next to one that falls below it. The number of plus signs $x$ has a binomial distribution (where $\Pi = 1/2$) if the null hypothesis is true. For large samples, the null hypothesis should be rejected if $x < n/2 - z_{\alpha/2}\sqrt{n/4}$ or if $x > n/2 + z_{\alpha/2}\sqrt{n/4}$, assuming that a two-tailed test is appropriate.

7. The *Mann-Whitney test* is used to test the null hypothesis that two samples come from the same population. The first step is to rank all observations in the two samples combined and to compute $R_1$, the sum of the ranks for the first sample. One then computes the test statistic

$$\frac{n_1 n_2 + \dfrac{n_1(n_1+1)}{2} - R_1 - \dfrac{n_1 n_2}{2}}{\sqrt{n_1 n_2(n_1+n_2+1)/12}},$$

where $n_1$ is the number of observations in the first sample and $n_2$ is the number of observations in the second sample. If this test statistic is greater than $z_{\alpha/2}$ or less than $-z_{\alpha/2}$ the null hypothesis should be rejected. (This assumes that $n_1$ and $n_2$ are at least 10.)

8. The *runs test* is designed to test the null hypothesis that a sequence of numbers, symbols, or objects is in random order. A *run* is a sequence of *identical* numbers, symbols, objects, or events preceded

and followed by *different* numbers, symbols, objects, or events, or by nothing at all. If a sequence is in random order, the expected number of runs $E_r$ equals $[(2n_1n_2)/(n_1 + n_2)] + 1$, and the standard deviation of the number of runs $\sigma_r$ equals

$$\sqrt{\frac{2n_1n_2(2n_1n_2 - n_1 - n_2)}{(n_1 + n_2)^2(n_1 + n_2 - 1)}},$$

where $n_1$ is the number of observations in the sequence of one type and $n_2$ is the number of the other type. If both $n_1$ and $n_2$ exceed 10, and the number of runs is $r$, the null hypothesis should be rejected if $(r - E_r)/\sigma_r$ exceeds $z_{\alpha/2}$ or is less than $-z_{\alpha/2}$.

## Chapter Review Exercises

**9.31** A taxi service has two taxis. It wants to determine whether one makes the trip from the center of town to the local airport more quickly than the other. Over a one-week period, the times (in minutes) that it takes each taxi to make this trip are shown below:

Taxi I: 6.3, 7.2, 5.2, 3.5, 4.8, 5.7, 5.2, 6.4, 5.9, 6.1, 6.3, 6.8
Taxi II: 7.4, 6.5, 7.4, 8.2, 5.6, 5.4, 6.3, 7.4, 6.2, 7.3, 4.6, 5.9

Use Minitab (if available) to answer the following questions. If you do not have Minitab, do the calculations by hand.

(a) Determine whether the median time for the two taxis is the same. (Let $\alpha = 0.05$.)
(b) Calculate and interpret the $p$-value for this test.

**9.32** A store has three branches. The store's managers want to find out whether the percentage of customers who are satisfied with the store's service is the same at each branch. A random sample of 100 customers is chosen from each branch's customers, and each person in the sample is asked whether he or she regards the store's service as satisfactory, the results being as follows:

|  | Branch A | Branch B | Branch C |
|---|---|---|---|
| Satisfactory | 81 | 72 | 60 |
| Not satisfactory | 19 | 28 | 40 |
| Total | 100 | 100 | 100 |

(a) Can the observed differences among branches be attributed to chance variation? (Set $\alpha = .01$.)
(b) If $\alpha$ is set equal to .05, can the observed differences be attributed to chance variation?

**9.33** The Department of Transportation wants to determine whether six types of cars differ with respect to the probability of stalling when the temperature is minus 40 degrees. The Department chooses a random sample of 100 cars of each type, and operates them under this set of circumstances, the results being:

|          | Type 1 | Type 2 | Type 3 | Type 4 | Type 5 | Type 6 |
|----------|--------|--------|--------|--------|--------|--------|
| Stall    | 31     | 22     | 29     | 33     | 23     | 24     |
| No stall | 69     | 78     | 71     | 67     | 77     | 76     |
| Total    | 100    | 100    | 100    | 100    | 100    | 100    |

Test whether these types of cars differ significantly with respect to the probability of stalling under these circumstances. (Set $\alpha = 0.10$.)

**9.34** The Alpha Corporation hires a market-research firm to interview 100 people between 20 and 30 years of age in each of four regions of the country who bought one of Alpha's competitors' bicycles in the past year. Each interviewee is asked whether he or she had ever heard of Alpha's bicycle, and if so, whether he or she felt it was overpriced. The results are as follows:

| Region | Never heard of Alpha's bicycle | Thought Alpha's bicycle was Overpriced | Thought Alpha's bicycle was Not overpriced | Total |
|--------|-------------------------------|-----------|---------------|-------|
| South  | 30  | 10 | 60 | 100 |
| North  | 41  | 21 | 38 | 100 |
| East   | 28  | 7  | 65 | 100 |
| West   | 32  | 14 | 54 | 100 |
| Total  | 131 | 52 | 217 | 400 |

Test whether there are regional differences in the probability that an interviewee who bought one of Alpha's competitors' bicycles in the past year (1) never heard of Alpha's bicycle; (2) thought it was overpriced: (3) did not think it was overpriced. (Let $\alpha = .01$.)

**9.35** A publisher is interested in determining which of three book covers is most attractive. It interviews 400 people in each of three states (California, Illinois, and New York) and asks each person which of the covers he or she prefers. The number preferring each cover is as follows:

|              | California | Illinois | New York |
|--------------|-----------|----------|----------|
| First cover  | 81        | 60       | 182      |
| Second cover | 78        | 93       | 95       |
| Third Cover  | 241       | 247      | 123      |
| Total        | 400       | 400      | 400      |

Do these data indicate that there are regional differences in people's preferences concerning these covers?
(a) Use the .05 level of significance.
(b) Use the .01 level of significance.

**9.36** Find a 95 percent confidence interval for the variance if $s = 8$ and if
(a) $n = 10$.
(b) $n = 20$.
(c) $n = 41$.

**9.37** Test the hypothesis that $\sigma = 10$, given that $s = 8$ for a sample of 15.
(a) The alternative hypothesis is $\sigma > 10$, and $\alpha = .05$.
(b) The alternative hypothesis is $\sigma < 10$, and $\alpha = .01$.

**9.38** A researcher claims that the median increase in output due to a new technique is 10 units of output per hour. A firm tries the new technique 100 times, and finds that in 41 of these trials the new technique results in an increase of more than 10 units of output per hour. In 59 of these trials there is an increase of less than 10 units of output per hour.

(a) Use the sign test to indicate whether the researcher's claim is correct, against a two-sided alternative. (Let $\alpha = .05$.)

(b) Use the sign test to indicate whether the researcher's claim is correct, against the alternative hypothesis that the median increase is less than 10 units of output per hour. (Let $\alpha = .05$.)

**9.39** The Crooked Arrow National Bank chooses a random sample of 36 depositors at its southern branch and finds that the 1986 income of each is as follows:

*Dollars*

| | | | | | |
|---|---|---|---|---|---|
| 20,100 | 8,200 | 13,200 | 5,200 | 11,100 | 51,100 |
| 19,400 | 8,900 | 14,300 | 10,100 | 10,800 | 9,600 |
| 10,100 | 10,300 | 15,800 | 12,300 | 9,100 | 10,900 |
| 23,000 | 26,000 | 16,100 | 14,000 | 7,200 | 12,000 |
| 24,200 | 11,400 | 17,200 | 15,100 | 4,300 | 13,200 |
| 25,100 | 12,900 | 18,900 | 16,000 | 38,000 | 15,100 |

Use the sign test to test the hypothesis that the median income of the depositors at the bank's southern branch equals $15,000, the alternative hypothesis being that it does not equal $15,000.

(a) Carry out this test at the 5 percent significance level.

(b) Carry out this test at the 1 percent significance level.

(c) Compare your results with those obtained in Exercise 8.29. Are the results consistent? Are the two tests aimed at testing the same hypothesis? Are the two tests based on the same assumptions? If not, what are the differences in the assumptions?

**\*9.40** In Minnesota, 400 small businesses were chosen at random. The sample was divided into three groups. Firms in the first group spent $100,000 for consulting services, firms in the second group spent $50,000 for consulting services, and firms in the third group spent nothing for consulting services. The number of firms in each group whose profits improved, deteriorated greatly, deteriorated slightly, or remained unchanged was as follows:

| Change in profits | Amount spent on consulting services | | |
|---|---|---|---|
| | $100,000 | $50,000 | $0 |
| Improved | 32 | 16 | 10 |
| Deteriorated greatly | 11 | 20 | 31 |
| Deteriorated slightly | 27 | 34 | 19 |
| Unchanged | 70 | 70 | 60 |
| Total | 140 | 140 | 120 |

Minitab was used to carry out a $\chi^2$ test to see whether the change in profits was independent of the amount spent on consulting services. Interpret the results, given below.

* Exercise requiring some familiarity with Minitab.

```
MTB > read into c10,c11,c12
DATA> 32 16 10
DATA> 11 20 31
DATA> 27 34 19
DATA> 70 70 60
DATA> end
     4 ROWS READ
MTB > chisquare c10,c11,c12
Expected counts are printed below observed counts
            C10        C11        C12      Total
     1       32         16         10        58
            20.3       20.3       17.4
     2       11         20         31        62
            21.7       21.7       18.6
     3       27         34         19        80
            28.0       28.0       24.0
     4       70         70         60       200
            70.0       70.0       60.0
Total      140        140        120       400
ChiSq =    6.74 +     0.91 +     3.15 +
           5.28 +     0.13 +     8.27 +
           0.04 +     1.29 +     1.04 +
           0.00 +     0.00 +     0.00 =     26.84
 df = 6
```

**9.41** Various financial institutions are asked whether they favor a tighter monetary policy by the Federal Reserve. The sample is divided into commercial banks, savings and loan associations, and savings banks. The results are as follows:

|  | Commercial banks | Savings and loan associations | Savings banks |
|---|---|---|---|
| Favor strongly | 20 | 12 | 19 |
| Favor slightly | 24 | 28 | 21 |
| Oppose strongly | 16 | 20 | 20 |
| Oppose slightly | 60 | 60 | 60 |
| Total | 120 | 120 | 120 |

Use Minitab (or some computer package) to test whether opinion on this issue is independent of whether a financial institution is a commercial bank, a savings and loan association, or a savings bank. (Let $\alpha = 0.05$.) If Minitab (or some other computer package) is not available, do the computations by hand.

# Appendix 9.1

## INTERPRETING COMPUTER OUTPUT FOR CHI-SQUARE TESTS

As pointed out in many previous chapters, computer programs are available to carry out most statistical calculations quickly and easily, and it is important that you be able to interpret the relevant computer output. Suppose, for example, that you want to test whether the industrial distribution of major American firms was the same in 1986 as in 1980. A random sample of 100 major American firms in 1986 shows that 51 were in manufacturing, 42 were in services, and 7 were in mining. A random sample of 100 major American firms in 1980 shows that 56 were in manufacturing, 32 were in services, and 12 were in mining. To test whether the difference between 1986 and 1980 in the percentage distribution of firms among these three industrial categories is statistically significant, Minitab is used to carry out the appropriate $\chi^2$ test, the computer output being shown below.

```
MTB > read into c16,c17
DATA> 56 51
DATA> 32 42
DATA> 12 7
DATA> end
      3 ROWS READ
MTB > chisquare c16,c17
Expected counts are printed below observed counts
                 C16              C17          Total
      1           56               51           107
                 53.5             53.5
      2           32               42            74
                 37.0             37.0
      3           12                7            19
                  9.5              9.5
  Total          100              100           200
  ChiSq =        0.12 +           0.12 +
                 0.68 +           0.68 +
                 0.66 +           0.66 = 2.90
  df = 2
```

As you can see, each expected frequency (or *count*, as it is called in the printout) in the contingency table is printed below the observed frequency. For example, the expected frequency corresponding to the observed frequency of 56 (for manufacturing firms in 1980) is 53.5. Below these frequencies, the value of

"total chi square," which is $\Sigma \frac{(f-e)^2}{e}$, is printed. In this case,

$\Sigma \frac{(f-e)^2}{e}$ equals 2.90. Finally, the number of degrees of freedom,

which is 2 in this case, is printed. Based on this computer output, we can easily test the null hypothesis that the percentage distribution of major American firms among these three industrial categories was the same in 1986 as in 1980. (Suppose that $\alpha$ is set equal to .05.) Since $\chi^2_{.05} = 5.99147$ when there are 2 degrees of

freedom (see Appendix Table 8), $\Sigma \frac{(f-e)^2}{e}$ does not exceed $\chi^2_{.05}$,

and there is no reason to reject the null hypothesis.

# 10 Experimental Design and the Analysis of Variance*

## 10.1 Introduction

As we have stressed often in previous chapters, one of the important functions of statistics is to provide principles for the proper design and analysis of experiments. Although the word *experiment* may evoke a vision of laboratory activities, many statistical investigations are experiments which have nothing whatever to do with a laboratory. Nonlaboratory experiments of various kinds are continually being carried out by business firms and government agencies for practical reasons. To make sure that these procedures are designed effectively and analyzed properly, you must know something about experimental design and the analysis of variance, the key topics of this chapter.

## 10.2 Design of Industrial Experiments

Although the neophyte commonly believes that it is a simple matter to design an experiment for testing a certain hypothesis, often a considerable amount of ingenuity is required to achieve a design that really tests the hypothesis one wants to test—and does so at something near minimum cost. As we know from Chapter 1, one of the commonest pitfalls is an experimental design in which the effect of the variable one wants to estimate is inextricably entangled with the effect of some other factor. (In other words, more than one factor may be responsible for a particular observed experimental result.) When this occurs, the effect of the variable one wants to estimate is said to be *confounded* with the effect of another factor or factors.

---

* Some instructors may want to skip the last sections of this chapter, which deal with the two-way analysis of variance. Others may want to take up only the first three sections, or skip this chapter entirely. Any of these options is feasible, since an understanding of this chapter is not required in subsequent chapters. Readers who skipped Chapter 9 should read Section 9.3 before reading Section 10.4.

| School grade | Treatment | Number of children | Number afflicted with polio | Number afflicted per 100,000 |
|---|---|---|---|---|
| Second grade | Salk vaccine | 222,000 | 38 | 17 |
| First and third grades | No vaccine | 725,000 | 330 | 46 |

**Table 10.1**
Incidence of Polio in Two Groups of Children

*Source:* W. Youden, "Chance, Uncertainty, and Truth in Science," *Journal of Quality Technology,* 1972.

To illustrate this, recall our discussion in Chapter 1 of the results of a nationwide test of the effectiveness of the Salk antipolio vaccine. These data (reproduced in Table 10.1) seem perfectly adequate for measuring the effects of the vaccine on the incidence of polio in children, but in fact they contain a major bias: Only those second-graders who received their parent's permission received the vaccine, whereas *all* first- and third-graders were taken into account in estimating what the incidence of polio would be without the vaccine. The problem was that the incidence of polio was substantially lower among nonvaccinated children who did not receive permission than among those who did. (Recall from Chapter 1 that children of lower-income parents are less likely to receive permission— and because they grow up in less hygienic conditions, they are less likely to get polio.) Table 10.2 shows the results of a much more adequate experimental design in which children who received permission are differentiated from those who did not. Those who received permission were assigned at random to either the group receiving the vaccine or the group receiving the placebo (similar in appearance to the vaccine, but of no medical significance). As you can see, the results indicate that the reduction in the incidence of polio due to the vaccine is much larger than is shown in Table 10.1[1]

In general, *an experiment dictates that one group of people (or machines, materials, or other experimental units) be treated differently from another group; and the effect of this difference in treatment is estimated by comparing certain measurable characteristics of the two groups.* However, unless the people (or machines, materials, or other experimental units) are assigned to one group or the other *at random,* all sorts of confounding can occur. For example, in the case of the Salk vaccine, if doctors had decided which children were to receive the vaccine and

*Randomization*

| Permission given | Treatment | Number of children | Number afflicted with polio | Number afflicted per 100,000 |
|---|---|---|---|---|
| Yes | Salk vaccine | 201,000 | 33 | 16 |
| Yes | Placebo | 201,000 | 115 | 57 |
| No | None | 339,000 | 121 | 36 |

**Table 10.2**
Randomized Experiment with Salk Vaccine

*Source:* See Table 10.1.

[1] This example is from W. Youden, "Chance, Uncertainty, and Truth in Science," *Journal of Quality Technology,* 1972.

which were not, there might have been a tendency to give the vaccine to children where the consent of the parents was easy to obtain, and such children might differ from the others in the likelihood that they would contract polio. *The importance of randomization cannot be overemphasized.* The experimental units or subjects should be assigned at random to the groups receiving different treatments.

### SHOULD THE EFFECTS OF SEVERAL FACTORS BE STUDIED ONE AT A TIME?

In many experiments, the statistician or decision maker is interested in estimating the effects of more than one factor on a certain characteristic of the the relevant experimental unit. For example, a thread manufacturer may be interested in estimating the effects of differences in raw materials and differences in types of machines on the strength of the thread produced. The traditional way of estimating these effects is the one-at-a-time method. That is, the firm would hold constant the raw material used and observe the effects of differences in the types of machine used. For example, if there are three types of raw materials that could be used (A, B, and C) and if there are three types of machines (I, II, and III), the firm might use one raw material, A, in all three types of machine in order to observe how the strength of the thread varies among the machines. Similarly, to estimate the effect of the type of raw material on the strength of the thread, the firm might use only one type of machine, I, and observe how the strength of the thread varies among the types of raw material.

An important disadvantage of the one-at-a-time approach is that the results may be too narrowly focused. For example, the experiments described in the previous paragraph will provide information concerning the effects of the type of machine on the strength of the thread *if raw material A is used.* However, the differences among machines in this regard may *not* be independent of the raw material used. Thus, although machine I may result in the strongest thread when raw material A is used, machine I may result in the weakest thread when raw material C is used. Similarly, the above experiments will provide information concerning the effects of the type of raw material on the strength of the thread when machine I is used, but the differences among raw materials in this regard may not be independent of the machine used. Thus, although raw material A may result in the strongest thread when machine I is used, raw material A may result in the weakest thread when machine III is used.

In general, modern statisticians tend to emphasize the advantages of not controlling an experiment too closely. Even if the effects of the factors are independent, it frequently is less expensive to conduct an experiment where the factors are allowed to vary, rather than hold one or more constant, because fewer observations often are required to obtain the same precision. For example, in the above case, a better experimental design may be to obtain data concerning the strength of the thread when each combination of raw material and machine is used. Thus, the firm might use raw material A with machine I, with machine II, and with machine III. Similarly, each of the other raw materials (B and C) might be used with machine I, with machine II, and with machine III. If the effects of the raw material are independent of the type of machine, and if the effects of the type of machine are independent of the raw material, the resulting nine observations could be used to estimate the effect of each raw material and each type of machine. (If they are not independent, more observations are needed, as we shall see below.) More will be said about this in subsequent sections of this chapter.

## RANDOMIZED BLOCKS

The type of experimental design recommended in the preceding paragraph is known as a randomized block design. In a *randomized block design* there are two kinds of effects, *treatment effects* and *block effects*. These terms are derived from agricultural research, where a field may be split into several blocks, and various treatments (such as fertilizers, pesticides, or some other factor whose effects the researcher is interested in) may be randomly assigned to plots in each block. Each block is constructed so that it contains relatively homogeneous experimental conditions. In an agricultural experiment each block may be a piece of land which has relatively homogeneous soil, sunlight, rainfall, and so forth. In the case of the thread manufacturer, either the raw materials or the machines can be regarded as treatments, and the other factor can be considered blocks. If each of the machines is regarded as a treatment, each raw material can then be regarded as a block because the use of this (and only this) raw material results in relatively homogeneous experimental conditions for comparing the effects of the treatments (the machines).

*Treatment Effects and Block Effects*

In a randomized block design, the statistician obtains data concerning the effect of each treatment in each block. Thus, in this case data are obtained concerning the strength of the thread resulting from the use of each type of machine with each raw material. The results might be as shown in Table 10.3. Note that only one observation is obtained concerning the effect of each treatment in each block. That is, no attempt is made to obtain more than one measurement of the strength of the thread resulting from the use of a particular type of machine with a particular raw material. In cases of this sort, the statistician must assume that the treatment effects and block effects are independent; if this assumption is violated, the *experimental errors*—the

| Raw material | Machine (strength of thread) | | |
| | I | II | III |
|---|---|---|---|
| A | 50 | 40 | 45 |
| B | 48 | 39 | 45 |
| C | 52 | 44 | 48 |

**Table 10.3**
Results of Experiment by Thread Manufacturer

**Table 10.4
Results of Experiment
by Thread
Manufacturer, where
Replication Occurs**

| Raw material | *Machine (strength of thread)* | | |
| | *I* | *II* | *III* |
|---|---|---|---|
| A | 50,52 | 40,38 | 45,46 |
| B | 48,47 | 39,39 | 45,44 |
| C | 52,53 | 44,45 | 48,47 |

errors due to chance variation in the effect of each treatment in each block—will tend to be overstated.

To avoid assuming that the treatment and block effects are independent, and to obtain more precise estimates of these effects, statisticians often specify that *replications* occur. In other words, they ask that the experiment be repeated so that more than one observation is obtained concerning the effect of each treatment in each block. Thus, in the case above, each type of machine might be used with each type of raw material to produce two pieces of thread, not one, and the results might be as shown in Table 10.4.

### LATIN SQUARES

Still another important concept in experimental design is *balance*. In either Table 10.3 or 10.4, we can estimate the net effect of each type of machine on the strength of the thread by comparing the mean strength of the pieces of thread made by this type of machine with the mean strength of the thread made by the other types of machines. Such a comparison is meaningful because each of these means pertains to the same set of raw materials. For example, in Table 10.3, each raw material is included once in calculating such a mean. In this sense, the design is balanced. In contrast, if the mean strength for machine I was based on the use of a different raw material than the mean strength for machine II, the design used by the thread manufacturer would not be balanced in this respect.

This concept of balance is used ingeniously in a commonly employed experimental design called a *Latin square*. Suppose that the thread manufacturer wants to estimate the effect of the type of raw material, the type of machine and the type of worker using them on the strength of the thread produced. (Note that an additional factor—type of worker—has been added to the two considered previously.) There are three types of raw material (A, B, and C), three types of machine (I, II, and III), and three types of worker (unskilled, semiskilled, and skilled). Although you may not believe it at first, the firm, by using a Latin square, can estimate the effects of all three of these factors on the basis of only nine observations.

Table 10.5 shows how this can be accomplished. Each vertical column signifies that a particular kind of raw material is used, and each horizontal row signifies that a particular type of machine is used. The letters in the body of the table indicate whether unskilled (U), semiskilled (E), or skilled (S) labor is used. There are nine entries in the

THE CASE OF FLUORESCEIN

A number of years ago, the claim was made by some agricultural scientists that plants grow better when watered with a dilute solution of fluorescein. To test this hypothesis, the agricultural scientists watered one group of plants with water and an adjacent group with fluorescein. The spatial distribution of plants was as shown below:

X X X X X X X X X   X X X X X X X X X X
X X X X X X X X X X   X X X X X X X X X X
X X X   Water only   X X X   X X X   Fluorescein   X X X
X X X X X X X X X X   X X X X X X X X X X
X X X X X X X X X X   X X X X X X X X X X

The results indicated a large difference in growth between the two groups. Unfortunately, however, these results were not supported by subsequent research because the experimental design was faulty. What fundamental mistake was made? How could the design have been improved?

SOLUTION: The biggest mistake was to put the plants watered with fluorescein on a different plot of land than the plants watered with water alone. The effects of fluorescein are confounded with whatever difference exists between the plots in fertility. In fact, the soil to which the fluorescein was applied contained more nutrients than the soil to which plain water was applied. This, rather than fluorescein, was responsible for the observed difference in plants' growth rates. A more effective test of the effect of fluorescein might have been accomplished by the spatial distribution of plants shown below:

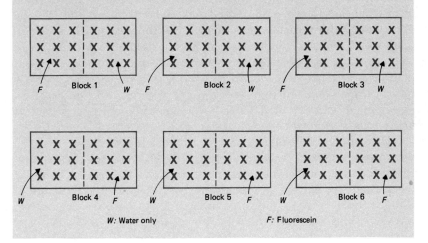

*W: Water only*          *F: Fluorescein*

TO ERR IS HUMAN ... BUT COSTLY

According to this design, the entire field is divided into six relatively homogeneous blocks, each of which contains 18 plants. Each block is then divided in half, and a coin is flipped to determine which half (that is, which nine plants) will be watered with fluorescein and which half will receive plain water. This is a randomized block with nine observations concerning the effect of each treatment in each block. The difference between the average rate of growth of the plants receiving flourescein and the average rate of growth of those receiving plain water is a measure of the net effect of the treatment. (The net block effects—that is, the differences in rate of growth among various blocks—are of subsidiary importance in this case.)[2]

[2] *Ibid.*

**Table 10.5**
A Latin-Square
Experimental Design

| Machine | Raw material A | B | C |
|---------|:---:|:---:|:---:|
| I | U | S | E |
| II | S | E | U |
| III | E | U | S |

table, each corresponding to an observation. For example, the entries in the first column show that in the first observation, raw material A will be used on machine I with unskilled labor; in the second observation, raw material A will be used on machine II with skilled labor; and in the third observation, raw material A will be used on machine III with semiskilled labor.

Several important points should be noted about the design shown in Table 10.5. First, each type of labor is used three times. Second, each type of labor shows up once and only once in each row, which means that each type of labor is used once and only once with each type of machine. Third, each type of labor shows up once only in each column, which means that each type of labor is used once and only once with each type of raw material. What these three points add up to is that *each type of labor is used once with each type of raw material and with each type of machine.* Further, *each type of raw material is used once with each type of labor and with each type of machine;* and *each type of machine is used once with each type of labor and with each type of raw material.*

Because of this balance, we can compare the strength of the thread resulting from various raw materials by comparing (1) the average of the three observations where raw material A was used, with (2) the average of the three observations where raw material B was used, with (3) the average of the three observations where raw material C was used. This is a valid comparison because each of these sets of observations is based on the one-time use of each type of machine and each type of labor. Similarly, we can compare the strength of the thread resulting from various types of machines by comparing (1) the average of the three observations where machine I was used with (2) the average of the three observations where machine II was used, with (3) the average of the three observations where machine III was used. Similar comparisons can be made among various types of labor.

## 10.3 Testing Textile Fabrics: A Case Study[3]

Latin squares are used often in industrial experimentation. To illustrate their use, consider the actual case of a chemical firm that wanted to test the durability of four types of rubber-covered fabric. The machine which is used for tests of this sort contains four rectangular brass plates, each of which is covered with a special emery paper. A mechanical device rubs samples of fabric over each of the four plates bearing the emery paper. Although the four plates, or positions in the machine, are much the same, they differ from one another slightly. Since a different type of fabric can be tested at each position, the machine can test four types of fabric simultaneously, each such testing being called a run. It is known that because of variations in the condition of the emery paper and in temperature, humidity, and other factors, the results of each test will vary from one run to the next.

The chemical firm used a Latin-square design for comparing the four types of fabric. As shown in Table 10.6, one factor considered in the experimental design was the position of the fabric in the machine (that is, the particular brass plate on which it was tested). This factor is shown by each vertical column beneath a number. A second factor was the run, which is shown by each horizontal row. The letters in the body of Table 10.6 indicate which of the four fabrics was tested in the indicated run, in the indicated position in the machine. The four types of fabric are designated by A, B, C, and D. As you can see, each type of fabric appears only four times in the table, once only in each row and once only in each column.

The numbers in parentheses in Table 10.6 show the results of the 16 tests, each of these numbers showing the loss in weight of the fabric after a certain amount of rubbing. The average result for fabric A was

| Run | Position in machine 4 | 2 | 1 | 3 | Mean |
|-----|------|------|------|------|------|
| 2 | A(251) | B(241) | D(227) | C(229) | 237 |
| 3 | D(234) | C(273) | A(274) | B(226) | 252 |
| 1 | C(235) | D(236) | B(218) | A(268) | 239 |
| 4 | B(195) | A(270) | C(230) | D(225) | 230 |
| Mean | 229 | 255 | 237 | 237 | |

**Table 10.6**
Results of a Latin-Square Design to Test Textile Fabrics

Source: O. L. Davies, *The Design and Analysis of Industrial Experiments* (London: Oliver and Boyd, 1956), p. 164.

[3] This example, is from O. L. Davies, *op cit.,* p. 163 ff.

266; for fabric B, 220; for fabric C, 242; and for fabric D, 231. These averages are comparable in the sense that each machine position and each run was included once in each average. Based on the results, it appeared that fabric B wore best, since it experienced the lowest weight loss. Also, based on the averages in Table 10.6, it appeared that position 2 in the machine tended to result in larger weight losses than the other positions, and that run 3 tended to produce higher weight losses than the other runs. Using techniques of the sort discussed in subsequent sections of this chapter, the firm was able to test whether these differences are statistically significant.

A final point should be noted concerning Latin squares. Underlying the use of this design is the assumption that the effect of each factor is independent of the other factors. In other words, it is assumed that the difference in weight loss between fabric A and fabric B will be the same, on the average, regardless of which machine position is chosen or which run is considered. Also, the difference in weight loss between machine positions is assumed to be the same, regardless of which type of fabric or run is considered. Frequently this assumption is perfectly reasonable, but in those cases where it is not, another type of experimental design should be used.

### EXERCISES

**10.1** Professor Malone assigns her students a new statistics textbook. She finds at the end of the year that 70 percent of the class do well on the final examination, and none fails it. Does this prove that the new book is effective?

**10.2** When studying the effects of a new drug, it is advisable to give the drug to some patients (the treatment group) but not others (the control group). Moreover, the experiment should be run *double-blind;* that is, neither the doctors who measure the results nor the subjects should know who is in which group. Why are these practices desirable?

**10.3** Many studies have been carried out to measure the effect of a particular surgical technique. Of the studies that showed marked enthusiasm for the technique, none had randomized controls. Of those that showed no enthusiasm for it, half had randomized controls. (By randomized controls, we mean that patients were assigned at random to this treatment rather than to another treatment.) How can you explain this?

**10.4** Experimental design reached the front page of the *Wall Street Journal* on August 25, 1977. The *Journal* defined a placebo as "any inactive substance or procedure used with a patient under the guise of an effective treatment." It reported that placebos often make people feel that they have recovered.[4] Given that this is the case, suppose that a new headache remedy is administered to 1,000 patients, and 700 say that it reduces the severity of their headaches. Is this proof of the effectiveness of this remedy? Why, or why not?

---

[4] See J. Critelli and K. Neumann, "The Placebo: Conceptual Analysis of a Construct in Transition," *American Psychologist,* January 1984.

**10.5** In the early 1960s, "gastric freezing" was used to treat duodenal ulcers. This treatment meant lowering a small balloon into the stomach and filling it with very cold alcohol to temporarily freeze the ulcer. Ulcer patients seemed to experience dramatic improvements after receiving this treatment.

(a) Is this proof of the effectiveness of this treatment? Why, or why not?

(b) To see whether "gastric freezing" really worked, researchers divided all patients entering a certain medical center into two groups: those who would undergo "gastric freezing," and those who would receive a placebo. (The latter patients were given the balloon treatment, but with alcohol that was not frigid enough to freeze the ulcer.) The latter group showed as much improvement as the former group. How do you account for this?

**10.6** An engineer wants to test whether process A results in higher output per worker in his plant than process B. To find out, he produces 10 percent of the plant's output, using process A. Since process A is still experimental, he is forced to use it only on the night shift, when supervision is less strict. Comment on this experimental design.

**10.7** In the 1950s, the U.S. Army carried out tests to determine whether large amounts of vitamins C and B complex would increase the physical performance of soldiers involved in a high-activity program in a cold environment.

(a) One way that this study could have been carried out is by picking a sample of 100 soldiers at random and sending them to a cold place to engage in calisthenics. All 100 would receive large amounts of the vitamins mentioned. Then their average calistenics performance might be compared with the average performance of all U.S. Army recruits. What problems can you detect in this design.

(b) Another way in which the experiment might be designed is as follows: Half of the 100 soldiers in the study might be given large amounts of vitamins C and B complex while the other half might receive a placebo. Then the average performance of one group might be compared with that of the other. In determining which group a soldier would join, the medical record of each individual would be inspected to see whether he had taken any vitamin supplements before. Only those who had not done so would be assigned to the group receiving the placebo. What problems can you detect in this design?

(c) As still another way of designing the experiment, half of the 100 soldiers in the study would be given large amounts of the vitamins, while the other half would receive nothing. Soldiers would be allocated at random to the two groups. Those receiving the vitamins would be told what they were being given, and those receiving nothing would be told that they were receiving nothing. Then the average performance of one group might be compared with that of the other. What problems can you detect in this design?

(d) The Army actually carried out this experiment in the following way: A random sample of soldiers was drawn from four platoons. The soldiers chosen from each platoon were randomly divided into two equal groups, one of which received large amounts of the vitamins, the other of which received a placebo. However, neither group knew that the latter was a placebo. In each platoon the average performance of one group was compared with that of the other.

(i) Is this a randomized block? Why, or why not?

(ii) If so, what are the blocks, and what are the treatments?

(iii) Is this a Latin square? Why, or why not?

(iv) If it is a Latin square, explain the balance of the design.

(v) Does this design contain replications?

**10.8** (a) In 1939, studies were conducted in which cases of lung cancer were identified, and some other persons comparable to the lung-cancer patients in sex, age, and other characteristics were selected. It was found that the persons with lung cancer smoked much more heavily than the other persons with similar characteristics. Was this a randomized experiment? What are the disadvantages of this experimental procedure?

(b) R. A. Fisher (the British statistician who will be discussed in the next section) argued that if people with a hereditary predilection for smoking also have a hereditary tendency to contract cancer, the results would be like those obtained in the studies in (a). How can one determine whether smoking is associated with lung cancer when heredity is held constant?[5]

## 10.4 The *F* Distribution

Earlier in this chapter we described particular experimental designs, but we have not yet indicated how one can test whether the observed differences among treatment means (that is, differences among the means of the observations resulting from various treatments) are statistically significant. For example, the results in Table 10.3 seem to indicate that the mean strength of thread produced by machine I is greater than that of machine II or machine III. But is this observed difference statistically significant? In other words, is it quite unlikely that it could have arisen due to chance? To answer this question, we must take up the *F* distribution, another of the most important probability distributions in statistics.

The *F* distribution is a continuous probability distribution. It was named for R. A. Fisher, the great British statistician who developed it in the early 1920s. Its definition is as follows.

> F DISTRIBUTION: *If a random variable* $Y_1$ *has a* $\chi^2$ *distribution with* $v_1$ *degrees of freedom, and if a random variable* $Y_2$ *has a* $\chi^2$ *distribution with* $v_2$ *degrees of freedom, then* $Y_1/v_1 \div Y_2/v_2$ *has an* **F distribution** *with* $v_1$ *and* $v_2$ *degrees of freedom (if* $Y_1$ *and* $Y_2$ *are independent).*

In other words, the ratio of two independent $\chi^2$ random variables, each divided by its number of degrees of freedom, is an *F* random variable.

Like the *t* and $\chi^2$ distributions, the *F* distribution is in reality a family of probability distributions, each corresponding to certain numbers of degrees of freedom. But unlike the *t* and $\chi^2$ distributions,

---

[5] For further discussion, see B. Brown, "Statistics, Scientific Method, and Smoking," in J. Tanur (ed.), *Statistics: A Guide to the Unknown* (San Francisco: Holden-Day, 1972).

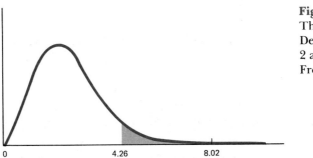

**Figure 10.1**
The *F* Probability
Density Function, with
2 and 9 Degrees of
Freedom

the *F* distribution has two numbers of degrees of freedom, not one.
Figure 10.1 shows the *F* distribution with 2 and 9 degrees of freedom.
As you can see, the *F* distribution is skewed to the right. However, as
both numbers of degrees of freedom become very large, the *F* distribution tends toward normality. As in the case of the $\chi^2$ distribution, the
probability that an *F* random variable is negative is zero. This must be
true since an *F* random variable is a ratio of two non-negative numbers.
($Y_1/v_1$ and $Y_2/v_2$ are both non-negative.) Once again, it should be emphasized that any *F* random variable has *two* numbers of degrees of
freedom. Be careful to keep these numbers of degrees of freedom in
*the correct order*, because an *F* distribution with $v_1$ and $v_2$ degrees of
freedom is *not* the same as an *F* distribution with $v_2$ and $v_1$ degrees of
freedom.

Tables are available which show the values of *F* that are exceeded
with certain probabilities, such as .05 and .01. Appendix Table 9
shows, for various numbers of degrees of freedom, the value of *F* that
is exceeded with probability equal to .05. For example, if the numbers
of degrees of freedom are 2 and 9, the value of *F* that is exceeded with
probability equal to .05 is 4.26. Similarly, Appendix Table 10 shows,
for various numbers of degrees of freedom the value of *F* that is exceeded with probability equal to .01. For example, if the numbers of
degrees of freedom are 2 and 9, the value of *F* exceeded with probability equal to .01 is 8.02.

In the following sections of this chapter, we shall show how the *F*
distribution can be used to test whether the differences among various
treatment means are statistically significant. For now, it is important to
become familiar with Appendix Tables 9 and 10. The following examples show how these tables are used.

EXAMPLE 10.1 A random variable has the *F* distribution with 15 and 12
degrees of freedom. What is the value of this random variable
that is exceeded with a probability of .05? With a probability of
.01?

SOLUTION: Appendix Table 9 shows that the answer to the
first question is 2.62; Appendix Table 10 shows that the answer to
the second question is 4.01.

EXAMPLE 10.2 A random variable has the $F$ distribution with 7 and 18 degrees of freedom. What is the value of this random variable that is exceeded with a probability of .05? With a probability of .01?

SOLUTION: Appendix Table 9 shows that the answer to the first question is 2.58; Appendix Table 10 shows that the answer to the second question is 3.84.

## 10.5 Analysis of a Completely Randomized Design

The procedure of one of the simplest experimental designs is to divide a set of people or other experimental units into groups, and subject each group to a different treatment. (The people or other experimental units are allocated to the groups at random.) Then the differences among the mean responses of the various groups are used to measure the net effects of the various treatments. This is called a *completely randomized design*. For example, suppose that a beer manufacturer picks a random sample of 20 beer drinkers whose favorite brand of beer is known to be brand W. These beer drinkers are then divided randomly into four groups of five individuals each. The first group is asked to rate a beer in an unmarked can which is in reality their favorite (brand W). The second group is asked to rate a beer in an unmarked can which is in reality brand X. The third group is asked to rate a beer in an unmarked can which is in reality brand Y. The fourth group is asked to rate a beer in an unmarked can which is in reality brand Z. If the results are as shown in Table 10.7, the mean ratings provided by the four groups are 60, 61, 58, and 61, respectively. The beer manufacturer would like to test the hypothesis that these differences among the four means are due to chance.

In this case, the null hypothesis is that $\mu_1 = \mu_2 = \mu_3 = \mu_4$, where $\mu_1$

**Table 10.7**
Results of Beer
Manufactuer's Survey

| | Respondent's favorite beer | Ratings Brand X | Ratings Brand Y | Ratings Brand Z |
|---|---|---|---|---|
| | 60 | 61 | 58 | 61 |
| | 59 | 60 | 54 | 57 |
| | 61 | 66 | 58 | 61 |
| | 55 | 62 | 58 | 61 |
| | 65 | 56 | 62 | 65 |
| $\bar{x}_j$ | 60 | 61 | 58 | 61 |
| $\sum_{i=1}^{5} (x_{ij} - \bar{x}_j)^2$ | 52 | 52 | 32 | 32 |
| $s_j^2$ | 13 | 13 | 8 | 8 |

is the mean rating in the population of the favorite beer, $\mu_2$ is the mean rating in the population of brand X, $\mu_3$ is the mean rating in the population of brand Y, and $\mu_4$ is the mean rating in the population of brand Z. The alternative hypothesis is that $\mu_1$, $\mu_2$, $\mu_3$, and $\mu_4$ are not all equal. Clearly, the null hypothesis tends to be supported if the four sample means are close together, whereas the alternative hypothesis tends to be supported if they are far apart. A reasonable measure of how close together or far apart the four sample means are is their variance—that is, the square of their standard deviation. Specifically, their variance equals

$$ s_{\bar{x}}^2 = \sum_{j=1}^{4} \frac{(\bar{x}_j - \bar{\bar{x}})^2}{3}, $$

where $\bar{x}_j$ is the mean of the $j$th sample and $\bar{\bar{x}}$ is the mean of the four sample means.[6] If the numerical values in Table 10.7 are substituted in this equation, we have

$$ s_{\bar{x}}^2 = \frac{(60 - 60)^2 + (61 - 60)^2 + (58 - 60)^2 + (61 - 60)^2}{3} = 2. $$

Suppose that each of the four populations of ratings can be approximated by a normal distribution, and that the standard deviation of each population $\sigma$ is the same. Then if the null hypothesis is true (that is, if $\mu_1 = \mu_2 = \mu_3 = \mu_4$), our four samples in Table 10.7 are four samples from the same population. Consequently, since the variance of the sample means drawn from the same population equals $\sigma^2/n$ (as we know from Chapter 6), it follows that $s_{\bar{x}}^2$ is an estimate of $\sigma^2/n$ where $n$ is the size of each sample. (In this case, $n = 5$.) Thus, $ns_{\bar{x}}^2$ is an estimate of $\sigma^2$, the common variance of the four populations, if the null hypothesis and our other assumptions are true. In particular, in the case in Table 10.7, 5(2), or 10, is an estimate of $\sigma^2$ if the null hypothesis is true (since $n = 5$ and $s_{\bar{x}}^2 = 2$).

To test whether the null hypothesis is true, we compare this estimate of $\sigma^2$ with another estimate of $\sigma^2$ that is valid regardless of whether or not the null hypothesis is true. This latter estimate of $\sigma^2$ is the mean of the four variances within the samples, $s_1^2$, $s_2^2$, $s_3^2$, and $s_4^2$. As we know from Chapter 7, each of these sample variances is an unbiased estimate of the population variance (assumed to be the same in the four populations). Thus, the mean of these variances is also an unbiased estimate of the population variance; and this is true regardless of whether or not $\mu_1$, $\mu_2$, $\mu_3$, and $\mu_4$ are equal. In the case shown in Table 10.7, the mean of the sample variances equals

$$ \frac{s_1^2 + s_2^2 + s_3^2 + s_4^2}{4} = \frac{13 + 13 + 8 + 8}{4} = \frac{42}{4} = 10.5. $$

---

[6] The denominator is 3 because this is a sample variance, where the denominator is the sample size minus 1. There are four sample means, so the denominator is $(4 - 1) = 3$.

The derivation of $s_1^2$, $s_2^2$, $s_3^2$, and $s_4^2$ is shown in Table 10.7.

At this point, we have two estimates of $\sigma^2$, the first being $ns_{\bar{x}}^2$ (in this case, 10), the second being $(s_1^2 + s_2^2 + s_3^2 + s_4^2)/4$ (in this case, 10.5). To test the null hypothesis, we form the ratio of these two estimates:

*The Variance Ratio*

$$F = \frac{ns_{\bar{x}}^2}{(s_1^2 + s_2^2 + s_3^2 + s_4^2)/4}$$

(10.1)

This ratio is called a *variance ratio*. If the null hypothesis is true, this ratio should be fairly close to 1, since both the numerator and the denominator should be approximately equal to $\sigma^2$. If the null hypothesis is not true, $ns_{\bar{x}}^2$ should be much greater than $\sigma^2$, since it will also reflect the variation among $\mu_1$, $\mu_2$, $\mu_3$, and $\mu_4$. Consequently, this variance ratio, $F$, is used as a test statistic. High values of $F$ are evidence that the null hypothesis should be rejected.

But how large can $F$ be expected to be by chance if the null hypothesis is true? The answer is given by the $F$ distribution, since the variance ratio has an $F$ distribution with $(k - 1)$ and $k(n - 1)$ degrees of freedom (where $k$ is the number of population means being compared and $n$ is the sample size in each population), if the null hypothesis is true. Thus, in the case of Table 10.7, the variance ratio has an $F$ distribution with 3 and 16 degrees of freedom, since $k = 4$ and $n = 5$. As shown in Appendix Table 9, there is a .05 chance that an $F$ variable with 3 and 16 degrees of freedom will exceed 3.24. Thus, if the .05 significance level is appropriate, the null hypothesis should be rejected if the variance ratio exceeds 3.24. In fact, since $ns_{\bar{x}}^2 = 10$ and $(s_1^2 + s_2^2 + s_3^2 + s_4^2)/4 = 10.5$, the variance ratio in Table 10.7 equals 0.95, and there is no reason to reject the null hypothesis.

In general, the decision rule for this test procedure is as follows.

*Test that $\mu_1 = \mu_2 = \ldots = \mu_k$*

DECISION RULE: *Reject the null hypothesis that the population means are all equal if*

$$\frac{ns_{\bar{x}}^2}{\sum\limits_{j=1}^{k} s_j^2/k} > F_\alpha,$$

*where $F_\alpha$ is the value of an F random variable—with degrees of freedom equal to $(k - 1)$ and $k(n - 1)$—that is exceeded with probability equal to $\alpha$ (the desired level of significance), where k is the number of population means being compared, and n is the sample size in each population.*

In the next section, we shall show how this test can be presented in another format which is used in the design and analysis of experiments.

# 10.6 One-Way Analysis of Variance

The analysis of variance is a technique designed to divide the total variation in a set of data into its component parts, each of which can be ascribed to a particular source. For example, in the problem presented in the previous section, the total variation might be split into two parts, one representing the differences among beer brands in their average ratings, the other representing the variation among consumers in their rating of a particular brand. In this section, we will show how an analysis of variance can be carried out for this type of problem. Although this is merely another way of presenting the test described in the previous section, it is useful to know this format since it is often employed and since it can be adapted for more complex problems, as we shall see in Section 10.8.

## NOTATION

In any one-way analysis of variance, the purpose is to test whether the means of $k$ populations are equal. Since $n$ observations are chosen from each population, the data can be arrayed as shown in Table 10.8. Each (vertical) column contains observations from the same population. Thus, $x_{11}$ is the first observation from the first population, $x_{21}$ is the second observation from the first population, $x_{12}$ is the first observation from the second population, and so on. In general, $x_{ij}$ is the $i$th observation from the $j$th population. We can apply this notation to Table 10.7, where $k = 4$ and $n = 5$. Clearly, in Table 10.7, $x_{11} = 60$, $x_{21} = 59$, $x_{12} = 61$, and so on.

The mean of the observations in the $j$th column is denoted by $\bar{x}_{.j}$. Thus, in Table 10.7, $\bar{x}_{.1} = 60$, $\bar{x}_{.2} = 61$, $\bar{x}_{.3} = 58$, and $\bar{x}_{.4} = 61$. On the other hand, the mean of the observations in the $i$th (horizontal) row is denoted by $\bar{x}_{i.}$. The dot indicates the subscript over which the averaging takes place. Since the first subscript designates the row of the observation, $\bar{x}_{.j}$ indicates that the averaging is over the observations in all rows in the $j$th column. Similarly, since the second subscript designates the column of the observation, $\bar{x}_{i.}$ indicates that the averaging is over the observations in all columns in the $i$th row. Thus, in Table 10.7, $\bar{x}_{1.} = (60 + 61 + 58 + 61)/4$, or 60.

| | | | | |
|---|---|---|---|---|
| $x_{11}$ | $x_{12}$ | $\cdot$ $x_{1j}$ | $\cdot$ | $x_{1k}$ |
| $x_{21}$ | $x_{22}$ | $\cdot$ $x_{2j}$ | $\cdot$ | $x_{2k}$ |
| $x_{31}$ | $x_{32}$ | $\cdot$ $x_{3j}$ | $\cdot$ | $x_{3k}$ |
| $\cdot$ | $\cdot$ | | | |
| $x_{i1}$ | $x_{i2}$ | $\cdot$ $x_{ij}$ | $\cdot$ | $x_{ik}$ |
| $\cdot$ | $\cdot$ | $\cdot$ $\cdot$ | $\cdot$ | |
| $x_{n1}$ | $x_{n2}$ | $\cdot$ $x_{nj}$ | $\cdot$ | $x_{nk}$ |
| Means $\quad \bar{x}_{.1}$ | $\bar{x}_{.2}$ | $\cdot$ $\bar{x}_{.j}$ | $\cdot$ | $\bar{x}_{.k}$ |

**Table 10.8**
Array of Data for
Analysis of Variance

BASIC IDENTITY

Using this notation, we can define the total variation in a set of data as follows:

***Total Sum of Squares***

$$TSS = \sum_{i=1}^{n} \sum_{j=1}^{k} (x_{ij} - \bar{x}..)^2,$$

(10.2)

where $\bar{x}..$ is the overall mean of all $n \times k$ observations. The total variation is generally referred to as the ***total sum of squares,*** which accounts for the *TSS* on the left-hand side of equation (10.2). With a bit of algebraic manipulation, it can be shown that the total sum of squares is identically equal to the sum of two terms:

***Total Sum of Squares = Between-Group Sum of Squares + Within-Group Sum of Squares***

$$\sum_{i=1}^{n} \sum_{j=1}^{k} (x_{ij} - \bar{x}..)^2 = n \sum_{j=1}^{k} (\bar{x}.j - \bar{x}..)^2 + \sum_{i=1}^{n} \sum_{j=1}^{k} (x_{ij} - \bar{x}.j)^2.$$

(10.3)

This is the basic identity underlying the one-way analysis of variance.

To understand the one-way analysis of variance, it is essential to interpret each of the terms in equation (10.3). The total sum of squares, which is on the left-hand side of this equation, is a measure of how much variation exists among all of the $n \times k$ observations in the sample. According to equation (10.3), this total variation can be split into two parts, one reflecting differences among the means of the observations taken from different populations, the other reflecting differences among the observations taken from the same population. Holding $n$ and $k$ constant, the larger the first part (relative to the second part), the less likely it would seem that the population means are all equal.

The first term on the right-hand side of equation (10.3) is often called the ***between-group sum of squares*** since it reflects differences between the populations in the sample means. To satisfy yourself that it does reflect such differences, note that it equals $(k-1)$ times $ns_{\bar{x}}^2$. Another frequently used name for this term is the ***treatment sum of squares*** since the differences among the population means are often called the net effects of different treatments. *Treatment* is a completely general term used to characterize each of the populations being compared. For example, each brand of beer can be called a different treatment in Table 10.7.

The second term on the right-hand side of equation (10.3) is often referred to as the ***within-group sum of squares,*** since it measures the variation within the populations. To see that it does measure such variation, note that it is equal to $k(n-1)$ times the denominator of the variance ratio in equation (10.1). (In equation (10.1), $k=4$ and $n=5$.) It is also frequently called the ***error sum of squares,*** since the within-group variation is often interpreted as being due to experimental error.

To illustrate that the total sum of squares is identically equal to the between-group sum of squares plus the within-group sum of squares, let's return to the data in Table 10.7. Clearly, the total sum of squares equals

$$TSS = (60 - 60)^2 + (59 - 60)^2 + (61 - 60)^2 + (55 - 60)^2$$
$$+ (65 - 60)^2 + (61 - 60)^2 + (60 - 60)^2 + (66 - 60)^2$$
$$+ (62 - 60)^2 + (56 - 60)^2 + (58 - 60)^2 + (54 - 60)^2$$
$$+ (58 - 60)^2 + (58 - 60)^2 + (62 - 60)^2 + (61 - 60)^2$$
$$+ (57 - 60)^2 + (61 - 60)^2 + (61 - 60)^2 + (65 - 60)^2$$
$$= 198.$$

The between-group sum of squares equals

$$BSS = 5[(60 - 60)^2 + (61 - 60)^2 + (58 - 60)^2 + (61 - 60)^2]$$
$$= 30.$$

And the within-group sum of squares equals

$$WSS = (60 - 60)^2 + (59 - 60)^2 + (61 - 60)^2 + (55 - 60)^2$$
$$+ (65 - 60)^2 + (61 - 61)^2 + (60 - 61)^2 + (66 - 61)^2$$
$$+ (62 - 61)^2 + (56 - 61)^2 + (58 - 58)^2 + (54 - 58)^2$$
$$+ (58 - 58)^2 + (58 - 58)^2 + (62 - 58)^2 + (61 - 61)^2$$
$$+ (57 - 61)^2 + (61 - 61)^2 + (61 - 61)^2 + (65 - 61)^2$$
$$= 168.$$

Since $198 = 30 + 168$, it is obvious that the total sum of squares does equal the sum of the between-group sum of squares and the within-group sum of squares.

> *In the one-way analysis of variance, the total variation in the sample is divided into two parts, the between-group sum of squares and the within-group sum of squares.*    **Basic Idea #19**

## ANALYSIS-OF-VARIANCE TABLE

The test procedure used in the one-way analysis of variance is precisely the same as that described in Section 10.5. What is different is that the computations are presented in the format of an analysis of variance table. The general form of a one-way analysis of variance table is shown in Table 10.9. The first column shows the source or type of variation, and the second column shows the corresponding sum of squares. The third column shows the number of degrees of freedom

**Table 10.9**

General One-Way Analysis-of-Variance Table

| Source of variation | Sum of squares | Degrees of freedom | Mean square | F |
|---|---|---|---|---|
| Between groups | BSS | $k-1$ | $\dfrac{BSS}{k-1}$ | $\dfrac{BSS}{k-1} \div \dfrac{WSS}{k(n-1)}$ |
| Within groups | WSS | $k(n-1)$ | $\dfrac{WSS}{k(n-1)}$ | |
| Total | TSS | $nk-1$ | | |

corresponding to each sum of squares, these numbers being the figures that were used in equation (10.1) to divide each sum of squares to obtain an estimate of $\sigma^2$, the variance of each population.[7] The fourth column shows each mean square, which is the sum of squares divided by the number of degrees of freedom. The last column shows the ratio of the two mean squares. If the null hypothesis is true, this ratio—which is precisely the same as the variance ratio in equation (10.1)—has an $F$ distribution with $(k-1)$ and $k(n-1)$ degrees of freedom. (Note that these are the numbers of degrees of freedom in the first and second rows of the table.) Thus, in a one-way analysis of variance table, the decision rule is as follows.

*Test that* $\mu_1 = \mu_2 = \ldots = \mu_k$

DECISION RULE: *Reject the null hypothesis that the population means are all equal if the ratio of the between-group mean square to the within-group mean square exceeds* $F_\alpha$, *where* $\alpha$ *is the desired significance level.*

Of course, this decision rule is essentially the same as the decision rule in Section 10.5.

To illustrate the application of a one-way analysis of variance, let's go back to the experiment carried out by the beer manufacturer. Based on the data in Table 10.7, we know from previous paragraphs that the between-group sum of squares equals 30 and that the within-group sum of squares equals 168. These figures constitute the second column of Table 10.9. And since $k = 4$ and $n = 5$, it is evident that the numbers of degrees of freedom are 3 and 16. Thus, dividing 30 by 3, we get the between-group mean square, which is 10; and dividing 168 by 16, we get the within-group mean square, which is 10.5. The ratio of the mean squares is $10 \div 10.5 = 0.95$. Since $F_{.05} = 3.24$ when there are 3 and 16 degrees of freedom, the ratio of the mean squares is not greater than $F_{.05}$, and there is no reason (at the .05 significance level) to reject the null hypothesis that the mean ratings for the four brands of beer are equal. The analysis-of-variance table is given in Table 10.10.

---

[7] In the numerator of the right-hand side of equation (10.1), the between-group sum of squares is divided by 3. (Note that $ns_{\bar{x}}^2 = BSS \div 3$.) In the denominator of the right-hand side of equation (10.1), the within-group sum of squares is divided by $4 \times (n-1)$. (Note that $(s_1^2 + s_2^2 + s_3^2 + s_4^2) \div 4 = WSS \div [4 \times (n-1)]$.) Since $k = 4$ in equation (10.1), the numbers of degrees of freedom are indeed the numbers used in equation (10.1) to divide the sums of squares to get estimates of $\sigma^2$.

| Source of variation | Sum of squares | Degrees of freedom | Mean square | F |
|---|---|---|---|---|
| Between groups | 30 | 3 | 10 | 0.95 |
| Within groups | 168 | 16 | 10.5 | |
| Total | 198 | 19 | | |

**Table 10.10**
One-Way Analysis-of-Variance Table, Beer Example

The following example illustrates further how the one-way analysis of variance is used.

EXAMPLE 10.3 The thread manufacturer (encountered earlier in this chapter) wants to determine whether the mean strength of thread produced by three different types of machine are different when raw material A is used on each machine. Four pieces of thread are produced on each type of machine, the results being as follows:

| | Machine | |
|---|---|---|
| *I* | *II* | *III* |
| 50 | 41 | 49 |
| 51 | 40 | 47 |
| 51 | 39 | 45 |
| 52 | 40 | 47 |

Use a one-way analysis of variance to test whether the mean strength of thread is equal for the three types of machine. (Let the significance level equal .05.)

SOLUTION: The mean strength for the first type of machine is 51; for the second type it is 40; and for the third method it is 47. The mean for all types of machine is 46. Thus, the between-group sum of squares equals $4 \times [(51-46)^2 + (40-46)^2 + (47-46)^2]$, or 4(62), or 248. The within-group sum of squares equals $(50-51)^2 + (51-51)^2 + (51-51)^2 + (52-51)^2 + (41-40)^2 + (40-40)^2 + (39-40)^2 + (40-40)^2 + (49-47)^2 + (47-47)^2 + (45-47)^2 + (47-47)^2 = 12$. Thus, the analysis-of-variance table is

| Source of variation | Sum of squares | Degrees of freedom | Mean square | F |
|---|---|---|---|---|
| Between groups | 248 | 2 | 124 | 93 |
| Within groups | 12 | 9 | 1.33 | |
| Total | 260 | 11 | | |

Since there are 2 and 9 degrees of freedom, $F_{.05} = 4.26$. Since the observed value of $F$ far exceeds this amount, the thread manufacturer should reject the null hypothesis that the mean strength is the same for the three types of machine.[8]

---

[8] See Appendix 10.2 for some formulas that are useful in calculating the required sums of squares in more complicated cases.

## 10.7 Confidence Intervals for Differences among Means

In the previous section, we saw how the one-way analysis of variance can be used to test whether several population means are all equal. However, in most statistical investigations the purpose is to find out the *extent* to which these means differ, not just *whether* they differ. For example, in the case of the experiment carried out by the thread manufacturer in Example 10.3, it is important to estimate the differences between the mean strengths of thread produced by various types of machines. The firm wants to answer questions like: What is the difference between the mean strength of thread produced on machine I and that produced on machine II? In this section we will show how confidence intervals can be constructed for *all* differences among the population means. The confidence coefficient attached to these intervals is the probability that *all* these intervals will include the respective differences among the population means *simultaneously*.

In the case of the thread manufacturer, there are three differences between the population means, since there are three population means. If $\mu_1$ is the mean strength of thread produced by machine I, $\mu_2$ is the mean strength of that produced by machine II, and $\mu_3$ is the mean strength of thread produced by machine III, the probability is $(1 - \alpha)$ that *all* the following statements hold true *simultaneously*:

*Confidence Intervals for Differences among Means*

$$\bar{x}_{.1} - \bar{x}_{.2} - \sqrt{F_\alpha}s_w \sqrt{\frac{2(k-1)}{n}} < \mu_1 - \mu_2 < \bar{x}_{.1} - \bar{x}_{.2} + \sqrt{F_\alpha}s_w \sqrt{\frac{2(k-1)}{n}}$$

$$\bar{x}_{.1} - \bar{x}_{.3} - \sqrt{F_\alpha}s_w \sqrt{\frac{2(k-1)}{n}} < \mu_1 - \mu_3 < \bar{x}_{.1} - \bar{x}_{.3} + \sqrt{F_\alpha}s_w \sqrt{\frac{2(k-1)}{n}}$$

$$\bar{x}_{.2} - \bar{x}_{.3} - \sqrt{F_\alpha}s_w \sqrt{\frac{2(k-1)}{n}} < \mu_2 - \mu_3 < \bar{x}_{.2} - \bar{x}_{.3} + \sqrt{F_\alpha}s_w \sqrt{\frac{2(k-1)}{n}}$$

$(10.4)$

where $s_w$ is the square root of the within-group mean square defined as $WSS/k(n-1)$, $F_\alpha$ is the value of an $F$ random variable [with $(k-1)$ and $k(n-1)$ degrees of freedom] exceeded with a probability of $\alpha$, $k$ is the number of means being compared, and $n$ is the size of the sample taken from each population.[9]

Since $\bar{x}_{.1} = 51, \bar{x}_{.2} = 40, \bar{x}_{.3} = 47, F_{.05} = 4.26, s_w = 1.15, k = 3$, and $n = 4$, the 95 percent confidence interval for all of the differences between the population means is as follows:

[9] This result and that in Section 10.9 are from H. Scheffe, *The Analysis of Variance* (New York: Wiley, 1959).

$$6.63 < \mu_1 - \mu_2 < 11.37$$

$$1.63 < \mu_1 - \mu_3 < 6.37$$

$$-9.37 < \mu_2 - \mu_3 < -4.63.$$

Clearly, machine I seems to result in stronger thread than machine II or machine III, and machine III seems to result in a stronger thread than machine II.

Expression (10.4) can be used to construct confidence intervals for the differences between the population means, no matter what the values of $k$ and $n$ may be. The following is an example of how these intervals are constructed.

---

EXAMPLE 10.4 Using the data in Table 10.7, construct a 95 percent confidence interval for the six differences between the mean ratings of the four beers (the favorite, brand X, brand Y, and brand Z).

SOLUTION: In this case, $\bar{x}_{.1} = 60$, $\bar{x}_{.2} = 61$, $\bar{x}_{.3} = 58$, $\bar{x}_{.4} = 61$, $F_{.05} = 3.24$ (since there are 3 and 16 degrees of freedom), $s_w = \sqrt{10.5}$, $k = 4$, and $n = 5$. Thus, $\sqrt{F_\alpha}\, s_w \sqrt{2(k-1)/n} = \sqrt{3.24} \times \sqrt{10.5} \times \sqrt{(3/5)} \times 2 = 6.39$, and the confidence intervals are as follows:

$$-7.39 < \mu_1 - \mu_2 < 5.39$$

$$-4.39 < \mu_1 - \mu_3 < 8.39$$

$$-7.39 < \mu_1 - \mu_4 < 5.39$$

$$-3.39 < \mu_2 - \mu_3 < 9.39$$

$$-6.39 < \mu_2 - \mu_4 < 6.39$$

$$-9.39 < \mu_3 - \mu_4 < 3.39$$

As you can see, all these confidence intervals include zero, which is what would be expected, given that the analysis of variance in the previous section concluded that none of these differences is statistically significant.

---

## EXERCISES

**10.9** If an $F$ random variable has 40 and 30 degrees of freedom, what is the probability that it will exceed (a) 1.79; (b) 2.30?

**10.10** Suppose that $X$ has a $\chi^2$ distribution with 4 degrees of freedom and $Y$ has a $\chi^2$ distribution with 7 degrees of freedom. (a) What is the distribution of $X/4 \div Y/7$? (b) What is the distribution of $Y/7 \div X/4$?

**10.11** Suppose that a random variable has an $F$ distribution with 10 and 12 degrees of freedom. (a) What is the value of this variable that is exceeded with a probability of .05? (b) What is the value of this variable that is exceeded with a probability of .01?

**10.12** Fill in the blanks in the following analysis of variance table:

| Source of variation | Sum of squares | Degrees of freedom | Mean square | F |
|---|---|---|---|---|
| Between groups | _____ | 2 | _____ | _____ |
| Within groups | _____ | 11 | 14 | |
| Total | 200 | _____ | | |

**10.13** An American automobile manufacturer runs an experiment in which four of its cars are chosen at random, four of another U.S. firm's cars are chosen at random, four of a German firm's cars are chosen at random, and four of a Japanese firm's cars are chosen at random. Each of the 16 cars is operated under identical conditions for a month, and the mileage per gallon of gasoline is determined. The results are as follows:

*Miles per gallon*

| Auto firm | U.S. competitor | German firm | Japanese firm |
|---|---|---|---|
| 18 | 22 | 25 | 29 |
| 20 | 21 | 27 | 28 |
| 19 | 24 | 26 | 24 |
| 17 | 20 | 28 | 25 |

(a) Test the hypothesis that the mean number of miles per gallon is the same for all four firms' cars (using $\alpha = .05$).
(b) Construct an analysis-of-variance table summarizing the results.
(c) Construct 95 percent confidence intervals for the differences among the four types of cars with respect to the mean number of miles per gallon.

**10.14** A medical researcher wants to compare the effects of four drugs. He picks a sample of 48 people, and gives drug $A$ to 12 of them, drug $B$ to the next 12 of them, and so on. The drugs are meant to induce weight loss. The following table shows the mean weight loss (in pounds) and the sample standard deviation of the weight loss for those given each drug:

*Drug*

| | A | B | C | D |
|---|---|---|---|---|
| Mean | 10.1 | 11.4 | 9.3 | 8.8 |
| Standard deviation | 3.6 | 3.2 | 2.9 | 2.8 |

(a) Which drug seems most effective? Which drug seems least effective?
(b) Construct the appropriate analysis of variance table.
(c) Can the observed differences in the mean effects of these drugs be attributed to chance? (Let $\alpha = 0.05$.)
(d) What assumptions underlie your answer to (c)?

**10.15** An industrial engineer identifies four ways that a certain job can be done. To determine how long it takes workers to do the job when each of these methods is used, he asks four workers to do the job using method A, another four workers to do this job using method B, and so on. Each worker's time (in seconds) is shown below:

| Method A | Method B | Method C | Method D |
|----------|----------|----------|----------|
| 19 | 18 | 21 | 22 |
| 17 | 16 | 20 | 23 |
| 22 | 15 | 19 | 21 |
| 20 | 14 | 19 | 20 |

(a) Which method seems fastest? Slowest?

(b) Construct the relevant analysis of variance table.

(c) Can the differences among methods in average time be attributed to chance? (Let $\alpha = 0.01$.)

(d) Construct 99 percent confidence intervals for the differences among the means.

(e) Criticize the design of this experiment.

(f) If the engineer finds that one of these methods is faster than the others, does this mean that only this method should be used?

# 10.8 Two-Way Analysis of Variance

The one-way analysis of variance is the simplest type; there are a variety of more complicated types which are taken up in more specialized textbooks. For present purposes, it is sufficient to discuss a single extension, the two-way analysis of variance. This technique varies from the one-way analysis because two, not one, sources of variation (other than the error sum of squares) are singled out for attention. In particular, two-way analysis is the technique used to test whether the differences among treatment means in a randomized block without replications are statistically significant.

To illustrate two-way analysis of variance, let's return to the beer manufacturer discussed previously. Suppose that, when the study described in Section 10.5 is presented to the firm's managers, they suggest to the firm's statisticians that a somewhat different experimental design be used. In particular, it is suggested that the same five people be asked to rate all four of the brands of beer, rather than employing a different sample for each brand. In this way, the differences among brands in sample means will not be clouded by differences in the composition of the sample. The result is a randomized block design in which each person constitutes a block.

Suppose that the statisticians carry out the experiment suggested and the results are as shown in Table 10.11. In this new experimental design, the firm's managers are interested in the differences among brands of beer and the differences among people. The differences among brands of beer in their mean ratings are called *differences among treatment means*, as in the previous section. The differences among persons in their mean ratings are called *differences among block means*, and they reflect the fact that some individuals may tend to rate all these brands of beer more highly than would other individuals. Our primary interest here is in whether or not the observed differences among treatment means are due to chance. The differences among the block

**Table 10.11**
**Results of Survey by**
**Beer Manufacturers**

| Person | Favorite brand | Rating | | | Block mean |
| | | Brand X | Brand Y | Brand Z | |
|---|---|---|---|---|---|
| Jones | 63 | 59 | 62 | 61 | 61.25 |
| Smith | 61 | 62 | 57 | 63 | 60.75 |
| Klein | 61 | 64 | 60 | 58 | 60.75 |
| Carlucci | 62 | 62 | 60 | 62 | 61.50 |
| Weill | 58 | 63 | 61 | 61 | 60.75 |
| Treatment mean | 61 | 62 | 60 | 61 | |

means are of only secondary interest in this case; but since they contribute to the total sum of squares, they must be included in the analysis.

As in the case of the one-way analysis of variance, the total sum of squares equals

**Total Sum of Squares**

$$TSS = \sum_{i=1}^{n} \sum_{j=1}^{k} (x_{ij} - \bar{x}..)^2. \tag{10.5}$$

In this case, however, the total sum of squares can be split into three parts in the following way:

$$\sum_{i=1}^{n} \sum_{j=1}^{k} (x_{ij} - \bar{x}..)^2 = n \sum_{j=1}^{k} (\bar{x}_{.j} - \bar{x}..)^2 + k \sum_{i=1}^{n} (\bar{x}_{i.} - \bar{x}..)^2 \tag{10.6}$$

$$+ \text{ error sum of squares,}$$

where the error sum of squares (denoted by *ESS*) can be obtained by subtraction. That is,

**Error Sum of Squares**

$$ESS = \sum_{i=1}^{n} \sum_{j=1}^{k} (x_{ij} - \bar{x}..)^2 - n \sum_{j=1}^{k} (\bar{x}_{.j} - \bar{x}..)^2 \tag{10.7}$$

$$- k \sum_{i=1}^{n} (\bar{x}_{i.} - \bar{x}..)^2.$$

The identity in equation (10.6) is the basis for the two-way analysis of variance. The first term on the right-hand side of equation (10.6) is called the ***treatment sum of squares,*** as in previous sections. Clearly, it reflects differences in the treatment means (for example, differences in the average rating of various brands of beer in Table 10.11). We denote this term by *BSS*. The second term on the right-hand side of equation (10.6) is called the ***block sum of squares,*** since it reflects differences in the block means (for example, differences among the persons in Table 10.11 in their average ratings of all beers). We denote the block sum of squares by *RSS*. The identity in equation (10.6) says that

> total sum of squares = treatment sum of squares + block
> sum of squares + error sum of squares

*The Basic Identity*

In other words, $TSS = BSS + RSS + ESS$.

> *In the two-way analysis of variance, the total variation in the sample is divided into three parts, the treatment sum of squares, the block sum of squares, and the error sum of squares.*

*Basic Idea #20*

This identity is used in the two-way analysis-of-variance table shown in Table 10.12. The first column of the table shows the source of variation (treatment, block, or error), and the second column shows the corresponding sum of squares. The third column shows the number of degrees of freedom for each sum of squares—$(k-1)$ for the treatment sum of squares, $(n-1)$ for the block sum of squares, and $(k-1)(n-1)$ for the error sum of squares. The fourth column shows the mean square for treatments, blocks, and errors, each being the relevant sum of squares divided by the relevant degrees of freedom (that is, the second column divided by the third column). The last column shows (1) the treatment mean square divided by the error mean square and (2) the block mean square divided by the error mean square. As explained in the next paragraph, these two ratios are the pay dirt of the entire analysis.

**Table 10.12**
General Two-Way Analysis-of-Variance Table

| Source of variation | Sum of squares | Degrees of freedom | Mean square | F |
|---|---|---|---|---|
| Treatments | BSS | $k-1$ | $\dfrac{BSS}{k-1}$ | $\dfrac{BSS}{k-1} \div \dfrac{ESS}{(k-1)(n-1)}$ |
| Blocks | RSS | $n-1$ | $\dfrac{RSS}{n-1}$ | $\dfrac{RSS}{n-1} \div \dfrac{ESS}{(k-1)(n-1)}$ |
| Error | ESS | $(k-1)(n-1)$ | $\dfrac{ESS}{(k-1)(n-1)}$ | |
| Total | TSS | $nk-1$ | | |

In a two-way analysis of variance, we can test two different null hypotheses, not just one. The first null hypothesis is that *the treatment means are all equal.* In terms of our example, this hypothesis says that the mean rating for each brand of beer is the same. To test this hypothesis, we use the ratio of the treatment mean square to the error mean square. If $\alpha$ is the significance level, the decision rule is the following.

*Test that the
Treatment Means are
All Equal*

DECISION RULE: *Reject the above null hypothesis if the ratio of the
treatment mean square to the error mean square exceeds* $F_\alpha$ *where there
are* (k − 1) *and* (k − 1)(n − 1) *degrees of freedom.*

The second null hypothesis is that *the block means are all equal.* In terms
of our example, this hypothesis says that the mean rating is the same
for all people. To test this second hypothesis, we use the ratio of the
block mean square to the error mean square. If $\alpha$ is the significance
level, the decision rule is the following.

*Test that the Block
Means are All Equal*

DECISION RULE: *Reject the above null hypothesis if the ratio of the block
mean square to the error mean square exceeds* $F_\alpha$ *where there are*
(n − 1) *and* (k − 1)(n − 1) *degrees of freedom.*

To illustrate the application of two-way analysis of variance, consider the data concerning the beers in Table 10.11. As shown in Table
10.13, the total sum of squares equals 66, the treatment sum of squares
equals 10, the block sum of squares equals 2, and the error sum of
squares equals 54. (Each of these numbers is derived under the table.)
Since $k = 4$ and $n = 5$, the degrees of freedom for each sum of squares is
as shown in the third column of the table; and dividing each sum of
squares by its number of degrees of freedom, we get the mean squares
in the fourth column. Finally, dividing the treatment mean square by
the error mean square, we get 0.74. This is less than 3.49, which is $F_{.05}$
when there are 3 and 12 degrees of freedom. Thus, there is no reason
to reject the hypothesis that the treatment means are equal. Dividing
the block mean square by the error mean square, we get 0.11. This is
less than 3.26, which is $F_{.05}$ when there are 4 and 12 degrees of freedom. Thus, we should not reject the hypothesis that the block means
are equal. Overall, the results of this analysis indicate that the individuals in the sample do not exhibit different average ratings for all included beer brands, and there is no evidence of differences in the
average ratings of the four brands of beer.

The two-way analysis of variance is of great practical importance
because, as we know from earlier sections, randomized blocks are a
commonly used experimental design. The following is a further example of how two-way analysis of variance is carried out.

EXAMPLE 10.5 Use the data in Table 10.3 to construct a two-way analysis of variance to test whether the mean strength of thread differs
among the three types of machine. Then test whether the mean
strength of thread differs among the three types of raw material.
Set the significance level of each test equal to .05.

SOLUTION: Let $x_{ij}$ be the strength of the thread made from
the $i$th type of raw material on the $j$th type of machine, where A
is the first type of raw material. B the second type, and so on.

**Table 10.13**
Analysis of Results
of Beer Survey

| Source of variation | Sum of squares | Degrees of freedom | Mean square | F |
|---|---|---|---|---|
| Treatments (beers) | 10 | 3 | 3.33 | $\dfrac{3.33}{4.50} = 0.74$ |
| Blocks (people) | 2 | 4 | 0.50 | $\dfrac{0.50}{4.50} = 0.11$ |
| Error | 54 | 12 | 4.50 | |
| Total | 66 | 19 | | |

As shown in Table 10.11, the treatment means are 61, 62, 60, and 61. Thus, since $\bar{x}.. = 61$,

$$\text{Treatment sum of squares} = 5[(61-61)^2+(62-61)^2+(60-61)^2+(61-61)^2]$$
$$= 5(2) = 10$$

As shown in Table 10.11, the block means are 61.25, 60.75, 60.75, 61.50, and 60.75. Thus,

Block sum of squares =

$$4[(61.25 - 61)^2 + (60.75 - 61)^2 + (60.75 - 61)^2$$
$$+ (61.50 - 61)^2 + (60.75 - 61)^2] = 4(.5) = 2.$$

Based on the data in Table 10.11,

$$\text{Total sum of squares} = (63-61)^2+(59-61)^2+(62-61)^2+(61-61)^2$$
$$+(61-61)^2+(62-61)^2+(57-61)^2+(63-61)^2$$
$$+(61-61)^2+(64-61)^2+(60-61)^2+(58-61)^2$$
$$+(62-61)^2+(62-61)^2+(60-61)^2+(62-61)^2$$
$$+(58-61)^2+(63-61)^2+(61-61)^2+(61-61)^2$$
$$= 66.$$

Using the previous results,

Error sum of squares $= 66 - 10 - 2 = 54$.

Based on Table 10.3, it is clear that

$$\bar{x}_{.1} = 50 \qquad \bar{x}_{1.} = 45$$
$$\bar{x}_{.2} = 41 \qquad \bar{x}_{2.} = 44$$
$$\bar{x}_{.3} = 46 \qquad \bar{x}_{3.} = 48.$$

If the various types of machines are regarded as treatments, and if the raw materials are regarded as blocks, the treatment sum of squares is

$$3 \times \left[\left(50 - 45\tfrac{2}{3}\right)^2 + \left(41 - 45\tfrac{2}{3}\right)^2 + \left(46 - 45\tfrac{2}{3}\right)^2\right] = 122,$$

and the block sum of squares is

$$3 \times \left[ \left(45 - 45\frac{2}{3}\right)^2 + \left(44 - 45\frac{2}{3}\right)^2 + \left(48 - 45\frac{2}{3}\right)^2 \right] = 26,$$

since $\bar{x}.. = 45\ 2/3$. Thus, the error sum of squares is

$$\left(50 - 45\frac{2}{3}\right)^2 + \left(48 - 45\frac{2}{3}\right)^2 + \left(52 - 45\frac{2}{3}\right)^2$$

$$+ \left(40 - 45\frac{2}{3}\right)^2 + \left(39 - 45\frac{2}{3}\right)^2 + \left(44 - 45\frac{2}{3}\right)^2$$

$$+ \left(45 - 45\frac{2}{3}\right)^2 + \left(45 - 45\frac{2}{3}\right)^2 + \left(48 - 45\frac{2}{3}\right)^2$$

$$- 122 - 26 = 150 - 122 - 26 = 2.$$

The analysis-of-variance table is as follows:

| Source of variation | Sum of squares | Degrees of freedom | Mean square | F |
|---|---|---|---|---|
| Machines (treatments) | 122 | 2 | 61 | $61/0.5 = 122$ |
| Raw materials (blocks) | 26 | 2 | 13 | $13/0.5 = 26$ |
| Error | 2 | 4 | 0.5 | |
| Total | 150 | 8 | | |

Since $F_{.05} = 6.94$ when there are 2 and 4 degrees of freedom, both values of $F$ in the table exceed $F_{.05}$. Thus, the firm should reject both the null hypothesis that the mean strength is the same for all machines and the null hypothesis that the mean strength is the same for all raw materials.[10]

## 10.9 Confidence Intervals for Differences among Means

In the previous section, we saw how the two-way analysis of variance can be used to test whether the treatment means are all equal. In most experiments, however, the purpose is to find out the *extent* to which the treatment means differ, not just *whether* they differ. In this section, we will show how confidence intervals can be constructed for all differences between treatment means. The confidence coefficient attached to these intervals is the probability that *all* these intervals will include the true differences among treatment means *simultaneously*.

---

[10] See Appendix 10.2 for some formulas that are useful in calculating the required sums of squares in more complicated cases.

# Industrial Experimentation in Cotton Textiles

The cotton textile industry is extremely important, both in the United States and in many other countries. To illustrate how cotton textile firms have used the statistical techniques taken up in this part of the book, consider an experiment carried out by the British Cotton Industry Research Association to determine the effect of the type of flyer and the number of turns per inch on the number of breaks per 100 pounds of materials.[11] (A "flyer" is a rotary guide which by rotation puts twist into the strand and guides it into the bobbin.) There are six treatments: (1) an ordinary flyer with 1.69 turns per inch; (2) an ordinary flyer with 1.78 turns per inch; (3) an ordinary flyer with 1.90 turns per inch; (4) a special flyer with 1.63 turns per inch; (5) a special flyer with 1.69 turns per inch; and (6) a special flyer with 1.78 turns per inch.

To estimate the differences among these treatments in the mean number of breaks per 100 pounds of material, a randomized block was used. There were 13 blocks, each being a different machine or a different set of experimental conditions. Thus, since there were 6 treatments and 13 blocks, there were 6 x 13, or 78 observations in all. The analysis of variance table is shown in Table 1. Since $F_{.05}$ with 5 and 60 degrees of freedom is 2.37, the differences among the treatment means are statistically significant. Of the six treatments included in the experiment, a special flyer with 1.78 turns per inch turned out to have the smallest mean number of breaks per 100 pounds of material.

| Source of variation | Sum of squares | Degrees of freedom | Mean square | F |
|---|---|---|---|---|
| Treatments | 462 | 5 | 92.4 | 9.1 |
| Blocks | 354 | 12 | 29.5 | 2.9 |
| Error | 612 | 60 | 10.2 | |
| Total | 1,428 | 77 | | |

Table 1

Analysis of Variance for Randomized Block

As an example of the use of a one-way analysis of variance, consider a textile plant that was interested in determining whether some types of cotton are more likely than others to break in the weaving process.[12] (The fewer breaks, the higher the weaving quality of the cotton.) In particular, the plant wanted to know whether two growths of cotton (A and B) differed in weaving quality. Also, the plant's statisticians wanted to know whether there was a difference in this respect among the number of turns per inch in the yarn: low (L), medium (M), and high (H). Thus, there are six combinations of growth of cotton and

[11] R. Peake, "Planning an Experiment in a Cotton Spinning Mill," *Applied Statistics*, 1953.

[12] L. Tippett, *Technological Applications of Statistics* (New York: Wiley, 1950).

number of turns in the yarn that the plant wanted to study: (1) low number of turns with cotton A; (2) medium number of turns with cotton A; (3) high number of turns with cotton A; (4) low number of turns with cotton B; (5) medium number of turns with cotton B; and (6) high number of turns with cotton B.

The firm chose at random nine warps of each of these six types. (A *warp* is a quantity of warp yarn that goes into one loom as a unit.) The number of warp threads that broke during the weaving of each warp was counted and expressed as a particular number of breaks per unit-length of warp. The results were as follows:

| Low number of turns with cotton A | Medium number of turns with cotton A | High number of turns with cotton A | Low number of turns with cotton B | Medium number of turns with cotton B | High number of turns with cotton B |
|---|---|---|---|---|---|
| 26 | 18 | 36 | 27 | 42 | 20 |
| 30 | 21 | 21 | 14 | 26 | 21 |
| 54 | 29 | 24 | 29 | 19 | 24 |
| 25 | 17 | 18 | 19 | 16 | 17 |
| 70 | 12 | 10 | 29 | 39 | 13 |
| 52 | 18 | 43 | 31 | 28 | 15 |
| 51 | 35 | 28 | 41 | 21 | 15 |
| 26 | 30 | 15 | 20 | 39 | 16 |
| 67 | 36 | 26 | 44 | 29 | 28 |

The firm used a one-way analysis of variance to test the observed differences among the mean breakage rates.

The lowest mean breakage rate (18.78) occurred when there were a high number of turns with cotton B, and the highest mean breakage rate (44.56) occurred when there were a low number of turns with cotton A. The analysis of variance table was as follows:

| Source of variation | Sum of Squares | Degrees of freedom | Mean square | F |
|---|---|---|---|---|
| Between types | 3487.7 | 5 | 698 | 5.82 |
| Within types | 5745.1 | 48 | 120 | |
| Total | 9232.8 | 53 | | |

Since $F_{.01}$ equals about 3.4, it is clear that the observed value of $F$ is much greater than $F_{.01}$, and that the observed differences in the mean breakage rates are statistically significant.

The foregoing illustrations are only a small sample of the ways in which the analysis of variance has been used in the cotton textile industry. $\chi^2$ tests and nonparametric techniques have also been used very frequently. In cotton textiles, as in a broad array of other industries, these are bread-and-butter techniques that are used over and over again to obtain needed information.

PROBING DEEPER

1. Based on the results in Table 1, can you determine whether a cotton mill should use a special flyer with 1.78 turns per inch?

2. Does the fact that the differences among the treatment means in Table 1 are statistically significant tell us that the differences are of practical importance?

3. If there is reason to believe that the number of breaks per 100 pounds of material is much more variable when a special flyer, rather than an ordinary flyer, is used, will this affect your interpretation of the results in Table 1. If so, how?

4. Can the experiment involving the two growths of cotton (A and B) and the number of turns per inch in the yarn be viewed as a randomized block? If so, explain how.

5. An engineer claims that, for a low number of turns with cotton A, the standard deviation of the number of breaks per unit-length of warp is 10. Do you agree? Why, or why not?

6. Can you construct a 95 percent confidence interval for the standard deviation in the previous question? If you can, do so, and state your assumptions.

In the case of the thread manufacturer there are three differences among the treatment (that is, machine) means. Assuming that the raw material is held constant, let $\mu_1$ once again be the mean strength of thread produced by machine I, $\mu_2$ the mean strength for machine II, and so on. The probability is $(1 - \alpha)$ that *all* the following statements hold true *simultaneously*:

*Confidence Intervals for Differences among Treatment Means*

$$\bar{x}_{.1} - \bar{x}_{.2} - \sqrt{F_\alpha} s_E \sqrt{\frac{2(k-1)}{n}} < \mu_1 - \mu_2 < \bar{x}_{.1} - \bar{x}_{.2} + \sqrt{F_\alpha} s_E \sqrt{\frac{2(k-1)}{n}}$$

$$\bar{x}_{.1} - \bar{x}_{.3} - \sqrt{F_\alpha} s_E \sqrt{\frac{2(k-1)}{n}} < \mu_1 - \mu_3 < \bar{x}_{.1} - \bar{x}_{.3} + \sqrt{F_\alpha} s_E \sqrt{\frac{2(k-1)}{n}}$$

$$\bar{x}_{.2} - \bar{x}_{.3} - \sqrt{F_\alpha} s_E \sqrt{\frac{2(k-1)}{n}} < \mu_2 - \mu_3 < \bar{x}_{.2} - \bar{x}_{.3} + \sqrt{F_\alpha} s_E \sqrt{\frac{2(k-1)}{n}}$$

$$(10.8)$$

where $k$ is the number of treatments, $n$ the number of blocks, $F_\alpha$ is the value of $F$ with $(k-1)$ and $(k-1)(n-1)$ degrees of freedom exceeded with a probability of $\alpha$, and $s_E$ is the square root of the error mean square. Since $\bar{x}_{.1} = 50$, $\bar{x}_{.2} = 41$, $\bar{x}_{.3} = 46$, $F_{.05} = 6.94$, $k = 3$, $n = 3$, and $s_E = .71$, the 95 percent confidence interval for all the differences between treatment (machine) means when the block (raw material) is held constant, is as follows:

$$6.84 < \mu_1 - \mu_2 < 11.16$$

$$1.84 < \mu_1 - \mu_3 < \phantom{0}6.16$$

$$-7.16 < \mu_2 - \mu_3 < -2.84.$$

It is also possible to construct confidence intervals of this sort for the differences among the block means when the treatment is held constant. Thus, in this case, when the type of machine is held constant, one can construct confidence intervals for the differences among the means corresponding to various raw materials. (Since this entails only a slight modification of the procedures covered in the previous paragraph, we describe this procedure in a footnote.)[13]
The following is a further illustration of the construction of confidence intervals of this sort.

---

[13] To obtain confidence intervals for the block means, all that one has to do is substitute $k$ for $n$ and vice versa in expression (10.8). Of course, $F_\alpha$ must also differ from that in expression (10.8) because there are $(n-1)$ and $(k-1)(n-1)$ degrees of freedom in this case.

EXAMPLE 10.6 Use the data in Table 10.11 to construct a 95 percent confidence interval for the six differences between the mean ratings of the four beers, when the person making the ratings is held constant.

SOLUTION: In this case, $\bar{x}_{.1} = 61, \bar{x}_{.2} = 62, \bar{x}_{.3} = 60, \bar{x}_{.4} = 61,$ $F_{.05} = 3.49,$   $s_E = \sqrt{4.50},$   $k = 4,$   and   $n = 5.$   Thus, $\sqrt{F_\alpha} s_E \sqrt{2 \times [(k-1)/n]} = \sqrt{3.49} \times \sqrt{4.50} \times \sqrt{6/5} = 4.34$, and the confidence intervals are as follows:

$$-5.34 < \mu_1 - \mu_2 < 3.34$$

$$-3.34 < \mu_1 - \mu_3 < 5.34$$

$$-4.34 < \mu_1 - \mu_4 < 4.34$$

$$-2.34 < \mu_2 - \mu_3 < 6.34$$

$$-3.34 < \mu_2 - \mu_4 < 5.34$$

$$-5.34 < \mu_3 - \mu_4 < 3.34.$$

All the above confidence intervals include zero, which would be expected, since the analysis of variance in the previous section concluded that none of the differences among the treatment means is statistically significant.

# 10.10  A Final Caution

Before concluding this chapter, it is important to note the assumptions underlying the analysis of variance. If these assumptions are not met, the use of the analysis of variance may be misleading. One assumption is that the populations being compared are normally distributed. Fortunately, this assumption is not as stringent as it sounds, since studies have shown that the validity of the analysis of variance is not significantly affected by moderate departures from normality. In the jargon of the statistician, the analysis of variance is a "robust test" in this regard. Another assumption underlying the analysis of variance is that the variances of the populations are equal. (Appendix 10.1 shows how the $F$ distribution can be used to test the hypothesis that the variances of two populations are equal.) If this assumption is not met, trouble can result. Statisticians have devised techniques for handling this problem in some cases; these techniques are more properly discussed in more advanced texts. Still another important assumption is that the observations are statistically independent. To repeat, the analysis of variance should not be used unless the relevant assumptions are at least approximately fulfilled.

EXERCISES

**10.16** Fill in the blanks in the following analysis of variance table:

| Source of variation | Sum of squares | Degrees of freedom | Mean square | F |
|---|---|---|---|---|
| Treatments | _____ | 2 | 19 | _____ |
| Blocks | 122 | _____ | _____ | _____ |
| Error | _____ | _____ | 40 | |
| Total | 400 | 11 | | |

**10.17** A consumer research organization tests three brands of tires to see how many miles they can be driven before they should be replaced. One tire of each brand is tested in each of five types of cars. The results (in thousands of miles) are as follows:

| Type of car | Brand A | Brand B | Brand C |
|---|---|---|---|
| I | 26 | 29 | 24 |
| II | 30 | 32 | 27 |
| III | 27 | 30 | 26 |
| IV | 28 | 28 | 25 |
| V | 29 | 31 | 28 |

(a) Which brand seems to wear best? Which seems to wear most poorly?
(b) Do tires seem to wear better in one type of car than in another type?
(c) Construct the relevant analysis of variance table.
(d) Can the differences among brands be attributed to chance? (Let $\alpha = 0.05$.)
(e) Can the differences among types of cars be attributed to chance? (Let $\alpha = 0.05$.)
(f) Construct 95 percent confidence intervals for the differences among the brand means.

**10.18** A psychologist wants to determine which of three teaching methods results in the highest scores on a particular test. She picks three girls and three boys at random. One of each sex is taught by each method, after which all are given the test. Their grades are as follows:

| | Method A | Method B | Method C |
|---|---|---|---|
| Boys | 80 | 78 | 64 |
| Girls | 91 | 85 | 70 |

(a) Which method seems to result in the highest scores? In the lowest scores?
(b) Do boys seem to do better on this test than girls?
(c) Construct the relevant analysis of variance table.
(d) Can the differences in average score among methods be attributed to chance? (Let $\alpha = 0.05$.)
(e) Can the difference in average score between boys and girls be attributed to chance? (Let $\alpha = 0.05$.)
(f) Criticize the design of the experiment.

(g) If one method can be shown to result in a higher average score than the other methods, should this method replace the others?

**10.19** Four machines are used with three different types of raw materials to make a particular metal part. The number of satisfactory parts produced per day with each machine and type of raw material is:

|  | Machine 1 | Machine 2 | Machine 3 | Machine 4 |
|---|---|---|---|---|
| Raw material A | 69 | 88 | 63 | 74 |
| Raw material B | 73 | 85 | 60 | 71 |
| Raw material C | 75 | 86 | 59 | 69 |

Use Minitab (or some other computer package) to determine whether the different types of raw materials produce equivalent results, on the average. (Let $\alpha = .05$.) If Minitab (or some other computer package) is not available, do the calculations by hand.

**10.20** Suppose that an automobile manufacturer designs the following experiment: It picks at random four of its cars, four of its U.S. competitor's cars, four of a German firm's cars, and four of a Japanese firm's cars. Then it has one of each firm's cars drive in city traffic; one of each firm's cars drive under suburban conditions; one of each firm's cars drive under mountainous conditions; and one of each firm's cars drive in flat, open country. The results are as follows:

| Manufacturer | City | Suburbs | Miles per gallon Mountains | Flat country |
|---|---|---|---|---|
| Auto firm | 14 | 20 | 21 | 24 |
| U.S. competitor | 15 | 21 | 25 | 26 |
| German firm | 18 | 24 | 26 | 28 |
| Japanese firm | 19 | 25 | 25 | 27 |

(a) Construct an analysis-of-variance table summarizing the results.
(b) What are the treatment effects here? What are the block effects?
(c) Does it appear that each make of car gets the same number of miles per gallon? If not, which seems to get the most, and which seems to get the least?
(d) Does it appear that the number of miles per gallon is the same for all types of driving conditions? If not, which type of driving condition seems to result in the largest number of miles per gallon, and which type seems to result in the smallest number of miles per gallon?
(e) Construct 95 percent confidence intervals for the differences among the four types of cars with respect to the mean number of miles per gallon when the type of driving condition is held constant.

## Chapter Review

1. An *experiment* dictates that one group of people, machines, or other experimental units be treated differently from another group; and the effect of this difference in treatment is estimated by comparing certain measurable characteristics of the two groups. In an attempt to

prevent confounding, statisticians recommend that the people, machines, or other experimental units be assigned to one group or the other *at random.*

2. A frequently used experimental design is the *randomized block design,* in which the experimental units are classified into blocks, and the statistician obtains data concerning the effect of each *treatment* in each *block.* In this design it is possible to estimate both the differences among the treatment means (holding the block constant) and the differences among the blocks means (holding the treatment constant).

3. Another commonly used experimental design is the *Latin square.* In a Latin square design there are three factors: (1) the treatments being compared: (2) a factor corresponding to the horizontal rows of the Latin square: and (3) a factor corresponding to the vertical columns of the Latin square. In each horizontal row of a Latin square, each treatment appears only once: and in each vertical column each treatment appears only once.

4. If $Y_1$ has a $\chi^2$ distribution with $v_1$ degrees of freedom, and if $Y_2$ has a $\chi^2$ distribution with $v_2$ degrees of freedom, then $Y_1/v_1 \div Y_2/v_2$ has an $F$ distribution with $v_1$ and $v_2$ degrees of freedom (if $Y_1$ and $Y_2$ are independent). The $F$ distribution is generally skewed to the right, but as both degrees of freedom become very large, the $F$ distribution tends toward normality.

5. In a *completely randomized design,* experimental units are classified at random into groups, and each group is subjected to a different treatment. Then the sample mean in each group is used as an estimate of the population mean, and a test is made of the null hypothesis that the population means are all equal. The null hypothesis should be rejected if $ns_{\bar{x}}^2 \div \Sigma s_j^2/k > F_\alpha$, where $k$ is the number of population means being compared, $n$ is the sample size in each population, $s_{\bar{x}}^2$ is the variance of the $k$ sample means, $s_j^2$ is the variance of the sample from the $j$th population, and $F_\alpha$ is the value of an $F$ random variable with $(k-1)$ and $k(n-1)$ degrees of freedom that is exceeded with a probability of $\alpha$.

6. An *analysis of variance* is a technique designed to analyze the total variation in a set of data, the object being to split up this total variation into component parts, each of which can be ascribed to a particular source. The *one-way analysis of variance* splits the total sum of squares into two parts, the *between-group* (or *treatment*) *sum of squares* and the *within-group* (or *error*) *sum of squares.* Using this breakdown of the total sum of squares, the one-way analysis of variance tests the null hypothesis that the population means in a completely randomized design are all equal. The test procedure is precisely the same as that given in the previous paragraph. What is different is that the computations are presented in the format of an analysis-of-variance table.

7. In most experiments, the purpose is to find out the *extent to which* the population means (corresponding to the various treatments) differ, not just *whether* they differ. Confidence intervals can be constructed for all differences among the population means. The *confidence coefficient* attached to these intervals is the probability that all these intervals will include the respective differences among the population means simultaneously.

8. The *two-way analysis of variance* recognizes not one, but two sources of variation other than the error sum of squares. The two-way technique is used to test whether the differences among treatment means in a randomized block (without replications) are statistically significant. The two-way analysis of variance splits the total sum of squares into three parts, the *treatment sum of squares,* the *block sum of squares,* and the *error sum of squares.* The null hypothesis that the treatment means are equal should be rejected if the ratio of the treatment mean square to the error mean square exceeds $F_\alpha$, where there are $(k - 1)$ and $(k - 1)(n - 1)$ degrees of freedom. The null hypothesis that the block means are equal should be rejected if the ratio of the block mean square to the error mean square exceeds $F_\alpha$, where there are $(n - 1)$ and $(k - 1)(n - 1)$ degrees of freedom.

9. Confidence intervals can be constructed for all differences among the treatment means (holding the block constant). The confidence coefficient attached to these intervals is the probability that all these intervals will include the respective differences among the treatment means simultaneously. Confidence intervals of this sort can also be constructed for the differences among the block means (holding the treatment constant).

## Chapter Review Exercises

**10.21** A chemical firm wants to determine how four catalysts differ in yield. The firm runs the experiment in eight of its plants. In each plant, the yield is measured with each catalyst. The yields are as follows:

| | | Catalyst | | |
|---|---|---|---|---|
| *Plant* | *1* | *2* | *3* | *4* |
| A | 35 | 40 | 41 | 40 |
| B | 36 | 37 | 39 | 41 |
| C | 38 | 41 | 42 | 40 |
| D | 38 | 40 | 43 | 39 |
| E | 39 | 41 | 40 | 39 |
| F | 37 | 36 | 39 | 38 |
| G | 35 | 37 | 38 | 37 |
| H | 36 | 37 | 36 | 36 |

(a) Which catalyst seems to have the highest yield? The lowest yield?
(b) Which plant seems to have the highest yield?
(c) Construct the relevant analysis of variance table.
(d) Can the differences among catalysts be attributed to chance? (Let $\alpha = 0.01$.)
(e) Can the difference in average yield among the plants be attributed to chance? (Let $\alpha = 0.01$.)
(f) Calculate 99 percent confidence intervals for the differences among catalysts in mean yield.

**10.22** An agricultural experiment station treats three plots of ground with fertilizer I, another three plots of ground with fertilizer II, and still another three plots of ground with fertilizer III. Then wheat is grown on the nine plots, the yield on each plot being as follows:

| Fertilizer I | Fertilizer II | Fertilizer III |
|---|---|---|
| 9.1 | 8.2 | 9.3 |
| 8.3 | 5.9 | 8.8 |
| 6.4 | 7.0 | 7.9 |

(a) Which fertilizer seems to result in the highest average yield? The lowest average yield?

(b) Construct the relevant analysis of variance table.

(c) Can the observed differences among fertilizers in the mean yield be attributed to chance? (Let $\alpha = .05$)

(d) What assumptions underlie your answer to (c)?

(e) Construct 95 percent confidence intervals for the differences among the means.

(f) Criticize the design of this experiment.

(g) If the station determines that the mean yield with one of these fertilizers is higher than with the others, does this mean that farmers should replace the other fertilizers with this one?

**10.23** An electronics firm has three plants, one in Germany, one in Singapore, and one in the United States. The firm's vice president of manufacturing is interested in whether or not output per hour of labor differs among the plants. Fifteen workers are chosen at random from each plant, and the output per hour of each worker is measured. The mean and sample standard deviation of these workers' output per hour of labor are shown below:

|  | Germany | Singapore | United States |
|---|---|---|---|
| Mean | 15.1 | 16.2 | 13.4 |
| Standard Deviation | 3.2 | 4.0 | 3.1 |

(a) In which plant does output per hour of labor seem highest? In which one does it seem lowest?

(b) Construct the relevant analysis of variance table.

(c) Can the observed differences among the plants in the mean output per hour of labor be attributed to chance? (Let $\alpha = 0.05$.)

(d) What assumptions underlie your answer to (c)?

(e) Construct 95 percent confidence intervals for the differences among the means.

(f) To what extent are these data trustworthy indicators of the relative quality of the work force at each plant?

**10.24** The following table shows the results of nine different measurements of the viscosity of silicone gum rubber.[14] As can be seen, three of the measurements pertain to the first batch of this material, three pertain to the second batch, and three to the third batch. Also, three of the measurements were made by William Jones, three were made by John Beam, and three were made by Joan Read. Three of the measurements were made with type A viscosity measuring jars, three were made with type B measuring jars, and three were made with type C measuring jars.

---

[14] These data are from A. Duncan, *op. cit.*, p. 747. They are coded; that is, a constant amount is subtracted from each measurement to make the data easier to work with.

| Batch | Jones | Measurer Beam | Read |
|-------|-------|---------------|------|
| I | 9 (A) | 8 (B) | −3 (C) |
| II | 17 (B) | −2 (C) | 7 (A) |
| III | −2 (C) | 41 (A) | 2 (B) |

(a) Is this a randomized block? Why, or why not?
(b) If so, what are the blocks, and what are the treatments?
(c) Is this a Latin square? Why, or why not?
(d) If it is a Latin square, explain the balance of the design.
(e) Does the design contain replications?
(f) Which of the batches has the highest average rating?
(g) Which type of measuring jar seems to result in the highest average rating?
(h) Without carrying out the appropriate tests of significance, can we be sure that these differences among batches and among types of measuring jars are not due to chance?

*10.25 The following data show the number of words per minute which a secretary types (on a number of occasions) on five different typewriters.

Typewriter A:  67, 72, 64, 70, 67
Typewriter B:  68, 73, 70, 73, 69
Typewriter C:  71, 63, 72, 75, 72
Typewriter D:  70, 68, 74, 69, 74
Typewriter E:  64, 59, 68, 66, 75

Minitab is used to carry out an analysis of variance to determine whether the differences among the typewriters in this regard can be attributed to chance. Interpret the results, shown below.

```
MTB > read into c21,c22,c23,c24,c25
DATA > 67 68 71 70 64
DATA > 72 73 63 68 59
DATA > 64 70 72 74 68
DATA > 70 73 75 69 66
DATA > 67 69 72 74 75
DATA > end
      5 ROWS READ
MTB > aovoneway in c21,c22,c23,c24,c25

ANALYSIS OF VARIANCE
SOURCE   DF      SS      MS      F
FACTOR    4    81.8    20.5   1.32
ERROR    20   309.6    15.5
TOTAL    24   391.4              INDIVIDUAL 95 PCT CI'S FOR
                                 MEAN BASED ON POOLED STDEV
LEVEL     N    MEAN   STDEV    ----+---------+---------+---------+--
C21       5  68.000   3.082       (----------*----------)
C22       5  70.600   2.302           (----------*----------)
C23       5  70.600   4.506           (----------*----------)
C24       5  71.000   2.828            (---------*----------)
C25       5  66.400   5.857     (-----------*----------)
                                 ----+---------+---------+---------+--
POOLED STDEV = 3.934            64.0      68.0      72.0      76.0
```

---

* Exercise requiring some familiarity with Minitab or with Appendix 10.3.

# Appendix 10.1

COMPARING TWO POPULATION VARIANCES

The $F$ distribution can be used to test the hypothesis that the variance of one normal population equals the variance of another normal population. This test is often useful because a decision maker wants to determine whether one population is more variable than another. For example, a production manager may want to determine whether the variability of the errors made by one measuring instrument is less than the variability of those made by another measuring instrument. In addition, this test is often used to determine whether the assumptions underlying other statistical tests are valid. For example, in carrying out the $t$ test to determine whether the means of two populations are the same, we assume that the variances of the two populations are equal. (Recall Chapter 8.) This assumption can be checked by carrying out the test described below.

The null hypothesis is that the variance of one normal population $\sigma_1^2$ equals the variance of the other normal population $\sigma_2^2$. To test this hypothesis, a sample of $n_1$ observations is taken from the first population and a sample of $n_2$ observations is taken from the second population. The test statistic is $s_1^2 \div s_2^2$, where $s_1^2$ is the sample variance of the observations taken from the first population and $s_2^2$ is the sample variance of the observations taken from the second population. If the null hypothesis (that $\sigma_1^2 = \sigma_2^2$) is true, this test statistic has the $F$ distribution with $(n_1 - 1)$ and $(n_2 - 1)$ degrees of freedom.

DECISION RULE: *When the alternative hypothesis[15] is $\sigma_1^2 > \sigma_2^2$, reject the null hypothesis if the test statistic exceeds $F_\alpha$. When the alternative hypothesis is $\sigma_1^2 \neq \sigma_2^2$, let the population with the larger sample variance be the first population (that is, the one whose sample variance is in the numerator of the test statistic); and reject the null hypothesis if the test statistic exceeds $F_{\alpha/2}$.*

To illustrate the use of this test, suppose that a financial analyst wants to test whether the variance of the prices paid for 100 shares of common stock of firm A equals the variance of the prices paid for 100 shares of common stock of firm B. (The significance level is set at .02.) A random sample of 25 prices paid for firm A's stock is selected, as is a random sample of 25 prices paid for firm B's stock, and it is found that the sample variance is 50 (dollars squared) for firm A's stock and 80 (dollars squared) for firm B's stock. Since the alternative hypothesis is that the two variances are unequal (regardless of which is bigger), we let the prices of firm B's stock constitute the first population (since its sample variance is larger than the sample variance of the prices of firm A's stock). Thus, $s_1^2 \div s_2^2 = 80 \div 50$, or 1.6. The null hypothesis should be rejected if this test statistic exceeds $F_{.01}$ (the number of degrees of freedom being 24 and 24). According to Appendix Table 10, $F_{.01} = 2.66$. Since the test statistic does not exceed $F_{.01}$, the null hypothesis should not be rejected.

---

[15] When the alternative hypothesis is one-sided (that is, when the variance of one population is larger than that of the other population, according to the alternative hypothesis), the population with the larger variance according to the alternative hypothesis should be designated as the first population.

# Appendix 10.2

## FORMULAS FOR COMPUTATIONS IN THE ANALYSIS OF VARIANCE

In calculating the sums of squares that are required by the analysis of variance, it is usually best to use formulas that require finding only sums and sums of squares of the observations. Such formulas are given below for the sums of squares in Table 10.9:

$$BSS = \frac{1}{n} \sum_{j=1}^{k} T_j^2 - \frac{1}{nk} T^2$$

$$TSS = \sum_{i=1}^{n} \sum_{j=1}^{k} x_{ij}^2 - \frac{1}{kn} T^2,$$

where $T_j$ is the total of the observations from the $j$th population (or the $j$th treatment), and $T$ is the total of all observations. Once these two sums of squares are calculated, we can obtain $WSS$ by subtraction. That is,

$$WSS = TSS - BSS.$$

An additional formula that applies to Table 10.12 is

$$RSS = \frac{1}{k} \sum_{i=1}^{n} T_i^2 - \frac{1}{kn} T^2,$$

where $T_i$ is the total of the observations in the $i$th block.

To illustrate the use of these formulas, consider once more the data in Example 10.3. Clearly, $T = 552$, $T_1 = 204$, $T_2 = 160$, and $T_3 = 188$. Thus,

$$BSS = \frac{1}{4} (204^2 + 160^2 + 188^2) - \frac{1}{4(3)} 552^2 = 25{,}640 - 25{,}392 = 248$$

$$TSS = 25{,}652 - \frac{1}{4(3)} 552^2 = 25{,}652 - 25{,}392 = 260$$

$$WSS = 260 - 248 = 12.$$

Comparing these results with those in the solution to Example 10.3, we find that they are identical. The advantage of the formulas given here is that they are easier and more efficient to calculate. Although this is not obvious in this case (since the numbers were intentionally chosen so that the calculations would be simple), this is generally true.

# Appendix 10.3

## INTERPRETING COMPUTER OUTPUT FOR THE ANALYSIS OF VARIANCE

This appendix illustrates how computer output for the analysis of variance can be interpreted. Suppose that an umbrella manufacturer wants to determine whether there are significant differences among six months (January, March, May, July, September, and November) in the average number of days of rainfall. For each month, data are obtained concerning the number of days of rainfall in 1986 in seven randomly chosen areas in the firm's market. Minitab is used to calculate the one-way analysis of variance, the results being shown below:

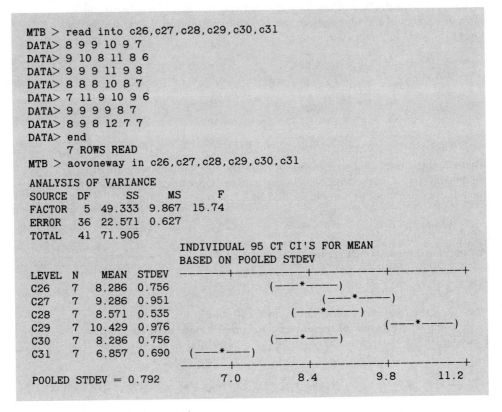

```
MTB > read into c26,c27,c28,c29,c30,c31
DATA> 8 9 9 10 9 7
DATA> 9 10 8 11 8 6
DATA> 9 9 9 11 9 8
DATA> 8 8 8 10 8 7
DATA> 7 11 9 10 9 6
DATA> 9 9 9 9 8 7
DATA> 8 9 8 12 7 7
DATA> end
      7 ROWS READ
MTB > aovoneway in c26,c27,c28,c29,c30,c31

ANALYSIS OF VARIANCE
SOURCE  DF     SS      MS      F
FACTOR   5  49.333   9.867  15.74
ERROR   36  22.571   0.627
TOTAL   41  71.905
                        INDIVIDUAL 95 CT CI'S FOR MEAN
                        BASED ON POOLED STDEV
LEVEL   N    MEAN   STDEV  ----+---------+---------+---------+--
C26     7   8.286   0.756               (----*-----)
C27     7   9.286   0.951                    (----*-----)
C28     7   8.571   0.535               (----*-----)
C29     7  10.429   0.976                        (----*-----)
C30     7   8.286   0.756               (----*-----)
C31     7   6.857   0.690      (----*-----)
                        ----+---------+---------+---------+--
POOLED STDEV = 0.792       7.0       8.4       9.8      11.2
```

The data for January are put in column C26, the data for March are put in column C27, . . . , and the data for November are put in column C31. The line after the data contains the Minitab command (ao-voneway) for a one-way analysis of variance. The next three lines of figures that are printed out show the analysis-of-variance table. The next six lines show the mean and standard deviation of the number of days of rainfall for each of the six months. (Level C26 refers to January, level C27 to March, . . . , and level C31 to November.) The "pooled stdev" is an estimate of the standard deviation among the areas in the number of days of rainfall in a given month. Finally, each horizontal bar of asterisks shows a 95 percent confidence interval for the relevant month's mean number of days of rainfall. Thus, for January (level C26), the confidence interval is approximately 7.8–9.0 days.

Based on this printout, it is clear that the differences among these months in the average number of days of rainfall are statistically significant. According to Appendix Table 9, $F_{.05}$ equals about 2.5, since there are 5 and 36 degrees of freedom. The observed value of $F$, which is 15.74, is far greater than 2.5.

PART V

# Regression and Correlation

# Regression and Correlation Techniques

# 11

## 11.1 Introduction

Statisticians frequently must estimate how one variable is related to, or affected by, another variable. A firm may need to determine how its sales are related to the gross national product; or it may need to determine how its total production costs are related to its output rate. To estimate such relationships, statisticians use regression techniques; and to determine how strong such relationships are, they use correlation techniques. Regression and correlation are among the most important and most frequently used methods of statistics. In this chapter, we begin the study of regression and correlation. The next chapter will provide further information concerning these topics.

## 11.2 Relationship among Variables, and the Scatter Diagram

DETERMINISTIC AND STATISTICAL RELATIONSHIPS

The statistical techniques presented in previous chapters were concerned with a single variable, $X$. For example, in Chapter 7 we described how to estimate the mean of $X$; and in Chapter 8 we described how to test whether this mean equals a specified value. In many important practical situations, statisticians must be concerned with more than a single variable; in particular, they must be concerned with the relationships among variables. For example, they may want to determine whether changes in one variable, $X$, tend to be associated with changes in another variable, $Y$. (For example, does $Y$ tend to increase when $X$ increases?) The techniques presented in previous chapters are powerless to handle such a problem.

*Deterministic Relationship*

When we say that statisticians are interested in the relationships among variables, it is important to note at the outset that these relationships seldom are deterministic. To see what we mean by a *deterministic relationship*, suppose that $Y$ is the variable we want to estimate and $X$ is the variable whose value will be used to make this estimate. If the relationship between $Y$ and $X$ is *exact*, we say that it is a *deterministic relationship*. For example, if $Y$ is the perimeter of a square and $X$ is the length of a side of the square, it is evident that $Y = 4X$. This is a deterministic relationship since, once we are given $X$, we can predict $Y$ *exactly*. (For example, in this case if $X$ equals 2, we know that $Y$ must equal 8—no more, or less.)

*Statistical Relationship*

Statisticians are generally interested in statistical, not deterministic, relationships. If a *statistical relationship* exists between $Y$ and $X$, the average value of $Y$ tends to be related to the value of $X$, but it is impossible to predict with certainty the value of $Y$ on the basis of the value of $X$. For example, suppose that $X$ is a family's annual income and $Y$ is the amount the family saves per year. On the average, the amount saved by a family tends to increase as its income increases; and this relationship can be used to predict how much a family will save, if we know the amount of its income. However, this relationship is far from exact. Since families with the same income do not all save the same amount, it is impossible to predict with certainty the amount a family will save on the basis of its income alone.

## REGRESSION ANALYSIS

*Regression analysis* describes the way in which one variable is related to another. (As we shall see in Chapter 12, regression and correlation techniques can handle more than two variables, but only two are considered in this chapter.) Regression analysis derives an equation which can be used to estimate the unknown value of one variable on the basis of the known value of the other variable. For example, suppose that a hosiery mill is scheduled to produce 4 tons of output next month and wants to estimate how much its costs will be. In this case, although the mill's output is known, its costs are unknown. Regression analysis can be used to estimate the value of the costs on the basis of the known value of output. Regression analysis can also be used to estimate the level of capital expenditures required to establish a plant with a certain capacity. In the case of the hosiery mill, if the plant's capacity were known, regression analysis could be used to predict the firm's level of expenditure.

The term *regression analysis* comes from studies carried out by the English statistician Francis Galton about 80 years ago. Galton compared the heights of parents with the heights of their offspring and found that very tall parents tended to have offspring who were shorter than their parents, while very short parents tended to have offspring who were taller than their parents. In other words, the heights of the offspring of unusually tall or unusually short parents tended to "regress" toward the mean height of the population. Because Galton used the height of the parent to predict the height of the offspring, this type of analysis came to be called regression analysis, even though subsequent applications actually had very little to do with Galton's "regression" of heights toward the mean.

## SCATTER DIAGRAM

Since regression analysis is concerned with how one variable is related to another, an analysis of this sort generally begins with data concerning the two variables in question. Again, suppose that the hosiery mill wants to estimate how its monthly costs are related to its monthly output rate. Clearly, a sensible first step for the firm is to obtain data regarding its costs and output during a sample of months in the past. Suppose that the firm collects such data for a sample of nine months, the results being shown in Table 11.1. It is convenient to plot data of this sort in a so-called **scatter diagram.** In this diagram, the known variable—in this case the monthly output rate—is plotted along the horizontal axis and is called the *independent variable.* The unknown variable—the monthly cost, in this case—is plotted along the vertical axis and is called the *dependent variable.* Of course, during the period to which the data pertain, *both* variables are known, but when the regression analysis is used to estimate how much it will cost to produce a particular output, only output, rather than cost, will be known.

Figure 11.1 shows the scatter diagram based on the data in Table 11.1. Clearly, this diagram provides a useful visual portrait of the relationship between the dependent and independent variables. Based on such a diagram, one can form an initial impression concerning the following three important questions.

| Output (tons) | Production cost (thousands of dollars) |
|---|---|
| 1 | 2 |
| 2 | 3 |
| 4 | 4 |
| 8 | 7 |
| 6 | 6 |
| 5 | 5 |
| 8 | 8 |
| 9 | 8 |
| 7 | 6 |

**Table 11.1**
Cost and Output of a Hosiery Mill, Sample of Nine Months

IS THE RELATIONSHIP DIRECT OR INVERSE? The relationship between $X$ and $Y$ is *direct* if increases in $X$ tend to be associated with increases in $Y$, and decreases in $X$ tend to be associated with decreases in $Y$. On the other hand, the relationship between $X$ and $Y$ is *inverse* if increases in $X$ tend to be associated with decreases in $Y$, and decreases in $X$ tend to be associated with increases in $Y$. Figure 11.1 indicates that the

**Figure 11.1**
Scatter Diagram of
Data on Cost and
Output of Hosiery Mill

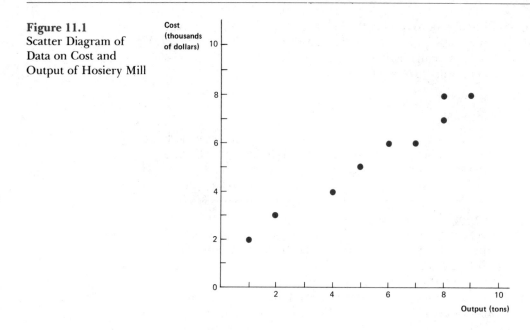

relationship between the firm's output and its costs is direct, as would be expected. Panel A of Figure 11.2 shows a case where the relationship between $X$ and $Y$ is inverse. Not all scatter diagrams indicate either a direct or inverse relationship. Some, like panel C, indicate no correlation at all between $X$ and $Y$. That is, changes in $X$ do not seem to have any effect on the value of $Y$.

IS THE RELATIONSHIP LINEAR OR NONLINEAR? A relationship between $X$ and $Y$ is *linear* if a straight line provides an adequate representation of the average relationship between the two variables. On the other hand, if the points in the scatter diagram fall along a curved line or depart in some other way from a linear relationship, the relationship between $X$ and $Y$ is *nonlinear*. Figure 11.1 suggests that the relationship between cost and output is linear (at least in this range). Panel D of Figure 11.2 shows a case where the relationship between $X$ and $Y$ is nonlinear.

HOW STRONG IS THE RELATIONSHIP? The relationship between $X$ and $Y$ is relatively strong if the points in the scatter diagram lie close to the line of average relationship. For example, panel E of Figure 11.2 shows a case where the relationship is strong enough so that one can predict the value of $Y$ quite accurately on the basis of the value of $X$. This is evidenced by the fact that all the points lie very close to the line. On the other hand, panel F of Figure 11.2 shows a case where the relationship is so weak that one cannot predict the value of $Y$ at all well on the basis of the value of $X$. This follows from the fact that the points are scattered widely around the line.

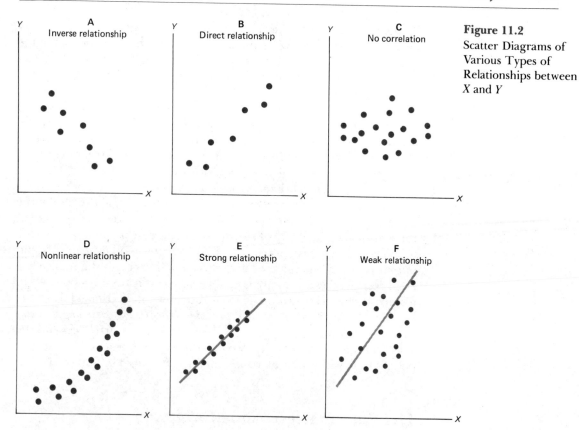

**Figure 11.2**
Scatter Diagrams of
Various Types of
Relationships between
$X$ and $Y$

## CORRELATION ANALYSIS

*Correlation analysis* is concerned with the strength of the relationship between two variables. As we have seen, some relationships among variables are much stronger than others. For example, the size of a person's left foot is ordinarily very strongly related to the size of his or her right foot. On the other hand, there is some relationship between a firm's size and how rapidly it adopts new techniques, but this relationship may be rather weak. Correlation analysis is an important and useful complement to regression analysis. Whereas regression analysis describes the *type* of relationship between the two variables, correlation analysis describes the *strength* of this relationship.

# 11.3 Aims of Regression and Correlation Analysis

Basically, there are four principal goals of regression and correlation analysis. First, *regression analysis provides estimates of the dependent variable for given values of the independent variable.* If the hosiery mill in Figure 11.1 wants to estimate its monthly cost of producing 4 tons of output per month, regression analysis provides such an estimate, based on the *regression line*. This line (which is fitted to the data by a method described in Section 11.6) estimates the mean value of $Y$ for each value of $X$. Thus, in the case of the hosiery mill the regression line would estimate the mean value of cost for each value of output.

Second, *regression analysis provides measures of the errors that are likely to be involved in using the regression line to estimate the dependent variable.* For example, in the case of the hosiery mill it clearly would be useful to know how much faith one can put in the cost estimate based on the regression line. To answer such questions, statisticians constuct confidence intervals which are described in Section 11.9.

Third, *regression analysis provides an estimate of the effect on the mean value of $Y$ of a one-unit change in $X$.* For example, in the case of the hosiery mill, the management might well be interested in the value of the marginal cost, which, of course, is the increase in the total cost due to a one-unit increase in output. If the relationship between cost and output is linear, the slope of the regression line equals the mill's marginal cost. Regression analysis enables us to estimate this slope and to test hypotheses concerning its value, as shown in Section 11.13.

Fourth, *correlation analysis provides estimates of how strong the relationship is between the two variables.* The *coefficient of correlation* and the *coefficient of determination* are two measures generally used for this purpose. These will be discussed in detail in Sections 11.10 and 11.11.

## 11.4 Linear Regression Model

A *model* is a simplified or idealized representation of the real world. All scientific inquiry is based to some extent on the use of models. In this section, we describe the model—that is, the set of simplifying assumptions—on which regression analysis is based. To begin with, the statistician visualizes a population of all relevant pairs of observations of the independent and dependent variables. For example, in the case of the hosiery mill, the statistician would visualize a population of pairs of observations concerning output and cost. This population would include all the levels of cost corresponding to all the output rates in the history of the mill.

Holding constant the value of $X$ (the independent variable), the statistician assumes that each corresponding value of $Y$ (the dependent variable) is drawn at random from the population. For example, the second pair of observations in Table 11.1 is a case where output equals 2 tons and cost $3,000. The statistician views this pair of observations as arising in the following way. The value of output (the independent variable) is fixed at 2 tons. The value of cost (the dependent variable) is the result of a random choice of all levels of cost corresponding to an output of 2 tons. Thus, the value of the dependent variable is a random variable, which happens in this case to equal $3,000.

What determines the shape of the probability distribution of the dependent variable $Y$ when the value of the independent variable is fixed at its specified value? For example, suppose that the probability

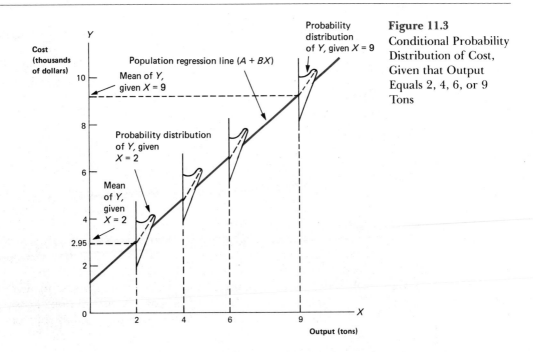

**Figure 11.3**
Conditional Probability
Distribution of Cost,
Given that Output
Equals 2, 4, 6, or 9
Tons

distribution of cost, given that output equals 2 tons, is as shown in Figure 11.3. According to this figure, this probability distribution is bell-shaped with a mean of $2,950. Why does the probability distribution have this shape? Because in the population as a whole the values of cost (the dependent variable), when output is fixed at 2 tons, have a bell-shaped distribution with a mean of $2,950.

The probability distribution of $Y$, given a specified value of $X$, is called the **conditional probability distribution of $Y$**. Thus, Figure 11.3 shows the conditional probability distribution of cost, given that output equals 2 tons (in other words, under the *condition* that output equals 2 tons).[1] Figure 11.3 also shows the conditional probability distribution of cost, given that output equals 4 tons, 6 tons, and 9 tons. The conditional probability distribution of $Y$, given the specified value of $X$, is denoted by

$$P(Y \mid X),$$

where $Y$ is the value of the dependent variable and $X$ is the specified value of the independent variable. The mean of this conditional probability distribution is denoted by $\mu_{Y \cdot X}$, and the standard deviation of this probability distribution is denoted by $\sigma_{Y \cdot X}$. For example, in Figure 11.3, $\mu_{Y \cdot X}$ equals $2,950 if output equals 2 tons. (That is, $\mu_{Y.2} = 2.95$ thousands of dollars.)

---

[1] The concept of a conditional probability distribution has already been touched on in Appendix 4.3. The present discussion is self-contained and does not presume any knowledge of the material in Appendix 4.3.

Regression analysis makes the following assumptions about the conditional probability distribution of $Y$. First, it assumes that *the mean value of* Y, *given the value of* X, *is a linear function of* X. In other words, the mean value of the dependent variable is assumed to be a linear function of the independent variable. Put still differently, the means of the conditional probability distributions are assumed to lie on a straight line, the equation of this line being

$$\mu_{Y \cdot X} = A + BX.$$

**Population Regression Line**

Figure 11.3 shows a case of this sort, as evidenced by the fact that the means lie on a straight line. This straight line is called the ***population regression line*** or the ***true regression line***.

Second, regression analysis assumes that *the standard deviation of the conditional probability distribution is the same, regardless of the specified value of the independent variable*. Thus, in Figure 11.3, the spread of each of the conditional probability distributions is the same. For example, the standard deviation of the probability distribution of cost, given that output is 2 tons, is the same as the standard deviation of cost, given that output is 9 tons. This characteristic (of equal standard deviations) is called *homoscedasticity*.

Third, regression analysis assumes that *the values of* Y *are independent of one another*. For example, if one observation lies below the mean of its conditional probability distribution, it is assumed that this will not affect the chance that some other observation in the sample will lie below the mean of its conditional probability distribution. Obviously, this assumption need not be true. For example, in the case of the hosiery mill, if one month's costs are below average, the next month's costs may also be below average because the same factors may be at work for an extended period of time.

Fourth, regression analysis assumes that the *conditional probability distribution of* Y *is normal*. Actually, as pointed out below, not all aspects of regression analysis require this assumption, but some do. It is also worth noting that in regression analysis only $Y$ is regarded as a random variable. The values of $X$ are assumed to be fixed. Thus, when regression analysis is used to estimate $Y$ on the basis of $X$, the true value of $Y$ is subject to error, but the value of $X$ is known. For example, if regression analysis is used to estimate the hosiery mill's costs when its output is 4 tons, the true cost of producing this output can be predicted only subject to error; but the output (4 tons) is known precisely.

The four assumptions underlying regression analysis can be stated somewhat differently. Together they imply that

$$Y_i = A + BX_i + e_i, \tag{11.1}$$

where $Y_i$ is the $i$th observed value of the dependent variable, $X_i$ is the $i$th observed value of the independent variable, and $e_i$ is a normally distributed random variable with a mean of zero and a standard deviation equal to $\sigma_e$. Essentially, $e_i$ is an *error term*, that is, a random amount that is added to $A + BX_i$ (or subtracted from it if $e_i$ is negative). Because of

**Error Term**

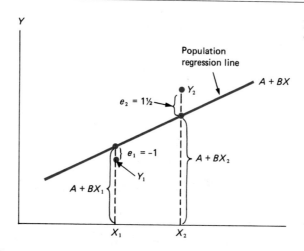

**Figure 11.4**
The Regression Model

the presence of this error term, the observed values of $Y_i$ fall around the population regression line, not on it. Thus, as shown in Figure 11.4, if $e_1$ (the value of the error term for the first observation) is $-1$, $Y_1$ will lie 1 below the population regression line. And if $e_2$ (the value of the error term for the second observation) is $+1.50$, $Y_2$ will lie 1.50 above the population regression line. Regression analysis assumes that the values of $e_i$ are independent.[2]

Although the assumptions underlying regression analysis are unlikely to be met completely, they are close enough to the truth in a sufficiently large number of cases so that regression analysis is a powerful technique. Nonetheless, it is important to recognize at the start that if these assumptions are not at least approximately valid, the results of a regression analysis can be misleading. In the next chapter we shall indicate how tests can be carried out to check these assumptions, and we will describe how violations of these assumptions may affect the results.

## 11.5 Sample Regression Line

To carry out a regression analysis, we must obtain the mathematical equation for a line that describes the average relationship between the dependent and independent variable. This line is calculated from the sample observations and is called the *sample* or *estimated regression line*. It should not be confused with the *population regression line* discussed in the previous section. Whereas the population regression line is based on the entire population, the sample regression line is based only on the sample.

*Sample Regression Line*

The general expression for the sample regression line is

$$\hat{Y} = a + bX,$$

---

[2] In this chapter (and in Chapter 12), we use capital letters $X$ and $Y$ to denote both random variables and realized values. Which is referred to should be obvious from the context. (Practically all texts use the same symbols for both at this point. There seems to be general agreement that there is little danger of confusion.)

**Figure 11.5**
Sample Regression
Line

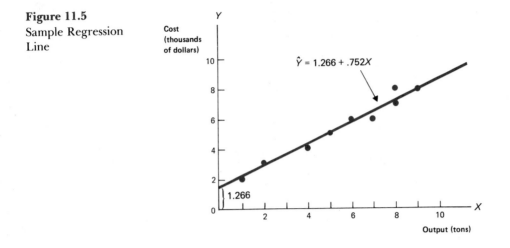

where $\hat{Y}$ is the value of the dependent variable predicted by the regression line, and $a$ and $b$ are estimators of $A$ and $B$, respectively. Since this equation implies that $\hat{Y} = a$ when $X = 0$, it follows that $a$ is the value of $Y$ at which the line intersects the $Y$ axis. Thus, $a$ is often called the $Y$ *intercept* of the regression line. And $b$, which clearly is the *slope* of the line, measures the change in the predicted value of $Y$ associated with a one-unit increase in $X$.

Figure 11.5 shows the estimated regression line for the data concerning cost and output of the hosiery mill. The equation for this regression line is

$$\hat{Y} = 1.266 + 0.752X,$$

where $\hat{Y}$ is monthly cost in thousands and $X$ is monthly output in tons. What is 1.266? It is the value of $a$, the estimator of $A$. What is 0.752? It is the value of $b$, the estimator of $B$. We are not interested here in how this equation was determined. (The methods used and their rationale are described in detail in the next section.) What we do want to consider is how this equation should be interpreted.

To begin with, note the difference between $Y$ and $\hat{Y}$. Whereas $Y$ denotes an *observed* value of monthly cost, $\hat{Y}$ denotes the *computed* or *estimated* value of monthly cost, based on the regression line. For example, the first row of Table 11.1 shows that in the first month the actual value of cost was $2,000 when output was 1 ton. Thus, $Y = 2.0$ thousands of dollars when $X = 1$. In contrast, the regression line indicates that $\hat{Y} = 1.266 + 0.752(1)$, or 2.018 thousands of dollars when $X = 1$. In other words, while the regression line predicts that cost will equal $2,018 when output equals 1 ton, the actual cost under these circumstances (in the first month) was $2,000.

It is important to be able to identify and interpret the $Y$ intercept and slope of a regression line. What is the $Y$ intercept of the regression line in the case of the hosiery mill? It is 1.266 thousands of dollars. This means that if the monthly output is zero, the estimated monthly cost would be $1,266. (As shown in Figure 11.5, $1,266 is the value of the dependent variable at which the regression line intersects the vertical axis.) What is the slope of the regression line in this case? It is 0.752 thousands of dollars. This means that the estimated monthly cost increases by $752 when the monthly output increases by 1 ton.

## 11.6 Method of Least Squares

In this section, we describe how a sample regression line is calculated. To illustrate how this is done, suppose that we want to estimate a regression line to represent the relationship beween cost and output in Fig. 11.1. Since the equation for this line is

$$\hat{Y} = a + bX,$$

the estimation of a regression line really amounts to the choice of numerical values of $a$ and $b$. There are an infinite number of possible values of $a$ and $b$ we could choose. One possibility, resulting in the line shown in panel A of Figure 11.6, is that $a = 1/2$ and $b = 1$. Another possibility, resulting in the line in panel B of Figure 11.6, is that $a = 0$ and $b = 1$. Still another possibility, resulting in the line in panel C of Figure 11.6, is that $a = -1$ and $b = 1$. How can we figure out which is best?

The *method of least squares* answers this question in the following way: We should take each possible line (that is, each possible value of $a$ and $b$) and measure the deviation of each point in the sample from this line. Thus, in panel A of Figure 11.6 the deviation of each point from the line is measured by the broken vertical line from the point to the line. Then, according to the method of least squares, we square each of these deviations and add them up. Thus, in panel A of Figure 11.6 the sum of the squared deviations of the points from the line shown there is 8.25. In panel B the sum of squared deviations is 5, and in panel C the sum of squared deviations is 12.[3] Then *the method of least squares dictates that we choose the line where the sum of the squared deviations of the points from the line is a minimum.*

Certainly, on an intuitive level, it makes sense to choose the line (that is, the values of $a$ and $b$) that minimizes the sum of squared deviations of the data in the sample from the line. Why? Because the bigger the sum of squared deviations of the data from the line, the poorer the line fits the data. Thus, one should minimize the sum of squared deviations if one wants to obtain a line that fits the data as well as possible. This is illustrated in Figure 11.6. Clearly, the line in panel C does not fit the data as well as the line in panel A, which in turn does not fit the data as well as the line in panel B. This fact is reflected in the differences in

---

[3] In panel A the deviations of the points from the line are $1/2, 1/2, -1/2, -1/2, -1/2, -3/2, -3/2, -1/2, -3/2$. Thus, the sum of squared deviations is 8.25. In panel B the deviations are $1, 1, 0, 0, 0, -1, -1, 0, -1$. Thus, the sum of squared deviations is 5. In panel C the deviations are $2, 2, 1, 1, 1, 0, 1, 0, 0$. Thus, the sum of squared deviations is 12. Since the deviations are in units of thousands of dollars, the squared deviations are in units of millions of dollars squared.

**Figure 11.6**
Alternative Values of *a*
and *b*

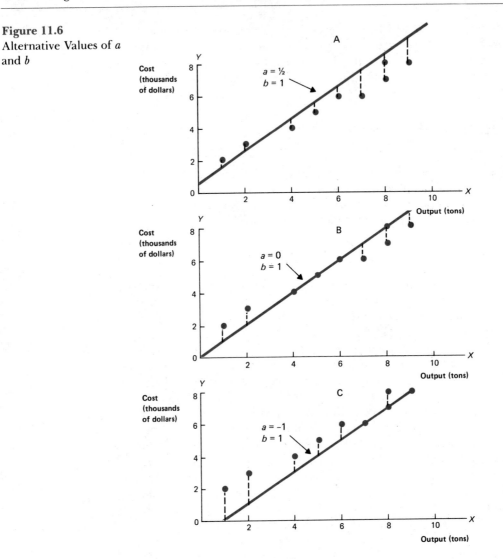

the sum of the squared deviations. As would be expected, the sum of squared deviations is higher for the line in panel c than for the line in panel A, and higher for the line in panel A than for the line in panel B.

Mathematically, it can be shown that if $Y_i$ and $X_i$ are the *i*th pair of observations concerning the dependent and independent variables, the values of *a* and *b* that result in the minimization of the sum of squared deviations from the regression line satisfy the following equations:

*Least-Squares
Estimators of A
and B*

$$b = \frac{\sum_{i=1}^{n} (X_i - \overline{X})(Y_i - \overline{Y})}{\sum_{i=1}^{n} (X_i - \overline{X})^2} \qquad (11.2a)$$

$$a = \overline{Y} - b\overline{X}. \qquad (11.2b)$$

where $n$ is the number of values of $X_i$ (and $Y_i$) on which the calculation of the sample regression line is based. The value of $b$ in equation (11.2a) is often called the **estimated regression coefficient.**

---

*If you are interested in the relationship between two variables, X and Y, a good way to begin is to plot Y against X in a scatter diagram. The method of least squares dictates that we choose a regression line where the sum of the squared deviations of the points from the line is a minimum.*

*Basic Idea #21*

---

From the standpoint of computational ease, it frequently is preferable to use a somewhat different formula for $b$ than the one given in equation (11.2a). This alternate formula, which yields the same answer as equation (11.2a), is

$$b = \frac{n \sum_{i=1}^{n} X_i Y_i - \left(\sum_{i=1}^{n} X_i\right)\left(\sum_{i=1}^{n} Y_i\right)}{n \sum_{i=1}^{n} X_i^2 - \left(\sum_{i=1}^{n} X_i\right)^2}.$$

In the case of the hosiery mill, Table 11.2 shows the calculation of $\Sigma X_i Y_i$, $\Sigma X_i^2$, $\Sigma X_i$, and $\Sigma Y_i$. Based on these calculations,

$$b = \frac{9(319) - (50)(49)}{9(340) - 50^2} = \frac{2871 - 2450}{3060 - 2500}$$

$$= \frac{421}{560} = .752.$$

**Table 11.2**
Computation of $\Sigma X_i$, $\Sigma Y_i$, $\Sigma X_i^2$, $\Sigma Y_i^2$, and $\Sigma X_i Y_i$

| | $X_i$ | $Y_i$ | $X_i^2$ | $Y_i^2$ | $X_i Y_i$ |
|---|---|---|---|---|---|
| | 1 | 2 | 1 | 4 | 2 |
| | 2 | 3 | 4 | 9 | 6 |
| | 4 | 4 | 16 | 16 | 16 |
| | 8 | 7 | 64 | 49 | 56 |
| | 6 | 6 | 36 | 36 | 36 |
| | 5 | 5 | 25 | 25 | 25 |
| | 8 | 8 | 64 | 64 | 64 |
| | 9 | 8 | 81 | 64 | 72 |
| | 7 | 6 | 49 | 36 | 42 |
| Total | 50 | 49 | 340 | 303 | 319 |

$$\bar{X} = \frac{50}{9} = 5.556$$

$$\bar{Y} = \frac{49}{9} = 5.444,$$

Thus, the value of *b*, generally referred to as the *least-squares estimator of B*, is .752 thousands of dollars, which is the result given in the previous section. In other words, an increase in output of 1 ton results in an increase in estimated cost of about $752.

Having calculated *b*, we can readily determine the value of *a*, generally referred to as the *least-squares estimator of A*. According to equation (11.2b),

$$a = \bar{Y} - b\bar{X},$$

where $\bar{Y}$ is the mean of the values of *Y*, and $\bar{X}$ is the mean of the values of *X*. Since, as shown in Table 11.2, $\bar{Y} = 5.444$ and $\bar{X} = 5.556$, it follows that

$$a = 5.444 - .752(5.556)$$

$$= 1.266.$$

Thus, the least-squares estimate of *A* is 1.266 thousands of dollars. Recall that this is the result given in the previous section.

Given *a* and *b*, it is a simple matter to specify the average relationship in the sample between cost and output for the hosiery firm. This relationship is

$$\hat{Y} = 1.266 + 0.752X. \tag{11.3}$$

where $\hat{Y}$ is measured in thousands of dollars and *X* is measured in tons. As we know, this line is often called the *sample regression line* or the *regression of* Y *on* X. It is the line that we presented in the previous section and that we plotted in Figure 11.5. Now we have shown how this line is derived.

A regression line of this sort can be of great practical importance. For example, suppose that the managers of the hosiery mill want to predict the firm's monthly costs if they decide to produce 4 tons per month. Using equation (11.3), the firm's statisticians would predict that its costs would be

$$1.266 + 0.752(4) = 4.274. \tag{11.3a}$$

Since costs are measured in thousands of dollars, this means that total costs would be expected to be 4.274 thousands of dollars, or $4,274.

The following example illustrates further how one calculates a least-squares regression line.

EXAMPLE 11.1 An economist wants to estimate the relationship in a small Appalachian community between a family's annual income and the amount that the family saves. The following data from nine families are obtained:

| Annual income (thousands of dollars) | Annual savings (thousands of dollars) |
|---|---|
| 12 | 0.0 |
| 13 | 0.1 |
| 14 | 0.2 |
| 15 | 0.2 |
| 16 | 0.5 |
| 17 | 0.5 |
| 18 | 0.6 |
| 19 | 0.7 |
| 20 | 0.8 |

Calculate the least-squares regression line, where annual savings is the dependent variable and annual income is the independent variable.

SOLUTION: Letting $X_i$ be the income (in thousands of dollars) of the $i$th family, and $Y_i$ be the saving (in thousands of dollars) of the $i$th family, we find that

$$\sum_{i=1}^{9} X_i Y_i = 63.7 \quad \sum_{i=1}^{9} Y_i = 3.6 \quad \bar{Y} = 0.4,$$

$$\sum_{i=1}^{9} X_i^2 = 2364 \quad \sum_{i=1}^{9} X_i = 144 \quad \bar{X} = 16.$$

Thus, substituting these values in the alternate formula for $b$, we obtain

$$b = \frac{9(63.7) - (144)(3.6)}{9(2364) - 144^2} = \frac{573.3 - 518.4}{21,276 - 20,736} = .1017.$$

Consequently,

$$a = \bar{Y} - b\bar{X} = 0.4 - .1017(16) = -1.2272.$$

Thus, the regression line is

$$\hat{Y} = -1.2272 + .1017X,$$

where both $X$ and $Y$ are measured in thousands of dollars.

**EXERCISES**

**11.1** In aircraft manufacture, rivets are used to join parts. The following table, taken from a study published in *Industrial Quality Control*, shows the number of oversize rivet holes and the number of minor repairs on 10 sections of an airplane.

| Oversize rivet holes | Minor repairs |
|:---:|:---:|
| 45 | 22 |
| 52 | 26 |
| 49 | 21 |
| 60 | 28 |
| 67 | 33 |
| 61 | 32 |
| 70 | 33 |
| 54 | 25 |
| 52 | 34 |
| 67 | 35 |

(a) Construct a scatter diagram of these data.

(b) Calculate the sample regression line, where the number of minor repairs is the dependent variable and the number of oversize rivet holes is the independent variable.

(c) Use this regression line to predict the average number of minor repairs if the number of oversize rivet holes is 50.

(d) Predict the average number of minor repairs if the number of oversize rivet holes is 70.

**11.2** For 15 chemical firms, the percent of value-added spent on research and development in the United States ($X$) and the percent of value-added spent on research and development outside the United States ($Y$) are shown below.[4]

| Firm | X | Y |
|:---:|:---:|:---:|
| 1 | 5.00 | 0.00 |
| 2 | 8.90 | 0.43 |
| 3 | 7.15 | 0.00 |
| 4 | 6.10 | 0.24 |
| 5 | 7.70 | 0.00 |
| 6 | 8.20 | 0.91 |
| 7 | 1.01 | 0.00 |
| 8 | 0.61 | 0.03 |
| 9 | 0.72 | 0.01 |
| 10 | 0.68 | 0.00 |
| 11 | 1.14 | 0.00 |
| 12 | 1.18 | 0.01 |
| 13 | 0.73 | 0.00 |
| 14 | 0.87 | 0.20 |
| 15 | 1.47 | 0.00 |

(a) Construct the scatter diagram of these data.

(b) Calculate the sample regression line, where $Y$ is the dependent variable and $X$ is the independent variable.

[4] E.Mansfield, "R and D and Innovation: Some Empirical Findings," in Z. Griliches ed., *R and D, Patents, and Productivity* (Chicago: University of Chicago Press, 1984).

(c) Predict the percent of value-added spent on research and development outside the United States if a chemical firm spends 6 percent of value-added on research and development in the United States.

(d) Is the relationship between $X$ and $Y$ a causal relationship? Why, or why not?

**11.3** A psychologist obtains the IQ of 14 individuals, each of which is asked to take a particular test. The results are as follows:

| IQ | Test score |
|----|-----------|
| 99 | 54 |
| 110 | 70 |
| 141 | 100 |
| 108 | 61 |
| 123 | 83 |
| 129 | 86 |
| 132 | 98 |
| 88 | 51 |
| 101 | 64 |
| 105 | 68 |
| 97 | 52 |
| 96 | 55 |
| 118 | 71 |
| 113 | 76 |

(a) Construct a scatter diagram of these data.

(b) Calculate the sample regression line, where the test score is the dependent variable and the IQ is the independent variable.

(c) Predict the average test score if a person's IQ is 130.

(d) If the maximum test score is 100, do you think that the relationship between IQ and test score is linear regardless of the level of IQ? Explain.

**11.4** The following data are obtained concerning the R and D expenditures and sales of six telecommunications firms:

| Firm | Sales | R and D |
|------|-------|---------|
| | (millions of dollars) | |
| A T & T | 50,790 | 419 |
| Comsat | 300 | 12 |
| GTE | 9,980 | 162 |
| Rolm | 201 | 13 |
| United | 1,904 | 3 |
| Western Union | 794 | 5 |

(a) Contruct a scatter diagram of these data.

(b) Compute the sample regression line, where R and D expenditure is the dependent variable and sales is the independent variable.

(c) Predict the average R and D expenditure if sales equal $10 billion.

(d) Do you think that this regression line can be applied to other industries? Explain.

**11.5** In August 1985, *Business Week* presented the profits and sales (in millions of dollars) for the second quarter of 1985 of six appliance firms:

|                         | Sales | Profit |
|-------------------------|-------|--------|
| Allegheny International  | 500   | 10     |
| Hoover                  | 167   | 7      |
| Magic Chef              | 313   | 19     |
| Maytag                  | 172   | 19     |
| Whirlpool               | 967   | 49     |
| White                   | 591   | 16     |

(a) Construct a scatter diagram of these data.

(b) Which variable (profit or sales) do you regard as the dependent variable?

(c) Compute the sample regression line.

## 11.7 Characteristics of Least-Squares Estimates

Optimally, in their calculations the decision maker and the statistician would like to know the population regression line. For example, the hosiery mill would like to know the population regression line relating its costs to its output—the regression based on *all* the possible observations of cost and output. However, this regression line cannot be calculated because the statistician has only the sample of observations to work with. Therefore, the best the statistician can do is to calculate the sample regression line and to use it as an estimate of the population regression line. The statistics—$a$ and $b$—defined in the previous section are estimators of $A$ and $B$, the constants in the population regression. *Whether or not the conditional probability distribution of the dependent variable is normal,* these estimators have the following desirable properties (if the other assumptions in Section 11.4 are met):

1. *Unbiasedness.* It can be shown that $a$ is an unbiased estimator of $A$, and that $b$ is an unbiased estimator of $B$. In other words, if we were to draw one sample after another, and calculate the least-squares estimators $a$ and $b$ from each sample, the mean value of $a$ would equal $A$, and the mean value of $b$ would equal $B$, if we were to draw a very large number of samples. (Recall from Chapter 7 that unbiasedness is one of the criteria statisticians use for choosing among estimators.)

2. *Efficiency.* It can also be shown that of all estimators which are unbiased (and which are linear functions of the dependent variables), $a$ and $b$ have the smallest standard deviation. In other words, they are the most efficient estimators of this type. This very important result was proved by the so-called *Gauss-Markov theorem.* (Recall from Chapter 7 that efficiency is one of the criteria used by statisticians to choose among estimators.)

3. *Consistency.* We also find that $a$ is a consistent estimator of $A$, and that $b$ is a consistent estimator of $B$. In other words, as the sample size becomes larger and larger, the value of $a$ homes in on $A$, and the value of $b$ homes in on $B$. (Recall from Chapter 7 that consistency is one of the criteria used by statisticians for choosing among estimators.)

*Because of these very desirable properties, the least-squares estimators a and b are the standard estimators used by statisticians to estimate the con-*

*stants in the population regression line,* A *and* B. The fact that *a* and *b* have these desirable properties provides fundamental and strong support for the intuitive judgment in the previous section that the method of least squares is a good way to fit a sample regression line.

## 11.8 Standard Error of Estimate

In previous sections we have shown how regression analysis provides estimates of the dependent variable for given values of the independent variable (the first goal cited in Section 11.3). We will now describe how regression analysis provides measures of the errors that are likely to be involved in this estimation procedure (the second goal cited in Section 11.3). First, it is essential to recall that the standard deviation of the conditional probability distribution of the dependent variable is assumed to be the same, regardless of the value of the independent variable. *This standard deviation, which we denote by* $\sigma_e$, *is a measure of the amount of scatter about the regression line in the population.* If $\sigma_e$ is large, there is much scatter; if $\sigma_e$ is small, there is little scatter.

The sample statistic used to estimate $\sigma_e$ is the standard error of estimate. It is defined as

*Formula for $s_e$*

$$s_e = \sqrt{\frac{\sum_{i=1}^{n}(Y_i - \hat{Y}_i)^2}{n-2}}, \qquad (11.4)$$

where $Y_i$ is the *i*th value of the dependent variable, *n* is the sample size, and $\hat{Y}_i$ is the estimate of $Y_i$ from the regression line (that is, $\hat{Y}_i = a + bX_i$). Clearly, the value of $s_e$ rises with increases in the amount of scatter about the regression line in the sample. If there is no scatter at all (which means that all the points are on the regression line), $s_e$ equals zero, since $Y_i$ and $\hat{Y}_i$ are always the same. But if there is much scatter, $Y_i$ often differs greatly from $\hat{Y}_i$, with the result that $s_e$ will be large.

Another formula used frequently for the standard error of estimate is

$$s_e = \sqrt{\frac{\sum_{i=1}^{n} Y_i^2 - a \sum_{i=1}^{n} Y_i - b \sum_{i=1}^{n} X_i Y_i}{n-2}}. \qquad (11.5)$$

This expression is often easier to calculate than the one given in equation (11.4). Of course, equations (11.4) and (11.5) will always give the same result. In each of these equations $(n-2)$ is the denominator because this results in $s_e^2$ being an unbiased estimate of $\sigma_e^2$.

To illustrate the use of equation (11.5), let's return to the case of the hosiery mill. Since Table 11.2 shows that $\Sigma Y_i^2 = 303$, $\Sigma Y_i = 49$, and

**Figure 11.7**
Conditional Probability
Distribution of Cost,
Given that Output
Equals 2, 4, 6, or 9
Tons

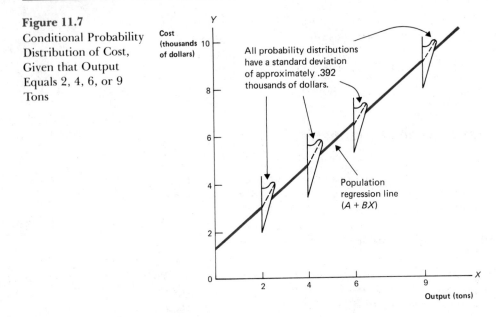

$\Sigma X_iY_i = 319$, it follows that

$$s_e = \sqrt{\frac{303 - (1.266)(49) - (0.752)(319)}{7}} = \sqrt{.154} = 0.392.$$

Thus, if we knew the true values of $A$ and $B$, the standard deviation of the errors in prediction based on this true regression line would be about .392 thousands of dollars, or $392. In other words, the situation would be as shown in Figure 11.7.

The standard error of estimate will be used frequently in subsequent sections. The example below further illustrates how it is calculated.

EXAMPLE 11.2 Based on the data in Example 11.1, the economist wants to calculate the standard error of estimate. What is its value?

SOLUTION: From equation (11.5), it follows that

$$s_e = \sqrt{\frac{\sum_{i=1}^{9} Y_i^2 - a \sum_{i=1}^{9} Y_i - b \sum_{i=1}^{9} X_iY_i}{7}}$$

$$= \sqrt{\frac{2.08 + (1.2272)(3.6) - (.1017)(63.7)}{7}}$$

$$= \sqrt{\frac{2.08 + 4.4179 - 6.4783}{7}} = \sqrt{.0028}$$

$$= .053.$$

Thus, the standard error of estimate is .053 thousands of dollars, or $53.

## 11.9 Estimators of (1) the Conditional Mean and (2) an Individual Value of $Y$

ESTIMATING THE CONDITIONAL MEAN

In this section we show how one can estimate the conditional mean of $Y$. In contrast to the previous section, it is not assumed that we know $A$ and $B$; instead, we assume that both the regression line and the estimate are based on least-squares estimates of $A$ and $B$. To be specific, let's return again to the example of the hosiery mill. Suppose that the managers of the mill are interested in predicting the *mean* monthly cost of the mill if it were to achieve an output of 4 tons per month *over and over again*. In other words, the firm's managers are interested in estimating the vertical coordinate of the point on the population regression line corresponding to an output of 4 tons per month. In Figure 11.8, this conditional mean is denoted by $\mu_{y.4}$.

Since the population regression line is unknown, the best that the mill's managers can do is to substitute the sample regression line. That is, to estimate $A + B(4)$, one uses $a + b(4)$, or $\$4,274$. This, of course, is the point on the sample regression line corresponding to $X = 4$, as shown in Figure 11.8. In general, in order to estimate the conditional mean of $Y$ (that is, the vertical coordinate of the point on the true regression line) when the independent variable equals $X^*$, one should use

$a + bX^*$.

A confidence interval for the conditional mean of $Y$ (given that the independent variable equals $X^*$) is

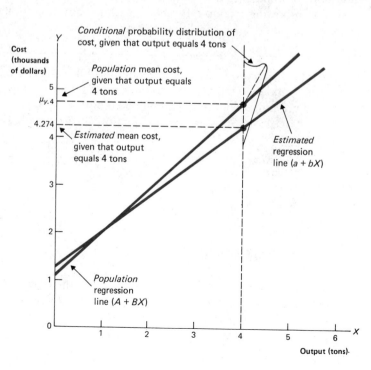

Conditional probability distribution of cost, given that output equals 4 tons

Cost (thousands of dollars)

Population mean cost, given that output equals 4 tons

$\mu_{y.4}$

Estimated mean cost, given that output equals 4 tons

Estimated regression line $(a + bX)$

Population regression line $(A + BX)$

Output (tons)

**Figure 11.8**
Estimated (and Population) Mean Cost and Conditional Probability Distribution of Cost, If Output Equals 4 Tons

*Confidence Interval
for Conditional
Mean*

$$(a + bX^*) \pm t_{\alpha/2} s_e \sqrt{\frac{1}{n} + \frac{(X^* - \overline{X})^2}{\sum_{i=1}^{n} X_i^2 - n\overline{X}^2}} \qquad (11.6)$$

*If the conditional probability distribution of the dependent variable is normal, the probability is* $(1 - \alpha)$ *that this interval will include this conditional mean.* Note that this confidence interval becomes wider as the value of $X^*$ lies farther and farther from $\overline{X}$. This makes sense because as one moves farther and farther from $\overline{X}$, an error in the estimated slope of the regression line will result in an increasingly larger error in the location of the regression line.

To illustrate the use of the confidence interval in expression (11.6), let's consider the prediction of the mean cost that would be incurred if the hosiery mill were to achieve repeatedly an output of 4 tons per month. In this case, the 95 percent confidence interval for this mean cost would be

$$4.274 \pm 2.365(0.392) \sqrt{\frac{1}{9} + \frac{(4 - 5.556)^2}{62.2}}$$

or

$$4.274 \pm .359.$$

(Note that $t_{\alpha/2}$ is based on $(n - 2)$, or 7, degrees of freedom.) Thus, if the firm is interested in estimating the point on the population regression line corresponding to an output of 4 tons per month, the 95 percent confidence interval for this conditional mean is 3.915 to 4.633 thousands of dollars (that is, $3,915 to $4,633).

The following example further illustrates how a confidence interval is computed for a conditional mean.

---

EXAMPLE 11.3 Based on the data in Example 11.1, calculate a 95 percent confidence interval for the mean amount of family savings among families with an income of $20,000.

SOLUTION: Substituting in equation (11.6), we obtain

$$-1.2272 + .1017(20) \pm 2.365(.053) \sqrt{\frac{1}{9} + \frac{(20 - 16)^2}{2364 - 9(16)^2}},$$

or

$$.8068 \pm 2.365(.053)(.615),$$

or

$$0.730 \text{ to } 0.884.$$

Thus, the 95 percent confidence interval for the mean amount of savings among families with an income of $20,000 is .730 to .884 thousands of dollars, or $730 to $884.

---

## PREDICTING AN INDIVIDUAL VALUE OF Y

Sample regression lines are often used to predict an individual value of $Y$. The hosiery mill's managers might want to predict the mill's costs for next month if it produces 4 tons of output then. This kind of prediction can be made with the use of the sample regression line. The vertical coordinate of the point on the sample regression line corresponding to $X = 4$ can be used as a predictor. As shown in Figure 11.8 (and as we already know from Section 11.6), the prediction in this case would be that $Y = 4.274$ thousands of dollars, or $4,274. In general, to predict an individual value of $Y$ when the independent variable equals $X^*$, one should use

$$a + bX^*.$$

A confidence interval for the value of $Y$ that will occur if the independent variable is set at $X^*$ is

$$(a + bX^*) \pm t_{\alpha/2} s_e \sqrt{\frac{n+1}{n} + \frac{(X^* - \overline{X})^2}{\sum_{i=1}^{n} X_i^2 - n\overline{X}^2}} \qquad (11.7)$$

*Confidence Interval for an Individual Value of Y*

*If the conditional probability distribution of the dependent variable is normal, the probability is $(1 - \alpha)$ that this interval will include the true value of* $Y$. A comparison of this confidence interval with the one shown in expression (11.6) indicates that this one is wider than the other. This, of course, is reasonable because the sampling error in predicting an *individual value* of the dependent variable will be greater than the sampling error in estimating the *conditional mean value* of the dependent variable.

To illustrate the use of this confidence interval, suppose that the hosiery mill is interested in predicting the cost that will be incurred if next month's output rate is set at 4 tons. Based on expression (11.7), the 95 percent confidence interval for this cost is

$$4.274 \pm 2.365(0.392) \sqrt{\frac{10}{9} + \frac{(4 - 5.556)^2}{62.2}}$$

or

$$4.274 \pm 0.994.$$

(Again, $t_{\alpha/2}$ is based on $(n - 2)$, or 7, degrees of freedom.) Thus, if the firm is interested in estimating the cost in a particular month when the output rate is 4 tons, the 95 percent confidence interval for this cost is 3.280 to 5.268 thousands of dollars, or $3,280 to $5,268.

It is important to recognize that the estimate of the *conditional mean* of $Y$ covered above is quite different from the prediction of an *in-*

*dividual value* of Y being discussed here. An estimate of the *average* cost that would occur if ouput were set *repeatedly* at 4 tons per month is not the same as a prediction of the cost that would occur if output were set at 4 tons next month only. In the latter case, we are trying to predict the value of the random variable Y whose distribution is shown in Figure 11.8; in the former case, we are attempting to estimate the conditional mean of this distribution ($\mu_{Y \cdot 4}$), also shown in Figure 11.8. Put still differently, in the latter case we are attempting to predict an individual value of Y, while in the former we are trying to estimate a point on the population regression line. Although these two things are not the same, we use the same point estimate—$a + b(4)$—for both. However, as we have seen, we use a different confidence interval for each.

Before leaving the topic of the confidence intervals in expressions (11.6) and (11.7), it is important to spell out the underlying assumptions once more. First, *it is assumed that the population regression is linear; that is, that the conditional mean of the dependent variable is a linear function of the independent variable*. Second, *it is assumed that the conditional probability distribution of the dependent variable is normal and that its standard deviation is the same, regardless of the value of the independent variable*. If any one of these assumptions is violated, the results will be in error to at least some degree. It is also worth noting that the confidence interval in expresssion (11.7) relates only to a single prediction. In other words, the confidence interval pertains to the entire process of gathering a sample of size n and making a single prediction. It does not pertain to more than one prediction based on a single sample.

Example 11.4 shows how a confidence interval is computed for a predicted value of the dependent variable.

---

EXAMPLE 11.4 Based on the data in Example 11.1, predict the savings of a family chosen at random from among families with an income of $20,000, and calculate a 95 percent confidence interval for this prediction.

SOLUTION: The prediction is $-1.2272 + .1017(20)$, or .807 thousands of dollars (that is, $807). Using expression (11.7), the confidence interval is

$$-1.2272 + .1017(20) \pm 2.365(.053) \sqrt{\frac{10}{9} + \frac{(20-16)^2}{2364 - 9(16^2)}},$$

or

$$.8068 \pm 2.365(.053)1.174,$$

or

$$.660 \text{ to } .954.$$

Thus, the 95 percent confidence interval for the savings of a family chosen at random from among families with an income of $20,000 is .660 to .954 thousands of dollars, or $660 to $954.

---

## EXERCISES

**11.6** A sample of 10 economics textbooks shows the following relationship between their price and length (prices are rounded off):

| Price (dollars) | Length (pages) |
| --- | --- |
| 16 | 520 |
| 20 | 680 |
| 21 | 740 |
| 9 | 200 |
| 10 | 400 |
| 22 | 800 |
| 20 | 750 |
| 15 | 500 |
| 10 | 300 |
| 10 | 350 |

(a) Construct a scatter diagram of these data.
(b) Compute the sample regression line, where price is the dependent variable and length is the independent variable.
(c) Calculate the standard error of estimate.
(d) Compute a 95 percent confidence interval for the conditional mean price, if the book's length is 600 pages.
(e) Compute a 95 percent confidence interval for the price of a book, if its length is 600 pages.
(f) Is length causally related to price?

**11.7** For eight families, the amount spent annually on food, and their annual incomes, are given below:

| Food expenditure | Income |
| --- | --- |
| (thousands of dollars) | |
| 4 | 20 |
| 6 | 40 |
| 3 | 11 |
| 5 | 30 |
| 2 | 9 |
| 2 | 12 |
| 3 | 15 |
| 3 | 21 |

(a) Compute the sample regression line, where income is the independent variable and amount spent on food is the dependent variable.
(b) Calculate the standard error of estimate.
(c) Compute a 90 percent confidence interval for the conditional mean food expenditure if family income is $20,000.
(d) Compute a 90 percent confidence interval for a family's food expenditure if its income is $20,000.

**11.8** (a) In Exercise 11.1, what is the standard error of estimate?
(b) Compute a 99 percent confidence interval for the conditional mean number of minor repairs if the number of oversize rivets is 50.

**11.9** (a) In Exercise 11.2, what is the standard error of estimate?
  (b) Compute a 50 percent confidence interval for a chemical firm's per-cent of value-added spent on R and D outside the United States if it spends 6 percent of value-added on R and D in the United States.

**11.10** (a) In Exercise 11.3, what is the standard error of estimate?
  (b) Compute a 90 percent confidence interval for a person's test score if his IQ is 130.

**11.11** The U. S. Department of Agriculture has published data concerning the strength of cotton yarn and the length of the cotton fibers that make up the yarn.[5] Results for 10 pieces of yarn are as follows:

| Strength of yarn (pounds) | Fiber length (hundredths of an inch) |
|---|---|
| 99 | 85 |
| 93 | 82 |
| 99 | 75 |
| 97 | 74 |
| 90 | 76 |
| 96 | 74 |
| 93 | 73 |
| 130 | 96 |
| 118 | 93 |
| 88 | 70 |

  (a) Construct a scatter diagram of these data.
  (b) Based on this scatter diagram, does the relationship between these two variables seem to be direct or inverse? Is this in accord with common sense? Why, or why not? Does the relationship seem to be linear?
  (c) Assume that the conditional mean value of yarn strength is a linear function of fiber length. Calculate the least-squares estimates of the parameters (*A* and *B*) of this linear function.
  (d) What is the sample regression line for these data? Use this regression line to predict the average stength of yarn made from fibers of length equal to 0.80 inches. Use this regression line to predict the average strength of yarn made from fibers of length equal to 0.90 inches.
  (e) Calculate the standard error of estimate.
  (f) Compute a 90 percent confidence interval for the conditional mean strength of yarn corresponding to a fiber length of 0.80 inches.
  (g) Compute a 90 percent confidence interval for the strength of a piece of yarn, if the fiber length is 0.80 inches. Why is this confidence interval wider than the confidence interval in (f)?

## 11.10 Coefficient of Determination

In previous sections, we have shown how a regression line can be cal-culated. Once the regression line has been found, the statistician wants

[5] U.S. Department of Agriculture, *Results of Fiber and Spinning Tests for Some Vari-eties of Upland Cotton Grown in the U.S.*

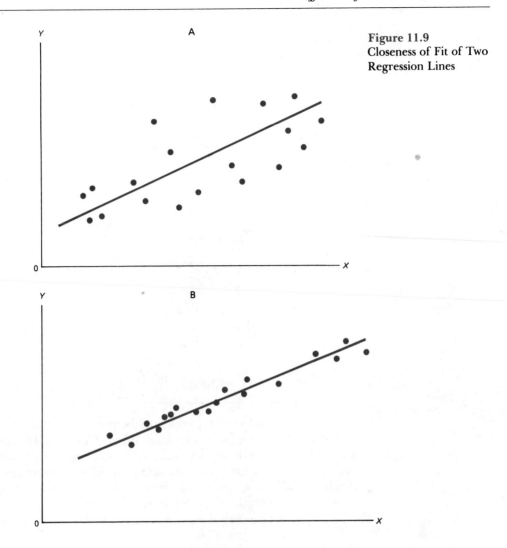

**Figure 11.9**
Closeness of Fit of Two
Regression Lines

to know how well this line fits the data. There can be vast differences in
how well a regression line fits a set of data, as shown in Figure 11.9.
Clearly, the regression line in panel B of Figure 11.9 provides a better
fit than the regression line in panel A of the same figure. How can we
measure how well a regression line fits the data?

As a first step toward answering this question, we must discuss the
concept of *variation,* which refers to a sum of squared deviations. The
total variation in the dependent variable $Y$ equals

$$\sum_{i=1}^{n} (Y_i - \bar{Y})^2. \tag{11.8}$$

In other words, the total variation equals the sum of the squared deviations of $Y$ from its mean. (In Chapter 10, this was often called the
total sum of squares.)

To measure how well a regression line fits the data, we divide the
total variation in the dependent variable into two parts: (1) the varia-

**Figure 11.10**
Division of $(Y_i - \overline{Y})$ into
Two Parts: $(Y_i - \hat{Y}_i)$ and
$(\hat{Y}_i - \overline{Y})$

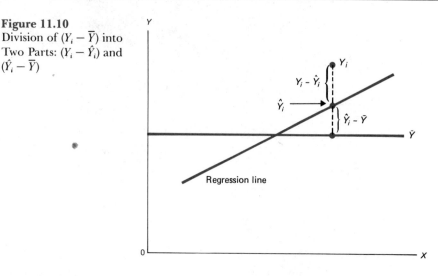

tion that *can* be explained by the regression line; and (2) the variation
that *cannot* be explained by the regression line. To divide the total variation in this way, we must note that for the *i*th observation,

$$(Y_i - \overline{Y}) = (Y_i - \hat{Y}_i) + (\hat{Y}_i - \overline{Y}), \tag{11.9}$$

where $\hat{Y}_i$ is the value of $Y_i$ that would be predicted on the basis of the
regression line. In other words, as shown in Figure 11.10, the discrepancy between $Y_i$ and the mean value of $Y$ can be split into two parts: the
discrepancy between $Y_i$ and the point on the regression line directly
below (or above) $Y_i$ and the discrepancy between the point on the regression line directly below (or above) $Y_i$ and $\overline{Y}$.

If we square both sides of equation (11.9) and sum the result over
all values of *i*, we find that[6]

*Total Variation =*
*Unexplained*
*Variation +*
*Explained Variation*

$$\sum_{i=1}^{n} (Y_i - \overline{Y})^2 = \sum_{i=1}^{n} (Y_i - \hat{Y}_i)^2 + \sum_{i=1}^{n} (\hat{Y}_i - \overline{Y})^2.$$

The term on the left-hand side of this equation shows the *total variation*
in the dependent variable. The first term on the right-hand side measures the *variation in the dependent variable that is not explained by the re-*
*gression.* This is a reasonable interpretation of this term since it is the
sum of squared deviations of the actual observations from the regres-

[6] To derive this result, note that

$$\sum_{i=1}^{n} (Y_i - \overline{Y})^2 = \sum_{i=1}^{n} [(Y_i - \hat{Y}_i) + (\hat{Y}_i - \overline{Y})]^2$$

$$= \sum_{i=1}^{n} (Y_i - \hat{Y}_i)^2 + \sum_{i=1}^{n} (\hat{Y}_i - \overline{Y})^2 + 2 \sum_{i=1}^{n} (Y_i - \hat{Y}_i)(\hat{Y}_i - \overline{Y}).$$

The last term on the right-hand side equals zero, so this result follows.

sion line. Clearly, the larger the value of this term, the poorer the regression equation fits the data.

The second term on the right-hand side of the equation measures the *variation in the dependent variable that is explained by the regression.* This is a reasonable interpretation of this term since it shows how much the dependent variable would be expected to vary on the basis of the regression alone. Putting it differently, this second term shows the reduction in unexplained variation due to the use of the regression instead of $\bar{Y}$ as an estimator of $Y$. Using $\bar{Y}$ as an estimator, the total variation is unexplained, whereas the first term on the right is unexplained when predictions based on the regression are used. Thus, the second term on the right shows the reduction in unexplained variation due to the use of predictions based on the regression instead of $\bar{Y}$.

To measure the closeness of fit of a regression line, statisticians use the **coefficient of determination:**

*Formula for $r^2$*

$$r^2 = 1 - \frac{\sum_{i=1}^{n} (Y_i - \hat{Y}_i)^2}{\sum_{i=1}^{n} (Y_i - \bar{Y})^2}.$$    (11.10)

In other words,

$$r^2 = 1 - \frac{\text{Variation not explained by regression}}{\text{Total variation}}$$

$$= \frac{\text{Variation explained by regression}}{\text{Total variation}}.$$    (11.11)

Clearly, the coefficient of determination is a reasonable measure of the closeness of fit of the regression line, since it equals *the proportion of the total variation in the dependent variable that is explained by the regression line.*[7]

In practical work, a more convenient formula for the coefficient of determination is

*Alternative Formula for $r^2$*

$$r^2 = \frac{\left[ n \sum_{i=1}^{n} X_i Y_i - \left( \sum_{i=1}^{n} X_i \right) \left( \sum_{i=1}^{n} Y_i \right) \right]^2}{\left[ n \sum_{i=1}^{n} X_i^2 - \left( \sum_{i=1}^{n} X_i \right)^2 \right] \left[ n \sum_{i=1}^{n} Y_i^2 - \left( \sum_{i=1}^{n} Y_i \right)^2 \right]}.$$    (11.12a)

To illustrate the computation of the coefficient of determination, Table 11.2 shows the various quantities needed in equation (11.12a) in the case of the hosiery mill's cost function. Substituting these quantities

[7] However, if random sampling is not used, the coefficient of determination can be difficult to interpret because its value may depend heavily on the set of values of the independent variable that is chosen. See Sanford Weisberg, *Applied Linear Regression* (New York: Wiley, 1980), pp.64–67.

into equation (11.12a), we have

$$r^2 = \frac{[9(319) - 50(49)]^2}{[9(340) - 50^2][9(303) - 49^2]} = \frac{421^2}{560(326)}$$

$$= 0.97.$$

Thus, the coefficient of determination between cost and output for the hosiery mill is 0.97. In other words, the regression line in Figure 11.5 can explain about 97 percent of the variation in cost.

Still another formula that is sometimes even more convenient is

$$r^2 = \frac{a \sum_{i=1}^{n} Y_i + b \sum_{i=1}^{n} X_i Y_i - \frac{1}{n}\left(\sum_{i=1}^{n} Y_i\right)^2}{\sum_{i=1}^{n} Y_i^2 - \frac{1}{n}\left(\sum_{i=1}^{n} Y_i\right)^2}.$$

*(11.12b)*

The advantage of this formula over equation (11.12a) is that if one has already calculated the regression line, the values of $a$ and $b$ are already available. If this formula is used in the case of the hosiery mill, the result is

$$r^2 = \frac{1.266(49) + .752(319) - 266.778}{36.222} = \frac{62.034 + 239.888 - 266.778}{36.222}$$

$$= .97,$$

which, of course, is the answer we obtained in the previous paragraph.

## 11.11 The Correlation Coefficient

As pointed out at the beginning of this chapter, the purpose of correlation analysis is to measure the strength of the relationship between two variables, $X$ and $Y$. The assumptions (or model) underlying correlation analysis are as follows: First, both $X$ and $Y$ are assumed to be normally distributed random variables. This is different from regression analysis where $Y$ is assumed to be a random variable but $X$ is not. Second, the standard deviation of the $Y$s is assumed to be constant for all values of $X$, and the standard deviation of the $X$s is assumed to be constant for all values of $Y$.

*Basic Idea #22*    *If two variables are statistically independent, the correlation coefficient between them equals zero. If there is a perfect linear relationship between them, the correlation coefficient will be +1 or −1 depending on whether the relationship is direct or inverse.*

The correlation coefficient is commonly used as a measure of the strength of the relationship between two variables. The ***correlation coefficient*** *r* is simply the square root of the coefficient of determination. That is,

$$r = \sqrt{r^2}.$$

The sign of *r* must equal the sign of the slope of the regression line. Thus, the positive square root of $r^2$ is taken if $b > 0$, and the negative square root is taken if $b < 0$. For ease of computation, the following formula is often used:[8]

*Formula for r*

$$r = \frac{n \sum_{i=1}^{n} X_i Y_i - \sum_{i=1}^{n} X_i \sum_{i=1}^{n} Y_i}{\sqrt{n \sum_{i=1}^{n} X_i^2 - \left(\sum_{i=1}^{n} X_i\right)^2} \sqrt{n \sum_{i=1}^{n} Y_i^2 - \left(\sum_{i=1}^{n} Y_i\right)^2}}$$

(11.13)

The correlation coefficient cannot be greater than 1 or less than −1. *If* r = 1, *there is a perfect linear relationship between the independent and dependent variables, and the relationship is direct.* In other words, the situation is like that shown in panel A of Figure 11.11. On the other hand, *if* r = −1, *there is a perfect linear relationship between the independent and dependent variables, and the relationship is inverse.* The situation is like that shown in panel B of Figure 11.11. In each case the relationship is perfect in the sense that the regression explains all the variation in the dependent variable (since all the points fall on the regression line). Why does the correlation coefficient equal either +1 or −1 if all the variation in *Y* is explained by the regression line? Because under these circumstances, the actual value of *Y* must always be equal to the value computed from the regression, which means that $\Sigma(Y_i - \hat{Y}_i)^2 = 0$. In other words, none of the variation in *Y* is unexplained by the regression. Hence, it follows from equation (11.11) that $r^2$ must equal 1, which means that *r* must equal +1 or −1.

*If* r = 0, *there is zero correlation between the independent and dependent variables.* In this case, the least-squares estimate of *B* will turn out to be zero, indicating that on the average, changes in the independent variable have no effect on the dependent variable. Under such cir-

---

[8] The correlation coefficient in the text is the correlation coefficient *unadjusted* for degrees of freedom, and is a biased estimate of the population correlation coefficient. (It is biased away from zero.) An unbiased estimate is the *adjusted* correlation coefficient, which is

$$r = \sqrt{1 - \frac{\sum_{i=1}^{n} (Y_i - \hat{Y}_i)^2 \div (n - 2)}{\sum_{i=1}^{n} (Y_i - \overline{Y})^2 \div (n - 1)}},$$

where *n* is the number of observations. (*Adjusted* here means adjusted for degrees of freedom.)

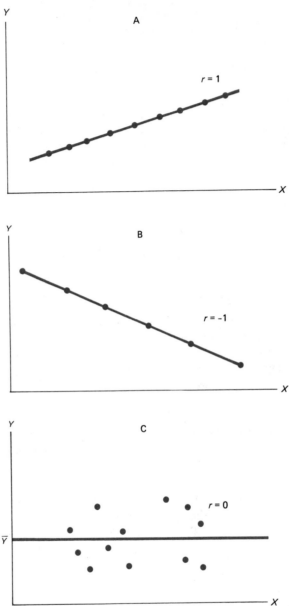

**Figure 11.11**
Cases where $r = 1, -1,$ and 0

cumstances, the correlation coefficient is zero because the regression explains none of the variation in the dependent variable. In other words, $\Sigma(\hat{Y}_i - \bar{Y})^2 = 0$ since the mean value of $Y$ is always equal to the value computed from the regression. Also, $\Sigma(Y_i - \hat{Y}_i)^2 = \Sigma(Y_i - \bar{Y})^2$ since $\hat{Y}_i$ always equals $\bar{Y}$. Hence, it follows from equation (11.10) that $r^2$ must equal 0, which means that $r$ must equal 0. In a situation of this sort, the best estimate of the dependent variable is $\bar{Y}$; the value of the independent variable provides no additional useful information on this score. The situation is like that shown in panel c of Figure 11.11.

The calculation and interpretation of the sample correlation coefficient are illustrated in the following example.

EXAMPLE 11.5 An industrial psychologist obtains the IQ score and productivity of 10 workers, the results being

| IQ score | Productivity (output per hour) |
|---|---|
| 110 | 5.2 |
| 120 | 6.0 |
| 130 | 6.3 |
| 126 | 5.7 |
| 122 | 4.8 |
| 121 | 4.2 |
| 103 | 3.0 |
| 98 | 2.9 |
| 80 | 2.7 |
| 97 | 3.2 |

Compute the correlation coefficient between IQ score and productivity. Is the relationship direct or inverse?

SOLUTION: Letting the productivity of the $i$th worker equal $Y_i$ and the IQ score of the $i$th worker equal $X_i$,

$$\sum_{i=1}^{10} X_i Y_i = 5042.6 \quad \sum_{i=1}^{10} X_i = 1107 \quad \sum_{i=1}^{10} Y_i = 44.0$$

$$\sum_{i=1}^{10} Y_i^2 = 210.84 \quad \sum_{i=1}^{10} X_i^2 = 124{,}823 \quad n = 10.$$

Thus, from equation (11.13), it follows that

$$r = \frac{10(5{,}042.6) - (1107)(44)}{\sqrt{10(124{,}823) - (1107)^2} \ \sqrt{10(210.84) - (44)^2}}$$

$$= \frac{50{,}426 - 48{,}708}{\sqrt{1{,}248{,}230 - 1{,}225{,}449} \ \sqrt{2108.4 - 1936}} = \frac{1718}{\sqrt{22{,}781} \ \sqrt{172.4}}$$

$$= \frac{1718}{(150.93)(13.13)} = \frac{1718}{1982} = .867.$$

The value of the correlation coefficient is about .87. Since this value is positive, the relationship seems to be direct.

THE CASE OF COPPER AND LEAD

An economist is interested in whether a country's copper output is related to its lead output. He is particularly interested in three countries, A, B, and C. To normalize for differences among the sizes of the countries, he uses copper output *per capita* as one variable and lead output per capita as the other. The scatter diagram, shown below, indicates that there is a strong positive correlation between these variables.

TO ERR IS
HUMAN ... BUT
COSTLY

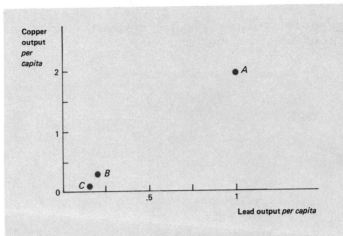

He shows this scatter diagram to his boss, who says that it is extremely misleading. What was his mistake?

SOLUTION: A spurious correlation can be created by dividing both the independent and dependent variables by the same quantity. In this case, suppose that the populations of countries A, B, and C are 1 million, 10 million, and 20 million, respectively. Then this procedure (of using copper output per capita and lead output per capita) results in a high correlation, as shown above. But in fact there is no relationship whatever between a country's copper output and its lead output, as shown in the graph below. The moral is that one should be very careful to avoid creating a spurious correlation by dividing both the independent and dependent variables by the same quantity.

In addition, one must be very careful about drawing conclusions from such a small sample. Even if the sample correlation coefficient is quite high, it may be due to chance. (For a relevant test procedure, see Section 11.12.)

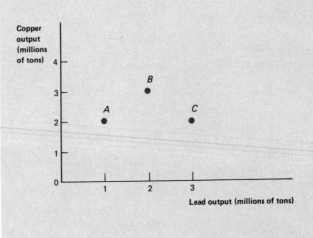

# 11.12 Inference Concerning the Population Correlation Coefficient[9]

The correlation coefficient described in the previous section is a *sample* correlation coefficient, and it varies from one sample to another. In contrast, the **population correlation coefficient,** $\rho$ (rho), pertains to the entire population. The definition of $\rho$ is the same as that of $r$, the only difference being that $r$ is based on the sample data whereas $\rho$ is based on all the data in the population. In most cases, statisticians are much more interested in the correlation coefficient in the population than in the sample, and they use $r$ as an estimate of $\rho$.

Frequently, statisticians are interested in testing whether the population correlation coefficient is zero. *If one variable is independent of another variable, the population correlation coefficient equals zero.* For example, if an employee's productivity on a particular job is completely unrelated to his or her age, the correlation coefficient in the population between a person's productivity and age will be zero. The personnel director of a firm may want to determine whether productivity and age really are uncorrelated for a particular job. Having measured productivity and age for a sample of workers, the director may want to test the null hypothesis that $\rho = 0$.

To carry out such a test, the first step is to specify the null hypothesis and the alternative hypothesis. The null hypothesis is that $\rho = 0$, and the alternative hypothesis is that $\rho \neq 0$. To test the null hypothesis, we compute

*Test that $\rho = 0$*

$$t = \frac{r}{\sqrt{(1 - r^2)/(n - 2)}},$$

(11.14)

which has the $t$ distribution with $(n - 2)$ degrees of freedom if the null hypothesis is true. Thus, the decision rule used to test this null hypothesis is: *Reject the null hypothesis that $\rho = 0$ if* t *is greater than* $t_{\alpha/2}$ *or less than* $-t_{\alpha/2}$.

To illustrate the way in which this test is carried out, suppose that in a sample of 18 workers the sample correlation coefficient between productivity and age is 0.42. To test the null hypothesis that the population correlation coefficient equals zero, we compute

$$t = \frac{0.42}{\sqrt{(1 - .42^2)/16}} = \frac{0.42}{\sqrt{.2059}} = 0.93.$$

---

[9] Some instructors may prefer to take up Chapter 11 before Chapter 8. Chapter 11 has been written so this can be done; but Sections 11.12 and 11.13 should be taken up after Chapter 8 has been covered.

**GETTING DOWN TO CASES**

PIG IRON AND LIME CONSUMPTION IN THE PRODUCTION OF STEEL

A number of years ago, the well-known British statistician L.H.C. Tippett conducted a study of the relationship between the percentage of pig iron in a cast of steel and the lime consumption per cast.[10] The basic data used were the following:

| Percentage of pig iron | Lime consumption (hundred weights) | Percentage of pig iron | Lime consumption (hundred weights) | Percentage of pig iron | Lime consumption (hundred weights) |
|---|---|---|---|---|---|
| 23 | 164 | 37 | 140 | 45 | 187 |
| 25 | 141 | 37 | 170 | 45 | 194 |
| 26 | 140 | 37 | 176 | 45 | 216 |
| 29 | 156 | 37 | 182 | 45 | 219 |
| 30 | 165 | 37 | 191 | 45 | 219 |
| 30 | 177 | 37 | 194 | 46 | 205 |
| 30 | 178 | 37 | 198 | 46 | 235 |
| 30 | 182 | 37 | 216 | 47 | 193 |
| 30 | 184 | 38 | 145 | 47 | 197 |
| 31 | 172 | 38 | 157 | 47 | 206 |
| 32 | 159 | 38 | 164 | 47 | 218 |
| 32 | 185 | 38 | 175 | 47 | 218 |
| 33 | 138 | 38 | 225 | 47 | 220 |
| 33 | 155 | 38 | 281 | 47 | 274 |
| 33 | 170 | 39 | 190 | 47 | 310 |
| 33 | 192 | 39 | 201 | 48 | 170 |
| 33 | 228 | 40 | 138 | 48 | 205 |
| 34 | 161 | 40 | 200 | 48 | 241 |
| 35 | 133 | 40 | 223 | 48 | 242 |
| 35 | 146 | 40 | 241 | 49 | 193 |
| 35 | 156 | 41 | 212 | 49 | 204 |
| 35 | 165 | 42 | 166 | 49 | 206 |
| 35 | 176 | 42 | 182 | 50 | 158 |
| 35 | 193 | 42 | 194 | 50 | 195 |
| 35 | 194 | 42 | 213 | 50 | 196 |
| 36 | 124 | 42 | 246 | 50 | 198 |
| 36 | 132 | 43 | 207 | 52 | 208 |
| 36 | 146 | 43 | 210 | 52 | 219 |
| 36 | 174 | 43 | 212 | 52 | 262 |
| 36 | 180 | 44 | 176 | 53 | 170 |
| 36 | 195 | 44 | 215 | 53 | 188 |
| 36 | 201 | 45 | 174 | 53 | 193 |
| 37 | 126 | 45 | 184 | 53 | 219 |
|  |  |  |  | 53 | 240 |

[10] See L.H.C. Tippett, *op. cit.*

> (a) Plot the above data in a scatter diagram.
> (b) Calculate the least-squares regression of lime consumption on percentage of pig iron. (That is, let lime consumption be the dependent variable and percentage of pig iron be the independent variable.)
> (c) Compute the correlation coefficient between lime consumption and percentage of pig iron.
> (d) Interpret the results of (a), (b), and (c) above.

If the significance level of the test is set at .05, the null hypothesis should be rejected if $t > 2.12$ or if $t < -2.12$, since Appendix Table 6 shows that $t_{.025} = 2.12$ if there are 16 degrees of freedom. Because the observed value of $t$ does not exceed 2.12 or fall below $-2.12$, the null hypothesis should not be rejected.

## 11.13 Inference Concerning the Value of *B*

In regression analysis, the slope of the sample regression line $b$ varies from one sample to another. Like any sample statistic, it has a sampling distribution; and an estimate of the standard deviation of this sampling distribution is

$$s_b = s_e \div \sqrt{\sum_{i=1}^{n} X_i^2 - n\overline{X}^2},$$

which is often called the *standard error of* b. There are many occasions when the statistician wants to use the observed value of $b$ to calculate a confidence interval for $B$, the slope of the population regression line. Such a confidence interval is

$$b \pm t_{\alpha/2}s_b. \qquad (11.15)$$

**Confidence Interval for B**

*If the conditional probability distribution of the dependent variable is normal, the probability is* $(1 - \alpha)$ *that these limits will include the true value of* B. To illustrate the use of this formula, we can calculate the 95 percent confidence interval for $B$ in the case of the hosiery mill. Since $b = .752$, $t_{.025} = 2.365$, $s_e = .392$, and $\sum X_i^2 - n\overline{X}^2 = 62.2$, it follows that the confidence interval is

$.752 \pm 2.365(.392) \div \sqrt{62.2},$

or

$.752 \pm .118.$

Thus, the 95 percent confidence interval for the cost of producing an extra ton of output is .634 to .870 thousands of dollars (that is, $634 to $870).

In addition to estimating the value of $B$, the statistician frequently wants to test the hypothesis that $B$ equals zero. If this hypothesis is true, the mean of the dependent variable is the same, regardless of the value of the independent variable. (Specifically, the mean equals $A$, since $BX = 0$.) Thus, a knowledge of the independent variable is of no use in predicting the dependent variable since the conditional probability distribution of the dependent variable is not influenced by the value of the independent variable. In other words, if this hypothesis is true (and if the assumptions given in Section 11.4 hold), there is *no relationship* between the dependent and the independent variable.

**Test that B = 0**

If $B$ equals zero, it does not follow that $b$ must equal zero. On the contrary, it is quite likely that $b$ will be non-zero because of random fluctuations. The decision rules for testing the null hypothesis that $B = 0$ are given below. These rules assume that the conditional probability distribution of the dependent variable is normal, and which rule is appropriate depends on the nature of the alternative hypothesis.

ALTERNATIVE HYPOTHESIS: B $>$ 0. *Reject the null hypothesis if* b $\div$ s$_b$ $>$ t$_\alpha$. *(The number of degrees of freedom is* n $-$ 2, *and* $\alpha$ *is the significance level.)*

ALTERNATIVE HYPOTHESIS: B $<$ 0. *Reject the null hypothesis if* b $\div$ s$_b$ $<$ $-$t$_\alpha$.

ALTERNATIVE HYPOTHESIS: B $\neq$ 0. *Reject the null hypothesis if* b $\div$ s$_b$ $>$ t$_{\alpha/2}$ *or* $<$ $-$t$_{\alpha/2}$.

As an illustration of how this test is carried out, suppose that the hosiery mill wants to test whether the slope of the population regression line relating cost to output equals zero (against the alternative hypothesis that $B \neq 0$.) The significance level is .05. From previous sections, we know that in this case $b = 0.752$, $s_e = 0.392$, $\Sigma X_i^2 - n\bar{X}^2 = 62.2$, and $n = 9$. Thus,

$$b \div s_b = \frac{.752}{.392 \div \sqrt{62.2}} = 15.1.$$

Since $n = 9$, there are 7 degrees of freedom, and $t_{.025} = 2.365$. Since the value of the test statistic (15.1) exceeds 2.365, the null hypothesis should be rejected.

Below is a further illustration of how one can test the hypothesis that $B = 0$.

EXAMPLE 11.6 Based on the data in Example 11.1, test the hypothesis that the slope of the true regression line relating savings to income equals zero. The alternative hypothesis is that this slope exceeds zero. Use the .05 significance level.

SOLUTION: From the results in Examples 11.1 and 11.2, we know that $b = .1017$, $s_e = .053$, $\Sigma X_i^2 - n\overline{X}^2 = 60$, and $n = 9$. Thus,

$$b \div s_b = \frac{.1017}{.053 \div \sqrt{60}} = 15.$$

Since $t_{.05} = 1.895$ when there are 7 degrees of freedom, the null hypothesis should be rejected. That is, the evidence seems to indicate that the slope of the true regression is positive, not zero.

## 11.14  Statistical Cost Functions: A Case Study

In preceding sections we have used some hypothetical data concerning the costs and output of a hosiery mill to illustrate regression and correlation techniques. Since these data are hypothetical, they may give the impression that the procedures employed are not of much practical use in estimating and analyzing cost functions. Nothing could be further from the truth. One of the classic studies carried out in industrial economics used precisely these techniques to estimate the relationship between cost and output for a hosiery mill. This study,[11] carried out by Joel Dean of the Columbia University Graduate School of Business, obtained the following regression line:

$$\hat{Y} = 2936 + 1.998X,$$

where $\hat{Y}$ is the computed monthly cost of production in dollars and $X$ is the monthly output of the mill in dozens of pairs of stockings. The scatter diagram and the above regression line are shown in Figure 11.12. The standard error of estimate was $6110. The correlation coefficient was 0.973.

This study is famous for several reasons, one being that it provided some early evidence concerning the shape of firms' cost functions. In particular, the study suggested that within the relevant range marginal cost (that is, the extra cost of an extra unit of output) did not vary with the output level of the firm. However, this may have been due in considerable part to the limited range of the observations. In other words, if data were obtained concerning much higher output levels than those shown in Figure 11.12, it is likely that costs would no longer conform to the linear relationship, due to capacity constraints. Since Dean's pioneer study, a large number of similar investigations have been made for other firms in a variety of industries. The results, based on the types of regression and correlation techniques described in this chapter, have proved valuable both to the firms themselves and to economists and government agencies.

---

[11] J. Dean, "Statistical Cost Functions of a Hosiery Mill," *Journal of Business*, 1941. Reprinted in part in E. Mansfield, *Statistics for Business and Economics: Readings and Cases* (New York: Norton, 1980).

**Figure 11.12**
Actual Regression of
Cost on Output,
Hosiery Mill

*Source:* J. Dean, "Statistical
Cost Functions of a
Hosiery Mill," *Journal of
Business,* 1941.

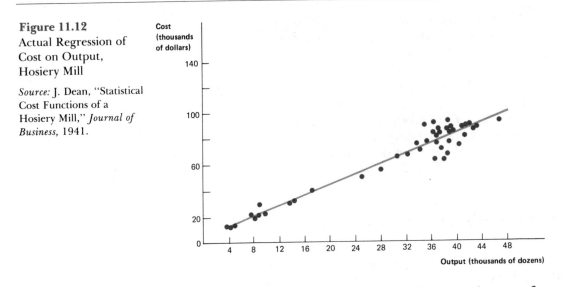

## 11.15 Hazards and Problems in Regression and Correlation

There are a number of pitfalls in regression and correlation analyses that should be emphasized. First, *it is by no means true that a high coefficient of determination (or correlation coefficient) between two variables means that one variable causes the other variable to vary.* For example, if one regresses the size of a person's left foot on the size of his or her right foot, this regression is bound to fit very well, since the size of a person's left foot is closely correlated with the size of his or her right foot. But this does not mean that the size of a person's right foot *causes* a person's left foot to be as large or small as it is. Two variables can be highly correlated without causation being implied.

Second, *even if an observed correlation is due to a causal relationship, the direction of causation may be the reverse of that implied by the regression.* For example, suppose that we regress a firm's profits on its R and D (research and development) expenditure, the firm's profits being the dependent variable and its R and D expenditures being the independent variable. If the correlation between these two variables turns out to be high, does this imply that high R and D expenditures produce high profits? Obviously not. The line of causation could run the other way: High profits could result in high R and D expenditures. Thus, in interpreting the results of regression and correlation studies, it is important to ask oneself whether the line of causation assumed in the studies is correct.

Third, *regressions are sometimes used to forecast values of the dependent variable corresponding to values of the independent variable lying beyond the sample range.* For example, in Figure 11.13, the scatter diagram shows that the data for the independent variable range from about 1 to 7. But the regression may be used to forecast the dependent variable when the independent variable assumes a value of 9, which is outside the sample range. This procedure, known as **extrapolation,** is dangerous because the available data provide no evidence that the true

*Hazards of
Extrapolation*

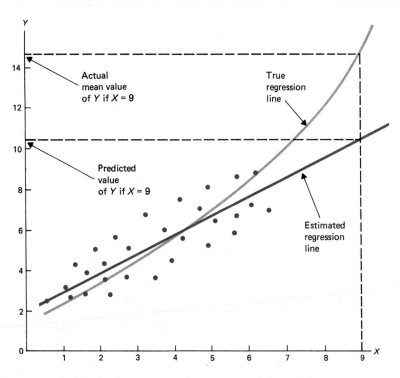

**Figure 11.13**
Dangers of
Extrapolation

*Note:* If the curved line is the true (that is, the population) regression line, and if the estimated regression line is extrapolated to forecast the value of Y when X = 9, the result will be very inaccurate. Whereas the forecast is that Y will equal about 10.5, the mean value of Y(if X = 9) is really about 14.5

regression is linear beyond the range of the sample data. For example, the true regression may be as shown in Figure 11.13, in which case a forecast based on the estimated regression may be very poor.

Fourth, *it is important to recognize that a regression based on past data may not be a good predictor, due to shifts in the regression line.* For example, suppose that the hosiery mill experiences considerable increases over time in wage rates and materials prices. If so, the regression line relating cost to output is likely to shift upward and to the left; and predictions based on historical data are likely to underestimate future costs. On the other hand, suppose that while input prices remain constant, the hosiery mill experiences considerable productivity growth due to new technology. If this happens, the regression line relating cost to output is likely to shift downward and to the right. Predictions based on historical data are likely to overestimate future costs in a situation of this sort.

Finally, *when carrying out a regression, it is important to try to make sure that the assumptions in Section 11.4 are met.* Statisticians often plot the regression line on the scatter diagram and look for evidence of departures from the assumptions. For example, the scatter diagram may indicate that the relationship between the variables is curvilinear, not linear. (See panel D of Figure 11.2.) If this is so, logarithmic regression techniques (described in Appendix 11.1) or multiple regression (described in the next chapter) may be more appropriate than the simple linear regression techniques discussed here. There are a considerable number of tests that can be carried out to determine whether the assumptions underlying the use of regression techniques are met. Some of the most commonly used tests are presented in the next chapter.

EXERCISES

**11.12** For a random sample of 10 students, the relationship between their college grade-point average (Y) and their high school grade-point average (X) is as follows:

| Y | X |
|---|---|
| 2.5 | 3.1 |
| 2.0 | 2.7 |
| 3.1 | 3.6 |
| 3.4 | 3.7 |
| 3.9 | 4.0 |
| 2.1 | 3.0 |
| 2.8 | 3.3 |
| 2.9 | 3.5 |
| 3.0 | 3.6 |
| 3.8 | 4.0 |

(a) Calculate the correlation coefficient.
(b) Test whether the population correlation coefficient is zero. (Let $\alpha = .01$.)

**11.13** According to data published in *Industrial Quality Control,* the relationship between detergency (Y) and concentration (X) was as follows for a sample of eight detergents:

| Y | X |
|---|---|
| 37 | 10 |
| 42 | 20 |
| 46 | 30 |
| 48 | 40 |
| 53 | 10 |
| 62 | 10 |
| 79 | 30 |
| 84 | 40 |

(a) Calculate the correlation coefficient.
(b) Test whether the population correlation coefficient is zero. (Let $\alpha = .05$.)

**11.14** (a) Based on the data in Exercise 11.2, calculate the coefficient of determination. Interpret your result.
(b) Test whether the slope of the true regression line in Exercise 11.2 is zero. (Let $\alpha = .05$, and use a one-tailed test where the alternative hypothesis is that the slope is positive.)

**11.15** (a) Based on the data in Exercise 11.1, calculate the coefficient of determination. Interpret your result.
(b) Test whether the slope of the true regression line in Exercise 11.1 is zero. (Let $\alpha = .05$, and use a two-tailed test.)

**11.16** (a) Based on the data in Exercise 11.7, calculate the standard error of the slope of the sample regression line.
(b) Calculate a 95 percent confidence interval for the slope of the true regression line in Exercise 11.7.

**11.17** (a) Based on the data in Exercise 11.6, calculate the standard error of the slope of the sample regression line.

(b) Calculate a 90 percent confidence interval for the slope of the true regression line in Exercise 11.6.

**11.18** Pamela Jackson of Rhode Island College and Leo Carroll of the University of Rhode Island carried out a study of the determinants of municipal police expenditures in 90 non-Southern cities.[12] Among other things, they found that the coefficient of correlation between a city's 1971 police expenditures (per capita) and its 1968–70 crime rate was 0.54. Test whether the population correlation coefficient equals zero. (Let $\alpha = .05$.)

**11.19** (a) Using the data in Exercise 11.11, calculate the sample correlation coefficient between fiber length and yarn strength.

(b) What proportion of the variation in yarn strength in the sample can be explained by fiber length?

**11.20** If the sample correlation coefficient is .30 and $n$ equals 20, test whether the population correlation coefficient is zero. (Let $\alpha = .05$, and use a two-tailed test.)

**11.21** Using the data in Exercise 11.11, test the hypothesis that the slope of the true regression line relating yarn strength (the dependent variable) and fiber length (the independent variable) equals zero. The alternative hypothesis is that the slope exceeds zero. (Use the .01 significance level.) Can we be sure that variation in fiber length *causes* variation in yarn strength? Why, or why not?

**11.22** If two variables $X$ and $Y$ are statistically independent, their correlation coefficient is zero. But is it true that if the population correlation coefficient between $X$ and $Y$ is zero they must be statistically independent? Explain.

**11.23** If the standard deviation of the conditional probability distribution of the dependent variable varies with the value of the independent variable, *heteroscedasticity* is said to occur.

(a) Can you think of any reasons why heteroscedasticity might occur in the case of the relationship between savings and income in Example 11.1?

(b) What factors might cause heteroscedasticity in the case of the relationship between a firm's costs and its level of output?

# Chapter Review

1. *Regression analysis* indicates how one variable is related to another. A first step in describing such a relationship is to plot a scatter diagram. The *regression line* shows the average relationship between the dependent variable and the independent variable. The method of least squares is the standard technique used to fit a regression line to a set of data. If the regression line is $\hat{Y}$

[12] P. Jackson and L. Carroll, "Race and the War on Crime: The Socio-political Determinants of Municipal Police Expenditures in 90 Non-Southern U.S. Cities," *American Sociological Review*, June 1981.

$= a + bX$, and if $a$ and $b$ are calculated by least squares,

$$b = \frac{\sum_{i=1}^{n} (X_i - \overline{X})(Y_i - \overline{Y})}{\sum_{i=1}^{n} (X_i - \overline{X})^2}$$

$$a = \overline{Y} - b\overline{X}.$$

This value of $b$ is often called the *estimated regression coefficient.*

2. The regression line calculated by the method of least squares is generally based on a sample, not on the entire population. Suppose that the means of the conditional probability distributions of the dependent variable fall on a straight line: $\mu_{Y \cdot X} = A + BX$. Also, suppose that the standard deviation of these conditional probability distributions is the same for all values of $X$, that the observations in the sample are statistically independent, and that the values of $X$ are known with certainty. Then $a$ is an unbiased and consistent estimator of $A$, and is the most efficient estimator of $A$ (among those unbiased estimators that are linear functions of the dependent variable). Similarly, $b$ is an unbiased and consistent estimator of $B$, and is the most efficient estimator of $B$ (among those unbiased estimators that are linear functions of the dependent variable).

3. The *standard deviation of the conditional probability distribution of the dependent variable is* $\sigma_e$. It is a measure of the amount of scatter about the regression line in the population. The sample statistic which is used to estimate $\sigma_e$ is the standard error of estimate, defined as

$$s_e = \sqrt{\frac{\sum_{i=1}^{n} (Y_i - \hat{Y}_i)^2}{n - 2}},$$

where $\hat{Y}_i$ is the estimate of $Y_i$ based on the regression line.

4. If we can assume that the conditional probability distribution of the dependent variable is normal, we can calculate a confidence interval for the conditional mean of the dependent variable when the independent variable equals $X^*$. This confidence interval is

$$(a + bX^*) \pm t_{\alpha/2} s_e \sqrt{\frac{1}{n} + \frac{(X^* - \overline{X})^2}{\sum_{i=1}^{n} X_i^2 - n\overline{X}^2}}$$

if the confidence coefficient is $(1 - \alpha)$. Also, we can calculate a confidence interval for the value of the dependent variable that will occur if the independent variable is set at $X^*$. This confidence interval is

$$(a + bX^*) \pm t_{\alpha/2} s_e \sqrt{\frac{n + 1}{n} + \frac{(X^* - \overline{X})^2}{\sum_{i=1}^{n} X_i^2 - n\overline{X}^2}}$$

if the confidence coefficient is $(1 - \alpha)$. The number of degrees of freedom is $n - 2$.

5. Assuming that the conditional probability distribution of the dependent variable is normal, a confidence interval for the slope of the population regression line $B$ is

$$b \pm t_{\alpha/2}s_b,$$

where

$$s_b = s_e \div \sqrt{\sum_{i=1}^{n} X_i^2 - n\overline{X}^2}$$

and where $(1 - \alpha)$ is the confidence coefficient. To test the null hypothesis that $B = 0$ (against the alternative hypothesis that $B \neq 0$), reject the null hypothesis if $b \div s_b > t_{\alpha/2}$ or $< -t_{\alpha/2}$. The number of degrees of freedom is $(n - 2)$, and $\alpha$ is the significance level.

6. To measure the closeness of fit of a regression line, statisticians often use the *coefficient of determination,* defined as

$$r^2 = \frac{\left[ n\sum_{i=1}^{n} X_i Y_i - \left( \sum_{i=1}^{n} X_i \right)\left( \sum_{i=1}^{n} Y_i \right) \right]^2}{\left[ n\sum_{i=1}^{n} X_i^2 - \left( \sum_{i=1}^{n} X_i \right)^2 \right]\left[ n\sum_{i=1}^{n} Y_i^2 - \left( \sum_{i=1}^{n} Y_i \right)^2 \right]}.$$

The *coefficient of determination* equals the proportion of the total variation in the dependent variable that is explained by the regression line.

7. *Correlation analysis* is concerned with measuring the strength of the relationship between two variables. The correlation coefficient, which is the square root of the coefficient of determination, is often used for this purpose. If $r = 1$, there is a perfect linear relationship between the two variables, and the relationship is direct. If $r = -1$, there is a perfect linear relationship between the two variables, and the relationship is inverse. If two variables are statistically independent, $r = 0$. To test whether the population correlation coefficient $\rho$ equals zero, one can calculate

$$t = \frac{r}{\sqrt{\dfrac{(1 - r^2)}{(n - 2)}}}$$

The null hypothesis that $\rho = 0$ should be rejected if this test statistic exceeds $t_{\alpha/2}$ or is less than $-t_{\alpha/2}$, where $\alpha$ is the significance level and the number of degrees of freedom equals $n - 2$. The alternative hypothesis is that $\rho \neq 0$.

8. There are many pitfalls in regression and correlation analysis. A high correlation between two variables does not necessarily mean that the variables are causally related. And even if they are causally related, the direction of causation may be different from that presumed in the analysis. It is extremely dangerous to extrapolate a regression line beyond the range of the data. One should be careful to avoid causing spurious correlation by dividing both the independent and the dependent variable by the same quantity. A regression based on past data may not be a good predictor, due to shifts in the regression line. It is important to try to determine whether the assumptions underlying regression analysis are met.

## Chapter Review Exercises

**11.24** A firm examines a random sample of 10 spot welds of steel. In each case, the shear strength of the weld and the diameter of the weld are determined, the results being as follows:[13]

| Shear strength (pounds) | Weld diameter (thousandths of a inch) |
|---|---|
| 680 | 190 |
| 800 | 200 |
| 780 | 209 |
| 885 | 215 |
| 975 | 215 |
| 1,025 | 215 |
| 1,100 | 230 |
| 1,030 | 250 |
| 1,175 | 265 |
| 1,300 | 250 |

(a) Construct a scatter diagram of these data.

(b) Based on this scatter diagram, does the relationship between these two variables seem to be direct or inverse? Does this accord with common sense? Why, or why not? Does the relationship seem to be linear?

(c) Assume that the conditional mean value of a weld's shear strength is a linear function of its diameter. Calculate the least-squares estimates of the parameters ($A$ and $B$) of this linear function.

(d) Plot the regression line. Use this regression line to predict the average shear strength of a weld $1/5$ inch in diameter. Use the regression line to predict the average shear strength of a weld $1/4$ inch in diameter.

(e) Compute the standard error of estimate. What does this number mean?

(f) Compute the 95 percent confidence interval for the mean shear strength among welds with a diameter of $1/4$ inch.

(g) Compute the 95 percent confidence interval for the shear strength of a weld if its diameter is $1/4$ inch. What assumptions underlie the calculation of this confidence interval?

**11.25** The 1980 sales and profits of seven steel companies were as follows:

| | (billions of dollars) | |
| Firm | Sales | Profit |
|---|---|---|
| Armco | 5.7 | 0.27 |
| Bethlehem | 6.7 | 0.12 |
| Bundy | 0.2 | 0.00 |
| Carpenter | 0.6 | 0.04 |
| Republic | 3.8 | 0.05 |
| U.S. Steel (now USX) | 12.5 | 0.46 |
| Westran | 0.5 | 0.00 |

[13] A. Duncan, *op. cit*, p. 651.

(a) Construct a scatter diagram of these data.

(b) Compute the sample regression line, where profit is the dependent variable and sales is the independent variable.

(c) Estimate the 1980 average profit of a steel firm with sales of $2 billion then.

(d) Can this regression line be used to predict a steel firm's profit in 1990? Explain.

**11.26** Based on the data in Example 11.5, test whether the population correlation coefficient between IQ score and productivity is zero. (Let $\alpha = .05$, and use a two-tailed test.)

**11.27** On the basis of the data in Table 11.1 and the regression equation derived therefrom, the hosiery mill estimates that if output is 10 tons per month, costs should be $8,786 per month. Comment on this estimate.

**11.28** If a regression line cannot explain 36 percent of the variation in the dependent variable, what is the correlation coefficient?

**11.29** (a) Using the data in Exercise 11.24, calculate the sample correlation coefficient between weld strength and weld diameter.

(b) What proportion of the variation in weld strength in the sample can be explained by weld diameter?

**11.30** Using the data in Exercise 11.24, test the hypothesis that the slope of the true regression line relating a weld's strength (the dependent variable) to its diameter (the independent variable) equals zero. The alternative hypothesis is that the slope is greater or less than zero. (Use the .05 significance level.) Can we be sure that variation in a weld's shear strength does *not* cause variation in its diameter? Why, or why not?

**11.31** For the following 10 countries, the number of telephones and television receivers are as follows:

| Country | Telephones (millions) | Televisions (millions) |
|---|---|---|
| Argentina | 2.6 | 5.1 |
| Belgium | 3.6 | 3.5 |
| Canada | 16.5 | 11.3 |
| France | 24.7 | 19.0 |
| Italy | 19.3 | 22.0 |
| Japan | 53.6 | 63.0 |
| Spain | 11.9 | 9.4 |
| Turkey | 1.9 | 3.3 |
| United Kingdom | 26.7 | 22.6 |
| West Germany | 28.6 | 20.8 |

*Source: Statistical Abstract of the U.S., 1982–83.*

Use Minitab (or some other computer package) to construct the scatter diagram. If no computer package is available, do it by hand.

**\*11.32** Based on the data in Exercise 11.31, Minitab is used to calculate the regression of the number of televisions in a country (in millions) on the number of telephones in the country (in millions). Interpret the results, shown below. (The dependent variable has been put in column thirty-five; the independent variable has been put in column thirty-four. See the answer to Exercise 11.31.)

* This exercise assumes some familiarity with Minitab or with Appendix 11.3.

```
MTB > read into c34,c35
DATA > 2.6 5.1
DATA > 3.6 3.5
DATA > 16.5 11.3
DATA > 24.7 19
DATA > 28.6 20.8
DATA > 19.3 22
DATA > 53.6 63
DATA > 11.9 9.4
DATA > 1.9 3.3
DATA > 26.7 22.6
DATA > end
      10 ROWS READ
MTB > regress c35 on 1 predictor in c34

The regression equation is
C35 = - 2.24 + 1.07 C34

Predictor        Coef        Stdev       t-ratio
Constant        -2.243       2.724        -0.82
C34              1.0688      0.1129        9.47
s = 5.337       R-sq = 91.8%        R-sq(adj) = 90.8%


Analysis of Variance
SOURCE              DF              SS            MS
Regression           1           2554.7        2554.7
Error                8            227.9          28.5
Total                9           2782.6


Unusual Observations

Obs.   C34    C35     Fit  Stdev.Fit  Residual  St.Resid
  7   53.6  63.00   55.05      4.26      7.95     2.47RX


R denotes an obs. with a large st. resid.
X denotes an obs. whose X value gives it large
influence.
```

# Appendix 11.1

### LOGARITHMIC REGRESSION

In this chapter we have dealt only with the case where the regression of $Y$ on $X$ is linear. That is, we have assumed that $\mu_{Y \cdot x}$, the conditional mean of $Y$, is a linear function of $X$. That is,

$$\mu_{Y \cdot x} = A + BX_i.$$

But in many cases, the mean value of $Y_i$ is not a linear function of $X_i$. Despite this fact, the methods described in this chapter may still be serviceable in some instances. In this appendix we describe two ways in which these methods can be modified to handle cases where the relationship between $Y_i$ and $X_i$ is nonlinear.

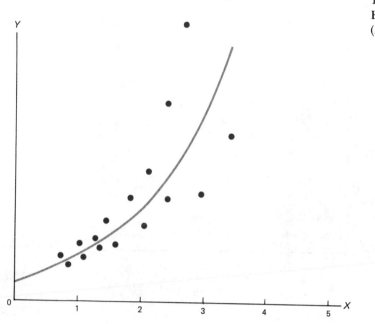

**Figure 11.14**
Exponential Function
(Assuming $\beta = 2$)

*Where* $Y_i$ *Is an Exponential Function of* $X_i$. Suppose that

$$Y_i = \alpha\beta^{X_i}z_i, \qquad\qquad (11.16)$$

where $z_i$ is an error term with a positive (+) value. It is assumed that the expected value of log $z_i$ equals zero, that the variance of log $z_i$ equals $\sigma_e^2$, and that the values of log $z_i$ are independent. Figure 11.14 shows this relationship between $Y_i$ and $X_i$. In economic statistics, this sort of relationship occurs frequently. For example, if an employee's salary increases at an approximately constant percentage rate, there will be an exponential relationship between the employee's years of experience and his or her salary. By taking the logarithm of both sides of equation (11.16) we obtain

$$\log Y_i = \log \alpha + \log \beta\, X_i + \log z_i.$$

If we let $\log \alpha = A$, $\log \beta = B$, and $\log z_i = e_i$, we have

$$\log Y_i = A + BX_i + e_i,$$

which is an ordinary linear equation. Thus, we can estimate $A \,(= \log \alpha)$ and $B$ $(= \log \beta)$ by regressing $\log Y_i$ on $X_i$. In other words, the trick here is to use $\log Y_i$, not $Y_i$, as the dependent variable in the regression.

*Where the Relationship Between* $Y_i$ *and* $X_i$ *is*

$$Y_i = \alpha X_i^B z_i. \qquad\qquad (11.17)$$

Figure 11.15 shows the relationship between $Y_i$ and $X_i$ when $B = .6$ and when $B = -1$. This sort of relationship occurs frequently in economics. For example,

**Figure 11.15**
Graph of $Y = \alpha X_i^B$, for $B = 0.6$ and $-1.0$

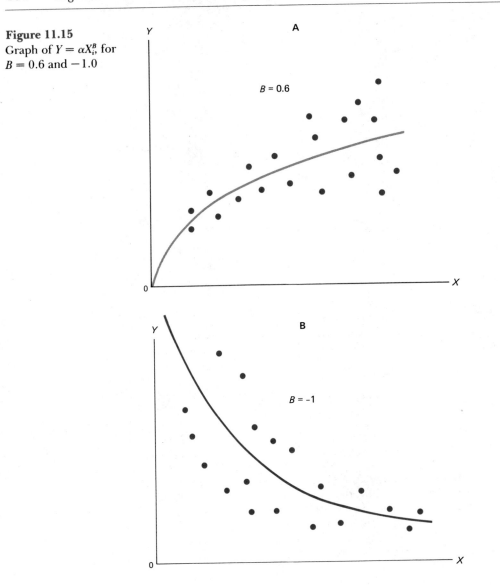

the relationship in panel B of Figure 11.15 is a demand curve of unit elasticity, if $Y_i$ is quantity demanded and $X_i$ is price. Once again, we can convert this equation to linear form by taking logarithms of both sides, the result being

$$\log Y_i = \log \alpha + B \log X_i + \log z_i.$$

If we let $\log \alpha = A$, and $\log z_i = e_i$, we have

$$\log Y_i = A + B \log X_i + e_i,$$

which is an ordinary linear equation. We can estimate $A$ $(= \log \alpha)$ and $B$ by regressing $\log Y_i$ on $\log X_i$. In other words, the trick is to use $\log Y_i$, not $Y_i$, as the dependent variable and to use $\log X_i$, not $X_i$, as the independent variable in the regression.

LOGARITHMIC REGRESSION AND ECONOMIES OF SCALE: A CASE STUDY. Economists and engineers have often used logarithmic regression to estimate economies of scale. For example, take the mundane, but important case of sewer pipelines. Based on engineering considerations, there is reason to believe that

$$C_i = \alpha Q_i^B z_i, \qquad (11.18)$$

where $C_i$ is the total construction cost of the $i^{th}$ pipeline, $Q_i$ is the capacity of the pipeline, and $\alpha$ and $B$ are unknown constants.

Steven Hanke and Roland Wentworth of Johns Hopkins University have used logarithmic regression to estimate $B$, which is the crucial parameter determining the extent of economies of scale.[14] To see why $B$ is crucial in this regard, note that a 1-percent increase in capacity results in a $B$-percent increase in total construction cost, according to equation (11.18). Thus, the lower the value of $B$, the greater the economies of scale.

To estimate $B$, Hanke and Wentworth took logarithms of both sides of equation (11.18), the result being

$$\log C_i = \log \alpha + B \log Q_i + e_i,$$

where $e_i = \log z_i$. Data concerning $\log C_i$ and $\log Q_i$ were obtained for 6,791 separate wastewater pipeline projects. The least-squares estimate of $B$ was calculated,[15] the formula being exactly like equation (11.2a), except that $\log C_i$ is substituted for $Y_i$ and $\log Q_i$ is substituted for $X_i$.

The resulting estimate of $B$ was 0.462. The sampling errors were relatively small, a 95 percent confidence interval for $B$ being 0.445 to 0.478. Based on these results, Hanke and Wentworth concluded that there are very substantial economies of scale in the construction of pipelines. "If pipeline capacity is increased by 100 percent, it is certain that, within a 95 percent reliability, total pipeline construction costs will increase by only 36.1 to 39.3 percent."[16]

# Appendix 11.2

## RANK CORRELATION

In Chapter 9 we pointed out that nonparametric tests are often available as analogues to parametric tests. An important nonparametric technique used for measuring the degree of correlation between two variables $Y$ and $X$ is *rank correlation*. This technique is particularly useful when one or more of the vari-

---

[14] S. Hanke and R. Wentworth, "Statistical Cost Function Developed for Sewer Lines," *Water and Sewer Works*, December 1980.

[15] Actually, the regression was carried out somewhat differently than described here, but for present purposes, this can be ignored.

[16] Hanke and Wentworth, *op. cit.*, p. 55. If $B = 0.445$, and if $Q_i$ doubles, $C_i$ will increase by 36.1 percent, according to equation (11.18). Similarly, if $B = 0.478$, and if $Q_i$ doubles, $C_i$ will increase by 39.3 percent, according to equation (11.18).

**Table 11.3**
Results of Psychologist's
Tests

| Person | (1) Persistence Rank | (2) IQ score | (3) IQ rank | (4) $d_i = (3) - (1)$ | (5) $d_i^2$ |
|---|---|---|---|---|---|
| A | 1 | 120 | 8 | 7 | 49 |
| B | 9 | 130 | 3 | −6 | 36 |
| C | 4 | 132 | 2 | −2 | 4 |
| D | 3 | 127 | 4 | 1 | 1 |
| E | 8 | 126 | 5 | −3 | 9 |
| F | 2 | 110 | 11 | 9 | 81 |
| G | 11 | 98 | 12 | 1 | 1 |
| H | 10 | 113 | 10 | 0 | 0 |
| I | 6 | 121 | 7 | 1 | 1 |
| J | 5 | 123 | 6 | 1 | 1 |
| K | 7 | 119 | 9 | 2 | 4 |
| L | 12 | 140 | 1 | −11 | 121 |
| $\Sigma d_i^2$ | | | | | 308 |

$$r_r = 1 - \frac{6(308)}{12(144-1)} = 1 - \frac{1848}{1716} = -.077$$

ables in question cannot be measured in ordinary ways, but can only be ranked. For example, a psychologist may want to determine how closely a person's persistence is associated with his or her IQ score. Since there is no objective measure of persistence, the psychologist asks a professor to rank 12 students on the basis of his perception of their persistence. Each student's persistence rank and IQ score are shown in Table 11.3.

The **rank correlation coefficient** *measures the closeness of the relationship between the two sets of rankings—that is, between the rankings of the one variable and the rankings of the other variable.* For example, column (3) in Table 11.3 shows each person's rank with regard to IQ score and column (1) shows the corresponding rank with regard to persistence. Are these ranks correlated? In other words, do people who rank highly with regard to IQ score also tend to rank highly with regard to persistence? If so, is this relationship close? To help answer such questions, statisticians have devised the rank correlation coefficient, which is defined as follows:

$$r_r = 1 - \frac{6 \sum_{i=1}^{n} d_i^2}{n(n^2 - 1)}, \qquad (11.19)$$

where $d_i$ is the difference between the two ranks of the $i$th observation, and $n$ is the number of observations. As shown in Table 11.3, the rank correlation coefficient in the case of the psychologist's study equals $-.077$.

The value of the rank correlation coefficient can vary from $+1$ to $-1$. A positive value of the rank correlation coefficient suggests that the two variables are directly related; a negative value suggests that they are inversely related. If the rank correlation coefficient is $+1$, there is a perfect direct relationship between the two rankings. If the rank correlation coefficient is $-1$, there is a perfect inverse relationship between the two rankings. If the two variables are statistically independent, the population rank correlation coefficient is zero. (However, due to chance, the *sample* rank correlation coefficient may not be zero under these circumstances.)

To test whether the population rank correlation coefficient is zero, we compute the following test statistic:

$$t = \frac{r_r}{\sqrt{(1 - r_r^2)/(n - 2)}}.$$

If the null hypothesis (that the population rank correlation coefficient is zero) is true, this statistic has the $t$ distribution with $(n - 2)$ degrees of freedom. Thus, if the alternative hypothesis is that the population rank correlation coefficient is either greater or less than zero the decision rule is as follows.

DECISION RULE: *Reject the null hypothesis that the population rank correlation coefficient is zero if* $t < -t_{\alpha/2}$ *or if* $t > t_{\alpha/2}$. *(There are* $(n - 2)$ *degrees of freedom, and* $\alpha$ *is the significance level.)*

To illustrate the application of this decision rule, let's return to the psychologist's results in Table 11.3. Since the sample rank correlation coefficient in this case equals $-.077$,

$$t = \frac{-.077}{\sqrt{.9941/10}} = \frac{-.077}{.315} = -.24.$$

If the significance level is set at .05, $t_{.025} = 2.228$ since there are 10 degrees of freedom. Since the observed value of $t$ is not less than $-2.228$ or greater than 2.228, the psychologist should not reject the null hypothesis that persistence and IQ score are unrelated.

The rank correlation coefficient is often used instead of the ordinary correlation coefficient when the dependent variable is far from normally distributed. Wide departures from normality have no effect on the test described in the two previous paragraphs because this test does not assume normality. Also, the rank correlation coefficient is sometimes preferred to the ordinary correlation coefficient because the former is easier to compute. As you can see from Table 11.3, the computations are relatively simple.

# Appendix 11.3

## INTERPRETING COMPUTER OUTPUT FOR REGRESSION AND CORRELATION

This appendix illustrates how computer output for regression and correlation can be interpreted. Suppose that a stock broker wants to estimate the relationship in a particular industry between a firm's profits and its sales in the first quarter of 1986. Based on data for 12 randomly chosen firms, the broker uses Minitab to calculate the regression. Profit is designated as C37, and sales are designated as C36. The printout is as follows:

```
The regression equation is
C37 = - 45.5 + 0.0920 C36

Predictor      Coef      Stdev    t-ratio
Constant      -45.49     19.33     -2.35
C36          0.092018  0.008251   11.15

s = 43.14    R-sq = 92.6%    R-sq(adj) = 91.8%
```

According to this printout, the regression equation is $\hat{Y} = -45.5 + 0.092X$. The standard error ($s_b$) of the regression coefficient (b) is .008251, and $b \div s_b$ equals 11.15. Thus, the regression coefficient differs significantly from zero at the .01 level, since $b \div s_b$ exceeds 3.169, the value of $t_{.005}$ when there are 10 degrees of freedom. (See Appendix Table 6.) The standard error of estimate is 43.14 millions of dollars. (Both profits and sales are expressed in millions of dollars.) The coefficient of determination is 0.926; adjusted for degrees of freedom (recall footnote 8), it is 0.918. Thus, the regression fits the data very well.

Not all of the printout is shown above. The results of other calculations, explained in the following chapter, are printed out as well.

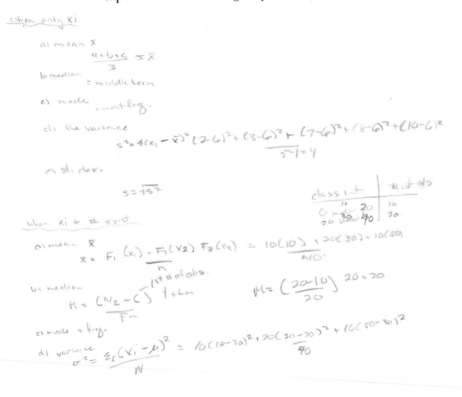

# Multiple Regression and Correlation*

12

## 12.1 Introduction

In the previous chapter we discussed regression and correlation techniques in the case where there is only one independent variable. In practical applications of regression and correlation techniques, it frequently is necessary and desirable to include two or more independent variables. In this chapter we will extend our treatment of regression and correlation to the case in which there is more than one independent variable. In addition, we will indicate how tests can be carried out to check some of the assumptions underlying regression analysis, and we will describe how violations of these assumptions may affect the results.

## 12.2 Multiple Regression: Nature and Purposes

Whereas a *simple regression* includes only one independent variable, a *multiple regression* includes two or more independent variables. Basically, there are two important reasons why a multiple regression must often be used instead of a simple regression. First, *one frequently can predict the dependent variable more accurately if more than one independent variable is used.* In the case of the hosiery mill (discussed in the previous chapter), the firm's statisticians may feel that factors other than the output rate have an important effect on the firm's cost. For example, it may seem likely that costs will tend to decrease if the mill's manager is relatively experienced. Thus, holding the output rate constant, the expected value of the firm's costs may be a linear function of the number of years of experience of the mill's manager. In other words, it may be

---

* Some instructors may prefer to omit the latter half of this chapter, which deals with dummy variables, multicollinearity, and serial correlation. Others may wish to skip this chapter completely. Either option is feasible since an understanding of subsequent chapters does not depend on coverage of this chapter.

reasonable to assume that

$$E(Y_i) = A + B_1 X_{1i} + B_2 X_{2i}, \qquad (12.1)$$

where $Y_i$ is the cost (in thousands of dollars) for the $i$th month, $X_{1i}$ is the output rate (in tons) for that month, and $X_{2i}$ is the number of years of experience of the manager in charge of the mill during the $i$th month. Of course, if increases in the experience of the manager result in lower costs, $B_2$ is negative.

According to equation (12.1), the expected value of cost in a certain month is dependent on the output rate for that month and on the amount of experience of the mill's manager during that month. The equation says that if the output rate increases by 1 ton, the expected value of cost increases by $B_1$ thousands of dollars, and that if the amount of experience of the manager increases by 1 year, the expected value of cost increases by $B_2$ thousands of dollars. (Since $B_2$ is presumed to be negative, this means that the expected value of cost will decrease if the amount of experience increases by 1 year.) If it is true that cost depends on both the output rate and the amount of managerial experience, then the firm's statisticians can predict cost more accurately by using a multiple regression—that is, an equation relating the dependent variable (cost) to both independent variables (output rate and the amount of managerial experience)—than by using the simple regression in the previous chapter. The latter relates the dependent variable (cost) to only one of the independent variables (output rate).

A second reason for using multiple regression instead of simple regression is that *if the dependent variable depends on more than one independent variable, a simple regression of the dependent variable on a single independent variable may result in a biased estimate of the effect of this independent variable on the dependent variable.* For example, suppose that, in accord with equation (12.1), cost is dependent on both output and the amount of managerial experience. Estimating the simple regression of cost on output (as we did in the previous chapter) may result in a biased estimate of $B_1$, which measures the cost of an extra ton of output. To understand why such a bias may arise, suppose that the mill has tended to produce greater outputs in those months when it has had relatively experienced managers than in months when it has had relatively inexperienced managers. If this is the case, the costs of the low-output months have tended to be high because the managers have tended to be inexperienced, and the costs of the high-output months have tended to be low because the managers have tended to be experienced. Thus, the estimated simple regression will result in an estimate of $B_1$ that will be biased downward.[1]

When a dependent variable is a function of more than one inde-

[1] Why will the estimate of $B_1$ be biased downward? Because the costs of the high-output months are lower and the costs of the low-output months are higher than if the amount of managerial experience were held constant at its average level. Thus, the extra cost resulting from a certain extra amount of output is underestimated, which is the same as saying that $B_1$ is biased downward.

pendent variable, the observed relationship between the dependent variable and any one of the independent variables may be misleading because the observed relationship may reflect the variation in the other independent variables. Since these other independent variables are totally uncontrolled, they may be varying in such a way as to make it appear that this independent variable has more effect or less effect on the dependent variable than in fact is true. To estimate the true effects of this independent variable on the dependent variable, we must include all the independent variables in the regression; that is, we must construct a multiple regression.

> *To predict a particular dependent variable, it often is useful to include several independent variables, not just one. Also, to obtain an accurate estimate of the effect of one independent variable on the dependent variable, it may be essential to include other independent variables in the analysis.*

*Basic Idea #23*

## 12.3 The Multiple-Regression Model

As pointed out in Chapter 11, the basic model underlying simple regression is

$$Y_i = A + BX_i + e_i,$$

(12.2)

where $Y_i$ is the $i$th observed value of the dependent variable, $X_i$ is the $i$th observed value of the independent variable, and $e_i$ is a normally distributed random variable with a mean of zero and a standard deviation of $\sigma_e$. Essentially, $e_i$ is an *error term*—that is, a random amount that is added to $A + BX_i$ (or subtracted from it, if $e_i$ is negative). The conditional mean of $Y_i$ is assumed to be a linear function of $X_i$—namely, $A + BX_i$. And the values of $e_i$ are assumed to be statistically independent.

The model underlying multiple regression is essentially the same as that above, the only difference being that the conditional mean of the dependent variable is assumed to be a linear function of more than one independent variable. If there are two independent variables $X_1$ and $X_2$, the model is

$$Y_i = A + B_1 X_{1i} + B_2 X_{2i} + e_i,$$

(12.3)

*Regression Model: Two Independent Variables*

where $e_i$ is an error term. As in the case of simple regression, it is assumed that the expected value of $e_i$ is zero, that $e_i$ is normally distributed, and that the standard deviation of $e_i$ is the same, regardless of the value of $X_{1i}$ or $X_{2i}$. Also, the values of $e_i$ are assumed to be statistically independent. In contrast to the case of simple regression, the conditional mean of $Y_i$ is a linear function of both $X_{1i}$ and $X_{2i}$. Specifically, the conditional mean equals $A + B_1 X_{1i} + B_2 X_{2i}$.

## 12.4 Least-Squares Estimates of the Regression Coefficients

*The first step in multiple-regression analysis is to identify the independent variables, and to specify the mathematical form of the equation relating the expected value of the dependent variable to these independent variables.* In the case of the hosiery mill this step is carried out in equation (12.1), which indicates that the independent variables are the output rate and the manager's years of experience. The relationship between the expected value of the dependent variable (costs) and these independent variables is linear. Having carried out this first step, *we next estimate the unknown constants A, $B_1$, and $B_2$ in the true regression equation.* Just as in the case of simple regression, these constants are estimated by finding the value of each that minimizes the sum of the squared deviations of the observed values of the dependent variable from the values of the dependent variable predicted by the regression equation.

To understand more precisely the nature of least-squares estimates of A, $B_1$, and $B_2$, suppose that $a$ is an estimator of A, $b_1$ an estimator of $B_1$, and $b_2$ an estimator of $B_2$. Then the value of the dependent variable $\hat{Y}_i$ predicted by the estimated regression equation is

$$\hat{Y}_i = a + b_1 X_{1i} + b_2 X_{2i},$$

and the deviation of this predicted value from the actual value of the dependent variable is

$$Y_i - \hat{Y}_i = Y_i - a - b_1 X_{1i} - b_2 X_{2i}.$$

Just as in the case of simple regression, the closeness of fit of the estimated regression equation to the data is measured by the sum of squares of these deviations:

$$\sum_{i=1}^{n} (Y_i - \hat{Y}_i)^2 = \sum_{i=1}^{n} (Y_i - a - b_1 X_{1i} - b_2 X_{2i})^2, \qquad (12.4)$$

where $n$ is the number of observations in the sample. The larger the sum of squares, the less closely the estimated regression equation fits; the smaller this sum of squares, the more closely it fits. Thus, it seems reasonable to choose the values of $a$, $b_1$, and $b_2$ that minimize the expression in equation (12.4). These estimates are least-squares estimates, as in the case of simple regression.

It can be shown that the values of $a$, $b_1$, and $b_2$ that minimize the sum of squared deviations in equation (12.4) are as follows:

$$b_1 = \frac{\sum_{i=1}^{n} (X_{2i} - \overline{X}_2)^2 \sum_{i=1}^{n} (X_{1i} - \overline{X}_1)(Y_i - \overline{Y}) - \sum_{i=1}^{n} (X_{1i} - \overline{X}_1)(X_{2i} - \overline{X}_2) \sum_{i=1}^{n} (X_{2i} - \overline{X}_2)(Y_i - \overline{Y})}{\sum_{i=1}^{n} (X_{1i} - \overline{X}_1)^2 \sum_{i=1}^{n} (X_{2i} - \overline{X}_2)^2 - \left[\sum_{i=1}^{n} (X_{1i} - \overline{X}_1)(X_{2i} - \overline{X}_2)\right]^2}$$

$$b_2 = \frac{\sum_{i=1}^{n} (X_{1i} - \overline{X}_1)^2 \sum_{i=1}^{n} (X_{2i} - \overline{X}_2)(Y_i - \overline{Y}) - \sum_{i=1}^{n} (X_{1i} - \overline{X}_1)(X_{2i} - \overline{X}_2) \sum_{i=1}^{n} (X_{1i} - \overline{X}_1)(Y_i - \overline{Y})}{\sum_{i=1}^{n} (X_{1i} - \overline{X}_1)^2 \sum_{i=1}^{n} (X_{2i} - \overline{X}_2)^2 - \left[\sum_{i=1}^{n} (X_{1i} - \overline{X}_1)(X_{2i} - \overline{X}_2)\right]^2}$$

$$a = \overline{Y} - b_1\overline{X}_1 - b_2\overline{X}_2.$$

*Least-Squares Estimators of A, $B_1$, and $B_2$*

(12.5)

To make these computations simpler, note that

$$\sum_{i=1}^{n} (X_{1i} - \overline{X}_1)^2 = \sum_{i=1}^{n} X_{1i}^2 - \frac{\left(\sum_{i=1}^{n} X_{1i}\right)^2}{n}$$

$$\sum_{i=1}^{n} (X_{2i} - \overline{X}_2)^2 = \sum_{i=1}^{n} X_{2i}^2 - \frac{\left(\sum_{i=1}^{n} X_{2i}\right)^2}{n}$$

$$\sum_{i=1}^{n} (X_{1i} - \overline{X}_1)(Y_i - \overline{Y}) = \sum_{i=1}^{n} X_{1i}Y_i - \frac{\left(\sum_{i=1}^{n} X_{1i}\right)\left(\sum_{i=1}^{n} Y_i\right)}{n}$$

$$\sum_{i=1}^{n} (X_{2i} - \overline{X}_2)(Y_i - \overline{Y}) = \sum_{i=1}^{n} X_{2i}Y_i - \frac{\left(\sum_{i=1}^{n} X_{2i}\right)\left(\sum_{i=1}^{n} Y_i\right)}{n}$$

$$\sum_{i=1}^{n} (X_{1i} - \overline{X}_1)(X_{2i} - \overline{X}_2) = \sum_{i=1}^{n} X_{1i}X_{2i} - \frac{\left(\sum_{i=1}^{n} X_{1i}\right)\left(\sum_{i=1}^{n} X_{2i}\right)}{n}$$

The following example illustrates how least-squares estimates of A, $B_1$, and $B_2$ are calculated. Although (as we shall see in subsequent sections) electronic computers are generally used for such calculations, it is worthwhile to work through at least one sample calculation of this sort by hand.

---

EXAMPLE 12.1 The hosiery mill's statisticians feel that equation (12.1) is true, and they want to obtain least-squares estimates of A, $B_1$, and $B_2$. Data are obtained concerning cost, output, and managerial experience for nine months, the results being shown in Table 12.1. Calculate the least-squares estimates of A, $B_1$, and $B_2$. Compare the least-squares estimate of $B_1$ with the estimate of B that we obtained in the previous chapter, based on the same data concerning cost and output. Why is the present estimate different from the latter? Which of the two estimates is likely to be better?

**Table 12.1**
Cost and Output of a
Hosiery Mill, with the
Number of Years of
Experience of Manager
in Charge during the
Month (Sample of
Nine Months)

| Output (tons) | Production cost (thousands of dollars) | Manager's years of experience |
|---|---|---|
| 2 | 3 | 0 |
| 1 | 2 | 1 |
| 8 | 8 | 2 |
| 5 | 5 | 3 |
| 6 | 6 | 4 |
| 4 | 4 | 5 |
| 7 | 6 | 6 |
| 9 | 8 | 7 |
| 8 | 7 | 8 |

SOLUTION: Based on the data in Table 12.1,

$$\sum_{i=1}^{9} X_{1i}^2 = 340 \qquad \sum_{i=1}^{9} X_{1i} = 50 \qquad \sum_{i=1}^{9} Y_i^2 = 303$$

$$\sum_{i=1}^{9} X_{2i}^2 = 204 \qquad \sum_{i=1}^{9} X_{2i} = 36 \qquad \sum_{i=1}^{9} Y_i = 49$$

$$\sum_{i=1}^{9} X_{1i}X_{2i} = 245 \qquad \sum_{i=1}^{9} X_{1i}Y_i = 319 \qquad \sum_{i=1}^{9} X_{2i}Y_i = 225.$$

Inserting these figures into the formulas given prior to this example,

$$\sum_{i=1}^{9} (X_{1i} - \overline{X}_1)^2 = 340 - \frac{50^2}{9} = 340 - 277.78 = 62.22$$

$$\sum_{i=1}^{9} (X_{2i} - \overline{X}_2)^2 = 204 - \frac{36^2}{9} = 204 - 144 = 60$$

$$\sum_{i=1}^{9} (X_{1i} - \overline{X}_1)(Y_i - \overline{Y}) = 319 - \frac{(50)(49)}{9} = 319 - 272.22 = 46.78$$

$$\sum_{i=1}^{9} (X_{2i} - \overline{X}_2)(Y_i - \overline{Y}) = 225 - \frac{(36)(49)}{9} = 225 - 196 = 29$$

$$\sum_{i=1}^{9} (X_{1i} - \overline{X}_1)(X_{2i} - \overline{X}_2) = 245 - \frac{(50)(36)}{9} = 245 - 200 = 45.$$

Then, inserting these figures into the equations in (12.5), we obtain

$$b_1 = \frac{(60)(46.78) - (45)(29)}{(62.22)(60) - 45^2} = 0.88$$

$$b_2 = \frac{(62.22)(29) - (45)(46.78)}{(62.22)(60) - 45^2} = -0.18$$

$$a = \frac{49}{9} - .88\left(\frac{50}{9}\right) + .18\left(\frac{36}{9}\right) = 1.28.$$

Consequently, the estimated regression equation is

$$\hat{Y}_i = 1.28 + 0.88\, X_{1i} - 0.18 X_{2i}.$$

The estimated value of $B_1$ is 0.88, as contrasted with the estimate of $B$ in the previous chapter, which was 0.75. In other words, a one-ton increase in output results in an increase in estimated costs of 0.88 thousands of dollars, as contrasted with .75 thousands of dollars in the previous chapter. The reason these estimates differ is that the present estimate of the effect of output on costs holds constant the manager's years of experience, whereas the earlier estimate did not hold this factor constant. Since this factor affects costs, the earlier estimate is likely to be a biased estimate of the effect of output on cost.[2]

## 12.5 Confidence Intervals and Tests of Hypotheses Concerning $B_1$ and $B_2$

As in the case of simple regression, least-squares estimators have many statistically desirable properties. Specifically, a, b$_1$, *and* b$_2$ *are unbiased and consistent.* Moreover, the Gauss-Markov theorem tells us that *of all unbiased estimators that are linear functions of the dependent variables,* a, b$_1$, *and* b$_2$ *have the smallest standard deviation.* As in the case of simple regression, these desirable properties hold if the observations are independent and if the standard deviation of the conditional probability distribution is the same, regardless of the value of the independent variables. It is *not* necessary that the conditional probability distribution of the dependent variable be normal. Or, stating the same thing differently, it is *not* necessary that the probability distribution of $e_i$ be normal.

---

[2] Of course, this regression is only supposed to be appropriate when $X_{1i}$ and $X_{2i}$ vary in a certain limited range. If $X_{2i}$ is large and $X_{1i}$ is small, the regression would predict a negative value of cost, which obviously is inadmissable. But as long as the regression is not used to make predictions for values of $X_{1i}$ and $X_{2i}$ outside the range of the data given in Table 12.1, this is no problem. For simplicity, we assume in equation (12.1) that the effect of the number of years of managerial experience on the expected value of cost (holding output constant) can be regarded as linear in the relevant range. Alternatively, we could have assumed that it was quadratic (or some other nonlinear forms could have been used). In Section 12.13 we discuss how multiple regression can be used to estimate quadratic equations.

If we are willing to assume that $e_i$ is normally distributed, we can calculate a confidence interval for $B_1$ or $B_2$. This frequently is an important purpose of a multiple-regression analysis. For example, in the case of the hosiery mill, an important purpose of the analysis may be to obtain a confidence interval for $B_1$, which is often called the **true regression coefficient of $X_1$**. (The least-squares estimator of $B_1$—that is, $b_1$—is often called the *estimated regression coefficient of* $X_1$). This true regression coefficient is of interest because it measures the effect of a one-unit increase in $X_1$ (tons of output) on the expected value of $Y$ (cost), when $X_2$ (years of experience of the mill's manager) is held constant. Another important purpose of the analysis may be to obtain a confidence interval for $B_2$, which is often called the **true regression coefficient of $X_2$**. (The least-squares estimator of $B_2$—that is, $b_2$—is often called the *estimated regression coefficient of* $X_2$). This true regression coefficient is of interest because it measures the effect of a one-unit increase in $X_2$ (the number of years of experience of the mill's manager) on the expected value of $Y$ (cost), when $X_1$ (output) is held constant.[3]

As pointed out in the previous section, multiple regressions are generally carried out by electronic computer, not by hand. Because of the importance of multiple-regression techniques, standard programs have been formulated for calculation by computer of the least-squares estimates of $A$, $B_1$, and $B_2$. Since these estimates are sample statistics, they are obviously subject to sampling error. Besides calculating the values of the least-squares estimators of $A$, $B_1$, and $B_2$ (that is, $a$, $b_1$, and $b_2$), these programs provide an estimate of the standard deviation of $b_1$ (often called the *standard error of* $b_1$), and an estimate of the standard deviation of $b_2$ (often called the *standard error of* $b_2$).

Given the computer printout, it is relatively simple to construct a confidence interval for $B_1$ or $B_2$. As noted above, in any standard computer printout the standard error of $b_1$ and the standard error of $b_2$ are shown. If the confidence coefficient is set at $(1 - \alpha)$, *a confidence interval for* $B_1$ *is*

**Confidence Interval for $B_1$**

$$b_1 \pm t_{\alpha/2} s_{b_1}, \qquad (12.6)$$

where $s_{b_1}$ is the standard error of $b_1$ and where $t$ has $n - k - 1$ degrees of freedom (where $k$ is the number of independent variables included in the regression—two in this case). If the confidence coefficient is set at $(1 - \alpha)$, *a confidence interval for* $B_2$ *is*

**Confidence Interval for $B_2$**

$$b_2 \pm t_{\alpha/2} s_{b_2}, \qquad (12.7)$$

where $s_{b_2}$ is the standard error of $b_2$.[4]

[3] Sometimes statisticians are also interested in obtaining a confidence interval for $A$, which is often called the intercept of the regression, and which measures the expected value of $Y$ when both $X_1$ and $X_2$ are zero. (See note 4.)

[4] Many computer printouts also show $s_a$, the standard error of $a$. If the confidence coefficient is set at $(1 - \alpha)$, a confidence interval for $A$ is $a \pm t_{\alpha/2} s_a$.

Given the computer printout, it is also easy to test the null hypothesis that $B_1$ or $B_2$ equals zero, if we assume once again that $e_i$—in equation (12.3)—is normally distributed. The computer printout shows the t *statistic (or t ratio) for* $b_1$, this statistic being defined as $b_1 \div s_{b_1}$. If $B_1$ equals zero, this $t$ statistic has the $t$ distribution with $(n - k - 1)$ degrees of freedom. Thus, if the alternative hypothesis is two-sided and $\alpha$ is the significance level, the decision rule is as follows.

DECISION RULE: *Reject the null hypothesis that* $B_1$ *equals zero if the* t *statistic for* $b_1$ *exceeds* $t_{\alpha/2}$ *or is less than* $-t_{\alpha/2}$. *(The number of degrees of freedom is* n − k − 1.)     ***Test that*** $B_1 = 0$

Similarly, if $B_2$ equals zero, the $t$ statistic (or $t$ ratio) for $b_2$—defined as $b_2 \div s_{b_2}$—has the $t$ distribution with $(n - k - 1)$ degrees of freedom; and the decision rule is as follows.

DECISION RULE: *Reject the null hypothesis that* $B_2$ *equals zero if the* t *statistic for* $b_2$ *exceeds* $t_{\alpha/2}$ *or is less than* $-t_{\alpha/2}$. *(The number of degrees of freedom is* n − k − 1.)     ***Test that*** $B_2 = 0$

The following examples illustrate how, with a computer printout of the results of a multiple regression, one can construct confidence intervals for some of the true regression coefficients, and test whether some of these coefficients are zero.[5]

EXAMPLE 12.2 The computer printout of the multiple regression of cost on (1) output and (2) the number of years of experience of the mill's manager is shown, in part, in Table 12.2. (The basic data were given in Table 12.1.) In this printout (based on Minitab), output is designated as variable C38, the number of years of experience of the mill's manager is designated as variable C40, and cost is designated as C39. As you can see, the computer prints out the value of $b_1$, $s_{b_1}$, and the $t$ statistic for $b_1$ in the row labeled C38, it prints out the value of $b_2$, $s_{b_2}$, and the $t$ statistic for $b_2$ in the row labeled C40, and it prints out the value of $a$, the standard error of $a$, and the $t$ statistic for $a$ in the row labeled *constant*. In all cases, $t$ statistics are referred to as $t$-ratios in this computer printout. (Because of rounding errors, the estimates of $a$, $b_1$, and $b_2$ obtained by the computer will differ slightly from those obtained by hand.) Use these results to calculate a 95 percent confidence interval for $B_1$; then use these results to test the hypothesis that $B_2$ equals zero. (The alternative hypothesis is two-sided, and the significance level should be set equal to .05.)

[5] It is also possible, since the printout shows the $t$ statistic (or $t$ ratio) for $a$, to test the null hypothesis that $A$ equals zero. If $A$ equals zero, the $t$ statistic (or $t$ ratio) for $a$—defined as $a \div s_a$—has the $t$ distribution with $(n - k - 1)$ degrees of freedom; and the decision rule is: Reject the null hypothesis that $A$ equals zero if the $t$ statistic for $a$ exceeds $t_{\alpha/2}$ or is less than $-t_{\alpha/2}$; otherwise, do not reject the null hypothesis. (The number of degrees of freedom is $n - k - 1$.) Note that $a$ is designated as "constant" in Table 12.2.

**Table 12.2**
Section of Minitab
Computer Printout,
Showing Results of
Multiple Regression of
Cost on Output and
Manager's Years of
Experience

```
The regression equation is
C39 = 1.26 + 0.879 C38 − 0.176 C40

Predictor          Coef        Stdev      t-ratio
Constant          1.2647      0.1442        8.77
C38               0.87902     0.03469      25.34
C40              -0.17593     0.03532      -4.98
```

SOLUTION: Since the standard error of $b_1$ equals .0347, a 95 percent confidence interval for $B_1$ is $0.879 \pm t_{.025}(.0347)$. Because $n - k - 1 = 9 - 2 - 1$, or 6, the $t$ distribution has 6 degrees of freedom, and $t_{.025} = 2.447$. Thus, the confidence interval for $B_1$ is $0.879 \pm .0849$. In other words, the 95 percent confidence interval for the increase in expected cost due to a one-ton increase in output is .794 to .964 thousands of dollars.

If $B_2$ were zero, the probability would be .05 that the $t$ statistic for $b_2$ would be greater than 2.447 or less than −2.447. (This is because $t_{.025} = 2.447$.) Since the $t$ statistic for $b_2$ equals −4.98, it is less than −2.447. Thus, we must reject the null hypothesis that $B_2$ equals zero.

EXAMPLE 12.3 The computer printout in Table 12.3 indicates the results of a multiple regression where the dependent variable is a firm's 1984 profit (in millions of dollars), designated as C7. Unlike the case in Example 12.2, there are three, not two, independent variables: the firm's sales (designated as C5), its rate of growth of sales (C6), and its expenditures on research and development (C8). This multiple regression is based on data for 22 major American chemical firms. Calculate a 95 percent confidence interval for the true regression coefficient of the firm's sales. Using a two-tailed test, test the null hypothesis that the true regression coefficient of the firm's rate of growth of sales is zero. (Set the significance level at .05.)

SOLUTION: Since the estimated regression coefficient of the firm's sales is 0.016887, and the standard error of this estimated regression coefficient is 0.004013, the 95 percent confidence interval is $0.016887 \pm 2.101 (0.004013)$, since $t_{.025} = 2.101$ when there are 22-3-1, or 18 degrees of freedom. In other words, the confidence interval is $0.016887 \pm .008431$, or 0.008456 to

**Table 12.3**
Section of Minitab
Printout, Showing
Results of Multiple
Regression of a
Chemical Firm's 1984
Profit (C7) on its Sales
(C5), Rate of Growth of
Sales (C6), and its
Expenditure on
Research and
Development (C8).

```
The regression equation is C7 = 32.0 + 0.0169 C5 −
1.42 C6 + 0.729 C8

Predictor          Coef        Stdev      t-ratio
Constant         31.95        11.86        2.69
C5               0.016887     0.004013     4.21
C6              -1.423        1.027       -1.39
C8               0.7295       0.1260       5.79
```

## THE CASE OF STATISTICAL DEMAND CURVES

The director of marketing research of a firm producing personal computers wants to estimate the market demand curve for such computers. In other words, he wants to estimate the relationship between the price of a personal computer and the quantity of such computers demanded per year in the United States. Based on existing studies, he knows that the demand curve for personal computers has shifted to the right in recent years. That is, when price is held constant, the quantity of such computers demanded in the United States has increased. In addition, he knows that the price of personal computers has fallen in recent years because the cost of producing them has fallen. To estimate the demand curve, he obtains data on the price and quantity of personal computers sold in each of the last 10 years. He calculates the regression of quantity sold on price. When he presents the results to the firm's board of directors, one board member says that the resulting regression equation is not a demand curve at all. Is the board member correct? If so, what mistake did the marketing research director make?

SOLUTION: The board member is correct. Suppose that the 1987 demand curve was $D_1$, that the 1986 demand curve was $D_2$, and so on. And suppose that the supply curve for such computers was $S_1$ in 1987, $S_2$ in 1986, and so on. Then the actual combination of price and quantity in each year is shown by each dot in the graph below (if the market is in equilibrium). As you can see, the regression line (shown by $D\,D'$) is a hybrid that resembles none of the true demand curves. Sophisticated econometric techniques have been developed to deal with this so-called *identification problem.* But for present purposes, the moral is that a regression equation of this sort is likely to be a poor estimate of a product's demand curve.[6]

**TO ERR IS HUMAN ... BUT COSTLY**

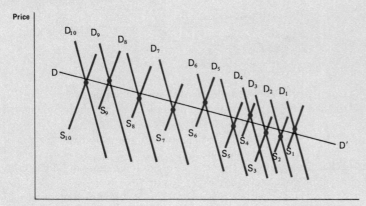

[6] For further discussion of this point, see E. Working, "What Do Statistical 'Demand Curves' Show?," in E. Mansfield, *Statistics for Business and Economics, op. cit.*

0.025318. To interpret this result, suppose that a firm's sales increase by \$1 million. Based on the regression, it appears that such an increase is associated, on the average, with an increase in profit of 0.008456 to 0.025318 millions of dollars (that is, of \$8,456 to \$25,318), according to the 95 percent confidence interval.[7]

The $t$ ratio of the estimated regression coefficient of the firm's rate of growth of sales is $-1.39$, which is not less than $-2.101$, the value of $-t_{.025}$ when there are 18 degrees of freedom. Thus, we should not reject the null hypothesis that the true regression coefficient of the firm's rate of growth of sales is zero.

---

[7] These numerical results, like many others in this text, are merely illustrative. That is, they are used to explain the nature of a statistical technique and should not be interpreted as an accurate or complete representation of the relevant economic phenomena. Indeed, some of the data used in other examples of this sort are hypothetical, as is evident from our treatment of them.

---

EXERCISES

**12.1** An economist wants to determine the relationship in San Francisco between the annual amount spent on clothing, on the one hand, and the number of family members and family income, on the other. She obtains the following data concerning 10 randomly selected San Francisco families:

| Annual amount spent on clothing (\$000) | Number of family members | Annual family income (\$000) |
|:---:|:---:|:---:|
| 0.8 | 1 | 21 |
| 1.4 | 2 | 25 |
| 0.3 | 1 | 10 |
| 2.3 | 2 | 37 |
| 3.8 | 2 | 48 |
| 2.1 | 3 | 27 |
| 3.3 | 4 | 35 |
| 2.9 | 3 | 35 |
| 3.0 | 5 | 25 |
| 4.0 | 3 | 51 |

(a) Calculate the sample multiple regression, clothing expenditure being the dependent variable and number of family members and family income being the independent variables.
(b) What is the estimated effect of an additional family member on annual clothing expenditure?
(c) What is the estimated effect of an additional thousand dollars of income on annual clothing expenditure?
(d) Would you expect that this relationship would be linear for all values of the independent variables?
(e) Predict a family's annual clothing expenditure if it contains two members and its income is \$25,000.
(f) Predict a family's annual clothing expenditure if it contains four members and its income is \$35,000.

(g) Do you think that the regression you calculated in (a) would be applicable to Minneapolis families? If not, what differences would you expect between this regression and that for Minneapolis families?

**12.2** A college wants to estimate the relationship between a student's grade-point average, on the one hand, and his or her high school grade-point average and SAT score. For 12 students, these variables are as follows:

| College grade-point average | High School grade-point average | SAT score |
|---|---|---|
| 2.5 | 3.0 | 1,000 |
| 2.9 | 3.2 | 1,100 |
| 2.9 | 3.5 | 980 |
| 3.1 | 3.4 | 1,120 |
| 3.2 | 3.8 | 1,010 |
| 3.6 | 4.0 | 1,140 |
| 3.4 | 3.6 | 1.180 |
| 3.6 | 3.8 | 1,250 |
| 3.6 | 3.8 | 1,200 |
| 3.9 | 3.9 | 1,260 |
| 3.8 | 4.0 | 1,210 |
| 4.0 | 4.0 | 1,300 |

(a) Calculate the sample multiple regression, college grade-point average being the dependent variable and high school grade-point average and SAT score being the independent variables.

(b) What is the estimated effect of an additional 1.0 in high school grade-point average on college grade-point average?

(c) What is the estimated effect of an additional 100 points in SAT score on college grade-point average?

(d) Do colleges actually use formulas of this sort? For what reason?

(e) Predict a student's grade-point average at this college if he has a high school grade-point average of 3.0 and an SAT score of 1150.

(f) Predict a student's grade-point average at this college if she has a high school grade-point average of 3.8 and an SAT score of 1240.

(g) Do you think that the multiple regression you calculated in (a) can be applied to any college, or just this one? Explain.

**12.3** An agricultural economist is interested in the effects on corn output of the amount of nitrogen fertilizer and of the amount of phosphate fertilizer. Using the number of bushels of corn per acre (designated as C1) as the dependent variable, he regresses this variable on the number of pounds of nitrogen per acre (designated as C2) and the number of pounds of phosphate per acre (designated as C3). There are 25 observations. Part of the computer printout is as follows:

| Predictor | Coef | Stdev | t-ratio |
|---|---|---|---|
| Constant | 19 | 6 | 3.17 |
| C2 | 0.29 | 0.11 | 2.64 |
| C3 | 0.67 | 0.23 | 2.91 |

(a) Compute a 90 percent confidence interval for the true regression coefficient of the number of pounds of nitrogen per acre.

(b) Compute a 95 percent confidence interval for the true regression coefficient of the number of pounds of phosphate per acre.

(c) Suppose that the printout did not include the standard errors of the regression coefficients, but that the other figures shown above were included. Could you figure out the standard errors? If so, how?

**12.4** For a sample of 15 students, the grade each obtained on the final examination and the two hour examinations of a finance course are given below:

| First hour examination | Second hour examination | Final examination |
|---|---|---|
| 68 | 73 | 74 |
| 75 | 78 | 81 |
| 84 | 82 | 85 |
| 93 | 98 | 94 |
| 95 | 91 | 90 |
| 87 | 89 | 85 |
| 86 | 82 | 81 |
| 72 | 65 | 63 |
| 61 | 70 | 80 |
| 54 | 46 | 53 |
| 87 | 91 | 86 |
| 78 | 75 | 76 |
| 72 | 78 | 77 |
| 75 | 62 | 68 |
| 84 | 80 | 83 |

(a) Calculate the sample multiple-regression equation, where the final examination grade is the dependent variable and the two hour examination grades are the independent variables.
(b) Is the regression coefficient for the first hour examination grade significantly different from zero?
(c) Is the regression coefficient for the second hour examination grade significantly different from zero?

## 12.6  Multiple Coefficient of Determination

In the previous chapter we described how the coefficient of determination can be used to measure how well a simple regression equation fits the data. When a multiple regression is calculated, the multiple coefficient of determination, rather than the simple coefficient of determination discussed in the previous chapter, is used for this purpose. The multiple coefficient of determination is defined as

*Formula for $R^2$*

$$R^2 = 1 - \frac{\sum_{i=1}^{n}(Y_i - \hat{Y}_i)^2}{\sum_{i=1}^{n}(Y_i - \overline{Y})^2},$$

(12.8)

where $\hat{Y}_i$ is the value of the dependent variable that is predicted from the regression equation.[8] Thus, as in the case of the simple coefficient of determination covered in the previous chapter,

$$R^2 = \frac{\text{Variation explained by regression}}{\text{Total variation}}, \qquad (12.9)$$

which means that $R^2$ *measures the proportion of the total variation in the dependent variable that is explained by the regression equation.* The positive square root of the multiple coefficient of determination is called the *multiple correlation coefficient* and is denoted by $R$. It, too, is sometimes used to measure how well a multiple-regression equation fits the data.

If there are only two independent variables in a multiple regression, as in equation (12.1), a relatively simple way to compute the multiple coefficient of determination is as follows:

*Alternative Formula for $R^2$*

$$R^2 = \frac{b_1 \sum_{i=1}^{n} (X_{1i} - \overline{X}_1)(Y_i - \overline{Y}) + b_2 \sum_{i=1}^{n} (X_{2i} - \overline{X}_2)(Y_i - \overline{Y})}{\sum_{i=1}^{n} Y_i^2 - \dfrac{\left(\sum_{i=1}^{n} Y_i\right)^2}{n}}. \qquad (12.10)$$

If there are more than two independent variables, a multiple regression is almost always carried out on an electronic computer, which is programmed to print out the value of the multiple coefficient of determination (or of the multiple correlation coefficient).

The following are examples of the calculation and interpretation of the multiple coefficient of determination.

---

EXAMPLE 12.4 Use the data in Table 12.1 to calculate the multiple coefficient of determination between the hosiery mill's costs, on the one hand, and its output and the number of years of experience of the mill's manager, on the other. Interpret your result.

---

[8] This is the *unadjusted* multiple coefficient of determination, which is biased away from zero. An unbiased estimate of the population multiple coefficient of determination is the *adjusted* multiple coefficient of determination, which is

$$\overline{R}^2 = 1 - \frac{\sum_{i=1}^{n} (Y_i - \hat{Y}_i)^2 \div (n - k - 1)}{\sum_{i=1}^{n} (Y_i - \overline{Y})^2 \div (n - 1)},$$

where $n$ is the number of observations and $k$ is the number of independent variables. (By *adjusted,* we mean adjusted for degrees of freedom.) See Section 12.4 for a more detailed definition of $\hat{Y}_i$.

SOLUTION: We know from Example 12.1 that $b_1 = 0.88$, $b_2 = -0.18$,

$$\sum_{i=1}^{9} (X_{1i} - \overline{X}_1)(Y_i - \overline{Y}) = 46.78, \sum_{i=1}^{9} (X_{2i} - \overline{X}_2)(Y_i - \overline{Y}) = 29, \text{ and}$$

$$\sum_{i=1}^{9} Y_i^2 - \frac{\left(\sum_{i=1}^{9} Y_i\right)^2}{9} = 303 - \frac{(49)^2}{9} = 303 - 266.78 = 36.22.$$

Thus it follows from equation (12.10) that

$$R^2 = \frac{.88(46.78) - .18(29)}{36.22} = .99.$$

This means that 99 percent of the variation in the firm's monthly costs during the period covered by the data can be explained by the multiple regression equation derived in Example 12.1.

---

EXAMPLE 12.5 Table 12.4 shows another part of the computer printout of the results of the multiple regression of a chemical firm's 1984 profit on the three independent variables described in Example 12.3. Interpret the figure labeled R-sq in this printout.

**Table 12.4**
Another Section of
Minitab Printout,
Showing Results of
Multiple Regression
Concerning Chemical
Firms' Profits

| s = 26.27 | R-sq = 99.4% | R-sq(adj) = 99.3% |
|-----------|--------------|-------------------|

SOLUTION: This figure—99.4%—is the multiple coefficient of determination. It means that the regression equation shown in the printout explains 99.4 percent of the variation among these firms in the level of their 1984 profits.

## 12.7 Geometrical Interpretation of Results of Multiple Regression and Correlation

As we have seen, the estimated multiple regression equation shows the average relationship between the dependent variable and the independent variables. If there are only two independent variables, it is possible to represent this average relationship by a plane rather than by a line (which was used for this purpose in the previous chapter). Figure 12.1 shows the plane corresponding to the regression equation relating the hosiery mill's costs to its output rate and the number of years of experience of the mill's manager:

$$\hat{Y}_i = 1.28 + 0.88X_{1i} - 0.18X_{2i}.$$

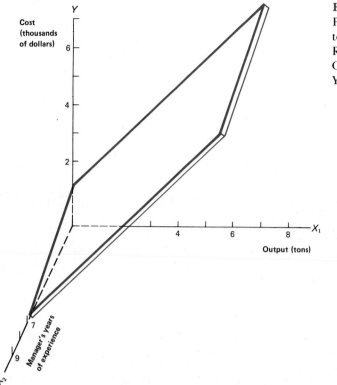

**Figure 12.1**
Plane Corresponding to Regression Equation Relating Cost to Output and Manager's Years of Experience

When both $X_{1i}$ and $X_{2i}$ are zero, Figure 12.1 shows that the predicted value of cost is 1.28 thousands of dollars, which is the intercept of the regression. Figure 12.1 shows that when $X_{1i}$ is held constant, if $X_{2i}$ increases by one unit (one year in this case), the average value of the dependent variable decreases by an amount equal to the regression coefficient of $X_{2i}$, which is .18 thousands of dollars (or \$180) in this case. Figure 12.1 also shows that when $X_{2i}$ is held constant, if $X_{1i}$ is increased by one unit (one ton in this case), the average value of the dependent variable increases by an amount equal to the regression coefficient of $X_{1i}$, which is .88 thousands of dollars (or \$880).

Just as each observation in a simple regression can be represented as a point in a scatter diagram, each observation in a multiple regression (with two independent variables) can be represented as a point in three-dimensional space. Figure 12.2 shows how we can represent an observation where the dependent variable equals 7 and the independent variables equal 3 and 5. *The multiple coefficient of determination is a measure of how well the plane representing the regression equation fits the points representing the individual observations.* For example, panel A of Figure 12.3 shows a case where the regression plane does not fit the points at all well; on the other hand, panel B of Figure 12.3 shows a case where the regression plane fits the points much more closely than in panel A. Clearly, the multiple coefficient of determination (as well as the multiple correlation coefficient) is higher in panel B than in panel A.

**Figure 12.2**
Geometrical Represen-
tation of an Observa-
tion (*Y* Is the Value of
the Dependent
Variable; $X_1$ and $X_2$
Are the Values of the
Independent Variables)

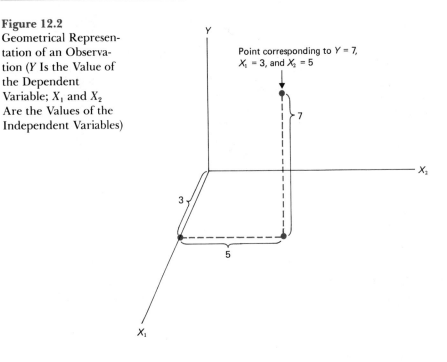

## 12.8 Analysis of Variance[9]

The analysis of variance, which we discussed at length in Chapter 10, is used to test the overall statistical significance of a regression equation. That is, it is used to test whether *all* the true regression coefficients in the equation equal zero. In the case of the hosiery mill, we might want to test whether both $B_1$ and $B_2$ are zero in order to see whether there is any relationship between the dependent variable and all the independent variables taken together. The analysis of variance can be used in this way in simple as well as multiple regressions. In simple regressions, the result is precisely the same as in the test of $B = 0$, described in Section 11.13.

In our discussion of the analysis of variance in Chapter 10, we defined the total variation in *Y* as the sum of the squared deviations of the values of *Y* from the mean of *Y*. Thus, the total variation in *Y* equals

$$\sum_{i=1}^{n} (Y_i - \overline{Y})^2.$$

As pointed out in Section 11.10,

$$\sum_{i=1}^{n} (Y_i - \overline{Y})^2 = \sum_{i=1}^{n} (\hat{Y}_i - \overline{Y})^2 + \sum_{i=1}^{n} (Y_i - \hat{Y}_i)^2.$$

[9] To understand this section, it is not essential that the reader be familiar with Chapter 10. However, it is essential that Section 10.4 (on the *F* distribution) be read prior to studying this section.

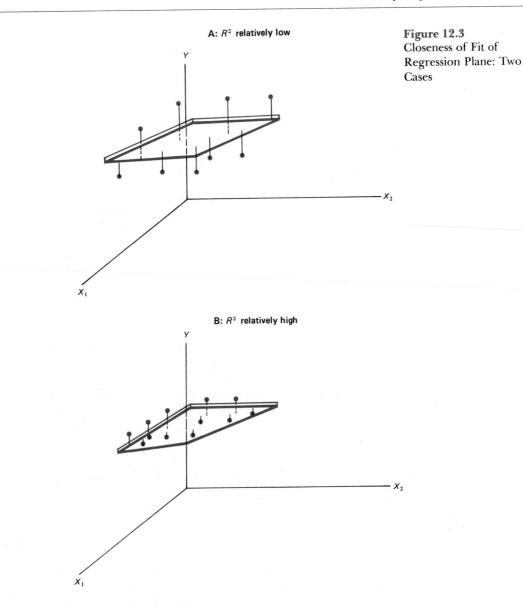

A: $R^2$ relatively low

B: $R^2$ relatively high

**Figure 12.3**
Closeness of Fit of
Regression Plane: Two
Cases

The first term on the right is the variation explained by the regression, and the second term on the right is the variation unexplained by the regression (also called the *error sum of squares* or the *residual sum of squares*). Thus

Total variation = Explained variation + Unexplained variation.

To carry out the analysis of variance, we construct a table of the general form of Table 12.5. The first column shows the source or type of variation, and the second column shows the corresponding sum of squares (that is, the corresponding variation). The third column shows the number of degrees of freedom corresponding to each sum of squares. For the explained variation, the number of degrees of freedom equals the number of independent variables $k$. For the unexplained variation, the number of degrees of freedom equals $(n - k - 1)$.

**Table 12.5**
Analysis of
Variance for
Regression

| Source of variation | Sum of squares | Degrees of freedom | Mean square | F ratio |
|---|---|---|---|---|
| Explained by regression | $\sum_{i=1}^{n} (\hat{Y}_i - \overline{Y})^2$ | $k$ | $\dfrac{\sum_{i=1}^{n} (\hat{Y}_i - \overline{Y})^2}{k}$ | $\dfrac{\sum_{i=1}^{n} (\hat{Y}_i - \overline{Y})^2 / k}{\sum_{i=1}^{n} (Y_i - \hat{Y}_i)^2 / (n - k - 1)}$ |
| Unexplained by regression | $\sum_{i=1}^{n} (Y_i - \hat{Y}_i)^2$ | $n - k - 1$ | $\dfrac{\sum_{i=1}^{n} (Y_i - \hat{Y}_i)^2}{(n - k - 1)}$ | |
| Total | $\sum_{i=1}^{n} (Y_i - \overline{Y})^2$ | $n - 1$ | | |

The fourth column shows the mean square, which is the sum of squares divided by the number of degrees of freedom. The fifth column shows the ratio of the two mean squares.[10]

If the null hypothesis is true (that is, if all the true regression coefficients are zero), this ratio has an $F$ distribution with $k$ and $(n - k - 1)$ degrees of freedom. Thus, to carry out the appropriate analysis of variance, the decision rule is as follows.

*Test that All True*
*Regression*
*Coefficients = 0*

DECISION RULE: *Reject the null hypothesis that the true regression coefficients are all zero if the ratio of the explained mean square to the unexplained mean square exceeds* $F_\alpha$, *where* $\alpha$ *is the desired significance level.*

Because multiple regressions are generally carried out on an electronic computer, it is seldom necessary to calculate the numbers in Table 12.5 by hand. The computer printout generally gives the analysis-of-variance table. Some printouts refer to the explained mean square as the *regression mean square* and to the unexplained mean square as the *error mean square* or the *residual mean square*. The ratio of the explained to the unexplained mean square is often referred to as the $F$ value or the $F$ ratio. The way in which the analysis-of-variance table is displayed on the computer printout may vary from that in Table 12.5, but the information imparted is the same. In Section 12.14 we will examine the printout of one computer program in some detail.

The following is an example of the application of the analysis of variance.

---

[10] In calculating the sum of squares explained by the regression, it is convenient to note that

$$\sum_{i=1}^{n} (\hat{Y}_i - \overline{Y})^2 = b_1 \sum_{i=1}^{n} (X_{1i} - \overline{X}_1)(Y_i - \overline{Y}) + b_2 \sum_{i=1}^{n} (X_{2i} - \overline{X}_2)(Y_i - \overline{Y}).$$

The right-hand side of this equation can be evaluated relatively easily. (This assumes that $k = 2$.)

EXAMPLE 12.6 The $F$ ratio is 937 for the multiple regression of a chemical firm's 1984 profit on the three independent variables described in Example 12.3. Use this $F$ ratio to test the hypothesis that all the true regression coefficients in this multiple regression are zero. (Set $\alpha = .05$.)

SOLUTION: Since $n = 22$ and $k = 3$, the number of degrees of freedom are 3 and 18. (Why? Because $k = 3$ and $(n - k - 1) = 18$.) Thus $F_{.05} = 3.16$. (See Appendix Table 9.) Since the $F$ ratio exceeds 3.16, we should reject the null hypothesis that all the true regression coefficients in this multiple regression equal zero.

## EXERCISES

**12.5** Based on the data in Exercise 12.1, calculate the multiple coefficient of determination. Interpret your results.

**12.6** Based on the data in Exercise 12.2, calculate the multiple coefficient of determination. Interpret your results.

**12.7** A multiple regression is carried out where there are 4 independent variables and 32 observations. The multiple coefficient of determination ($R^2$) equals 0.63.
  (a) What is $\overline{R}^2$, the multiple coefficient of determination adjusted for degrees of freedom?
  (b) In what way is $\overline{R}^2$ preferable to $R^2$?

**12.8** A multiple regression is carried out where there are three independent variables and 45 observations. The multiple coefficient of determination adjusted for degree of freedom, $\overline{R}^2$, equals 0.45.
  (a) What is $R^2$?
  (b) Must $R^2$ always be greater than $\overline{R}^2$?

**12.9** A statistician at an automobile plant carries out an analysis of variance, with the following results:

| Source of variation | Sum of squares | Degrees of freedom | Mean square | F |
|---|---|---|---|---|
| Explained by regression | ——— | 3 | 42 | ——— |
| Unexplained by regression | ——— | ——— | ——— | |
| Total | 542 | 31 | | |

  (a) Fill in the blanks.
  (b) How many independent variables are there?
  (c) How many observations are included?
  (d) What is the regression mean square?
  (e) What is the error mean square?
  (f) What is the residual mean square?
  (g) Interpret the results of the analysis of variance.

**12.10** Based on the data in Exercise 12.1, carry out an analysis of variance to test whether all of the true regression coefficients equal zero. (Let $\alpha = .05$.)

**12.11** Based on the data in Exercise 12.2, carry out an analysis of variance to test whether all of the true regression coefficients equal zero. (Let $\alpha = .01$.)

**12.12** An economist believes that there is a quadratic relationship between the interest rate in a particular year and the amount spent that year by a particular industry on plant and equipment. In other words, the economist believes that

$$E(S_t) = A + B_1\, i_t + B_2\, i_t^2,$$

where $S_t$ is the amount spent on plant and equipment by this industry during the $t$th year, and $i_t$ is the interest rate in the $t$th year. Can the economist use multiple regression techniques to estimate $A$, $B_1$, and $B_2$? If so, what assumptions must be made?

**12.13** An economist asks a research assistant to calculate a multiple regression on an electronic computer. The dependent variable is a particular industry's rate of productivity increase during 1948–66, and the independent variables are (1) the percent of the industry's employees belonging to unions in 1953 (designated as C2); (2) the percent of the industry's value-added spent on applied research and development in 1958 (designated as C4); and (3) the percentage of the firms in the industry reporting in 1958 that they expected their R and D (research and development) expenditures to pay out in no less than six years (designated as C8). Only 17 industries can be included because data are lacking concerning the third independent variable for 3 of the industries. The research assistant comes back with the Minitab printout, part of which is as follows:

| Predictor | Coef | Stdev | t-ratio |
|-----------|---------|--------|---------|
| Constant | 4.6595 | 1.5000 | 4.0518 |
| C2 | −0.0622 | 0.0090 | −6.9400 |
| C4 | 0.0963 | 0.0154 | 6.2700 |
| C8 | 0.0704 | 0.0098 | 7.2000 |

(a) What is the estimated regression equation?

(b) Calculate a 95 percent confidence interval for the true regression coefficient of the percent of the industry's value-added spent on applied research and development.

(c) Calculate a 90 percent confidence interval for the true regression coefficient of the percent of the industry's employees belonging to unions.

(d) The economist wants to test the hypothesis that whether an industry's R and D consists mainly of long-term or short-term projects has no effect on its rate of productivity increase. If the significance level is set at .05, and if the alternative hypothesis is two-sided, use the printout to test this hypothesis.

(e) Another part of the computer printout is shown below. What is the multiple coefficient of determination in this case? What does this result mean?

$$s = 0.3458 \qquad\qquad R\text{-sq} = 87.0\%$$

(f) The F-value is 29.01. Test the null hypothesis that all the true regression coefficients in the multiple regression are zero. (Set $\alpha = .05$.)

**\*12.14** Data on the detergency of a given detergent at various temperatures and concentrations are as follows:

| Detergency | Concentration | Temperature |
|---|---|---|
| 37 | 10 | 100 |
| 42 | 20 | 100 |
| 46 | 30 | 100 |
| 48 | 40 | 100 |
| 53 | 10 | 120 |
| 62 | 20 | 120 |
| 79 | 30 | 120 |
| 84 | 40 | 120 |
| 59 | 10 | 140 |
| 74 | 20 | 140 |
| 88 | 30 | 140 |
| 102 | 40 | 140 |

*Source:* A. J. Duncan, *op. cit.*, p. 700.

Minitab is used to estimate how detergency is related to concentration and temperature. Interpret the results, shown below.

```
MTB > read into c41,c42,c43
DATA>  37 10 100
DATA>  42 20 100
DATA>  46 30 100
DATA>  48 40 100
DATA>  53 10 120
DATA>  62 20 120
DATA>  79 30 120
DATA>  84 40 120
DATA>  59 10 140
DATA>  74 20 140
DATA>  88 30 140
DATA> 102 40 140
DATA> end
     12 ROWS READ
MTB > regress c41 on 2 predictors in c42 and c43

The regression equation is
C41 = -72.2 + 0.967 C42 + 0.938 C43


Predictor        Coef        Stdev      t-ratio
Constant       -72.17        16.17        -4.46
C42            0.9667       0.1868         5.18
C43            0.9375       0.1279         7.33


s = 7.233      R-sq = 90.0%      R-sq(adj) = 87.7%

Analysis of Variance

SOURCE          DF           SS           MS
Regression       2       4214.2       2107.1
Error            9        470.8         52.3
Total           11       4685.0
```

* Exercise requires some familiarity with Minitab.

## 12.9 Dummy-Variable Techniques

Multiple regression can be used to analyze the effects of *qualitative* variables (that is, variables that do not assume numerical values) as well as *quantitative* variables. For example, suppose that an economist is studying the saving behavior of families in a lower-middle-class Chicago neighborhood and that he wants to estimate the effect on a family's saving rate of two variables: (1) the family's annual income, and (2) whether the family owns its home or rents. The second independent variable is a qualitative, not a quantitative, variable; yet it can be included in a multiple regression as a so-called dummy variable, defined below.

DUMMY VARIABLE: *A dummy variable is a variable that equals 0 or 1.*

In the present case, the economist can construct a dummy variable $H_i$, which equals 1 if the family owns its own home and 0 if the $i$th family rents. Then, if the relationship between the dependent and independent variables is linear, it may be assumed that

$$S_i = A + B_1 I_i + B_2 H_i + e_i,$$ 

(12.11)

where $S_i$ is the annual amount (in thousands of dollars) saved by the $i$th family, $I_i$ is its annual income (in thousands of dollars), and $e_i$ is the difference between $S_i$ and $E(S_i)$. Using the procedures described in Section 12.4, least-squares estimates of $A$, $B_1$, and $B_2$ can be obtained in the ordinary way.

To understand more clearly what equation (12.11) means and what the use of dummy variables really amounts to, suppose that the economist obtains data from 20 families (6 of whom own their homes and 14 of whom rent), and that these data are as shown in Table 12.6. If we plot the amount that each family saves yearly against its annual income, we obtain the results shown in Figure 12.4. Clearly, the relationship between the amount saved yearly and annual income seems different for the home owners than for the renters: Specifically, the home owners seem to save more at each level of income than do the renters. In other words, there seem to be two regression lines, one for families who own their homes, and one for those who rent. These two regression lines have the same slope but different intercepts. Suppose that two regression lines (with the same slope but different intercepts) exist in the population, as well as in this sample. What equation (12.11) does is to compress these two regression lines into a single equation.

To see that equation (12.11) does this, suppose that the model representing the amounts saved by renting families is

$$S_i = A + B_1 I_i + e_i,$$

**Table 12.6**
Annual Savings and
Annual Income of 6
Home-Owning and 14
Home-Renting Families

| Name | Annual savings (thousands of dollars) | Annual income (thousands of dollars) | Owns/rents | Value of $H_i$ |
|------|------|------|------|------|
| Jones | 1.0 | 20 | Rents | 0 |
| Smith | 1.3 | 24 | Rents | 0 |
| Kargill | 0.7 | 12 | Rents | 0 |
| Mennon | 0.8 | 16 | Rents | 0 |
| Billings | 0.5 | 11 | Rents | 0 |
| Stratahan | 2.4 | 32 | Owns | 1 |
| Cohen | 0.3 | 10 | Rents | 0 |
| Lamb | 3.2 | 40 | Owns | 1 |
| Schmidt | 2.8 | 32 | Owns | 1 |
| Palucci | 0.0 | 7 | Rents | 0 |
| Chichester | 0.3 | 9 | Rents | 0 |
| LaRue | 0.0 | 6 | Rents | 0 |
| Liu | 1.0 | 18 | Rents | 0 |
| Armour | 2.0 | 20 | Owns | 1 |
| Christenson | 0.4 | 12 | Rents | 0 |
| Howe | 0.7 | 14 | Rents | 0 |
| Pitt | 1.5 | 15 | Owns | 1 |
| Drummond | 1.6 | 16 | Owns | 1 |
| Tracy | 0.6 | 15 | Rents | 0 |
| Holland | 0.6 | 14 | Rents | 0 |

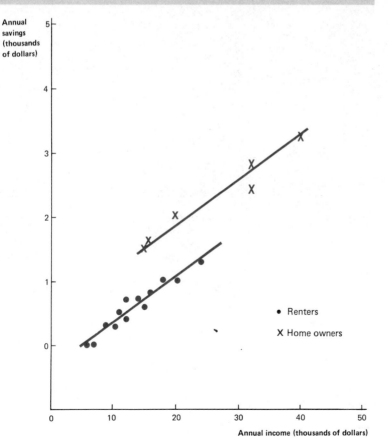

**Figure 12.4**
Relationship between
Amount Saved and
Income, Renters and
Home Owners

and suppose that the model representing the amounts saved by home-owning families is

$$S_i = (A + B_2) + B_1 I_i + e_i,$$

where $B_2$ is the *extra* amount (in thousands of dollars) that home owners save, on the average. Then it follows that equation (12.11) is valid both for families who own their homes and for those who rent. (Why? Because $B_2 H_i$ in equation (12.11) equals zero for renting families, while it equals $B_2$ for home-owning families.)

If equation (12.11) is accepted as a reasonable model, the least-squares estimates of $A$, $B_1$, and $B_2$ can be calculated in the usual way, as shown in Table 12.7. The resulting regression equation, shown by the two parallel lines in Figure 12.4, is

$$\hat{S}_i = -.3207 + .0675 I_i + .827 H_i, \qquad (12.12)$$

Table 12.7
Calculation of Least-Squares Estimates of $A$, $B_1$, and $B_2$, Based on Data in Table 12.6

$$\sum_{i=1}^{20}(H_i - \overline{H})^2 = \sum_{i=1}^{20}H_i^2 - \frac{\left(\sum_{i=1}^{20}H_i\right)^2}{20} = 6 - 1.8 = 4.2$$

$$\sum_{i=1}^{20}(I_i - \overline{I})^2 = \sum_{i=1}^{20}I_i^2 - \frac{\left(\sum_{i=1}^{20}I_i\right)^2}{20} = 7,377 - 5,882.4 = 1,494.6$$

$$\sum_{i=1}^{20}(S_i - \overline{S})^2 = \sum_{i=1}^{20}S_i^2 - \frac{\left(\sum_{i=1}^{20}S_i\right)^2}{20} = 39.27 - 23.544 = 15.726$$

$$\sum_{i=1}^{20}(I_i - \overline{I})(S_i - \overline{S}) = \sum_{i=1}^{20}I_i S_i - \frac{\left(\sum_{i=1}^{20}I_i\right)\left(\sum_{i=1}^{20}S_i\right)}{20} = 516.1 - 372.15 = 143.95$$

$$\sum_{i=1}^{20}(H_i - \overline{H})(S_i - \overline{S}) = \sum_{i=1}^{20}H_i S_i - \frac{\left(\sum_{i=1}^{20}H_i\right)\left(\sum_{i=1}^{20}S_i\right)}{20} = 13.5 - 6.51 = 6.99$$

$$\sum_{i=1}^{20}(I_i - \overline{I})(H_i - \overline{H}) = \sum_{i=1}^{20}I_i H_i - \frac{\left(\sum_{i=1}^{20}I_i\right)\left(\sum_{i=1}^{20}H_i\right)}{20} = 155 - 102.9 = 52.1$$

$$b_1 = \frac{(4.2)(143.95) - (52.1)(6.99)}{(1,494.6)(4.2) - (52.1)^2} = \frac{604.590 - 364.179}{6,277.32 - 2,714.41} = .0675$$

$$b_2 = \frac{(1,494.6)(6.99) - (52.1)(143.95)}{(1,494.6)(4.2) - (52.1)^2} = \frac{10,447.254 - 7,499.795}{6,277.32 - 2,714.41} = .827$$

$$a = \left(\frac{21.7}{20}\right) - .0675\left(\frac{343}{20}\right) - .827\left(\frac{6}{20}\right) = -.3207$$

(In deriving $a$, note that $\sum_{i=1}^{20}S_i = 21.7$, $\sum_{i=1}^{20}I_i = 343$, and $\sum_{i=1}^{20}H_i = 6$.)

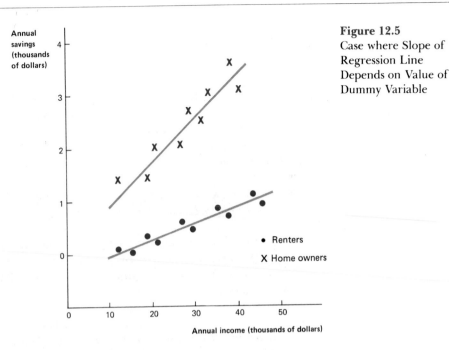

**Figure 12.5**
Case where Slope of Regression Line Depends on Value of Dummy Variable

where $\hat{S}_i$ is the amount of the $i$th family's savings (in thousands of dollars) predicted by the regression equation. According to this equation, a $1,000 increase in income results in an increase in estimated savings of .0675 thousands of dollars, or $67.50. Holding income constant, a home-owning family is estimated to save .827 thousands of dollars (or $827) more than a renting family.

Note that if $H_i$ were omitted from this equation and if the simple regression of $S_i$ on $I_i$ alone were used to estimate $B_1$, the result would be a biased estimate of $B_1$. As is evident from Figure 12.4, the simple regression of $S_i$ on $I_i$ would overestimate $B_1$ because families who rent (1) tend to have lower incomes than home owners, and (2) tend to save less than home owners even when their incomes equal those of home owners. This illustrates the fact (stressed at the beginning of this chapter) that if a dependent variable is influenced by more than one independent variable, a simple regression of the dependent variable on one of the independent variables is likely to result in a biased estimate of the effect of this independent variable on the dependent variable.

When this dummy variable technique is used, it is assumed that the values of the other regression coefficients in the regression equation are not affected by the value of the dummy variable. For example, in equation (12.11) it is assumed that the value of $B_1$ is the same regardless of whether $H_i$ equals zero or 1. In other words, it is assumed that the slope of the relationship between savings and income is the same among families who rent as among those who own their homes. This assumption may or may not be true. (Figure 12.5 shows a case where it is not true.) If this is not true, separate regression equations should be estimated for families who rent and for those who own their homes.[11]

[11] Alternatively, other more advanced techniques can be used in a case of this sort.

## 12.10 Multicollinearity

Regression analysis, like any tool, should not be applied blindly. It is important to check whether the assumptions underlying regression analysis are at least approximately correct, and to be aware of the problems that regression analysis can encounter. One important problem that can arise in multiple-regression studies is multicollinearity, which is defined as follows.

> MULTICOLLINEARITY: *Multicollinearity is a situation in which two or more of the independent variables are very highly correlated.*

In the case of the hosiery mill, suppose that the relationship in the past between output and the number of years of experience of the mill's manager has been as shown in Figure 12.6. If so, there has been a perfect linear relationship between the two independent variables in equation (12.1). In a case of this sort, it is impossible to estimate the regression coefficients of both independent variables $(X_1$ and $X_2)$ because the data provide no information concerning the effect of one independent variable, holding the other independent variable constant. All that can be observed is the effect of both independent variables together, given that they both move together in the way they have in the past.

To understand why it is impossible to estimate the regression coefficients of both independent variables in a case of this sort, let's look at the case of the hosiery mill in the situation shown in Figure 12.6. Regression analysis estimates the effect of each independent variable by seeing how much effect this one independent variable has on the dependent variable when other independent variables are held constant. However, in the situation in Figure 12.6 it is impossible for such an analysis to be carried out because one cannot separate the effects of the

**Figure 12.6**
Relationship between Independent Variables in Equation (12.1), Case of Complete Multicollinearity

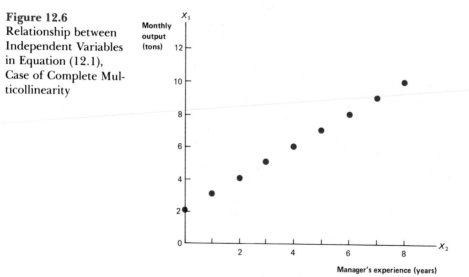

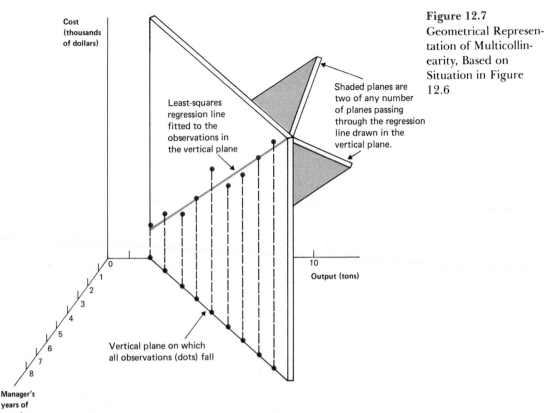

**Figure 12.7**
Geometrical Represen-
tation of Multicollin-
earity, Based on
Situation in Figure
12.6

Within the figure:

Cost
(thousands
of dollars)

Least-squares
regression line
fitted to the
observations in
the vertical plane

Shaded planes are
two of any number
of planes passing
through the regression
line drawn in the
vertical plane.

10
Output (tons)

0
1
2
3
4
5
6
7
8

Vertical plane on which
all observations (dots) fall

Manager's
years of
experience

output rate on cost from the effects of the years of experience of the
mill's manager. Whenever the output rate increased, so did the manager's number of years of experience. Given that the two independent
variables move together in this rigid, lockstep fashion, there is no way
to tell how much effect each has separately; all that we can observe is
the effect of both combined. Stated in terms of our earlier discussion of
experimental design, these effects are hopelessly confounded.

Another way of characterizing multicollinearity is by portraying
it geometrically. Figure 12.7 shows the relationship between cost on
the one hand and output and amount of managerial experience on the
other, given the situation in Figure 12.6. Because the two independent
variables are perfectly correlated (that is, the correlation coefficient
equals 1), there is no single plane that minimizes the sum of squared
deviations; instead, there are any number of planes that do equally well
in this regard. Thus, the method of least squares does not result in estimates of the effect of each independent variable and can only be used
to estimate their combined effects.

If there is good reason to believe that the independent variables
will continue to move in lockstep in the future as they have in the past,
multicollinearity does not prevent us from using regression analysis to
predict the dependent variable. Since the two independent variables
are perfectly correlated, one of them in effect stands for both; and we
therefore need use only one in the regression analysis. However, if the

independent variables cannot be counted on to continue to move in lockstep this procedure is dangerous, since it takes no account of the variation in the excluded independent variable.

In fact, one seldom encounters cases where independent variables are *perfectly* correlated, as they are in Figure 12.6. *But one often encounters cases where independent variables are so highly correlated that, although it is possible to estimate the regression coefficient of each variable, these regression coefficients cannot be estimated at all accurately.* How can one tell whether multicollinearity is so strong that it is responsible for the large standard errors of certain estimated regression coefficients? This is done by estimating the correlation coefficients among the independent variables. If some of these correlation coefficients are close to 1 (or − 1), multicollinearity is likely to be a problem.[12]

In cases where multicollinearity exists, it sometimes is possible to alter the independent variables in such a way as to reduce it. For example, suppose that a statistician wants to estimate a regression equation where the quantity demanded per year of a certain good is the dependent variable and the average price of this good and disposable income of American consumers are the independent variables. If disposable income is measured in money terms (that is, without adjustment for changes in the price level), there may be a high correlation between the independent variables. But if disposable income is measured in real terms (that is, with adjustment for changes in the price level), this correlation may be reduced considerably. Thus, this may be a good reason to measure disposable income in real rather than money terms.

If multicollinearity cannot be avoided in this way, there may be no alternative but to acquire new data which do not contain the high correlation among the independent variables. Just as bricks cannot be made without straw, there may be no way to estimate accurately the regression coefficient of a particular independent variable that is very highly correlated with some other independent variable.

### EXERCISES

**12.15** An economist wants to estimate a multiple-regression equation in which the amount saved by the $i$th family depends on the family's income, whether the family is white or nonwhite, and whether the family is headed by a male or a female. Explain how a regression equation of this sort can be estimated. What is the dependent variable? What are the independent variables? What assumptions must be made?

**12.16** A firm is interested in estimating the effect of its advertising expenditures and its number of salespeople on its sales. The firm calculates a multiple regression of its sales on its advertising expenditures and its number of salespeople. Are the independent variables in this regression

---

[12] Of course, if there are only two independent variables, there is only one simple correlation coefficient between the independent variables. But as the number of independent variables gets larger, the number of such correlation coefficients increases also. Besides looking at the simple correlation coefficients among pairs of independent variables, it is advisable to look at the multiple correlation coefficients of each of the independent variables on all of the others. See J. Johnston, *Econometric Methods*, 3rd. ed. (New York: McGraw-Hill, 1984), p. 163.

equation the only ones that should be included? If not, what are some other variables that might be considered?

**12.17** A government economist wants to estimate the relationship between a worker's productivity on a particular job, on the one hand, and the worker's sex and score on an aptitude test, on the other hand. He obtains the following data for 15 people:

| Productivity | Sex | Test score |
|---|---|---|
| 41 | Male | 70 |
| 52 | Female | 70 |
| 63 | Female | 75 |
| 74 | Male | 92 |
| 32 | Female | 48 |
| 48 | Female | 63 |
| 52 | Male | 75 |
| 61 | Female | 73 |
| 55 | Male | 80 |
| 54 | Female | 69 |
| 52 | Male | 73 |
| 39 | Male | 65 |
| 42 | Female | 60 |
| 45 | Female | 61 |
| 39 | Male | 62 |

(a) Calculate the multiple regression of productivity on sex and test score.
(b) Holding test score constant, what is the average difference in productivity between a male and female on this job?
(c) Holding sex constant, what is the effect of a 10-point increase in test score on productivity?
(d) If a male scores 82 on this test, what would you estimate his productivity to be?
(e) If a female scores 76 on this test, what would you estimate her productivity to be?

**12.18** Based on the data in Exercise 12.2, is there evidence of multicollinearity?

**12.19** A statistician working for a municipal government is interested in the determinants of a city's per capita expenditure on police salaries and operations. She calculates a multiple regression, where a city's per capita police expenditure is the dependent variable and its population size and population density are the independent variables. Are these the only independent variables that should be included? If not, what are some other variables that should be considered?

**12.20** The printout of the regression (based on the data in Table 12.6) of the annual amount saved by a family on its income (designated as C2) and the dummy variable showing whether it owns its home or rents (designated as C3) is shown (in part) below:

| Predictor | Coef | Stdev | t-ratio |
|---|---|---|---|
| C2 | 0.0675 | 0.0040 | 16.89 |
| C3 | 0.8273 | 0.0754 | 10.98 |

(a) Calculate a 95 percent confidence interval for the true regression coefficient of the family's income (which indicates the effect of income on the amount saved).

(b) Is there any evidence from the printout that multicollinearity is a problem? How would you determine whether multicollinearity is a problem in this case?

**12.21** A government agency wants to estimate the effect of a city's unemployment rate on its crime rate (as measured by the number of major crimes per million people). The agency believes that the crime rate (holding the unemployment rate constant) tended to be higher in 1987 than in 1985.

(a) Using the data below, calculate the multiple regression of the crime rate on the unemployment rate and a dummy variable representing the year to which the data pertain.

| City | Crime rate | Unemployment rate (percent) | Year |
|------|-----------|-----------------------------|------|
| A | 100 | 5 | 1985 |
| B | 120 | 7 | 1985 |
| C | 140 | 8 | 1985 |
| D | 170 | 9 | 1985 |
| E | 110 | 5 | 1987 |
| F | 160 | 8 | 1987 |
| G | 200 | 9 | 1987 |
| H | 120 | 6 | 1987 |
| I | 130 | 7 | 1987 |
| J | 150 | 7 | 1987 |

(b) Based on your results, what is the least-squares estimate of the mean difference between 1987 and 1985 in a city's crime rate, holding the unemployment rate constant? What is the effect of an increase of one percentage point in the unemployment rate, holding the year constant?

(c) Suppose that a noted criminologist were to publish a study indicating that the effect of a 1 percent increase in the unemployment rate on the crime rate was much greater in 1987 than in 1985. Would this make you skeptical of the results you obtained in (a)? Why, or why not? Can the data in (a) be used to determine whether the criminologist's hypothesis is true? If so, how?

# 12.11 Serial Correlation and the Durbin-Watson Test

SERIAL CORRELATION

Besides multicollinearity, another important problem that can arise in regression analysis is that the error terms (that is, the values of $e_i$) are not independent. This, of course, is a violation of one of the assumptions underlying regression analysis. (Recall Section 12.3.) Many regressions (both simple and multiple) in business and economics are based on time series; that is, the data consist of observations pertaining to various periods of time. For example, the data in Exercise 12.26 are time series because they pertain to 15 years of sales by the Rotunda Corporation. When regressions are based on time series, the error terms frequently are serially correlated.

> **SERIAL CORRELATION:** *If the value of a time series at time* t *is correlated with its value* h *periods before, the time series exhibits serial correlation (also known as autocorrelation). If no such correlation exists, the time series is said to be serially independent.*

To illustrate serial correlation, take the case of the Dow-Jones average of stock prices. If this average is relatively high (by historical standards) today, it is likely to be relatively high tomorrow as well. Not all economic time series are serially correlated. For example, in Chapter 8 the time series of the mean height of metal pieces (part of which is shown in Figure 8.13) shows no serial correlation.

As we pointed out above, the error terms in equation (12.3) may be serially correlated. Since this may be true for either simple or multiple regressions, let's assume that there is only one independent variable and that

$$Y_i = A + BX_i + e_i,$$

where $Y_i$ is the $i$th value of the dependent variable, $X_i$ is the $i$th value of the independent variable, and $e_i$ is the difference between $Y_i$ and the expected value of $Y_i$. Suppose that the error terms—that is, the $e_i$—are serially correlated. In particular, suppose that each error term is positively correlated with the subsequent error term. Thus, if $Y_1$ tends to be above its expected value (which means that $e_1 > 0$), $Y_2$ tends also to be above its expected value (which means that $e_2 > 0$).[13] If serial correlation of this sort is present, the least-squares method may still result in *unbiased estimates of A and B. The principal problem resulting from such serial correlation is that the standard errors of the estimators of* A *and* B *are really larger than would be estimated from the formulas that assume no serial correlation.* For example, the customary formula for the standard error of the estimator of $B$ is

$$s_b = \frac{s_e}{\sqrt{\sum_{i=1}^{n}(X_i - \overline{X})^2}} = \frac{s_e}{\sqrt{\sum_{i=1}^{n}X_i^2 - n\overline{X}^2}},$$

where $s_e$ is the standard error of estimate. This formula assumes that the $e_i$ (the error terms) are independent. But if serial correlation of this sort is present, this formula for the standard error of $b$ is an underestimate of the true standard error.

Another effect of serial correlation is that *it may make it appear that cycles exist in the data.* For example, if an observation is above the mean in one period, the next period's observation may also tend to be

---

[13] This is a case of *positive* serial correlation. (It is the sort of situation frequently encountered in economics.) If $Y_2$ tends to be *below* its expected value when $Y_1$ is *above* its expected value, this is a case of *negative* serial correlation. More is said about this below.

above the mean; and if an observation is below the mean in one period, the next period's observation may also tend to be below the mean. The result gives the appearance of a cycle but is actually due to serial correlation. Still another effect of serial correlation is that *it may induce a falsely high or falsely low agreement between two variables.* If both of the variables are serially correlated, what appears to be a large number of cases where the two variables are much the same or not the same may really amount to relatively few cases because the observations are not independent. (The observations are dependent on one another and tend to measure the same thing.)

DURBIN-WATSON TEST

To test whether serial correlation is present in the error terms in a regression, we can use the Durbin-Watson test. Let $\hat{e}_i$ be the difference between $Y_i$ and $\hat{Y}_i$, the value of $Y_i$ predicted by the sample regression; then

$$\hat{e}_i = Y_i - a - bX_i.$$

In order to apply the Durbin-Watson test, we must compute

$$d = \frac{\sum_{i=2}^{n} (\hat{e}_i - \hat{e}_{i-1})^2}{\sum_{i=1}^{n} \hat{e}_i^2}. \qquad (12.13)$$

Durbin and Watson have provided tables which show whether $d$ is so high or so low that the null hypothesis that there is no serial correlation should be rejected.

Suppose that we want to test this null hypothesis against the alternative hypothesis that there is *positive* serial correlation. (*Positive* serial correlation would mean that $e_i$ is *directly* related to $e_{i-1}$.) If so, we should reject the null hypothesis if $d < d_L$ and accept the null hypothesis if $d > d_u$. If $d_L \leq d \leq d_u$, the test is inconclusive. The values of $d_L$ and $d_u$ are shown in Appendix Table 11. (Note that these values depend on the sample size $n$ and on $k$, the number of independent variables in the regression.) On the other hand, suppose that the alternative hypothesis is that there is *negative* serial correlation. (*Negative* serial correlation would mean that $e_i$ is *inversely* related to $e_{i-1}$.) If so, we should reject the null hypothesis if $d > 4 - d_L$ and accept the null hypothesis if $d < 4 - d_u$. If $4 - d_u \leq d \leq 4 - d_L$, the test is inconclusive. Finally, for a two-tailed test of both positive and negative serial correlation, reject the null hypothesis if $d < d_L$ or if $d > 4 - d_L$, and accept the null hypothesis if $d_u < d < 4 - d_u$. Otherwise, the test is inconclusive. For a two-tailed test, note that the significance level is double that shown in Appendix Table 11.

The following example illustrates the use of the Durbin-Watson test to detect serial correlation.

EXAMPLE 12.7 An ice cream firm regresses its daily sales on two independent variables (namely, the mean temperature during the day and the price it charges). The residual from this regression for each of the 61 days included in the sample—that is, each value of $\hat{e}_i$—is shown in Table 12.8. Use the Durbin-Watson test to determine whether there is evidence of serial correlation in the error terms. (Set the significance level equal to .05, and use a two-tailed test.)

SOLUTION: As shown in Table 12.8,

$$d = \frac{13,908,400}{9,711,000} = 1.43.$$

The alternative hypothesis here is two-tailed, because we would like to detect either positive or negative serial correlation. Since

Table 12.8
Computation of Durbin-Watson Statistic in Example 12.7

| Day | $\hat{e}_i$ | $(\hat{e} - \hat{e}_{i-1})^2$ | $\hat{e}_i^2$ | Day | $\hat{e}_i$ | $(\hat{e}_i - \hat{e}_{i-1})^2$ | $\hat{e}_i^2$ |
|---|---|---|---|---|---|---|---|
| 1 | −190 | — | 36,100 | 31 | −400 | 10,000 | 160,000 |
| 2 | −200 | 100 | 40,000 | 32 | 350 | 562,500 | 122,500 |
| 3 | −220 | 400 | 48,400 | 33 | 50 | 90,000 | 2,500 |
| 4 | −100 | 14,400 | 10,000 | 34 | 50 | 0 | 2,500 |
| 5 | 800 | 810,000 | 640,000 | 35 | 50 | 0 | 2,500 |
| 6 | 20 | 608,400 | 400 | 36 | 150 | 10,000 | 22,500 |
| 7 | 300 | 78,400 | 90,000 | 37 | −900 | 1,102,500 | 810,000 |
| 8 | 400 | 10,000 | 160,000 | 38 | −800 | 10,000 | 640,000 |
| 9 | 0 | 160,000 | 0 | 39 | 300 | 1,210,000 | 90,000 |
| 10 | −440 | 193,600 | 193,600 | 40 | 300 | 0 | 90,000 |
| 11 | 800 | 1,537,600 | 640,000 | 41 | −30 | 108,900 | 900 |
| 12 | 300 | 250,000 | 90,000 | 42 | 50 | 6,400 | 2,500 |
| 13 | 500 | 40,000 | 250,000 | 43 | 0 | 2,500 | 0 |
| 14 | 500 | 0 | 250,000 | 44 | −50 | 2,500 | 2,500 |
| 15 | 200 | 90,000 | 40,000 | 45 | 300 | 122,500 | 90,000 |
| 16 | 210 | 100 | 44,100 | 46 | −300 | 360,000 | 90,000 |
| 17 | −50 | 67,600 | 2,500 | 47 | −700 | 160,000 | 490,000 |
| 18 | 0 | 2,500 | 0 | 48 | 300 | 1,000,000 | 90,000 |
| 19 | 50 | 2,500 | 2,500 | 49 | −150 | 202,500 | 22,500 |
| 20 | 200 | 22,500 | 40,000 | 50 | 100 | 62,500 | 10,000 |
| 21 | 600 | 160,000 | 360,000 | 51 | −150 | 62,500 | 22,500 |
| 22 | −100 | 490,000 | 10,000 | 52 | −150 | 0 | 22,500 |
| 23 | 450 | 302,500 | 202,500 | 53 | −100 | 2,500 | 10,000 |
| 24 | 500 | 2,500 | 250,000 | 54 | 0 | 10,000 | 0 |
| 25 | −300 | 640,000 | 90,000 | 55 | −500 | 250,000 | 250,000 |
| 26 | −600 | 90,000 | 360,000 | 56 | −1000 | 250,000 | 1,000,000 |
| 27 | 200 | 640,000 | 40,000 | 57 | −900 | 10,000 | 810,000 |
| 28 | 700 | 250,000 | 490,000 | 58 | −150 | 562,500 | 22,500 |
| 29 | 300 | 160,000 | 90,000 | 59 | 300 | 202,500 | 90,000 |
| 30 | −300 | 360,000 | 90,000 | 60 | −400 | 490,000 | 160,000 |
|  |  |  |  | 61 | −150 | 62,500 | 22,500 |
|  |  |  |  | Total | | 13,908,400 | 9,711,000 |

.05 is the significance level, we should look in Appendix Table 11 for $\alpha = .025$. Because $n = 61$, and $k = 2$, this table shows that $d_L = 1.44$ and $d_u = 1.57$ (since $n$ is approximately 60), which means that we should reject the null hypothesis if $d < 1.44$ or $d > 2.56$, and that we should accept the null hypothesis if $1.57 < d < 2.43$. Since $d = 1.43$, we should reject the null hypothesis. In other words, there seems to be statistically significant evidence of serial correlation.

## 12.12 Analyzing the Residuals

In the previous section we defined $\hat{e}_i$ as the difference between $Y_i$ and $\hat{Y}_i$. In other words, it is the difference between the actual value of $Y_i$ and the value predicted by the regression. Since it is a measure of the extent to which $Y_i$ *cannot* be explained by the regression, $\hat{e}_i$ is often called the **residual** for the $i$th observation. If the assumptions underlying regression analysis are met, the error terms should be successively independent, as we know from previous sections. If they are not independent, this is likely to show up in a plot of the residuals. For example, panel A in Figure 12.8 shows a case where successive residuals are positively correlated. (Why positively? Because if one residual is large, the next tends to be large; and if one residual is small, the next tends to be small.) Panel B of Figure 12.8 shows a case where successive residuals are negatively correlated. (Why negatively? Because if one residual is large, the next tends to be small; and if one residual is small, the next tends to be large.)

The printouts of many computer programs show the residuals, and statisticians frequently plot these to detect departures from the assumptions underlying regression analysis. To illustrate how an analysis of the residuals can be useful in detecting such departures, let's return to the ice cream firm in Example 12.7. After regressing the firm's daily sales on two independent variables (temperature and price), the firm's statistician calculates the residuals (shown in Table 12.8), plots them (Figure 12.9), and determines whether their pattern suggests that an important independent variable has been omitted. The statistician notices that many of the days when the residuals were large and positive were days when the firm had all three of its salespersons on the job, whereas many of the days when the residuals were large and negative were days when only one of the firm's salespeople worked (because the others were ill or on vacation). This pattern, shown in Figure 12.9, suggests that the regression does not contain all the important independent variables. (Statisticians often refer to this as a *specification error*.) To decide whether the number of salespeople working on a particular day should be used as an additional independent variable in the regression, the statistician can use this number as a third independent variable, and see whether its regression coefficient is statistically significant.

Besides being useful in detecting specification errors, the resid-

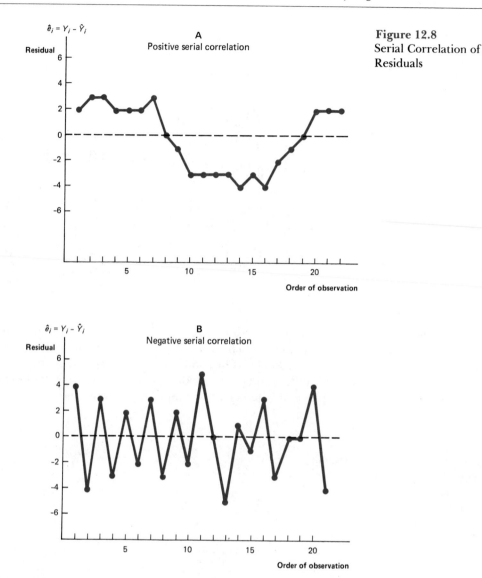

**Figure 12.8**
Serial Correlation of
Residuals

uals can also be plotted to detect departures from the assumption that
the standard deviation of the error terms in the regression is the same,
regardless of the values of the independent variables. For example,
suppose that the ice cream firm's statistician plots each day's residual
against the mean temperature for that day, the results being shown in
Figure 12.10. Apparently, the standard deviation of the residuals tends
to increase as the temperature increases. If this tendency is statistically
significant, and if the standard deviation of the residuals is a reasonably
good estimate of the standard deviation of the error terms, it indicates
a departure from one of the assumptions underlying regression analysis. (As emphasized, the form of regression analysis described here assumes that the standard deviation of the error terms is constant.)

Statisticians generally plot and analyze the residuals from their
regression equations; so do economists, as is indicated by Nobel laureate Paul A. Samuelson's well-known statement: "To the scientific

**Figure 12.9**
Residuals in Table
12.8: Detecting
Specification Errors

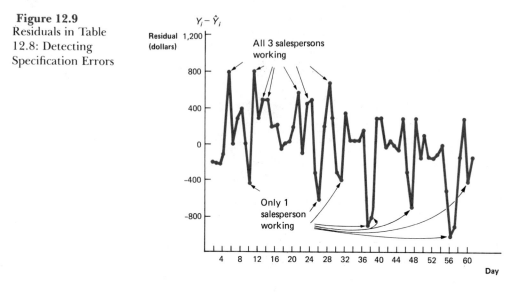

forecaster I say, 'Always study your residuals.' " What Samuelson
meant was that in evaluating any model it is useful to calculate the difference between each observation and what the model predicts this
observation will be. These differences—or residuals—are very useful
in indicating whether the model excludes some important explanatory
variables and whether its assumptions are valid.

**Figure 12.10**
Scatter Diagram of
Residual and Tempera-
ture: Departure from
Homoscedasticity

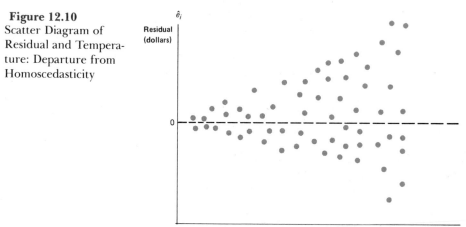

**Basic Idea #24** *Be sure to check whether the assumptions underlying regression
analysis are at least approximately correct. Multicollinearity,
serial correlation, and specification errors are among the things
to watch out for.*

## 12.13 Coping with Departures from the Assumptions

In previous sections and in Chapter 11, we have discussed ways of detecting certain departures from the assumptions underlying regression analysis. What can be done if a particular assumption is violated? Let's begin with the assumption that the conditional mean of the dependent variable is a linear function of the independent variables. If an examination of the data indicates that the relationship is curvilinear rather than linear, what can be done? One possibility is to use the logarithmic regression techniques described in Appendix 11.1 to fit a nonlinear relationship to the data. Another possibility is to fit a quadratic equation to the data. For example, suppose that a statistician is interested in the relationship between a family's income and the amount the family spends on food. To detect signs of nonlinearity, the statistician examines the scatter diagram and finds there is an obvious tendency for the relationship between these two variables to be curvilinear. He then assumes that

$$Y_i = A + B_1 X_i + B_2 X_i^2 + e_i,$$

where $Y_i$ is the amount spent on food by the $i$th family and $X_i$ is its income. Letting $X_i$ be the first independent variable and $X_i^2$ be the second independent variable, $A$, $B_1$, and $B_2$ are estimated in the way described in Section 12.4. In this way, a quadratic relationship is fitted to the data.

Next, let's turn to the assumption that the error terms in a regression are independent. If a statistician finds evidence of serial correlation in the error terms, what can be done about it? Specifically, suppose that a dependent variable $Y$ is regressed on an independent variable $X$, and the Durbin-Watson test indicates the presence of serial correlation. What can be done? One possible way to proceed is to perform a transformation on the data. In other words, rather than calculating the regression of $Y$ and $X$, one can calculate the regression of some function of the $Y$s on some function of the $X$s. The particular function that should be used is too complicated to be presented here,[14] but for present purposes, all that we want to convey is that if serial correlation is detected, there are methods for handling the problem.

Similarly, if other complications exist, such as departures from homoscedasticity, there are techniques for coping with these problems. This does not mean that the application of these techniques is cut-and-dried, or that it is possible to deal with any and all the departures from the assumptions. But anyone who uses regression techniques should appreciate the importance of testing the assumptions on which these techniques are based, while realizing that when these tests indicate that the assumptions are violated, a variety of methods exist that are aimed

---

[14] See J. Johnston, *op. cit.*

at handling the problem. Descriptions of these methods can be found in more advanced texts.[15]

## 12.14 Computer Programs and Multiple Regression

The advent of computer technology has caused a marked reduction in the amount of effort and expense required to calculate multiple regressions with large numbers of independent variables. Thirty years ago there was an enormous amount of drudgery in computing a multiple regression with more than a few independent variables; now such computations are relatively simple. It is important for you to be familiar with the kind of information printed out by computers and the form in which it appears. Since there is a wide variety of "canned" programs for calculating regressions, there is no single format or list of items which are printed out. However, the various sorts of computer printouts are sufficiently similar so that it is worthwhile looking at one illustration—Minitab—in some detail.

Table 12.9 shows the Minitab printout from a multiple regression of a chemical firm's 1984 profit on three independent variables: (1) the firm's sales (C5), (2) its rate of growth of sales (C6), and (3) its expenditures on research and development (C8). We are already familiar with the top half of this printout, which was reproduced in Table 12.3. As

**Table 12.9**
Minitab Printout of Results of Multiple Regression Concerning Chemical Firms' 1984 Profit

```
MTB > regress c7 on 3 predictors in c5,c6,c8

The regression equation is C7 = 32.0 + 0.0169 C5 -
1.42 C6 + 0.729 C8

    Predictor          Coef          Stdev        t-ratio
    Constant          31.95          11.86           2.69
    C5              0.016887       0.004013          4.21
    C6                -1.423          1.027          -1.39
    C8               0.7295         0.1260           5.79

    s = 26.27     R-sq = 99.4%     R-sq (adj) = 99.3%

    Analysis of Variance

    SOURCE               DF            SS             MS
    Regression            3       1940092         646697
    Error                18         12422            690
    Total                21       1952514
```

[15] In some cases, an equation is part of a system of equations. In such cases, the unwary statistician may obtain biased estimates of the population regression equation unless special techniques are used. For a simple introduction to this so-called identification problem (recall the case of statistical demand curves described earlier in this chapter), see E. Working, "What Do Statistical Demand Curves Show?" in E. Mansfield, *Statistics for Business and Economics: Readings and Cases* (New York: Norton. 1980). For a much more complete and advanced discussion, see J. Johnston, *op. cit.*

we already know, these figures show the value, standard error, and *t* statistic (or *t*-ratio) of each of the estimated regression coefficients (and of the intercept, which is designated as the "constant" in the printout).

The next line in Table 12.9 contains s, R-sq, and R-sq (adj). As pointed out in Example 12.5, R-sq is the multiple coefficient of determination. R-sq (adj) is the adjusted multiple coefficient of determination, discussed in footnote 8. Thus the only item left to be explained is s, which is the *standard error of estimate.* As in the case of simple regression, the standard error of estimate is an estimate of the standard deviation of the probability distribution of the dependent variable when the independent variables are all held constant. Thus, it is a measure of the amount of scatter of individual observations about the regression equation. In this case the standard error of estimate is 26.27 millions of dollars, which means that in the population as a whole, the standard deviation of the differences between a firm's profit and that predicted by the true regression is estimated to be $26.27 million.

The last part of Table 12.9 presents an analysis of variance for the multiple regression. The *variation explained by the regression* equals 1,940,092. The *error sum of squares* equals 12,422. The *regression mean square* is the variation explained by the regression divided by its degrees of freedom (which equal the number of independent variables). Thus, the regression mean square in this case equals 1,940,092 ÷ 3, or 646,697. The *error mean square* here is the error sum of squares divided by its degrees of freedom (which is one less than the number of observations minus the number of independent variables). Thus, the error mean square in this case equals 12,422 ÷ 18, or 690. The F value in Table 12.9 is 646,697 ÷ 690, or 937.[16]

## 12.15 Choosing the Best Form of a Multiple-Regression Equation

As emphasized throughout this chapter, the widespread availability and application of computer technology have facilitated greatly the use of multiple regression and correlation. Statisticians now can try various versions of a particular multiple regression equation to see which version fits best. These experiments can take various forms. For one thing, one can experiment with *various measures of a particular variable.* For example, the hosiery mill's statistician may want to try alternative output measures which take account of the fact that all tons of output are not the same. For another thing, one can experiment with *various forms of the regression equation.* Thus, in the case of the hosiery mill, the statis-

---

[16] The printout in Table 12.9 is not meant to be exhaustive. Besides the information shown there, the printout often shows the values of each of the dependent and independent variables (and their means and standard deviations); the simple correlation coefficient between the dependent variable and each independent variable and between each pair of independent variables; and the values of the residuals. Each of these items has been explained previously. (In the case of Minitab, observations that depart greatly from the regression and those with a large influence on the results are also printed out.)

tician might want to try the logarithm of the output, rather than the output, as the independent variable.

Still another way in which statisticians can experiment with a particular regression equation is to *add and omit various independent variables to see how the results are affected.* The hosiery mill's statisticians might be interested in adding another independent variable, the price of nylon, to the multiple regression to see whether it seems to have a significant effect on the firm's costs. Also, they may want to determine (1) whether the estimated regression coefficients of the independent variables currently used are statistically significant when this new independent variable is introduced, and (2) whether the values of these regression coefficients are altered considerably by its introduction.

*Stepwise Multiple Regression*

Statisticians sometimes use a procedure called *stepwise multiple regression* to specify which independent variables seem to provide the best explanation of the behavior of the dependent variable. Using commonly available programs, a computer can determine which of a set of independent variables is most highly correlated with the dependent variable $Y$. (The computer programs allow the inclusion of dozens of independent variables in the set.) Suppose that this independent variable is $X$. The computer then selects the independent variable from the remainder of the set that results in the greatest reduction in the variation unexplained by the regression of $Y$ on $X$. Suppose that this added independent variable is $R$. Then the computer selects the independent variable from the remainder of this set that results in the greatest reduction in the variation unexplained by the regression of $Y$ on $X$ and $R$. If this independent variable is $S$, then the computer selects the independent variable from the remainder of this set that results in the greatest reduction in the variation unexplained by the regression of $Y$ on $X$, $R$, and $S$. This process continues until all the independent variables in the set have been added to the regression, or until none of the remaining independent variables in the set reduces significantly the unexplained variation.

*Given the relative ease with which one can calculate alternative forms and versions of a multiple regression equation, it is extremely important that the statistician have good a priori reasons for including each of the independent variables in the regression equation being used.* Using stepwise multiple regression, it is not very difficult to select some independent variables that will explain much of the variation in practically any dependent variable, even if these independent variables really have little or no effect on the dependent variable. Why is this so? Because some variables are bound to be correlated with the dependent variable by chance, or they may be influenced by the same factors as the dependent variable, with the result that they are correlated with the dependent variable even though they do not influence the dependent variable. If one uses the computer to hunt around long enough, one is likely to find a combination of independent variables that explains much of the variation in the dependent variable. But if these variables are merely the result of an indiscriminate, mechanical quest for a good-fitting regression equation, the resulting equation is likely to be useless for purposes of prediction.

# How to Estimate the Mileage per Gallon of a New Car

Automobile buyers are interested in estimates of mileage per gallon because fuel economy is one way to save money. If one make of car gets many more miles per gallon than another make, this is something that car buyers want to know. Automobile manufacturers are required by law to provide new car buyers with Environmental Protection Agency (EPA) estimates of mileage per gallon, but there are questions concerning the accuracy of these estimates. The Environmental Protection Agency has stated: "While EPA tests provide a measure of how much fuel each vehicle uses under precisely controlled laboratory conditions, they cannot predict exactly what mileage you personally will achieve. . . . The mileage per gallon ratings are estimates, not guaranteed fuel economy claims. They are published to help you compare the relative mileage per gallon of each model in a vehicle class.[17] To obtain these estimates, the EPA uses a dynamometer to simulate driving conditions in a 23-minute laboratory test.

Another source of data concerning mileage per gallon is *Consumer Reports*, published periodically by Consumers Union. It provides estimates based on actual experience in city driving, expressway driving, and driving on a 195-mile test trip. (For some of the results, see Table 1.) However, because such testing takes a good deal of time and money, *Consumer Reports* can provide such estimates for only a limited number of car makes. Also, its estimates are not available until several months after the introduction of new models.

If estimates by *Consumer Reports* are not available for a particular make of car, can one somehow use other sorts of data to arrive at such an estimate? Michael Lovell of Wesleyan University has studied this question, using the regression techniques discussed in Chapters 11 and 12. Based on 1984 data for the 46 cars for which both EPA and *Consumer Reports* estimates are available, he found that one can predict a car's mileage per gallon, as estimated by *Consumer Reports*, reasonably well by the physical characteristics of the car. In particular, if the $i$th car's mileage per gallon ($Y_i$) is regressed on (1) its weight ($X_{1i}$), (2) a dummy variable that is 0 for standard transmissions and 1 for automatics ($X_{2i}$), and (3) a dummy variable that is 0 for a gasoline engine and 1 for a diesel engine ($X_{3i}$), the result is

$$\hat{Y}_i = 43.6 - 0.006X_{1i} - 3.75X_{2i} + 6.26\ X_{3i} \qquad (12.14)$$

[17] M. Lovell, "EPA and CU Mileage per Gallon Estimates," unpublished paper, July 1984. I am indebted to Professor Lovell for permission to quote from this paper.

**Table 1**
Mileage per Gallon
Estimates, Selected
Automobiles, 1985

| Automobile | Mileage per gallon* | Automobile | Mileage per gallon* |
|---|---|---|---|
| **Small Cars:** | | **Medium Cars:** | |
| Chevrolet Chevette | 28 | Audi 5000S | 20 |
| Dodge Omni/Charger | 33 | Buick Century | 27 |
| Ford Escort | 34 | Chrysler LeBaron | 26 |
| Honda Civic | 30 | Oldsmobile Cutlass Supreme | 20 |
| Mazda GLC | 33 | | |
| Nissan Sentra | 38 | **Large Cars:** | |
| Renault Alliance | 33 | | |
| Toyota Tercel | 34 | Mercury Grand Marquis | 19 |

Source: *Consumer Reports*, April 1985.

* Unweighted average of figures for city and expressway driving.

The multiple coefficient of determination is 0.74.

Interestingly enough, equation (12.14) can predict a car's mileage per gallon more accurately than its EPA estimate. In other words, if you calculate the difference between a car's mileage per gallon (according to *Consumer Reports*) and its predicted mileage per gallon based on equation (12.14), this difference tends to be smaller than the difference between its mileage per gallon (according to *Consumer Reports*) and the EPA estimate for this make of car.

However, this does not mean that the EPA estimate for this make of car is not of use in predicting mileage per gallon. On the contrary, if the EPA estimate for the $i$th make of car is introduced as an additional independent variable in equation (12.14), its regression coefficient turns out to be statistically significant. The new regression equation is:

$$\hat{Y}_i = 22.0 - 0.002X_{1i} - 2.76X_{2i} + 3.28X_{3i} + 0.415X_{4i} \qquad (12.15)$$

where $X_{4i}$ is the EPA estimate for the $i$th make of car. The standard error of the EPA estimate's regression coefficient is 0.097; thus its $t$-statistic is 4.28, and it differs significantly from 0. Holding constant the physical characteristics $(X_1, X_2, X_3)$ of a car, the EPA estimate is of use in predicting its mileage per gallon.

Equation (12.15) fits the data better than equation (12.14), as evidenced by the fact that the multiple coefficient of determination is 0.82 for equation (12.15), as compared with 0.74 for equation (12.14). Also, the standard error of estimate is smaller (2.04 versus 2.42) for equation (12.15) than for equation (12.14). The fact that equation (12.15) fits the data better than equation (12.14) is an indication that the EPA estimate should be included as an independent variable in the equation.

Lovell concludes his study by pointing out that, although the EPA estimates "do not offer precise estimates as to which vehicles are most likely to yield high mileage on the road," more precise estimates can be obtained by considering the physical characteristics $(X_1, X_2, X_3)$ of the car together with the EPA estimate. In other words, one can estimate mileage per gallon more accurately by using equation (12.15) than by simply assuming that mileage per gallon will equal the EPA figure. This is an interesting result, with potential importance to the many new car buyers in the United States. (Cautionary note: The results summarized here pertain to 1984 only and are only illustrative. Readers interested in pursuing this topic further should consult Lovell's entire body of work in this area.)

### PROBING DEEPER

1. According to equation (12.15), the estimated regression coefficient of the EPA estimate is 0.415. Does this mean that the actual mileage per gallon tends to be about 41.5 percent of the EPA estimate? Why, or why not?

2. Is serial correlation a problem in equation (12.15)? Why, or why not?

3. How would you determine whether multicollinearity is a problem in equation (12.15)?

4. Comparing equation (12.14) with equation (12.15), it appears that the estimated extra mileage per gallon due to a diesel engine (rather than a gasoline engine) is greater when the EPA estimate is not included in the regression equation. Does the estimated regression coefficient for $X_{3i}$ in equation (12.14) mean the same thing as the estimated regression coefficient for $X_{3i}$ in equation (12.15)? If not, how does the interpretation of the two coefficients differ?

5. According to the above discussion, the standard error of estimate is smaller for equation (12.15) than for equation (12.14). Does this mean that equation (12.15) is a better indication of the factors *causing* differences among cars in mileage per gallon? Explain.

6. According to the above discussion, the multiple coefficient of determination is higher for equation (12.15) than for equation (12.14). Does this imply that a prediction based on equation (12.15) will be more accurate than one based on equation (12.14)?

In choosing the best form of a multiple regression equation, the statistician sometimes must decide whether to include an independent variable which on *a priori* grounds is almost certain to influence the dependent variable, but which has an estimated regression coefficient that is not statistically significant. If there are very strong *a priori* grounds for believing that an independent variable affects the dependent variable, it is acceptable to include this variable in the regression equation, even if its regression coefficient is not statistically significant. After all, the fact that the estimated regression coefficient is not statistically significant does not prove that the true regression coefficient is zero. And the estimated regression coefficient would be likely to constitute a better estimate of the true regression coefficient than zero (which would be the value attributed to it if the independent variable were omitted from the equation).

## 12.16 Determinants of the Strength of Cotton Yarn: A Case Study

As a final illustration of the use of multiple regression, let's consider a study by Acheson Duncan of Johns Hopkins University of the determinants of the strength of cotton yarn.[18] The U.S. Department of Agriculture has presented data concerning the skein strength (in pounds) of cotton yarn, which is perhaps the most important single measure of spinning quality. Besides indicating good spinning and weaving performance, good yarn strength increases the range of usefulness of a given cotton.

In his analysis, Duncan hypothesizes that the strength of a particular variety of cotton depends on three qualities of cotton yarn: (1) fiber length, (2) fiber tensile strength, and (3) fiber fineness. Fiber length was determined by the fibrograph instrument, a type of photoelectric device. Fiber tensile strength, measured in 1,000 pounds per square inch, is a very important factor in determining yarn strength, and varieties of cotton with good fiber strength cause less trouble in spinning than do weak-fibered varieties. In general, fiber fineness (measured in 0.1 milograms per inch of fiber) is believed to contribute to yarn strength. However, the desirability of fiber fineness depends on the specific end product or use.

Based on a sample of 20 pieces of cotton yarn, Duncan calculated the following multiple regression equation:

$$\hat{Y}_i = 39.32 + 1.069X_{1i} + 0.164X_{2i} - 0.936X_{3i},$$

where $Y_i$ is the skein strength of the $i^{th}$ type of yarn, $X_{1i}$ is its fiber length, $X_{2i}$ is its fiber tensile strength, and $X_{3i}$ is its fiber fineness. The standard error of the regression coefficient for fiber length was 0.189, the standard error of the regression for fiber tensile strength was

---

[18] A. Duncan, *op. cit.* The basic data came from U.S. Department of Agriculture, *Results of Fiber and Spinning Tests for Some Varieties of Upland Cotton Grown in the United States.*

0.507, and the standard error of the regression coefficient for fiber fineness was 0.371. The multiple coefficient of determination was 0.70.

These results provide interesting information concerning the effects of each of the independent variables on the strength of cotton yarn, when the other independent variables are held constant. As expected, there seems to be a strong direct relationship between fiber length and yarn strength. On the other hand, fiber fineness seems to be inversely related to yarn strength. As for the effects of fiber tensile strength, the relevant regression coefficient (0.164) is not very reliable, since its standard error is 0.507. Throughout industry and government, multiple-regression analyses of this sort are being carried out by the thousands, and a great deal of valuable new knowledge and insight has resulted.

### EXERCISES

**12.22** A statistician regresses a dependent variable on four independent variables, based on 30 observations. The Durbin-Watson statistic equals 1.06. Test whether there is serial correlation in the deviations of the observations from the regression equation. Let $\alpha = .05$, and suppose that the alternative hypothesis is that there is positive serial correlation.

**12.23** Using 40 observations, an economist regresses a dependent variable on two independent variables. The Durbin-Watson statistic equals 2.71. Test whether there is serial correlation in the deviations of the observations from the regression equation. Let $\alpha = .01$, and assume the alternative hypothesis to be that there is negative serial correlation.

**12.24** A department store's manager asks her staff to calculate a multiple regression: the dependent variable is the store's sales in a particular year, the independent variables are gross national product and the population of the city in which the store is located. The residuals (in millions of dollars) from the multiple regression are as follows:

| Year | Residual | Year | Residual |
|------|----------|------|----------|
| 1959 | −0.4 | 1973 | −0.2 |
| 1960 | +0.2 | 1974 | +0.0 |
| 1961 | +0.3 | 1975 | +0.1 |
| 1962 | −0.1 | 1976 | +0.2 |
| 1963 | −0.2 | 1977 | −0.7 |
| 1964 | +0.3 | 1978 | −0.6 |
| 1965 | +0.2 | 1979 | −0.3 |
| 1966 | +0.2 | 1980 | +0.2 |
| 1967 | +0.3 | 1981 | +0.3 |
| 1968 | +0.4 | 1982 | +0.2 |
| 1969 | +0.5 | 1983 | +0.4 |
| 1970 | −0.6 | 1984 | 0.0 |
| 1971 | −0.7 | 1985 | +0.1 |
| 1972 | −0.1 | 1986 | 0.0 |

(a) Plot the residuals. Based on visual inspection, is there any evidence of serial correlation?

(b) Use the Durbin-Watson test to see whether there is significant evidence of positive serial correlation. (Let $\alpha = .025$.)

**12.25** Edwin Mansfield and Harold Wein[19] carried out a study for one of the nation's largest railroads to determine how the costs incurred in a freight yard are related to the output of the yard. The two most important services performed by a yard are switching and delivery, and it seems reasonable to use the number of cuts switched and the number of cars delivered during a particular period as a measure of output. (A *cut* is a group of cars that rolls as a unit onto the same classification track; it is often used as a unit of switching output.) The study assumed that

$$C_i = A + B_1 S_i + B_2 D_i + e_i,$$

where $C_i$ is the cost incurred in this freight yard on the $i$th day, $S_i$ is the number of cuts switched in this yard on the $i$th day, $D_i$ is the number of cars delivered in this yard on the $i$th day, and $e_i$ is the difference between $C_i$ and $E(C_i)$. Data were obtained regarding $C_i$, $S_i$, and $D_i$ for 61 days. Based on the procedures described in Section 12.4, these data were used to obtain estimates of $A$, $B_1$, and $B_2$. The resulting regression equation was

$$\hat{C}_i = 4{,}914 + 0.42 S_i + 2.44 D_i,$$

where $\hat{C}_1$ is the cost (in dollars) predicted by the regression equation for the $i$th day.
(a) If you were asked to evaluate this study, what steps would you take to determine whether the principal assumptions underlying regression analysis were met?
(b) If you were satisfied that the underlying assumptions were met, of what use might the above regression equation be to the railroad? Be specific.
(c) Before using the above regression equation, what additional statistics would you like to have? Why?

# Chapter Review

1. Whereas a *simple regression* includes only one independent variable, a *multiple regression* includes more than one independent variable. An advantage of multiple regression over simple regression is that one frequently can predict the dependent variable more accurately if more than one independent variable is used. Also, if the dependent variable is influenced by more than one independent variable, a simple regression of the dependent variable on a single independent variable may result in a biased estimate of the effect of this independent variable on the dependent variable.

2. The first step in *multiple-regression analysis* is to identify the independent variables, and then to specify the mathematical form of the equation relating the expected value of the dependent variable to the independent variables. For example, if $Y$ is the dependent variable and

[19] E. Mansfield and H. Wein, "A Regression Control Chart for Costs," *Applied Statistics*, March 1958.

$X_1$ and $X_2$ are identified as the independent variables, one might specify that

$$Y_i = A + B_1 X_{1i} + B_2 X_{2i} + e_i,$$

where $e_i$ is the difference between $Y_i$ and $E(Y_i)$. To estimate $B_1$ and $B_2$ (called the true regression coefficients of $X_1$ and $X_2$) as well as $A$ (the intercept of this true regression equation), we use the values that minimize the sum of squared deviations of $Y_i$ from $\hat{Y}_i$, the value of the dependent variable predicted by the estimated regression equation.

3. Multiple regressions are generally calculated by computers rather than by hand. The standard programs print out the *estimated standard deviations* of the least-squares estimators of $B_1$ and $B_2$ (these estimated standard deviations being called *standard errors*). Using the value of the least-squares estimator of $B_1$ or $B_2$ together with these standard errors, one can obtain a confidence interval for $B_1$ or $B_2$. For example, a confidence interval for $B_1$ is $b_1 \pm t_{\alpha/2} s_{b_1}$, where $b_1$ is the least-squares estimator of $B_1$ and $s_{b_1}$ is the standard error of $b_1$. The standard programs also print out the *t*-ratio (or *t* statistic) for each estimated regression coefficient, the *t*-ratio being the estimated regression coefficient divided by its standard error. To test whether a true regression coefficient equals zero, one should see whether its *t*-ratio exceeds $t_{\alpha/2}$ or is less than $-t_{\alpha/2}$; if so, one should reject the hypothesis that the true regression coefficient is zero. (This is a two-tailed test.)

4. The *multiple coefficient of determination $R^2$* equals the ratio of the variation explained by the multiple regression to the total variation in the dependent variable. The positive square root of the multiple coefficient of determination is called the *multiple correlation coefficient* and is denoted by $R$. Both $R^2$ and $R$ are measures of how well the regression equation fits the data: The closer they are to zero, the poorer the fit; the closer they are to 1, the better the fit.

5. Multiple regression can be used to analyze the effects of *qualitative variables* (that is, variables which do not assume numerical values) as well as *quantitative variables*. To represent a qualitative variable we use a *dummy variable* (that is, a variable that can equal 0 or 1). For example, to include whether or not a family owns its home in a regression, we construct a dummy variable which equals 1 if the family owns its own home and 0 if it rents.

6. An important problem that can occur in multiple regression is *multicollinearity*, a situation where two or more of the independent variables are highly correlated. If two independent variables are perfectly correlated, there is no way of estimating the effect of each, holding the other constant; all that we can observe is the effect of both combined. Even if two independent variables are highly (but not perfectly) correlated, there may be no way of estimating the regression coefficient of each with even minimal accuracy. In cases of this sort, it may be necessary to obtain additional data or to redefine the variables in order to reduce the correlation between the independent variables.

7. Another problem arises when the error terms in the equation (that is, the values of $e_i$) are serially correlated. If the basic data constitute a time series, the existence of serial correlation means that succes-

sive values of $e_i$ are not independent. For example, if the value of the error term for one period is positive, the value of the error term for the next period may tend to be positive also. The principal problem resulting from serial correlation of this sort is that the ordinary formulas underestimate the standard errors of the regression coefficients and intercept. To test whether serial correlation of this kind is present, we can use the *Durbin-Watson test*, which is based on the value of $d$,

$$d = \sum_{i=2}^{n} (\hat{e}_i - \hat{e}_{i-1})^2 \div \sum_{i=1}^{n} \hat{e}_i^2,$$

where $\hat{e}_i$ equals $Y_i - \hat{Y}_i$. (Each value of $\hat{e}_i$ is called a *residual*.) Tables are available which show whether $d$ is so high or so low that the null hypothesis (that there is no serial correlation) should be rejected.

8. The advent of computer technology has caused a marked reduction in the amount of effort and expense required to calculate multiple regressions. We have described and explained the items generally printed out by programs like Minitab, including the $F$ test for determining whether all true regression coefficients are zero. With existing computer technology, one can experiment with various forms of a regression equation. It is important that the statistician have good *a priori* reasons for including each of the independent variables in the equation. If independent variables are chosen indiscriminately in a mechanical quest for a good-fitting equation, the resulting equation is likely to be useless for purposes of prediction.

## Chapter Review Exercises

**12.26** The Rotunda Corporation believes that its annual sales depend on disposable income in its city of location and on the price of its product. Data for the past 15 years concerning these variables are given below:

| Year | Number of units sold annually (millions) | Disposable income (billions of dollars) | Price of product (dollars per unit) |
|---|---|---|---|
| 1987 | 8 | 4 | 9 |
| 1986 | 8 | 4 | 8 |
| 1985 | 5 | 3 | 8 |
| 1984 | 4 | 3 | 9 |
| 1983 | 6 | 3 | 7 |
| 1982 | 4 | 3 | 10 |
| 1981 | 2 | 2 | 8 |
| 1980 | 3 | 2 | 6 |
| 1979 | 4 | 2 | 5 |
| 1978 | 2 | 2 | 7 |
| 1977 | 2 | 2 | 8 |
| 1976 | 1 | 1 | 6 |
| 1975 | 1 | 1 | 5 |
| 1974 | 1 | 1 | 7 |
| 1973 | 1 | 1 | 5 |

(a) Calculate the multiple-regression equation, if Rotunda's sales are the dependent variable and disposable income and price of product are the independent variables.

(b) Based on your results in (a), predict the volume of annual sales for the Rotunda Corporation under each of the following sets of circumstances:

    (i) disposable income equals $3 billion and price equals $7;
    (ii) disposable income equals $1 billion and price equals $6;
    (iii) disposable income equals $2 billion and price equals $10.

(c) What is the estimated effect on the expected value of Rotunda's annual sales of an increase of $1 billion in disposable income? What is the estimated effect of an increase of $1 in price?

(d) The president of the Rotunda Corporation uses the multiple-regression equation you derived in (a) to predict the firm's annual sales for next year. Since he believes that disposable income will be $6 billion and product price will be $11, he bases his prediction on these values of the independent variables. What objections might legitimately be raised against these procedures?

(e) Use the data to calculate the multiple coefficient of determination between Rotunda's sales, on the one hand, and disposable income and price, on the other. Interpret your results. Does it appear that the regression equation you derived in (a) fits the data well? Why, or why not?

(f) Between 1973 and 1987, the population of the city in which the Rotunda Corporation is located grew steadily. By 1987, it was about four times its size in 1973. The marketing director of the Rotunda Corporation objects to the equation that you derived in (a) on the grounds that population, not disposable income, has been the variable that has influenced Rotunda's sales.

    (i) Is the marketing director correct in saying that if population and disposable income are perfectly correlated, the observed effect of disposable income in your equation may be due to population?

    (ii) Can you suggest some ways of altering the multiple regression so that the marketing director's hypothesis can be tested?

**12.27** (a) Based on the data in Exercise 12.1, is there evidence of multicollinearity? (b) If there were multicollinearity in this case, how would you go about dealing with it?

**12.28** A store wants to estimate the relationship between its sales, on the one hand, and the number of salespersons it hires and its advertising expenditures, on the other. Using its monthly sales as the dependent variable, the store's statistician regresses this variable on the number of salespersons employed by the store during the month (designated as C2) and the store's monthly advertising expenditure (designated as C3). There are 43 observations. Both sales and advertising expenditure are measured in millions of dollars. The computer printout is as follows:

| Predictor | Coef | Stdev | t-ratio |
|-----------|------|-------|---------|
| Constant  | 0.310 | 0.190 | 1.63 |
| C2        | 0.052 | 0.021 | 2.48 |
| C3        | 5.621 | 2.012 | 2.79 |

(a) Compute a 95 percent confidence interval for the true regression coefficient of the number of salespersons employed by the store.

(b) Compute a 90 percent confidence interval for the true regression coefficient of the store's advertising expenditure.

(c) If both the number of salespersons and advertising expenditure equal zero, wouldn't one expect that sales should equal zero? Given that the standard error of the intercept is 0.19, can you test whether this is true? How? (Let $\alpha = .05$.)

**12.29** (a) Calculate the deviation of each of the observations in Exercise 12.26 from the multiple-regression equation (when sales are regressed on disposable income and price).

(b) Test whether there is serial correlation in the error terms (that is, the deviations of the observations from the regression equation) when the Rotunda Corporation's sales are regressed on disposable income and price. Let $\alpha = .05$, and assume the alternative hypothesis to be that there is either positive or negative serial correlation.

**12.30** In *Bazemore vs. Friday*, the U.S. Court of Appeals ruled that, to challenge regression studies as evidence in a trial, one need only demonstrate that there are other variables (besides those included as independent variables) that may influence the dependent variable and that might be taken into account. Whether or not a listing of untested variables should be enough to discredit a regression analysis was taken to the Supreme Court in 1985.[20] What are some objections to such a ruling?

**12.31** For 20 working women (with children) in a particular community, their age at the birth of their first child, their score on an IQ test, and their current annual income are as follows:

| Age at birth of first child (years) | IQ test score | Current annual income ($000) |
|:---:|:---:|:---:|
| 18 | 105 | 16 |
| 17 | 108 | 17 |
| 20 | 110 | 17 |
| 22 | 109 | 18 |
| 24 | 110 | 17 |
| 26 | 120 | 21 |
| 27 | 125 | 20 |
| 19 | 111 | 17 |
| 18 | 85 | 9 |
| 20 | 96 | 11 |
| 21 | 94 | 10 |
| 24 | 130 | 26 |
| 30 | 134 | 21 |
| 28 | 103 | 11 |
| 25 | 102 | 12 |
| 24 | 89 | 8 |
| 22 | 92 | 12 |
| 20 | 95 | 13 |
| 19 | 100 | 14 |
| 17 | 91 | 9 |

Use Minitab (or some other computer package) to calculate the regression of income on IQ test score and age at birth of first child, as well as the multiple correlation coefficient. If no computer package is available, do the calculations by hand.

[20] "Statistical Evidence Becomes a Federal Case," *Business Week,* December 2, 1985.

# PART VI

# Time Series and Index Numbers

Let x be a binomial random variable with $n = 12$ and $E(x) = 3$. What is the probability that this random variable lies between its standard deviation and its expected value?

$E(x) = n \cdot \pi$
$3 = 12 \cdot \pi$
$\pi = \frac{3}{12} = .25$

$\sigma(x) = \sqrt{n \cdot \pi (1 - \pi)} = \sqrt{12 \times .25 \times .75}$
$= \sqrt{2.25} = 1.5$

$P(\sigma(x) < x < E(x)) = P(1.5 < x < 3) = P(x = 2) = .2323$

$P(\sigma(x) \leq x \leq E(x)) = \begin{array}{c} \text{or} \\ P \end{array} (1.5 \leq x \leq 3) = P(x = 2) + P(x = 3) = .2323 + .2581$
$= .4904$

---

Suppose no info about IQ in a country. 600 students gives .90 confidence int. $115.54 \leq \mu \leq 118.46$
as mean IQ of the 600 students?

① $\bar{x} - z_{\alpha/2} \frac{s}{\sqrt{n}} = 115.54$

② $\bar{x} + z_{\alpha/2} \frac{s}{\sqrt{n}} = 118.46$

① + ② $\quad 2\bar{x} = 115.54 + 118.46$
$\bar{x} = 117$

b) sample stand. dev.?

$\bar{x} - z_{\alpha/2} \frac{s}{\sqrt{n}} = 115.54$

$117 - 1.65 \frac{s}{\sqrt{600}} = 115.54$

$s = \left( \frac{115.54 - 117}{-1.65} \right) \sqrt{600} = s \frac{-1.46}{-1.65} (24.49) = 21.67$

c) if only 20 students of 600 does answers is a) & b) change?

a) does not

b) becomes $\bar{x} - t_{\alpha/2} \frac{s}{\sqrt{n}} =$
$117 - (1.729) \frac{s}{\sqrt{n}} = 115.54$
$s = 3.78$

Many semesters an instructor has recorded " "
students grades is 72. the current class of 25
that the current is superior.
mean $\bar{x} = 75.2$ present sufficient evidence incurrent class
superior $\alpha = .05$.

$H_0 : \mu = 72$

$H_1 : \mu > 72$

Decision
rej. $H_0$ if $z > z_\alpha$
Do not $H_0$ if $z < z_\alpha$

$$z = \frac{\bar{x} - \mu}{\sigma/\sqrt{n}} = \frac{75.2 - 72}{12/\sqrt{25}}$$

$$= \frac{3.2}{2.4} = 1.33$$

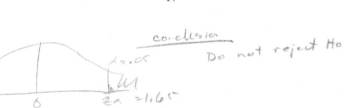

conclusion
Do not reject $H_0$

$\alpha = .05$

$\mu$

$z_\alpha = 1.65$

rej.

Do not.

What is the prob of type I error.

$\alpha = .05$

What is the prob of type II error.
cannot be calculated from the
information given. We need to know
what is the "true" mean.

# Introduction to Time Series

# 13

## 13.1 Introduction

Business and economic data generally are of two types: ***cross-section*** data and ***time-series*** data. *Cross-section data pertain to a variety of units or entities at a given point in time.* For example, the profits data regarding the petroleum firms exhibited in Table 1.8 are cross-section data. They show the profitability of various oil firms during a particular period of time, 1984. *Time series data, on the other hand, pertain to a given unit or entity at a number of points in time.* Thus, if we had obtained data concerning the profitability of Exxon during each year from 1955 to 1987, this would have been an example of a time series.

Decision makers in business and government are continually involved with time series. Month by month, business executives pore over the latest month's sales figures. They compare these figures with those for previous periods and with past forecasts in an attempt to gauge their current performance and determine their future moves. Moreover, sales figures are only one of a myriad of time series that are generated by most firms. The accounting department of the typical company generates time series on sales, costs, taxes, profits, assets, debts, dividends, and many other variables. Moreover, the engineering department, production department, marketing department, and other parts of a firm also contribute to the generation of time series.

Statisticians have developed certain techniques for describing and analyzing time series. For one thing, they have tried in various ways to break down a time series into such elements as its trend, seasonal variation, and cyclical variation. This chapter describes these procedures and indicates how they have been used in business and economic forecasting. In addition, we point out the limitations of these techniques. (As we shall see, they are based on a very rudimentary theory of the determinants of economic variables, and should be used with caution.) Finally, we will describe briefly some alternative types of forecasting techniques which are based on econometric models.

## 13.2 The Traditional Time-Series Model

The classical approach to the analysis of time series, devised primarily by economic statisticians, was essentially descriptive. It assumed that an economic time series could be decomposed into four components: trend, seasonal variation, cyclical variation, and irregular movements. More specifically, it assumed that the value of an economic variable at a certain point in time could be represented as the product of each of these four components. For example, the value of a company's sales in January 1987 was viewed as equal to

$$T \times S \times C \times I,$$

(13.1)

where $T$ is the trend value of the firm's sales during that month, $S$ is the seasonal variation attributable to January, $C$ is the cyclical variation occurring that month, and $I$ is the irregular variation that occurred then.[1] Each of these components is defined below.

**Trend:** *A trend is a relatively smooth long-term movement of a time series.* For example, the population of the United States increased rather steadily between 1940 and 1984, as shown in Figure 13.1.

**Figure 13.1**
Population of the United States, 1940–84

*Source: Economic Report of the President, 1985* (Washington, D.C.: Government Printing Office, 1985).

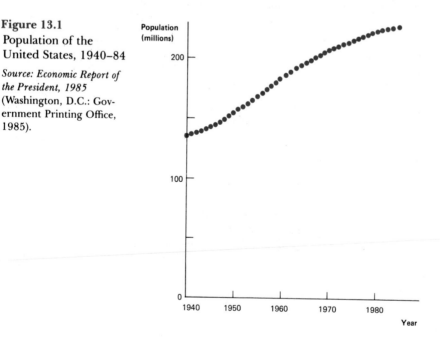

----

[1] In some versions of this model, the components are added rather than multiplied. That is, it is assumed that

$$Y = T + S + C + I,$$

where $Y$ is the value of the time series.

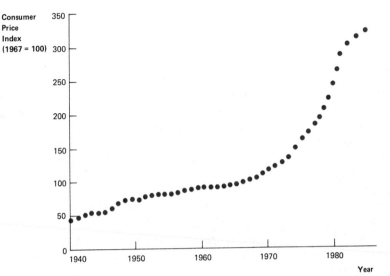

**Figure 13.2**
Consumer Price Index, 1940–84

*Source: Economic Report of the President, 1985* (Washington, D.C.: Government Printing Office, 1985).

Thus, there has been an upward trend in the U.S. population. Similarly, there has been an upward trend in the price level in the United States, as shown in Figure 13.2. Of course, not all trends are upward. For example, the trend in the average length of the marketable interest-bearing public debt in the United States has been downward, as shown in Figure 13.3.[2] Whether upward or downward, the trend of a time series is represented by a smooth curve. In equation (13.1) $T$ is the value of the firm's sales that would be predicted for January 1987, based on such a curve.

***Seasonal Variation:** In a particular month, the value of an economic variable is likely to differ from what would be expected on the basis of its*

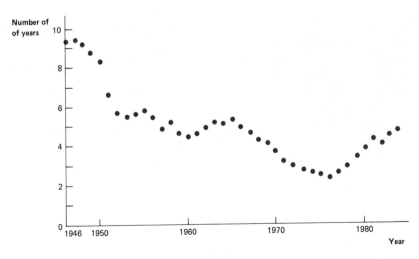

**Figure 13.3**
Average Length of Marketable Interest-Bearing U.S. Debt, 1946–84

*Source: Economic Report of the President, 1985* (Washington, D.C.: Government Printing Office, 1985).

---

[2] In still other cases the trend is horizontal; that is, there is no upward tendency or downward tendency in the time series. In these cases it it often said that there is *no trend*.

*trend, due to seasonal factors.* For example, consider the sales of a firm that produces air conditioners. Since the demand for air conditioners is much higher in the summer than in the winter, one would expect that the monthly time series of the firm's sales would show a pronounced and predictable seasonal pattern. Specifically, sales each year would tend to be higher from June through August than during the rest of the year. As we shall see, it is possible to calculate *seasonal indexes* which estimate how much each month departs from what would be expected on the basis of its trend. In equation (13.1) we must multiply the trend value *T* by the seasonal index *S* to allow for the effect of this seasonal variation.

*Cyclical Variation: Another reason why an economic variable may differ from its trend value is that it may be influenced by the so-called business cycle.* As is well known, the general tempo of economic activity in our society has exhibited a cyclical nature, with booms being followed by recessions, and recessions being followed by expansions. These cycles have not been regular or consistent (which is one reason why many economists prefer the term *business fluctuations* to *business cycles*); but unquestionably there has been a certain cyclical ebb and flow of economic activity, which has been reflected in a great many time series. For this reason, $T \times S$ is multiplied by $C$, which is supposed to indicate the effect of cyclical variation on the firm's sales in equation (13.1).

*Irregular Variation:* After having been multiplied by both $S$ and $C$, the trend value $T$ has been altered to reflect seasonal and cyclical forces. However, besides these forces, *a variety of short-term, erratic forces are also at work.* Their effects are represented by $I$. In effect, $I$ reflects the effects of all factors other than the trend, seasonal variation, and cyclical variation. According to the classical model, these irregular forces are too unpredictable to be useful for forecasting purposes.

---

*Basic Idea #25*    **The classical approach to the analysis of economic time series assumes that such a time series can be decomposed into four components: (1) trend, which is a relatively smooth, long-term movement, (2) seasonal variation, which is monthly or quarterly variation due to seasonal factors, (3) cyclical variation, which is due to the business cycle, and (4) irregular variation, which results from a variety of short-term, erratic forces.**

---

## 13.3 Estimation of a Least-Squares Linear Trend Line

Many studies have been carried out to estimate the trend, seasonal variation, and cyclical variation in particular economic time series. In Sections 13.3 to 13.7 of this chapter we discuss the methods used to es-

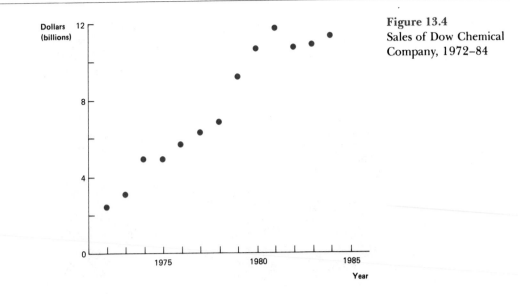

**Figure 13.4**
Sales of Dow Chemical
Company, 1972–84

timate a trend; in subsequent sections we shall take up seasonal and cyclical variation. First, we take up the case where the long-term overall
movement of the time series seems to be linear. For example, this
seems true for the sales of the Dow Chemical Company during 1972–
84. (These sales are plotted in Figure 13.4.) In a case where the trend
seems to be linear, statisticians frequently use the method of least
squares to calculate the trend. In other words, they assume that if the
long-term forces underlying the trend were the only ones at work the
time series would be approximately linear. Specifically, they assume
that

*Linear Trend*

$$Y_t = A + Bt, \qquad\qquad (13.2)$$

where $Y_t$ is the trend value of the variable at time $t$. (Note that $t$ assumes
values like 1987 or 1988 if time is measured in years.) The *trend value*
is the value of the variable that would result if only the trend were at
work. The deviation of $Y$, the actual value of the variable, from the
trend value is *the deviation from trend*.

If the deviations from trend—that is, $Y - Y_t$—can be regarded as
random variables with zero mean and constant standard deviation, and
if they are independent, we know from Chapter 11 that the most efficient linear unbiased estimates of $A$ and $B$ can be obtained by the
method of least squares. Actually, because the deviations from trend
are not independent (see Section 12.11) these assumptions frequently
are not met. Thus, one should be very cautious about the interpretation and use of such trend lines. As stressed in previous chapters, regression techniques can be quite misleading when the assumptions on
which they are based are not valid.[3]

---

[3] Another way of interpreting least-squares trend lines is as a purely descriptive
device, the values of the time series included in the analysis being viewed as the population.

**Table 13.1**
Sales of Dow Chemical
Company, 1972–84,
and Calculation of $\Sigma S_t$,
$\Sigma S_t t'$, and $\Sigma t'^2$

| Year (t) | Sales (billions of dollars) ($S_t$) | $t'$ | $S_t t'$ | $t'^2$ |
|---|---|---|---|---|
| 1972 | 2.40 | −6 | −14.40 | 36 |
| 1973 | 3.07 | −5 | −15.35 | 25 |
| 1974 | 4.94 | −4 | −19.76 | 16 |
| 1975 | 4.89 | −3 | −14.67 | 9 |
| 1976 | 5.65 | −2 | −11.30 | 4 |
| 1977 | 6.23 | −1 | − 6.23 | 1 |
| 1978 | 6.89 | 0 | 0 | 0 |
| 1979 | 9.26 | 1 | 9.26 | 1 |
| 1980 | 10.63 | 2 | 21.26 | 4 |
| 1981 | 11.87 | 3 | 35.61 | 9 |
| 1982 | 10.62 | 4 | 42.48 | 16 |
| 1983 | 10.95 | 5 | 54.75 | 25 |
| 1984 | 11.42 | 6 | 68.52 | 36 |
| Total | 98.82 | | 150.17 | 182 |

$$\bar{S} = \frac{98.82}{13} = 7.602$$

To illustrate the calculation of a linear trend, let's examine Dow Chemical Company's annual sales from 1972 to 1984, as shown in Table 13.1. Since sales in year $t$ is the dependent variable and $t$ is the independent variable, it follows from our discussion in Chapter 11 that

$$b = \frac{\sum_{t=t_0}^{t_0+n-1} (S_t - \bar{S})(t - \bar{t})}{\sum_{t=t_0}^{t_0+n-1} (t - \bar{t})^2}, \tag{13.3}$$

$$a = \bar{S} - b\bar{t}, \tag{13.4}$$

where $S_t$ is sales (in billions of dollars) in year $t$, $t_0$ is the earliest year in the time series (that is, 1972), and $t_0 + n - 1$ is the latest year in the time series (that is, 1984).

In equations (13.3) and (13.4), time is measured in ordinary calendar years. In other words, $t$ varies from 1972 to 1984. To carry out the calculations, it is advisable to convert time into a coded variable which has a mean of zero. If there are an odd number of years (as in this case), let $t' = 0$ for the middle year, and let the rest of the years be ..., −3, −2, −1, 0, 1, 2, 3, .... If there is an even number of years, let $t' = -1$ and $t' = 1$ for the two middle years, and let the rest of the years be ... −5, −3, −1, 1, 3, 5, .... The advantage of this coding is that if it is carried out,[4]

---

[4] Since the time scale is changed so that $\bar{t} = 0$, it is obvious that equation (13.3) simplifies to equation (13.5) and that equation (13.4) simplifies to equation (13.6).

$$b = \frac{\Sigma S_{t'} t'}{\Sigma t'^2} \qquad\qquad (13.5)$$

$$a = \overline{S}. \qquad\qquad (13.6)$$

These expressions are simpler to compute than those in the previous paragraph.

Table 13.1 shows the values of $\Sigma S_{t'} t'$, $\Sigma t'^2$, and $\overline{S}$ for the time series of Dow's sales. Inserting these values into equations (13.5) and (13.6), we have

$$b = \frac{150.17}{182} = 0.825$$

$$a = 7.602.$$

Thus, the trend line (shown in Figure 13.5) is

$$S_{t'} = 7.602 + 0.825 t'.$$

To interpret this equation, note that $t' = 0$ when $t = 1978$. Thus, this equation says that the estimated trend value of Dow's sales was 7.602 billions of dollars in 1978, and that it increased by 0.825 billions of dollars per year.

As pointed out in the previous paragraph, this trend line has 1978, the middle year in the time series, as its origin. This is because $t' = 0$ for 1978. Suppose that we want to obtain an equation for the trend where $t$, not $t'$, is the independent variable. In other words, suppose that we want to return to the original time scale where time is

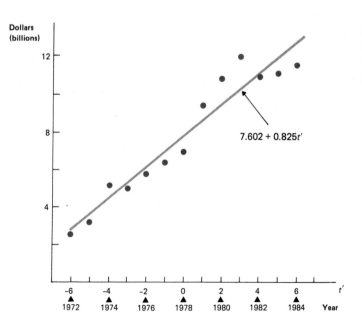

**Figure 13.5**
Linear Trend in Sales, Dow Chemical Company, 1972–84

measured in calendar years rather than in deviations from 1978. To change scale in this way, we need only change $a$, since $b$ will be unaffected. Why? Because regardless of whether time is measured in calendar years or in deviations from 1978, the annual increase in the trend value is $b$ ($= .825$) billions of dollars.

How is the value of $a$ changed if we return to the original time scale where time is measure in calendar years (like 1980 or 1981)? Letting $a'$ be the new value of $a$, it must be true that

$$a' + b(1978) = 7.602.$$

Why is this true? Because if time is measured in calendar years (and the origin is at year zero), the left-hand side of the above equation equals the trend value in 1978. Since this trend value must be the same regardless of which scale is used for measuring time, it must equal 7.602, the former value of $a$. Thus, since $b = .825$,

$$a' = 7.602 - .825(1978) = 7.602 - 1,631.850 = -1,624.248.$$

Consequently, if the original scale is used for time—that is, if $t$ is set equal to 1978 in 1978—the trend line is

$$S_t = -1,624.248 + 0.825t. \qquad (13.7)$$

The following example shows how, once a trend line of this sort has been obtained, it can be used to summarize and describe the long-term behavior of a time series.

---

EXAMPLE 13.1 Use the least-squares trend line to estimate Dow's 1980 sales which would have been expected on the basis of this trend alone. Use this trend line to estimate the average annual increase in Dow's sales from 1972 to 1984.

SOLUTION: According to equation (13.7), the trend value of Dow's sales in 1980 was $-1,624.248 + 0.825(1980)$, or 9.252 billions of dollars. According to this equation, the firm's sales tended to increase annually by 0.825 billions of dollars (that is, by $825 million) during this period.

---

## 13.4 Estimation of a Nonlinear Trend

Many time series exhibit nonlinear trends. In some such cases, a quadratic function of time provides an adequate trend. Such a trend can be represented as

$$Y_t = A + B_1 t + B_2 t^2.$$

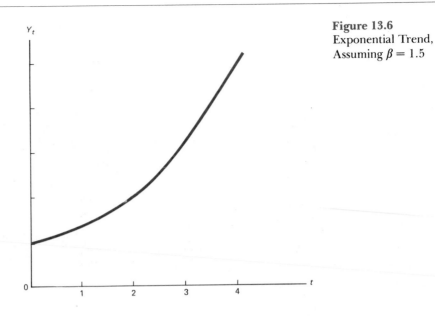

**Figure 13.6**
Exponential Trend,
Assuming $\beta = 1.5$

To estimate $A$, $B_1$, and $B_2$, we can use the multiple-regression techniques described in the previous chapter. (Recall Section 12.13.[5]) As indicated in the previous chapter, there are standard computer programs for making these computations.

For many variables, an exponential curve provides a better-fitting trend than a quadratic curve. The equation for such a trend (shown in Figure 13.6) is

*Exponential Trend*

$$Y_t = \alpha\beta^t, \qquad (13.8)$$

where $Y_t$ is the trend value of the time series at time $t$. A trend of this sort seems to fit many business and economic time series, including the capital-labor ratio and the savings-income ratio of the American economy. If this sort of trend is present, we can take logarithms of both sides of equation (13.8), the result being

$$\log Y_t = A + Bt, \qquad (13.9)$$

where $A = \log \alpha$ and $B = \log \beta$. Since equation (13.9) is linear, we can estimate $A$ and $B$ by the method of least squares.[6] Then we can take antilogs of $A$ and $B$ to estimate $\alpha$ and $\beta$, the unknown coefficients in equation (13.8). In this way, we can estimate the nonlinear trend shown in equation (13.8).

The following example illustrates the calculation of an exponential trend where $B$ is negative, not positive. That is, the trend is downward, not upward.

---

[5] For those readers who skipped Chapter 12, the only point necessary to grasp is that it is possible to calculate a quadratic trend by the method of least squares.

[6] This has been pointed out in Appendix 11.1.

EXAMPLE 13.2 An economist wants to estimate the trend in the ratio of U.S. consumption expenditures to gross national product. Table 13.2 shows this ratio (expressed as a percentage and given at 10-year intervals) from 1940 to 1980. Calculate an exponential trend for this ratio.

**Table 13.2**
Calculation of Exponential Trend for Consumption as a Percent of Gross National Product, United States, 1940–80

| Year | $Y$ | $\log Y$ | $t'$ | $\log Y(t')$ | $t'^2$ |
|------|------|----------|------|--------------|--------|
| 1940 | 71.0 | 1.8513 | −20 | −37.026 | 400 |
| 1950 | 67.0 | 1.8261 | −10 | −18.261 | 100 |
| 1960 | 64.1 | 1.8069 | 0 | 0 | 0 |
| 1970 | 62.6 | 1.7966 | 10 | 17.966 | 100 |
| 1980 | 63.4 | 1.8021 | 20 | 36.042 | 400 |
| Total | | 9.0830 | 0 | −1.279 | 1,000 |

*Source: Economic Report of the President* (Washington, D.C.: Government Printing Office, 1985).

SOLUTION: Taking logarithms of the ratios, we obtain the figures in the third column of the table. The least-squares estimator of $B$ is

$$b = \frac{\Sigma \log Y(t')}{\Sigma t'^2} \qquad (13.10)$$

and the least-squares estimator of $A$ is

$$a = \frac{\Sigma \log Y}{n}, \qquad (13.11)$$

if time is coded as shown in Table 13.2. (That is, time equals −20, −10, 0, 10, and 20, which means that 1960 is the new origin.) Inserting the figures from Table 13.2 into equations (13.10) and (13.11), we have

$$b = \frac{-1.279}{1000} = -.001279$$

$$a = \frac{9.083}{5} = 1.8166.$$

Since the antilog of $b$ is .9971 and the antilog of $a$ is 65.6, the regression line in equation (13.8) is estimated to be

$$Y_t = 65.6(0.9971)^{t'},$$

where $t'$ is measured in years from 1960.

## 13.5 The Trend in the Capital-Labor Ratio: A Case Study

An important characteristic of any economy is the ratio of capital to labor. Since capital includes the various means of production such as buildings, equipment, and inventories, the capital-labor ratio shows the average amount of such equipment and related items (in constant dollars) that can be utilized by a member of the labor force. All other things equal, the higher the capital-labor ratio, the more productive a laborer will be. There has been considerable interest in the changes over time in the capital-labor ratio in the American economy. Nobel laureate Simon Kuznets of Harvard University analyzed this ratio in some of his studies; his estimates are shown in part in Table 13.3. Apparently, there was a steady growth in this ratio from 1900 to 1953; it was about $5,000 per person in 1900 and about $6,000 per person after World War II.[7]

Nobel laureate Lawrence Klein of the University of Pennsylvania and Richard Kosobud of Wayne State University estimated the trend in this ratio, using Kuznets's data (shown in part in Table 13.3).[8] Klein and Kosobud assumed that the trend line conformed to equation (13.8). That is, they assumed that

$$Y_t = \alpha \beta^{t'},$$

where $Y_t$ in this case is the trend value of the capital-labor ratio at time $t'$, and $t'$ is measured in units of six months from the beginning of 1927. In accord with our discussion in the previous section, they regressed

**Table 13.3**
Capital per Worker, United States. Selected Years 1900–53

| Year | Capital per worker (dollars) | Year | Capital per worker (dollars) |
|---|---|---|---|
| 1900 | 4,820 | 1927 | 6,157 |
| 1903 | 4,843 | 1930 | 6,666 |
| 1906 | 4,970 | 1933 | 7,326 |
| 1909 | 5,231 | 1936 | 6,202 |
| 1912 | 5,309 | 1939 | 6,393 |
| 1915 | 5,580 | 1942 | 5,736 |
| 1918 | 5,090 | 1945 | 5,499 |
| 1921 | 5,995 | 1948 | 6,066 |
| 1924 | 5,877 | 1951 | 6,049 |
| | | 1953 | 6,812 |

[7] The data in Table 13.3 are from S. Kuznets, *Capital in the American Economy* (New York: National Bureau of Economic Research, 1959); and J. Kendrick, *Productivity Trends in the United States* (New York: National Bureau of Economic Research, 1960).
[8] L. Klein and R. Kosobud, "Great Ratios of Economics," reprinted in E. Mansfield, *Statistics for Business and Economics: Reading and Cases* (New York: Norton, 1980).

log $Y$ on $t'$ to obtain the least-squares estimates of log $\alpha$ and log $\beta$, the result being

$$\log Y_t = 3.76126 + 0.0010t'.$$

Taking antilogs of both sides, Klein and Kosobud obtained the following trend line:

$$Y_t = 5771 \ (1.0023)^{t'}, \qquad\qquad (13.12)$$

since the antilog of 3.76126 is 5771 and the antilog of .0010 is 1.0023. The trend line in equation (13.12) indicates that on the average, the capital-labor ratio grew at a compound rate of slightly under 1/4 of 1 percent semiannually. This result is of considerable interest to economists concerned with the long-term economic growth of the United States.

## 13.6 Moving Averages

In some cases, there is no relatively simple mathematical function that can adequately portray the long-term movement of a particular time series. For example, consider the annual price of the common stock of the IBM Corporation during the years 1967–84. Certainly, it is questionable whether this time series, shown in Figure 13.7, exhibits a simple linear, exponential, or quadratic trend. The price of the stock reached a peak in 1972, declined until 1975, increased again until

**Figure 13.7**
**Annual Price of the Common Stock of IBM Corporation, 1967–84.**

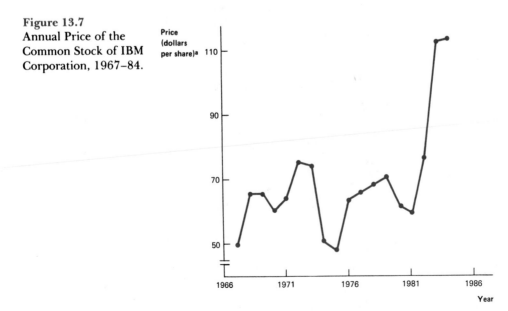

[a] Mean of annual high and low price (adjusted for stock splits)

| Year | Price (dollars per share)* | Five-year moving total | Five-year moving average | Seven-year moving total | Seven-year moving average |
|------|----------------------------|------------------------|--------------------------|-------------------------|---------------------------|
| 1967 | 50  |     |      |     |      |
| 1968 | 66  |     |      |     |      |
| 1969 | 66  | 308 | 61.6 |     |      |
| 1970 | 61  | 334 | 66.8 | 459 | 65.6 |
| 1971 | 65  | 343 | 68.6 | 460 | 65.7 |
| 1972 | 76  | 328 | 65.6 | 442 | 63.1 |
| 1973 | 75  | 315 | 63.0 | 440 | 62.9 |
| 1974 | 51  | 314 | 62.8 | 445 | 63.6 |
| 1975 | 48  | 304 | 60.8 | 448 | 64.0 |
| 1976 | 64  | 297 | 59.4 | 443 | 63.3 |
| 1977 | 66  | 317 | 63.4 | 430 | 61.4 |
| 1978 | 68  | 331 | 66.2 | 439 | 62.7 |
| 1979 | 71  | 327 | 65.4 | 468 | 66.9 |
| 1980 | 62  | 338 | 67.6 | 517 | 73.9 |
| 1981 | 60  | 383 | 76.6 | 565 | 80.7 |
| 1982 | 77  | 426 | 85.2 |     |      |
| 1983 | 113 |     |      |     |      |
| 1984 | 114 |     |      |     |      |

Table 13.4
Five-year and Seven-year Moving Averages of the Price of IBM Common Stock.

* Mean of annual high and low price.

1979, fell again during 1980 and 1981, and then reached new heights in 1983 and 1984. In cases of this sort, statisticians frequently use moving averages to **smooth** the time series—that is, *to generate a smooth curve showing the long-term movements of the series*. Thus, *moving averages are sometimes used to estimate trends where the trends do not lend themselves to the treatment described in previous sections.*

To illustrate what we mean by a moving average, consider Table 13.4, which shows the annual price of IBM common stock from 1967 to 1984. *To obtain a **smoothed** or trend value for a particular year, we average the figures for an interval of time centered on this year.* For example, suppose that we were to compute a *five-year* moving average. In this case, we would use the mean price during the period 1967–71 as the trend value for 1969; the mean price during 1968–72 as the trend value for 1970; the mean price during 1969–73 as the trend value for 1971; and so on. The computation of the five-year moving average is shown in Table 13.4, and the moving average itself is plotted in Figure 13.8.

Of course, a five-year moving average is not the only kind. Suppose that we want to compute a *seven-year moving average* for the data in Table 13.4. In this case, we would use the mean price during the period 1967–73 as the trend value for 1970; the mean price during 1968–74 as the trend value for 1971; the mean price during 1969–75 as the trend value for 1972; and so on. The computation of the seven-year moving average is shown in Table 13.4, and the moving average itself is plotted in Figure 13.8. Other possible kinds of moving averages are 9-year moving averages, 10-year moving averages, and so on.

**Figure 13.8**
Five-year and Seven-year Moving Average of Price of Common Stock of IBM Corporation, 1967–84.

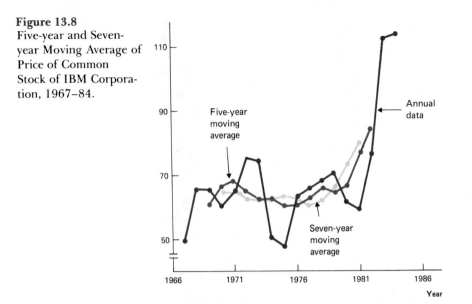

The basic idea underlying the use of moving averages is that, *if the time series contains certain fluctuations or cycles that tend to recur, the effect of these cycles can be eliminated by taking a moving average where the number of years in the average equals the period of the cycle.* For example, Figure 13.9 shows a time series where there is a simple four-year cycle superimposed on a linear trend. In this case, if a moving average is taken of the time series (the number of years in the average being equal to the period of the cycle, which is four years), the result will be the linear trend only. Obviously, however, practically no business or economic time series contains as simple or regular a cycle as that shown in Figure 13.9. Thus, all that the use of a moving average can do is to smooth a time series, not eliminate the short-term fluctuations entirely.

**Figure 13.9**
Time Series Composed of a Linear Trend plus a Simple Four-Year Cycle

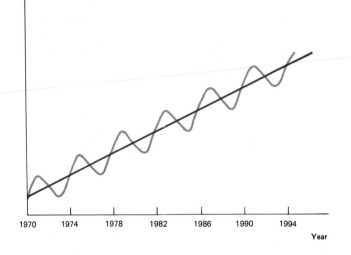

Finally, several points should be noted concerning moving averages. First, *if a time series is a purely random sequence of numbers, a moving average of this time series will tend to exhibit cyclical fluctuations.* (In terms of our discussion in the previous chapter, this is due to the fact that a moving average is serially correlated.) Thus, one must be careful to recognize that many apparent cycles in moving averages may be spurious. Second, *the peaks and troughs in the moving average may occur at different times than the peaks and troughs in the original time series.* (For example, as shown in Figure 13.8, the price of IBM common stock reached a trough in 1975, but the seven-year moving average reached a trough in 1977.) Third, *a moving average of this type cannot be calculated for the latest or earliest years in a time series,* since the average depends on numbers that precede or occur after the time series.

## 13.7 Exponential Smoothing

Another method of smoothing a time series is to use *exponential smoothing.* According to this method, *the trend value at time* t *is a weighted average of all available previous values, where the weights decline geometrically as one goes backward in time.* Using the moving average techniques described in the previous section, the trend value at time *t* is obtained by calculating an unweighted average of the observations centered at time *t,* the number of observations being 5, 7, or some other fixed number. In contrast, exponential smoothing uses *all* the previous observations to obtain the trend value at time *t,* and the weight attached to each observation declines *geometrically* with the age of the observation.

To illustrate the nature of exponential smoothing, let's assume that a firm has been in existence for five years and that its sales have been \$1 million, \$3 million, \$3 million, \$2 million, and \$4 million. (See Figure 13.10.) Then, the trend value in the fifth year would be a weighted average of \$1 million, \$3 million, \$3 million, \$2 million, and \$4 million, where the weights decline geometrically as we go backward in time. Specifically, the weight attached to the observation at time *t* equals $\theta$, the weight attached to the observation at time $t - 1$ equals $(1 - \theta)\theta$, the weight attached to the observation at time $t - 2$ equals $(1 - \theta)^2\theta$, the weight attached to the observation at time $t - 3$ equals $(1 - \theta)^3\theta., \ldots$, and the weight attached to the observation at the earliest relevant point in time (time 0) equals $(1 - \theta)^t$. Clearly, the weights decline geometrically as one goes backward in time; that is, the weight attached to the observation at time $t - 1$ is $(1 - \theta)$ times the weight attached to the observation at time *t;* the weight attached to the observation at time $t - 2$ is $(1 - \theta)$ times the weight attached to the observation at time $t - 1$; and so on.

To calculate an exponentially smoothed time series, it is necessary to choose a value of $\theta$, which is designated the ***smoothing constant.*** If          *Smoothing Constant*

**Figure 13.10**
Sales of Firm, Actual
and Exponentially
Smoothed*

* Year 1 is the firm's first
year in existence, year 2 is
its second year, and so on.

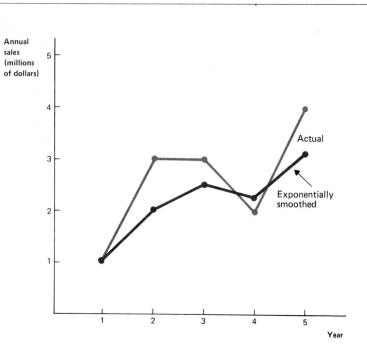

we choose a value of 0.5 for $\theta$, the exponentially smoothed value of the firm's sales in each of the five years is as follows:

$$S_0 = 1$$
$$S_1 = (.5)(3) + (1 - .5)(1) = 2$$
$$S_2 = (.5)(3) + (1 - .5)(.5)(3) + (1 - .5)^2(1) = 2.5$$
$$S_3 = (.5)(2) + (1 - .5)(.5)(3) + (1 - .5)^2(.5)(3)$$
$$\quad + (1 - .5)^3(1) = 2.25$$
$$S_4 = (.5)(4) + (1 - .5)(.5)(2) + (1 - .5)^2(.5)(3)$$
$$\quad + (1 - .5)^3(.5)(3) + (1 - .5)^4(1) = 3.125,$$

where $S_0$ is the exponentially smoothed value of the firm's sales in the first year of its existence, $S_1$ is this value in the second year, $S_2$ the value in the third year, and so on. Figure 13.10 shows both the original time series and the exponentially smoothed time series.

A noteworthy characteristic of an exponentially smoothed time series is that in order to calculate the value of such a smoothed time series at time $t$ all we need is the value of the smoothed time series at time $t - 1$ and the actual value of the time series at time $t$. This is because the smoothed value of the time series at time $t$ is a simple weighted average of the smoothed value at time $t - 1$ and the actual value at time $t$. If $S_t$ is the smoothed value at time $t$,

**Formula for Exponential Smoothing**

$$S_t = \theta Y(t) + (1 - \theta)S_{t-1}, \qquad\qquad (13.13)$$

where $Y(t)$ is the value of the time series at time $t$.[9] This means that in

---

[9] It is easy to prove that equation (13.13) is true. If $Y(t)$ is the actual value of the time series at time $t$, then equation (13.13) implies that

order to calculate an exponentially smoothed time series, one does not need to keep all the previous values of the actual time series. For example, the firm in the previous paragraph does not need to keep all its previous sales figures; *it only needs to keep the value of the exponentially smoothed sales in the previous year.* From this information alone (together with the present value of sales and the smoothing constant), the firm can calculate the smoothed current sales figure. For example, if the firm's sales in its sixth year of existence are $5 million, the smoothed value of sales for the sixth year is

$$(0.5)(5) + (1 - 0.5)(3.125) = 4.062,$$

or $4.062 million.

In choosing the value of the smoothing constant $\theta$, it is essential that a number between 0 and 1 be picked. In other words, it is essential that $0 \leqslant \theta \leqslant 1$. If $\theta$ is close to 1, past values of the time series are given relatively little weight (compared to recent values) in calculating smoothed values. If $\theta$ is close to 0, past values of the time series are given considerable weight (as compared to recent values) in calculating smoothed values. If the time series contains a great deal of random variation, it is often advisable to choose a relatively small value of $\theta$ since this results in relatively little weight being put on $Y(t)$, which is more influenced than $S_{t-1}$ by this variation. On the other hand, if one wants the smoothed time series to reflect relatively quickly whatever changes occur in the average level of the time series, the value of $\theta$ should be set at a high level.

Exponential smoothing has both theoretical and practical advantages. If the time series conforms to certain assumptions concerning its probability distribution, it can be shown (with more advanced mathematical techniques than are appropriate here) that exponential smoothing has desirable theoretical properties.[10] As for its practical advantages, it is obvious that exponential smoothing can be carried out very simply if the computations are based on equation (13.13). For these reasons, exponential smoothing is sometimes preferred over the use of simple moving averages. (For a discussion of the use of exponential smoothing for forecasting purposes, see Appendix 13.2.)

The following example illustrates the construction of an exponentially smoothed time series.

---

$S_t = \theta Y(t) + (1 - \theta)S_{t-1}$

$\quad = \theta Y(t) + (1 - \theta)[\theta Y(t-1) + (1 - \theta)S_{t-2}]$

$\quad = \theta Y(t) + (1 - \theta)\theta Y(t-1) + (1 - \theta)^2[\theta Y(t-2) + (1 - \theta)S_{t-3}]$

$\quad = \theta Y(t) + (1 - \theta)\theta Y(t-1) + (1 - \theta)^2\theta Y(t-2) + \ldots + (1 - \theta)^t Y(0).$

Since the last expression on the right-hand side is equivalent to the definition of an exponentially smoothed time series in the first paragraph of this section, it follows that equation (13.13) is true.

[10] For example, see R. G. Brown, *Smoothing, Forecasting, and Prediction of Discrete Time Series* (Englewood Cliffs: Prentice-Hall, 1963).

EXAMPLE 13.3 A bicycle shop sells the following number of bicycles from 1976 to 1987. Calculate an exponentially smoothed time series from these data. (Set $\theta$ equal to $1/3$.)

| Year | Number sold (thousands) | Year | Number sold (thousands) |
|---|---|---|---|
| 1976 | 3 | 1982 | 6 |
| 1977 | 3 | 1983 | 6 |
| 1978 | 3 | 1984 | 6 |
| 1979 | 3 | 1985 | 9 |
| 1980 | 6 | 1986 | 10 |
| 1981 | 6 | 1987 | 12 |

SOLUTION: Let $S_0$ be the exponentially smoothed sales (in thousands) in 1976, $S_1$ the exponentially smoothed sales in 1977, and so on. Then, from equation (13.13),

$S_0 = 3.00$  $S_6 = 1/3(6) + 2/3(4.67) = 5.11$
$S_1 = 1/3(3) + 2/3(3) = 3.00$  $S_7 = 1/3(6) + 2/3(5.11) = 5.41$
$S_2 = 1/3(3) + 2/3(3) = 3.00$  $S_8 = 1/3(6) + 2/3(5.41) = 5.61$
$S_3 = 1/3(3) + 2/3(3) = 3.00$  $S_9 = 1/3(9) + 2/3(5.61) = 6.74$
$S_4 = 1/3(6) + 2/3(3) = 4.00$  $S_{10} = 1/3(10) + 2/3(6.74) = 7.83$
$S_5 = 1/3(6) + 2/3(4) = 4.67$  $S_{11} = 1/3(12) + 2/3(7.83) = 9.22$

## EXERCISES

**13.1** "Trends are useless. By itself time does not cause a variable to change." Comment and evaluate.

**13.2** "Although a trend may be a useful summary description of the long-term movement in a time series, it tells us little or nothing about why the time series behaved in this way." Comment and evaluate.

**13.3** "Trends are inexorable, if they fit the data well. Once you find a trend that fits past data, you can be sure it will continue into the future." Comment and evaluate.

**13.4** "You can fit a trend by plotting the time series against time and by drawing a freehand curve that seems to fit well. You don't have to bother with least squares techniques, since the assumptions underlying the use of least squares are not met." Comment and evaluate.

**13.5** The Union Carbide Corporation's sales during the period 1960–75 are given below:

| Year | Sales (billions of dollars) | Year | Sales (billions of dollars) |
|---|---|---|---|
| 1960 | 1.5 | 1968 | 2.7 |
| 1961 | 1.6 | 1969 | 2.9 |
| 1962 | 1.6 | 1970 | 3.0 |
| 1963 | 1.7 | 1971 | 3.0 |
| 1964 | 1.9 | 1972 | 3.3 |
| 1965 | 2.1 | 1973 | 3.9 |
| 1966 | 2.2 | 1974 | 5.3 |
| 1967 | 2.5 | 1975 | 5.7 |

(a) Fit a linear (least-squares) trend line to these data.
(b) What is the equation for the trend line if the origin is set at 1968? What is the equation for the trend line if the origin is set at 1965?
(c) Fit an exponential trend line to the data.
(d) Plot the linear trend line (from (a)) against the time series. Plot the exponential trend line (from (c)) on the same graph. Which trend line —the linear or the exponential—seems to provide a better fit to the data?
(e) In 1980, Union Carbide's sales were $9.994 billion. Suppose that in 1976 both the linear trend line and the exponential trend line had been used to forecast the firm's 1980 sales. Which forecast would have been more accurate?
(f) In 1984, Union Carbide's sales were $9.508 billion. Suppose that in 1976 both the linear trend line and the exponential trend line had been used to forecast the firm's 1984 sales. Which forecast would have been more accurate?

**13.6** The gross national product (GNP) of the United States (in billions of dollars) during the period 1961–81 is given below:

| Year | GNP | Year | GNP | Year | GNP |
|------|-----|------|-----|------|-----|
| 1961 | 523 | 1968 | 868 | 1975 | 1,529 |
| 1962 | 564 | 1969 | 936 | 1976 | 1,706 |
| 1963 | 595 | 1970 | 982 | 1977 | 1,918 |
| 1964 | 636 | 1971 | 1,063 | 1978 | 2,156 |
| 1965 | 688 | 1972 | 1,171 | 1979 | 2,414 |
| 1966 | 753 | 1973 | 1,307 | 1980 | 2,626 |
| 1967 | 796 | 1974 | 1,413 | 1981 | 2,922 |

(a) Calculate a linear (least-squares) trend for these data.
(b) On the average, how much did GNP increase annually during this period?
(c) If time is measured in years from 1970, what is the equation for the trend line?
(d) If time is measured in years from 1965, what is the equation for the trend line?
(e) Plot the above data. Does it appear that a linear trend is adequate?
(f) In 1985, GNP was $3,992 billion. Suppose that this trend line had been used to forecast GNP in 1985. How large would the forecasting error have been?
(g) Can you suggest a type of trend line that might fit the GNP time series better than a linear trend?

**13.7** The percent of gross national product devoted to research and development in the United States during the period 1961–83 was as follows:

| Year | Percent | Year | Percent | Year | Percent | Year | Percent |
|------|---------|------|---------|------|---------|------|---------|
| 1961 | 2.74 | 1967 | 2.91 | 1973 | 2.34 | 1979 | 2.27 |
| 1962 | 2.73 | 1968 | 2.83 | 1974 | 2.32 | 1980 | 2.38 |
| 1963 | 2.87 | 1969 | 2.74 | 1975 | 2.30 | 1981 | 2.45 |
| 1964 | 2.97 | 1970 | 2.64 | 1976 | 2.27 | 1982 | 2.58 |
| 1965 | 2.91 | 1971 | 2.50 | 1977 | 2.27 | 1983 | 2.65 |
| 1966 | 2.90 | 1972 | 2.43 | 1978 | 2.25 | | |

(a) Calculate a three-year moving average.
(b) Calculate a five-year moving average.
(c) Plot both moving averages against the original data.

**13.8** A textile firm's statistician uses least-squares techniques to obtain the following trend:

$$Y_t = 100 + 8.12t + 0.42t^2,$$

where $Y_t$ equals the firm's sales (in millions of dollars) in period $t$, where $t$ is measured in years from 1984.
(a) What statements can you make about the changes over time, as measured in *absolute* terms, that occurred during the period 1979–88 in the firm's sales rate, as indicated by this trend?
(b) What statements can you make about the changes over time, as measured in *percentage* terms, that have occurred in the firm's sales rate, as indicated by this trend?
(c) To interpret this trend properly, what additional information do you need?

**13.9** (a) Using the data in Exercise 13.5, calculate the exponentially smoothed value of Union Carbide's annual sales during the period 1960–64, if the smoothing constant equals 1/5.
(b) Calculate the exponentially smoothed value of Union Carbide's annual sales during the period 1960–64, if the smoothing constant equals 1/10.

**13.10** Fixed investment in nonresidential structures and producers' durable equipment (in billions of dollars) in the United States during the period 1958–84 were as follows:

| Year | Expenditure | Year | Expenditure | Year | Expenditure |
|---|---|---|---|---|---|
| 1958 | 42 | 1967 | 84 | 1976 | 174 |
| 1959 | 46 | 1968 | 91 | 1977 | 206 |
| 1960 | 49 | 1969 | 101 | 1978 | 242 |
| 1961 | 48 | 1970 | 104 | 1979 | 280 |
| 1962 | 52 | 1971 | 108 | 1980 | 296 |
| 1963 | 55 | 1972 | 121 | 1981 | 327 |
| 1964 | 61 | 1973 | 143 | 1982 | 350 |
| 1965 | 73 | 1974 | 157 | 1983 | 353 |
| 1966 | 83 | 1975 | 158 | 1984 | 426 |

(a) Calculate a three-year moving average.
(b) Calculate a five-year moving average.

**13.11** An electronics firm's sales (in millions of dollars) during the period 1972–86 were as follows:

| Year | Sales | Year | Sales | Year | Sales |
|---|---|---|---|---|---|
| 1972 | 1 | 1977 | 7 | 1982 | 14 |
| 1973 | 2 | 1978 | 8 | 1983 | 16 |
| 1974 | 4 | 1979 | 6 | 1984 | 13 |
| 1975 | 6 | 1980 | 10 | 1985 | 19 |
| 1976 | 5 | 1981 | 12 | 1986 | 20 |

(a) Calculate an exponentially smoothed time series from these data. (Set $\theta = .4$.)

(b) Calculate an exponentially smoothed time series from these data, but set $\theta$ equal to .2 rather than .4.

**13.12** The unemployment rate (that is, unemployment as a percent of the civilian labor force) in the United States during the period 1966–84 was as follows:

| Year | Unemployment rate | Year | Unemployment rate |
|---|---|---|---|
| 1966 | 3.8 | 1976 | 7.7 |
| 1967 | 3.8 | 1977 | 7.1 |
| 1968 | 3.6 | 1978 | 6.1 |
| 1969 | 3.5 | 1979 | 5.8 |
| 1970 | 4.9 | 1980 | 7.1 |
| 1971 | 5.9 | 1981 | 7.6 |
| 1972 | 5.6 | 1982 | 9.7 |
| 1973 | 4.9 | 1983 | 9.6 |
| 1974 | 5.6 | 1984 | 7.5 |
| 1975 | 8.5 | | |

(a) Calculate an exponentially smoothed time series from these data. (Set $\theta = .3$.)

(b) Calculate an exponentially smoothed time series from these data, but set $\theta$ equal to .1 rather than .3.

**13.13** During the 1980s there has been widespread concern and controversy over the extent to which American producers are competitive in world markets. The data presented in Exercise 13.7 have played a role in this discussion.

(a) In what way, and to what extent, are these data relevant?

(b) In two brief sentences, characterize what these data show.

(c) Do these data exhibit a linear or exponential trend?

**\*13.14** (a) Based on the data in Exercise 13.6, calculate a quadratic trend for gross national product.

(b) Plot the resulting quadratic trend against the data. Does it seem to fit reasonably well?

## 13.8 Seasonal Variation

Many times series are composed of monthly or quarterly rather than annual data. For such time series, decision makers and statisticians must recognize that seasonal variation is likely to be present in the series. As pointed out in Section 13.2, seasonal variation in many economic time series is due to the weather. However, this is not always the case, as illustrated by a firm that sells ornaments for Christmas trees. There is likely to be a pronounced seasonal variation in this firm's sales, but this is due to the location of a specific holiday (Christmas) on

* Exercise 13.14 presumes that the reader has covered Chapter 12. If not, this exercise should be skipped.

the calendar, not to the weather. Still other reasons for seasonal variation are the fact that some industries tend to grant vacations at a particular time of year, or that taxes have to be paid at particular times of the year, or that schools tend to open at particular times of the year.

As we stated early in this chapter, statisticians have devised methods for estimating the pattern of seasonal variation in a particular time series. In other words, they can determine the extent to which a particular month or quarter is likely to differ from what would be expected on the basis of the trend and cyclical variation in the same series. (In terms of the traditional model in equation (13.1), statisticians can determine the value of $S$ for each month or quarter.) For example, the statistician for a manufacturer of soft drinks may tell the company's managers that U.S. production of soft drinks tends in June to be 5.9 percent higher than what the trend and cyclical variation in soft drink production would indicate. Or the statistician may find that U.S. production of soft drinks in December tends to be 7.0 percent lower than the trend and cyclical variation would indicate.

*Seasonal Index*

*The seasonal variation in a particular time series is described by a figure for each month, the **seasonal index**, which shows the way in which that month tends to depart from what would be expected on the basis of the trend and cyclical variation in the time series.* For example, Table 13.5 shows the seasonal variation in U.S. production of soft drinks. January's production tends to be about 93.4 percent of the amount expected on the basis of trend and cyclical variation; February's production tends to be about 89.3 percent of this amount; March's production tends to be about 90.7 percent of this amount; and so on. Figures of this sort can be used in a number of ways. *One important application is to forecast what the time series will be in the future.* For example, suppose that, based on the trend and cyclical variation, it appears likely that about 30 million gallons of soft drinks will be produced next January. If this is the case, a reasonable forecast of actual January production is .934 (30 million) = 28.02 million gallons, since January's production tends to be 93.4 percent of the amount expected on the basis of trend and cyclical variation.

**Table 13.5**
Seasonal Variation in Production of Soft Drinks in the United States

| Month | Seasonal index |
|---|---|
| January | 93.4 |
| February | 89.3 |
| March | 90.7 |
| April | 94.9 |
| May | 99.0 |
| June | 105.9 |
| July | 112.4 |
| August | 113.4 |
| September | 108.3 |
| October | 103.9 |
| November | 95.8 |
| December | 93.0 |

In addition, *a knowledge of seasonal variation is often useful in interpreting current or recent developments.* For example, suppose that it was announced in June 1987 that soft drink output in the United States had increased by about 4 percent between April and May. Firms and government agencies interested in the soft drink industry were likely to ask: To what extent is this increase due to seasonal factors? In other words, even if no increase occurred between April and May in the trend value, and if cyclical factors remained constant, would one expect an increase of this magnitude on the basis of seasonal factors alone? Looking at Table 13.5, since May's output is generally about 99.0 percent of what would be expected on the basis of trend and cyclical variation, and since April's output is generally about 94.9 percent of this amount, one would expect May's output to be 4.3 percent higher than April's, even if the trend value and cyclical variation were the same for each month.[11] Thus, the observed increase was no more than would be expected on the basis of seasonal factors alone.

Another way of handling this problem is to **deseasonalize** the data, a procedure that is often carried out in business and government. *By deseasonalizing we mean removing the seasonal element from the data.* Recall that according to the traditional model, the value of a time series at time *t* is equal to the seasonal variation at *t* times a combination of trend, cyclical, and irregular factors. *By dividing the value of the time series at time* t *by the seasonal index for time* t *(divided by 100), we obtain the deseasonalized value at time* t. This deseasonalized value should be purged of the effects of seasonal variation.

For example, if each month's soft drink output in 1987 was as shown in Table 13.6, we can use the seasonal indexes in Table 13.5 to remove the seasonal element from each month. For example, since January is generally 93.4 percent of the deseasonalized amount, the

*Deseasonalizing Data*

**Table 13.6**
Deseasonalizing Production Figures for Soft Drinks in 1987

| Month | Output (millions of gallons) (1) | Seasonal index (2) | Deseasonalized output (millions of gallons) (1) ÷ [(2)/100] |
|---|---|---|---|
| January | 25 | 93.4 | 26.8 |
| February | 24 | 89.3 | 26.9 |
| March | 24 | 90.7 | 26.5 |
| April | 25 | 94.9 | 26.3 |
| May | 26 | 99.0 | 26.3 |
| June | 28 | 105.9 | 26.4 |
| July | 30 | 112.4 | 26.7 |
| August | 30 | 113.4 | 26.5 |
| September | 30 | 108.3 | 27.7 |
| October | 31 | 103.9 | 29.8 |
| November | 30 | 95.8 | 31.3 |
| December | 30 | 93.0 | 32.3 |

[11] If May's output is 99.0 percent of what would be expected on the basis of trend and cyclical variation, and if April's output is 94.9 percent of the same amount, May's output is 99.0/94.9 × April's output, or 104.3 percent of April's output.

deseasonalized output in January 1987 was 25 million ÷ .934, or 26.8 million gallons. Similarly, since July is generally 112.4 percent of the deseasonalized amount, the deseasonalized output in July 1987 was 30 million ÷ 1.124, or 26.7 million gallons. Table 13.6 shows the result for each month of 1987. To see whether the 4 percent increase between April and May of 1987 was greater than would be expected on the basis of seasonal variation, we can compare the deseasonalized figures for April and May. Since May's deseasonalized output was no greater than April's, the increase between April and May was no greater than would be expected due to seasonal factors alone. Of course, this is precisely the same result that we obtained in the next-to-last paragraph.

The example below further illustrates the construction and use of deseasonalized data.

EXAMPLE 13.4 A statistician estimates the seasonal index for each month for a firm's sales, the result being

| January | 90 | May | 100 | September | 110 |
|---------|----|-----|-----|-----------|-----|
| February | 90 | June | 100 | October | 110 |
| March | 90 | July | 100 | November | 110 |
| April | 90 | August | 100 | December | 110 |

If actual sales in January of this year are $20 million, and if each month's sales rose this year by $100,000, what was the deseasonalized value of sales for each month? Taking account of seasonal factors, did sales generally tend to rise or fall during this year?

SOLUTION: The deseasonalized value of sales each month (in millions of dollars) was

| January | $20.0 \div .90 = 22.2$ | July | $20.6 \div 1.00 = 20.6$ |
|---------|------------------------|------|-------------------------|
| February | $20.1 \div .90 = 22.3$ | August | $20.7 \div 1.00 = 20.7$ |
| March | $20.2 \div .90 = 22.4$ | September | $20.8 \div 1.10 = 18.9$ |
| April | $20.3 \div .90 = 22.6$ | October | $20.9 \div 1.10 = 19.0$ |
| May | $20.4 \div 1.00 = 20.4$ | November | $21.0 \div 1.10 = 19.1$ |
| June | $20.5 \div 1.00 = 20.5$ | December | $21.1 \div 1.10 = 19.2$ |

It is therefore evident that when seasonal factors are taken into account, sales generally tended to fall during this year.

## 13.9 Calculation of a Seasonal Index: Ratio-to-Moving-Average Method

How can we calculate a seasonal index like that shown in Table 13.5? In other words, how do we know that January's output tends to be 93.4 percent of the amount that would be expected on the basis of trend and cyclical variation, that February's output tends to be 89.3 percent of this amount, and so on? The most commonly used method is to begin by computing a 12-month moving average of the monthly time series. Table 13.7 shows the 12-month moving average computed for the time series for the U.S. output of soft drinks. Note, however, that this moving average, shown in the second column of Table 13.7, is centered between two months rather than at the middle of a month. Thus, to center the moving average at the *middle* of each month, we take the average of two moving averages: (1) the one centered at the *beginning* of the month in question, and (2) the one centered at the *end* of the month in question. The results are shown in the third column of Table 13.7.

*The moving average pertaining to the middle of a given month is viewed as a good approximation of what would be expected on the basis of trend and cyclical variation.* Why? Because, as pointed out in Section 13.6, a moving average should eliminate the effect of a cycle if the length of the moving average is equal to the period of the cycle.

**Table 13.7**
Computation of Centered 12-Month Moving Average for Production of Soft Drinks

| Month | Soft drink output (1982 = 100.0) | 12-Month moving average[a] | Centered 12-month moving average[b] | Actual output as a ratio of centered moving average[c] |
|---|---|---|---|---|
| January 1982 | 88.0 | | | |
| February | 86.1 | | | |
| March | 91.8 | | | |
| April | 96.4 | | | |
| May | 100.1 | | | |
| June | 107.5 | 100.0 | | |
| July | 111.0 | 100.8 | 100.4 | 1.106 |
| August | 114.1 | 101.6 | 101.2 | 1.127 |
| September | 108.9 | 102.0 | 101.8 | 1.070 |
| October | 103.4 | 102.5 | 102.3 | 1.011 |
| November | 95.7 | 103.2 | 102.9 | .930 |
| December | 96.7 | 104.1 | 103.7 | .932 |
| January 1983 | 97.9 | 105.6 | 104.8 | .934 |
| February | 95.5 | 106.8 | 106.2 | .899 |
| March | 97.0 | 108.2 | 107.5 | .902 |
| April | 102.7 | 109.6 | 108.9 | .943 |
| May | 108.5 | 111.0 | 110.3 | .984 |
| June | 118.0 | 111.8 | 111.4 | 1.059 |
| July | 128.5 | 113.0 | 112.4 | 1.143 |
| August | 129.3 | 114.1 | 113.5 | 1.139 |
| September | 125.4 | 115.1 | 114.6 | 1.094 |
| October | 120.0 | 116.4 | 115.7 | 1.037 |
| November | 112.9 | 117.7 | 117.0 | .965 |
| December | 106.2 | 118.9 | 118.3 | .898 |
| January 1984 | 111.5 | 120.2 | 119.5 | .933 |
| February | 108.9 | 121.7 | 120.9 | .901 |
| March | 108.7 | 123.4 | 122.5 | .887 |
| April | 118.7 | 124.6 | 124.0 | .957 |
| May | 124.2 | 125.4 | 125.0 | .994 |

| Month | Soft drink output (1982 = 100.0) | 12-Month moving average[a] | Centered 12-month moving average[b] | Actual output as a ratio of centered moving average[c] |
|---|---|---|---|---|
| June | 132.1 | | 126.1 | 1.048 |
| July | 144.7 | 126.7 | 127.0 | 1.139 |
| August | 146.5 | 127.3 | 127.6 | 1.148 |
| September | 145.9 | 128.0 | 128.5 | 1.135 |
| October | 134.3 | 129.0 | 129.3 | 1.039 |
| November | 123.5 | 129.5 | 130.0 | .950 |
| December | 121.4 | 130.4 | 130.9 | .927 |
| January 1985 | 118.9 | 131.4 | 131.6 | .903 |
| February | 117.0 | 131.8 | 132.0 | .886 |
| March | 121.2 | 132.1 | 132.1 | .917 |
| April | 124.7 | 132.1 | 132.5 | .941 |
| May | 134.3 | 132.9 | 133.3 | 1.008 |
| June | 144.9 | 133.6 | 134.0 | 1.081 |
| July | 149.3 | 134.4 | 134.8 | 1.108 |
| August | 150.2 | 135.3 | 135.5 | 1.108 |
| September | 145.7 | 135.7 | 136.0 | 1.071 |
| October | 143.6 | 136.3 | 136.7 | 1.050 |
| November | 132.4 | 137.2 | 137.4 | .964 |
| December | 130.3 | 137.6 | 137.9 | .945 |
| January 1986 | 129.9 | 138.2 | 138.8 | .936 |
| February | 121.4 | 139.5 | 139.9 | .868 |
| March | 128.4 | 140.3 | 141.0 | .911 |
| April | 135.7 | 141.8 | 142.3 | .954 |
| May | 139.1 | 142.8 | 143.4 | .970 |
| June | 152.9 | 144.0 | 144.6 | 1.057 |
| July | 163.9 | 145.2 | | |
| August | 159.8 | | | |
| September | 164.3 | | | |
| October | 156.0 | | | |
| November | 146.8 | | | |
| December | 144.3 | | | |

*Source:* These are not the actual data for these years, although they are based on Federal Reserve Board data. However, they are adequate for present purposes, which are purely pedagogical and illustrative.

[a] To obtain each figure in this column, the first step is to calculate the 12-month moving sum of the output figures. Thus, the sum of the first 12 output figures on page 567 is 88.0 + 86.1 + . . . + 95.7 + 96.7 = 1199.7. Dividing this sum by 12, we get 100.0, the first figure in this column. The sum of the second through thirteenth output figures is 86.1 + 91.8 + . . . + 96.7 + 97.9 = 1209.6. Dividing this sum by 12, we get 100.8, the second figure in this column. Continuing in this way, we derive each figure in this column.

[b] To obtain each figure in this column, average (1) the figure that is 1/2 line *above* it in the previous column and (2) the figure that is 1/2 line *below* it in the previous column. Thus, the top figure on page 567 is the average of 100.0 and 100.8, or 100.4. And the next figure is the average of 100.8 and 101.6, or 101.2.

[c] To obtain each figure in this column, divide the figure in the first column by the figure in the third column.

Clearly, the period of the seasonal variation is 12 months; thus, a 12-month moving average should eliminate the seasonal variation if this variation is completely regular from year to year. Stated differently, the 12-month moving average, since it is 1/12 of the annual total (for the year centered at the middle of the month), should be essentially free of seasonal variation, because seasonal variation is defined so that it averages out over a one-year period. Besides eliminating most of the seasonal variation, the 12-month moving average eliminates much of the irregular variation, with the result that it is viewed as a good approximation of what would be expected on the basis of trend and cyclical variation.

The next step in constructing the monthly seasonal index is to *divide the actual value for each month by the moving average.* According to the traditional model in equation (13.1), the actual value equals $T \times S \times C \times I$, and the moving average is approximately equal to $T \times C$. Thus, the ratio of the actual value to the moving average, shown in the last column of Table 13.7, is an estimate of $S \times I$. Because these ratios reflect both seasonal and irregular variation, we want to eliminate the irregular variation to the extent possible. To do this, *we calculate the median of the ratios for each month,* which should be relatively free from the effects of irregular variation. Table 13.8 shows that (after all ratios are multiplied by 100) the median of the January ratios is 93.35, the median of the February ratios is 89.25, and so on.

Finally, *we adjust each month's median so that the mean value of the 12 monthly indexes equals 100.* To understand why, suppose that the deseasonalized value of a time series is the same for each month of the year. Due to seasonal variation, the actual value varies from month to month. However, the seasonal variation cancels out *for the year as a whole.* In this case, if we sum up the values for the year as a whole, the sum of the actual values must equal the sum of the deseasonalized values. That is, the mean value of the monthly indexes must be 100, and the sum of the indexes for all 12 months must be 1200. To be sure

**Table 13.8**
Computation of Seasonal Index for Soft Drink Production

| Month | Ratios (times 100) | | | | Median | Seasonal Index |
|---|---|---|---|---|---|---|
| January | 93.4 | 93.3 | 90.3 | 93.6 | 93.35 | 93.4 |
| February | 89.9 | 90.1 | 88.6 | 86.8 | 89.25 | 89.3 |
| March | 90.2 | 88.7 | 91.7 | 91.1 | 90.65 | 90.7 |
| April | 94.3 | 95.7 | 94.1 | 95.4 | 94.85 | 94.9 |
| May | 98.4 | 99.4 | 100.8 | 97.0 | 98.90 | 99.0 |
| June | 105.9 | 104.8 | 108.1 | 105.7 | 105.80 | 105.9 |
| July | 110.6 | 114.3 | 113.9 | 110.8 | 112.35 | 112.4 |
| August | 112.7 | 113.9 | 114.8 | 110.8 | 113.30 | 113.4 |
| September | 107.0 | 109.4 | 113.5 | 107.1 | 108.25 | 108.3 |
| October | 101.1 | 103.7 | 103.9 | 105.0 | 103.80 | 103.9 |
| November | 93.0 | 96.5 | 95.0 | 96.4 | 95.70 | 95.8 |
| December | 93.2 | 89.8 | 92.7 | 94.5 | 92.95 | 93.0 |
| Total | | | | | 1199.15 | 1200.0 |

that this holds true for the monthly indexes we are constructing, we must multiply each median (in the next-to-last column of Table 13.8) by 1200 and divide the result by the sum of the medians, 1199.15. The results are shown in the last column of Table 13.8.

The figures in the last column are the monthly indexes we want. As pointed out in the previous section, these indexes have a variety of uses in business and government. However, now that we have shown how they are calculated, it is evident that they are based on a number of simplifying assumptions. In particular, we have assumed implicitly that the pattern of seasonal variation has and will remain relatively constant over time. This can be a dangerous assumption. For example, the seasonal variation in electric power consumption has been affected by the advent of air conditioning. There are ways to take account of changes of this sort in seasonal variation, but the simple methods described here make no attempt to do so. Nevertheless, these procedures are used very extensively in business and economics and are a standard part of the statistician's equipment.

## 13.10 Calculation of Seasonal Variation: Dummy-Variable Method[12]

Another way of calculating the seasonal variation in a time series is to use the dummy-variable technique discussed in the previous chapter on multiple regression. Suppose, for example, that a statistician has a time series composed of quarterly values; that is, each observation pertains to the first, second, third, or fourth quarter of a year. If the statistician believes that the time series has a linear trend, he or she may assume that the value of the observation at time $t$ equals

$$Y(t) = A + B_1 t + B_2 Q_1 + B_3 Q_2 + B_4 Q_3 + e_t, \qquad (13.14)$$

where $Q_1$ equals 1 if time $t$ is the first quarter and 0 otherwise, $Q_2$ equals 1 if time $t$ is the second quarter and zero otherwise, $Q_3$ equals 1 if time $t$ is the third quarter and 0 otherwise, and $e_t$ equals $Y(t) - E(Y(t))$.

It is important to understand the meaning of $B_1$, $B_2$, $B_3$, and $B_4$ in equation (13.14). Clearly, $B_1$ is the slope of the linear trend, but what are $B_2$, $B_3$, and $B_4$? The answer is that $B_2$ *is the difference between the expected value of an observation in the first quarter and the expected value of an observation in the fourth quarter, holding the trend value constant.* To see that this is true, note that if an observation pertains to the first quarter, its expected value equals

$$A + B_1 t + B_2,$$

according to equation (13.14). Similarly, if an observation pertains to the fourth quarter, its expected value equals

$$A + B_1 t,$$

according to equation (13.14). Thus, if the effect of the trend could be held constant (and if $t$ were the same), the difference between the expected value of an observation in the first quarter and the expected value of an observation in the fourth quarter would be

$$(A + B_1 t + B_2) - (A + B_1 t) = B_2,$$

which is what we set out to prove. Holding constant the trend value, one can show in the same way that $B_3$ *is the difference between the expected value of an observation in the second quarter and the expected value of an observation in the fourth quarter; and* $B_4$ *is the difference between the expected value of an observation in the third quarter and the expected value of an observation in the fourth quarter.*

Thus, if equation (13.14) is valid, the statistician can represent the seasonal variation in the time series by the three numbers $B_2$, $B_3$, and $B_4$. To estimate each of these numbers, ordinary multiple regression techniques can be used.[13] The dependent variable is $Y(t)$, and the independent variables are $t, Q_1, Q_2$, and $Q_3$. The latter three independent variables—$Q_1$, $Q_2$, and $Q_3$—are dummy variables (as we know from the previous chapter); that is, each of these can assume only two values: 0 or 1. As indicated in the previous chapter, the constants in equation (13.14)—$A, B_1, B_2, B_3$, and $B_4$—can be estimated by the ordinary least-squares procedure.

When using this dummy-variable technique, the statistician assumes that seasonal effects are *added* to the trend value, as shown in equation (13.14). This differs from the traditional model in equation (13.1), where it is assumed that seasonal effects *multiply* the trend value. (See footnote 1.) The former assumption is appropriate in some cases, while the latter assumption is appropriate in others. Both techniques are useful.

The following example shows how the dummy-variable technique can be used to estimate the seasonal variation in monthly data.

---

EXAMPLE 13.5 A statistician has monthly data concerning the sales of a particular firm. Indicate how he can estimate the seasonal variation (from month to month) in sales if there is a linear trend.

SOLUTION: The statistician can assume that

$$S(t) = A + B_1 t + B_2 M_1 + B_3 M_2 + \ldots + B_{12} M_{11} + e_t, \qquad (13.15)$$

---

[13] Of course, it should be recalled that multiple regression assumes that the error terms are independent and that the standard deviation of the error terms is constant. These assumptions may not be met. For example, the error terms may be serially correlated. See Chapter 12.

where $S(t)$ is the firm's sales in the month $t$, $M_1$ equals 1 if month $t$ is January and 0 otherwise, . . . , $M_{11}$ equals 1 if month $t$ is November and 0 otherwise, and $e_t$ equals $S(t) - E(S(t))$. Using ordinary multiple regression techniques, the statistician can estimate $A, B_1, B_2, . . . , B_{12}$. The estimates of $B_2, B_3, . . . , B_{11}$, and $B_{12}$ indicate the seasonal variation in the firm's sales. In particular, $B_2$ is the difference between the expected value of sales for January and December, $B_3$ is the difference between the expected value of sales for February and December, and so on, until $B_{12}$ is the difference between the expected value of sales for November and December (holding constant the trend value for each case).

## EXERCISES

**13.15** The Bona Fide Washing Machine Company's statistician calculates a seasonal index for the firm's sales, the results being shown in the second column below. The firm's monthly 1986 sales are shown in the third column.

| Month | Seasonal index | 1986 Sales (millions of dollars) |
|---|---|---|
| January | 97 | 2.5 |
| February | 96 | 2.4 |
| March | 97 | 2.7 |
| April | 98 | 2.9 |
| May | 99 | 3.0 |
| June | 100 | 3.1 |
| July | 101 | 3.2 |
| August | 103 | 3.1 |
| September | 103 | 3.2 |
| October | 103 | 3.1 |
| November | 102 | 3.0 |
| December | 101 | 2.9 |

Calculate deseasonalized sales figures for 1986.

**13.16** A firm claims that the seasonal index for its sales is 110 for each of the first six months of the year, and 95 for each of the last six months of the year. Comment on this claim.

**13.17** The unemployment rate in the United States was as follows during the period 1979–81:

| Month | 1979 | 1980 | 1981 |
|---|---|---|---|
| January | 5.8 | 6.3 | 7.4 |
| February | 5.9 | 6.2 | 7.4 |
| March | 5.8 | 6.3 | 7.3 |
| April | 5.8 | 6.9 | 7.3 |
| May | 5.6 | 7.5 | 7.5 |
| June | 5.6 | 7.5 | 7.4 |
| July | 5.6 | 7.8 | 7.2 |
| August | 5.9 | 7.7 | 7.3 |
| September | 5.8 | 7.5 | 7.6 |
| October | 5.9 | 7.5 | 8.0 |
| November | 5.9 | 7.5 | 8.3 |
| December | 6.0 | 7.3 | 8.8 |

(a) Calculate a seasonal index for this unemployment rate.

(b) A colleague of yours informs you that the above unemployment rates have already been seasonally adjusted. Why do any of the monthly indexes that you computed in (a) differ from 100.0?

**13.18** The Acme Corporation calculates a 12-month moving average of its sales. After centering this moving average at the middle of the month, Acme divides each month's sales by this centered moving average. The results (multiplied by 100) are as follows:

| | | Ratios | | |
|---|---|---|---|---|
| *Month* | *1983* | *1984* | *1985* | *1986* |
| January | 91.0 | 90.0 | 90.5 | 94.0 |
| February | 93.0 | 92.0 | 92.5 | 95.0 |
| March | 95.0 | 94.0 | 94.5 | 96.0 |
| April | 97.0 | 96.0 | 96.5 | 97.0 |
| May | 99.0 | 98.0 | 98.5 | 98.0 |
| June | 101.0 | 100.0 | 100.5 | 99.0 |
| July | 103.0 | 102.0 | 102.5 | 101.0 |
| August | 105.0 | 104.0 | 104.5 | 102.0 |
| September | 107.0 | 106.0 | 106.5 | 103.0 |
| October | 109.0 | 108.0 | 108.5 | 104.0 |
| November | 100.0 | 110.0 | 105.0 | 105.0 |
| December | 100.0 | 100.0 | 100.0 | 106.0 |

Calculate a seasonal index for the Acme Corporation's sales.

**13.19** A sheet metal firm has sales of $4.0 million in January and $3.8 million in February. For this firm's sales, the seasonal index is 110 in January and 101 in February.

(a) In a meeting of the board of directors, the firm's president says that January's sales were good but February's were disappointing. Do you agree? Explain.

(b) The firm's president also predicts on the basis of January's sales that the firm's sales during the entire year will be $48 million. How do you think he got this figure? Do you agree with it? Explain.

**13.20** (a) In equation (13.14), suppose that the statistician estimates $B_2$ to be 2, $B_3$ to be 3, and $B_4$ to be $-2$. Holding constant the trend value, what is the expected difference in the value of $Y$ between the first and third quarters?

(b) In equation (13.14), suppose that the statistician believes the trend to be exponential rather than linear. How should equation (13.14) be changed to reflect this?

(c) In equation (13.15), suppose that the statistician estimates $B_2$ to be 2 and $B_3$ to be 3. Holding constant the trend value, what is the expected difference between January and February in the value of $S$?

# 13.11 Cyclical Variation

As pointed out at the beginning of this chapter, time series in business and economics frequently exhibit cyclical variation, such variation often being termed the *business cycle*. To illustrate what we mean by the

**Figure 13.11**
Gross National Product
(1958 Dollars), United
States 1918–84
(Excluding World War
II)

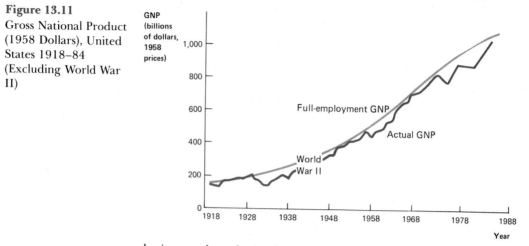

*Business
Fluctuations*

*business cycle* or *business fluctuations,* let's look at how national output has grown in the United States since World War I. Figure 13.11 shows the behavior of gross national product (GNP) in constant dollars in the United States since 1919. It is clear that output has grown considerably during this period; indeed, GNP is more than five times what it was 50 years ago. It is also clear that this growth has not been steady. On the contrary, although the long-term trend has been upward, there have been periods—1919–21, 1929–33, 1937–38, 1944–46, 1948–49, 1953–54, 1957–58, 1969–70, 1973–75, 1980, and 1981–82—when national output has declined.

Let's define the *full-employment level* of GNP as the total amount of goods and services that could have been produced if there had been full employment. Figure 13.11 shows that national output tends to rise and approach its full-employment level for a while, then falters and falls below this level, then rises to approach it once more, then falls below it again, and so on. For example, output remained close to the full-employment level in the prosperous mid-1920s, fell far below this level in the depressed 1930s, and rose again to this level once we entered World War II. This movement of national output is sometimes called the business cycle, but it must be recognized that these "cycles" are far from regular or consistent. (On the contrary, they are very irregular.)

Each cycle can be divided by definition into four phases, as shown in Figure 13.12. *The **trough** is the point where national output is lowest relative to its full-employment level.* ***Expansion*** *is the subsequent phase during which national output rises. The **peak** occurs when national output is highest relative to its full-employment level. Finally,* ***recession*** *is the subsequent phase during which national output falls.*[14] Besides these four phases, two other terms are frequently used to describe stages of the business cycle. A *depression* is a period when national output is well below its full-employment level; it is a severe recession. (Depressions are, of course, periods of excessive unemployment.) *Prosperity* is a period when national output is close to its full-employment level. Prosperity, if total spending

[14] The peak and trough may also be defined in terms of deviations from the long-term trend of GNP rather than in terms of deviations from the full-employment level of GNP.

CHANGES IN EMPLOYMENT IN FIVE NATIONS

Every nation of the world is concerned with the problem of promoting the employment of its citizens and keeping the unemployment rate at a minimum. In August 1985, the U.S. Bureau of Labor Statistics published the following data concerning the number of people employed and the unemployment rate in five countries during the period 1975–84.[15]

| | United States | Japan | France | West Germany | Great Britain |
|---|---|---|---|---|---|
| **Employment (millions)** | | | | | |
| 1975 | 85.8 | 51.5 | 20.7 | 25.2 | 24.0 |
| 1976 | 88.8 | 52.0 | 20.9 | 25.0 | 23.8 |
| 1977 | 92.0 | 52.7 | 21.0 | 25.0 | 23.8 |
| 1978 | 96.0 | 53.4 | 21.1 | 25.1 | 24.0 |
| 1979 | 98.8 | 54.0 | 21.1 | 25.5 | 24.4 |
| 1980 | 99.3 | 54.6 | 21.1 | 25.7 | 24.1 |
| 1981 | 100.4 | 55.1 | 21.0 | 25.5 | 23.2 |
| 1982 | 99.5 | 55.6 | 21.0 | 25.1 | 22.8 |
| 1983 | 100.8 | 56.6 | 20.8 | 24.6 | 22.6 |
| 1984 | 105.0 | 56.9 | 20.7 | 24.6 | 23.0 |
| **Unemployment rate (percent)** | | | | | |
| 1975 | 8.5 | 1.9 | 4.2 | 3.4 | 4.5 |
| 1976 | 7.7 | 2.0 | 4.5 | 3.4 | 5.9 |
| 1977 | 7.1 | 2.0 | 4.8 | 3.5 | 6.3 |
| 1978 | 6.1 | 2.3 | 5.3 | 3.4 | 6.2 |
| 1979 | 5.8 | 2.1 | 6.1 | 3.0 | 5.3 |
| 1980 | 7.1 | 2.0 | 6.4 | 2.9 | 6.8 |
| 1981 | 7.6 | 2.2 | 7.5 | 4.1 | 10.4 |
| 1982 | 9.7 | 2.4 | 8.4 | 5.9 | 11.8 |
| 1983 | 9.6 | 2.7 | 8.6 | 7.5 | 12.8 |
| 1984 | 7.5 | 2.8 | 10.1 | 7.8 | 13.0 |

(a) How useful are these data in estimating the trend of employment in each country? If they are useful in this regard, does this trend seem to be much the same in each country?

(b) Does there seem to be cyclical variation in employment? If so, is the cyclical variation much the same in each country?

(c) Does there seem to be cyclical variation in unemployment rates? If so, is the cyclical variation much the same in each country?

(d) During the mid-1980s, concern was expressed in many quarters regarding Europe's economic performance relative to that of the United States and Japan. Do the above data help to indicate the source and nature of this concern? If so, how?

[15] J. Moy, "Recent Trends in Unemployment and the Labor Force," *Monthly Labor Review*, August 1985.

**Figure 13.12**
Four Phases of
Business Fluctuations

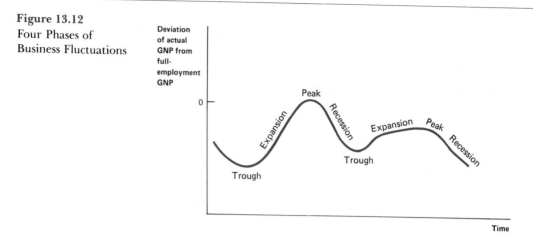

is too high relative to potential output, can be a time of rapid inflation. Of course, in some business cycles the peak may not be a period of prosperity because output may be below its full-employment level; in the same way, the trough may not be a period of depression because output may not be far below its full-employment level.

*Many business and economic time series go up and down with the business cycle.* For example, industrial output tends to be above its trend line at the peak of the business cycle and tends to fall below its trend line at the trough. Similarly, such diverse series as the money supply, industrial employment, and stock prices reflect the business cycle. However, not all series go up and down at exactly the same time. Some turn upward before others at a trough, and some turn downward before others at a peak. As we shall see in Section 13.13, the fact that some time series tend to precede others in cyclical variation sometimes is used to forecast the pace of economic activity.

## 13.12 Elementary Forecasting Techniques

One of the most difficult and important tasks that statisticians confront is forecasting. (As a well-known politician once pointed out with extraordinary incisiveness: "Today the real problem is the future.") In general, all forecasting techniques are extremely fallible, and all forecasts should be treated with caution. Nonetheless, businesses and government agencies have no choice but to make forecasts, however crude. Since governments, firms, and private individuals must continually make decisions that hinge on what they expect will happen, they must make implicit forecasts even if they do not make explicit ones. Thus, the central question is how best to forecast, not whether to forecast. In this section we shall present some elementary forecasting techniques; but these should be viewed as crude first approximations rather than as highly sophisticated methods. (More sophisticated techniques are taken up briefly in Section 13.14.)

*Trend Extrapolation*        The simplest type of forecasting method is a straightforward extrapolation of a trend. For example, let's return to the Dow Chemical

Company. At the end of 1984, suppose that Dow wanted to forecast its 1985 sales. During the period of 1972–84, we know from Section 13.3 that the firm's sales could be represented by the following trend line:

$$S_{t'} = 7.602 + 0.825t',$$

where $t'$ equals the year in question minus 1978. To forecast its 1985 sales, Dow could simply insert 7 (that is, 1985–1978) for $t'$ in the above equation. Thus, the forecast for 1985 is

$$7.602 + 0.825(7) = 13.377,$$

or 13.377 billions of dollars. As shown in Figure 13.13, this forecast is a simple extension, or extrapolation, of the trend line into the future.[16]

   In many cases, firms and government agencies want to forecast monthly rather than annual amounts. In such cases, it is necessary to recognize that seasonal variation, as well as trend, is likely to affect the value for a particular month. To see how a forecast can be made under such circumstances, consider a beer manufacturer that wants to forecast its sales during each month of 1988. On the basis of data for each month during the period 1960–87, the firm determines that its sales seem to conform to the following trend:

$$S_t = 1,410 + 33t,$$

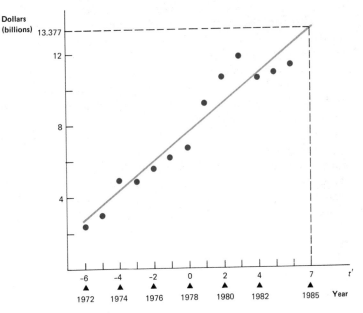

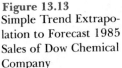

**Figure 13.13**
Simple Trend Extrapolation to Forecast 1985 Sales of Dow Chemical Company

[16] How accurate would this forecast have been? Dow's sales for 1985 were $11.5 billion, so this forecast was in error by about 14 percent, a relatively large amount. This illustrates the fact that simple mechanical forecasting techniques of this sort should be viewed with caution.

**Table 13.9**
Forecasted Trend
Value of Sales, Seasonal
Index, and Forecasted
Monthly Sales of Beer
Manufacturer, 1988

| Month | Forecasted trend value of sales* | Seasonal index | Forecasted sales (reflecting both trend and seasonal variation)* |
|---|---|---|---|
| January | 1,806 | 80 | 1,445 |
| February | 1,839 | 80 | 1,471 |
| March | 1,872 | 90 | 1,685 |
| April | 1,905 | 90 | 1,715 |
| May | 1,938 | 110 | 2,132 |
| June | 1,971 | 120 | 2,365 |
| July | 2,004 | 120 | 2,405 |
| August | 2,037 | 120 | 2,444 |
| September | 2,070 | 110 | 2,277 |
| October | 2,103 | 100 | 2,103 |
| November | 2,136 | 100 | 2,136 |
| December | 2,169 | 80 | 1,735 |

* Expressed in units of $1,000.

*Including Seasonal Variation*

where $S_t$ is the trend value of the firm's monthly sales (in thousands of dollars) and $t$ is time measured in months from January 1987. Thus, if this trend continues, the expected sales for each month in 1988 would be as shown in the second column of Table 13.9. But this ignores whatever seasonal variation exists in the firm's sales. In order to include seasonal variation, suppose that the beer manufacturer's statisticians analyze past sales data and, using the method described in Section 13.9, find that the monthly seasonal index for sales is as shown in the third column of Table 13.9. Thus, if this seasonal pattern continues in 1988 as in the past, we would expect that actual sales each month would equal the trend value (in the second column) times the seasonal index (in the third column) divided by 100. The result, which is shown in the fourth column of Table 13.9, is a forecast that includes both the trend and the seasonal variation.

Of course, it must be recognized that this entire procedure is simply a mechanical extrapolation of the firm's sales data into the future. Essentially, the assumption is made that the past trend and the past seasonal variation will continue. The hazards involved in simple extrapolation of this sort have been emphasized in the previous two chapters. Moreover, it is assumed that the trend and seasonal variation are the predominantly important factors that will determine sales in the coming months. The validity of this assumption depends on many considerations, including the extent to which the time series in question (in this case, sales) is affected by cyclical factors and the extent to which the economy is likely to change its cyclical position. In the next section, we will turn our attention to a particular method of forecasting business fluctuations.

# 13.13 Leading Indicators

Firms often want to modify their forecasts in order to take account of prospective overall changes in economic activity. For example, if the beer manufacturer in Table 13.9 is convinced that a serious depression will occur in 1988, it is likely to modify the forecasts in Table 13.9 accordingly. But how does the beer manufacturer—or anyone else—predict whether there is going to be a depression? There are a variety of ways of doing this, all of which are very imperfect. In this section, we discuss an essentially empirical approach, reserving a discussion of more sophisticated techniques for the next section.

*Perhaps the simplest way to forecast business fluctuations is to use leading indicators, which are certain economic series that typically go down or up before gross national product does.* The National Bureau of Economic Research, founded by Wesley C. Mitchell (1874–1948), has carried out detailed and painstaking examinations of the behavior of various economic variables over a long period of time, in some cases over as long as 100 years. The Bureau has attempted to find out whether each variable turns downward before, at, or after the peak of the business cycle, and whether it turns upward before, at, or after the trough. *Variables that go down before the peak and up before the trough are called* **leading series.** *Variables that go down at the peak and up at the trough are called* **coincident series.** *Variables that go down after the peak and up after the trough are called* **lagging series.**

It is worthwhile examining the kinds of variables that fall into each of these three categories, since they provide important facts about the anatomy of the cycle. According to the Bureau, some important leading series are business failures, new orders for durable goods, the average work week, building contracts, stock prices, certain wholesale prices, the money supply, the amount of business and consumer borrowing, and new incorporations. These are the variables that tend to turn downward before the peak and upward before the trough.[17] Coincident series include employment, industrial production, corporate profits, and gross national product, among many others. Some typical lagging series are retail sales, manufacturers' inventories, and personal income.

Economists sometimes use leading series as forecasting devices. There are sound economic reasons why these series turn downward before a peak or upward before a trough: In some cases leading series indicate changes in spending in strategic areas of the economy, while in others they indicate changes in businessmen's and investors' expectations. In order to guide the government in setting economic policies and to guide firms in their planning, it is important to try to spot turning points—peaks and troughs—in advance. (This, of course, is the

---

[17] Of course, business failures turn upward before the peak and downward before the trough.

toughest part of economic forecasting.) Economists sometimes use leading indicators as evidence that a turning point is about to occur. If a large number of leading indicators turn down, this is viewed as a sign of a coming peak. The upturn of a large number of leading indicators is thought to signal an impending trough.

Unfortunately, leading indicators are not very reliable. It is true that the economy has seldom turned downward in recent years without a warning from these indicators, but unfortunately these indicators have turned down on several occasions—in 1952 and 1962, for example—when the economy did *not* turn down subsequently. Thus, leading indicators sometimes provide false signals. Also, in periods of expansion they sometimes turn downward too far ahead of the real peak. And in periods of recession they sometimes turn upward only a very short while before the trough, so that we've turned the corner before anything can be done. Nonetheless, leading indicators are not worthless; they are watched closely and used to supplement other more sophisticated forecasting techniques.

## 13.14 Econometric Models as Forecasting Techniques

In recent years, statisticians and economists have tended to base their forecasts less on simple trend extrapolation of the sort described in Section 13.12 and more on multiple regression techniques and multi-equation models of the sort described in Chapter 12. *The emphasis has shifted toward the construction and estimation of an equation or system of equations that will show the effects of various independent variables on the variable or variables one wants to forecast.* For example, we may want to estimate the quantity of onions that will be supplied by American producers next year. Based on elementary economic theory, one would expect this quantity to depend on this year's price of onions and this year's level of production costs. To make such a forecast, modern statisticians and economists would be likely to estimate an equation relating these independent variables to the quantity of onions supplied; then they would use this equation to make the forecast.

To illustrate this, let's consider the study by D. B. Suits and S. Koizumi[18] which determined that the quantity of onions supplied in year $t$ (designated as $Y_t$) was related in the following way to the price of onions in year $t - 1$ (designated as $P_{t-1}$) and the level of costs of producing onions in year $t - 1$ (designated as $C_{t-1}$):

$$\log Y_t = 0.134 + 0.0123(t - 1924) + 0.324 \log P_{t-1} - 0.512 \log C_{t-1}$$

[18] D. B. Suits and S. Koizumi, "The Dynamics of the Onion Market." *Journal of Farm Economics*, 38, 1956, pp. 475–84. This study is cited for illustrative purposes only. It was carried out a number of years ago and would have to be updated if used today. However, this does not reduce its usefulness in the present context, since we are citing it only as a well-known example of equations of this sort.

Given the known price of onions this year and the known level of this year's production costs, this equation may be used to forecast the quantity of onions that will be supplied next year. If we insert the relevant values of *P, C,* and *t* into the right-hand side of this equation, the resulting value of *Y* is the desired forecast. (The procedure for estimating an equation of this sort has already been described in considerable detail in Chapter 12.)[19]

To forecast many economic variables (such as gross national product) statisticians often use multi-equation models. One of these is the Wharton Econometric Model, described in Appendix 13.1. The Wharton Model contains hundreds of equations variously intended to explain the level of expenditures by households, the level of business investment, aggregate output and employment, and wages, prices, and interest rates. The forecasts produced by the Wharton Model and other large ones like it are followed closely by major business firms and government agencies. Indeed, some firms and agencies have constructed their own multi-equation models. Of course, this does not mean that these large models have an unblemished forecasting record; on the contrary, they, like all other forecasting techniques, are quite fallible. However, these models continue to be widely used in business and government. (Table 13.10 describes the accuracy of four leading econometric models during the late 1970s and early 1980s.)

Both the single-equation model used to forecast the quantity of onions supplied and the Wharton Model, with its hundreds of equations, are examples of econometric models. Although it would be inappropriate in an elementary course to go far into econometrics, it is important that you know the definition of an econometric model.

ECONOMETRIC MODEL: *An econometric model is a system of equations (or a single equation) estimated from past data that is used to forecast economic and business variables.*

*Econometric Model*

Table 13.10
Average Error (in percentage points) of the Forecasts of Four Econometric Models, 1976–80*

| Econometric model | Variable forecasted | | |
|---|---|---|---|
| | Rate of increase of GNP (in real terms) | Unemployment rate | Rate of price increase |
| | *Average error (in percentage points)* | | |
| Bureau of Economic Analysis | 0.5 | 0.4 | 0.9 |
| Chase Econometrics | 1.4 | 1.0 | 1.2 |
| Data Resources, Inc. | 0.8 | 0.5 | 1.0 |
| Wharton | 0.7 | 0.3 | 1.0 |

*Source:* Stephen McNees, "The Recent Record of Thirteen Forecasters," *New England Economic Review,* September 1981.

* All forecasts included here were for one year ahead.

[19] Readers who skipped Chapter 12 can simply take on faith the fact that equations of this sort can be estimated by statisticians.

The essence of any econometric model is that it blends economic theory with statistical techniques. Since econometric models have come to play such an important role in business and economic forecasting, students who want to delve more deeply into forecasting procedures should consider taking the specialized econometrics courses available at many colleges and universities.

> *Basic Idea #26*    **Economic and business statisticians use a variety of forecasting techniques. Econometric models have gained increasing popularity in recent years. However, it is important to bear in mind that all forecasting techniques are extremely fallible and that all forecasts should be treated with caution.**

## 13.15 Forecasting Company Sales: A Case Study[20]

To illustrate the ways in which the simple forecasting techniques described in previous sections have been used, let's consider the case of the Long Island Lighting Company, a major New York public utility. This company has used techniques of this sort to forecast monthly sales for each of the next 18 months and annual sales for each of the next five years. Obviously, these forecasts have played an important role in the firm's planning and decision making.

The first step in Long Island Lighting's forecasting procedure has been to use the trend of population in Nassau and Suffolk counties to forecast the population in the area served by the firm in the relevant month or year in the future. The next step has been to multiply the forecasted population by the number of meters per person (which remains relatively constant) in order to obtain a forecast of the number of meters in the relevant month or year in the future. Next, the 12-month moving average of energy use per meter has been projected into the future on the basis of a trend line of the sort discussed in previous sections. The relationship between energy use per meter and sales per meter has then been used to forecast the sales per meter corresponding to the forecasted energy use per meter. Finally, to obtain a sales forecast, the predicted level of sales per meter has been multiplied by the forecasted number of meters.

This first sales forecast has been checked against another which

[20] This case study is from the Conference Board, *Studies in Business Policy,* and is reprinted in E. Mansfield, *Managerial Economics and Operations Research,* 3d ed. (New York: Norton, 1975). Needless to say, it pertains only to the firm's practices at the time of the Conference Board's publication. For an account of the sales forecasting procedures of the Timken Company, Cummins Engine Company, and RCA Corporation, see the Conference Board, *Sales Forecasting* (New York: Conference Board, 1978); reprinted in E. Mansfield, "Sales Forecasting: Three Case Studies," *Statistics for Business and Economics: Readings and Cases* (New York: Norton, 1980).

THE CEA's FORECASTS FOR 1984

**TO ERR IS
HUMAN ... BUT
COSTLY**

One of the functions of the President's Council of Economic Advisers (CEA) is to forecast the state of the American economy. At the beginning of 1984, President Reagan's Council of Economic Advisers published a forecast of what would occur in 1984. Its forecast of the economic outlook, shown below, called for the growth of real GNP to fall to about 4.5 percent in 1984, for the unemployment rate to be about 7.6 percent in the fourth quarter of 1984, and for the inflation rate to rise slightly to about 5 percent.

| Percent change (1983 fourth quarter to 1984 fourth quarter): | Forecast |
|---|---|
| Real GNP | 4.5 |
| GNP implicit price deflator | 5.0 |
| Output per hour of labor | 2.1 |
| *Level in fourth quarter:* | |
| Unemployment rate (percent) | 7.6 |

Other private forecasts were not vastly different. Data Resources, Inc., predicted a 4.4 percent increase in real GNP, a 7.8 percent unemployment rate, and a 5 percent inflation rate. Chase Econometrics forecasted a 4 percent increase in real GNP, a 8.1 percent unemployment rate, and a 5 percent inflation rate. Wharton Econometrics came up with a 6.6 percent increase in real GNP, a 7.8 percent unemployment rate, and a 3.9 percent inflation rate.

At the end of 1984, a number of major magazines and newspapers ran articles criticizing economic forecasters for their performance. Why?

SOLUTION: In December 1984, the U.S. Department of Commerce reported that real GNP had increased by about 6.7 percent during 1984. According to the U.S. Department of Labor, the civilian unemployment rate in November 1984 was 7.2 percent, and the Consumer Price Index in November 1984 was 4 percent higher than in November 1983. *Time* magazine (and other publications) pointed out that most forecasters had underestimated the rate of growth of real GNP and overestimated the rate of inflation.

The most important moral here is that it is unrealistic to expect economic forecasters to hit the target on the nose. No reputable forecaster would claim that he or she could do so. Moreover, to be useful, forecasts frequently do not have to be very accurate. For many purposes, all that is required is a rough estimate of what will occur, a partial dissipation of the fog that surrounds the economic future. (Furthermore, some forecasts for 1984, such as Wharton's, seemed quite accurate.)

has been derived in a different way but which has also been produced in a number of steps. The firm's engineering department has begun by estimating the peak demand for the previous year and has then estimated the increments in peak demand for the next five years. Next, the trend in the ratio of average kilowatt demand to peak demand has been projected into the future. (According to the firm, this is reasonably accurate, since the ratio moves slowly.) Then the forecast of the peak demand has been multiplied by the forecast of the ratio of average demand to peak demand, and the product has been multiplied by 8,760 (the number of hours in a year). The result (after the amounts of energy lost and company consumption have been deducted) has represented another forecast of the company's sales.

A planning committee has reconciled these two forecasts, and the result was the forecast used in the firm's planning. This method has been used by Long Island Lighting for a number of years. During an eight-year period, "the margin of error has been slightly above 1 percent for forecasts made one year in advance and between 5 percent and 10 percent for the most distant years."[21] Of course, it should not be assumed that the Long Island Lighting Company's techniques will work this well for other firms. But because there was a rather well-defined and stable trend in energy use per meter, in the ratio of average kilowatt demand to peak demand, and in other key variables, these techniques seem to have been relatively effective. For firms facing quite different conditions, such techniques might give much less accurate results.

### EXERCISES

**13.21** On the basis of a simple extrapolation of a linear trend, the predicted value of the Bona Fide Washing Machine Company's sales in February 1989 is $4 million. Adjust this figure to include seasonal variation, on the basis of the figures in Exercise 13.15.

**13.22** On the basis of a simple extrapolation of a linear trend, the predicted value of the Acme Corporation's sales in February 1989 is $25 million. Adjust this figure to include seasonal variation on the basis of the figures in Exercise 13.18.

**13.23** Explain why new orders for durable goods and building contracts would be expected to be leading series rather than lagging series.

**13.24** Explain why the level of stock prices would be expected to be a leading series rather than a lagging series.

**13.25** (a) Based on the data in Exercise 13.12, does it appear that the unemployment rate is subject to cyclical variation?
(b) 1981, 1980, and 1975 were all years of recession. Did the unemployment rate tend to be relatively high in these years?
(c) Explain your results in (b).

**13.26** On October 1, 1985, the U.S. Department of Commerce announced that the index of leading indicators rose seven-tenths of 1 percent in August 1985, the fourth consecutive monthly gain.

[21] *Ibid.*

(a) During August, the money supply increased. Is the money supply among the leading indicators? If so, did its increase help to raise the index?

(b) During August, the average work week rose. Is the average work week among the leading indicators? If so, did its increase help to raise the index?

(c) During August, stock prices fell. Is the level of stock prices among the leading indicators? If so, did its fall help to raise the index?

(d) During August, business and consumer borrowing fell. Is the level of business and consumer borrowing among the leading indicators? If so, did its fall help to raise the index?

**13.27** (a) Use a simple extrapolation of the linear trend in Exercise 13.5 to forecast the Union Carbide Corporation's sales in 1985.

(b) Use a simple extrapolation of the exponential trend in Exercise 13.5 to do the same thing.

(c) How close to one another are the results?

**13.28** Suits and Koizumi estimated an equation to explain the quantity of onions supplied. Do you think that an equation could be estimated to explain the quantity of onions demanded? If so, what independent variables would you propose to include in such an equation?

**13.29** The equation describing the sales trend of a chemical firm is

$$S_t = 21.3 + 1.3t,$$

where $S_t$ is the sales (in millions of dollars per month) of the firm, and $t$ is time measured in months from January 1984. The firm's seasonal index of sales is

| | | | | | |
|---|---|---|---|---|---|
| January | 103 | May | 101 | September | 121 |
| February | 80 | June | 104 | October | 101 |
| March | 75 | July | 120 | November | 75 |
| April | 103 | August | 139 | December | 78 |

Construct a monthly sales forecast for the firm for 1990.

**13.30** A copper firm's monthly sales have the following trend:

$$C_t = 4.12 + 0.32t,$$

where $C_t$ is the sales (in millions of dollars per month) of the firm, and $t$ is time measured in months from July 1985. The firm's seasonal index of sales is

| | | | | | |
|---|---|---|---|---|---|
| January | 81 | May | 137 | September | 79 |
| February | 98 | June | 122 | October | 101 |
| March | 102 | July | 104 | November | 74 |
| April | 76 | August | 101 | December | 125 |

Construct a monthly sales forecast for the firm in 1991.

## Chapter Review

1. The traditional approach to the analysis of time series, devised primarily by economic statisticians, was essentially descriptive. It assumed that an economic time series can be decomposed into four components: trend, seasonal variation, cyclical variation, and irregular movements. A *trend* is a relatively smooth long-term movement of a time series. *Seasonal variation,* which results in one month's (or one season's) being consistently higher or lower than another, is frequently due to the weather, holidays, and other such factors. *Cyclical variation* reflects the effects of business fluctuations, or the so-called business cycle. *Irregular variation* is attributable to short-term, erratic forces that are too unpredictable to be of use for forecasting purposes.

2. If the trend in a time series is *linear,* simple regression may be used to estimate an equation representing the trend. If it seems to be *nonlinear,* a quadratic equation may be estimated by multiple regression, or an exponential trend may be fitted. In some cases, there is no relatively simple mathematical function that can adequately portray the long-term movement of a particular time series. In cases of this sort, statisticians frequently use *moving averages* to smooth the time series. If the time series contains certain fluctuations or cycles that tend to recur, the effect of these cycles can be eliminated by taking a moving average where the number of years in the average is equal to the period of the cycle. Another method of smoothing a time series is to use *exponential smoothing.* According to this method, the smoothed value at a given point in time is a weighted average of all previous values where the weights decline geometrically as one goes backward in time.

3. The seasonal variation in a particular time series is described by a figure for each month (the *seasonal index*) that shows the extent to which that month's value typically departs from what would be expected on the basis of trend and cyclical variation. Such seasonal indexes can be used to *deseasonalize* a time series, that is, to remove the seasonal element from the data. To calculate a seasonal index for a given month, we compute a 12-month moving average centered at the beginning of this month, as well as a 12-month moving average centered at the end of this month. Then we express this month's actual value as a ratio of the mean of these two moving averages (and multiply by 100). The median value of this ratio (adjusted so that the mean value of all months' seasonal indexes is 100) is the seasonal index for this month. Another way to calculate a seasonal index is to use dummy variables in a multiple regression.

4. Many business and economic time series go up and down with the fluctuations of the economy as a whole. This cyclical variation, as well as trend and seasonal variation, is reflected in many time series. It is customary to divide business fluctuations into four phases: *trough, expansion, peak,* and *recession.* Variables that go down before the peak and up before the trough are called *leading series.* Some important leading series are business failures, new orders for durable goods, average work week, building contracts, stock prices, certain wholesale prices, and new incorporations. Economists sometimes use these series, which are often called *leading indicators,* to forecast whether a turning point is about to occur. If a large number of leading indicators turn downward, this is viewed as a sign of a coming peak. If a large number turn upward, this is thought to signal an impending trough. Although these indicators are not very reliable, they are watched closely and are used to supplement other more sophisticated forecasting techniques.

5. One of the most difficult and important tasks that statisticians con-

front is forecasting. In general, all forecasting techniques are extremely fallible, and all forecasts should be treated with caution. The simplest kind of forecasting method is a straightforward extrapolation of a trend. To allow for seasonal variation, such an extrapolation can be multiplied by the seasonal index (divided by 100) for the month to which the forecast applies. This entire procedure is simply a mechanical extrapolation of the time series into the future. In recent years, statisticians and economists have tended to base their forecasts less on simple extrapolations of this sort and more on equations (or systems of equations) showing the effects of various independent variables on the variable (or variables) one wants to forecast. These equations (or systems of equations), which are estimated using the techniques described in the previous chapter, are often called *econometric models.*

## Chapter Review Exercises

**13.31** The capacity utilization rate in all U.S. manufacturing is shown below for the years 1947–74.[22]

| Year | Percent utilized of all manufacturing capacity | Year | Percent utilized of all manufacturing capacity |
|------|------|------|------|
| 1947 | 95 | 1961 | 78 |
| 1948 | 92 | 1962 | 81 |
| 1949 | 81 | 1963 | 83 |
| 1950 | 90 | 1964 | 86 |
| 1951 | 91 | 1965 | 91 |
| 1952 | 88 | 1966 | 96 |
| 1953 | 92 | 1967 | 93 |
| 1954 | 83 | 1968 | 94 |
| 1955 | 91 | 1969 | 96 |
| 1956 | 90 | 1970 | 88 |
| 1557 | 86 | 1971 | 85 |
| 1958 | 76 | 1972 | 90 |
| 1959 | 82 | 1973 | 96 |
| 1960 | 80 | 1974 | 91 |

(a) Calculate a three-year moving average of this time series.
(b) Calculate a five-year moving average of this time series.
(c) Calculate a seven-year moving average of this time series.
(d) Plot the time series, together with the three-year moving average (derived in (a)). Based on this diagram, does it appear that this time series has a strong upward trend?
(e) For 1980 to 1984, the capacity utilization rate in all U.S. manufacturing was as follows:

| | |
|------|------|
| 1980 | 80 |
| 1981 | 79 |
| 1982 | 71 |
| 1983 | 75 |
| 1984 | 82 |

[22] See *Economic Report of the President, 1976* (Washington, D.C.: Government Printing Office, 1976), p. 211. This is the Wharton series, whereas the figures in part (e) are from the Federal Reserve. They are not entirely comparable, but that is of no importance here.

Do these more recent figures change your view as to whether or not this time series has a strong upward trend?

**13.32** According to the Federal Reserve Board, the capacity utilization rate in manufacturing (in percent) in each quarter during the period 1972–77 was as follows:

| Quarter | 1972 | 1973 | 1974 | 1975 | 1976 | 1977 |
|---------|------|------|------|------|------|------|
| First   | 80.9 | 87.1 | 85.7 | 70.9 | 79.1 | 81.2 |
| Second  | 82.4 | 87.6 | 85.8 | 71.3 | 80.3 | 82.7 |
| Third   | 83.4 | 87.8 | 85.5 | 75.3 | 80.8 | 83.0 |
| Fourth  | 85.8 | 87.7 | 79.7 | 76.9 | 80.6 | 82.8 |

Suppose that a statistician calculates a four-quarter moving average of the data, centers the four-quarter moving average, and computes the ratio (times 100) of each quarter's capacity utilization rate to the centered moving average. Then he calculates the median of the ratios for (1) the first quarter, (2) the second quarter, (3) the third quarter, and (4) the fourth quarter. Finally, he multiplies each of the four medians by 400 and divides the result by the sum of the medians for the four quarters.

(a) Is this result a seasonal index?

(b) If the above figures have already been seasonally adjusted by the Federal Reserve Board, can you guess what his results are likely to be?

**13.33** Civilian employment (in millions of persons) in the United States during 1961–81 was as follows:

| Year | Employment | Year | Employment | Year | Employment |
|------|-----------|------|-----------|------|-----------|
| 1961 | 66 | 1968 | 76 | 1975 | 86 |
| 1962 | 67 | 1969 | 78 | 1976 | 89 |
| 1963 | 68 | 1970 | 79 | 1977 | 92 |
| 1964 | 69 | 1971 | 79 | 1978 | 96 |
| 1965 | 71 | 1972 | 82 | 1979 | 99 |
| 1966 | 73 | 1973 | 85 | 1980 | 99 |
| 1967 | 74 | 1974 | 87 | 1981 | 100 |

(a) Calculate a linear (least-squares) trend for these data.

(b) On the average, how much did civilian employment increase annually during this period?

(c) If time is measured from 1961, what is the equation for the trend line?

(d) If time is measured from 1975, what is the equation for the trend line?

(e) Plot the above data. Does a linear trend appear to be adequate?

(f) In 1984, civilian employment was 105 million persons. Suppose that the linear trend you computed in (a) had been used to forecast civilian employment in 1984. How large would the forecasting error have been?

**13.34** Suppose that a statistician decides to change the smoothing constant from 1/3 to 1/5. He has the exponentially smoothed value of a time series for the previous period, $S_{t-1}$, based on the old smoothing constant. Is it true that $S_t = 1/5Y + (1 - 1/5)S_{t-1}$, where $S_{t-1}$ is based on $\theta = 1/3$? Why, or why not?

**\*13.35** (a) Using the data in Exercise 13.33, calculate a quadratic trend for civilian employment.

(b) Plot the resulting quadratic trend against the data. Does it seem to fit reasonably well?

**13.36** (a) Based on the data in Exercise 13.32, does it appear that the capacity utilization rate in manufacturing is subject to cyclical variation?

(b) On *a priori* grounds, would you expect the capacity utilization rate to vary cyclically?

**13.37** Based on the linear trend you calculated in Exercise 13.6, estimate gross national product in 1986.

**\*13.38** Based on the quadratic trend you calculated in Exercise 13.14, estimate gross national product in 1986.

**13.39** Based on the linear trend you calculated in Exercise 13.33, estimate civilian employment in 1985.

**\*13.40** Based on the quadratic trend you calculated in Exercise 13.35, estimate civilian employment in 1985.

**13.41** (a) Calculate the difference between each year's sales of the Union Carbide Corporation and the least-squares trend you derived in Exercise 13.5.

(b) Do these deviations from trend show a cyclical pattern? If so, when are some of the most pronounced peaks and troughs?

**13.42** (a) Does the time series of capacity utilization rates in Exercise 13.31 show cyclical variation?

(b) Do the peaks and troughs tend to coincide with those in the economy as a whole?

# Appendix 13.1

## THE WHARTON ECONOMETRIC MODEL

One of the most widely used econometric models is the Wharton model. Composed of hundreds of equations, this model has been in operation for over 20 years. In addition to being used to forecast gross national product, it attempts to forecast the composition of gross national product, the price level, unemployment, and other major variables pertaining to the national economy. With the help of high-speed electronic computers, the coefficients of the many equations in this model can be estimated and revised from time to time.

Due to the model's large size it is not feasible to list all its equations. But it is possible to describe the nature of some of the more important equations, thus providing some feel for the structure of the model. First, there are equations to explain the level of personal consumption expenditures, one equation

---

\* Exercise 13.35 assumes that the reader has covered Chapter 12. If not, it should be skipped.

\* Exercises 13.38 and 13.40 assume that the reader has covered Chapter 12. If not, they should be skipped.

pertaining to autos, one to nonautomotive durables, one to nondurables, and one to services. In each of these equations, the level of expenditure is dependent on disposable income and prices, among other things. This, of course, is in keeping with basic economic theory.

Second, there are equations which are used to explain the level of investment, one equation pertaining to business purchases of plant and equipment, one pertaining to construction of houses and apartments, and one pertaining to changes in inventories. In the case of expenditures on plant and equipment, the relevant equation makes these expenditures dependent on changes in output or the extent of capacity utilization as well as on the cost of capital. In the case of residential investment, the tightness of monetary markets is an important explanatory variable in the relevant equation.

Third, there are equations to explain output and employment. These variables are determined in the model primarily by the level of aggregate demand, which depends on consumption expenditure, investment, imports and exports, and government expenditures. Government expenditures are an exogenous variable in the model. Equations are included to explain imports and exports, these variables being functions of things like the prices of imported goods relative to domestic goods and the competitiveness of U.S. export prices relative to world prices. To estimate employment, production functions are estimated.[23] Given the estimates of output, these production functions can be used to forecast employment. Unemployment can also be forecasted, based on information concerning the size of the labor force.

Fourth, there are equations to explain wages, prices, and interest rates. The wage equations make the rate of increase of wages depend on the rate of unemployment and the rate of increase of prices. The price level is determined by the extent of capacity utilization, among other things. A number of equations relate to the monetary sector of the economy. These equations attempt to explain interest rates, the money supply, and other financial variables.

Every three months the Wharton Model is used to produce forecasts of gross national product, the price level, the unemployment rate, and other such variables for the next two years. During the 1960s its forecasting performance was particularly impressive; the average error of its forecasts of gross national product during the period 1959–67 was only about $3 billion (less than 1/2 of 1 percent). During the mid-1970s, its forecasting performance was less impressive, due in part to political events such as the behavior of the OPEC oil producers. (Obviously, such events are difficult to predict and incorporate into such a model.) Nonetheless, this and other multi-equation models are widely used by business firms and government agencies. They have an important influence on both government policy and business decision making here and abroad.

# Appendix 13.2

### FORECASTING BASED ON EXPONENTIAL SMOOTHING

In Section 13.7 we described how exponential smoothing can be used to smooth a time series. It is worth noting that exponential smoothing can also be used to forecast a time series. When used for forecasting purposes, the basic

---

[23] A *production function* is defined by economists as the relationship between input and output for a firm, industry, or the economy as a whole.

equation for exponential smoothing is

$$F_t = \theta A(t-1) + (1-\theta)F_{t-1},\qquad (13.16)$$

where $A(t-1)$ is the actual value of the time series at time $(t-1)$, and $F_t$ is the forecast for time $t$. We suppose that the forecast is being made at time $(t-1)$, so the actual value of the time series at this time is known. Given the smoothing constant $\theta$, the forecast for time $t$ is simply a weighted average of the *actual* value at time $(t-1)$ and the *forecasted* value for time $(t-1)$, where the actual value is weighted by $\theta$ and the forecasted value is weighted by $(1-\theta)$. It can easily be shown that the forecast for time $t$ is the weighted sum of the actual values prior to time $t$, where the weight attached to each value declines geometrically with the age of the observation.

To illustrate the use of exponential smoothing for forecasting purposes, let's return to the firm in Section 13.7 that had been existence for five years. Sales during the first year were $1 million, and we assume that the firm's sales forecast for the first year was also $1 million. What will be its sales forecast for the second year? To make such a forecast, the firm begins by choosing a value for the smoothing constant $\theta$. Suppose that a value of .2 is chosen. Then the forecast for the second year is $.2(1) + .8(1) = 1$, or $1 million. The firm's actual sales in the second year turn out to be $3 million. Hence, its sales forecast for the third year will be $.2(3) + .8(1) = 1.4$, or $1.4 million. The firm's actual sales in the third year turn out to be $3 million. Hence, its sales forecast for the fourth year will be $.2(3) + .8(1.4) = 1.72$, or $1.72 million. And so on.

Exponential smoothing is often used in this way to make forecasts, particularly where there is a need for a cheap, fast, and rather mechanical method to make forecasts for a large number of items. For example, to implement various kinds of inventory control models, demand forecasts for hundreds or thousands of items may be required. A major advantage of this technique is that the only number that must be stored until the next time period is the forecast for the next period. Why? Because this number, together with the actual value for the next period and $\theta$, determines the forecast for the period following the next period.

In accord with our discussion in Section 13.7, it is essential that $\theta$ be between 0 and 1. If $\theta$ is close to 1, past values of the time series are given relatively little weight (compared to recent values) in determining the forecast. If $\theta$ is close to zero, past values are given relatively heavy weight (compared to recent values) in determining the forecast. In practice, values of .3 or less are often used. Finally, a variety of smoothing techniques can be used as the basis for forecasting methods. The technique given in this appendix is certainly among the simplest, and thus has been chosen as the logical introduction to methods of this kind.

# 14

# Index Numbers*

## 14.1 Introduction

Decision makers and analysts in business and government must frequently try to summarize the differences that have occurred in a number of economic variables over a particular period of time. For example, between 1970 and 1987 changes occurred in the prices of practically all goods and services; and these price changes were by no means uniform. For many purposes, it is useful to construct a price index to summarize this variety of price changes into a single number which indicates the extent to which the cost of living in 1987 exceeded that in 1970. Similarly, the price of a given item is different in Los Angeles than in Chicago, and this intercity difference varies from one item to another. Again, it is useful to summarize this variety of price differences into a single number which tells whether (and to what extent) the cost of living is higher in Los Angeles than in Chicago. Statisticians and economists have devoted considerable attention to the formulation of index numbers of this kind; the purpose of this chapter is to describe and illustrate the methods used to construct such numbers.

At the outset, let's define somewhat more precisely what we mean by an index number.

> INDEX NUMBER: *An index number is a ratio (generally expressed in percentage terms) of one quantity to another where one of the quantities summarizes a given group of items and the other quantity summarizes a base group of items. The "base" group is used as a basis of comparison with the "given" group.*

For example, a government statistician might be asked to construct an index number showing changes in steel prices between 1970 and 1987.

* Some instructors may wish to skip Section 14.8, which deals with quantity indexes. This section can be omitted without loss of continuity.

In such an index number, 1970 would be the *base* year, and 1987 would be the *given* year. The 1987 prices of the various types of steel (sheets, bars, tin plate, stainless, and so on) would be compared with those in 1970, and this index number would indicate how much higher these prices were in 1987 than in 1970. Thus, if this index number equals 280, this means that on the average, steel prices were 180 percent higher in 1987 than in 1970.

## 14.2 Unweighted Index Numbers

As a first step toward understanding how index numbers are constructed, we consider so-called unweighted index numbers. Although unweighted index numbers are very seldom used (because of flaws discussed below), they are the simplest kind and thus provide a good beginning for our discussion. To see what we mean by unweighted index numbers, suppose that the government of a small agricultural country wants to construct an index number showing changes over time in the price of its major food crops. Assume that the price per bushel of four major crops—corn, wheat, oats, and soybeans—is shown in Table 14.1 for 1970, 1980, 1985, and 1987. Using 1970 as the base period, an index of these prices can be constructed by summing up the prices of each of these commodities in a given year and then dividing the sum by the corresponding sum in the base year. For example, since $1.00 is the price of corn in 1970, since $1.74 is the price of wheat in 1970, and so on, the sum for the base year is

$$\$1.00 + 1.74 + 0.60 + 2.13 = \$5.47.$$

Similarly, the corresponding sum for 1987 is

$$\$1.29 + 1.67 + 0.67 + 3.49 = \$7.12.$$

Thus, the value of the index number for 1987 is

$$\frac{\$7.12}{\$5.47}(100) = 130,$$

which indicates that these crop prices rose by 30 percent between 1970 and 1987.

| Crop | 1970 (dollars) | 1980 (dollars) | 1985 (dollars) | 1987 (dollars) |
|---|---|---|---|---|
| Corn | 1.00 | 1.16 | 1.33 | 1.29 |
| Wheat | 1.74 | 1.35 | 1.33 | 1.67 |
| Oats | 0.60 | 0.62 | 0.62 | 0.67 |
| Soybeans | 2.13 | 2.54 | 2.85 | 3.49 |

Table 14.1
Price per Bushel of Four Major Food Crops, 1970–87*

* These figures were chosen for illustrative purposes; the important point here is to facilitate our exposition of index numbers, not to describe the actual behavior of farm prices in any particular country during this period.

In general, the simple unweighted index we have constructed can be expressed as follows: *If $P_{01}$ is the price of the first commodity (corn) in the base period and $P_{11}$ is its price in year 1, $P_{02}$ is the price of the second commodity (wheat) in the base period and $P_{12}$ is its price in year 1, and so forth, then the simple unweighted price index for year 1 is*

$$I_1 = \frac{P_{11} + P_{12} + \ldots + P_{1m}}{P_{01} + P_{02} + \ldots + P_{0m}}(100),$$

*where m is the number of commodities included in the index.* Thus, we can express this index as

**Simple Aggregative Index**

$$I_1 = \frac{\sum_{i=1}^{m} P_{1i}}{\sum_{i=1}^{m} P_{0i}}(100). \qquad (14.1)$$

One major difficulty with this **simple aggregative index,** as it is called, is that *the result one obtains depends on the units in which each of the commodities is expressed.* For example, suppose that we measure soybeans in units of 100 bushels rather than bushels. Then the price of soybeans would be $213 in 1970 and $349 in 1987. Thus, the simple aggregative price index for 1987 would be

$$\frac{\$1.29 + 1.67 + 0.67 + 349.00}{\$1.00 + 1.74 + 0.60 + 213.00}(100) = \frac{352.63}{216.34}(100) = 163,$$

which indicates that these crop prices rose by 63 percent between 1970 and 1987. Clearly, this result is quite different from that obtained when the price of a single bushel of soybeans was used.

One way of avoiding this difficulty is to use an index which is an **average of relatives.** For example, to obtain an index of food crop prices, one could begin by obtaining relatives for each price, a *relative* being *the ratio of the price in the given year to that in the base year.* Table 14.2 shows the price relatives for each commodity. Then to obtain the index, we can compute the mean of these relatives. For example, the value of the index number for 1987 can be computed from the relatives shown in the last column of Table 14.2:

$$\frac{1.29 + 0.96 + 1.12 + 1.64}{4}(100) = 125.$$

Note that we use the mean of the price relatives; the median or mode could have been used instead, but the mean is the kind of average that is typically employed.

In general, an index number based on a simple average of relatives can be expressed in the following way. Using the notation de-

| Crop | $\dfrac{1980\ price}{1970\ price}$ | $\dfrac{1985\ price}{1970\ price}$ | $\dfrac{1987\ price}{1970\ price}$ |
|------|------|------|------|
| Corn | $\dfrac{1.16}{1.00} = 1.16$ | $\dfrac{1.33}{1.00} = 1.33$ | $\dfrac{1.29}{1.00} = 1.29$ |
| Wheat | $\dfrac{1.35}{1.74} = 0.78$ | $\dfrac{1.33}{1.74} = 0.76$ | $\dfrac{1.67}{1.74} = 0.96$ |
| Oats | $\dfrac{0.62}{0.60} = 1.03$ | $\dfrac{0.62}{0.60} = 1.03$ | $\dfrac{0.67}{0.60} = 1.12$ |
| Soybeans | $\dfrac{2.54}{2.13} = 1.19$ | $\dfrac{2.85}{2.13} = 1.34$ | $\dfrac{3.49}{2.13} = 1.64$ |

**Table 14.2**
Price Relatives for
Four Major Food
Crops, 1970–87*

\* See note, Table 14.1.

scribed above, the price index for year 1 equals

$$I_1 = \frac{\dfrac{P_{11}}{P_{01}} + \dfrac{P_{12}}{P_{02}} + \dfrac{P_{13}}{P_{03}} + \ldots + \dfrac{P_{1m}}{P_{0m}}}{m} (100).$$

Or using the summation symbol,

$$I_1 = \frac{\sum_{i=1}^{m} \dfrac{P_{1i}}{P_{0i}}}{m} (100). \qquad (14.2)$$

*Simple Average of
Relatives*

## 14.3 Weighted Index Numbers

Unweighted relative price indexes have a very important disadvantage which results in their seldom being used: *The various prices included in the index are given equal importance.* Consider the four commodities in Tables 14.1 and 14.2. As shown in Table 14.3, corn is produced and consumed in much greater quantities than wheat, oats, or soybeans in the country for which the price index is being computed. Consequently, it seems obvious that changes in the price of corn should have a greater weight in a price index than changes in the price of the other three crops. However, unweighted relative price indexes would treat them all as equally important.

To remedy this deficiency, statisticians generally use weighted index numbers. In the case of price indexes, the weights most often used for price relatives are value weights. For example, in the case of Table 14.3, the price relative for corn would be weighted by the value of corn produced, the price relative for wheat would be weighted by the value of wheat produced, and so on. Using the values in the base

**Table 14.3**
Production and Value
of Output of Major
Food Crops, 1970–87*

| Crop | 1970 | 1980 | 1985 | 1987 |
|---|---|---|---|---|
| | | Production (millions of bushels) | | |
| Corn | 3.907 | 4,084 | 4,099 | 5,553 |
| Wheat | 1,355 | 1,316 | 1,370 | 1,545 |
| Oats | 1,153 | 927 | 909 | 695 |
| Soybeans | 555 | 846 | 1,124 | 1,283 |
| | | Value of output (millions of dollars) | | |
| Corn | 3,907 | 4,737 | 5,452 | 7,163 |
| Wheat | 2,358 | 1,777 | 1,822 | 2,580 |
| Oats | 692 | 575 | 564 | 466 |
| Soybeans | 1,182 | 2,149 | 3,203 | 4,478 |

\* See note, Table 14.1.

year as weights, it is clear from Table 14.2 (which contains the price relatives) and Table 14.3 (which contains the value weights) that the value of the price index for 1987 is

$$\frac{3{,}907(1.29) + 2{,}358(0.96) + 692(1.12) + 1{,}182(1.64)}{3{,}907 + 2{,}358 + 692 + 1{,}182}(100)$$

$$= \frac{10{,}017}{8{,}139}(100) = 123.$$

In other words, the price relative for each commodity is weighted by its value in the base year, and the weighted sum of the price relatives is divided by the sum of the weights.

In symbols, this weighted relative price index can be expressed as

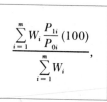

$$\frac{\sum_{i=1}^{m} W_i \frac{P_{1i}}{P_{0i}}(100)}{\sum_{i=1}^{m} W_i}, \tag{14.3}$$

where $W_i$ is the weight attached to the price relative of the $i$th commodity. When, as in this case, the weight equals $Q_{0i}P_{0i}$, where $Q_{0i}$ is the amount consumed of the $i$th commodity in the base period, the resulting price index is

$$\frac{\sum_{i=1}^{m} (Q_{0i}P_{0i})\left(\frac{P_{1i}}{P_{0i}}\right)}{\sum_{i=1}^{m} Q_{0i}P_{0i}}(100).$$

This is the Laspeyres index, defined below.

LASPEYRES INDEX: *The Laspeyres price index is the ratio (expressed as a percentage) of the total cost in the given year of the quantity of each commodity consumed in the base year to what was the total cost of these quantities in the base year. In symbols, the Laspeyres index is*

*Laspeyres Price Index*

$$\frac{\sum\limits_{i=1}^{m} Q_{0i} P_{1i}}{\sum\limits_{i=1}^{m} Q_{0i} P_{0i}} (100).$$

(14.4)

Another frequently encountered type of weighted relative price index is the Paasche index, defined as follows.

PAASCHE INDEX: *The Paasche price index is the ratio (expressed as a percentage) of the total cost in the given year of the quantity of each commodity consumed in the given year to what would have been the total cost of these quantities in the base year. In symbols, the Paasche index is*

*Paasche Price Index*

$$\frac{\sum\limits_{i=1}^{m} Q_{1i} P_{1i}}{\sum\limits_{i=1}^{m} Q_{1i} P_{0i}} (100).$$

(14.5)

Both the Laspeyres and Paasche indexes measure the change, from the base year to the given year, in the total cost of the commodities consumed; but while the Laspeyres index uses the quantities consumed in the *base year,* the Paasche index uses the quantities consumed in the *given year* (that is, year 1).

Table 14.4 shows the value of the Paasche index for our four commodities in 1985 and 1987. In 1987, for example, since the amounts consumed of each commodity were as shown in the last column of Table 14.3, it would have cost

5,553($1.00) + 1,545($1.74) + 695($0.60) + 1,283($2.13),

or $11,391 million, to buy these amounts in the base period. Thus, since it actually cost $14,687 million to buy these amounts in 1987,[1] the value of the Paasche index for 1987 is

$$\frac{14,687}{11,391} (100) = 129.$$

[1] Table 14.3 shows that the value of output in 1987 was 7, 163 + 2,580 + 466 + 4,478 = 14,687 millions of dollars. The same thing is shown in the latter part of Table 14.4.

**Table 14.4**
1985 and 1987
Paasche Price Indexes

| | Computation of 1985 Paasche price index | |
| Crop | 1985 output × 1985 price | 1985 output × 1970 price |
| --- | --- | --- |
| Corn | 5,452 | 4,099 |
| Wheat | 1,822 | 2,384 |
| Oats | 564 | 545 |
| Soybeans | 3,203 | 2,394 |
| Total | 11,041 | 9,422 |

Value of index $= \dfrac{11,041}{9,422} (100) = 117.$

| | Computation of 1987 Paasche price index | |
| Crop | 1987 output × 1987 price | 1987 output × 1970 price |
| --- | --- | --- |
| Corn | 7,163 | 5,553 |
| Wheat | 2,580 | 2,688 |
| Oats | 466 | 417 |
| Soybeans | 4,478 | 2,733 |
| Total | 14,687 | 11,391 |

Value of index $= \dfrac{14,687}{11,391} (100) = 129.$

Although the Paasche index is potentially useful, it has important practical disadvantages in cases where an index is calculated for one period after another. For example, the Consumer Price Index and the Producer Price Index (formerly the Wholesale Price Index) are calculated on a monthly and annual basis. To use the Paasche index for a series of this sort, one would have to obtain fresh data each period concerning the quantity weights, $Q_{1i}$. This would be expensive and time-consuming, making the use of the Paasche index inappropriate on practical grounds.

*Basic Idea # 27*    **Whereas the Laspeyres price index shows how much more the set of goods and services consumed in the base period cost in the given period than in the base period, the Paasche price index shows how much more the set of goods and services consumed in the given period cost in the given period than in the base period.**

## 14.4 Other Weighting Schemes and Base Periods

Although the Laspeyres index weights individual prices by quantity consumed or produced in the *base period* and the Paasche index weights them by quantity consumed or produced in the *given period* (that is, the period for which the index is computed), these two periods are not the only ones which could be used. For example, consider our index of crop prices. The Laspeyres 1987 index calculated in the previous section weighted individual prices by output in 1970; and the Paasche index for 1987 weighted individual prices by output in 1987. *But there is nothing to prevent us from using as weights output in some year other than 1970 or 1987.* Indeed, many important price indexes published by the federal government do just that.

One reason for using output for some period other than the base year or current year is that it is desirable for the weights to refer to a "normal" year. Suppose that 1970 and 1987 were both rather abnormal years for agriculture, whereas 1980 seemed much freer of major abnormalities. Then rather than using 1970 quantity weights (as in the Laspeyres index) or 1987 weights (as in the Paasche index), we might use 1980 quantity weights. In other words, to obtain the index for 1987, we might calculate the amount it would cost in 1987 to buy the amount of each commodity consumed in 1980. The result is

$$4{,}084(\$1.29) + 1{,}316(\$1.67) + 927(\$0.67) + 846(\$3.49) = \$11{,}040.$$

Then we might calculate the amount it would have cost in 1970 to buy the amount of each commodity consumed in 1980. The result is

$$4{,}084(\$1.00) + 1{,}316(\$1.74) + 927(\$0.60) + 846(\$2.13) = \$8{,}732.$$

Thus, the index for 1987 is

$$\frac{11{,}040}{8{,}732}(100) = 126$$

In symbols, this index, sometimes called a *fixed-weight aggregative index,* can be expressed as

*Fixed-Weight Aggregative Price Index*

$$I_1 = \frac{\sum_{i=1}^{m} Q_{wi}P_{1i}}{\sum_{i=1}^{m} Q_{wi}P_{0i}}(100), \qquad (14.6)$$

where $Q_{wi}$ is the amount consumed of the $i$th commodity in year $w$. Year $w$ is, of course, the period (chosen because it is freer of abnormalities or for some other reason) to which the weights pertain. Note that the only difference between this index and the Laspeyres index or the Paasche index is that $Q_{wi}$ is used as a weight here, whereas $Q_{0i}$ is used in the Laspeyres index and $Q_{1i}$ is used in the Paasche index.

Going a step further, there is no reason why the quantity weights must pertain to a single year. Instead, *one can use as weights the average level of output or consumption in a number of years.* For example, in the

**Table 14.5**
Fixed-Weight
Aggregative Index of
Crop Prices, Quantity
Weights Being Averages
of 1980 and 1985

| Crop | (1) 1980 output | (2) 1985 output | (3) Quantity weight [average of (1) and (2)] | (4) Quantity weight × 1970 price (dollars) | (5) Quantity weight × 1987 price (dollars) |
|---|---|---|---|---|---|
| Corn | 4,084 | 4,099 | 4,092 | 4,092 | 5,279 |
| Wheat | 1,316 | 1,370 | 1,343 | 2,337 | 2,243 |
| Oats | 927 | 909 | 918 | 551 | 615 |
| Soybeans | 846 | 1,124 | 985 | 2,098 | 3,438 |
| Total | | | | 9,078 | 11,575 |

$$\text{Index for 1987} = \frac{11,575}{9,078}(100) = 128.$$

case of the index of crop prices, we might feel that the best set of weights is an average of 1980 and 1985 quantities, as given in column three of Table 14.5. Based on these weights, the index for 1987 is computed (in Table 14.5) in the ordinary way. The resulting index number for 1987 is 128. Among the many well-known indexes which have used quantity weights of this sort is the U.S. Department of Agriculture's index of prices received by farmers. For example, during the 1960s this index used as quantity weights the average quantities during the period 1953–57.

Finally, *the base period for prices (as well as output) can also be an average of several years.* Suppose that we decide that an average of 1970 and 1980 is a good base period for crop prices. Assuming that we use 1980 quantity weights, Table 14.6 shows how the index for 1987 is computed. Government price indexes frequently use an average of several years as the base price. For example, during the 1960s, the Producer Price Index (then called the Wholesale Price Index) used 1957–59 as the base period for prices. Of course, there is no reason why the period to which the quantity weights pertain (the *weight base*) must be the same as the period from which price changes are measured (the *reference base*). For example, the weight base in Table 14.6 is 1980, whereas the reference base is the average of 1970 and 1980.

**Table 14.6**
Fixed-Weight
Aggregative Index of
Crop Prices, Using
1980 Quantity Weights
and Average of 1970
and 1980 as Price Base

| Crop | 1980 output | 1970 price (dollars) | 1980 price (dollars) | Average of 1970 and 1980 prices (dollars) | 1980 output × average price (dollars) | 1980 output × 1987 price (dollars) |
|---|---|---|---|---|---|---|
| Corn | 4,084 | 1.00 | 1.16 | 1.08 | 4,411 | 5,268 |
| Wheat | 1,316 | 1.74 | 1.35 | 1.55 | 2,040 | 2,198 |
| Oats | 927 | 0.60 | 0.62 | 0.61 | 565 | 621 |
| Soybeans | 846 | 2.13 | 2.54 | 2.34 | 1,980 | 2,953 |
| Total | | | | | 8,996 | 11,040 |

$$\text{Index for 1987} = \frac{11,040}{8,996}(100) = 123.$$

| Year | (1) Price of platinum (dollars per troy ounce) | (2) Price index (1979 = 100) | (3) Price index (Previous year = 100) |
|---|---|---|---|
| 1979 | 352 | $\frac{352}{352}(100) = 100$ | — |
| 1980 | 439 | $\frac{439}{352}(100) = 125$ | $\frac{439}{352}(100) = 125$ |
| 1981 | 475 | $\frac{475}{352}(100) = 135$ | $\frac{475}{439}(100) = 108$ |
| 1982 | 475 | $\frac{475}{352}(100) = 135$ | $\frac{475}{475}(100) = 100$ |
| 1983 | 475 | $\frac{475}{352}(100) = 135$ | $\frac{475}{475}(100) = 100$ |

**Table 14.7**
Index of Price of
Platinum, 1979–83

*Source: Statistical Abstract of the United States* (Washington, D.C.: Government
Printing Office, 1985).

## 14.5 Chain Index Numbers

To complete our survey of the various techniques of index number
construction, we must describe chain index numbers. To understand
the basic principles behind chain index numbers, one need only consider Table 14.7, which shows the price of platinum during the period
1979–83. Based on these data, it is a simple matter to make up an index
of the price of platinum during this period. All we need to do is divide
the price for each year by the price in 1979 and multiply by 100.

The resulting index in column (2) of Table 14.7 shows how the
price of platinum has varied relative to the 1979 base price. Suppose,
however, that we are interested in year-to-year changes in price. In this
case, it makes sense to use as a base the price in the previous year, not
the price in 1979. In other words, *we can use a shifting price base rather
than a fixed base.* For example, the index for 1981 relative to a 1980
base is

$$I_{80}^{81} = \frac{475}{439}(100) = 108,$$

and the index for 1982 relative to a 1981 base is

$$I_{81}^{82} = \frac{475}{475}(100) = 100.$$

These are chain index numbers.

From chain index numbers one can easily obtain the ordinary
fixed-base index numbers described in previous sections. To show this,

let $P_{83}$ be the price of platinum in 1983, $P_{82}$ its price in 1982, and so on. Then if the 1979 price is used as a base, the ordinary index number for 1983 is

$$I_{83} = \frac{P_{83}}{P_{79}} (100).$$

But this can be written

$$I_{83} = \left(\frac{P_{83}}{P_{82}}\right)\left(\frac{P_{82}}{P_{81}}\right)\left(\frac{P_{81}}{P_{80}}\right)\left(\frac{P_{80}}{P_{79}}\right) (100). \qquad (14.7)$$

Since $P_{83}/P_{82}$ is the chain index (divided by 100) indicating the change from 1982 to 1983, $\dfrac{P_{82}}{P_{81}}$ is the chain index (divided by 100) indicating the change from 1981 to 1982, and so forth, we can substitute $I\,^{83}_{82} \div 100$ for $P_{83}/P_{82}$, $I\,^{82}_{81} \div 100$ for $P_{82}/P_{81}$, and so on, the result being

$$I_{83} = \left(\frac{I\,^{83}_{82}}{100}\right)\left(\frac{I\,^{82}_{81}}{100}\right)\left(\frac{I\,^{81}_{80}}{100}\right)\left(\frac{I\,^{80}_{79}}{100}\right) (100)$$

Clearly, the chain index technique can be used to compute the value of an index number, given that one knows the value of the index number in the previous period. To see how this can be done, let's return to our example of the platinum price index. It is apparent that the value of the index for 1982 was

$$I_{82} = \left(\frac{P_{82}}{P_{81}}\right)\left(\frac{P_{81}}{P_{80}}\right)\left(\frac{P_{80}}{P_{79}}\right) (100).$$

Thus, if we substitute $I_{82}$ for its above equivalent expression in equation (14.7), we have

$$I_{83} = \frac{P_{83}}{P_{82}} I_{82}.$$

The meaning of this formula is as follows: To get the value of the price index for 1983, all that we have to do is multiply the 1982 value of the price index by the ratio of 1983 price to 1982 price. This latter ratio is, of course, the chain index (divided by 100) indicating the price change from 1982 to 1983. This is the way that many important price indexes, such as the Consumer Price Index, are calculated.

**EXERCISES**

**14.1** According to the Department of Agriculture, the prices received by farmers during the period 1910–74 could be represented by the following index numbers:

| Year | All farm products | Livestock | Cotton | Tobacco | Fruit |
|------|-------------------|-----------|--------|---------|-------|
| 1910–14 | 100 | 100 | 100 | 100 | 100 |
| 1930 | 125 | 134 | 104 | 140 | 149 |
| 1950 | 258 | 280 | 282 | 402 | 194 |
| 1970 | 280 | 326 | 183 | 604 | 233 |
| 1974 | 467 | 453 | 433 | 821 | 349 |

(a) For each of these index numbers, specify the base period and the given period.

(b) Did the price of farm products increase at a more rapid or less rapid annual rate during 1970–74 than during 1950–70?

(c) Did the price of farm products increase at a more rapid or less rapid annual rate during 1970–74 than during 1930–50?

(d) Which of the specific types of farm products (livestock, cotton, tobacco, and fruit) in the table above experienced the greatest percentage price increase from 1910–14 to 1974?

**14.2** The prices per pound of beef cattle and veal calves were as follows in 1940, 1950, 1960, and 1970:

|  | 1940 | 1950 | 1960 | 1970 |
|------|------|------|------|------|
| Beef | $0.08 | $0.23 | $0.20 | $0.27 |
| Veal | 0.09 | 0.26 | 0.23 | 0.35 |

The U.S. outputs of beef and veal (in billions of pounds) in 1940, 1950, 1960, and 1970 were as follows:

|  | 1940 | 1950 | 1960 | 1970 |
|------|------|------|------|------|
| Beef | 7.2 | 9.5 | 14.8 | 21.7 |
| Veal | 1.0 | 1.2 | 1.1 | 0.6 |

(a) A meat-packing firm wants to construct a price index for beef cattle and veal calves. Letting 1940 be the base year, calculate an average of relatives for 1950, 1960, and 1970.

(b) Letting 1940 be the base year, calculate a Laspeyres price index for 1950, 1960, and 1970.

(c) Letting 1940 be the base year, calculate a Paasche price index for 1950, 1960, and 1970.

(d) Letting 1940 be the base year, calculate a fixed-weight aggregative price index for 1950, 1960, and 1970, where the quantity weights pertain to 1950.

(e) Using nontechnical language, explain exactly what the results of (a)-(d) mean.

**14.3** A department store sells three models of television sets. Average unit selling prices and quantities sold (in thousands) in 1985 and 1987 were as follows:

| Model | 1985 Price | Quantity | 1987 Price | Quantity |
|-------|-----------|----------|-----------|----------|
|       | (dollars) |          | (dollars) |          |
| A     | 360       | 2.1      | 505       | 2.0      |
| B     | 480       | 1.7      | 520       | 1.6      |
| C     | 610       | 0.9      | 622       | 0.7      |

(a) The store wants to construct a price index for all the television sets it sells. If 1985 is the base year, calculate a simple unweighted price index for 1987.

(b) If 1985 is the base year, compute a Laspeyres price index for 1987.

(c) If 1985 is the base year, compute a Paasche price index for 1987.

(d) Using nontechnical language, explain exactly what the results of parts (a)-(c) mean.

**14.4** A chair manufacturer makes easy chairs and rocking chairs. Average unit selling prices and quantities produced (and sold) in 1984 and 1987 were as follows:

| Year | Easy chairs Price | Quantity | Rocking chairs Price | Quantity |
|------|------------------|----------|---------------------|----------|
|      | (dollars)        |          | (dollars)           |          |
| 1984 | 173              | 18,256   | 58                  | 9,433    |
| 1987 | 203              | 21,304   | 79                  | 8,243    |

(a) The chair manufacturer wants to construct a price index for all the chairs it sells. If 1984 is the base year, calculate a Laspeyres price index for 1987.

(b) If 1984 is the base year, calculate a Paasche price index for 1987.

**14.5** Irving Fisher, a well-known American economist, suggested that an "ideal" index number would equal

$$\sqrt{(\text{Laspeyres index})(\text{Paasche index})}.$$

(Some reasons why Fisher felt this way are touched on in subsequent exercises.) Using the symbols in equations (14.4) and (14.5), construct a general formula for Fisher's "ideal" price index.

## 14.6 Basic Considerations in Index Number Construction

Thus far we have focused exclusively on the types of formulas used to construct an index number. However, in many cases the practical problems of index number construction outweigh in importance the choice of a formula. This section discusses four fundamental questions that must be faced in constructing any index number.

> *In constructing and using index numbers, you should constantly be*    **Basic Idea #28**
> *on guard against lack of comparability of the items being com-*
> *pared over time. Concepts and definitions as well as techniques of*
> *measurement frequently differ from source to source and from*
> *period to period.*

First, *what is the purpose of the index number? What is it intended to represent or measure?* Unless this question is answered properly, there is a danger that the index number may not really measure what it is supposed to measure. In our example of crop prices, does the index number represent changes in the price level of *all* crops, or just food crops—in a number of countries, or only in the small agricultural country where the index number is being computed? Does the index number measure short-term changes—weekly or monthly price movements—or long-term changes? All these questions must be answered before the statistician can construct a proper index number.

Second, assuming that a particular index number is a price index, the statistician must decide *which prices should be included.* In our index of crop prices, should potatoes be included? Should rice be included? Such questions can only be answered on the basis of the *objectives* of the index, and the statistician must specify very carefully the nature and characteristics of the items involved. In establishing our index of crop prices, the analyst must be careful to specify precisely the *kind* of wheat being included. If such specifications are lacking, the price data may not be comparable, thereby rendering the index quite erroneous. Thus, if the 1987 wheat price was lower than the 1970 price because it pertained to a poorer grade of wheat, our index is misleading.

Third, the statistician must determine *the time period to which the index number should pertain.* Given the objectives of our index number of crop prices, should we be comparing recent price levels with 1970, or should we be comparing them with 1980, or 1985? In other words, what period should be used as the base period? Naturally, the answer depends on the purposes of the index number. Often, a base period is chosen in order to avoid a period of abnormality such as war, severe depression, hyperinflation, and so on. The reasons for this are evident. For example, an index of prices with 1933 (the bottom of the Great Depression) as a base year is bound to exaggerate the extent of recent price increases. The federal government has established a standard base period for most indexes computed by government agencies. During the early 1970s, this base period was 1967; during the late 1970s and 1980s, it was 1972 for some (but not all) indexes.

Fourth, the statistican must determine *the weights to be used.* As we have seen, there is a considerable amount of choice in this regard. The weights may pertain to the base period (as in the Laspeyres index), to the given period (as in the Paasche index), or to some other period (as in the fixed-weight aggregative index). But this is only part of the range of choice. In addition, there is often some question about the nature of the weights. For example, in a price index, should the quantity weights be the amounts consumed or the amounts produced? And

**TO ERR IS
HUMAN ... BUT
COSTLY**

THE CASE OF INFLATION IN R AND D

There are very great difficulties in measuring the rate of inflation in research and development (R and D). For the lack of anything better, the official government R and D statistics use the GNP deflator to deflate R and D expenditures. The GNP deflator is the price index used to deflate gross national product. The relevant government agencies are well aware that the GNP deflator is only a rough approximation, and work has begun on the construction of price indexes for R and D inputs. According to one study, the price index for R and D inputs equaled 291.4 in the chemical and petroleum industries in 1983 (1969 = 100). In 1983, the GNP deflator equaled 215.34 (1969 = 86.79).

A newspaper article says that the use of the GNP deflator tended to overstate real R and D expenditure in the chemical and petroleum industries in 1983. Specifically, it says that, if the above price index for R and D inputs had been used instead of the GNP deflator, the figure for 1983 R and D expenditure (in 1983 dollars) in these industries would have been divided by 2.914, rather than 2.1534, to obtain 1983 R and D expenditures (in 1969 dollars). Thus, it says that the figure for 1983 R and D expenditures (in 1969 dollars) would have been 26 percent lower (since 2.1534 ÷ 2.914 = .74). Is this article correct? If not, why not?

SOLUTION: Let $X$ be the 1983 R and D expenditure of these industries (in 1983 dollars). In 1969 dollars, this expenditure would equal $X \div 2.914$ if the above price index for R and D inputs is used. In 1969 dollars, this expenditure would equal

$$X \div \left(\frac{215.34}{86.79}\right) = X \div 2.48$$

if the GNP deflator is used. Thus, the ratio of the figure if the price index for R and D inputs is used to that if the GNP deflator is used equals

$$\frac{X}{2.914} \div \frac{X}{2.48} = \frac{2.48}{2.914} = 0.85.$$

In other words, the figure for 1983 R and D expenditure (in 1969 dollars) in these industries would be 15 percent lower if the price index for R and D inputs is used than if the GNP deflator is used. Thus, the article is wrong: The figure would be 15 percent lower, not 26 percent lower.

the amounts consumed or produced by whom? Like so many of the questions taken up in this section, these questions must be answered by the good judgment of the statistician in response to the particular objectives of the index number under construction. There are no mechanical rules or formulas for solving these problems.

## 14.7 The Consumer Price Index: A Case Study

Perhaps the most widely heralded and most influential index in the United States is the Consumer Price Index, computed by the U.S. Department of Labor's Bureau of Labor Statistics. Until 1978, the purpose of this index was to measure changes in prices of goods and services purchased by urban wage earners and clerical workers and their families. In 1978 the index was expanded to include all urban consumers (although the narrower index was not discontinued). The index includes the price of food, automobiles, clothing, homes, furniture, home supplies, drugs, fuel, medical fees, legal fees, rents, repairs, transportation fares, recreational goods, and so forth. Prices, as defined in the index, include sales taxes and excise taxes. Also, real estate taxes—but not income taxes or personal property taxes—are included in the index. As the *New York Times* put it, ". . . of all the torrent of statistics pouring out of Washington, none exceeds in importance the monthly Consumer Price Index issued by the Bureau of Labor Statistics."[2]

The base period for the Consumer Price Index has been changed from time to time. During the 1960s, the average of 1957–59 was used as a base period; during the 1970s and 1980s, the base period was 1967. Undoubtedly, the base period will be changed again in the future. The Consumer Price Index is computed monthly. A separate index is calculated for each of 28 Standard Metropolitan Statistical Areas as well as for all the urban places in the United States. Thus, the index provides information on inter-city differences in the rate of price increases. Table 14.8 shows the behavior of the Consumer Price Index during 1984. Table 14.9 provides index numbers for prices of various kinds of goods and services included in the Consumer Price Index during 1965-84. Clearly, the rate of increase of prices has varied from one kind of good to another. For example, note the large jump in food prices from 1972 to 1974.

The Consumer Price Index is widely used by industry and government. Labor contracts often stipulate that wages must increase in accord with changes in the Consumer Price Index. Also, other contracts, such as long-term leases, sometimes call for automatic adjustments based on the Index. In addition, it is often used by economists and government officials to measure changes in the purchasing power of the dollar. Pensions, welfare payments, royalties, and even alimony payments are often related to the Consumer Price Index.

The formula used to compute the Consumer Price Index is of the

---

[2] *New York Times,* June 22, 1974.

**Table 14.8**
Consumer Price Index,
January 1984–
December 1984

| Month | Consumer Price Index (1967 = 100) |
|---|---|
| January | 305.2 |
| February | 306.6 |
| March | 307.3 |
| April | 308.8 |
| May | 309.7 |
| June | 310.7 |
| July | 311.7 |
| August | 313.0 |
| September | 314.5 |
| October | 315.3 |
| November | 315.3 |
| December | 315.5 |

*Source:* U.S. Bureau of Labor Statistics.

**Table 14.9**
Consumer Price Index,
by Commodity and
Service Groups,
1965–84

| Year | All items | Food | Durable commodities | Nondurable commodities* (1967 = 100) | Services |
|---|---|---|---|---|---|
| 1965 | 94.5 | 94.4 | 98.4 | 94.8 | 92.2 |
| 1966 | 97.2 | 99.1 | 98.5 | 97.0 | 95.8 |
| 1967 | 100.0 | 100.0 | 100.0 | 100.0 | 100.0 |
| 1968 | 104.2 | 103.6 | 103.1 | 104.1 | 105.2 |
| 1969 | 109.8 | 108.9 | 107.0 | 108.8 | 112.5 |
| 1970 | 116.3 | 114.9 | 111.8 | 113.1 | 121.6 |
| 1971 | 121.3 | 118.4 | 116.5 | 117.0 | 128.4 |
| 1972 | 125.3 | 123.5 | 118.9 | 119.8 | 133.3 |
| 1973 | 133.1 | 141.4 | 121.9 | 124.8 | 139.1 |
| 1974 | 147.7 | 161.7 | 130.6 | 140.9 | 152.1 |
| 1975 | 161.2 | 175.4 | 145.5 | 151.7 | 166.6 |
| 1976 | 170.5 | 180.8 | 154.3 | 158.3 | 180.4 |
| 1977 | 181.5 | 192.2 | 163.2 | 166.5 | 194.3 |
| 1978 | 195.4 | 211.4 | 173.9 | 174.3 | 210.9 |
| 1979 | 217.4 | 234.5 | 191.4 | 198.7 | 234.2 |
| 1980 | 246.8 | 254.6 | 210.4 | 235.2 | 270.3 |
| 1981 | 272.4 | 274.6 | 227.1 | 257.5 | 305.7 |
| 1982 | 289.1 | 285.7 | 241.1 | 261.6 | 333.3 |
| 1983 | 298.4 | 291.7 | 253.0 | 266.3 | 344.9 |
| 1984 | 311.1 | 302.9 | 266.5 | 270.8 | 363.0 |

*Source: Economic Report of the President, 1985* (Washington, D.C.: Government Printing Office, 1985).
* Food is excluded from this and the previous column.

fixed-weight aggregative type. During more than 60 years of existence, the Consumer Price Index has experienced several revisions of its quantity weights. There has also been some change in the list of items in the Index in order to include new products like synthetic fibers and television sets and drop obsolete products such as buggy whips. At present the Consumer Price Index contains approximately

400 items. Prices are obtained by representatives of the Bureau of Labor Statistics from a sample of about 25,000 retail stores and service establishments. To make sure that quality and quantity do not vary over time, the Bureau has formulated detailed specifications for each item.

Finally, it is important to recognize the limitations of the Consumer Price Index. Despite warnings to the contrary, many people seem to believe that this index measures a family's living costs. This clearly is not so, since a family's living costs depend on how much the family consumes as well as on the level of prices it pays. All that the Consumer Price Index measures is the change in prices. Another important misconception is that this index applies to all Americans, whereas it applies only to urban consumers. Also, it is important to recognize that the Index does into take into account changes in income taxes and personal property taxes; nor is it free from errors due to sampling or incorrect reporting of prices. Still and all, the Consumer Price Index provides a reasonably accurate indication of changes over time in the prices paid by urban consumers, although it may not be a very good measure for a particular worker or occupation.[3]

The following example illustrates how the Consumer Price Index is used by industry and labor.

EXAMPLE 14.1 In many situations, ranging from the formulation of government economic policy to a particular labor negotiation between management and a union, it is important to distinguish between *money* wages and *real* wages. Money wages are wages expressed in ordinary dollars, whereas real wages are corrected to take into account changes in the purchasing power of the dollar. In other words, real wages are expressed in dollars of constant purchasing power. Suppose that your money wage during selected years from 1971 to 1984 was as shown in Table 14.10, and that the Consumer Price Index (divided by 100) during each

**Table 14.10**
Your Annual Wage and the Consumer Price Index

| Year | Your money wage (dollars) | Consumer Price Index ÷ 100 (1967 = 1.00) |
|---|---|---|
| 1971 | 19,000 | 1.213 |
| 1972 | 20,000 | 1.253 |
| 1973 | 21,000 | 1.331 |
| 1977 | 25,000 | 1.815 |
| 1978 | 27,000 | 1.954 |
| 1979 | 29,000 | 2.174 |
| 1980 | 31,000 | 2.468 |
| 1981 | 33,000 | 2.724 |
| 1984 | 35,000 | 3.111 |

[3] This discussion omits many important considerations that are too detailed to be taken up here. For further details, see U.S. Bureau of Labor Statistics, "The Consumer Price Index," in E. Mansfield, *Statistics for Business and Economics: Readings and Cases* (New York: Norton, 1980).

year was shown in the final column of this table. What was your real wage in 1967 dollars in each year?

SOLUTION: To compute real wages, one must divide money wages by the appropriate price index (divided by 100). Assuming that the Consumer Price Index is appropriate, your real wage was

```
1971:   19,000 ÷ 1.213 = $15,633
1972:   20,000 ÷ 1.253 =  15,961
1973:   21,000 ÷ 1.331 =  15,777
1977:   25,000 ÷ 1.815 =  13,774
1978:   27,000 ÷ 1.954 =  13,818
1979:   29,000 ÷ 2.174 =  13,340
1980:   31,000 ÷ 2.468 =  12,561
1981:   33,000 ÷ 2.724 =  12,114
1984:   35,000 ÷ 3.111 =  11,250
```

## 14.8 Quantity Indexes

Our discussion up to this point has been concerned solely with price indexes. In this section, we consider another important kind of index number, quantity indexes. Just as price indexes measure the change over time in prices, quantity indexes measure the change over time in the quantity produced. For example, to construct a quantity index for the production of food crops in the small agricultural country discussed in previous sections, we can use formulas that are entirely analogous to those used for price indexes. Basically, we only need to substitute price terms for quantity terms and vice versa, since prices are used as weights in quantity indexes.

Specifically, a *Laspeyres quantity index* can be computed, the formula being:

**Laspeyres Quantity Index**

$$I_1 = \frac{\sum_{i=1}^{m} P_{0i}Q_{1i}}{\sum_{i=1}^{m} P_{0i}Q_{0i}} (100).$$

(14.8)

The weights are the prices in the base year (year 0). To illustrate the computations, Table 14.11 shows how a Laspeyres quantity index for 1987 can be computed for the food crops, given that 1970 is the base year.

Similarly, a *Paasche quantity index* can be computed, the formula being

**Paasche Quantity Index**

$$I_1 = \frac{\sum_{i=1}^{m} P_{1i}Q_{1i}}{\sum_{i=1}^{m} P_{1i}Q_{0i}} (100).$$

(14.9)

Table 14.11
Computation of
Laspeyres Quantity
Index for 1987

| Copy | (1) 1987 quantity | (2) 1970 quantity | (3) Weight (1970 price) | (3) × (1) | (3) × (2) |
|---|---|---|---|---|---|
| Corn | 5,553 | 3,907 | 1.00 | 5,553 | 3,907 |
| Wheat | 1,545 | 1,355 | 1.74 | 2,688 | 2,358 |
| Oats | 695 | 1,153 | 0.60 | 417 | 692 |
| Soybeans | 1,283 | 555 | 2.13 | 2,733 | 1,182 |
| Total | | | | 11,391 | 8,139 |

$$\text{Index} = \frac{11,391}{8,139}(100) = 140.$$

The weights are the prices in the given year (year 1). To illustrate the computations, Table 14.12 shows how a Paasche quantity index for 1987 can be computed for the food crops, given that 1970 is the base year.

Table 14.12
Computation of
Paasche Quantity
Index for 1987

| Copy | (1) 1987 quantity | (2) 1970 quantity | (3) Weight (1987 price) | (3) × (1) | (3) × (2) |
|---|---|---|---|---|---|
| Corn | 5,553 | 3,907 | 1.29 | 7,163 | 5,040 |
| Wheat | 1,545 | 1,355 | 1.67 | 2,580 | 2,263 |
| Oats | 695 | 1,153 | 0.67 | 466 | 773 |
| Soybeans | 1,283 | 555 | 3.49 | 4,478 | 1,937 |
| Total | | | | 14,687 | 10,013* |

$$\text{Index} = \frac{14,687}{10,013}(100) = 147.$$

* Due to rounding errors, this differs from 10,017 in the first equation in Section 14.3.

Also, *a fixed-weight aggregative quantity index* can be computed, the formula being

$$I_1 = \frac{\sum_{i=1}^{m} P_{wi}Q_{1i}}{\sum_{i=1}^{m} P_{wi}Q_{0i}} (100).$$

(14.10)

where $P_{wi}$ is the price of the *i*th commodity in year *w*. The weights are the prices in year *w*. To illustrate the computations, Table 14.13 shows how an index of this sort can be computed for 1987 food crops, given that 1970 is the base (reference) year and the weights are 1980 prices.

# The Index of Industrial Production and Seasonal Variation

One of America's most important economic index numbers is the Index of Industrial Production, computed by the Federal Reserve Board. This is a quantity index which measures changes in the output of American manufacturing, mining, and electric and gas industries, as well as in individual parts of each of these three industrial sectors. For example, in manufacturing, individual indexes are computed for such industries as primary metals, fabricated metal products, machinery, transportation equipment, instruments, ordnance, lumber, furniture, textiles, paper, chemicals, and food.

Although the Index of Industrial Production pertains to only a part of the economy (such activities as wholesale and retail trade, finance, other services, transportation, construction, and agriculture being omitted), it nevertheless is an important indicator of the pace of economic activity in the United States. Table 1 shows the values of the index during the period 1953–84. In general, the index rose during this period, but, as one would expect, it fell during recessions. (For example, it dropped from 1953 to 1954, from 1957 to 1958, from 1969 to 1970, from 1973 to 1975, and from 1979 to 1980.)

The Index of Industrial Production, which is computed monthly, has used 1967 as a base (reference) period. As with the Consumer Price Index and other government indexes, its base period is revised periodi-

**Table 1**
Index of Industrial Production, 1953–84 (1967 = 100)

| Year | Index | Year | Index |
|------|-------|------|-------|
| 1953 | 54.8 | 1969 | 110.7 |
| 1954 | 51.9 | 1970 | 106.6 |
| 1955 | 58.5 | 1971 | 106.8 |
| 1956 | 61.1 | 1972 | 115.2 |
| 1957 | 61.9 | 1973 | 125.6 |
| 1958 | 57.9 | 1974 | 124.8 |
| 1959 | 64.8 | 1975 | 117.8 |
| 1960 | 66.2 | 1976 | 129.8 |
| 1961 | 66.7 | 1977 | 137.1 |
| 1962 | 72.2 | 1978 | 146.1 |
| 1963 | 76.5 | 1979 | 152.5 |
| 1964 | 81.7 | 1980 | 147.0 |
| 1965 | 89.2 | 1981 | 151.0 |
| 1966 | 97.9 | 1982 | 138.6 |
| 1967 | 100.0 | 1983 | 147.6 |
| 1968 | 105.7 | 1984* | 163.5 |

*Source:* Federal Reserve Board.

\* Preliminary estimate by Council of Economic Advisers.

cally, and the weights used are also updated from time to time. These weights are somewhat different from those described in Section 14.8. Instead of using the price of a commodity as a weight, the value-added per unit of output is used since this provides a better measure of the worth of a unit of an industry's output.

To understand why this is so, let us consider what is meant by *value-added*. In briefest terms, it can be defined as *the amount of value added by the industry to the total value of the product*. For example, suppose that $170 million of bread was produced in the United States in 1987. In order to produce it, flour mills turned out $90 million worth of flour. Thus, the value of the bread industry's production was the extra value it added to the $90 million of flour it bought—that is, $80 million. (In other words, the bread industry's value-added is $80 million.) Similarly, in the automobile industry the total value of the automobile is not attributable to this industry alone since the steel, tires, glass, and many other components of cars were not produced by the auto manufacturers. What the auto makers produce is the extra value that is added to the value of the steel, tires, glass, and other products that they purchase.

If we let $V_{67,i}$ be the value-added per unit of output of the $i$th commodity in 1967, the formula for the Index of Industrial Production is

$$I_1 = \frac{\sum_{i=1}^{m} V_{67,i} Q_{1i}}{\sum_{i=1}^{m} V_{67,i} Q_{67,i}} \ (100).$$

From this formula, it is clear that (as we have pointed out) the value-added per unit of output is used to weight each commodity's output.

The Index of Industrial Production is broken into separate indexes for various types of goods. As the Federal Reserve Board points out in the following quotation, the seasonal variation for these various types of goods can be quite different.

[Page 614] shows the seasonal patterns of production for several major market groups . . . . Output of consumer goods reaches a seasonal peak in the autumn months primarily as a result of the processing of food crops and a seasonal accumulation of business inventories of non-food products to supply cold weather and holiday demands. There is a less pronounced seasonal high in consumer goods output in February and March to meet spring demands for clothing and home goods. The subsequent decline and a marked rise in June, as shown in the top panel in the chart, are followed by sharp curtailments in July production schedules. In recent years these reductions have amounted to about 7.5 per cent.

Total equipment output shows less seasonal variation than consumer goods, and the defense and space component shows virtually none. Output of business equipment in the new index reaches an advanced level before seasonal adjustment during the first half of the year as production of construction and farm machinery reaches a peak in the spring months. A second high period is evident during the autumn.

Output of construction products has a marked seasonal rise—about 12 percent—during the spring and early summer, as shown in the bottom panel [below]. The seasonal decline in such output during the . . . month of July is more than that for business equipment and about the same as that for consumer goods.

Output of materials is at an advanced level from February to June. This has usually been true for total products too.[4]

The pattern of seasonal variation shown in the graphs below is reasonable. For example, one would expect that output of all of these types of goods would be relatively low in December because of the Christmas holidays and in July because of vacations. To interpret properly changes in the index, these seasonal factors must be taken into account. For example, if the output of consumer goods is 3 percent lower in November than in September, one might be inclined to think that a recession had begun. However, as shown below, the seasonal index for September is over 105, whereas it is about 100 in November; thus, more than a 3 percent reduction in output would be expected purely on the basis of seasonal variation.

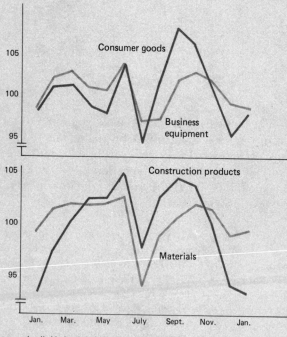

Implied indexes derived by dividing indexes without seasonal adjustment by seasonally adjusted indexes.

[4] Federal Reserve Board, *Industrial Production, 1971* (Washington, D.C.: Government Printing Office, 1972), pp. 47–48.

PROBING DEEPER

1. Based on Table 1, do you think that a linear regression would provide an adequate description of the trend in the Index of Industrial Production? Why, or why not?

2. Do you think that the Index of Industrial Production would be a useful leading indicator? Explain.

3. If the reference base for the Index of Industrial Production was changed from 1967 to 1977, what would be the new value of the index for 1983?

4. Does the seasonal variation in the production of construction products seem more severe than that for business equipment? If so, why?

5. In recent years, provisions for a longer Christmas vacation were written into labor contracts in the auto industry. What effect do you think they had on the seasonal variation in auto production?

6. If you wanted to construct a measure of total output in the economy, would you add up the value added of all firms? Or would it be better to add up the sales of all firms? Explain.

**Table 14.13**
Computation of
Fixed-Weight
Aggregate Quantity
Index for 1987 with
1980 Prices as Weights

| Copy | (1) 1987 quantity | (2) 1970 quantity | (3) Weight (1980 price) | (3) × (1) | (3) × (2) |
|------|------|------|------|------|------|
| Corn | 5,553 | 3,907 | 1.16 | 6,441 | 4,532 |
| Wheat | 1,545 | 1,355 | 1.35 | 2,086 | 1,829 |
| Oats | 695 | 1,153 | 0.62 | 431 | 715 |
| Soybeans | 1,283 | 555 | 2.54 | 3,259 | 1,410 |
| Total | | | | 12,217 | 8,486 |

$$\text{Index} = \frac{12,217}{8,486}(100) = 144.$$

EXERCISES

**14.6**  It is frequently asserted that the Laspeyres quantity index overstates changes, whereas the Paasche quantity index understates them. Can you think of an important reason why this is likely to be true?

**14.7**  Based on the data in Exercise 14.2, compute a Laspeyres quantity index for U.S. output of beef and veal, 1940 being the base year and 1950, 1960, and 1970 being the given years.

**14.8**  Based on the data in Exercise 14.2, compute a Paasche quantity index for U.S. output of beef and veal, 1940 being the base year and 1970 being the given year.

**14.9**  Using the data in Exercise 14.3, compute a Laspeyres quantity index for the number of television sets sold by the department store. Let 1985 be the base year and 1987 be the given year.

**14.10**  Using the data in Exercise 14.3, compute a Paasche quantity index for the number of television sets sold by the department store. Let 1985 be the base year and 1987 be the given year.

**14.11**  Based on the data in Exercise 14.4, calculate a Laspeyres quantity index for the number of chairs produced by the chair manufacturer. In this index, 1984 is the base year and 1987 is the given year.

**14.12**  Based on the data in Exercise 14.4, calculate a Paasche quantity index for the number of chairs produced by the chair manufacturer. In this index, 1984 is the base year, and 1987 is the given year.

**14.13**  Comparing 1987 with 1986, the sales of a firm's chemical division were down 6 percent while the sales of its drug division were up 5 percent. Can we conclude that the 1987 sales of its drug division were 11 percent higher than those of its chemical division? Explain.

**14.14**  The price of a unit of good X was $3 in 1987 and $1 in 1985. The price of a unit of good Y was $1 in 1987 and $2 in 1985. In 1985, 10 units of good X and 3 units of good Y were sold. In 1987, 2 units of good X and 11 units of good Y were sold.

(a) "Since good X's price increased by 200 percent and good Y's price fell by 50 percent, the percentage increase, on the average, of the prices of these two goods during the period 1985–87 was about $(200 - 50) \div 2$, or 75." Comment and evaluate this statement.

(b) "Since good Y's price was 100 percent higher in 1985 than in 1987, and good X's price was 67 percent lower in 1985 than in 1987, the prices of these two goods were higher, on the average, in 1985 than in 1987 by about (100 percent – 67 percent) $\div$ 2, or approximately 17 percent." Comment and evaluate.

**14.15** It seems desirable that when a price index (divided by 100) is multiplied by a quantity index (divided by 100), the result should equal the value index, which is

$$\sum_{i=1}^{m} Q_{1i} P_{1i} \div \sum_{i=1}^{m} Q_{0i} P_{0i}.$$

Basically, the reason why this is desirable is that

$$\text{Price} \times \text{Quantity} = \text{Value}.$$

Consequently, if the price index for a commodity doubles and if the quantity index for the commodity also doubles, the value index should quadruple. (This is the so-called *factor-reversal* test.)

(a) If both the price index and the quantity index are Laspeyres indexes, does this desirable result occur?

(b) If both the price index and the quantity index are Paasche indexes, does this desirable result occur?

(c) If both the price index and the quantity index are of Fisher's "ideal" type (see Exercise 14.5), does this desirable result occur?

**14.16** It seems desirable that if an index for year 1 with base year 0 equals 40, then an index for year 0 with base year 1 should equal 250. After all, if the item in question is 40 percent as big in year 1 as in year 0, it should be 250 percent as big in year 0 as in year 1. Whether or not an index number has this property determines whether it passes the *time-reversal* test.

(a) Does the Laspeyres index pass this test?

(b) Does Fisher's "ideal" index pass this test?

# Chapter Review

1. An *index number* is a ratio (generally expressed as a percentage) of one quantity to another, where one of the quantities summarizes a given group of items and the other quantity summarizes a base group of items. The *base* group is used as a standard of comparison with the *given* group. For example, an index number might be constructed to indicate the level of steel prices in 1987 (the given period) relative to 1970 (the base period).

2. If $P_{0i}$ is the price of the $i$th commodity in the base period and $P_{1i}$ is its price in the given period, the *simple aggregative price index* is

$$\frac{\sum_{i=1}^{m} P_{1i}}{\sum_{i=1}^{m} P_{0i}}(100),$$

and the *average of relatives* is

$$\frac{\sum_{i=1}^{m}\frac{P_{1i}}{P_{0i}}}{m}(100).$$

These indexes are seldom used because the simple aggregative price index is influenced by the units in which the quantities of various commodities are expressed, and the average of relatives does not weight prices by their importance.

3. In constructing price indexes, statisticians generally weight each commodity's *price relative* by its value. If the value in the base period is used as a weight, the result is the *Laspeyres price index,* which is

$$\frac{\sum_{i=1}^{m}Q_{0i}P_{1i}}{\sum_{i=1}^{m}Q_{0i}P_{0i}}(100),$$

where $Q_{0i}$ is the amount consumed of the *i*th commodity in the base period. If the value in the given period is used as a weight, the result is the *Paasche price index,* which is

$$\frac{\sum_{i=1}^{m}Q_{1i}P_{1i}}{\sum_{i=1}^{m}Q_{1i}P_{0i}}(100),$$

where $Q_{1i}$ is the amount consumed of the *i*th commodity in the given period. If the quantity weights refer to a period other than the base year or the given year, the result is a *fixed-weight aggregative index,* which is

$$\frac{\sum_{i=1}^{m}Q_{wi}P_{1i}}{\sum_{i=1}^{m}Q_{wi}P_{0i}}(100),$$

where $Q_{wi}$ is the amount consumed of the *i*th commodity in the period to which the quantity weights pertain.

4. There are a host of practical problems that must be met in the construction of an index number. It is particularly important that the purpose of the index number be kept in mind, and that a proper decision be made concerning which prices should be included, the time period to which the index number should pertain (including which base period to use), and the nature of the weights to be used. To a considerable extent proper decisions depend on the good judgment and integrity of the statistician, since there are no mechanical rules insuring good results.

5. Just as price indexes measure the change over time (or from place to place) in prices, so *quantity indexes* measure the change over time (or from place to place) in the quantity produced. Among the possible indexes of this sort are the *Laspeyres quantity index* (which uses base-period prices as weights), the

*Paasche quantity index* (which uses given-period prices as weights), and the *fixed-weight aggregative quantity index* (which uses prices in a period other than the base or given period as weights).

6. Probably the most famous and important price index is the Consumer Price Index, which is issued monthly by the Bureau of Labor Statistics. It is an important measure of the rate of inflation, and is used widely as a basis for escalator clauses that raise wages, pensions, and other payments when the price level rises. An important quantity index is the Index of Industrial Production, computed monthly by the Federal Reserve Board. Although it excludes such parts of the economy as wholesale and retail trade, services, finance, transportation, construction, and agriculture, it is regarded as an important indicator of the pace of economic activity in the United States.

## Chapter Review Exercises

**14.17** Suppose that the price of a trip from New York to London changed as follows during the period 1900–85:

| Year | Price |
|------|-------|
| 1900 | $100  |
| 1940 | 200   |
| 1985 | 400   |

Are these prices of a ship or airplane ticket an adequate measure of the cost of the trip? Why, or why not? If not, what additional factors should be taken into account? (Hint: What quality changes have occurred?)

**14.18** The Acme Corporation manufactures three types of machine tools. The price of each of these tools in 1960, 1970, and 1987 is given below:

| Year | Type A  | Type B  | Type C  |
|------|---------|---------|---------|
| 1960 | $1,000  | $ 500   | $2,000  |
| 1970 | 1,800   | 1,000   | 4,200   |
| 1987 | 2,500   | 1,500   | 6,100   |

(a) Calculate a simple aggregate index—that is, a simple unweighted index—of this firm's prices. (Use 1960 as the base year, and use 1970 and 1987 as the given years.) Indicate the limitations of this kind of price index, and state how it can be improved.

(b) Calculate index numbers based on a simple average of relatives to represent the changes over time in the Acme Corporation's prices. Indicate the limitations of this kind of price index, and state how it can be improved.

(c) Suppose that the quantity of each type of machine sold each year by the Acme Corporation was as follows:

| Year | Type A | Type B | Type C |
|------|--------|--------|--------|
| 1960 | 100    | 200    | 10     |
| 1970 | 200    | 300    | 20     |
| 1987 | 300    | 400    | 20     |

Calculate a Laspeyres price index for 1970 and 1987 for this firm. (Again, use 1960 as the base period.)

(d) Calculate a Paasche price index for this firm, assuming that 1987 is the given period and 1960 is the base period.

(e) Suppose that the Acme Corporation considers 1970 as a much more normal year than either 1960 or 1987. (i) Use the data to calculate a fixed-weight aggregative price index for 1960, where the weights pertain to 1970, with 1960 as the base period. (ii) Calculate a fixed-weight aggregative price index for 1987, where the weights pertain to 1970, with 1960 as the base period.

(f) Let $I_{60}^{70}$ be the 1970 index (relative to a 1960 base) of the price of type A machines. What is the value of $I_{60}^{70}$? If $I_{70}^{71}$ was 110, what was $I_{60}^{71}$?

(g) Calculate Fisher's "ideal" price index for 1987, using 1960 as the base period. (See Exercise 14.5.)

**14.19** In the table below are three important price indexes for the years, 1970–75.

(a) How do the objectives of these index numbers differ?

(b) Which includes the widest range of goods and services?

(c) Which of these index numbers do most economists prefer as a measure of general inflation? Why?

| Year | Consumer Price Index (1967 = 100) | Producer Price Index (formerly Wholesale Price Index) (1967 = 100) | Price deflator for GNP (1972 = 100) |
|---|---|---|---|
| 1970 | 116.3 | 110.4 | 91.36 |
| 1971 | 121.3 | 113.9 | 96.02 |
| 1972 | 125.3 | 119.1 | 100.00 |
| 1973 | 133.1 | 134.7 | 105.92 |
| 1974 | 147.7 | 160.1 | 116.20 |
| 1975 | 161.2 | 174.9 | 126.35 |

(d) "Since the data show that the Consumer Price Index for 1975 was 161.2, while the price deflator for gross national product (GNP) for 1975 was 126.35, it follows that consumer prices rose more rapidly during 1970–75 than did the prices of gross national product generally." Do you agree? Why, or why not?

**14.20** The average weekly earnings in manufacturing during the period 1970–75 were as follows.

| Year | Average weekly earnings |
|---|---|
| 1970 | $133.73 |
| 1971 | 142.44 |
| 1972 | 154.69 |
| 1973 | 166.06 |
| 1974 | 176.40 |
| 1975 | 189.51 |

(a) Using the Consumer Price Index (given in Exercise 14.19), compute real weekly earnings during each of these years.

(b) Based on your results, by what percentage did real average weekly earnings in manufacturing increase between 1970 and 1975? Describe in detail what the word *real* means in this context.

**14.21** A vegetarian couple of ages 72 and 70 are considering moving to Atlanta, Georgia, from Boston, Massachusetts. They read that the Consumer Price Index is 1 percent higher for Boston than for Atlanta. From this they conclude that it will cost them 1 percent less to maintain their existing standard of living in Atlanta than in Boston. Do you agree? Why, or why not?

**14.22** In 1984, the index of prices received by farmers was 142 (1977 = 100), while the index of prices paid by farmers was 164 (1977 = 100). What were the implications, if any, of these indexes for the U.S. farm economy?

**14.23** The Acme Corporation wants to construct a quantity index for its production of machine tools.
   (a) Using the data in Exercise 14.18, construct a Laspeyres index of this kind for 1970 and 1987. (Use 1960 as the base period.)
   (b) Using the data in Exercise 14.18, construct a Paasche quantity index for the Acme Corporation, where 1987 is the given year and 1960 is the base period.

PART VII

# Statistical Decision Theory

# Decision Theory: Prior Analysis and Posterior Analysis

## 15.1 Introduction

One of the most interesting developments in statistics over the past 30 years has been the growing application of decision theory. Decision makers in business and government are continually faced with the problem of making choices under conditions of uncertainty. For example, a business frequently must decide whether or not to introduce a new product while still uncertain of the size of the potential market for the product. And firms must decide on the location, size, and staffing of new plants when they are uncertain of the future pattern of demand, costs, and a variety of other relevant factors. Statistical decision theory provides useful ways of analyzing such problems and helps decision makers arrive at rational choices. In this and the following chapter we will describe this theory and its application.

## 15.2 Decision Trees

Any problem of decision making under uncertainty has the following two characteristics. First, *the decision maker must make a choice, or perhaps a series of choices, among alternative courses of action.* Second, *this choice leads to some consequence, but the decision maker cannot tell in advance the exact nature of this consequence because it depends on some unpredictable event, or series of events, as well as on the choice itself.* For example, let's consider the case of a marketing manager who must decide whether or not to adopt a new label for a product. In this case, the choice is between two alternatives: adopt the new label or stick with the old one. And the consequence of adopting the new label is uncertain since the marketing manager cannot be sure whether the new label is superior to the old one. If the new label is a better label the firm will gain $800,000, but if it is not the firm will lose $500,000. The marketing manager believes that if he adopts the new label there is a 50-50 chance that it will be superior and a 50-50 chance that it will not.

To represent any such problem of decision making under uncertainty, a decision tree is useful. Its definition is as follows:

*Decision Forks and Chance Forks*

> DECISION TREE: *A decision tree represents a decision problem as a series of choices, each of which is depicted by a fork (sometimes called a juncture or branching point). A **decision fork** is a juncture representing a choice where the decision maker is in control of the outcome; a **chance fork** is a juncture where "chance" controls the outcome.*

To differentiate between a decision fork and a chance fork, we shall place a small square at the former juncture but not at the latter. Figure 15.1 shows the decision tree for the problem facing the marketing manager. Beginning at the left-hand side of the diagram, the first choice is up to the marketing manager, who can either follow the branch representing the adoption of the new label or the branch representing the continued use of the old label. Since this fork is a decision fork, it is represented by a square. If the branch representing the continued use of the old label is followed, the consequence is certain: The firm will have no extra profits or losses. Thus zero extra profit is shown at the end of this branch. If the branch representing the adoption of the new label is followed, we come to a chance fork since it is uncertain whether the new label is superior to the old one. The upper branch following this chance fork represents the consequence that the new label is superior, in which case the extra profit to the firm is $800,000, shown at the end of this branch. The lower branch following this chance fork represents the consequence that the new label is not superior, in which case the outcome is −$500,000 (a loss), shown at the end of this branch. The probability that "chance" will choose each of these branches is shown above the branch.

*Backward Induction*

Given such a decision tree, it is easy to determine which branch the marketing manager should choose in order to maximize expected extra profit to the firm. The process by which we solve this problem, known as *backward induction,* requires that we begin at the right-hand side of the decision tree, where the monetary payoff figures are located. The first step is to calculate the expected monetary value of

**Figure 15.1**
Decision Tree for Marketing Manager's Problem of Whether or Not to Adopt New Label

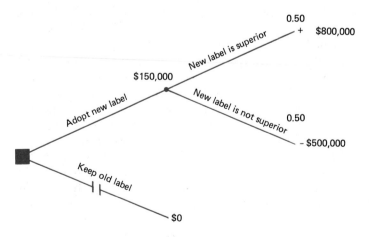

being situated at the chance fork immediately to the left of these payoff figures. In other words, this is the expected extra profit to the firm given that "chance" will choose which subsequent branch will be followed. Since there is a 0.50 probability that the branch culminating in a profit increase of $800,000 will be followed, and a 0.50 probability that the branch culminating in a profit decrease of $500,000 will be followed, the expected monetary value of being situated at this chance fork is

$$0.50(\$800,000) + 0.50(-\$500,000) = \$150,000.$$

This number is written above the chance fork in question to show that this is the expected monetary outcome of being located at that fork. Moving further along the decision tree to the left, it is evident that the marketing manager has a choice of two branches, one of which leads to an expected extra profit for the firm of $150,000, the other of which leads to a zero extra profit. If the marketing manager wants to maximize expected monetary value (in this case, extra profit), he should choose the former branch. In other words, he should adopt the new label. Since the latter branch (Keep old label) is nonoptimal, we place two vertical lines through it.

At this point, it is worth noting that this graphic procedure for analyzing the marketing manager's problem amounts to precisely the same thing as the calculations we made in Example 4.6 where this same problem was presented. Recall that in order to solve this problem then, we compared the expected extra profit if the new label was adopted ($150,000) with the extra profit if it was not adopted ($0) and followed the course of action that resulted in the larger of the two. Our procedure in Figure 15.1 is exactly the same.

## 15.3: Whether to Seed Hurricanes: A Case Study[1]

To illustrate how decision trees can be used in actual circumstances, let's consider the problem of whether or not the federal government should seed hurricanes. In the early 1970s statisticians at Stanford Research Institute (now SRI International) analyzed this decision for the U.S. Department of Commerce, and used decision trees in their work. To construct the decision tree they began by noting that the government (the decision maker in this case) has a choice of two courses of ac-

[1] This case study is based on R. Howard, J. Matheson, and D. North, "The Decision to Seed Hurricanes," *Science*, June 1972. It is reprinted in part in E. Mansfield, *Statistics for Business and Economics: Readings and Cases* (New York: Norton, 1980). Note that only a small portion of the authors' findings can be discussed here and that we must necessarily oversimplify their methods and results. The interested reader is referred to the original paper for a fuller discussion. Also see Roscoe Braham, "Field Experimentation in Weather Modification" (and the comments by William Kruskal, Frederick Mosteller, Jerzy Neyman, and others), *Journal of the American Statistical Association*, 74, March 1979, pp. 52-104. For further discussion, see Section 5.9.

**Figure 15.2**
Decision Tree for the
Government's Decision
to Seed or Not to Seed
a Hurricane

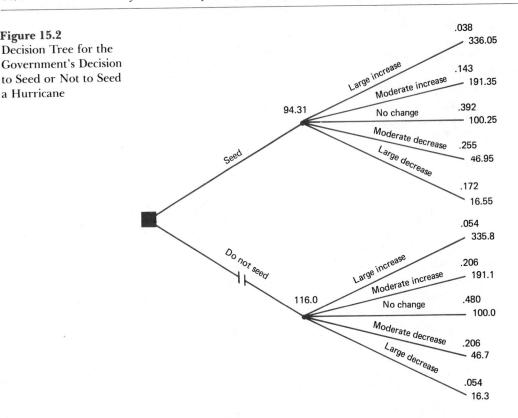

tion—to seed or not to seed. Thus, in Figure 15.2 at the left-hand end of the decision tree, there is a decision fork with two branches, one representing seeding, the other representing not seeding.

If the government follows the branch corresponding to seeding, it moves to a chance fork where there are five branches corresponding respectively to a large increase, a moderate increase, no change, a moderate decrease, and a large decrease in sustained wind speed. The property damage (in millions of dollars) corresponding to each of these consequences (plus the cost of seeding, which is $250,000) is shown at the right-hand end of each branch. (For example, the total cost is $336.05 million if there is a large increase in sustained wind speed.) The choice of which of these consequences will occur is up to "chance." The probability that each one will occur is given above each branch.

Continuing the method above, if the government follows the lower branch in Figure 15.2—the one corresponding to *not* seeding—it moves to a chance fork where there also are five branches corresponding respectively to a large increase, a moderate increase, no change, a moderate decrease, and a large decrease in sustained wind speed. The property damage resulting from each of these consequences is shown at the right-hand end of the relevant branch, and the probability that each will occur is given above its branch.

To determine whether the government should decide to seed or not to seed under these circumstances, the Stanford statisticians went through the procedure described in the previous section. That is, they calculated the expected cost (in millions of dollars) of being at the chance fork resulting from seeding, which is

$$.038(336.05) + .143(191.35) + .392(100.25)$$
$$+ .255(46.95) + .172(16.55) = 94.31.$$

They entered this figure (94.31) above the relevant chance fork in Figure 15.2. They then calculated the expected cost (in millions of dollars) of being at the chance fork resulting from not seeding, which is

$$.054(335.8) + .206(191.1) + .480(100.0) + .206(46.7) + .054(16.3)$$
$$= 116.0.$$

And they then entered the result (116.0) above its chance fork in Figure 15.2. If the government wants to minimize the expected cost, it should choose the branch that leads to the situation where the expected cost is lower.[2] Since the branch corresponding to seeding leads to an expected cost of $94.31 million and the one corresponding to not seeding leads to an expected cost of $116.00 million, it follows that the optimal decision is to seed. (And two vertical lines are placed through the nonoptimal branch.)

The following example, based on the data in Figure 15.2, provides further practice in using decision trees.

EXAMPLE 15.1 Suppose that the property damage corresponding to a large decrease in sustained wind speed is zero, rather than $16.3 million (the amount in Figure 15.2). If the government wants to minimize expected cost, should it decide to seed?

SOLUTION: The expected cost (in millions of dollars) of being at the chance fork resulting from seeding is

$$.038(336.05) + .143(191.35) + .392(100.25)$$
$$+ .255(46.95) + .172(.25) = 91.51.$$

(Under these circumstances, the cost when there is a large decrease in sustained wind speed is .25, since this is the cost of seeding.) The expected cost (in millions of dollars) of being at the chance fork resulting from not seeding is

$$.054(335.8) + .206(191.1) + .480(100.0)$$
$$+ .206(46.7) + .054(0) = 115.11.$$

Since seeding leads to an expected cost of $91.51 million and not seeding leads to an expected cost of $115.11 million, the government should seed in order to minimize expected cost.

---

[2] Whereas the monetary values in the marketing manager's problem were profits, the monetary values in this case are costs. Consequently, the decision maker wants to *minimize* the expected value of the monetary values in this case, whereas he wants to *maximize* the expected value of the monetary values in the previous section.

## 15.4 Maximization of Expected Utility

In discussing both the marketing manager's labeling decision and the U.S. Government's problem of whether or not to seed hurricanes, we have assumed that the decision maker wants to maximize expected monetary gain (or to minimize expected monetary loss). In this and the following section we will explain in greater detail why this may not be the right criterion, and will discuss how a more appropriate criterion can be formulated.[3] To understand why a decision maker may not want to maximize expected monetary gain, consider a situation where you are given a choice between (1) receiving $1,000,000 for certain, and (2) a gamble in which a fair coin is tossed, and you will receive $2,100,000 if it comes up heads, or you will lose $50,000 if it comes up tails. The expected monetary gain for the gamble is

$$0.50(\$2,100,000) + 0.50(-\$50,000) = \$1,025,000.$$

so you should choose the gamble over the certainty of $1,000,000 if you want to maximize expected monetary gain. However, it seems likely that many persons would prefer the certainty of $1,000,000 since the gamble entails a 50-50 chance that you will lose $50,000, a very substantial sum. Moreover, many people may feel that they can do almost as much with $1,000,000 as with $2,100,000, and therefore the extra amount is not worth the risk of losing $50,000.

Clearly, *whether or not you will want to maximize expected monetary gain in this situation depends on your attitude toward risk.* If you are a widow of modest means, you will probably be overwhelmed at the thought of taking a 50-50 chance of losing $50,000. On the other hand, if you are the president of a big corporation the prospect of a $50,000 loss may be not the least bit unsettling, and you may prefer the gamble to the certainty of a mere $1,000,000. And if you are the sort of person who enjoys danger and risk, you may prefer the gamble even though a $50,000 loss may wipe you out completely.

Fortunately, there is no need to assume that the decision maker wants to maximize expected monetary gain. Instead, as John von Neumann and Oskar Morgenstern pointed out several decades ago,[4] we

*Utility Function*

can construct a **utility function** for the decision maker based on his or her attitudes toward risk; from this, we can then go on to choose the alternative that offers the decision maker the maximum expected utility. The procedure used to construct a utility function is described in the following section, but before proceeding to this discussion, we must

*Assumptions Underlying Utility Function*

present the underlying assumptions of a utility function.

First, we must assume that *the decision maker's preferences are transitive.* That is, if he or she prefers Budweiser beer to Pabst beer,

---

[3] We assume here that the relevant probabilities can be formulated. If not, still other criteria of the sort discussed in more advanced texts on decision theory may have to be used.

[4] J. von Neumann and O. Morgenstern, *The Theory of Games and Economic Behavior* (Princeton: Princeton University Press, 1944).

and Pabst to Coors, he or she must therefore prefer Budweiser to Coors. Similarly, if the decision maker is indifferent between a hot dog and a hamburger and indifferent between a hamburger and a salami sandwich, he or she must therefore be indifferent between a hot dog and a salami sandwich. The assumption of transitivity plays an important role in the theory of consumer behavior. Students who have taken a course in microeconomics are likely to have encountered this assumption before.

Second, we must assume that *if there are three outcomes, A, B, and C, and if the decision maker prefers A to B and B to C, then there must be some probability P such that the decision maker will be indifferent between the certainty of B and a gamble where there is a probability of P that A will occur and a probability of (1 − P) that C will occur.* This probability may be big or small: that doesn't matter. What is important is that some value of *P* exists so that the decision maker is indifferent between the gamble that *A* or *C* will occur and the certainty of *B*.

Third, we must assume that *if the decision maker is indifferent between a hot dog and a hamburger, then he or she will be indifferent between two lottery tickets that are identical except that one offers a hot dog as a prize while the other offers a hamburger.* This is known as the *independence axiom.*

Fourth, we must assume that *the decision maker, faced with two lottery tickets for identical prizes, will always choose the one with the higher probability of winning.* Also, we must assume that if the decision maker is offered a lottery ticket whose prize is another lottery ticket, his attitude toward it will be the same as if he had computed the *ultimate* odds of winning or losing that are involved in this compound lottery ticket.

Although questions have been raised by some statisticians and economists concerning a few of these assumptions, most people seem to regard them as quite reasonable foundations on which to build a theory of choice under uncertainty. It is important to note, however, that we are not assuming that individuals actually conform to all these assumptions in their actual decision-making processes. Even if a person agreed with all the axioms involved, he or she might make mistakes or act irrationally at times. Our theory is designed to indicate how people *should* make choices if their decisions are to be in accord with their own preferences. This does not necessarily indicate how they *do* make choices.

## 15.5 Construction of a Utility Function

If the four assumptions listed in the previous section are met, it can be shown that *a rational decision maker will maximize expected utility.* In other words, the decision maker should choose the course of action with the highest expected utility. But what is a *utility*? It is a number that is attached to a possible outcome of the decision. Each outcome has a utility. The decision maker's *utility function* shows the utility that he or she attaches to each possible outcome. This utility function, as we shall see, shows the decision maker's preferences with respect to risk.

How can we know the utility that the decision maker attaches to each possible outcome? In other words, how can we construct a utility function for the decision maker? In order for us to construct such a function the decision maker must respond to a series of questions which indicate his or her preferences with regard to risk. For example, consider the case of the decision to seed or not to seed a hurricane. As indicated in Figure 15.2, the property damages corresponding to a large increase, a moderate increase, no change, a moderate decrease, and a large decrease in sustained wind speed are 335.8, 191.1, 100.0, 46.7, and 16.3 millions of dollars, respectively. To tell whether the expected utility resulting from seeding exceeds the expected utility of not seeding a hurricane, we must know the utility that the decision maker attaches to a monetary loss of $335.8 million—which we will represent by $U(-335.8)$—and the utility that he or she attaches to a monetary loss of $191.1 million—that is, $U(-191.1)$—and $U(-100.0)$, $U(-46.7)$, and $U(-16.3)$. Why? Because the expected utility of not seeding equals $.054U(-335.8) + .206U(-191.1) + .480U(-100.0) + .206U(-46.7) + .054U(-16.3)$. Also, we must know $U(-336.05)$, $U(-191.35)$, $U(-100.25)$, $U(-46.95)$, and $U(-16.55)$, since the expected utility of seeding equals

$$.038U(-336.05)+ .143U(-191.35) + .392U(-100.25)$$

$$+ .255U(-46.95) + .172U(-16.55).$$

(For the source of these numbers, see Figure 15.2).

**Step One**

We can find each of the required utilities in two steps. The first step is simple: *We set the utility attached to two monetary values arbitrarily.* The utility of the better consequence is set higher than the utility of the worse one. Often, the worst consequence involved is given a utility of zero, and the best consequence is given a utility of 1. Consequently, let's set $U(-336.05)$ equal to zero and $U(-16.3)$ equal to 1. It turns out that the ultimate results of the analysis do not depend on which two numbers we choose, as long as the utility of the better consequence is set higher than the utility of the worst one. Thus, we could set $U(-336.05)$ equal to 4 and $U(-16.3)$ equal to 10. It would make no difference to the ultimate outcome of the analysis.

**Step Two**

The second step is somewhat more complicated: *In this step we present the decision maker with a choice between the certainty of one of the other monetary values and a gamble where the possible outcomes are the two monetary values whose utilities we set arbitrarily.* For example, suppose that we want to find $U(-191.1)$. To do so, we ask the decision maker whether he or she would prefer the certainty of a $191.1 million loss to a gamble where there is a probability of $P$ that the loss is $16.3 million and a probability of $(1 - P)$ that the loss is $336.05 million. We then try various values of $P$ until we find the one where the decision maker is indifferent between the certainty of a $191.1 million loss and this gamble. Suppose that this value of $P$ is 0.6.

If the decision maker is indifferent between the certain loss of $191.1 million and this gamble, it must be that the expected utility of the certain loss of $191.1 million equals the expected utility of the gamble. (Why? Because under the assumptions in the previous section, the decision maker maximizes expected utility.) Thus.

$$U(-191.1) = 0.4U(-336.05) + 0.6U(-16.3).$$

And since we set $U(-336.05)$ equal to zero and $U(-16.3)$ equal to 1, it follows that $U(-191.1)$ must equal 0.6. In other words, the utility attached to a loss of $191.1 million is 0.6.

Similarly, we can find $U(-100.0)$, $U(-46.7)$, and the other utilities needed to compute the expected utility of seeding and the expected utility of not seeding. For example, to obtain $U(-100.0)$, we ask the decision maker whether he or she would prefer the certainty of a $100.0 million loss to a gamble where there is a probability of $P$ that the loss is $16.3 million and a probability of $(1 - P)$ that the loss is $336.05 million. Then we try various values of $P$ until we find the one where the decision maker is indifferent between the certainty of a $100.0 million loss and this gamble. Suppose that this value of $P$ is .85. Then the expected utility of a certain loss of $100.0 million must equal the expected utility of this gamble, which means that

$$U(-100.0) = .15U(-336.05) + .85U(-16.3).$$

And since $U(-336.05)$ equals zero and $U(-16.3)$ equals one, it follows that $U(-100.0)$ equals .85.

*The decision maker's utility function is the graph or relationship showing the utility he or she attaches to each amount of monetary gain or loss.* For example, in the case of the seeding example, the decision maker's utility function may be as shown in Figure 15.3. Through the repeated use of the procedure described above, we can obtain as many points on this utility function as we like—or as the decision maker's patience permits. But to see whether the decision maker, in the light of his or her own preferences toward risk, should seed or not, all that we need are the utilities shown in Table 15.1. Based on these utilities, each of which is

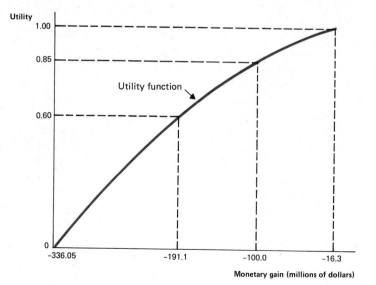

**Figure 15.3**
Decision Maker's
Utility Function,
Hurricane-Seeding
Example

**Table 15.1**
**Expected Utility from Seeding or Not Seeding a Hurricane**

| Seed | | | | Do not seed | | | |
|---|---|---|---|---|---|---|---|
| Monetary loss (millions of dollars) | (1) Utility | (2) Probability | (1) × (2) | Monetary loss (millions of dollars) | (1) Utility | (2) Probability | (1) × (2) |
| 336.05 | 0.000 | .038 | .000000 | 335.8 | 0.001 | .054 | .000054 |
| 191.35 | 0.599 | .143 | .085657 | 191.1 | 0.600 | .206 | .123600 |
| 100.25 | 0.849 | .392 | .332808 | 100.0 | 0.850 | .480 | .408000 |
| 46.95 | 0.949 | .255 | .241995 | 46.7 | 0.950 | .206 | .195700 |
| 16.55 | 0.999 | .172 | .171828 | 16.3 | 1.000 | .054 | .054000 |
| | | Expected utility: | .832288 | | | Expected utility: | .781354 |

obtained by the above procedure, it is clear that the decision maker should decide to seed, since the expected utility if a hurricane is seeded (.832288) exceeds that if it is not seeded (.781354).

## 15.6 Characteristics of Utility Functions

Not all utility functions look like the one in Figure 15.3. Although one can expect that utility increases with monetary gain, the shape of the utility function can vary greatly, depending on the preferences of the decision maker. Figure 15.4 shows three general types of utility functions. The one in panel A is like that in Figure 15.3 in the sense that utility increases with monetary value, but *at a decreasing rate*. In other words, an increase in monetary gain of $1 is associated with *smaller and smaller* increases in utility as the monetary gain increases in size. People *Risk Averters* with utility functions of this sort are **risk averters.** That is, when confronted with gambles with equal expected monetary gains, they prefer a gamble with a more certain outcome to one with a less certain outcome.

The utility function in panel B is one where utility increases with monetary value, but *at an increasing rate*. In other words, an increase in monetary gain of $1 is associated with *larger and larger* increases in utility as the monetary gain increases in size. People with utility func- *Risk Lovers* tions of this sort are **risk lovers.** That is, when confronted with gambles with equal expected monetary gains, they prefer a gamble with a less certain outcome to one with a more certain outcome.

Finally, the utility function in panel C is one where utility increases with monetary value and *at a constant rate*. In other words, an increase of $1 in monetary gain is associated with a *constant* increase in utility as the monetary gain grows larger and larger. Stated differently, utility in this case is a linear function of monetary gain:

$$U = a + bM$$

*(15.1)*

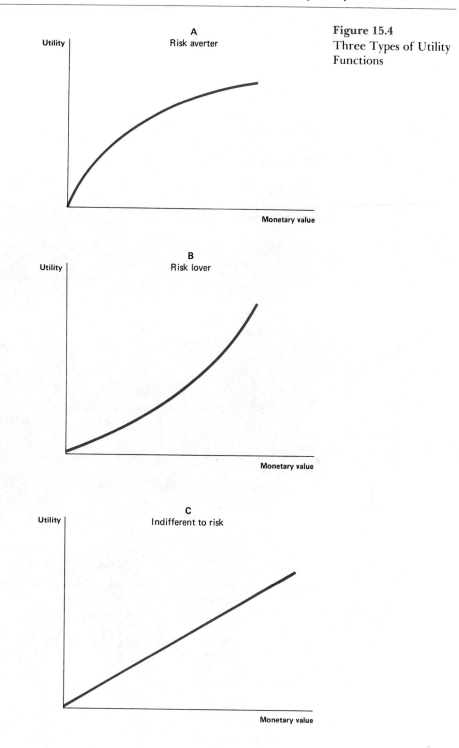

A
Utility | Risk averter

Monetary value

B
Utility | Risk lover

Monetary value

C
Utility | Indifferent to risk

Monetary value

**Figure 15.4**
Three Types of Utility
Functions

where $U$ is utility, $M$ is monetary gain, and $a$ and $b$ are constants. (Of course, $b > 0$.) People with utility functions of this sort are ***indifferent to risk.*** That is, they maximize expected monetary gain or value, regardless of risk. It is easy to show that this is true. Clearly, if equation (15.1)

*Indifference to Risk*

holds,

$$E(U) = a + bE(M),$$

(15.2)

where $E(U)$ is expected utility and $E(M)$ is expected monetary gain. (For a proof, see Appendix 4.1). Thus, since expected utility is directly related to expected monetary gain, it can only be a maximum when expected monetary gain is a maximum.

**GETTING DOWN TO CASES**

### SHOULD CUTLER-HAMMER PURCHASE AN OPTION?

In 1974, Cutler-Hammer, Inc., was offered the option to obtain a license to produce and sell a new flight-safety system. The inventor said that he had a strong patent position and that the new system was technically superior to other such systems; however, the market for the product was very uncertain because of pending legislative action. A team of Cutler-Hammer personnel and outside analysts carried out an analysis based on decision theory to help Cutler-Hammer to decide whether or not to purchase the option to obtain this license.[5]

According to this team, if Cutler-Hammer purchased the option, there was a 0.29 probability that it would not obtain the license, in which case it would lose $125,000, and a 0.71 probability that it would obtain the license. If it obtained such a license, the team estimated that there was a 0.85 probability that it would not obtain a defense contract, in which case it would lose $700,000, and a 0.15 probability that it would obtain a defense contract, in which case it would gain $5.25 million.

(a) Construct the decision tree.
(b) If Cutler-Hammer wanted to maximize expected monetary value, should it have purchased the option?
(c) Cutler-Hammer also analyzed the consequences of another course of action: waiting and seeking a sublicense. The team estimated that such a course of action would result in the following probability distribution of profit:

| Probability | Profit (thousands of dollars) |
|---|---|
| 0.94 | 0 |
| 0.06 | 830 |

After considerable discussion, there was a unanimous decision by the firm's decision-making group (the president and his vice presidents for business development and operations) to adopt this course of action. Were they averse to risk? Why or why not?

[5] J. Ulvila and R. Brown, "Decision Analysis Comes of Age," *Harvard Business Review*, September-October 1982. This account has been simplified somewhat for expository purposes.

> *Your best course of action is not necessarily to maximize expected*     **Basic Idea #29**
> *monetary value. Only if you are indifferent to risk is it true that*
> *you should maximize expected monetary value.*

In the remainder of this book, we shall assume that the decision maker's utility function is of the sort shown in panel c. If this is so, we can assume that the decision maker wants to maximize expected monetary gain. The reason for this assumption is that it simplifies the discussion. (Also, in many actual cases it may be a satisfactory approximation). In cases where this assumption does not hold, it is easy to alter our results accordingly. All that is required is that we substitute the decision maker's utilities for the monetary values in all subsequent sections.

The following example provides a simple illustration of the concepts discussed in this section.

---

EXAMPLE 15.2 John Jones and Cynthia Brown are each offered a choice of a certain gain of $200 or a gamble where there is a 50-50 chance of gaining $400 and a 50-50 chance of losing $100. John Jones chooses the gamble; Cynthia Brown chooses the certain gain of $200. Can we deduce from this that Jones is a risk lover and that Brown is a risk averter?

SOLUTION: Since the expected value of the gamble equals $0.5(\$400) - 0.5(\$100)$, or $150, John Jones must be a risk lover because he chose the alternative with the less certain outcome, even though it had a lower expected monetary value than the certain gain of $200. On the other hand, Cynthia Brown may or may not be a risk averter. If she were a risk averter, she would choose the certain gain because it has both a higher expected monetary gain and less uncertainty than the gamble. But if she were indifferent to risk, she would also choose the certain $200 gain, since it has a higher expected monetary gain than the gamble.

---

EXERCISES

**15.1** James McGuire's utility function is shown at the top of page 638.

(a) Is he a risk averter at all levels of monetary gain?
(b) Is he a risk lover at all levels of monetary gain?
(c) Does he prefer the certainty of gaining $50,000 over a gamble where there is a 0.5 probability of gaining $40,000 and a 0.5 probability of gaining $60,000? Explain.
(d) Does he prefer the certainty of gaining $70,000 over a gamble where there is a 0.5 probability of gaining $60,000 and a 0.5 probability of gaining $80,000? Explain.

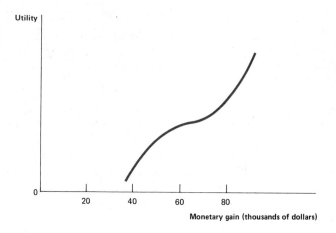

**15.2** Mill owner John Brown says that he is indifferent to risk. Suppose that we let 0 be the utility he attaches to $100,000 and 1 be the utility he attaches to $200,000. If what he says is true, what is the utility he attaches to (a) $400,000; (b) $40,000; (c) −$20,000.

**15.3** The owner of a sporting goods store says that he is indifferent between the certainty of receiving $7,500 and a gamble where there is a 0.5 chance of receiving $5,000 and a 0.5 chance of receiving $10,000. Also, he says that he is indifferent between the certainty of receiving $10,000 and a gamble where there is a 0.5 chance of receiving $7,500 and a 0.5 chance of receiving $12,500.

(a) In the following graph, draw four points on the utility function of the owner of the store.

(b) Does the owner of the sporting goods store seem to be a risk averter, a risk lover, or indifferent to risk? Explain.

**15.4** The Bona Fide Washing Machine Company is considering the purchase of a small firm that produces clocks. Bona Fide's management feels that there is a 50-50 chance, if Bona Fide buys the firm, that it can mold the firm into an effective producer of washing machine parts. If the firm can be transformed in this way, Bona Fide believes that it will make $500,000 if it buys the firm; if it cannot be transformed in this way, Bona Fide believes that it will lose $400,000.

(a) Construct a decision tree to represent Bona Fide's problem.

(b) What are the decision forks? (Are there more than one?)

(c) What are the chance forks? (Are there more than one?)

(d) Use the decision tree to solve Bona Fide's problem. In other words, assuming that the firm wants to maximize expected extra profit, should Bona Fide buy the firm or not?

(e) Before Bona Fide makes a decision concerning the purchase of the firm, Bona Fide's president learns that if the clock producer cannot be made into an effective producer of washing machine parts, there is a 0.2 probability that it can be resold to a Saudi Arabian syndicate at a profit of $100,000. (If the firm canot be resold, Bona Fide will lose $400,000.)

   (i) Does this information alter the decision tree?

   (ii) Can you think of three mutually exclusive outcomes if Bona Fide buys the firm?

   (iii) What is the probability of each of these outcomes?

   (iv) What is the monetary value to Bona Fide of each of these outcomes?

(f) Use your results in (e) to solve Bona Fide's problem under this new set of conditions. In other words, on the basis of this new information, should Bona Fide buy the firm or not?

(g) Bona Fide's executive vice president discovers an error in the estimate of how much Bona Fide will gain if it buys the clock manufacturer and turns it into an effective producer of washing machine parts.

   (i) Under the circumstances in (d), how big would this error have to be to reverse the indicated decision?

   (ii) Under the circumstances in (e), how big would the error have to be to reverse the indicated decision?

**15.5** The president of the Uphill Corporation, a maker of bicycle pedals, is indifferent between a certain gain of $10,000 and a gamble where there is a 0.50 chance of winning $100,000 and a 0.50 chance of losing $90,000.

(a) Draw three points on his utility function.

(b) Does it appear that Uphill's president is (i) indifferent to risk; (ii) a risk lover; (iii) a risk averter?

(c) Does it appear that Uphill's president's utility function is linear?

(d) Suppose that his utility function is linear between −$90,000 and $10,000 and linear between $10,000 and $100,000 (but that the slopes of these two line segments are not necessarily equal). Does this mean he maximizes expected monetary gain?

**15.6** A local bank president says that he is indifferent to risk.

(a) Suppose that we let 0 be the utility he attaches to $100,000 and 1 be the utility he attaches to $200,000. If what he says is true, what is the utility he attaches to (i) $300,000; (ii) $50,000; (iii) −$10,000?

(b) Suppose that the bank president must decide whether or not to make a loan where he perceives the probability to be 0.6 that he will gain $20,000 and 0.4 that he will lose $25,000. If he does not make the loan, he will incur neither a loss nor a gain. If what he says is true, will he make the loan?

(c) In contrast to the bank president in (a) and (b), the president of the Crooked Arrow National Bank is a risk lover. (i) Is it possible that he would make the loan in (b)? (ii) Can you be sure he would make such a loan? Why, or why not?

## 15.7 Posterior Analysis

*Prior Analysis vs.*
*Posterior Analysis*

Up to this point, we have been concerned with **prior analysis**—that is, decision making based solely on the basis of prior probabilities. A *prior probability* is a probability based on whatever information is available prior to the gathering of new experimental or sample evidence. As you will recall from Chapter 3, a prior probability may be a subjective probability that a certain hypothesis is true. For example, the probability that the new label (in Section 15.2) will be superior to the old one is a prior probability, just as the probability of a large increase in sustained speed if a hurricane is seeded (in Section 15.3) is a prior probability. In the next few sections, we turn to **posterior analysis,** in which decision making is based both on prior probabilities and on new experimental or sample evidence. Using Bayes' theorem, which we discussed at length in Chapter 3, we revise the prior probabilities to take account of this new, additional evidence.[6]

To illustrate the nature of posterior analysis, let's return to the marketing manager who must decide whether or not to adopt a new label for the firm's product. In contrast to Section 15.2 (and Example 4.6), suppose that the marketing manager does not want to rely solely on his prior probabilities that the new label will or will not be superior to the old one.[7] Instead, he sets in motion a survey of consumers. This survey is designed so that *if the new label is really superior,* the probability that the survey results will be favorable to the new label is 0.8, the probability that they will be neither favorable nor unfavorable to the new label is 0.1, and the probability that they will be unfavorable to the new label is 0.1. On the other hand, *if the new label is really not superior,* the probability that the survey results will be unfavorable to the new label is 0.7, the probability that they will be neither favorable nor unfavorable to to the new label is 0.1, and the probability that they will be favorable to the new label is 0.2.[8]

Suppose that the survey results turn out to be favorable to the new label. Using Bayes' theorem, we can compute the posterior probabilities that the new label is or is not superior to the old one. These posterior probabilities, which take account of the survey evidence, are

$$P(B|b) = \frac{P(b|B)P(B)}{P(b|B)P(B) + P(b|N)P(N)} \qquad (15.3)$$

---

[6] Note that *prior* and *posterior* are relative terms. For example, a prior probability may be derived, at least in part, from a previous sample. The essential point is that it is the probability prior to the availability of the evidence in question, not all evidence.

[7] Recall that the marketing manager's prior probabilities were .5 that the new label would be superior to the old one and .5 that it would not be superior to the old one.

[8] The reader may recall that this survey was discussed in Section 3.8, where we derived the posterior probabilities used in this section.

and

$$P(N|b) = \frac{P(b|N)P(N)}{P(b|B)P(B) + P(b|N)P(N)} \qquad (15.4)$$

where $P(B|b)$ is the posterior probability that the new label is really superior and $P(N|b)$ is the posterior probability that it is not. The prior probability that the new label is superior is $P(B)$, and the prior probability that it is not is $P(N)$. The probability that the survey results are favorable to the new label when it in fact is superior is $P(b|B)$, while the probability that they are favorable to the new label when it in fact is not superior is $P(b|N)$.

Inserting the numerical values into equations (15.3) and (15.4), we obtain

$$P(B|b) = \frac{0.8(0.5)}{0.8(0.5) + 0.2(0.5)} = 0.8$$

$$P(N|b) = \frac{0.2(0.5)}{0.8(0.5) + 0.2(0.5)} = 0.2.$$

Using these posterior probabilties, we can calculate the expected increase in profit if the marketing manager adopts the new label and if he does not. If he adopts the new label, the expected increase in profit, based on these posterior probabilities, is

$$0.8(\$800,000) + 0.2(-\$500,000) = \$540,000.$$

If he does not adopt the new label, the increase in profit is zero. Thus, if he wants to maximize expected profit, he should adopt the new label.

The effect of the evidence obtained from the survey has been to enlarge the expected increase in profit, should the new label be adopted. Before the survey the expected profit increase (if the new label were adopted) was $150,000. This figure was based only on the marketing manager's prior probabilities, and is called the *prior expected value*. After the survey the expected profit increase was $540,000, should the new label be adopted. This was based on the posterior probabilities and is called the *posterior expected value*. The reason why the posterior expected value exceeded the prior expected value in this case was that the sample produced evidence that increased the probability that the new label is in fact superior to the old one. Needless to say, the posterior expected value does not always exceed the prior expected value.

---

*In posterior analysis, Bayes' theorem is used to revise prior probabilities to take account of new evidence. The revised probabilities are called posterior probabilities, and they are used in essentially the same way as prior probabilities are used in prior analysis.*

*Basic Idea #30*

**TO ERR IS
HUMAN ... BUT
COSTLY**

THE CASE OF THE SUNKEN GOLD

Three similar boats sink off the coast of Florida. Each has two compartments where valuable articles are concealed. Boat $A$ contains a bar of gold in one compartment and a painting in the other compartment. Boat $B$ has a bar of gold in each compartment, and boat $C$ has a painting in each compartment. A diver tries to recover some of this treasure. He finds one of the boats, but cannot determine which one it is. He finds a bar of gold in one of its compartments. Thus, it must be either boat $A$ or boat $B$. Figuring that there is a 50–50 chance that it is boat $B$, he feels that there is a 50–50 chance that, if he can get the second compartment open, he will obtain a second bar of gold. (And since there is a 50–50 chance that it is boat $A$, he feels that there also is a 50–50 chance that, if he can get the compartment open, it will contain a painting.) Because any of these paintings will be destroyed by the water, it will be worthless. Thus, his decision concerning whether or not to risk his life to get the second compartment open depends heavily on the probability that it will contain gold. Has he calculated this probability correctly? That is, is this probability equal to 0.50?

SOLUTION: No. Assuming that it is equally likely that he has found each of the boats, Bayes' theorem says that the probability that it is boat $B$ (given that one of its compartments contains gold) is

$$\frac{P(g|B)P(B)}{P(g|B)P(B) + P(g|A)P(A)}$$

where $P(g|B)$ is the probability that a compartment contains gold, given that the boat is boat $B$, $P(g|A)$ is the probability that a compartment contains gold, given that the boat is boat $A$, $P(B)$ is the probability that the diver finds boat $B$, and $P(A)$ is the probability that he finds boat $A$. Since $P(g|B) = 1$, $P(g|A) = 1/2$, and $P(B) = P(A)$, it follows that, given that the one compartment contains gold, the probability that the other one contains gold is $2/3$, not $1/2$. This may seem hard to accept, but it is true.

## 15.8 Posterior Analysis in a Drug Firm: A Case Study [9]

To illustrate the use of posterior analysis, let's examine an actual situation. In 1968, the Canadian subsidiary of an American drug firm was considering introducing a certain kind of throat lozenge to the Canadian market. John Stonier of the firm's Consumer Products Division was charged with the responsibility of recommending whether or not to introduce the new product. To simplify the analysis, he used three profit levels—$200,000 per year, 0, and −$200,000 per year—to represent the possible profits from the new product. The prior probabilities he attached to these outcomes were 0.5, 0.3, and 0.2, respectively. Thus, on the basis of the prior probabilities, the expected profit increase resulting from the introduction to the new product was

$$0.5(\$200,000) + 0.3(0) + 0.2(-\$200,000) = \$60,000.$$

And since if the new product was not introduced the profit increase would be zero, Stonier was inclined to recommend introducing the product.

But Stonier also had the opportunity of carrying out a test-market study before making the final decision. Table 15.2 shows his estimate that the test market would indicate each level of increased profit from the introduction of the throat lozenges, when the actual increased profit from their introduction was as given at the top of each column in the table. For example, the first column of this table shows that if the actual profit increase would be $200,000, he felt that there was a 0.6 probability that the test-market study would indicate this fact, a 0.3 probability that it would indicate a zero profit increase, and a 0.1 probability that it would indicate a $200,000 loss.

If Stonier's test-market study indicated that the profit increase was −$200,000, should he recommend the introduction of this new

| Increased profit, test-market study (dollars) | Actual increased profit from new product (dollars) | | |
|---|---|---|---|
| | 200,000 | 0 | −200,000 |
| 200,000 | .6 | .2 | .1 |
| 0 | .3 | .6 | .3 |
| −200,000 | .1 | .2 | .6 |
| | 1.00 | 1.00 | 1.00 |

Table 15.2
Probability of Each Outcome of the Test-Market Study, Given Actual Level of Increased Profit from the New Product

[9] See C. H. von Lanzenauer, *Cases in Operations Research* (London and Canada: University of Western Ontario, 1975). This example is based on an actual case, but the names and numbers have been changed, and the circumstances and outcome have been altered and simplified considerably.

product? Let $P_1(200,000)$ be the posterior probablity that the profit increase will be $200,000, $P_1(0)$ be the posterior probability that it will be zero, and $P_1(-200,000)$ be the posterior probability that it will be $-\$200,000$. Under these circumstances,[10]

$$P_1(200,000) = \frac{.1(.5)}{.1(.5) + .2(.3) + .6(.2)} = 5/23,$$

$$P_1(0) = \frac{.2(.3)}{.1(.5) + .2(.3) + .6(.2)} = 6/23,$$

$$P_1(-200,000) = \frac{.6(.2)}{.1(.5) + .2(.3) + .6(.2)} = 12/23.$$

Thus, if the new product is introduced, the posterior expected value of the increased profit is

$$(5/23)(\$200,000) + (6/23)(0) + (12/23)(-\$200,000) = -\$60,870.$$

Since the expected increased profit is negative (and consequently less than the expected increased profit of zero if the new product is not introduced), it appears that Stonier should recommend that the product not be introduced. In contrast to the situation in the previous section, this is a case where the sample evidence recommends a change in the decision.

The following example is a further illustration of posterior analysis, based on the data in Table 15.2.

---

EXAMPLE 15.3 Suppose that the test-market study indicated an increased profit of zero. Should Stonier recommend the introduction of the new throat lozenge?

SOLUTION: Let $P_1(200,000)$ be the posterior probability that the profit increase will be $200,000, $P_1(0)$ be the posterior probability that it will be zero, and $P_1(-200,000)$ be the posterior probability that it will be $-\$200,000$. Since the test-market study

---

[10] To illustrate how the posterior probabilities $(5/23, 6/23, 12/23)$ are derived, consider $P_1(200,000)$. Based on Bayes' theorem, it equals

$$\frac{P(-200,000|200,000)P_0(200,000)}{P(-200,000|200,000)P_0(200,000)+P(-200,000|0)P_0(0)+P(-200,000|-200,000)P_0(-200,000)}$$

where $P_0(200,000)$, $P_0(0)$, and $P_0(-200,000)$ are the prior probabilities of an increased profit of $200,000, 0, and $-\$200,000$, respectively. $P(-200,000|200,000)$ is the probability that the test-market study will indicate an increased profit of $-\$200,000$, given that the profit increase will be $200,000; $P(-200,000|0)$ is the probability that the test-market study will indicate an increased profit of $-\$200,000$, given that the profit increase will be zero; and so forth. Inserting into this expression the prior probabilities in the first paragraph of this section and the conditional probabilities in Table 15.2, we get the equation for $P_1(200,000)$ in the text.

The equations in the text for $P_1(0)$ and $P_1(-200,000)$ can be derived in an analogous way. (Do this as an exercise.)

indicates that the profit increase will be zero, these posterior probabilities are as follows:[11]

$$P_1(200,000) \quad = \frac{.3(.5)}{.3(.5) + .6(.3) + .3(.2)} = .384$$

$$P_1(0) \qquad\qquad = \frac{.6(.3)}{.3(.5) + .6(.3) + .3(.2)} = .462,$$

$$P_1(-200,000) = \frac{.3(.2)}{.3(.5) + .6(.3) + .3(.2)} = .154.$$

Thus, if the new product is introduced, the posterior expected value of increased profit is

$$.384(\$200,000) + .462(0) + .154(-\$200,000) = \$46,000.$$

Since this exceeds zero, it appears that Stonier should recommend the introduction of the new product.

## 15.9 Sample Size and the Relationship between Prior and Posterior Probability Distributions

It is important to recognize that the relationship between the prior and posterior probability distributions of a random variable is influenced by the size of the sample. *As the sample size increases, the posterior distribution depends more and more on the sample results and less and less on the prior distribution.* To illustrate this, suppose that the marketing manager in Section 15.7 takes a second sample to indicate whether the new label is superior and that it, like the first sample, is favorable to the new label. Given that the surveys are independent and designed so that if the new label is superior, the probability that *each survey* will be favorable to it is 0.8, the probability that *both* surveys will be favorable to it

---

[11] To illustrate how the posterior probabilities (.384, .462, and .154) are derived, consider $P_1(200,000)$. Based on Bayes' theorem, it equals

$$\frac{P(0 \mid 200,000)P_0(0)}{P(0 \mid 200,000)P_0(200,000) + P(0 \mid 0)P_0(0) + P(0 \mid -200,000)P_0(-200,000)}$$

where $P_0(200,000)$, $P_0(0)$, and $P_0(-200,000)$ are the prior probabilities of an increased profit of $200,000, 0, and −$200,000, respectively. $P(0 \mid 200,000)$ is the probability that the test-market study will indicate an increased profit of zero, given that the profit increase will be $200,000; $P(0 \mid 0)$ is the probability that the test-market study will indicate an increased profit of zero, given that the profit increase will be zero; and so forth. Inserting the prior probabilities given in the first paragraph of this section and the conditional probabilities in Table 15.2 into this expression, we get the equation for $P_1(200,000)$ in the text.

The equations in the text for $P_1(0)$ and $P_1(-200,000)$ can be derived in an analogous way. (Do this as an exercise.)

under these circumstances is $0.8(0.8) = 0.64$. And given that the surveys are designed so that if the new label is not superior, the probability that *each* will be favorable to it is 0.2, the probability that *both* surveys will be favorable to the new label when in fact it is not superior is $0.2(0.2) = 0.04$.

Using Bayes' theorem, we can compute the posterior probabilities that the new label is or is not superior to the old one. These posterior probabilities are

$$P(B|bb) = \frac{P(bb|B)P(B)}{P(bb|B)P(B) + P(bb|N)P(N)} \tag{15.5}$$

and

$$P(N|bb) = \frac{P(bb|N)P(N)}{P(bb|B)P(B) + P(bb|N)P(N)}, \tag{15.6}$$

where $P(bb|B)$ is the probability that both surveys are favorable to the new label, given that it is in fact superior, and $P(bb|N)$ is the probability that both surveys are favorable to it, given that it is in fact not superior.

Inserting the numerical values into equations (15.5) and (15.6), we obtain

$$P(B|bb) = \frac{0.64(0.5)}{0.64(0.5) + .04(0.5)} = 16/17 = .941,$$

$$P(N|bb) = \frac{0.04(0.5)}{0.64(0.5) + .04(0.5)} = 1/17 = .059.$$

Given the larger sample size, the posterior probability that the new label is superior is much higher than it was after the first sample (0.941 vs. 0.8). And the posterior probability that the new label is not superior is much lower than after the first sample (0.059 vs. 0.2). This is what we would expect. As the sample size increases, the sample becomes increasingly trustworthy, and purely subjective feelings must be given relatively less weight. Thus, in this case, although the decision maker's prior probability that the new label was superior was only 0.5, the posterior probability increased to 0.8 after the first sample indicated that it was superior, and moved even further from 0.5 (to 0.941) after the second sample also indicated that it was superior.

## 15.10 Expected Value of Perfect Information, Prior and Posterior

An important concept in decision theory is the expected value of perfect information, which is defined as follows.

EXPECTED VALUE OF PERFECT INFORMATION: *The expected value of perfect information is the increase in expected profit if the decision maker could obtain completely accurate information concerning the random variable in question (but if he or she does not yet know what this information will be).*

In the case of the marketing manager, this expected value is the increase in expected profit if the manager could obtain perfectly accurate information indicating whether or not the new label is superior.

To illustrate how the expected value of perfect information can be computed, consider the marketing manager (before the survey described in Section 15.7 was carried out). To determine the expected value of perfect information, we must evaluate the expected gain to the company if he can obtain access to perfectly accurate information. Then we must calculate the extent to which this expected gain exceeds the expected gain based on the information actually available to the marketing manager.

If the manager obtains perfect information, he will be able to make the correct decision, regardless of whether the new label is superior. In other words, if the new label is superior the manager will be aware of this fact, and he will adopt the label. If the new label is not superior he will be aware of this fact also, and he will not adopt the label. Thus, given that the manager has access to perfect information, the expected value of the gain to the company is

$$0.5(\$800,000) + 0.5(0) = \$400,000.$$

To see why this is the expected gain if he has access to perfect information, it is important to recognize that although it is assumed that he has access to perfect information, *he does not yet know what this information will be.* Based on his prior probabilities, there is a .5 probability that this information will show that the new label really is superior, in which case he will adopt it and the gain will be $800,000. There is also a .5 probability that the information will show that the new label is really not superior, in which case he will not adopt it and the gain will be zero. Thus, as shown above, the expected value of the company's gain if the manager has access to perfect information (that is not yet revealed to him) is $400,000.

In contrast, the expected value of the company's gain if he bases his decision on existing information is $150,000, as we saw in Section 15.2. The difference between these two figures—$400,000 minus $150,000, or $250,000—is the expected value of perfect information. *It shows the amount by which the expected value of the firm's gains increases as a consequence of having access to perfect information.* Put differently, *it is the maximum amount that the decision maker should pay to obtain perfect information.*

In the case above, the $250,000 is the *prior* expected value of perfect information. In other words, it is the expected value of perfect information before the survey in Section 15.7 was carried out, and is based entirely on the decision maker's prior probabilities. After the survey has been carried out, we can compute the *posterior* expected value of perfect information—that is, the expected value of perfect information based on posterior probabilities. In the case of the marketing manager, the posterior expected value of perfect information is

$$0.8(\$800,000) + 0.2(0) - \$540,000 = \$100,000.$$

The posterior expected value of perfect information differs from the prior expected value of perfect information for two reasons. First, the posterior probabilities that the new label is or is not superior (0.8 and 0.2) are used in place of the prior probabilities (0.5 and 0.5). Second, the expected profit based on the posterior probabilities ($540,000) is used in place of the expected profit based on the prior probabilities ($150,000).

It is interesting to compare the prior and posterior expected values of perfect information. Before the survey was taken, the marketing manager should have been willing to pay up to $250,000 for perfect information concerning whether or not the new label was superior since this was the expected value of perfect information. But after the survey was taken and the results indicated that the new label was in fact superior, the manager should have been willing to pay only up to $100,000 for perfect information since this was then the expected value of perfect information. The survey, by reducing the decision maker's doubts concerning the superiority of the new label, has consequently reduced the expected value of perfect information.

Note, however, that the posterior expected value of perfect information can be either greater or less than the prior expected value of perfect information. For example, in the case of the Canadian drug executive in Section 15.8, the prior expected value of perfect information was [12]

$$0.5(\$200,000) + 0.3(0) + 0.2(0) - \$60,000 = \$40,000,$$

while the posterior expected value of perfect information was[13]

$$(5/23)(\$200,000) + (6/23)(0) + (12/23)(0) - (0) = \$43,478.$$

---

[12] According to Section 15.8, there is a 0.5 prior probability that the profit increase will be $200,000, in which case the new product will be introduced (and the profit increase will be $200,000). There is a 0.3 prior probability that the profit increase will be zero, in which case the new product will not be introduced (and the profit increase will be zero). And there is a 0.2 prior probability that the profit increase would be −$200,000, in which case the new product will not be introduced (and the profit increase will be zero). Thus, the expected value of the firm's increased profit if it has access to perfect information is 0.5 ($200,000) + 0.3(0) + 0.2(0). Subtracting $60,000, which (as we know from Section 15.8) is the expected profit increase with existing information, the expected value of perfect information is as shown in the equation in the text.

[13] According to Section 15.8, there is a 5/23 posterior probability that the profit increase will be $200,000, in which case the new product will be introduced (and the profit increase will be $200,000). There is a 6/23 posterior probability that the profit increase will be zero, in which case the new product will not be introduced (and the profit increase will be zero). And there is a 12/23 posterior probability that the profit increase would be −$200,000, in which case the new product will not be introduced (and the profit increase will be zero). Thus, the expected value of the firm's increased profit if it

The reason why the latter exceeds the former is that the posterior probabilities were quite different from the prior probabilities, and the optimal action was altered by the test-market results. In effect, the test-market results increased doubt concerning the proper decision and increased the expected value of perfect information.

## 15.11 Expected Opportunity Loss

In both prior analysis and posterior analysis, a frequently used concept is that of opportunity loss, defined as follows.

> OPPORTUNITY LOSS: *Given that the decision maker takes a particular action, the opportunity loss if a particular event (or state of nature) occurs is the difference between the profit actually achieved and the profit that would have been achieved if the action had been the best possible one for this event.*

For each event that can occur, one can compute the opportunity loss. For example, in the case of the marketing manager, there are two possible states of nature: that the new label is superior, or that it is not. If the former is the true state of nature, the best action that can be taken is to adopt the new label. Since the profit resulting from this act is $800,000 higher than if the old label is kept, the opportunity loss of keeping the old label is $800,000, and the opportunity loss of adopting the new label is zero. If, on the other hand, the true state of nature is that the new label is not superior, the best action that can be taken is to keep the old label. Since the profit resulting from this act is $500,000 higher than if the new label is adopted, the opportunity loss of adopting the new label is $500,000, and the opportunity loss of keeping the old label is zero. Table 15.3 shows, for each state of nature, both the profit and the opportunity loss resulting from each action the decision maker can take.

**Table 15.3**
Profit and Opportunity Loss for Each Action and Each Possible State of Nature*

| State of nature | Profit (dollars) | | Opportunity loss (dollars) | |
|---|---|---|---|---|
| | Adopt new label | Keep old label | Adopt new label | Keep old label |
| New label is superior | 800,000* | 0 | 0 | 800,000 |
| New label is not superior | −500,000 | 0* | 500,000 | 0 |

* The profit for the best act corresponding to each state of nature is marked with an asterisk. This informs the reader that for each state of nature opportunity losses are measured as differences from the figure marked with an asterisk.

has access to perfect information is (5/23)($200,000) + (6/23)(0) + (12/23)(0). Subtracting zero, which is the expected profit increase with existing information (since, as we know from Section 15.8, the new product will not be introduced), the expected value of perfect information is as shown in the text.

*If the decision maker wants to maximize expected monetary value, it can be shown that he or she should choose the action that minimizes the expected opportunity loss. This is an alternative way of taking the action that maximizes expected monetary value.* To demonstrate that the minimization of expected opportunity loss leads to the same choice as the maximization of expected monetary gain, consider the case of the marketing manager. Before the survey, the expected opportunity loss *if the marketing manager adopts the new label* is

$$0.5(0) + 0.5(\$500,000) = \$250,000.$$

Why? Because there is a 0.5 prior probability that the new label will be superior to the old one, in which case there is a zero opportunity loss (see Table 15.3); and because there is a 0.5 prior probability that the new label will not be superior to the old one, in which case there is a $500,000 opportunity loss (see Table 15.3). The expected opportunity loss *if he does not adopt the new label* is

$$0.5(\$800,000) + 0.5(0) = \$400,000.$$

Why? Because there is a 0.5 prior probability that the new label will be superior to the old one, in which case there is an $800,000 opportunity loss (see Table 15.3); and because there is a 0.5 prior probability that the new label will not be superior to the old one, in which case there is a zero opportunity loss (see Table 15.3).

Thus, if he minimizes the expected opportunity loss, he will choose the same action—that is, he will adopt the new label—as if he maximized expected profit. As stated above, this will be true in general.

Before concluding this section, two additional points should be noted. First, if you compare the results of this section with those of the last one, you will find that *the expected opportunity loss corresponding to the best action ($250,000) equals the expected value of perfect information (also $250,000).* This, too, will always be true. Thus, an alternative way of calculating the expected value of perfect information is to find the expected opportunity loss corresponding to the best action.

Second, as stated at the beginning of this section, *the concepts discussed here are relevant for posterior as well as prior analysis.* To illustrate their application in a posterior analysis, consider the case of the marketing manager after he received the survey results indicating that the new label was in fact superior. Using the resulting posterior probabilities, the expected opportunity loss if the marketing manager adopts the new label is[14]

$$.8(0) + .2(\$500,000) = \$100,000.$$

[14] According to Section 15.7, under these circumstances there is a 0.8 posterior probability that the new label will be superior to the old, in which case there is a zero opportunity loss (see Table 15.3); and there is a 0.2 posterior probability that the new label will not be superior to the old, in which case there is a $500,000 opportunity loss (see Table 15.3). Thus, the expected opportunity loss is given by the equation in the text.

The expected opportunity loss if he does not adopt it is[15]

.8($800,000) + .2(0) = $640,000.

In accord with our previous discussion, the action that minimizes expected opportunity loss—namely, adopting the new label—is the same as the action that maximizes expected profit. Also, the expected opportunity loss corresponding to the best action ($100,000) equals what was shown in the previous section to be the expected value of perfect information.

## EXERCISES

**15.7** An appliance firm must decide whether or not to offer Warren Whelan a job. If Whelan turns out to be a success, the firm will increase its profits by $100,000; if he turns out not to be a success, the firm's profits will decrease by $80,000. The firm feels that the chances are 50–50 that he will be a success.
(a) How can the firm obtain information concerning whether or not Whelan will be a success?
(b) What is the expected value of perfect information concerning whether or not he will be a success?

**15.8** A speculator must decide whether or not to invest $100,000 in a new Broadway musical. If the musical receives good reviews from the critics, the speculator will net $300,000 if he invests in it. If it does not receive good reviews, he will lose his $100,000. He thinks that the probability is 0.3 that it will receive good reviews.
(a) What is the expected opportunity loss if he invests in the musical?
(b) What is the expected opportunity loss if he does not invest in the musical?
(c) What is the expected value of perfect information?

**15.9** A steel firm is interested in merging with an oil firm. If the oil firm's current drilling activities are a success, the steel firm will gain $20 million from the merger. If the oil firm's current drilling activities are not a success, the steel firm will lose $10 million from the merger. The steel firm believes that there is a 0.4 probability that the oil firm's current drilling activities will be a success.
(a) What is the expected opportunity loss if the steel firm carries out the merger?
(b) What is the expected opportunity loss if the steel firm does not carry out the merger?
(c) What is the expected value of perfect information?

**15.10** Is it possible for the posterior expected value of perfect information to exceed the prior expected value of perfect information? If so, what would this mean? (Answer this question with one or two sentences of simple English.)

---

[15] According to Section 15.7, under these circumstances there is a 0.8 posterior probability that the new label will be superior to the old, in which case there is an $800,000 opportunity loss (see Table 15.3); and there is a 0.2 posterior probability that the new label will not be superior to the old, in which case there is a zero opportunity loss (see Table 15.3). Thus, the expected opportunity loss is given by the equation in the text.

**15.11** The Acme Corporation is considering the introduction of a new, lighter type of bicycle. The firm thinks that there is a 30 percent probability that the bicycle will be a major success (in which case the firm will make $1 million), a 40 percent probability that it will be no better (and no worse) than the firm's existing line of bicycles (in which case the firm will make nothing), and a 30 percent probability that it will be a flop (in which case the firm will lose $800,000). Acme hires a market-research firm to study the new bicycle's potential market, and this firm concludes that the new bike will be no better and no worse than the firm's existing line. Based on previous experience, the probability of coming to this conclusion, given that the new bicycle would be a major success, is thought to be 0.3. The probability of coming to this conclusion, given that the new bicycle is no better and no worse than the existing line, is thought to be 0.5. And the probability of coming to this conclusion, given that the new bicycle would be a flop, is thought to be 0.2.

(a) Calculate the posterior probabilities that the new bicycle will be a major success: no better or worse than the existing line; a flop.

(b) Based on these posterior probabilities, construct a decision tree representing Acme's decision problem.

(c) If Acme wants to maximize expected monetary gain, should it introduce the new bicycle?

(d) In this posterior analysis, what is the expected value of perfect information?

(e) Suppose that Acme had based its decision solely on the prior probabilities. What was the prior expected value of perfect information? How does it compare with the posterior expected value of perfect information? Why isn't the prior expected value of perfect information equal to the posterior expected value of perfect information?

(f) Construct a table showing the opportunity loss corresponding to each action, given each state of nature.

(g) Do a prior analysis based on the minimization of expected opportunity loss. Show that the expected opportunity loss corresponding to the best action is equal to the expected value of perfect information.

(h) Do a posterior analysis based on the minimization of expected opportunity loss. Show that the expected opportunity loss corresponding to the best action is equal to the expected value of perfect information.

# Chapter Review

1. Any problem of decision making under uncertainty has two characteristics. First, the decision maker must make a choice (or perhaps a series of choices) among alternative courses of action. Second, this choice (or series of choices) leads to some consequence, but the decision maker cannot tell in advance the exact nature of this consequence because it depends on some unpredictable event (or series of events) as well as on his or her choice (or series of choices). A *decision tree* represents such a problem as a series of choices, each of which is depicted by a fork. A *decision fork* is a juncture representing a choice where the decision maker is in control; a *chance fork* is a juncture where "chance" is in control of the outcome. By the process of *backward induction,* one can work one's way from the right-hand end of a decision tree to the left-hand end to solve the problem.

2. Whether a decision maker wants to maximize expected monetary

gain depends on his or her preferences with regard to risk. To reflect these preferences, we can construct a *utility function* for the decision maker (if the necessary assumptions are met). The first step in constructing a utility function is to establish arbitrarily the utilities attached to two monetary values. The second step is to present the decision maker with a choice between a gamble where the possible outcomes are the two monetary values whose utilities were arbitrarily set and the certainty of some third monetary value. After finding the probabilities in this gamble that will make the decision maker indifferent between these two choices, we can calculate the utility of the third monetary value. Next, the latter steps can be repeated over and over in order to obtain the utilities attached to as many monetary values as we need. If the assumptions discussed in this chapter are met, *the decision maker, if he or she is rational, will maximize expected utility.*

3. *Posterior analysis* is concerned with decisions based both on prior probabilities and on new experimental or sample evidence, whereas *prior analysis* is based on prior probabilities alone. To carry out a posterior analysis, we use Bayes' theorem to compute posterior probabilities which reflect the decision maker's prior probabilities and the sample results. As the sample size increases, the posterior probabilities depend more and more on the sample results and less and less on the prior probabilities. Once the posterior probabilities have been computed, the optimal choice is the one that maximizes expected monetary value if the decision maker is indifferent toward risk. In both posterior and prior analyses, we can compute the *expected value of perfect information.*

4. An *opportunity loss* is the loss incurred by the decision maker if he or she fails to take the best action possible. For each event that can occur, one can compute the opportunity loss. If the decision maker wants to maximize expected monetary value, it can be shown that he or she should choose the action that minimizes the expected opportunity loss. This is an alternative way of finding the action that maximizes expected monetary value. It can be applied in either a prior or a posterior analysis. An interesting point concerning this approach is that the expected opportunity loss corresponding to the best action is equal to the expected value of perfect information.

## Chapter Review Exercises

**15.12** The owner of a dress shop says that she is indifferent between the certainty of receiving \$10,000 and a gamble where there is a 0.5 chance of receiving \$25,000 and a 0.5 chance of receiving nothing. In the graph below, plot three points on her utility function.

Utility

Monetary gain

**15.13** A New York tavern owner must decide whether or not to expand his tavern. He thinks that the probability is 0.6 that the expansion will prove successful and 0.4 that it will not be successful. If it is successful, he will gain $100,000. If it is not successful, he will lose $80,000.

(a) Construct the decision tree for this problem, and use it to solve the problem, assuming that the tavern owner is indifferent to risk.

(b) List all forks in the decision tree you constructed, and indicate whether each is a decision fork or a chance fork. State why.

(c) Would the tavern owner's decision be altered if he felt that the probability that the expansion will prove successful is 0.5, not 0.6?

(d) What value of the probability that the expansion will prove successful will make the tavern owner indifferent between expanding and not expanding the tavern?

**15.14** The Internal Revenue Service claims that Mr. Hughes owes $50,000 in back taxes. If Mr. Hughes disputes this claim, it will cost him $5,000 in legal fees. He believes that the probability that he will win (and have to pay nothing in taxes) is 0.15. To get a better idea of the strength of his case, Mr. Hughes can ask a friend who is a tax expert. If he will win, Mr. Hughes believes that there is 0.7 probability that his friend will recognize this to be true. Mr. Hughes is indifferent to risk.

(a) If his friend's opinion is that Mr. Hughes will not win the case, should Mr. Hughes dispute the claim by the Internal Revenue Service?

(b) If his friend's opinion is that Mr. Hughes will win the case, should Mr. Hughes dispute the claim?

**15.15** A firm must decide whether or not to build a new warehouse. If its advertising campaign is successful, it will make $1 million by building the warehouse. If its advertising campaign is not successful, it will lose $0.5 million by building the warehouse. Based on previous experience, the firm's managers believe that there is a 0.5 probability that its advertising campaign will be successful. To obtain further information on this score, they carry out a survey. If the advertising campaign will be successful, there is a 0.6 probability that the survey will indicate that this is true. Similarly, if it will not be successful, there is a 0.6 probability that the survey will indicate that this is true.

(a) Suppose that the survey indicates that the advertising campaign will be successful. If the firm is indifferent to risk, should it build the new warehouse?

(b) Suppose that the survey indicates that the advertising campaign will not be successful. If the firm is indifferent to risk, should it build the new warehouse?

**15.16** In Exercise 15.13, what is the expected value of perfect information concerning whether or not the expansion of the tavern will be successful?

**15.17** In Exercise 15.15, what is the prior expected value of perfect information? What is the posterior expected value of perfect information, if the survey indicates that the advertising campaign will not be a success?

**15.18** In Exercise 15.15, what is the posterior expected value of perfect information, if the survey indicates that the advertising campaign will be a success?

# Appendix 15.1

## TWO-ACTION PROBLEMS WITH LINEAR PROFIT FUNCTIONS

In this appendix we will look at a special kind of problem which is encountered rather frequently and which can be handled simply and effectively by the methods discussed above. *This kind of problem is characterized by only two possible actions that the decision maker can take, and by the fact that the profit received is a linear function of the random variable representing the relevant chance event.* As an illustration, suppose that a firm is considering whether or not to establish a plant for producing and selling television sets in a particular foreign country. There are 5 million potential buyers in that country annually, and $p$ is the proportion of these potential buyers who would buy this firm's television set if it were available. The fixed cost of the new plant would be $1 million per year and the gross profit from each television set sold—that is, the price less the unit variable cost—is $20.

The firm must choose between two possible actions: to establish or not to establish the plant. If it establishes the plant, the firm's profit can easily be shown to be a linear function of $p$, which is a random variable. To see that this is the case, note that the number of television sets sold (in millions per year) will equal $5p$ so that profit (in millions of dollars per year) will be

$$\pi = -1 + 20(5)(p),$$

since annual fixed cost equals $1 million, and each television set sold results in a gross profit of $20. Simplifying terms, we have

$$\boxed{\pi = -1 + 100p,} \tag{15.7}$$

which shows that profit is a linear function of $p$. If the firm does not establish the plant, the profit clearly is zero.

*In a problem of this sort, there is a break-even value of the random variable which makes the decision maker indifferent between the two actions. If the random variable falls below this value, one action is optimal; if it falls above this value, the other action is optimal.* In the case at hand, the break-even value of $p$ is .01, since if $p = .01$ the profit obtained by establishing the plant is

$$-1 + 100(.01) = 0,$$

which is the same as the profit obtained by not establishing the plant. Figure 15.5 shows the profit functions if the firm establishes the plant and if it does not. As you can see, these functions intersect at the break-even value of $p$, which is .01. If $p$ is less than .01, the optimal action is not to establish the plant; if it is greater than .01, the optimal action is to establish it.

Suppose that the firm has prior probabilities concerning $p$ of the following sort: It believes that there is a .2 probability that $p$ will equal .006, a .3 probability that it will equal .009, a .4 probability that it will equal .012, and a .1 probability that it will equal .015. Under these circumstances, if the firm is indifferent to risk (and thus wants to maximize expected profit) which action should it take? As we shall see in the following section, there is a simple way of answering this question.

**Figure 15.5**
**Profit Functions if**
**Firm Does or Does Not**
**Establish Plant**

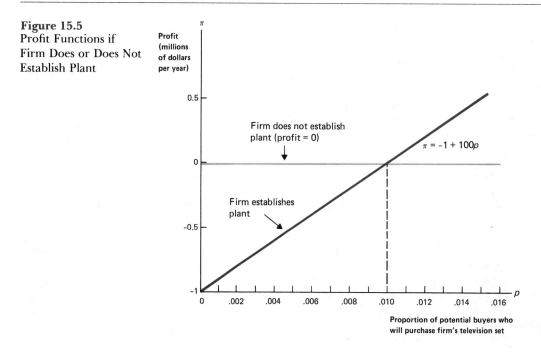

SOLVING A TWO-ACTION PROBLEM. Before we describe the simple way to solve this problem, let's solve it in the way discussed in previous sections. Figure 15.6 shows the decision tree in this case. Based on the firm's prior probabilities, the expected profit if it establishes the plant is

$$.2(-\$400,000) + .3(-\$100,000) + .4(\$200,000)$$

$$+.1(\$500,000) = \$20,000.$$

**Figure 15.6**
**Decision Tree for**
**Firm's Decision to**
**Establish or Not**
**Establish Plant**

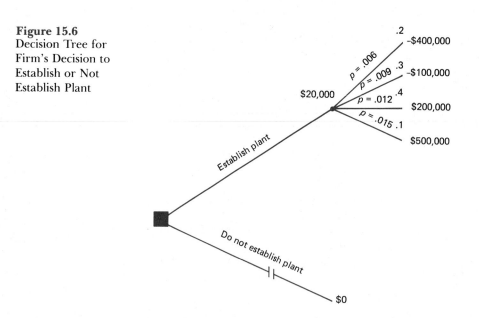

In other words, the profit[16] is −$400,000 if $p = .006$ (and the probability that this is the case is .2), the profit is −$100,000 if $p = .009$ (and the probability that this is the case is .3), the profit is $200,000 if $p = .012$ (and the probability that this is the case is .4), and the profit is $500,000 if $p = .015$ (and the probability that this is the case is .1). Because the result—$20,000—exceeds zero (the expected profit if the firm does not establish the plant), the firm should establish the plant.

A very simple way of solving this type of problem is to recognize that if profit is a linear function of $p$ (as shown in equation 15.7), then the expected value of profit is the same linear function of the expected value of $p$. In other words, if $A$ and $B$ are any two constants, and if

$$\pi = A + Bp,$$

then

$$E(\pi) = A + BE(p).$$

(The proof of this is given in Appendix 4.1.) Using this proposition, it follows from equation (15.7) that

$$E(\pi) = -1 + 100E(p) \tag{15.8}$$

if the firm establishes the plant. Thus, all we have to do to find the expected profit under these circumstances is determine the expected value of $p$, which is

$$.2(.006) + .3(.009) + .4(.012) + .1(.015) = .0102,$$

and substitute it in equation (15.8) to get

$$E(\pi) = -1 + 100(.0102) = .02.$$

which equals $20,000, the same result we got at the beginning of this section.[17]

An important point is that *the only essential information required to solve this type of problem is the expected value of the random variable—E(p) in this case.* We do not need to know the variance of $p$, or whether the probability distribution of $p$ is skewed, or the shape of the probability distribution. We only need to know the expected value. This is fortunate because prior probability distributions frequently cannot be specified with a great deal of confidence, and the decision maker may be more confident of the expected value than of the entire distribution. Based on equation (15.8), it is clear that the expected profit from establishing the plant will exceed zero (the expected profit if it does *not* establish the plant) if

$$-1 + 100E(p) > 0,$$

[16] To obtain each of these profit figures, we insert the relevant value of $p$ into equation (15.7). For example, when $p = .006$, equation (15.7) tells us that $\pi = -1 + 100(.006) = -.4$. Since $\pi$ is expressed in millions of dollars, this means that the profit is −$400,000 if $p = .006$.

[17] Since $\pi$ is expressed in millions of dollars, we must multiply .02 by $1 million to get the expected profit, which is $20,000. This explains how the figure in the text was derived.

or if

$$E(p) > .01.$$

*Thus, the decision boils down to whether or not the expected value of p exceeds .01 (the break-even value of p discussed in the previous section.) If it does, the firm should establish the plant; if not, it shouldn't.*

### EXERCISES

**15.19** There are 1 million potential buyers of hi-fi sets of a particular kind. The Electro Company is trying to decide whether to introduce a new type of hi-fi set in this market. It believes that the probability distribution of the proportion of these potential buyers who would buy its new product is as follows:

| Proportion of potential buyers that would buy its new product | Probability |
|---|---|
| 0.018 | 0.2 |
| 0.019 | 0.3 |
| 0.020 | 0.2 |
| 0.021 | 0.2 |
| 0.022 | 0.1 |

The fixed cost per year of the new product would be $1 million. The gross profit (price less average variable cost) from each of the new hi-fi sets would be $50.
(a) Is this a two-action problem?
(b) Is the Electro Company's annual profit a linear function of a random variable? If so, what is the random variable? What is the linear function?
(c) If the Electro Company wants to maximize expected profit, should it introduce the new product?

**15.20** A firm is considering whether or not to establish a plant to produce clocks. The fixed cost per year of the plant would be $100,000. The gross profit (price less average variable cost) from each clock sold would be $30. The firm believes that the probability distribution of the number of clocks that would be sold each year is

| Number of clocks sold | Probability |
|---|---|
| 2,000 | 0.2 |
| 3,000 | 0.4 |
| 4,000 | 0.2 |
| 5,000 | 0.2 |

(a) Is this a two-action problem?
(b) Is the firm's profit a linear function of a random variable? If so, what is the random variable? What is the linear function?
(c) If the firm wants to maximize expected profit, should it establish the plant?

# Decision Theory: Preposterior Analysis and Sequential Analysis*

## 16.1 Introduction

In the previous chapter, we discussed how decisions can be made on the basis of prior probabilities alone (prior analysis) and how they can be made after additional evidence has been obtained through sampling or experimentation (posterior analysis). However, nothing has yet been said about how the decision maker can determine whether or not it is worthwhile to obtain additional evidence of this sort. Preposterior analysis and sequential analysis are concerned with this latter problem, as well as with the problem of how much additional evidence to obtain and what action to take on the basis of it.

Whereas prior analysis and posterior analysis are concerned with the final choice of a best course of action, preposterior analysis and sequential analysis recognize that the decision maker may decide not to make a final choice until additional information has been obtained. For this reason, preposterior analysis and sequential analysis are richer and more interesting than prior analysis and posterior analysis. The former techniques are also more realistic, since executives in business and government continually must make choices in a situation where an important option is to postpone a final decision and buy further information.

## 16.2 Preposterior Analysis of the Marketing Manager's Problem

To show the nature of preposterior analysis, let's return to the case of the marketing manager who must decide whether or not to put a new label on his firm's product. At the beginning of the previous chapter we described how the manager could make this decision on the basis of his prior probabilities alone. However, this prior analysis assumed that

---

*Some instructors may want to skip this chapter entirely. The book is written so that this can easily be done.

the marketing manager could not—or did not want to—carry out a survey to obtain empirical evidence concerning the likelihood that the proposed new label was superior (or not superior) to the old label. Later, we described how the manager could make his decision after he had obtained evidence of this sort—but we did not indicate how he could decide whether it was worthwhile to carry out a survey to provide such evidence.

In contrast to the previous chapter, we now must recognize that the marketing manager really has the choice of whether or not to carry out a survey before making a final decision to use the new label or keep the old one. If he decides *not to carry out a survey,* the problem boils down to the one handled by the prior analysis in the previous chapter. Figure 16.1 shows the decision tree for this analysis. You will recall from the previous chapter that the monetary gain is $800,000 if the new label is introduced and is successful, and is −$500,000 if it is introduced and is unsuccessful. The prior probability is 0.50 that the new label is superior and 0.50 that it is not superior. As is evident from the decision tree, the optimal act, under these circumstances, is to introduce the new label since the expected profit is $150,000 if the manager does so and zero if he does not.

But what if the manager decides to *carry out a survey* in order to obtain additional information concerning whether or not the new label is superior? In accord with the previous chapter, suppose that he can carry out a survey which will result in three possible outcomes: (1) favorable to the new label; (2) ambiguous; and (3) unfavorable to the new label. Suppose that the probability that each outcome will occur, given that the new label is superior, is as shown in the first horizontal row of Table 16.1. (For example, the probability that the survey results will be favorable to the new label, given that it is really superior, is 0.8.) The probability that each outcome will occur, given that the new label is not superior, is as shown in the second horizontal row in Table 16.1. (For example, the probability that the survey results will be unfavorable to the new label, given that it really is not superior, is 0.7.) In addition, suppose that the cost of carrying out this survey is $100,000.

Given these characteristics of the sample, it is clear that if the survey is carried out the manager incurs a cost of $100,000, and

**Figure 16.1**
Decision Tree for the Marketing Manager's Problem if He Decides Not to Carry Out a Survey

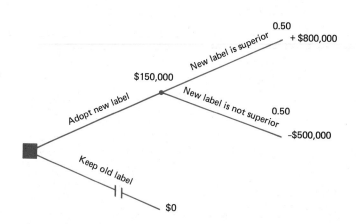

| State of nature | Sample outcome | | |
|---|---|---|---|
| | *Favorable to new label* | *Neither favorable nor unfavorable* | *Unfavorable to new label* |
| New label is superior | 0.8 | 0.1 | 0.1 |
| New label is not superior | 0.2 | 0.1 | 0.7 |

**Table 16.1**
Probability of Each
Survey Outcome,
Given the Actual State
of Nature

"chance" determines which of the three possible outcomes of the survey will result. Can the marketing manager estimate the probability of each of these three outcomes? Based on the information provided above, the answer is yes. *The probability that the sample results will be favorable to the new label equals*

$$P(b) = P(b|B)P(B) + P(b|N)P(N),$$

where $P(b|B)$ is the probability that the sample results are favorable, given that the new label is in fact superior, $P(b|N)$ is the probability that the sample results are favorable, given that the new label is not superior, $P(B)$ is the prior probability that the new label is superior, and $P(N)$ is the prior probability that it is not superior. Inserting numerical values into this equation,[1] we find that

$$P(b) = 0.8(0.5) + 0.2(0.5) = 0.5.$$

*The probability that the sample results are ambiguous*—that is, neither favorable nor unfavorable to the new label—is

$$P(a) = P(a|B)P(B) + P(a|N)P(N),$$

where $P(a|B)$ is the probability that the sample results are neither favorable nor unfavorable to the new label, given that the new label is in fact superior, and $P(a|N)$ is the probability that the sample results are neither favorable nor unfavorable to the new label, given that the new label is not superior. Inserting numerical values into this equation,[2] we find that

$$P(a) = 0.1(0.5) + 0.1(0.5) = 0.1.$$

Similarly, *the probability that the sample results are unfavorable to the new label* is

$$P(n) = P(n|B)P(B) + P(n|N)P(N),$$

---

[1] Recall from an earlier paragraph in this section that both $P(B)$ and $P(N)$ equal 0.5. Table 16.1 shows that $P(b|B) = 0.8$ and $P(b|N) = 0.2$. These are the required numerical values.

[2] As pointed out earlier in this section, both $P(B)$ and $P(N)$ equal 0.5. Table 16.1 shows that $P(a|B) = 0.1$ and $P(a|N) = 0.1$. These are the required numerical values.

where $P(n|B)$ is the probability that the sample results are unfavorable to the new label, given that the new label is in fact superior, and $P(n|N)$ is the probability that the sample results are unfavorable to the new label, given that the new label is in fact not superior. Inserting numerical values into this equation,[3] we find that

$$P(n) = 0.1(0.5) + 0.7(0.5) = 0.4.$$

Thus, if the marketing manager decides to carry out a survey, the cost will be $100,000, and the probability is 0.5 that the survey results will be favorable to the new label, 0.1 that they will be ambiguous (neither favorable nor unfavorable), and 0.4 that they will be unfavorable to a new label.

## 16.3 Completing the Decision Tree

Up to this point what we have said can be summarized by the decision tree in Figure 16.2. As shown there, the marketing manager has the

**Figure 16.2**
Portion of Decision
Tree for Marketing
Manager's Problem

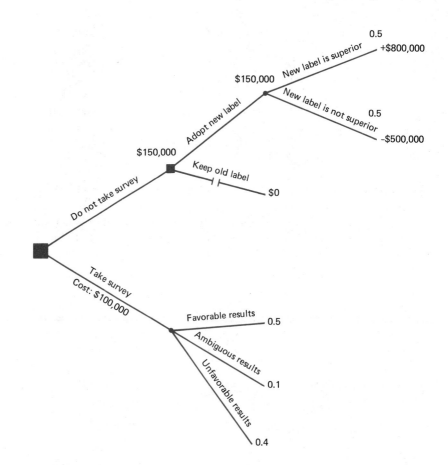

[3] As we have already pointed out, both $P(B)$ and $P(N)$ equal 0.5. Table 16.1 shows that $P(n|B) = 0.1$ and $P(n|N) = 0.7$. These are the required numerical values.

choice of carrying out a survey or not, this decision being shown at the left-hand end of the tree. *If he takes the upper branch* (that is, if he does not carry out the survey), he proceeds to a decision fork where he must choose whether or not to introduce the new label. The expected profit resulting from introducing the new label is $150,000 and the expected profit resulting from not doing so is zero; thus, the expected monetary value of taking the upper branch is $150,000. *If the manager takes the lower branch* (that is, if he does carry out the survey), he proceeds to a chance fork where "chance" is in control. From that point, the probability that survey results will be favorable to the new label is 0.5, the probability that ambiguous survey results will occur is 0.1, and the probability that unfavorable results will occur is 0.4.

To complete this tree we must trace out the consequences of each of the three possible survey results. First, let's take *the case where the survey results are favorable to the new label.* If this is the case, "chance" has taken the upper branch from the chance fork at the bottom of Figure 16.2. Now the next move is up to the marketing manager who must decide whether to adopt the new label or keep the old one. If he keeps the old label, the profit is zero. If he adopts the new label, the expected profit is

$$0.8(\$800,000) + 0.2(-\$500,000) = \$540,000$$

because the posterior probability that the new label is superior is 0.8, and the posterior probability that the new label is not superior is 0.2. (Recall the discussion in Section 15.7 of the previous chapter.) Thus, in this case the best action is to adopt the new label.

Next, let's take *the case where the survey results are neither favorable nor unfavorable to the new label.* If this is the case, "chance" has taken the middle branch from the chance fork at the bottom of Figure 16.2. Again, the next move is up to the marketing manager who must decide whether to adopt the new label or keep the old one. If he keeps the old label, the profit is zero. If he adopts the new label, the expected profit is

$$0.5(\$800,000) + 0.5(-\$500,000) = \$150,000,$$

because the posterior probability that the new label is superior is 0.5, and the posterior probability that the new label is not superior is 0.5.[4] Thus, in this case, the best action is to adopt the new label.

---

[4] The posterior probability that the new label is superior is

$$P(B|a) = \frac{P(a|B)P(B)}{P(a|B)P(B) + P(a|N)P(N)} = \frac{(0.1)(0.5)}{(0.1)(0.5) + (0.1)(0.5)} = 0.5.$$

The posterior probability that the new label is not superior is

$$P(N|a) = \frac{P(a|N)P(N)}{P(a|N)P(N) + P(a|B)P(B)} = \frac{(0.1)(0.5)}{(0.1)(0.5) + (0.1)(0.5)} = 0.5.$$

The values of $P(a|B)$ and $P(a|N)$ come from Table 16.1.

Finally, let's take *the case where the survey results are unfavorable to the new label.* If this is the case, "chance" has taken the bottom branch from the chance fork at the bottom of Figure 16.2, and once again the next move is up to the marketing manager, who must decide whether to adopt the new label or keep the old one. If he keeps the old label the profit is zero, whereas if he adopts the new label the expected profit is

$$(1/8)(\$800,000) + (7/8)(-\$500,000) = -\$337,500,$$

because the posterior probability that the new label is superior is $1/8$, and the posterior probability that the new label is not superior is $7/8$.[5] Thus, in this case, the best action is not to adopt the new label.

Having analyzed the consequences of each of the three possible survey results, we can complete the decision tree which is shown in its entirety in Figure 16.3. Working backward from the right-hand side of the tree, we can use *backward induction* (described in Chapter 15) to solve the problem. As noted above, the expected monetary value of being situated at decision fork (a) is $540,000, the expected monetary value of being at decision fork (b) is $150,000, and the expected monetary value of being at decision fork (c) is zero. Thus, since, if the marketing manager decides to carry out the survey, the probabilities of arriving at decision forks (a), (b), and (c) are 0.5, 0.1, and 0.4, respectively, it follows that the expected profit, if he carries out the survey, is

$$0.5(\$540,000) + 0.1(\$150,000) + 0.4(0) = \$285,000.$$

**Expected Value of Sample Information**

**Expected Net Gain of Sample Information**

Comparing this amount with the $150,000 expected profit that results when no survey is taken, it is clear that the marketing manager would be warranted in spending up to $285,000 − $150,000, or $135,000, for the survey. This is the *expected value of the sample information.* Note, however, that this is a gross figure since it takes no account of the amount that the firm must pay for the survey—$100,000 in this case. The *expected net gain of sample information* is the difference between the expected value of the sample information and the cost of the survey. The expected net gain shows the amount by which the survey increases or reduces the firm's expected profit when account is taken of the cost of the survey. Since the expected net gain of sample information is $135,000 − $100,000, or $35,000, it follows that the marketing manager should carry out the survey. Why? Because, even after we deduct the cost of the survey, the expected profit is greater if the survey is taken than if it is not.

[5] The posterior probability that the new label is superior is

$$P(B|n) = \frac{P(n|B)P(B)}{P(n|B)P(B) + P(n|N)P(N)} = \frac{(0.1)(0.5)}{(0.1)(0.5) + (0.7)(0.5)} = 1/8.$$

The posterior probability that the new label is not superior is

$$P(N|n) = \frac{P(n|N)P(N)}{P(n|N)P(N) + P(n|B)P(B)} = \frac{(0.7)(0.5)}{(0.7)(0.5) + (0.1)(0.5)} = 7/8.$$

The values of $P(n|B)$ and $P(n|N)$ come from Table 16.1.

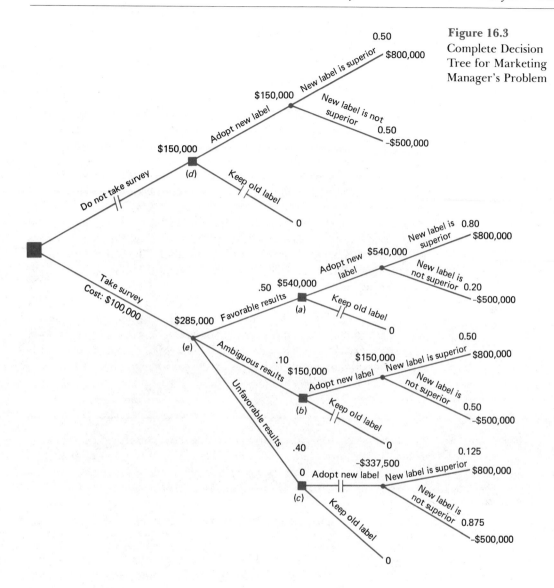

**Figure 16.3**
Complete Decision Tree for Marketing Manager's Problem

## 16.4 Extensive Form Analysis and Normal Form Analysis

In the previous two sections we have carried out a type of preposterior analysis known as *extensive form analysis*. More specifically, in terms of the decision tree in Figure 16.3, the upper part of the tree represents a prior analysis which results in the prior expected profit shown at fork (d), and the lower part of the tree represents an extensive form analysis which results in the preposterior expected profit shown at fork (e). To carry out an extensive form analysis, one begins by constructing a decision tree, by characterizing the states of nature at the right-hand ends of the tree, and by assigning payoffs to each action assuming that each state of nature is true. To determine whether the survey should be carried out, the preposterior expected payoff should be reduced by the cost of the survey, and the result should be compared with the prior expected payoff if no survey were carried out. If the former exceeds the latter, the survey is worthwhile.

Another type of preposterior analysis is *normal form analysis.* In this type of analysis we break down the problem in quite a different way, the first step being to list all the possible *strategies* that might be employed.

**Strategy**

STRATEGY: *A strategy is a decision rule that specifies which action the decision maker will take if each survey result occurs.*

For example, one strategy is the following: Adopt the new label if the survey results are favorable to it; keep the old label if the survey results are ambiguous or unfavorable to the new label. There are eight possible strategies that the marketing manager could choose, all of which are listed in Table 16.2. *In normal form analysis, one compares the value of all these strategies to find the one that is optimal. The result, of course, is the same as that arrived at by extensive form analysis, but, for reasons explained in a later section, it is sometimes advantageous to use a normal form analysis rather than an extensive form analysis.* (On the other hand, it is sometimes advantageous to use an extensive form analysis; the advantages of each are given in Section 16.5.)

**Basic Idea #31**    ***Preposterior analysis is concerned with decision making where one of the options open to the decision maker is to obtain more information. Preposterior analysis is of two types, extensive form analysis and normal form analysis. Both result in precisely the same answer.***

To illustrate how a normal form analysis is carried out, let's use this type of analysis to solve the marketing manager's problem. To add some variety to the illustration, let's minimize expected opportunity loss rather than maximize expected profit. You will recall from Section 15.11 of the previous chapter that a decision maker, if he or she minimizes expected opportunity loss, will maximize expected profit, so that the result should be the same. For each strategy, we must calculate the expected opportunity loss, given that a particular state of nature prevails. Then we can use these conditional expected losses to compute the expected opportunity loss from the strategy.

**Strategy 1**

Let's begin with **strategy 1** in Table 16.2. *Given that the new label in fact is superior,* the expected opportunity loss with this strategy is

$$0.8(0) + 0.1(0) + 0.1(0) = 0.$$

Why? Because, if the new label really is superior, the probability that the sample results will be favorable to the new label is 0.8, the probabil-

Table 16.2
Possible Strategies in Marketing Manager's Problem

| Survey outcome | Strategy* | | | | | | | |
|---|---|---|---|---|---|---|---|---|
| | 1 | 2 | 3 | 4 | 5 | 6 | 7 | 8 |
| Favorable to new label | Adopt new | Keep old | Adopt new | Adopt new | Adopt new | Keep old | Adopt new | Keep old |
| Ambiguous | Adopt new | Adopt new | Keep old | Adopt new | Keep old | Adopt new | Keep old | Keep old |
| Unfavorable to new label | Adopt new | Adopt new | Adopt new | Keep old | Adopt new | Keep old | Keep old | Keep old |

* Each strategy specifies the course of action to be taken if each survey outcome occurs. Thus, strategy 1 specifies that the new label be adopted regardless of the survey outcome. Strategy 4 specifies that the new label be adopted if the survey outcome is favorable to the new label or ambiguous, but that the old label be kept if the survey outcome is unfavorable to the new label. Strategy 8 specifies that the old label be kept regardless of the survey outcome.

**Table 16.3**
Opportunity Loss for
Each Action and Each
State of Nature

| State of nature | Action | |
| --- | --- | --- |
| | Adopt new label (dollars) | Keep old label (dollars) |
| New label is superior | 0 | 800,000 |
| New label is not superior | 500,000 | 0 |

*Source:* Table 15.3

ity that they will be ambiguous is 0.1, and the probability that they will be unfavorable is 0.1. (See Table 16.1.) Regardless of which of these sample outcomes occurs, strategy 1 dictates that the new label should be accepted. If the new label is in fact superior the opportunity loss associated with accepting it is zero (as shown in Table 16.3). Thus, the expected opportunity loss in this case is zero.

*If the new label is in fact not superior,* the expected opportunity loss with this strategy is

$$0.2(\$500,000) + 0.1(\$500,000) + 0.7(\$500,000) = \$500,000.$$

The probability of each sample outcome is as given in Table 16.1. Regardless of which of these sample outcomes occurs, strategy 1 dictates that the new label should be accepted. If the new label really is not superior, the opportunity loss associated with accepting it is $500,000 (as shown in Table 16.3). Thus, the expected opportunity loss in this case is $500,000.

To find the expected opportunity loss from this strategy, we must multiply each of these conditional expected opportunity losses—zero and $500,000—by their probability of occurrence and sum the results. Since the prior probability that the new label is superior is 0.5 and the prior probability that it is not superior is 0.5, the expected opportunity loss for this strategy is

$$0.5(0) + 0.5(\$500,000) = \$250,000.$$

*Strategy 7*

Next, let's consider **strategy 7.** What is the expected opportunity loss with this strategy? *If the new label is in fact superior,* the expected opportunity loss with this strategy is

$$0.8(0) + 0.1(\$800,000) + 0.1(\$800,000) = \$160,000.$$

To see why, note the following. If the sample results are favorable to the new label (a probability of 0.8), this strategy calls for acceptance of the new label, which results in an opportunity loss of zero if the new label is really superior. If the sample results are ambiguous (a probability of 0.1), this strategy calls for keeping the old label, which results in an opportunity loss of $800,000 if the new label is really superior. And if the sample results are unfavorable to the new label (a probability of 0.1), this strategy calls for keeping the old label, which results in an opportunity loss of $800,000 if the new label is really superior.

*If the new label really is not superior,* the expected opportunity loss with this strategy is

0.2($500,000) + 0.1(0) + 0.7(0) = $100,000.

To see why, note the following. If the sample results are favorable to the new label (a probability of 0.2), this strategy calls for acceptance of the new label, which results in an opportunity loss of $500,000 if the new label is in fact not superior. If the sample results are ambiguous (a probability of 0.1), this strategy calls for keeping the old label, which results in an opportunity loss of zero if the new label is in fact not superior. And if the sample results are unfavorable to the new label (a probability of 0.7), this strategy calls for keeping the old label, which results in an opportunity loss of zero if the new label is in fact not superior.

To find the expected opportunity loss with this strategy, we must multiply each of these conditional expected opportunity losses—$160,000 and $100,000 —by their probability of occurrence and sum the results. Since the prior probability that the new label is superior is 0.5, and the prior probability that it is not superior is 0.5, the expected opportunity loss with this strategy is

$$0.5(\$160,000) + 0.5(\$100,000) = \$130,000. \qquad (16.1)$$

Since this is considerably less than the expected opportunity loss with strategy 1, strategy 7 seems preferable to strategy 1.

Finally, let's consider **strategy 4.** What is the expected opportunity loss with this strategy? *If the new label is in fact superior,* the expected opportunity loss with this strategy is    ***Strategy 4***

0.8(0) + 0.1(0) + 0.1($800,000) = $80,000.

This is the same as for strategy 7, except that zero is substituted for $800,000 in the second term on the left-hand side because this strategy, unlike strategy 7, calls for accepting the new label when the sample results are ambiguous, and the opportunity loss associated with doing so is zero, not $800,000. *If the new label really is not superior,* the expected opportunity loss is

0.2($500,000) + 0.1($500,000) + 0.7(0) = $150,000.

This is the same as for strategy 7, except that $500,000 is substituted for zero in the second term on the left-hand side because this strategy, unlike strategy 7, calls for accepting the new label when the sample results are ambiguous, and the opportunity loss associated with doing so is $500,000, not zero.

To find the expected opportunity loss with strategy 4, we must multiply each of these conditional expected opportunity losses—$80,000 and $150,000—by their probability of occurrence and sum the results. We obtain

**Table 16.4**
Expected Opportunity
Loss for Each Possible
Strategy Listed in
Table 16.2

| Strategy | Expected opportunity loss (dollars) |
|---|---|
| 1 | 250,000 |
| 2 | 520,000 |
| 3 | 265,000 |
| 4 | 115,000 |
| 5 | 535,000 |
| 6 | 385,000 |
| 7 | 130,000 |
| 8 | 400,000 |

$$0.5(\$80,000) + 0.5(\$150,000) = \$115,000, \qquad (16.2)$$

which is lower than the expected opportunity loss with strategy 7. Table 16.4 lists the expected opportunity loss associated with each of the eight possible strategies that the marketing manager could adopt in handling this problem. As you can see, strategy 4 has the lowest expected opportunity loss and thus is the *optimal strategy*.

## 16.5 Comparison of Results, and Advantages of Each Type of Preposterior Analysis

To carry out a normal form analysis, one begins by listing all possible strategies (as in Table 16.2), each strategy being a specification of the action to be taken if each sample outcome arises. The next step is to calculate the expected opportunity loss associated with each strategy. (To carry out this step, it is convenient to compute first the conditional expected opportunity loss that corresponds to each state of nature, and then to weight each of these conditional figures by the prior probability that the relevant state of nature will occur.) Finally, one selects the strategy which has the minimum expected opportunity loss. This strategy is called **Bayes' strategy.**

Let's compare the results of the normal form analysis carried out in the previous section with the results of the extensive form analysis of the same problem in Sections 16.2 and 16.3. The first thing to note is that both types of analysis lead to the choice of the same strategy. As you can see from the bottom part of the decision tree in Figure 16.3, the extensive form analysis resulted in the decision to accept the new label if the survey results are favorable to the new label or if they are ambiguous, and to keep the old label if they are unfavorable to the new label. This is precisely the same as strategy 4 which, as we saw in the previous section, is the optimal strategy derived through the normal form analysis.

Another important fact is that if we had carried out the extensive form analysis in terms of opportunity losses rather than profits, the expected opportunity loss of the optimal strategy would have been the

# Decision Making at Maxwell House Concerning the Quick-Strip Can*

In October 1962, Folger's began test-marketing its coffee in Stockton, California, in a keyless container. Whereas earlier coffee cans had to be opened with a key, the new can did not. In November 1963, Maxwell House, the nation's largest producer of coffee, surveyed 125 Stockton users of Folger's coffee in the new can, and found that 86 percent of the consumers who had tried it preferred the new can to the old can. Other pieces of marketing intelligence also indicated that a keyless can would be popular among customers.

Maxwell House, together with the American Can Company, had helped to develop its own keyless container which operated on the tear-strip opening principle. Based on the reception of Folger's new can, Maxwell House executives were optimistic about consumer reaction to their own new can. One important decision that had to be made before introducing this new can was whether or not to raise the per-pound price of coffee in the new can by 2 cents. Coffee in the quick-strip can was expected to cost an average of 0.7 cents per pound more than that in the old container. According to Joseph Newman, who studied this case, if Maxwell House raised its price by 2 cents per pound it might have been reasonable to expect (1) a .25 probability that its market share would decline by 1.5 percentage points; (2) a .25 probability that its market share would remain constant; (3) a .25 probability that its market share would increase by 1.0 percentage points; and (4) a .25 probability that its market share would increase by 2.5 percentage points. The change in Maxwell House's profits corresponding to each change in its market share is given in Table 1.

According to Newman, if Maxwell House did not raise its price, it might have been reasonable to expect (1) a 0.1 probability that its market share would decline by 0.6 percentage points; (2) a 0.2 probability that its market share would remain constant; (3) a 0.5 probability that

**Table 1**

Changes in Profit Corresponding to Selected Changes in Market Share

| Price per pound held constant | | Price per pound increased by 2 cents | |
|---|---|---|---|
| Change in market share (percentage points) | Change in profit (thousands of dollars) | Change in market share (percentage points) | Change in profit (thousands of dollars) |
| +2.8 | 4,104 | 2.5 | 11,939 |
| +1.0 | − 591 | 1.0 | 6,489 |
| 0 | − 840 | 0 | 2,856 |
| −0.6 | −1,218 | −1.5 | −1,050 |

* This case is based on a section from Joseph Newman's *Management Applications of Decision Theory* (New York: Harper and Row, 1971).

its market share would increase by 1.0 percentage points; and (4) a 0.2 probability that its market share would increase by 2.8 percentage points. The change in Maxwell House's profits corresponding to each of these market-share changes is provided in Table 1.

Based on the above data, the decision tree representing Maxwell House's pricing problem was as follows:

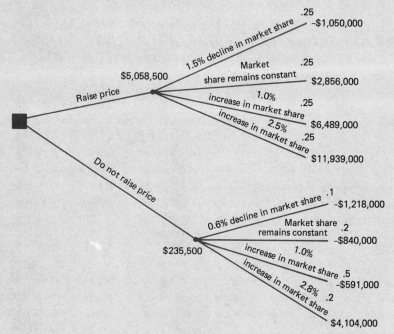

If Maxwell House wanted to maximize expected monetary value, it should have increased the price per pound of coffee in its new can by 2 cents.

In 1963, Maxwell House had to decide how quickly to introduce its new quick-strip can. One possible action at that time was to conduct a three-month sales test of the quick-strip can in Muncie and Stockton, rather than go ahead immediately with adoption of the new can (as visualized in the analysis above). A three-month sales test of this sort would have cost about $6,000. The results of such a test might indicate that the new can was (1) very successful; (2) successful; (3) no better than the old; or (4) worse than the old. According to Joseph Newman's study of this case, the probability that each market-test result would occur, given the true effect of the quick-strip can on Maxwell House's market share, is as follows:

| True effect of new can on Maxwell House's market share (percentage points) | Results of three-month sales test (probabilities) | | | |
|---|---|---|---|---|
| | Very successful | Successful | No better than old | Worse than old |
| +2.5 | 0.5 | 0.3 | 0.2 | 0 |
| +1.0 | 0.2 | 0.5 | 0.2 | 0.1 |
| 0 | 0.1 | 0.2 | 0.5 | 0.2 |
| −1.5 | 0 | 0.2 | 0.3 | 0.5 |

Since it was assumed in this analysis that the price per pound would be increased by 2 cents, the change in Maxwell House's profits corresponding to each of these changes in market share is as follows:

| Change in market share (percentage points) | Change in profit (thousands of dollars) |
|:---:|:---:|
| +2.5 | 11,939 |
| +1.0 | 6,489 |
| 0 | 2,856 |
| −1.5 | −1,050 |

According to Newman, if Maxwell House adopted the new can without the three-month sales test, it might have been reasonable to expect a 0.25 probability that its market share would increase by 2.5 percentage points, a 0.25 probability that its market share would increase by 1.0 percentage points, a 0.25 probability that its market share would remain constant, and a 0.25 probability that its market share would decrease by 1.5 percentage points.

Based on the above information, the decision tree to determine whether or not Maxwell House should have carried out the three-month sales test is as shown on the following page.

Thus, it is clear that, if Maxwell House wanted to maximize expected profit, it should not carry out the sales test, because the expected profit is only about $3,000 higher with the sales test than without it. This is less than the cost of the sales test, which was $6,000.

### PROBING DEEPER

1. According to Table 1, Maxwell House's profits would fall by $840,000 if no change occurred in either its price or its market share. How can this be?

2. In this analysis, what is the assumed shape of Maxwell House's utility function? Explain.

3. According to the discussion above, it might have been reasonable to expect that, if Maxwell House raised its price by 2 cents per pound, there would have been a 0.25 probability that its market share would decline by 1.5 percentage points. How can one determine such a prior probability? How can you tell whether it is in fact "reasonable"?

4. Suppose that Maxwell House was certain that its market share would increase by 2.8 percentage points if it did not raise its price. If all other aspects of the situation were unchanged, should Maxwell House have increased the price by 2 cents? Explain.

5. If Maxwell House had carried out the sales test, would its decision concerning whether or not to adopt the new can have depended on the sales test's outcome?

6. If Maxwell House did not want to maximize expected monetary value, how would you analyze this problem?

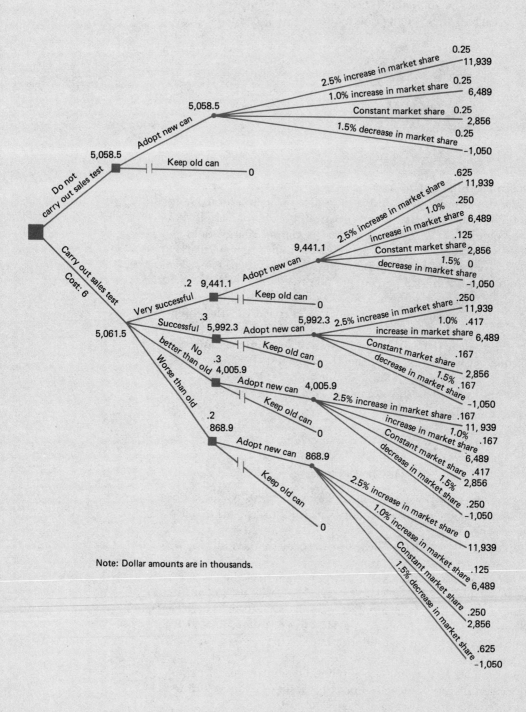

Note: Dollar amounts are in thousands.

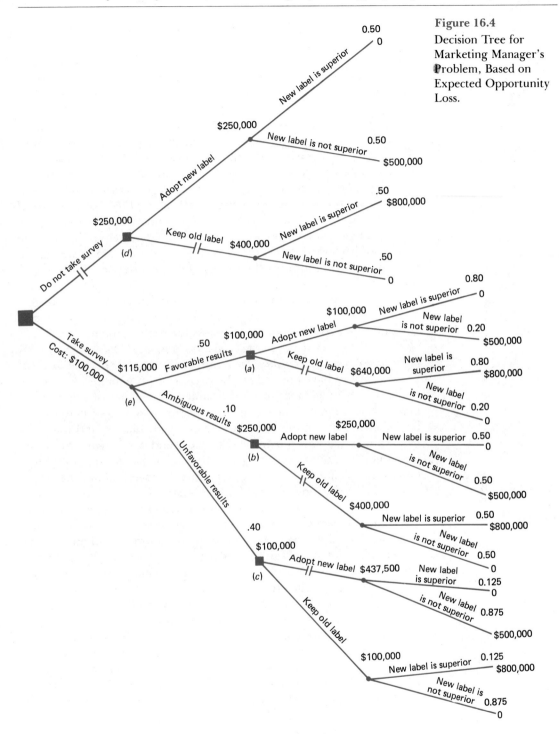

**Figure 16.4**
Decision Tree for Marketing Manager's Problem, Based on Expected Opportunity Loss.

figure at fork (e) in Figure 16.3. To prove that this is the case, Figure 16.4 shows the decision tree for this problem if opportunity losses rather than profits are used. As you can see, the figure at fork (e) is $115,000, the expected opportunity loss corresponding to strategy 4. To decide whether it is worthwhile to carry out the survey, this figure must be compared with the expected opportunity loss if no survey were taken—which, as Figure 16.4 shows, is $250,000. If we subtract the former figure from the latter, we get the expected value of sample information: $250,000 − $115,000 = $135,000. Reassuringly, this is precisely the amount that we obtained in Section 16.3 when we derived the expected value of sample information based on profits rather than on opportunity losses.

Since the extensive form analysis and the normal form analysis result in precisely the same answer, it may not seem a matter of importance as to which one is used. But this overlooks the fact that each type of analysis has certain advantages; for this reason it is worthwhile to learn both. *Perhaps the most important advantage of extensive form analysis is that the computations are much simpler and quicker than in normal form analysis.* In extensive form analysis, it is not necessary to compute the expected opportunity loss for each possible strategy. Through the process of backward induction we recognize that certain strategies are nonoptimal, and thus only the expected opportunity loss for the optimal strategy has to be calculated. This is an important advantage because real problems often involve a very large number of possible strategies, and it would be expensive and time-consuming to have to compute the expected opportunity loss for each of them.

On the other hand, *normal form analysis has the advantage of making it easier to see how sensitive the results are to the prior probabilities that are used.* As we have pointed out before, prior probabilities frequently are based on judgement or on the subjective feelings of the decision maker. It is understandable that both decision makers and analysts may be somewhat uncomfortable about these probabilities, and that they may want to see how much the results depend upon them. (Since the probabilities of survey results, given the state of nature, are often based on relative frequencies, there is usually less concern about them.)

*Strategy 4 Versus Strategy 7*

To illustrate how normal form analysis can be used to determine the sensitivity of the results to the prior probabilities, let's return to the marketing manager's problem in the previous section. A close inspection of Table 16.4 shows that two strategies—strategies 4 and 7—are the leading contenders, given that the prior probability that the new label is superior is 0.5 (and that the prior probability that it is not superior is 0.5). Most of the other strategies in Table 16.2 do not seem very sensible. For example, strategy 5 would dictate that the decision maker keep the old label if the survey results are favorable to the new label. (One hardly needs a course in statistics to reject such a strategy!) But the choice between strategies 4 and 7 is by no means obvious. As we know from Table 16.4, strategy 4 beats out strategy 7 if the prior probability that the new label is superior is 0.5. However, what if this prior probability is somewhat higher or lower than 0.5? Will strategy 4 continue to beat out strategy 7? Based on the results of a normal form analysis, this question is easy to answer.

The first thing to note is that the expected opportunity loss with strategy 4 is

$$P(\$80,000) + (1 - P)(\$150,000), \qquad (16.3)$$

where $P$ is the prior probability that the new label is superior. This follows directly from equation (16.2), where $P$ was set equal to 0.5. The next thing to note is that the expected opportunity loss with strategy 7 is

$$P(\$160,000) + (1 - P)(\$100,000). \qquad (16.4)$$

This follows directly from equation (16.1), where $P$ was set equal to 0.5. Clearly, the expected opportunity loss with strategy 7 goes up as $P$ increases, while the expected opportunity loss with strategy 4 goes down as $P$ increases.[6] To see whether there is some value of $P$ which makes the expected opportunity losses of these two strategies equal, we can set the expressions in equation (16.3) and equation (16.4) equal to one another:

$$P(\$80,000) + (1 - P)(\$150,000) = P(\$160,000) + (1 - P)(\$100,000).$$

Then we can solve for $P$, the result being

$$150,000 - 70,000P = 100,000 + 60,000P$$

$$130,000P = 50,000$$

$$P = 5/13.$$

Thus, if P is greater than 5/13, the expected opportunity loss with strategy 4 is less than that with strategy 7, while if $P$ is less than 5/13, the reverse is true.

What does this result tell the marketing manager? It tells him that strategy 4, which he knows to be better than strategy 7 when $P = 0.5$, will continue to be better than strategy 7 when $P$ is greater than 0.5. Also, it tells him that strategy 4 will beat out strategy 7 when $P$ is between 5/13 and 0.5, but that strategy 4 will be poorer than strategy 7 if $P$ is less than 5/13. On the basis of this information, the marketing manager knows precisely how sensitive his choice of a strategy is to variations in $P$.

Finally, it should be recognized that some strategies are better than others, regardless of the value of $P$. One strategy is said to **dominate**    *Domination* *another if its conditional expected opportunity loss is less than that of the*

---

[6]It follows from equation (16.4) that the expected opportunity loss with strategy 7 equals

$$\$100,000 + \$60,000(P).$$

Thus, the expected opportunity loss increases with increases in $P$. Similarly, it follows from equation (16.3) that the expected opportunity loss with strategy 4 equals

$$\$150,000 - \$70,000(P).$$

Thus, the expected opportunity loss decreases with increases in $P$.

*other strategy for some possible states of nature, and is no greater than that of the other strategy for the other possible states of nature.* The following example shows how this concept of domination can be applied to the marketing manager's problem.

---

EXAMPLE 16.1 Using the results in this section, show that strategies 4 and 7 dominate strategy 5. What implications does this have for the marketing manager?

SOLUTION: The expected opportunity loss with strategy 5 is[7].

$$P(\$720{,}000) + (1 - P)(\$350{,}000) = \$350{,}000 + \$370{,}000(P).$$

It follows from equation (16.3) that the expected opportunity loss with strategy 4 equals

$$\$150{,}000 - \$70{,}000(P).$$

From equation (16.4) it follows that the expected opportunity loss with strategy 7 equals

$$\$100{,}000 + \$60{,}000(P).$$

Thus, if $P$ is the same for all strategies, the expected loss with strategy 5 is greater than with strategy 4 or strategy 7, *regardless of the value of P.*[8] This means that strategy 5 must be dominated by both strategy 4 and strategy 7. To the marketing manager, this means that strategy 5 is a poor strategy, regardless of the value of $P$. In other words, there is no value of $P$ which would make this strategy a good one.

---

[7] To see why this is the expected opportunity loss with strategy 5, note that if the new label is in fact superior, the expected loss with this strategy is

$$0.8(\$800{,}000) + 0.1(\$800{,}000) + 0.1(0) = \$720{,}000.$$

If the new label is in fact not superior, the expected opportunity loss with this strategy is

$$0.2(0) + 0.1(0) + 0.7(\$500{,}000) = \$350{,}000.$$

Thus, if $P$ is the probability that the new label is in fact superior, the expected opportunity loss with strategy 5 is

$$P(\$720{,}000) + (1 - P)(\$350{,}000),$$

as indicated above.

[8] To prove that this is the case, note that the difference between the expected opportunity loss with strategy 5 and that with strategy 4 equals

$$[\$350{,}000 + \$370{,}000(P)] - [\$150{,}000 - \$70{,}000(P)]$$

$$= \$200{,}000 + \$440{,}000(P).$$

# 16.6 The Use of Prior Probabilities: A Final Warning

Finally, a warning is in order. While preposterior analysis (and prior and posterior analysis) is a powerful procedure for helping to solve problems faced by business firms and government agencies, it is not magic. In particular, the statistician should recognize that if analyses of this sort are based on distorted unreliable values of the prior probabilities, the results are likely to be incorrect and misleading. For example, in the case of the marketing manager, if his prior probability of the superiority of the new label is merely an optimistic figment of his imagination, and if the probabilities of each survey outcome (given the actual state of nature) reflect major biases on the part of the survey designers, the results may be worse than useless. Unless these and other basic data are meaningful, the results are likely to be misleading, no matter how scientific the analysis may appear. Although preposterior analysis (like prior and posterior analysis) is useful, it, like all statistical techniques, must be applied with proper caution.

## EXERCISES

**16.1** A large Chicago bank must decide whether or not to merge with a small Chicago bank. If the government allows the merger, the bank feels that it will gain $10 million by merging with the small bank. If the government does not allow the merger, the bank feels that it will lose $1 million in legal fees and other costs. The bank feels that there is a 0.1 chance that the government will allow the merger (and a 0.9 chance that it will not allow it). The bank can hire an antitrust expert for $50,000 to provide advice as to what the government's response will be. The probability that the expert will be correct is as follows:

| | Expert's opinion | |
| | Government | Government |
| State of nature | will approve | will not approve |
| --- | --- | --- |
| Government allows merger | 0.9 | 0.1 |
| Government does not allow merger | 0.1 | 0.9 |

(a) Construct the decision tree representing the bank's decision if the expert is not hired.

(b) Construct the decision tree when account is taken of the possibility that the bank can hire the expert.

---

Since $P \geqslant 0$, this difference must be positive regardless of the value of $P$. Similarly, the difference between the expected opportunity loss with strategy 5 and that with strategy 7 equals

$$[\$350,000 + \$370,000(P)] - [\$100,000 + \$60,000(P)]$$

$$= \$250,000 + \$310,000(P).$$

Since $P \geqslant 0$ this difference, too, must be positive, regardless of the value of $P$.

(c) Should the bank hire the expert?
(d) If the expert's opinion is that the government will allow the merger, should the bank go ahead with the merger?
(e) If the expert's opinion is that the government will not allow it, should the bank go ahead with the merger?

**16.2** A petroleum exploration firm must decide whether or not to drill a well in a particular location. If it strikes oil, the firm will gain $2 million. If it does not strike oil, it will lose $0.2 million. The firm feels that the probability of its striking oil is 0.1. The firm can hire a geologist to study the relevant location and give his advice concerning the presence or absence of oil there. The geologist's fee is $10,000. Given each state of nature, the probability that the geologist will be correct is the following:

| | Geologist's conclusion | |
| State of nature | Firm will strike oil | Firm will not strike oil |
| --- | --- | --- |
| Firm will strike oil | 0.6 | 0.4 |
| Firm will not strike oil | 0.4 | 0.6 |

(a) Construct the decision tree representing the firm's decision if it does not hire the geologist.
(b) Construct the decision tree when account is taken of the possibility that the firm can hire the geologist.
(c) Should the firm hire the geologist?
(d) If the geologist concludes that the firm will strike oil, should the firm drill the well?
(e) If the geologist concludes that the firm will not strike oil, should the firm drill the well?

**16.3** A firm must decide whether or not to expand its plant. If its current advertising campaign is successful, it will gain $5 million from such an expansion. If its current advertising campaign is not successful, it will lose $2 million from such an expansion. The firm believes that the probability that the current advertising campaign will be successful is 0.5. The firm can carry out a survey for $100,000 to determine whether its current advertising campaign will be successful. Given each state of nature, the probability of each survey outcome is as follows:

| | Survey outcome | |
| State of nature | Campaign will be successful | Campaign will not be successful |
| --- | --- | --- |
| Campaign will be successful | 0.8 | 0.2 |
| Campaign will not be successful | 0.2 | 0.8 |

(a) Construct a table showing the opportunity loss for each action and each state of nature.
(b) List all of the possible strategies. Are most of them sensible?
(c) Calculate the expected opportunity loss with each strategy.
(d) Let $P$ be the prior probability that the firm's advertising campaign will be successful. Obtain an expression (involving $P$) for the expected opportunity loss with each strategy.
(e) Do some strategies dominate others? Do those that are sensible dominate those that are not sensible?

**16.4** A publisher is considering a manuscript of a novel for publication. If it publishes the novel, the publisher will gain $2 million if the novel can be sold to some motion picture firm as the basis for a movie. If it cannot be sold to a motion picture firm, the publisher will lose $100,000 if it publishes the novel. The publisher feels that there is a 0.05 probability that the novel can be sold to a motion picture firm. To get a better idea of this probability, the publisher can pay a Hollywood agent $5,000 to read the manuscript and indicate whether or not, in his judgment, the novel, if published, can be sold to a motion picture firm. Given each state of nature, the probabilities that the agent will be correct or incorrect are as follows:

| State of nature | Agent's feeling | |
| --- | --- | --- |
| | *Novel can be sold* | *Novel cannot be sold* |
| Novel can be sold | 0.75 | 0.25 |
| Novel cannot be sold | 0.15 | 0.85 |

(a) Construct a table indicating the opportunity loss for each action and each state of nature.
(b) List all of the possible strategies.
(c) Calculate the expected opportunity loss with each strategy.
(d) In this case, what is the Bayes' strategy?

# 16.7 Sequential Sampling

In previous sections of this chapter we have been concerned with the choice of whether or not to gather information before making a decision. Now we turn to sequential analysis, which is closely related to the sort of analysis presented in the previous sections. *In sequential analysis, it is recognized that if the decision maker chooses to gather information before making a decision, he or she has the option, after this information has been gathered, of obtaining still more information before making the decision.* For example, the marketing manager, once he obtains the results of the survey, has the option of carrying out another survey of perhaps a different type before deciding whether to adopt the new label or stick with the old one.

In this section, we describe the use of sequential techniques to test hypotheses. Abraham Wald, a well-known statistician at Columbia University, did much of the pioneering work in this field during World War II. He showed how, given certain fixed probabilities of Type I and Type II errors, one can establish a sequential procedure for testing a particular hypothesis. In contrast to the test procedures in Chapters 8, 9, and 10, *no fixed sample size is specified* in sequential procedures. Instead, one computes a test statistic after each observation is chosen, and when this statistic falls beyond certain predetermined limits, the null hypothesis is rejected or not. Thus, it is possible that the statistician might either reject or not reject the null hypothesis after only a few observations have been chosen; or it might take many observations,

**Table 16.5
Sequential Sampling
Plan[a]**

| n | $A_n$ | $R_n$ | n | $A_n$ | $R_n$ | n | $A_n$ | $R_n$ | n | $A_n$ | $R_n$ |
|---|---|---|---|---|---|---|---|---|---|---|---|
|   | * |   | 16 | * | 2 | 31 | 0 | 3 | 46 | 0 | 3 |
| 2 | * | 2 | 17 | * | 2 | 32 | 0 | 3 | 47 | 0 | 3 |
| 3 | * | 2 | 18 | * | 2 | 33 | 0 | 3 | 48 | 0 | 3 |
| 4 | * | 2 | 19 | * | 2 | 34 | 0 | 3 | 49 | 0 | 4 |
| 5 | * | 2 | 20 | * | 3 | 35 | 0 | 3 | 50 | 0 | 4 |
| 6 | * | 2 | 21 | * | 3 | 36 | 0 | 3 | 51 | 0 | 4 |
| 7 | * | 2 | 22 | * | 3 | 37 | 0 | 3 | 52 | 0 | 4 |
| 8 | * | 2 | 23 | * | 3 | 38 | 0 | 3 | 53 | 0 | 4 |
| 9 | * | 2 | 24 | * | 3 | 39 | 0 | 3 | 54 | 0 | 4 |
| 10 | * | 2 | 25 | * | 3 | 40 | 0 | 3 | 55 | 0 | 4 |
| 11 | * | 2 | 26 | * | 3 | 41 | 0 | 3 | 56 | 0 | 4 |
| 12 | * | 2 | 27 | * | 2 | 42 | 0 | 3 | 57 | 0 | 4 |
| 13 | * | 2 | 28 | * | 3 | 43 | 0 | 3 | 58 | 0 | 4 |
| 14 | * | 2 | 29 | * | 3 | 44 | 0 | 3 | 59 | 0 | 4 |
| 15 | * | 2 | 30 | * | 3 | 45 | 0 | 3 | 60 | 0 | 4 |

[a]See footnotes 9 and 10.

*$A_n$ is less than zero.

*Source: A Duncan, Quality Control and Industrial Statistics* (Homewood, Ill.: Irwin, 1959), p. 154.

depending on what the observations turn out to be. *The advantage of sequential sampling is that on the average, it requires a smaller number of observations than a sample of fixed size in order to attain the same probability of Type I error (and of Type II error) as a fixed-size sample.*

To illustrate how such a sequential sampling scheme works, suppose that a firm wants to test whether the proportion of defective items in an incoming lot of goods is 0.01, the alternative hypothesis being that this proportion is 0.08. Suppose that the firm wants the probability of Type I error—that is, the probability that it will reject the hypothesis that the proportion defective is .01 when in fact this is true—to equal 0.05. Suppose also that it wants the probability of Type II error—that is, the probability that it will not reject the hypothesis that the proportion defective is .01 when in fact it is .08—to equal 0.10. Then Table 16.5 shows a sequential sampling scheme that will satisfy these requirements.[9] In this table, $n$ is the number of items that have been sampled. If $x_n$ is the number of these items that are defective, then the firm should not reject the null hypothesis if $x_n \leq A_n$, and it should reject the null hypothesis in favor of the alternative hypothesis if $x_n \geq R_n$. If $A_n < x_n < R_n$, the firm should sample another item from the shipment.

For example, suppose that the first 20 items sampled by the firm are not defective. Since Table 16.5 indicates that this result does not call for a decision one way or the other, the firm should sample a 21st item. If this item is defective, no decision is yet possible. If the next 10

[9] Table 16.5, although complete enough for present purposes, only provides information up to $n = 60$. For a more complete table, see A. Duncan, *op. cit.*, p. 154.

items are not defective one still cannot come to a decision since the cumulative number of defectives (1) is above $A_{31}$ (zero) and below $R_{31}$ (which is 3). Suppose that the 32nd item chosen is also defective. Still no decision can be made since the cumulative number of defectives (which is 2) is above $A_{32}$ (which is zero) and below $R_{32}$ (which is 3). Thus, another item must be sampled. Finally, if the 33rd item to be chosen is also defective, the firm should reject the hypothesis that only 1 percent of the items in the shipment are defective since the cumulative number of defectives (3) equals $R_{33}$ (also 3).

---

*In sequential sampling no fixed sample size is specified. A test statistic is computed after each observation is chosen, and when this test statistic falls outside predetermined limits, the null hypothesis is rejected or not.*

*Basic Idea # 32*

---

To repeat, the principal advantage of a sequential sampling scheme of this sort is that it allows the decision maker to test fewer items, on the average, before coming to a decision. This can be a very important advantage, particularly where testing involves the destruction of the item that is being tested. In some cases of this sort, each item tested is very expensive, and the decision maker can save large amounts of money by using sequential sampling rather than sampling based on a fixed sample size.[10]

The following example is a further illustration of how sequential sampling is carried out.

---

EXAMPLE 16.2 A manufacturer of rockets wants to test whether the probability of failure of its rockets equals .01, the alternative hypothesis being that it equals .08. The manufacturer uses the sampling scheme shown in Table 16.5. The first rocket tested is not defective, but the next two are. Should the firm continue sampling? If not, should it reject the null hypothesis?

SOLUTION: Since $n = 3$, Table 16.5 shows that $R_n = 2$. Thus, since two out of the three rockets are defective, $x_n = 2$, which means that $x_n = R_n$. Consequently, the firm should reject the null hypothesis that the probability of failure equals .01.

---

[10] Tables are available for various kinds of sequential sampling plans. For example, in the case of plans designed to test the proportion defective, one can choose the null hypothesis, the alternative hypothesis, the probability of Type I error, and the probability of Type II error. Given these choices, the tables show the sorts of information contained in Table 16.5. See *ibid.*, and A. Wald, *Sequential Analysis* (New York: Wiley, 1947).

EXERCISES

**16.5** A firm adopts the sequential sampling plan in Table 16.5. It tests one item after another, with the following results:

*N N N N N N N N N N N N N N N N N N D N N N N N N N N*
*D N N N N N N N N N N N D N N N N N N N N N N N N D N*

An *N* means that the item was not defective; a *D* means that it was defective. Each letter indicates the results of testing a particular item. Thus, the first letter is an *N* because the first item tested was not defective; the second letter is an *N* because the second item tested was not defective; and so on.

(a) Was it necessary to test as many items as the firm tested, according to the above sequence of *N*s and *D*s?

(b) At what point should the firm have reached a decision?

(c) What decision should the firm have reached?

**16.6** An automobile firm receives a shipment of components from a supplier. The automobile firm uses the sequential sampling plan in Table 16.5 to test whether the proportion of components that are defective exceeds the the level called for by the contract between the automobile firm and the component manufacturer.

(a) What level of the proportion defective was called for by this contract?

(b) After accepting the shipment, the automobile firm's inspector says that the minimum number of items had to be tested. How many items were tested?

# Chapter Review

1. In a *preposterior analysis* the decision maker has the option of gathering new information rather than making a final decision immediately. If it is decided not to obtain this information, the resulting expected profit can be calculated by the sort of prior analysis described in the previous chapter. If the decision maker chooses to obtain additional information, the probability distribution of the outcome of the survey or experiment can be computed. Also, using Bayes' theorem, the probability distribution of the states of nature, given the outcome of the survey or experiment, can be computed. The expected profit for each outcome can then be calculated, weighted by its probability of occurrence, and summed in order to obtain the expected profit. The difference between this value and the expected profit if it is decided not to obtain this additional information is the *expected value of the sample information*. The *expected net gain of sample information* is the expected value of the sample information less the cost of the survey or experiment. If the expected net gain of sample information is greater than zero, it is worthwhile obtaining the information.

2. There are two types of preposterior analysis: *extensive form analysis* and *normal form analysis*. To carry out an *extensive form analysis* one begins by constructing a decision tree, by characterizing the states

of nature at the right-hand ends of the tree, and by assigning payoffs to
each action, assuming that each state of nature is true. The next step is
to calculate the probabilities that the survey or experiment will result
in various outcomes, and to carry out a posterior analysis, assuming
each outcome occurs, in order to find the expected payoff corresponding to each outcome. Then, as described in the previous para-

The following is a set of true false questions.
Please, check the appropriate box.

a) The number of automobile accidents that you were
involved in as a driver last year is an example of
a discrete random variable.
2 pts        ☒ True        ☐ False

b) sample statistics are represented by letters from
the Greek alphabet.
2 pts        ☐ True        ☒ False

c) The standard error of the sample mean increases
as the sample size increases.
2 pts        ☐ True        ☒ False    $\bar{X}, n$   $\sigma_{\bar{x}} = \dfrac{\sigma}{\sqrt{n}}$

d) Rejection of a null hypothesis that is false
is a type II error.
2 pts        ☐ True        ☒ False

e) when the test statistic is t and the number
of degrees of freedom exceeds 30, the critical
value of t is very close to that of z.
2 pts        ☒ True        ☐ False

Europeans to the firm's products. If this substantial resistance occurs, the
firm will lose $2.5 million. One possibility is that the firm may carry out a
survey of European consumers to see what their response to the firm's
products might be. A polling organization says that for a fee of $75,000 it

can carry out a survey which will have the following probabilities of indicating various outcomes, given the actual state of nature:

| | Sample outcome | |
| State of nature | No resistance (probability) | Substantial resistance (probability) |
| --- | --- | --- |
| No resistance to Energetic's products | 0.7 | 0.3 |
| Substantial resistance to Energetic's products | 0.2 | 0.8 |

(a) Construct the decision tree representing the firm's decision if no survey is carried out.

(b) What is the probability that the survey will indicate no resistance? What is the probability that it will indicate substantial resistance?

(c) Assuming that the survey indicates no resistance, construct the subsequent branches of the decision tree.

(d) Assuming that the survey indicates substantial resistance, construct the subsequent branches of the decision tree.

(e) Construct the entire decision tree representing the Energetic Corporation's problem. Should the firm elect to have the survey carried out? Given that it elects to do so, what decision should the firm take if the results indicate no resistance? What decision should it take if the results indicate substantial resistance?

(f) What is the expected value of the sample information? What is the expected net gain of sample information?

(g) List all possible strategies that might be employed.

(h) For each of the strategies listed in (g), calculate the expected opportunity loss.

(i) Is the strategy with the lowest expected opportunity loss, as shown by your calculations in (h), the same as the result obtained in (e)? If so, explain why. If not, explain why the differences exist.

(j) One of the possible strategies is the following: Establish the plant if the survey indicates no resistance, and do not establish it if the survey indicates substantial resistance. Letting $P$ be the prior probability of no resistance, express the expected opportunity loss with this strategy as a function of $P$.

(k) For the other strategies besides that presented in (j), express the expected opportunity loss with each one as a function of $P$, the prior probability of no resistance. Do some strategies dominate others? If so, which ones are dominated by which others?

(l) Are there any values of $P$ for which the strategy in (j) is not as good as some other strategies? If so, what are these values of $P$, and why are the other strategies superior?

**16.8** The Uphill Corporation establishes the sequential sampling scheme shown in Table 16.5 to test whether the bolts it receives from a particular supplier are of the proper size. The supplier maintains that only 1 percent of the bolts Uphill receives in any given shipment should be more than .01 inches larger or smaller than the size specified by the design. Uphill uses the sequential sampling scheme to test whether this is true.

(a) Uphill samples a given shipment and finds that 1 out of the first 30 bolts is more than .01 inches larger or smaller than specified. Should Uphill reject the shipment?

(b) Uphill continues sampling and finds that 3 out of the next 30 bolts are more than .01 inches bigger or smaller than specified. Should Uphill reject the shipment?

# Appendix 16.1

SEQUENTIAL DECISION MAKING UNDER UNCERTAINTY

In this appendix, we return now to sequential decision-making procedures viewed from the point of view of Bayesian decision theory. To show how such procedures work, suppose that the Ellsworth Manufacturing Company is considering whether or not to install a major new process. If the process is introduced and is successful, the firm will reap a $5 million profit. On the other hand, if the process is introduced and is not successful, the firm will incur a $2 million loss. The firm attaches a prior probability of 0.5 to success and a prior probability of 0.5 to its being unsuccessful. Thus, on the basis of a prior analysis the firm would conclude, if it wants to maximize expected profit, that it should introduce the new process since the expected profit resulting from this action is

$$0.5(\$5,000,000) + 0.5(-\$2,000,000) = \$1,500,000,$$

while the profit resulting from its not introducing the new process is zero.

However, this prior analysis is based on the supposition that the firm gathers no evidence before making its decision. In fact, let us assume that the firm can carry out a research project to obtain information concerning the new process, and that the cost of this project is $100,000. If the new process is in fact successful, the probability that the research project will indicate this is 0.7, while the probability that it will indicate that the new process is unsuccessful is 0.3. If the new process is in fact unsuccessful, the probability that the research project will indicate this is 0.8, while the probability that it will indicate that the new process is successful is 0.2.

If the firm decides to carry out the research project, then it can decide at the end of the project whether or not to install the new process, or it can gather still more information by building and studying a pilot plant for the new process. If the new process is in fact successful, the probability that the pilot plant will indicate this is 0.8, while the probability that it will indicate that the new process is unsuccessful is 0.2. If the new process is in fact unsuccessful, the probability that the pilot plant will indicate this is 0.7, while the probability that it will indicate that the new process is successful is 0.3. The cost of building and studying the pilot plant is $100,000.

Figure 16.5 shows the decision tree for this problem. As you can see, the first decision the Ellsworth Manufacturing Company must make is whether or not to carry out the research project. If it decides not to do so, it must decide whether or not to install the new process. If it decides to install it, "chance" decides whether or not it is successful. On the other hand, if the company decides to carry out the research project, "chance" decides whether the project indicates success or not. Clearly, the probability that the research project will indicate success is $0.5(0.7) + 0.5(0.2) = 0.45$ since the prior probability that the process is successful is 0.5 and the probability that the research project will indicate that it is successful when this in fact is the case is 0.7, while the prior probability that the process is not successful is 0.5 and the probability that the research project will indicate that it is successful when in fact it is not successful is 0.2. Similarly, the probability that the research project will not indicate success is 0.55.

**Figure 16.5**
Decision Tree for
Problem Facing
Ellsworth Manufactur-
ing Company*

* All monetary values are
expressed in millions of
dollars.

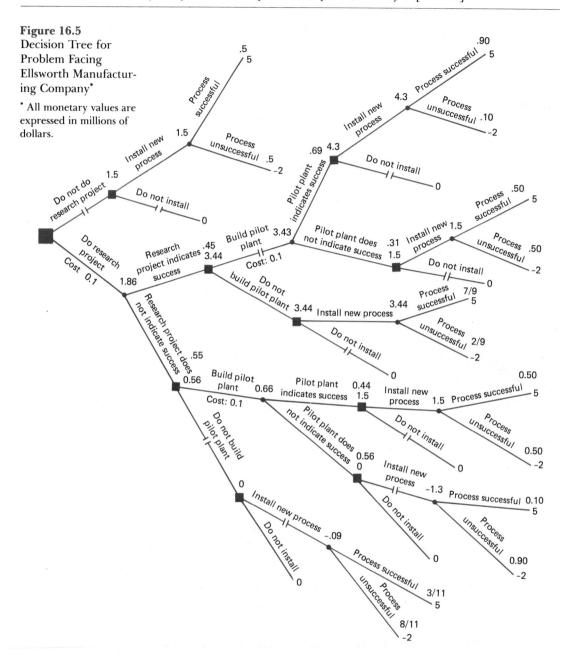

***Computation of
Posterior
Probabilities,
Based on Results of
Research Project
Alone***

If the research project above is carried out and *the results indicate that the new
process is a success,* what is the posterior probability that it will be a success?
Applying Bayes' theorem, this probability equals

$$P(S|s) = \frac{P(S)P(s|S)}{P(S)P(s|S) + P(N)P(s|N)},$$

where $P(S)$ is the prior probability that the new process is a success, $P(s|S)$ is the
probability that the research project indicates success when the new process is
in fact successful, $P(N)$ is the prior probability that the new process is not a suc-
cess, and $P(s|N)$ is the probability that the research project indicates success

when the new process is in fact not successful. Inserting numerical values in this expression,[11] we have

$$P(S|s) = \frac{0.5(0.7)}{0.5(0.7) + 0.5(0.2)} = 7/9.$$

Letting $P(N|s)$ be the posterior probability that the new process will not be a success, given that the research project indicates that it is a success, it follows that

$$P(N|s) = 1 - P(S|s) = 2/9.$$

If the research project is carried out and *the results indicate that the new process is not a success,* what is the posterior probability that it will be a success? Applying Bayes' theorem, this probability equals

$$P(S|n) = \frac{P(S)P(n|S)}{P(S)P(n|S) + P(N)P(n|N)},$$

where $P(n|S)$ is the probability that the research project does not indicate success when the new process is in fact successful, and $P(n|N)$ is the probability that the research project does not indicate success when the new process is in fact not successful. Inserting numerical values in this expression,[12] we have

$$P(S|n) = \frac{0.5(0.3)}{0.5(0.3) + 0.5(0.8)} = 3/11.$$

Letting $P(N|n)$ be the posterior probability that the new process will not be a success, given that the research project indicates that it is not a success, it follows that

$$P(N|n) = 1 - P(S|n) = 8/11.$$

With these results in hand, we can derive the expected profit if the firm carries out the research project and it indicates success, whereupon the firm decides to install the new process without building a pilot plant. This expected profit is[13]

$$(7/9)(5) + (2/9)(-2) = 3.44,$$

the figure shown in Figure 16.5. Since this expected profit exceeds zero, it is better to install the new process under these circumstances than not to do so. Also, we can derive the expected profit if the firm carries out the research project and it does not indicate success, whereupon the firm decides to install

[11] Recall from the previous section that both $P(S)$ and $P(N)$ equal 0.5, and that $P(s|S) = 0.7$ and $P(s|N) = 0.2$. These are the required numerical values.

[12] As pointed out in the previous section, both $P(S)$ and $P(N)$ equal 0.5, $P(n|S) = 0.3$, and $P(n|N) = 0.8$. These are the required numerical values.

[13] To obtain the expected profit, we multiply the posterior probability of success $(7/9)$ times 5, the profit (in millions of dollars) if the process is successful. To this we add the posterior probability of lack of success $(2/9)$ times $-2$, the profit (in millions of dollars) if the process is not successful. The same procedure is used in the next equation in the text, but the posterior probabilities are different because they are based on the assumption that the research project does not indicate success.

the new process without building a pilot plant. This expected profit is

$$(3/11)(5) + (8/11)(-2) = -.09,$$

which is the figure shown in Figure 16.5. Since this expected profit is less than zero, it is better not to install the new process under these circumstances.

***Computation of Posterior Probabilities, Based on Results of Research Project and Pilot Plant Together***

To derive the rest of the figures in Figure 16.5, we must compute the posterior probabilities that the new process will or will not be a success, based on the results of *both* the research project and the pilot plant. First, suppose that *both the research project and the pilot plant indicate success.* What is the posterior probability that the new process will be successful? Clearly, this probability equals

$$P(S\,|\,ss) = \frac{P(S)P(ss\,|\,S)}{P(S)P(ss\,|\,S) + P(N)P(ss\,|\,N)},$$

where $P(ss\,|\,S)$ is the probability that both the research project and the pilot plant indicate that the new process will be a success, given that the new process in fact is a success, and $P(ss\,|\,N)$ is the probability that both the research project and the pilot plant indicate that the new process is a success, given that the new process is not a success. If the results of the pilot plant and the research project are independent, it follows that[14]

$$P(S\,|\,ss) = \frac{0.5[(0.7)(0.8)]}{0.5[(0.7)(0.8)] + 0.5[(0.3)(0.2)]} = 0.90.$$

And we can also conclude that

$$P(N\,|\,ss) = 1 - 0.90 = 0.10,$$

where $P(N\,|\,ss)$ is the posterior probability that the new process is not a success, given that both the research project and the pilot plant indicate that it is a success.

Next, let's consider the case where *the research project indicates that the new process is a success and the pilot plant indicates that it is not a success.* Under these circumstances, what is the posterior probability that the new process is a success? Clearly, this probability equals

$$P(S\,|\,sn) = \frac{P(S)P(sn\,|\,S)}{P(S)P(sn\,|\,S) + P(N)P(sn\,|\,N)},$$

where $P(sn\,|\,S)$ is the probability that the research project will indicate that the new process is a success and that the pilot plant will indicate that it is not a success, given that the new program in fact is a success, and $P(sn\,|\,N)$ is the proba-

---

[14] Since we assume that the results of the research project and the pilot plant are independent, the probability that the research project and the pilot plant will both indicate success is the product of the probability that the research project indicates success and the probability that the pilot plant indicates success. According to page 687, the probability that the research project will indicate success (if the process is successful) is 0.7, and the probability that the pilot plant will indicate success (if the process is successful) is 0.8. Thus, $P(ss\,|\,S) = 0.7(0.8)$. According to page 687, the probability that the research project will indicate success (if the process is not successful) is 0.2, and the probability that the pilot plant will indicate success (if the process is not successful) is 0.3. Thus $P(ss\,|\,N) = 0.3(0.2)$.

bility that the research project will indicate that the new process is a success and that the pilot plant will indicate that it is not a success, given that the new process in fact is not a success. Inserting numerical values,[15] we have

$$P(S|sn) = \frac{0.5[(0.7)(0.2)]}{0.5[(0.7)(0.2)] + 0.5[(0.2)(0.7)]} = 0.5$$

And we can also conclude that

$$P(N|sn) = 1 - 0.5 = 0.5,$$

where $P(N|sn)$ is the posterior probability that the new process is not a success, given that the research project indicates that it is a success although the pilot plant indicates that it is not a success.

Next, let's consider the case where *the research project indicates that the new process is not a success although the pilot plant indicates that it is a success.* Under these circumstances, what is the posterior probability that the new process is a success? Clearly, this probability equals

$$P(S|ns) = \frac{P(N)P(ns|S)}{P(S)P(ns|S) + P(N)P(ns|N)},$$

where $P(ns|S)$ is the probability that the research project will indicate that the new process is not a success and that the pilot plant will indicate that it is a success, given that the new process is a success, and $P(ns|N)$ is the probability that the research project will indicate that the new process is not a success and that the pilot plant will indicate that it is a success, given that in fact it is not a success. Inserting numerical values,[16] we have

$$P(S|ns) = \frac{0.5[(0.3)(0.8)]}{0.5[(0.3)(0.8)] + 0.5[(0.8)(0.3)]} = 0.5$$

And we can also conclude that

$$P(N|ns) = 1 - 0.5 = 0.5,$$

where $P(N|ns)$ is the posterior probability that the new process is not a success if the research project indicates it is not a success and the pilot plant indicates it is a success.

Finally, suppose that *both the research project and the pilot plant indicate*

[15] Recall from page 687 that both $P(S)$ and $P(N)$ equal 0.5. Also $P(sn|S) = 0.7(0.2)$ since the probability that the research project will indicate success (if the process is successful) is 0.7 and the probability that the pilot plant will not indicate success (if the process is successful) is 0.2, according to page 687. Further, $P(sn|N) = 0.2(0.7)$ since the probability that the research project will indicate success (if the process is not successful) is 0.2, and the probability that the pilot plant will not indicate success (if the process is not successful) is 0.7 according to page 687. These are the required numerical values.

[16] From page 687, we know that both $P(S)$ and $P(N)$ equal 0.5. Also, $P(ns|S) = 0.3(0.8)$ since the probability that the research project will not indicate success (if the process is successful) is 0.3 and the probability that the pilot plant will indicate success (if the process is successful) is 0.8, according to page 687. Further, $P(ns|N) = 0.8(0.3)$ since the probability that the research project will not indicate success (if the process is not successful) is 0.8 and the probability that the pilot plant will indicate success (if the process is not successful) is 0.3, according to page 687. These are the required numerical values.

*that the new process is not a success.* What is the posterior probability that the new process is successful? This probability equals

$$P(S \mid nn) = \frac{P(S)P(nn \mid S)}{P(S)P(nn \mid S) + P(N)P(nn \mid N)},$$

where $P(nn \mid S)$ is the probability that the research project and the pilot plant both indicate that the new process is not a success when in fact it is a success, and $P(nn \mid N)$ is the probability that the research project and the pilot plant both indicate that the new process is not a success when in fact it is not a success. Inserting numerical values,[17] we have

$$P(S \mid nn) = \frac{0.5[(0.3)(0.2)]}{0.5[(0.3)(0.2)] + 0.5[(0.8)(0.7)]} = .10.$$

And we can also conclude that

$$P(N \mid nn) = 1 - .10 = 90,$$

where $P(N \mid nn)$ is the posterior probability that the new process is not a success if the research project and the pilot plant both indicate that this is the case.

**Solution to the Problem**

Having derived the posterior probabilities, based on the results of the research project alone and on the basis of both the research project and the pilot plant, we must now compute still another set of probabilities before being able to determine what the Ellsworth Manufacturing Company should decide. The first probability we need is the probability that the pilot plant will indicate success, given that the research project has indicated success. This probability is[18]

$$P(s \mid s) = \frac{P(ss)}{P(s)} = \frac{P(ss \mid S)P(S) + P(ss \mid N)P(N)}{P(s \mid S)P(S) + P(s \mid N)P(N)},$$

where $P(ss)$ is the probability that both the research project and the pilot plant indicate that the new process will be a success, and $P(s)$ is the probabability that the research project indicates success. Inserting numerical values,[19] we have

$$P(s \mid s) = \frac{[(0.7)(0.8)(0.5)] + [(0.3)(0.2)](0.5)}{(0.7)(0.5) + (0.2)(0.5)} = 0.69.$$

---

[17] As pointed out above, both $P(S)$ and $P(N)$ equal 0.5. Also, $P(nn \mid S) = 0.3(0.2)$ since the probability that the research project will not indicate success (if the process is successful) is 0.3 and the probability that the pilot plant will not indicate success (if the process is successful) is 0.2, according to page 687. Further, $P(nn \mid N) = 0.8(0.7)$ since the probability that the research project will not indicate success (if the process is not successful) is 0.8 and the probability that the pilot plant will not indicate success (if the process is not successful) is 0.7, according to page 687. These are the required values.

[18] To prove that $P(s \mid s) = P(ss) \div P(s)$, recall the definition of a conditional probability, given in equation (3.3).

[19] The values of $P(ss \mid S)$ and $P(ss \mid N)$ are derived in footnote 14 above. $P(s \mid S) = 0.7$ and $P(s \mid N) = 0.2$, according to page 687.

From this we can also conclude that the probability that the pilot plant will not indicate success, given that the research project has indicated success, equals

$$P(n|s) = 1 - .69 = .31.$$

The next probability that we need is the probability that the pilot plant will indicate success, given that the research project has indicated that the new process will not be a success. This probability is

$$P(s|n) = \frac{P(ns)}{P(n)} = \frac{P(ns|S)P(S) + P(ns|N)P(N)}{P(n|S)P(S) + P(n|N)P(N)},$$

where $P(ns)$ is the probability that the research project will not indicate success but that the pilot plant will indicate success, and $P(n)$ is the probability that the research project will not indicate success. Inserting numerical values,[20] we have

$$P(s|n) = \frac{[(0.3)(0.8)](0.5) + [(0.8)(0.3)](0.5)}{(0.3)(0.5) + (0.8)(0.5)} = 0.44.$$

And from this it follows that the probability that the pilot plant will not indicate success, given that the research project has not indicated success, equals

$$P(n|n) = 1 - 0.44 = .56.$$

Now we have all the probabilities set forth in Figure 16.5. To solve the problem, we begin by calculating the expected profit if the firm does not carry out the research project and goes ahead and installs the new process. Since (as shown in Figure 16.5) the expected profit is $1.5 million, the firm, if it does not carry out the research project, is better off installing the new process than not installing it. Thus, *the expected profit if the firm does not carry out the research project is $1.5 million* (as we saw at the beginning of this Appendix).

Next, suppose that the firm carries out the research project and builds the pilot plant, and that both indicate the new process will be successful. Under these circumstances, the expected profit is $4.3 million if the firm introduces the new process, which compares with a zero profit if it does not introduce it. (See Figure 16.5)[21] Thus, assuming that the research project indicates success,

[20] The values of $P(ns|S)$ and $P(ns|N)$ are derived in footnote 16 above. $P(n|S) =$ 0.3 and $P(n|N) = 0.8$, according to page 687.

[21] In Figure 16.5, the expected profit (in millions of dollars) when the decision maker is situated at each fork is shown above the fork, in accord with our discussion in Section 15.2. Thus, 4.3 is shown above the fork corresponding to the situation where both the research project and the pilot plant indicate that the new process will be successful, and the firm installs the new process. Why is the expected profit equal to $4.3 million? Because (as derived on page 690 and shown in Figure 16.5) there is a 0.9 probability of success (in which case the profit is $5 million) and a 0.1 probability of lack of success (in which case the profit is −$2 million). Thus, the expected profit is $0.9(5) + 0.1(-2) = 4.3$.

Each of the expected profit figures in Figure 16.5 can be calculated in a similar way. For example, what is the expected profit if the firm introduces the new process after the research project indicates success and the pilot plant does not indicate success? Using the values of $P(S|sn)$ and $P(N|sn)$ derived on page 691 and shown in Figure 16.5, the answer is $0.5(5) + 0.5(-2) = 1.5$. And as you can see, this is the number shown above this fork in Figure 16.5.

the expected profit if the pilot plant is built and indicates success is $4.3 million. On the other hand, suppose that whereas the research project indicates success, the pilot plant does not. In this case the expected profit if the new process is introduced is $1.5 million, which is obviously greater than the zero profit to be obtained if the new process is not introduced. (See Figure 16.5.) Thus, the better decision under these circumstances is to install the new process. Since the probability is 0.69 that the pilot plant will indicate success and 0.31 that it will not do so, *the expected profit if the firm builds the pilot plant (after the research project indicates success) is $3.43 million.* To compare this with the expected profit if the firm does *not* build a pilot plant (after the research project indicates success), look at the relevant section of Figure 16.5, which shows that the expected profit under the latter circumstances is $3.44 million. Thus, *if the research project indicates success, the firm will do better not to build the pilot plant. (The expected profit if it is built is $.01 million less than if it is not built, and there is an additional $.10 million cost of building it.)*

Now suppose that the firm carries out the research project and it does not indicate success. If the firm builds a pilot plant under these circumstances, and if the plant indicates success, the expected profit if the firm installs the new process is $1.5 million. Since this is more than the profit from not installing the process, the firm's best action if the pilot plant indicates success is to install the new process. If the pilot plant does not indicate success under these circumstances, the expected profit if the firm installs the new process is −$1.3 million. Thus, the firm's best action under the latter circumstances is not to introduce the new process. Since the probability is 0.44 that the pilot plant will indicate success and 0.56 that it will not do so under these circumstances, *the expected profit if the firm builds the pilot plant (after the research project does not indicate success) is $0.66 million.* If the firm does not build the pilot plant the expected profit is zero since its best action then is not to install the new process. (The expected profit if it installs the new process would be −$.09 million). Thus, *if the research project does not indicate success, the firm's best action is to build the pilot plant since (after subtracting the pilot plant's cost of $100,000) the expected profit is $0.56 million if the pilot plant is built and zero if it is not.*

At this point we can calculate the expected profit if the firm carries out the research project. Since there is a 0.45 probability that the research project will indicate success and a 0.55 probability that it will not, the expected profit is $1.86 million, as shown in Figure 16.5. Allowing for the $100,000 cost of the research project, the expected profit is nevertheless higher than $1.5 million, the expected profit if the research project is not carried out. Thus, *the firm should carry out the research project.*

To summarize, the solution to the problem is as follows: The Ellsworth Manufacturing Company, if it wants to maximize expected profit, should carry out the research project. If the research project indicates that the new process will be successful, the firm should not build the pilot plant, but go ahead and introduce the new process. If the research project does not indicate success, the firm should build the pilot plant. If the pilot plant indicates success, the firm should install the new process, but if the pilot plant does not indicate success, it should not install the new process.

THE CASE OF R AND D COST ESTIMATES

Some firms use statistical decision theory to help decide whether particular research and development (R and D) projects should be carried out. The analysis frequently is similar to that described here in the case of the Ellsworth Manufacturing Company. To carry out such an analysis, the firm's personnel must estimate how much the R and D project would cost, if it were carried out. In one major drug firm, the frequency distribution of 49 projects by the ratio of actual to estimated cost is as shown below: [22]

| Actual cost divided by estimated cost | Number of projects |
| --- | --- |
| Less than 1.01 | 6 |
| 1.01 and under 2.01 | 24 |
| 2.01 and under 3.01 | 16 |
| 3.01 and under 4.01 | 3 |

If this firm were using statistical decision theory to help determine whether particular R and D projects should be carried out, would there be hazards in adopting the analysis used in connection with the Ellsworth Manufacturing Company? What would they be?

SOLUTION: This firm seems to underestimate the cost of its R and D projects. (This is true in many firms, both because of the inherent uncertainties and because R and D personnel sometimes try to sell projects by underestimating their costs.) If the cost estimates are biased in this way, the analysis used in connection with the Ellsworth Manufacturing Company will be biased somewhat in favor of carrying out the R and D project in question. One possible way around this problem is to attempt to adjust the firm's cost estimates upward in accord with past experience. But unless one is careful, this may result in still further biases downward, as the R and D personnel adapt to this new environment. If scientists and engineers are rewarded for reasonably accurate estimates, the adjustment process may be unnecessary. As stressed in Section 16.6, unless the basic data are meaningful and relatively free from bias, the results may be misleading.

[22] E. Mansfield, J. Rapoport, J. Schnee, S. Wagner, and M. Hamburger, *Research and Innovation in the Modern Corporation* (New York: Norton, 1971).

EXERCISES

**16.9** A firm is interested in determining whether or not the market for a new product is big enough so that the product can be produced and sold profitably. The firm believes that there is a 0.5 probability that this is the case. To obtain further information on this score, the firm can hire a marketing consultant to give his opinion. Given each state of nature, the probability that the consultant will be correct is as follows:

| State of nature | Consultant's report | |
| --- | --- | --- |
| | Market is big enough | Market is not big enough |
| Market is big enough | 0.75 | 0.25 |
| Market is not big enough | 0.25 | 0.75 |

The firm also can carry out a survey to determine the size of the market. Given each state of nature, the probability that the survey will be correct is as follows:

| State of nature | Results of survey | |
| --- | --- | --- |
| | Market is big enough | Market is not big enough |
| Market is big enough | 0.8 | 0.2 |
| Market is not big enough | 0.2 | 0.8 |

Given each state of nature, the results of the survey and the consultant's report are statistically independent.
(a) If the firm hires the consultant and he reports that the market is big enough, what is the probability that this is the case?
(b) If the firm hires the consultant and he reports that the market is not big enough, what is the probability that this is the case?
(c) If both the consultant and the survey indicate that the market is big enough, what is the probability that this is the case?
(d) If both the consultant and the survey indicate that the market is not big enough, what is the probability that this is the case?
(e) If the consultant thinks that the market is big enough but the survey indicates that the market is not big enough, what is the probability that the market is big enough?
(f) If the consultant thinks that the market is not big enough but the survey indicates that the market is big enough, what is the probability that the market is big enough?

**16.10** The Energetic Corporation is studying the desirability of buying some very large and expensive pieces of equipment in order to be eligible to bid on a particular government contract. Based on its current information, the firm believes that if it buys the equipment the probability of getting the contract is 0.55 and the probability of not getting it is 0.45. If it buys the equipment and gets the contract, the firm foresees additional profits of $2 million; if it buys the equipment and does not get the contract, the firm foresees a loss of $1.5 million. If the firm does not buy the equipment profits will be unaffected.

Before deciding whether to buy these pieces of equipment, Energetic can hire a consulting firm to look closely at the factors determining whether or not it will get the contract. Based on past experience, Energetic feels that the probability that the consulting firm will conclude that Energetic will get the contract (if it has the equipment), given that this actually is the case, is 0.9, while the probability that it will conclude that Energetic will not get the contract under these circumstances is 0.1. Energetic also feels that the probability that the consulting firm will conclude that it will get the contract (if it has the equipment), given that this actually is not the case, is 0.3, while the probability of the consultant's concluding that Energetic will not get the contract under these circumstances is 0.7. The consulting firm will charge Energetic $150,000 for its study.

Once the study has been carried out (if it is carried out) by the consulting firm, Energetic can have its report gone over by a panel of experts in order to judge whether Energetic will or will not get the contract. If Energetic will get the contract, the probability that this panel will conclude that this is so is 0.9, while the probability that it will reach the reverse conclusion is 0.1. If Energetic will not get the contract, the probability that the panel will conclude that this is so is 0.9, while the probability that it will reach the reverse conclusion is 0.1. Given each state of nature, the results of the panel and of the consulting firm are regarded as being statistically independent. The panel of experts will charge Energetic $50,000 for providing a judgment of this sort.

(a) If Energetic hires the consulting firm, and if the firm concludes that Energetic will get the contract, what is the probability that Energetic will in fact get the contract? Under these circumstances, what is the probability that Energetic will in fact not get the contract?

(b) If Energetic hires the consulting firm, and if the firm concludes that Energetic will not get the contract, what is the probability that Energetic will in fact get the contract? Under these circumstances, what is the probability that Energetic will in fact not get the contract?

(c) If both the consulting firm and the panel conclude that Energetic will get the contract, what is the probability that it will in fact get the contract? Under these circumstances, what is the probability that it will in fact not get the contract?

(d) If both the consulting firm and the panel conclude that Energetic will not get the contract, what is the probability that it will in fact get the contract? Under these circumstances, what is the probability that it will in fact not get the contract?

(e) If the consulting firm concludes that Energetic will get the contract whereas the panel of experts concludes the reverse, what is the probability that it will in fact get the contract? Under these circumstances, what is the probability that it will in fact not get the contract?

(f) If the panel of experts concludes that Energetic will get the contract whereas the consulting firm concludes the reverse, what is the probability that it will in fact get the contract? Under these circumstances, what is the probability that it will in fact not get the contract?

(g) What is the probability that if the consulting firm concludes that Energetic will get the contract, the panel of experts will conclude the same thing? What is the probability that if the consulting firm concludes that Energetic will get the contract, the panel of experts will come to the opposite conclusion? Suppose that the consulting firm concludes that Energetic will not get the contract. Under these circumstances, what is the probability that the panel of experts will come to the opposite conclusion? What is the probability that they will reach the same conclusion?

(h) Construct the decision tree representing the Energetic Corporation's problem. If Energetic does not hire the consulting firm, is it better to buy the equipment or not? Why, or why not?

(i)  If Energetic hires the consulting firm, and if the consulting firm says Energetic will get the contract, should Energetic hire the panel of experts? Why, or why not?

(j)  If Energetic hires the consulting firm, and if the consulting firm says Energetic will not get the contract, should Energetic hire the panel of experts? Why, or why not?

(k)  What is the complete solution to Energetic's problem?

# Appendix

---

## Appendix Table 1

BINOMIAL PROBABILITY DISTRIBUTION

This table shows the value of

$$P(x) = \frac{n!}{(n-x)!x!} \, \Pi^x(1-\Pi)^{n-x}$$

for selected values of $\Pi$ and for $n = 1$ to 20. For values of $\Pi$ exceeding 0.5, the value of $P(x)$ can be obtained by substituting $(1-\Pi)$ for $\Pi$ and by finding $P(n-x)$. (See Section 4.7.)

| | | | | | | $\Pi$ | | | | | |
|---|---|------|------|------|------|------|------|------|------|------|------|
| $n$ | $x$ | .05 | .10 | .15 | .20 | .25 | .30 | .35 | .40 | .45 | .50 |
| 1 | 0 | .9500 | .9000 | .8500 | .8000 | .7500 | .7000 | .6500 | .6000 | .5500 | .5000 |
|   | 1 | .0500 | .1000 | .1500 | .2000 | .2500 | .3000 | .3500 | .4000 | .4500 | .5000 |
| 2 | 0 | .9025 | .8100 | .7225 | .6400 | .5625 | .4900 | .4225 | .3600 | .3025 | .2500 |
|   | 1 | .0950 | .1800 | .2550 | .3200 | .3750 | .4200 | .4550 | .4800 | .4950 | .5000 |
|   | 2 | .0025 | .0100 | .0225 | .0400 | .0625 | .0900 | .1225 | .1600 | .2025 | .2500 |
| 3 | 0 | .8574 | .7290 | .6141 | .5120 | .4219 | .3430 | .2746 | .2160 | .1664 | .1250 |
|   | 1 | .1354 | .2430 | .3251 | .3840 | .4219 | .4410 | .4436 | .4320 | .4084 | .3750 |
|   | 2 | .0071 | .0270 | .0574 | .0960 | .1406 | .1890 | .2389 | .2880 | .3341 | .3750 |
|   | 3 | .0001 | .0010 | .0034 | .0080 | .0156 | .0270 | .0429 | .0640 | .0911 | .1250 |
| 4 | 0 | .8145 | .6561 | .5220 | .4096 | .3164 | .2401 | .1785 | .1296 | .0915 | .0625 |
|   | 1 | .1715 | .2916 | .3685 | .4096 | .4219 | .4116 | .3845 | .3456 | .2995 | .2500 |
|   | 2 | .0135 | .0486 | .0975 | .1536 | .2109 | .2646 | .3105 | .3456 | .3675 | .3750 |
|   | 3 | .0005 | .0036 | .0115 | .0256 | .0469 | .0756 | .1115 | .1536 | .2005 | .2500 |
|   | 4 | .0000 | .0001 | .0005 | .0016 | .0039 | .0081 | .0150 | .0256 | .0410 | .0625 |

# Appendix Table 1 (Continued)

|       |       |       |       |       | $\Pi$ |       |       |       |       |       |
|-------|-------|-------|-------|-------|-------|-------|-------|-------|-------|-------|
| *n* *x* | *.05* | *.10* | *.15* | *.20* | *.25* | *.30* | *.35* | *.40* | *.45* | *.50* |
| 5  0 | .7738 | .5905 | .4437 | .3277 | .2373 | .1681 | .1160 | .0778 | .0503 | .0312 |
|    1 | .2036 | .3280 | .3915 | .4096 | .3955 | .3602 | .3124 | .2592 | .2059 | .1562 |
|    2 | .0214 | .0729 | .1382 | .2048 | .2637 | .3087 | .3364 | .3456 | .3369 | .3125 |
|    3 | .0011 | .0081 | .0244 | .0512 | .0879 | .1323 | .1811 | .2304 | .2757 | .3125 |
|    4 | .0000 | .0004 | .0022 | .0064 | .0146 | .0284 | .0488 | .0768 | .1128 | .1562 |
|    5 | .0000 | .0000 | .0001 | .0003 | .0010 | .0024 | .0053 | .0102 | .0185 | .0312 |
| 6  0 | .7351 | .5314 | .3771 | .2621 | .1780 | .1176 | .0754 | .0467 | .0277 | .0156 |
|    1 | .2321 | .3543 | .3993 | .3932 | .3560 | .3025 | .2437 | .1866 | .1359 | .0938 |
|    2 | .0305 | .0984 | .1762 | .2458 | .2966 | .3241 | .3280 | .3110 | .2780 | .2344 |
|    3 | .0021 | .0146 | .0415 | .0819 | .1318 | .1852 | .2355 | .2765 | .3032 | .3125 |
|    4 | .0001 | .0012 | .0055 | .0154 | .0330 | .0595 | .0951 | .1382 | .1861 | .2344 |
|    5 | .0000 | .0001 | .0004 | .0015 | .0044 | .0102 | .0205 | .0369 | .0609 | .0938 |
|    6 | .0000 | .0000 | .0000 | .0001 | .0002 | .0007 | .0018 | .0041 | .0083 | .0516 |
| 7  0 | .6983 | .4783 | .3206 | .2097 | .1335 | .0824 | .0490 | .0280 | .0152 | .0078 |
|    1 | .2573 | .3720 | .3960 | .3670 | .3115 | .2471 | .1848 | .1306 | .0872 | .0547 |
|    2 | .0406 | .1240 | .2097 | .2753 | .3115 | .3177 | .2985 | .2613 | .2140 | .1641 |
|    3 | .0036 | .0230 | .0617 | .1147 | .1730 | .2269 | .2679 | .2903 | .2918 | .2734 |
|    4 | .0002 | .0026 | .0109 | .0287 | .0577 | .0972 | .1442 | .1935 | .2388 | .2734 |
|    5 | .0009 | .0002 | .0012 | .0043 | .0115 | .0250 | .0466 | .0774 | .1172 | .1641 |
|    6 | .0000 | .0000 | .0001 | .0004 | .0013 | .0036 | .0084 | .0172 | .0320 | .0547 |
|    7 | .0000 | .0000 | .0000 | .0000 | .0001 | .0002 | .0006 | .0016 | .0037 | .0078 |
| 8  0 | .6634 | .4305 | .2725 | .1678 | .1001 | .0576 | .0319 | .0168 | .0084 | .0039 |
|    1 | .2793 | .3826 | .3847 | .3355 | .2670 | .1977 | .1373 | .0896 | .0548 | .0312 |
|    2 | .0515 | .1488 | .2376 | .2936 | .3115 | .2965 | .2587 | .2090 | .1569 | .1094 |
|    3 | .0054 | .0331 | .0839 | .1468 | .2076 | .2541 | .2786 | .2787 | .2568 | .2188 |
|    4 | .0004 | .0046 | .0815 | .0459 | .0865 | .1361 | .1875 | .2322 | .2627 | .2734 |
|    5 | .0000 | .0004 | .0026 | .0092 | .0231 | .0467 | .0808 | .1239 | .1719 | .2188 |
|    6 | .0000 | .0000 | .0002 | .0011 | .0038 | .0100 | .0217 | .0413 | .0703 | .1094 |
|    7 | .0000 | .0000 | .0000 | .0001 | .0004 | .0012 | .0033 | .0079 | .0164 | .0312 |
|    8 | .0000 | .0000 | .0000 | .0000 | .0000 | .0001 | .0002 | .0007 | .0017 | .0039 |
| 9  0 | .6302 | .3874 | .2316 | .1342 | .0751 | .0404 | .0207 | .0101 | .0046 | .0020 |
|    1 | .2985 | .3874 | .3679 | .3020 | .2253 | .1556 | .1004 | .0605 | .0339 | .0176 |
|    2 | .0629 | .1722 | .2597 | .3020 | .3003 | .2668 | .2162 | .1612 | .1110 | .0703 |
|    3 | .0077 | .0446 | .1069 | .1762 | .2336 | .2668 | .2716 | .2508 | .2119 | .1641 |
|    4 | .0006 | .0074 | .0283 | .0661 | .1168 | .1715 | .2194 | .2508 | .2600 | .2461 |
|    5 | .0000 | .0008 | .0050 | .0165 | .0389 | .0735 | .1181 | .1672 | .2128 | .2461 |
|    6 | .0000 | .0001 | .0006 | .0028 | .0087 | .0210 | .0424 | .0743 | .1160 | .1641 |
|    7 | .0000 | .0000 | .0000 | .0003 | .0012 | .0039 | .0098 | .0212 | .0407 | .0703 |
|    8 | .0000 | .0000 | .0000 | .0000 | .0001 | .0004 | .0013 | .0035 | .0083 | .0716 |
|    9 | .0000 | .0000 | .0000 | .0000 | .0000 | .0000 | .0001 | .0003 | .0008 | .0020 |

# Appendix Table 1 (Continued)

| | | | | | | $\Pi$ | | | | | |
|---|---|---|---|---|---|---|---|---|---|---|---|
| n | x | .05 | .10 | .15 | .20 | .25 | .30 | .35 | .40 | .45 | .50 |
| 10 | 0 | .5987 | .3487 | .1969 | .1074 | .0563 | .0282 | .0135 | .0060 | .0025 | .0010 |
| | 1 | .3151 | .3874 | .3474 | .2684 | .1877 | .1211 | .0725 | .0403 | .0207 | .0098 |
| | 2 | .0746 | .1937 | .2759 | .3020 | .2816 | .2335 | .1757 | .1209 | .0763 | .0439 |
| | 3 | .0105 | .0574 | .1298 | .2013 | .2503 | .2668 | .2522 | .2150 | .1665 | .1172 |
| | 4 | .0010 | .0112 | .0401 | .0881 | .1460 | .2001 | .2377 | .2508 | .2384 | .2051 |
| | 5 | .0001 | .0015 | .0085 | .0264 | .0584 | .1029 | .1536 | .2007 | .2340 | .2461 |
| | 6 | .0000 | .0001 | .0012 | .0055 | .0162 | .0368 | .0689 | .1115 | .1596 | .2051 |
| | 7 | .0000 | .0000 | .0001 | .0008 | .0031 | .0090 | .0212 | .0425 | .0746 | .1172 |
| | 8 | .0000 | .0000 | .0000 | .0001 | .0004 | .0014 | .0043 | .0106 | .0229 | .0439 |
| | 9 | .0000 | .0000 | .0000 | .0000 | .0000 | .0001 | .0005 | .0016 | .0042 | .0098 |
| | 10 | .0000 | .0000 | .0000 | .0000 | .0000 | .0000 | .0000 | .0001 | .0003 | .0010 |
| 11 | 0 | .5688 | .3138 | .1673 | .0859 | .0422 | .0198 | .0088 | .0036 | .0014 | .0005 |
| | 1 | .3293 | .3835 | .3248 | .2362 | .1549 | .0932 | .0518 | .0266 | .0125 | .0054 |
| | 2 | .0867 | .2131 | .2866 | .2953 | .2581 | .1998 | .1395 | .0887 | .0513 | .0269 |
| | 3 | .0137 | .0710 | .1517 | .2215 | .2581 | .2568 | .2254 | .1774 | .1259 | .0806 |
| | 4 | .0014 | .0158 | .0536 | .1107 | .1721 | .2201 | .2428 | .2365 | .2060 | .1611 |
| | 5 | .0001 | .0025 | .0132 | .0388 | .0803 | .1321 | .1830 | .2207 | .2360 | .2256 |
| | 6 | .0000 | .0003 | .0023 | .0097 | .0268 | .0566 | .0985 | .1471 | .1931 | .2256 |
| | 7 | .0000 | .0000 | .0003 | .0017 | .0064 | .0173 | .0379 | .0701 | .1128 | .1611 |
| | 8 | .0000 | .0000 | .0000 | .0002 | .0011 | .0037 | .0102 | .0234 | .0462 | .0806 |
| | 9 | .0000 | .0000 | .0000 | .0000 | .0001 | .0005 | .0018 | .0052 | .0126 | .0269 |
| | 10 | .0000 | .0000 | .0000 | .0000 | .0000 | .0000 | .0002 | .0007 | .0021 | .0054 |
| | 11 | .0000 | .0000 | .0000 | .0000 | .0000 | .0000 | .0000 | .0000 | .0002 | .0005 |
| 12 | 0 | .5404 | .2824 | .1422 | .0687 | .0317 | .0138 | .0057 | .0022 | .0008 | .0002 |
| | 1 | .3413 | .3766 | .3012 | .2062 | .1267 | .0712 | .0368 | .0174 | .0075 | .0029 |
| | 2 | .0988 | .2301 | .2924 | .2835 | .2323 | .1678 | .1088 | .0639 | .0339 | .0161 |
| | 3 | .0173 | .0852 | .1720 | .2362 | .2581 | .2397 | .1954 | .1419 | .0923 | .0537 |
| | 4 | .0021 | .0213 | .0683 | .1329 | .1936 | .2311 | .2367 | .2128 | .1700 | .1208 |
| | 5 | .0002 | .0038 | .0193 | .0532 | .1032 | .1585 | .2039 | .2270 | .2225 | .1934 |
| | 6 | .0000 | .0005 | .0040 | .0155 | .0401 | .0792 | .1281 | .1766 | .2124 | .2256 |
| | 7 | .0000 | .0000 | .0006 | .0033 | .0115 | .0291 | .0591 | .1009 | .1489 | .1934 |
| | 8 | .0000 | .0000 | .0001 | .0005 | .0024 | .0078 | .0199 | .0420 | .0762 | .1208 |
| | 9 | .0000 | .0000 | .0000 | .0001 | .0004 | .0015 | .0048 | .0125 | .0277 | .0537 |
| | 10 | .0000 | .0000 | .0000 | .0000 | .0000 | .0002 | .0008 | .0025 | .0068 | .0161 |
| | 11 | .0000 | .0000 | .0000 | .0000 | .0000 | .0000 | .0001 | .0003 | .0010 | .0029 |
| | 12 | .0000 | .0000 | .0000 | .0000 | .0000 | .0000 | .0000 | .0000 | .0001 | .0002 |
| 13 | 0 | .5133 | .2542 | .1209 | .0550 | .0238 | .0097 | .0037 | .0013 | .0004 | .0001 |
| | 1 | .3512 | .3672 | .2774 | .1787 | .1029 | .0540 | .0259 | .0113 | .0045 | .0016 |
| | 2 | .1109 | .2448 | .2937 | .2680 | .2059 | .1388 | .0836 | .0453 | .0220 | .0095 |
| | 3 | .0214 | .0997 | .1900 | .2457 | .2517 | .2181 | .1651 | .1107 | .0660 | .0349 |
| | 4 | .0028 | .0277 | .0838 | .1535 | .2097 | .2337 | .2222 | .1845 | .1350 | .0873 |

## Appendix Table 1 (Continued)

| | | | | | | $\Pi$ | | | | | |
|---|---|---|---|---|---|---|---|---|---|---|---|
| n | x | .05 | .10 | .15 | .20 | .25 | .30 | .35 | .40 | .45 | .50 |
| | 5 | .0003 | .0055 | .0266 | .0691 | .1258 | .1803 | .2154 | .2214 | .1989 | .1571 |
| | 6 | .0000 | .0008 | .0063 | .0230 | .0559 | .1030 | .1546 | .1968 | .2169 | .2095 |
| | 7 | .0000 | .0001 | .0011 | .0058 | .0186 | .0442 | .0833 | .1312 | .1775 | .2095 |
| | 8 | .0000 | .0000 | .0001 | .0011 | .0047 | .0142 | .0336 | .0656 | .1089 | .1571 |
| | 9 | .0000 | .0000 | .0000 | .0001 | .0009 | .0034 | .0101 | .0243 | .0495 | .0873 |
| | 10 | .0000 | .0000 | .0000 | .0000 | .0001 | .0006 | .0022 | .0065 | .0162 | .0349 |
| | 11 | .0000 | .0000 | .0000 | .0000 | .0000 | .0001 | .0003 | .0012 | .0036 | .0095 |
| | 12 | .0000 | .0000 | .0000 | .0000 | .0000 | .0000 | .0000 | .0001 | .0005 | .0016 |
| | 13 | .0000 | .0000 | .0000 | .0000 | .0000 | .0000 | .0000 | .0000 | .0000 | .0001 |
| 14 | 0 | .4877 | .2288 | .1028 | .0440 | .0178 | .0068 | .0024 | .0008 | .0002 | .0001 |
| | 1 | .3593 | .3559 | .2539 | .1539 | .0832 | .0407 | .0181 | .0073 | .0027 | .0009 |
| | 2 | .1229 | .2570 | .2912 | .2501 | .1802 | .1134 | .0634 | .0317 | .0141 | .0056 |
| | 3 | .0259 | .1142 | .2056 | .2501 | .2402 | .1943 | .1366 | .0845 | .0462 | .0222 |
| | 4 | .0037 | .0348 | .0998 | .1720 | .2202 | .2290 | .2022 | .1549 | .1040 | .0611 |
| | 5 | .0004 | .0078 | .0352 | .0860 | .1468 | .1963 | .2178 | .2066 | .1701 | .1222 |
| | 6 | .0000 | .0013 | .0093 | .0322 | .0734 | .1262 | .1759 | .2066 | .2088 | .1833 |
| | 7 | .0000 | .0002 | .0019 | .0092 | .0280 | .0618 | .1082 | .1574 | .1952 | .2095 |
| | 8 | .0000 | .0000 | .0003 | .0020 | .0082 | .0232 | .0510 | .0918 | .1398 | .1833 |
| | 9 | .0000 | .0000 | .0000 | .0003 | .0018 | .0066 | .0183 | .0408 | .0762 | .1222 |
| | 10 | .0000 | .0000 | .0000 | .0000 | .0003 | .0014 | .0049 | .0136 | .0312 | .0611 |
| | 11 | .0000 | .0000 | .0000 | .0000 | .0000 | .0002 | .0010 | .0033 | .0093 | .0222 |
| | 12 | .0000 | .0000 | .0000 | .0000 | .0000 | .0000 | .0001 | .0005 | .0019 | .0056 |
| | 13 | .0000 | .0000 | .0000 | .0000 | .0000 | .0000 | .0000 | .0001 | .0002 | .0009 |
| | 14 | .0000 | .0000 | .0000 | .0000 | .0000 | .0000 | .0000 | .0000 | .0000 | .0001 |
| 15 | 0 | .4633 | .2059 | .0874 | .0352 | .0134 | .0047 | .0016 | .0005 | .0001 | .0000 |
| | 1 | .3658 | .3432 | .2312 | .1319 | .0668 | .0305 | .0126 | .0047 | .0016 | .0005 |
| | 2 | .1348 | .2669 | .2856 | .2309 | .1559 | .0916 | .0476 | .0219 | .0090 | .0032 |
| | 3 | .0307 | .1285 | .2184 | .2501 | .2252 | .1700 | .1110 | .0634 | .0318 | .0139 |
| | 4 | .0049 | .0428 | .1156 | .1876 | .2252 | .2186 | .1792 | .1268 | .0780 | .0417 |
| | 5 | .0006 | .0105 | .0449 | .1032 | .1651 | .2061 | .2123 | .1859 | .1404 | .0916 |
| | 6 | .0000 | .0019 | .0132 | .0430 | .0917 | .1472 | .1906 | .2066 | .1914 | .1527 |
| | 7 | .0000 | .0003 | .0030 | .0138 | .0393 | .0811 | .1319 | .1771 | .2013 | .1964 |
| | 8 | .0000 | .0000 | .0005 | .0035 | .0131 | .0348 | .0710 | .1181 | .1647 | .1964 |
| | 9 | .0000 | .0000 | .0001 | .0007 | .0034 | .0116 | .0298 | .0612 | .1048 | .1527 |
| | 10 | .0000 | .0000 | .0000 | .0001 | .0007 | .0030 | .0096 | .0245 | .0515 | .0916 |
| | 11 | .0000 | .0000 | .0000 | .0000 | .0001 | .0006 | .0024 | .0074 | .0191 | .0417 |
| | 12 | .0000 | .0000 | .0000 | .0000 | .0000 | .0001 | .0004 | .0016 | .0052 | .0139 |
| | 13 | .0000 | .0000 | .0000 | .0000 | .0000 | .0000 | .0001 | .0003 | .0010 | .0032 |
| | 14 | .0000 | .0000 | .0000 | .0000 | .0000 | .0000 | .0000 | .0000 | .0001 | .0005 |
| | 15 | .0000 | .0000 | .0000 | .0000 | .0000 | .0000 | .0000 | .0000 | .0000 | .0000 |

# Appendix Table 1 (Continued)

|  |  |  |  |  |  | *Π* |  |  |  |  |  |
|---|---|---|---|---|---|---|---|---|---|---|---|
| *n* | *x* | *.05* | *.10* | *.15* | *.20* | *.25* | *.30* | *.35* | *.40* | *.45* | *.50* |
| 16 | 0 | .4401 | .1853 | .0743 | .0281 | .0100 | .0033 | .0010 | .0003 | .0001 | .0000 |
|  | 1 | .3706 | .3294 | .2097 | .1126 | .0535 | .0228 | .0087 | .0030 | .0009 | .0002 |
|  | 2 | .1463 | .2745 | .2775 | .2111 | .1336 | .0732 | .0353 | .0150 | .0056 | .0018 |
|  | 3 | .0359 | .1423 | .2285 | .2463 | .2079 | .1465 | .0888 | .0468 | .0215 | .0085 |
|  | 4 | .0061 | .0514 | .1311 | .2001 | .2252 | .2040 | .1553 | .1014 | .0572 | .0278 |
|  | 5 | .0008 | .0137 | .0555 | .1201 | .1802 | .2099 | .2008 | .1623 | .1123 | .0667 |
|  | 6 | .0001 | .0028 | .0180 | .0550 | .1101 | .1649 | .1982 | .1983 | .1684 | .1222 |
|  | 7 | .0000 | .0004 | .0045 | .0197 | .0524 | .1010 | .1524 | .1889 | .1969 | .1746 |
|  | 8 | .0000 | .0001 | .0009 | .0055 | .0197 | .0487 | .0923 | .1417 | .1812 | .1964 |
|  | 9 | .0000 | .0000 | .0001 | .0012 | .0058 | .0185 | .0442 | .0840 | .1318 | .1746 |
|  | 10 | .0000 | .0000 | .0000 | .0002 | .0014 | .0056 | .0167 | .0392 | .0755 | .1222 |
|  | 11 | .0000 | .0000 | .0000 | .0000 | .0002 | .0013 | .0049 | .0142 | .0337 | .0667 |
|  | 12 | .0000 | .0000 | .0000 | .0000 | .0000 | .0002 | .0011 | .0040 | .0115 | .0278 |
|  | 13 | .0000 | .0000 | .0000 | .0000 | .0000 | .0000 | .0002 | .0008 | .0029 | .0085 |
|  | 14 | .0000 | .0000 | .0000 | .0000 | .0000 | .0000 | .0000 | .0001 | .0005 | .0018 |
|  | 15 | .0000 | .0000 | .0000 | .0000 | .0000 | .0000 | .0000 | .0000 | .0001 | .0002 |
|  | 16 | .0000 | .0000 | .0000 | .0000 | .0000 | .0000 | .0000 | .0000 | .0000 | .0000 |
| 17 | 0 | .4181 | .1668 | .0631 | .0225 | .0075 | .0023 | .0007 | .0002 | .0000 | .0000 |
|  | 1 | .3741 | .3150 | .1893 | .0957 | .0426 | .0169 | .0060 | .0019 | .0005 | .0001 |
|  | 2 | .1575 | .2800 | .2673 | .1914 | .1136 | .0581 | .0260 | .0102 | .0035 | .0010 |
|  | 3 | .0415 | .1556 | .2359 | .2393 | .1893 | .1245 | .0701 | .0341 | .0144 | .0052 |
|  | 4 | .0076 | .0605 | .1457 | .2093 | .2209 | .1868 | .1320 | .0796 | .0411 | .0182 |
|  | 5 | .0010 | .0175 | .0668 | .1361 | .1914 | .2081 | .1849 | .1379 | .0875 | .0472 |
|  | 6 | .0001 | .0039 | .0236 | .0680 | .1276 | .1784 | .1991 | .1839 | .1432 | .0944 |
|  | 7 | .0000 | .0007 | .0065 | .0267 | .0668 | .1201 | .1685 | .1927 | .1841 | .1484 |
|  | 8 | .0000 | .0001 | .0014 | .0084 | .0279 | .0644 | .1134 | .1606 | .1883 | .1855 |
|  | 9 | .0000 | .0000 | .0003 | .0021 | .0093 | .0276 | .0611 | .1070 | .1540 | .1855 |
|  | 10 | .0000 | .0000 | .0000 | .0004 | .0025 | .0095 | .0263 | .0571 | .1008 | .1484 |
|  | 11 | .0000 | .0000 | .0000 | .0001 | .0005 | .0026 | .0090 | .0242 | .0525 | .0944 |
|  | 12 | .0000 | .0000 | .0000 | .0000 | .0001 | .0006 | .0024 | .0021 | .0215 | .0472 |
|  | 13 | .0000 | .0000 | .0000 | .0000 | .0000 | .0001 | .0005 | .0021 | .0068 | .0182 |
|  | 14 | .0000 | .0000 | .0000 | .0000 | .0000 | .0000 | .0001 | .0004 | .0016 | .0052 |
|  | 15 | .0000 | .0000 | .0000 | .0000 | .0000 | .0000 | .0000 | .0001 | .0003 | .0010 |
|  | 16 | .0000 | .0000 | .0000 | .0000 | .0000 | .0000 | .0000 | .0000 | .0000 | .0001 |
|  | 17 | .0000 | .0000 | .0000 | .0000 | .0000 | .0000 | .0000 | .0000 | .0000 | .0000 |
| 18 | 0 | .3972 | .1501 | .0536 | .0180 | .0056 | .0016 | .0004 | .0001 | .0000 | .0000 |
|  | 1 | .3763 | .3002 | .1704 | .0811 | .0338 | .0126 | .0042 | .0012 | .0003 | .0001 |
|  | 2 | .1683 | .2835 | .2556 | .1723 | .0958 | .0458 | .0190 | .0069 | .0022 | .0006 |
|  | 3 | .0473 | .1680 | .2406 | .2297 | .1704 | .1046 | .0547 | .0246 | .0095 | .0031 |
|  | 4 | .0093 | .0700 | .1592 | .2153 | .2130 | .1681 | .1104 | .0614 | .0291 | .0117 |

## Appendix Table 1 (Continued)

| n | x | .05 | .10 | .15 | .20 | Π .25 | .30 | .35 | .40 | .45 | .50 |
|---|---|---|---|---|---|---|---|---|---|---|---|
| | 5 | .0014 | .0218 | .0787 | .1507 | .1988 | .2017 | .1664 | .1146 | .0666 | .0327 |
| | 6 | .0002 | .0052 | .0301 | .0816 | .1436 | .1873 | .1941 | .1655 | .1181 | .0708 |
| | 7 | .0000 | .0010 | .0091 | .0350 | .0820 | .1376 | .1792 | .1892 | .1657 | .1214 |
| | 8 | .0000 | .0002 | .0022 | .0120 | .0376 | .0811 | .1327 | .1734 | .1864 | .1669 |
| | 9 | .0000 | .0000 | .0004 | .0033 | .0139 | .0386 | .0794 | .1284 | .1694 | .1855 |
| | 10 | .0000 | .0000 | .0001 | .0008 | .0042 | .0149 | .0385 | .0771 | .1248 | .1669 |
| | 11 | .0000 | .0000 | .0000 | .0001 | .0010 | .0046 | .0151 | .0374 | .0742 | .1214 |
| | 12 | .0000 | .0000 | .0000 | .0000 | .0002 | .0012 | .0047 | .0145 | .0354 | .0708 |
| | 13 | .0000 | .0000 | .0000 | .0000 | .0000 | .0002 | .0012 | .0044 | .0134 | .0327 |
| | 14 | .0000 | .0000 | .0000 | .0000 | .0000 | .0000 | .0002 | .0011 | .0039 | .0117 |
| | 15 | .0000 | .0000 | .0000 | .0000 | .0000 | .0000 | .0000 | .0002 | .0009 | .0031 |
| | 16 | .0000 | .0000 | .0000 | .0000 | .0000 | .0000 | .0000 | .0000 | .0001 | .0006 |
| | 17 | .0000 | .0000 | .0000 | .0000 | .0000 | .0000 | .0000 | .0000 | .0000 | .0001 |
| | 18 | .0000 | .0000 | .0000 | .0000 | .0000 | .0000 | .0000 | .0000 | .0000 | .0000 |
| 19 | 0 | .3774 | .1351 | .0456 | .0144 | .0042 | .0011 | .0003 | .0001 | .0000 | .0000 |
| | 1 | .3774 | .2852 | .1529 | .0685 | .0268 | .0093 | .0029 | .0008 | .0002 | .0000 |
| | 2 | .1787 | .2852 | .2428 | .1540 | .0803 | .0358 | .0138 | .0046 | .0013 | .0003 |
| | 3 | .0533 | .1796 | .2428 | .2182 | .1517 | .0869 | .0422 | .0175 | .0062 | .0018 |
| | 4 | .0112 | .0798 | .1714 | .2182 | .2023 | .1491 | .0909 | .0467 | .0203 | .0074 |
| | 5 | .0018 | .0266 | .0907 | .1636 | .2023 | .1916 | .1468 | .0933 | .0497 | .0222 |
| | 6 | .0002 | .0069 | .0374 | .0955 | .1574 | .1916 | .1844 | .1451 | .0949 | .0518 |
| | 7 | .0000 | .0014 | .0122 | .0443 | .0974 | .1525 | .1844 | .1797 | .1443 | .0961 |
| | 8 | .0000 | .0002 | .0032 | .0166 | .0487 | .0981 | .1489 | .1797 | .1771 | .1442 |
| | 9 | .0000 | .0000 | .0007 | .0051 | .0198 | .0514 | .0980 | .1464 | .1771 | .1762 |
| | 10 | .0000 | .0000 | .0001 | .0013 | .0066 | .0220 | .0528 | .0976 | .1449 | .1762 |
| | 11 | .0000 | .0000 | .0000 | .0003 | .0018 | .0077 | .0233 | .0532 | .0970 | .1442 |
| | 12 | .0000 | .0000 | .0000 | .0000 | .0004 | .0022 | .0083 | .0237 | .0529 | .0961 |
| | 13 | .0000 | .0000 | .0000 | .0000 | .0001 | .0005 | .0024 | .0085 | .0233 | .0518 |
| | 14 | .0000 | .0000 | .0000 | .0000 | .0000 | .0001 | .0006 | .0024 | .0082 | .0222 |
| | 15 | .0000 | .0000 | .0000 | .0000 | .0000 | .0000 | .0001 | .0005 | .0022 | .0074 |
| | 16 | .0000 | .0000 | .0000 | .0000 | .0000 | .0000 | .0000 | .0001 | .0005 | .0018 |
| | 17 | .0000 | .0000 | .0000 | .0000 | .0000 | .0000 | .0000 | .0000 | .0001 | .0003 |
| | 18 | .0000 | .0000 | .0000 | .0000 | .0000 | .0000 | .0000 | .0000 | .0000 | .0000 |
| | 19 | .0000 | .0000 | .0000 | .0000 | .0000 | .0000 | .0000 | .0000 | .0000 | .0000 |
| 20 | 0 | .3585 | .1216 | .0388 | .0115 | .0032 | .0008 | .0002 | .0000 | .0000 | .0000 |
| | 1 | .3774 | .2702 | .1368 | .0576 | .0211 | .0068 | .0020 | .0005 | .0001 | .0000 |
| | 2 | .1887 | .2852 | .2293 | .1369 | .0669 | .0278 | .0100 | .0031 | .0008 | .0002 |
| | 3 | .0596 | .1901 | .2428 | .2054 | .1339 | .0716 | .0323 | .0123 | .0040 | .0011 |
| | 4 | .0133 | .0898 | .1821 | .2182 | .1897 | .1304 | .0738 | .0350 | .0139 | .0046 |
| | 5 | .0022 | .0319 | .1028 | .1746 | .2023 | .1789 | .1272 | .0746 | .0365 | .0148 |
| | 6 | .0003 | .0089 | .0454 | .1091 | .1686 | .1916 | .1712 | .1244 | .0746 | .0370 |
| | 7 | .0000 | .0020 | .0160 | .0545 | .1124 | .1643 | .1844 | .1659 | .1221 | .0739 |
| | 8 | .0000 | .0004 | .0046 | .0222 | .0609 | .1144 | .1614 | .1797 | .1623 | .1201 |
| | 9 | .0000 | .0001 | .0011 | .0074 | .0271 | .0654 | .1158 | .1597 | .1771 | .1602 |

## Appendix Table 1 (Continued)

| n | x | $\Pi$ | | | | | | | | | |
|---|---|-----|-----|-----|-----|-----|-----|-----|-----|-----|-----|
|   |   | .05 | .10 | .15 | .20 | .25 | .30 | .35 | .40 | .45 | .50 |
|   | 10 | .0000 | .0000 | .0002 | .0020 | .0099 | .0308 | .0686 | .1171 | .1593 | .1762 |
|   | 11 | .0000 | .0000 | .0000 | .0005 | .0030 | .0120 | .0336 | .0710 | .1185 | .1602 |
|   | 12 | .0000 | .0000 | .0000 | .0001 | .0008 | .0039 | .0136 | .0355 | .0727 | .1201 |
|   | 13 | .0000 | .0000 | .0000 | .0000 | .0002 | .0010 | .0045 | .0146 | .0366 | .0739 |
|   | 14 | .0000 | .0000 | .0000 | .0000 | .0000 | .0002 | .0012 | .0049 | .0150 | .0370 |
|   | 15 | .0000 | .0000 | .0000 | .0000 | .0000 | .0000 | .0003 | .0013 | .0049 | .0148 |
|   | 16 | .0000 | .0000 | .0000 | .0000 | .0000 | .0000 | .0000 | .0003 | .0013 | .0046 |
|   | 17 | .0000 | .0000 | .0000 | .0000 | .0000 | .0000 | .0000 | .0000 | .0002 | .0011 |
|   | 18 | .0000 | .0000 | .0000 | .0000 | .0000 | .0000 | .0000 | .0000 | .0000 | .0002 |
|   | 19 | .0000 | .0000 | .0000 | .0000 | .0000 | .0000 | .0000 | .0000 | .0000 | .0000 |
|   | 20 | .0000 | .0000 | .0000 | .0000 | .0000 | .0000 | .0000 | .0000 | .0000 | .0000 |

*Source:* This table is taken from National Bureau of Standards, *Tables of the Binomial Probability Distribution,* Applied Mathematics Series, U.S. Department of Commerce, 1950.

## Appendix Table 2

### AREAS UNDER THE STANDARD NORMAL CURVE

This table shows the area between zero (the mean of a standard normal variable) and $z$. For example, if $z = 1.50$, this is the shaded area shown below which equals .4332.

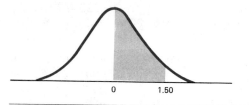

| z | .00 | .01 | .02 | .03 | .04 | .05 | .06 | .07 | .08 | .09 |
|---|-----|-----|-----|-----|-----|-----|-----|-----|-----|-----|
| 0.0 | .0000 | .0040 | .0080 | .0120 | .0160 | .0199 | .0239 | .0279 | .0319 | .0359 |
| 0.1 | .0398 | .0438 | .0478 | .0517 | .0557 | .0596 | .0636 | .0675 | .0714 | .0753 |
| 0.2 | .0793 | .0832 | .0871 | .0910 | .0948 | .0987 | .1026 | .1064 | .1103 | .1141 |
| 0.3 | .1179 | .1217 | .1255 | .1293 | .1331 | .1368 | .1406 | .1443 | .1480 | .1517 |
| 0.4 | .1554 | .1591 | .1628 | .1664 | .1700 | .1736 | .1772 | .1808 | .1844 | .1879 |
| 0.5 | .1915 | .1950 | .1985 | .2019 | .2054 | .2088 | .2123 | .2157 | .2190 | .2224 |
| 0.6 | .2257 | .2291 | .2324 | .2357 | .2389 | .2422 | .2454 | .2486 | .2517 | .2549 |
| 0.7 | .2580 | .2611 | .2642 | .2673 | .2704 | .2734 | .2764 | .2794 | .2823 | .2852 |
| 0.8 | .2881 | .2910 | .2939 | .2967 | .2995 | .3023 | .3051 | .3078 | .3106 | .3133 |
| 0.9 | .3159 | .3186 | .3212 | .3238 | .3264 | .3289 | .3315 | .3340 | .3365 | .3389 |
| 1.0 | .3413 | .3438 | .3461 | .3485 | .3508 | .3531 | .3554 | .3577 | .3599 | .3621 |
| 1.1 | .3643 | .3665 | .3686 | .3708 | .3729 | .3749 | .3770 | .3790 | .3810 | .3830 |
| 1.2 | .3849 | .3869 | .3888 | .3907 | .3925 | .3944 | .3962 | .3980 | .3997 | .4015 |
| 1.3 | .4032 | .4049 | .4066 | .4082 | .4099 | .4115 | .4131 | .4147 | .4162 | .4177 |
| 1.4 | .4192 | .4207 | .4222 | .4236 | .4251 | .4265 | .4279 | .4292 | .4306 | .4319 |
| 1.5 | .4332 | .4345 | .4357 | .4370 | .4382 | .4394 | .4406 | .4418 | .4429 | .4441 |
| 1.6 | .4452 | .4463 | .4474 | .4484 | .4495 | .4505 | .4515 | .4525 | .4535 | .4545 |
| 1.7 | .4554 | .4564 | .4573 | .4582 | .4591 | .4599 | .4608 | .4616 | .4625 | .4633 |
| 1.8 | .4641 | .4649 | .4656 | .4664 | .4671 | .4678 | .4686 | .4693 | .4699 | .4706 |
| 1.9 | .4713 | .4719 | .4726 | .4732 | .4738 | .4744 | .4750 | .4756 | .4761 | .4767 |
| 2.0 | .4772 | .4778 | .4783 | .4788 | .4793 | .4798 | .4803 | .4808 | .4812 | .4817 |
| 2.1 | .4821 | .4826 | .4830 | .4834 | .4838 | .4842 | .4846 | .4850 | .4854 | .4857 |
| 2.2 | .4861 | .4864 | .4868 | .4871 | .4875 | .4878 | .4881 | .4884 | .4887 | .4890 |
| 2.3 | .4893 | .4896 | .4898 | .4901 | .4904 | .4906 | .4909 | .4911 | .4913 | .4916 |
| 2.4 | .4918 | .4920 | .4922 | .4925 | .4927 | .4929 | .4931 | .4932 | .4934 | .4936 |
| 2.5 | .4938 | .4940 | .4941 | .4943 | .4945 | .4946 | .4948 | .4949 | .4951 | .4952 |
| 2.6 | .4953 | .4955 | .4956 | .4957 | .4959 | .4960 | .4961 | .4962 | .4963 | .4964 |
| 2.7 | .4965 | .4966 | .4967 | .4968 | .4969 | .4970 | .4971 | .4972 | .4973 | .4974 |
| 2.8 | .4974 | .4975 | .4976 | .4977 | .4977 | .4978 | .4979 | .4979 | .4980 | .4981 |
| 2.9 | .4981 | .4982 | .4982 | .4983 | .4984 | .4984 | .4985 | .4985 | .4986 | .4986 |
| 3.0 | .4987 | .4987 | .4987 | .4988 | .4988 | .4989 | .4989 | .4989 | .4990 | .4990 |

*Source:* This table is adapted from National Bureau of Standards, *Tables of Normal Probability Functions,* Applied Mathematics Series 23, U.S. Department of Commerce, 1953.

# Appendix Table 3

POISSON PROBABILITY DISTRIBUTION

This table shows the value of

$$P(x) = \frac{\mu^x e^{-\mu}}{x!}$$

for selected values of $x$ and for $\mu = .005$ to $8.0$.

| | | | | | $\mu$ | | | | | |
|---|---|---|---|---|---|---|---|---|---|---|
| $x$ | .005 | .01 | .02 | .03 | .04 | .05 | .06 | .07 | .08 | .09 |
| 0 | .9950 | .9900 | .9802 | .9704 | .9608 | .9512 | .9418 | .9324 | .9231 | .9139 |
| 1 | .0050 | .0099 | .0192 | .0291 | .0384 | .0476 | .0565 | .0653 | .0738 | .0823 |
| 2 | .0000 | .0000 | .0002 | .0004 | .0008 | .0012 | .0017 | .0023 | .0030 | .0037 |
| 3 | .0000 | .0000 | .0000 | .0000 | .0000 | .0000 | .0000 | .0001 | .0001 | .0001 |

| | | | | | $\mu$ | | | | | |
|---|---|---|---|---|---|---|---|---|---|---|
| $x$ | 0.1 | 0.2 | 0.3 | 0.4 | 0.5 | 0.6 | 0.7 | 0.8 | 0.9 | 1.0 |
| 0 | .9048 | .8187 | .7408 | .6703 | .6065 | .5488 | .4966 | .4493 | .4066 | .3679 |
| 1 | .0905 | .1637 | .2222 | .2681 | .3033 | .3293 | .3476 | .3595 | .3659 | .3679 |
| 2 | .0045 | .0164 | .0333 | .0536 | .0758 | .0988 | .1217 | .1438 | .1647 | .1839 |
| 3 | .0002 | .0011 | .0033 | .0072 | .0126 | .0198 | .0284 | .0383 | .0494 | .0613 |
| 4 | .0000 | .0001 | .0002 | .0007 | .0016 | .0030 | .0050 | .0077 | .0111 | .0153 |
| 5 | .0000 | .0000 | .0000 | .0001 | .0002 | .0004 | .0007 | .0012 | .0020 | .0031 |
| 6 | .0000 | .0000 | .0000 | .0000 | .0000 | .0000 | .0001 | .0002 | .0003 | .0005 |
| 7 | .0000 | .0000 | .0000 | .0000 | .0000 | .0000 | .0000 | .0000 | .0000 | .0001 |

| | | | | | $\mu$ | | | | | |
|---|---|---|---|---|---|---|---|---|---|---|
| $x$ | 1.1 | 1.2 | 1.3 | 1.4 | 1.5 | 1.6 | 1.7 | 1.8 | 1.9 | 2.0 |
| 0 | .3329 | .3012 | .2725 | .2466 | .2231 | .2019 | .1827 | .1653 | .1496 | .1353 |
| 1 | .3662 | .3614 | .3543 | .3452 | .3347 | .3230 | .3106 | .2975 | .2842 | .2707 |
| 2 | .2014 | .2169 | .2303 | .2417 | .2510 | .2584 | .2640 | .2678 | .2700 | .2707 |
| 3 | .0738 | .0867 | .0998 | .1128 | .1255 | .1378 | .1496 | .1607 | .1710 | .1804 |
| 4 | .0203 | .0260 | .0324 | .0395 | .0471 | .0551 | .0636 | .0723 | .0812 | .0902 |
| 5 | .0045 | .0062 | .0084 | .0111 | .0141 | .0176 | .0216 | .0260 | .0309 | .0361 |
| 6 | .0008 | .0012 | .0018 | .0026 | .0035 | .0047 | .0061 | .0078 | .0098 | .0120 |
| 7 | .0001 | .0002 | .0003 | .0005 | .0008 | .0011 | .0015 | .0020 | .0027 | .0034 |
| 8 | .0000 | .0000 | .0001 | .0001 | .0001 | .0002 | .0003 | .0005 | .0006 | .0009 |
| 9 | .0000 | .0000 | .0000 | .0000 | .0000 | .0000 | .0001 | .0001 | .0001 | .0002 |

| | | | | | $\mu$ | | | | | |
|---|---|---|---|---|---|---|---|---|---|---|
| $x$ | 2.1 | 2.2 | 2.3 | 2.4 | 2.5 | 2.6 | 2.7 | 2.8 | 2.9 | 3.0 |
| 0 | .1225 | .1108 | .1003 | .0907 | .0821 | .0743 | .0672 | .0608 | .0550 | .0498 |
| 1 | .2572 | .2438 | .2306 | .2177 | .2052 | .1931 | .1815 | .1703 | .1596 | .1494 |
| 2 | .2700 | .2681 | .2652 | .2613 | .2565 | .2510 | .2450 | .2384 | .2314 | .2240 |
| 3 | .1890 | .1966 | .2033 | .2090 | .2138 | .2176 | .2205 | .2225 | .2237 | .2240 |
| 4 | .0992 | .1082 | .1169 | .1254 | .1336 | .1414 | .1488 | .1557 | .1622 | .1680 |

## Appendix Table 3 (Continued)

| | | | | | | | | | |
|---|---|---|---|---|---|---|---|---|---|
| 5 | .0417 | .0476 | .0538 | .0602 | .0668 | .0735 | .0804 | .0872 | .0940 | .1008 |
| 6 | .0146 | .0174 | .0206 | .0241 | .0278 | .0319 | .0362 | .0407 | .0455 | .0504 |
| 7 | .0044 | .0055 | .0068 | .0083 | .0099 | .0118 | .0139 | .0163 | .0188 | .0216 |
| 8 | .0011 | .0015 | .0019 | .0025 | .0031 | .0038 | .0047 | .0057 | .0068 | .0081 |
| 9 | .0003 | .0004 | .0005 | .0007 | .0009 | .0011 | .0014 | .0018 | .0022 | .0027 |
| 10 | .0001 | .0001 | .0001 | .0002 | .0002 | .0003 | .0004 | .0005 | .0006 | .0008 |
| 11 | .0000 | .0000 | .0000 | .0000 | .0000 | .0001 | .0001 | .0001 | .0002 | .0002 |
| 12 | .0000 | .0000 | .0000 | .0000 | .0000 | .0000 | .0000 | .0000 | .0000 | .0001 |

$\mu$

| $x$ | 3.1 | 3.2 | 3.3 | 3.4 | 3.5 | 3.6 | 3.7 | 3.8 | 3.9 | 4.0 |
|---|---|---|---|---|---|---|---|---|---|---|
| 0 | .0450 | .0408 | .0369 | .0334 | .0302 | .0273 | .0247 | .0224 | .0202 | .0183 |
| 1 | .1397 | .1304 | .1217 | .1135 | .1057 | .0984 | .0915 | .0850 | .0789 | .0733 |
| 2 | .2165 | .2087 | .2008 | .1929 | .1850 | .1771 | .1692 | .1615 | .1539 | .1465 |
| 3 | .2237 | .2226 | .2209 | .2186 | .2158 | .2125 | .2087 | .2046 | .2001 | .1954 |
| 4 | .1734 | .1781 | .1823 | .1858 | .1888 | .1912 | .1931 | .1944 | .1951 | .1954 |
| 5 | .1075 | .1140 | .1203 | .1264 | .1322 | .1377 | .1429 | .1477 | .1522 | .1563 |
| 6 | .0555 | .0608 | .0662 | .0716 | .0771 | .0826 | .0881 | .0936 | .0989 | .1042 |
| 7 | .0246 | .0278 | .0312 | .0348 | .0385 | .0425 | .0466 | .0508 | .0551 | .0595 |
| 8 | .0095 | .0111 | .0129 | .0148 | .0169 | .0191 | .0215 | .0241 | .0269 | .0298 |
| 9 | .0033 | .0040 | .0047 | .0056 | .0066 | .0076 | .0089 | .0102 | .0116 | .0132 |
| 10 | .0010 | .0013 | .0016 | .0019 | .0023 | .0028 | .0033 | .0039 | .0045 | .0053 |
| 11 | .0003 | .0004 | .0005 | .0006 | .0007 | .0009 | .0011 | .0013 | .0016 | .0019 |
| 12 | .0001 | .0001 | .0001 | .0002 | .0002 | .0003 | .0003 | .0004 | .0005 | .0006 |
| 13 | .0000 | .0000 | .0000 | .0000 | .0001 | .0001 | .0001 | .0001 | .0002 | .0002 |
| 14 | .0000 | .0000 | .0000 | .0000 | .0000 | .0000 | .0000 | .0000 | .0000 | .0001 |

$\mu$

| $x$ | 4.1 | 4.2 | 4.3 | 4.4 | 4.5 | 4.6 | 4.7 | 4.8 | 4.9 | 5.0 |
|---|---|---|---|---|---|---|---|---|---|---|
| 0 | .0166 | .0150 | .0136 | .0123 | .0111 | .0101 | .0091 | .0082 | .0074 | .0067 |
| 1 | .0679 | .0630 | .0583 | .0540 | .0500 | .0462 | .0427 | .0395 | .0365 | .0337 |
| 2 | .1393 | .1323 | .1254 | .1188 | .1125 | .1063 | .1005 | .0948 | .0894 | .0842 |
| 3 | .1904 | .1852 | .1798 | .1743 | .1687 | .1631 | .1574 | .1517 | .1460 | .1404 |
| 4 | .1951 | .1944 | .1933 | .1917 | .1898 | .1875 | .1849 | .1820 | .1789 | .1755 |
| 5 | .1600 | .1633 | .1662 | .1687 | .1708 | .1725 | .1738 | .1747 | .1753 | .1755 |
| 6 | .1093 | .1143 | .1191 | .1237 | .1281 | .1323 | .1362 | .1398 | .1432 | .1462 |
| 7 | .0640 | .0686 | .0732 | .0778 | .0824 | .0869 | .0914 | .0959 | .1002 | .1044 |
| 8 | .0328 | .0360 | .0393 | .0428 | .0463 | .0500 | .0537 | .0575 | .0614 | .0653 |
| 9 | .0150 | .0168 | .0188 | .0209 | .0232 | .0255 | .0280 | .0307 | .0334 | .0363 |

## Appendix Table 3 (Continued)

$\mu$

| x | 4.1 | 4.2 | 4.3 | 4.4 | 4.5 | 4.6 | 4.7 | 4.8 | 4.9 | 5.0 |
|---|------|------|------|------|------|------|------|------|------|------|
| 10 | .0061 | .0071 | .0081 | .0092 | .0104 | .0118 | .0132 | .0147 | .0164 | .0181 |
| 11 | .0023 | .0027 | .0032 | .0037 | .0043 | .0049 | .0056 | .0064 | .0073 | .0082 |
| 12 | .0008 | .0009 | .0011 | .0014 | .0016 | .0019 | .0022 | .0026 | .0030 | .0034 |
| 13 | .0002 | .0003 | .0004 | .0005 | .0006 | .0007 | .0008 | .0009 | .0011 | .0013 |
| 14 | .0001 | .0001 | .0001 | .0001 | .0002 | .0002 | .0003 | .0003 | .0004 | .0005 |
| 15 | .0000 | .0000 | .0000 | .0000 | .0001 | .0001 | .0001 | .0001 | .0001 | .0002 |

$\mu$

| x | 5.1 | 5.2 | 5.3 | 5.4 | 5.5 | 5.6 | 5.7 | 5.8 | 5.9 | 6.0 |
|---|------|------|------|------|------|------|------|------|------|------|
| 0 | .0061 | .0055 | .0050 | .0045 | .0041 | .0037 | .0033 | .0030 | .0027 | .0025 |
| 1 | .0311 | .0287 | .0265 | .0244 | .0225 | .0207 | .0191 | .0176 | .0162 | .0149 |
| 2 | .0793 | .0746 | .0701 | .0659 | .0618 | .0580 | .0544 | .0509 | .0477 | .0446 |
| 3 | .1348 | .1293 | .1239 | .1185 | .1133 | .1082 | .1033 | .0985 | .0938 | .0892 |
| 4 | .1719 | .1681 | .1641 | .1600 | .1558 | .1515 | .1472 | .1428 | .1383 | .1339 |
| 5 | .1753 | .1748 | .1740 | .1728 | .1714 | .1697 | .1678 | .1656 | .1632 | .1606 |
| 6 | .1490 | .1515 | .1537 | .1555 | .1571 | .1584 | .1594 | .1601 | .1605 | .1606 |
| 7 | .1086 | .1125 | .1163 | .1200 | .1234 | .1267 | .1298 | .1326 | .1353 | .1377 |
| 8 | .0692 | .0731 | .0771 | .0810 | .0849 | .0887 | .0925 | .0962 | .0998 | .1033 |
| 9 | .0392 | .0423 | .0454 | .0486 | .0519 | .0552 | .0586 | .0620 | .0654 | .0688 |
| 10 | .0200 | .0220 | .0241 | .0262 | .0285 | .0309 | .0334 | .0359 | .0386 | .0413 |
| 11 | .0093 | .0104 | .0116 | .0129 | .0143 | .0157 | .0173 | .0190 | .0207 | .0225 |
| 12 | .0039 | .0045 | .0051 | .0058 | .0065 | .0073 | .0082 | .0092 | .0102 | .0113 |
| 13 | .0015 | .0018 | .0021 | .0024 | .0028 | .0032 | .0036 | .0041 | .0046 | .0052 |
| 14 | .0006 | .0007 | .0008 | .0009 | .0011 | .0013 | .0015 | .0017 | .0019 | .0022 |
| 15 | .0002 | .0002 | .0003 | .0003 | .0004 | .0005 | .0006 | .0007 | .0008 | .0009 |
| 16 | .0001 | .0001 | .0001 | .0001 | .0001 | .0002 | .0002 | .0002 | .0003 | .0003 |
| 17 | .0000 | .0000 | .0000 | .0000 | .0000 | .0001 | .0001 | .0001 | .0001 | .0001 |

$\mu$

| x | 6.1 | 6.2 | 6.3 | 6.4 | 6.5 | 6.6 | 6.7 | 6.8 | 6.9 | 7.0 |
|---|------|------|------|------|------|------|------|------|------|------|
| 0 | .0022 | .0020 | .0018 | .0017 | .0015 | .0014 | .0012 | .0011 | .0010 | .0009 |
| 1 | .0137 | .0126 | .0116 | .0106 | .0098 | .0090 | .0082 | .0076 | .0070 | .0064 |
| 2 | .0417 | .0390 | .0364 | .0340 | .0318 | .0296 | .0276 | .0258 | .0240 | .0223 |
| 3 | .0848 | .0806 | .0765 | .0726 | .0688 | .0652 | .0617 | .0584 | .0552 | .0521 |
| 4 | .1294 | .1249 | .1205 | .1162 | .1118 | .1076 | .1034 | .0992 | .0952 | .0912 |
| 5 | .1579 | .1549 | .1519 | .1487 | .1454 | .1420 | .1385 | .1349 | .1314 | .1277 |
| 6 | .1605 | .1601 | .1595 | .1586 | .1575 | .1562 | .1546 | .1529 | .1511 | .1490 |
| 7 | .1399 | .1418 | .1435 | .1450 | .1462 | .1472 | .1480 | .1486 | .1489 | .1490 |
| 8 | .1066 | .1099 | .1130 | .1160 | .1188 | .1215 | .1240 | .1263 | .1284 | .1304 |
| 9 | .0723 | .0757 | .0791 | .0825 | .0858 | .0891 | .0923 | .0954 | .0985 | .1014 |

# Appendix Table 3 (Continued)

$\mu$

| x | 6.1 | 6.2 | 6.3 | 6.4 | 6.5 | 6.6 | 6.7 | 6.8 | 6.9 | 7.0 |
|---|---|---|---|---|---|---|---|---|---|---|
| 10 | .0441 | .0469 | .0498 | .0528 | .0558 | .0588 | .0618 | .0649 | .0679 | .0710 |
| 11 | .0245 | .0265 | .0285 | .0307 | .0330 | .0353 | .0377 | .0401 | .0426 | .0452 |
| 12 | .0124 | .0137 | .0150 | .0164 | .0179 | .0194 | .0210 | .0227 | .0245 | .0264 |
| 13 | .0058 | .0065 | .0073 | .0081 | .0089 | .0098 | .0108 | .0119 | .0130 | .0142 |
| 14 | .0025 | .0029 | .0033 | .0037 | .0041 | .0046 | .0052 | .0058 | .0064 | .0071 |
| 15 | .0010 | .0012 | .0014 | .0016 | .0018 | .0020 | .0023 | .0026 | .0029 | .0033 |
| 16 | .0004 | .0005 | .0005 | .0006 | .0007 | .0008 | .0010 | .0011 | .0013 | .0014 |
| 17 | .0001 | .0002 | .0002 | .0002 | .0003 | .0003 | .0004 | .0004 | .0005 | .0006 |
| 18 | .0000 | .0001 | .0001 | .0001 | .0001 | .0001 | .0001 | .0002 | .0002 | .0002 |
| 19 | .0000 | .0000 | .0000 | .0000 | .0000 | .0000 | .0000 | .0001 | .0001 | .0001 |

$\mu$

| x | 7.1 | 7.2 | 7.3 | 7.4 | 7.5 | 7.6 | 7.7 | 7.8 | 7.9 | 8.0 |
|---|---|---|---|---|---|---|---|---|---|---|
| 0 | .0008 | .0007 | .0007 | .0006 | .0006 | .0005 | .0005 | .0004 | .0004 | .0003 |
| 1 | .0059 | .0054 | .0049 | .0045 | .0041 | .0038 | .0035 | .0032 | .0029 | .0027 |
| 2 | .0208 | .0194 | .0180 | .0167 | .0156 | .0145 | .0134 | .0125 | .0116 | .0107 |
| 3 | .0492 | .0464 | .0438 | .0413 | .0389 | .0366 | .0345 | .0324 | .0305 | .0286 |
| 4 | .0874 | .0836 | .0799 | .0764 | .0729 | .0696 | .0663 | .0632 | .0602 | .0573 |
| 5 | .1241 | .1204 | .1167 | .1130 | .1094 | .1057 | .1021 | .0986 | .0951 | .0916 |
| 6 | .1468 | .1445 | .1420 | .1394 | .1367 | .1339 | .1311 | .1282 | .1252 | .1221 |
| 7 | .1489 | .1486 | .1481 | .1474 | .1465 | .1454 | .1442 | .1428 | .1413 | .1396 |
| 8 | .1321 | .1337 | .1351 | .1363 | .1373 | .1382 | .1388 | .1392 | .1395 | .1396 |
| 9 | .1042 | .1070 | .1096 | .1121 | .1144 | .1167 | .1187 | .1207 | .1224 | .1241 |
| 10 | .0740 | .0770 | .0800 | .0829 | .0858 | .0887 | .0914 | .0941 | .0967 | .0993 |
| 11 | .0478 | .0504 | .0531 | .0558 | .0585 | .0613 | .0640 | .0667 | .0695 | .0722 |
| 12 | .0283 | .0303 | .0323 | .0344 | .0366 | .0388 | .0411 | .0434 | .0457 | .0481 |
| 13 | .0154 | .0168 | .0181 | .0196 | .0211 | .0227 | .0243 | .0260 | .0278 | .0296 |
| 14 | .0078 | .0086 | .0095 | .0104 | .0113 | .0123 | .0134 | .0145 | .0157 | .0169 |
| 15 | .0037 | .0041 | .0046 | .0051 | .0057 | .0062 | .0069 | .0075 | .0083 | .0090 |
| 16 | .0016 | .0019 | .0021 | .0024 | .0026 | .0030 | .0033 | .0037 | .0041 | .0045 |
| 17 | .0007 | .0008 | .0009 | .0010 | .0012 | .0013 | .0015 | .0017 | .0019 | .0021 |
| 18 | .0003 | .0003 | .0004 | .0004 | .0005 | .0006 | .0006 | .0007 | .0008 | .0009 |
| 19 | .0001 | .0001 | .0001 | .0002 | .0002 | .0002 | .0003 | .0003 | .0003 | .0004 |
| 20 | .0000 | .0000 | .0001 | .0001 | .0001 | .0001 | .0001 | .0001 | .0001 | .0002 |
| 21 | .0000 | .0000 | .0000 | .0000 | .0000 | .0000 | .0000 | .0000 | .0001 | .0001 |

# Appendix Table 4

VALUES OF $e^{-x}$

| $x$ | $e^{-x}$ | $x$ | $e^{-x}$ | $x$ | $e^{-x}$ |
|-----|----------|-----|----------|-----|----------|
| .0  | 1.000    | 1.5 | .223     | 3.0 | .050     |
| .1  | .905     | 1.6 | .202     | 3.1 | .045     |
| .2  | .819     | 1.7 | .183     | 3.2 | .041     |
| .3  | .741     | 1.8 | .165     | 3.3 | .037     |
| .4  | .670     | 1.9 | .150     | 3.4 | .033     |
| .5  | .607     | 2.0 | .135     | 3.5 | .030     |
| .6  | .549     | 2.1 | .122     | 3.6 | .027     |
| .7  | .497     | 2.2 | .111     | 3.7 | .025     |
| .8  | .449     | 2.3 | .100     | 3.8 | .022     |
| .9  | .407     | 2.4 | .091     | 3.9 | .020     |
| 1.0 | .368     | 2.5 | .082     | 4.0 | .018     |
| 1.1 | .333     | 2.6 | .074     | 4.5 | .011     |
| 1.2 | .301     | 2.7 | .067     | 5.0 | .007     |
| 1.3 | .273     | 2.8 | .061     | 6.0 | .002     |
| 1.4 | .247     | 2.9 | .055     | 7.0 | .001     |

# Appendix Table 5

RANDOM NUMBERS

---

```
11850 53535 04260 77609 93799 92171 45524 10968 30231 70864 29908
11851 41292 15201 66342 59155 46163 69248 31029 62034 21855 27863
11852 07320 22682 09595 44805 54593 53350 61354 14029 10195 18644
11853 77676 67772 45072 08940 02592 45976 82099 90739 77072 42081
11854 43227 20568 16309 23841 53173 39475 27282 82699 00022 96419

11855 90712 41695 67474 27567 93269 10163 94190 36188 41491 71217
11856 88103 21514 60787 33170 58215 89951 01634 98155 05154 08971
11857 72252 35791 84125 31962 81093 93068 41197 57779 88515 48002
11858 51702 49516 69510 19678 47298 11355 68459 96360 13436 66314
11859 63055 86998 22187 59898 96371 61370 35937 34292 00678 33505

11860 32373 57889 85880 66515 37489 37854 72926 23437 62233 38651
11861 71996 16525 25618 56577 69130 25035 93551 54394 81572 90624
11862 26912 70619 22576 22780 99118 18487 58801 36063 32886 60453
11863 74589 82677 13353 67658 17080 43212 34585 17179 86980 81899
11864 56041 53072 19912 47466 32585 41414 07564 80712 27286 07966

11865 09286 68067 84883 10023 78195 84711 85988 31545 39904 14984
11866 33610 84843 07145 38437 06148 06094 89601 96751 49124 55092
11867 14113 06396 59084 02534 09360 81918 77118 91640 92978 24815
11868 56302 89765 63857 42747 28592 41784 00822 60356 96389 11728
11869 06362 94540 29532 09994 55277 43897 63268 40481 00312 46039

11870 48568 34412 84939 54850 84317 92032 60430 49071 68962 28953
11871 65975 60965 77679 95782 67541 50654 09482 56111 98710 35803
11872 66686 32977 48472 30226 54226 72490 18395 37338 88279 79089
11873 51610 13000 73849 46654 30324 78000 72852 28934 83197 59003
11874 47600 86103 25788 08774 72020 04543 25849 88887 41159 30131

11875 34860 67572 83116 99579 81303 41889 56577 64142 51596 25329
11876 76649 50908 67006 29332 29689 68786 98987 34815 53512 20620
11877 78321 54309 85956 04976 37863 06711 72679 03405 28770 08515
11878 35775 21295 39621 02339 16537 42246 06571 81193 94930 05376
11879 06783 21338 89886 78826 02303 37886 70453 11021 62887 36855

11880 25887 53024 71881 51208 95739 98572 01903 68043 62661 71273
11881 37784 42100 70838 78963 10927 05448 25759 74051 47577 30196
11882 02120 59536 82996 22671 89267 65924 46725 69179 15182 59158
11883 55292 03836 28883 71134 08547 93204 09656 11671 29735 59573
11884 66186 43648 97926 80469 66412 73647 36779 84688 96862 51937

11885 55010 11479 55036 82146 37120 62328 56276 28906 45311 61818
11886 02322 18679 18478 30052 05666 84405 47513 09244 78978 91819
11887 78056 67836 82582 25809 20198 37222 62629 75733 77420 58746
11888 69812 88260 83519 10062 60865 35038 14665 18163 59351 25794
11889 84904 66864 26982 37928 32988 87652 81415 24416 93778 20391
```

## Appendix Table 5 (Continued)

```
11890 83143 47631 79772 08576 10311 17597 71049 63326 47168 05737
11891 44423 71197 91081 40781 72403 76245 31881 55716 89255 71997
11892 59882 58479 59609 80115 91569 23152 51781 85744 78640 80172
11893 74890 90405 75945 31645 61008 24448 42249 84909 29013 12529
11894 52174 64334 77631 19855 17723 02897 80427 20700 92210 92091

11895 41361 24347 53420 33639 83765 97935 83630 33765 21502 15589
11896 94585 84798 98480 08335 08728 60428 22282 76784 37316 08624
11897 36020 71966 61443 12554 67446 08676 46177 22422 87471 27283
11898 08112 59807 28404 60316 49676 52901 90604 48379 85233 52060
11899 05853 69681 52034 77617 78644 57321 14162 01849 94684 14628
```

*Source:* Rand Corporation, *A Million Random Digits with 100,000 Normal Deviates* (New York: Free Press, 1955), p. 238.

# Appendix Table 6

VALUES OF *t* THAT WILL BE EXCEEDED WITH SPECIFIED
PROBABILITIES

    This table shows the value of *t* where the area under the *t* distribution exceeding this value of *t* equals the specified amount. For example, the probability that a *t* variable with 14 degrees of freedom will exceed 1.345 equals .10.

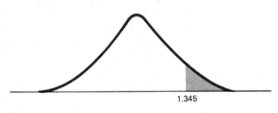

| Degrees of freedom | Probability |||||||
|---|---|---|---|---|---|---|---|
| | 0.40 | 0.25 | 0.10 | 0.05 | 0.025 | 0.01 | 0.005 |
| 1 | 0.325 | 1.000 | 3.078 | 6.314 | 12.706 | 31.821 | 63.657 |
| 2 | .289 | 0.816 | 1.886 | 2.920 | 4.303 | 6.965 | 9.925 |
| 3 | .277 | .765 | 1.638 | 2.353 | 3.182 | 4.541 | 5.841 |
| 4 | .271 | .741 | 1.533 | 2.132 | 2.776 | 3.747 | 4.604 |
| 5 | 0.267 | 0.727 | 1.476 | 2.015 | 2.571 | 3.365 | 4.032 |
| 6 | .265 | .718 | 1.440 | 1.943 | 2.447 | 3.143 | 3.707 |
| 7 | .263 | .711 | 1.415 | 1.895 | 2.365 | 2.998 | 3.499 |
| 8 | .262 | .706 | 1.397 | 1.860 | 2.306 | 2.896 | 3.355 |
| 9 | .261 | .703 | 1.383 | 1.833 | 2.262 | 2.821 | 3.250 |
| 10 | 0.260 | 0.700 | 1.372 | 1.812 | 2.228 | 2.764 | 3.169 |
| 11 | .260 | .697 | 1.363 | 1.796 | 2.201 | 2.718 | 3.106 |
| 12 | .259 | .695 | 1.356 | 1.782 | 2.179 | 2.681 | 3.055 |
| 13 | .259 | .694 | 1.350 | 1.771 | 2.160 | 2.650 | 3.012 |
| 14 | .258 | .692 | 1.345 | 1.761 | 2.145 | 2.624 | 2.977 |
| 15 | 0.258 | 0.691 | 1.341 | 1.753 | 2.131 | 2.602 | 2.947 |
| 16 | .258 | .690 | 1.337 | 1.746 | 2.120 | 2.583 | 2.921 |
| 17 | .257 | .689 | 1.333 | 1.740 | 2.110 | 2.567 | 2.898 |
| 18 | .257 | .688 | 1.330 | 1.734 | 2.101 | 2.552 | 2.878 |
| 19 | .257 | .688 | 1.328 | 1.729 | 2.093 | 2.539 | 2.861 |
| 20 | 0.257 | 0.687 | 1.325 | 1.725 | 2.086 | 2.528 | 2.845 |
| 21 | .257 | .686 | 1.323 | 1.721 | 2.080 | 2.518 | 2.831 |
| 22 | .256 | .686 | 1.321 | 1.717 | 2.074 | 2.508 | 2.819 |
| 23 | .256 | .685 | 1.319 | 1.714 | 2.069 | 2.500 | 2.807 |
| 24 | .256 | .685 | 1.318 | 1.711 | 2.064 | 2.492 | 2.797 |

## Appendix Table 6 (Continued)

| Degrees of freedom | 0.40 | 0.25 | 0.10 | Probability 0.05 | 0.025 | 0.01 | 0.005 |
|---|---|---|---|---|---|---|---|
| 25 | 0.256 | 0.684 | 1.316 | 1.708 | 2.060 | 2.485 | 2.787 |
| 26 | .256 | .684 | 1.315 | 1.706 | 2.056 | 2.479 | 2.779 |
| 27 | .256 | .684 | 1.314 | 1.703 | 2.052 | 2.473 | 2.771 |
| 28 | .256 | .683 | 1.313 | 1.701 | 2.048 | 2.467 | 2.763 |
| 29 | .256 | .683 | 1.311 | 1.699 | 2.045 | 2.462 | 2.756 |
| 30 | 0.256 | 0.683 | 1.310 | 1.697 | 2.042 | 2.457 | 2.750 |
| 40 | .255 | .681 | 1.303 | 1.684 | 2.021 | 2.423 | 2.704 |
| 60 | .254 | .679 | 1.296 | 1.671 | 2.000 | 2.390 | 2.660 |
| 120 | .254 | .677 | 1.289 | 1.658 | 1.980 | 2.358 | 2.617 |
| ∞ | .253 | .674 | 1.282 | 1.645 | 1.960 | 2.326 | 2.576 |

*Source: Biometrika Tables for Statisticians* (Cambridge, England: Cambridge University, 1954).

# Appendix Table 7a

### CHART PROVIDING 95 PERCENT CONFIDENCE INTERVAL FOR POPULATION PROPORTION, BASED ON SAMPLE PROPORTION*

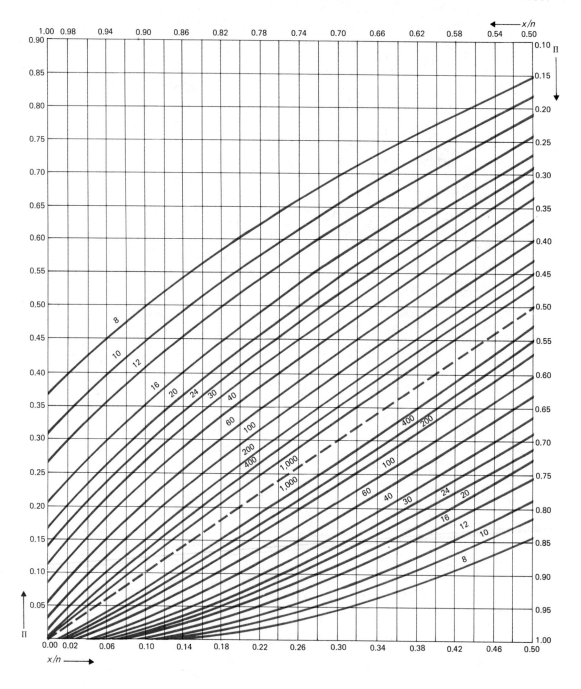

\* The sample proportion equals $x/n$; the population proportion equals $\Pi$.

# Appendix Table 7b

CHART PROVIDING 99 PERCENT CONFIDENCE INTERVAL FOR
POPULATION PROPORTION, BASED ON SAMPLE PROPORTION*

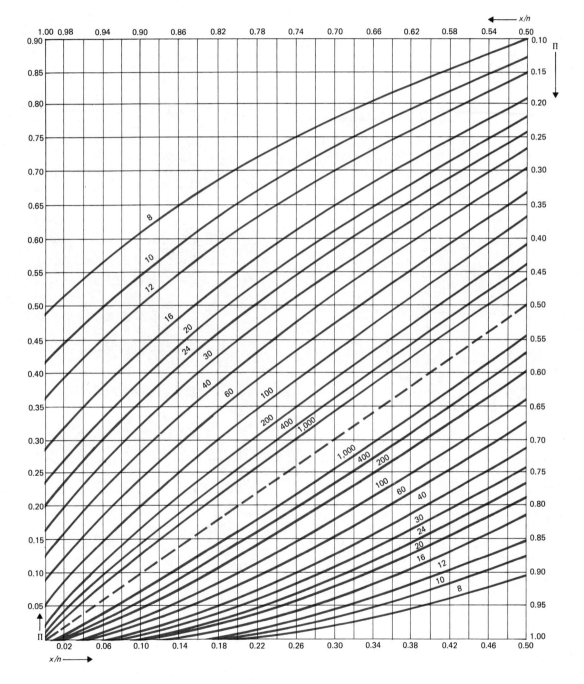

SOURCE: See Appendix Table 6.
* See footnote, Appendix Table 7a.

# Appendix Table 8

VALUES OF $\chi^2$ THAT WILL BE EXCEEDED WITH SPECIFIED PROBABILITIES

This table shows the value of $\chi^2$ where the area under the $\chi^2$ distribution exceeding this value of $\chi^2$ equals the specified amount. For example, the probability that a $\chi^2$ variable with 8 degrees of freedom will exceed 13.3616 equals 0.10.

| Degrees of freedom | Probabilities | | | | | | | |
|---|---|---|---|---|---|---|---|---|
| | 0.990 | 0.975 | 0.950 | 0.900 | 0.100 | 0.050 | 0.025 | 0.010 |
| 1 | $157088.10^{-9}$ | $982069.10^{-9}$ | $393214.10^{-8}$ | 0.0157908 | 2.70554 | 3.84146 | 5.02389 | 6.63490 |
| 2 | 0.0201007 | 0.0506356 | 0.102587 | 0.210720 | 4.60517 | 5.99147 | 7.37776 | 9.21034 |
| 3 | 0.114832 | 0.215795 | 0.351846 | 0.584375 | 6.25139 | 7.81473 | 9.34840 | 11.3449 |
| 4 | 0.297110 | 0.484419 | 0.710721 | 1.063623 | 7.77944 | 9.48773 | 11.1433 | 13.2767 |
| 5 | 0.554300 | 0.831211 | 1.145476 | 1.61031 | 9.23635 | 11.0705 | 12.8325 | 15.0863 |
| 6 | 0.872085 | 1.237347 | 1.63539 | 2.20413 | 10.6446 | 12.5916 | 14.4494 | 16.8119 |
| 7 | 1.239043 | 1.68987 | 2.16735 | 2.83311 | 12.0170 | 14.0671 | 16.0128 | 18.4753 |
| 8 | 1.646482 | 2.17973 | 2.73264 | 3.48954 | 13.3616 | 15.5073 | 17.5346 | 20.0902 |
| 9 | 2.087912 | 2.70039 | 3.32511 | 4.16816 | 14.6837 | 16.9190 | 19.0228 | 21.6660 |
| 10 | 2.55821 | 3.24697 | 3.94030 | 4.86518 | 15.9871 | 18.3070 | 20.4831 | 23.2093 |
| 11 | 3.05347 | 3.81575 | 4.57481 | 5.57779 | 17.2750 | 19.6751 | 21.9200 | 24.7250 |
| 12 | 3.57056 | 4.40379 | 5.22603 | 6.30380 | 18.5494 | 21.0261 | 23.3367 | 26.2170 |
| 13 | 4.10691 | 5.00874 | 5.89186 | 7.04150 | 19.8119 | 22.3621 | 24.7356 | 27.6883 |
| 14 | 4.66043 | 5.62872 | 6.57063 | 7.78953 | 21.0642 | 23.6848 | 26.1190 | 29.1413 |

# Appendix Table 8 (Continued)

| Degrees of freedom | Probabilities | | | | | | | | |
|---|---|---|---|---|---|---|---|---|---|
| | 0.990 | 0.975 | 0.950 | 0.900 | 0.100 | 0.050 | 0.025 | 0.010 |
| 15 | 5.22935 | 6.26214 | 7.26094 | 8.54675 | 22.3072 | 24.9958 | 27.4884 | 30.5779 |
| 16 | 5.81221 | 6.90766 | 7.96164 | 9.31223 | 23.5418 | 26.2962 | 28.8454 | 31.9999 |
| 17 | 6.40776 | 7.56418 | 8.67176 | 10.0852 | 24.7690 | 27.5871 | 30.1910 | 33.4087 |
| 18 | 7.01491 | 8.23075 | 9.39046 | 10.8649 | 25.9894 | 28.8693 | 31.5264 | 34.8053 |
| 19 | 7.63273 | 8.90655 | 10.1170 | 11.6509 | 27.2036 | 30.1435 | 32.8523 | 36.1908 |
| 20 | 8.26040 | 9.59083 | 10.8508 | 12.4426 | 28.4120 | 31.4104 | 34.1696 | 37.5662 |
| 21 | 8.89720 | 10.28293 | 11.5913 | 13.2396 | 29.6151 | 32.6705 | 35.4789 | 38.9321 |
| 22 | 9.54249 | 10.9823 | 12.3380 | 14.0415 | 30.8133 | 33.9244 | 36.7807 | 40.2894 |
| 23 | 10.19567 | 11.6885 | 13.0905 | 14.8479 | 32.0069 | 35.1725 | 38.0757 | 41.6384 |
| 24 | 10.8564 | 12.4011 | 13.8484 | 15.6587 | 33.1963 | 36.4151 | 39.3641 | 42.9798 |
| 25 | 11.5240 | 13.1197 | 14.6114 | 16.4734 | 34.3816 | 37.6525 | 40.6465 | 44.3141 |
| 26 | 12.1981 | 13.8439 | 15.3791 | 17.2919 | 35.5631 | 38.8852 | 41.9232 | 45.6417 |
| 27 | 12.8786 | 14.5733 | 16.1513 | 18.1138 | 36.7412 | 40.1133 | 43.1944 | 46.9630 |
| 28 | 13.5648 | 15.3079 | 16.9279 | 18.9392 | 37.9159 | 41.3372 | 44.4607 | 48.2782 |
| 29 | 14.2565 | 16.0471 | 17.7083 | 19.7677 | 39.0875 | 42.5569 | 45.7222 | 49.5879 |
| 30 | 14.9535 | 16.7908 | 18.4926 | 20.5992 | 40.2560 | 43.7729 | 46.9792 | 50.8922 |
| 40 | 22.1643 | 24.4331 | 26.5093 | 29.0505 | 51.8050 | 55.7585 | 59.3417 | 63.6907 |
| 50 | 29.7067 | 32.3574 | 34.7642 | 37.6886 | 63.1671 | 67.5048 | 71.4202 | 76.1539 |
| 60 | 37.4848 | 40.4817 | 43.1879 | 46.4589 | 74.3970 | 79.0819 | 83.2976 | 88.3794 |
| 70 | 45.4418 | 48.7576 | 51.7393 | 55.3290 | 85.5271 | 90.5312 | 95.0231 | 100.425 |
| 80 | 53.5400 | 57.1532 | 60.3915 | 64.2778 | 96.5782 | 101.879 | 106.629 | 112.329 |
| 90 | 61.7541 | 65.6466 | 69.1260 | 73.2912 | 107.565 | 113.145 | 118.136 | 124.116 |
| 100 | 70.0648 | 74.2219 | 77.9295 | 82.3581 | 118.498 | 124.342 | 129.561 | 135.807 |

*Source:* See Appendix Table 6.

# Appendix Table 9

VALUE OF AN *F* VARIABLE THAT IS EXCEEDED WITH
PROBABILITY EQUAL TO .05

*Degrees of freedom for numerator*

| | | 1 | 2 | 3 | 4 | 5 | 6 | 7 | 8 | 9 |
|---|---|---|---|---|---|---|---|---|---|---|
| | 1 | 161.4 | 199.5 | 215.7 | 224.6 | 230.2 | 234.0 | 236.8 | 238.9 | 240.5 |
| | 2 | 18.51 | 19.00 | 19.16 | 19.25 | 19.30 | 19.33 | 19.35 | 19.37 | 19.38 |
| | 3 | 10.13 | 9.55 | 9.28 | 9.12 | 9.01 | 8.94 | 8.89 | 8.85 | 8.81 |
| | 4 | 7.71 | 6.94 | 6.59 | 6.39 | 6.26 | 6.16 | 6.09 | 6.04 | 6.00 |
| | 5 | 6.61 | 5.79 | 5.41 | 5.19 | 5.05 | 4.95 | 4.88 | 4.82 | 4.77 |
| | 6 | 5.99 | 5.14 | 4.76 | 4.53 | 4.39 | 4.28 | 4.21 | 4.15 | 4.10 |
| | 7 | 5.59 | 4.74 | 4.35 | 4.12 | 3.97 | 3.87 | 3.79 | 3.73 | 3.68 |
| | 8 | 5.32 | 4.46 | 4.07 | 3.84 | 3.69 | 3.58 | 3.50 | 3.44 | 3.39 |
| | 9 | 5.12 | 4.26 | 3.86 | 3.63 | 3.48 | 3.37 | 3.29 | 3.23 | 3.18 |
| | 10 | 4.96 | 4.10 | 3.71 | 3.48 | 3.33 | 3.22 | 3.14 | 3.07 | 3.02 |
| | 11 | 4.84 | 3.98 | 3.59 | 3.36 | 3.20 | 3.09 | 3.01 | 2.95 | 2.90 |
| *Degrees of freedom for denominator* | 12 | 4.75 | 3.89 | 3.49 | 3.26 | 3.11 | 3.00 | 2.91 | 2.85 | 2.80 |
| | 13 | 4.67 | 3.81 | 3.41 | 3.18 | 3.03 | 2.92 | 2.83 | 2.77 | 2.71 |
| | 14 | 4.60 | 3.74 | 3.34 | 3.11 | 2.96 | 2.85 | 2.76 | 2.70 | 2.65 |
| | 15 | 4.54 | 3.68 | 3.29 | 3.06 | 2.90 | 2.79 | 2.71 | 2.64 | 2.59 |
| | 16 | 4.49 | 3.63 | 3.24 | 3.01 | 2.85 | 2.74 | 2.66 | 2.59 | 2.54 |
| | 17 | 4.45 | 3.59 | 3.20 | 2.96 | 2.81 | 2.70 | 2.61 | 2.55 | 2.49 |
| | 18 | 4.41 | 3.55 | 3.16 | 2.93 | 2.77 | 2.66 | 2.58 | 2.51 | 2.46 |
| | 19 | 4.38 | 3.52 | 3.13 | 2.90 | 2.74 | 2.63 | 2.54 | 2.48 | 2.42 |
| | 20 | 4.35 | 3.49 | 3.10 | 2.87 | 2.71 | 2.60 | 2.51 | 2.45 | 2.39 |
| | 21 | 4.32 | 3.47 | 3.07 | 2.84 | 2.68 | 2.57 | 2.49 | 2.42 | 2.37 |
| | 22 | 4.30 | 3.44 | 3.05 | 2.82 | 2.66 | 2.55 | 2.46 | 2.40 | 2.34 |
| | 23 | 4.28 | 3.42 | 3.03 | 2.80 | 2.64 | 2.53 | 2.44 | 2.37 | 2.32 |
| | 24 | 4.26 | 3.40 | 3.01 | 2.78 | 2.62 | 2.51 | 2.42 | 2.36 | 2.30 |
| | 25 | 4.24 | 3.39 | 2.99 | 2.76 | 2.60 | 2.49 | 2.40 | 2.34 | 2.28 |
| | 26 | 4.23 | 3.37 | 2.98 | 2.74 | 2.59 | 2.47 | 2.39 | 2.32 | 2.27 |
| | 27 | 4.21 | 3.35 | 2.96 | 2.73 | 2.57 | 2.46 | 2.37 | 2.31 | 2.25 |
| | 28 | 4.20 | 3.34 | 2.95 | 2.71 | 2.56 | 2.45 | 2.36 | 2.29 | 2.24 |
| | 29 | 4.18 | 3.33 | 2.93 | 2.70 | 2.55 | 2.43 | 2.35 | 2.28 | 2.22 |
| | 30 | 4.17 | 3.32 | 2.92 | 2.69 | 2.53 | 2.42 | 2.33 | 2.27 | 2.21 |
| | 40 | 4.08 | 3.23 | 2.84 | 2.61 | 2.45 | 2.34 | 2.25 | 2.18 | 2.12 |
| | 60 | 4.00 | 3.15 | 2.76 | 2.53 | 2.37 | 2.25 | 2.17 | 2.10 | 2.04 |
| | 120 | 3.92 | 3.07 | 2.68 | 2.45 | 2.29 | 2.17 | 2.09 | 2.02 | 1.96 |
| | $\infty$ | 3.84 | 3.00 | 2.60 | 2.37 | 2.21 | 2.10 | 2.01 | 1.94 | 1.88 |

# Appendix Table 9 (Continued)

*Degrees of freedom for numerator*

| | 10 | 12 | 15 | 20 | 24 | 30 | 40 | 60 | 120 | ∞ |
|---|---|---|---|---|---|---|---|---|---|---|
| 1 | 241.9 | 243.9 | 245.9 | 248.0 | 249.1 | 250.1 | 251.1 | 252.2 | 253.3 | 254.3 |
| 2 | 19.40 | 19.41 | 19.43 | 19.45 | 19.45 | 19.46 | 19.47 | 19.48 | 19.49 | 19.50 |
| 3 | 8.79 | 8.74 | 8.70 | 8.66 | 8.64 | 8.62 | 8.59 | 8.57 | 8.55 | 8.53 |
| 4 | 5.96 | 5.91 | 5.86 | 5.80 | 5.77 | 5.75 | 5.72 | 5.69 | 5.66 | 5.63 |
| 5 | 4.74 | 4.68 | 4.62 | 4.56 | 4.53 | 4.50 | 4.46 | 4.43 | 4.40 | 4.36 |
| 6 | 4.06 | 4.00 | 3.94 | 3.87 | 3.84 | 3.81 | 3.77 | 3.74 | 3.70 | 3.67 |
| 7 | 3.64 | 3.57 | 3.51 | 3.44 | 3.41 | 3.38 | 3.34 | 3.30 | 3.27 | 3.23 |
| 8 | 3.35 | 3.28 | 3.22 | 3.15 | 3.12 | 3.08 | 3.04 | 3.01 | 2.97 | 2.93 |
| 9 | 3.14 | 3.07 | 3.01 | 2.94 | 2.90 | 2.86 | 2.83 | 2.79 | 2.75 | 2.71 |
| 10 | 2.98 | 2.91 | 2.85 | 2.77 | 2.74 | 2.70 | 2.66 | 2.62 | 2.58 | 2.54 |
| 11 | 2.85 | 2.79 | 2.72 | 2.65 | 2.61 | 2.57 | 2.53 | 2.49 | 2.45 | 2.40 |
| 12 | 2.75 | 2.69 | 2.62 | 2.54 | 2.51 | 2.47 | 2.43 | 2.38 | 2.34 | 2.30 |
| 13 | 2.67 | 2.60 | 2.53 | 2.46 | 2.42 | 2.38 | 2.34 | 2.30 | 2.25 | 2.21 |
| 14 | 2.60 | 2.53 | 2.46 | 2.39 | 2.35 | 2.31 | 2.27 | 2.22 | 2.18 | 2.13 |
| 15 | 2.54 | 2.48 | 2.40 | 2.33 | 2.29 | 2.25 | 2.20 | 2.16 | 2.11 | 2.07 |
| 16 | 2.49 | 2.42 | 2.35 | 2.28 | 2.24 | 2.19 | 2.15 | 2.11 | 2.06 | 2.01 |
| 17 | 2.45 | 2.38 | 2.31 | 2.23 | 2.19 | 2.15 | 2.10 | 2.06 | 2.01 | 1.96 |
| 18 | 2.41 | 2.34 | 2.27 | 2.19 | 2.15 | 2.11 | 2.06 | 2.02 | 1.97 | 1.92 |
| 19 | 2.38 | 2.31 | 2.23 | 2.16 | 2.11 | 2.07 | 2.03 | 1.98 | 1.93 | 1.88 |
| 20 | 2.35 | 2.28 | 2.20 | 2.12 | 2.08 | 2.04 | 1.99 | 1.95 | 1.90 | 1.84 |
| 21 | 2.32 | 2.25 | 2.18 | 2.10 | 2.05 | 2.01 | 1.96 | 1.92 | 1.87 | 1.81 |
| 22 | 2.30 | 2.23 | 2.15 | 2.07 | 2.03 | 1.98 | 1.94 | 1.89 | 1.84 | 1.78 |
| 23 | 2.27 | 2.20 | 2.13 | 2.05 | 2.01 | 1.96 | 1.91 | 1.86 | 1.81 | 1.76 |
| 24 | 2.25 | 2.18 | 2.11 | 2.03 | 1.98 | 1.94 | 1.89 | 1.84 | 1.79 | 1.73 |
| 25 | 2.24 | 2.16 | 2.09 | 2.01 | 1.96 | 1.92 | 1.87 | 1.82 | 1.77 | 1.71 |
| 26 | 2.22 | 2.15 | 2.07 | 1.99 | 1.95 | 1.90 | 1.85 | 1.80 | 1.75 | 1.69 |
| 27 | 2.20 | 2.13 | 2.06 | 1.97 | 1.93 | 1.88 | 1.84 | 1.79 | 1.73 | 1.67 |
| 28 | 2.19 | 2.12 | 2.04 | 1.96 | 1.91 | 1.87 | 1.82 | 1.77 | 1.71 | 1.65 |
| 29 | 2.18 | 2.10 | 2.03 | 1.94 | 1.90 | 1.85 | 1.81 | 1.75 | 1.70 | 1.64 |
| 30 | 2.16 | 2.09 | 2.01 | 1.93 | 1.89 | 1.84 | 1.79 | 1.74 | 1.68 | 1.62 |
| 40 | 2.08 | 2.00 | 1.92 | 1.84 | 1.79 | 1.74 | 1.69 | 1.64 | 1.58 | 1.51 |
| 60 | 1.99 | 1.92 | 1.84 | 1.75 | 1.70 | 1.65 | 1.59 | 1.53 | 1.47 | 1.39 |
| 120 | 1.91 | 1.83 | 1.75 | 1.66 | 1.61 | 1.55 | 1.50 | 1.43 | 1.35 | 1.25 |
| ∞ | 1.83 | 1.75 | 1.67 | 1.57 | 1.52 | 1.46 | 1.39 | 1.32 | 1.22 | 1.00 |

*Degrees of freedom for denominator*

*Source:* See Appendix Table 6.

# Appendix Table 10

### VALUE OF AN *F* VARIABLE EXCEEDED WITH PROBABILITY EQUAL TO .01

*Degrees of freedom for numerator*

| | 1 | 2 | 3 | 4 | 5 | 6 | 7 | 8 | 9 |
|---|---|---|---|---|---|---|---|---|---|
| 1 | 4052 | 4999.5 | 5403 | 5625 | 5764 | 5859 | 5928 | 5982 | 6022 |
| 2 | 98.50 | 99.00 | 99.17 | 99.25 | 99.30 | 99.33 | 99.36 | 99.37 | 99.39 |
| 3 | 34.12 | 30.82 | 29.46 | 28.71 | 28.24 | 27.91 | 27.67 | 27.49 | 27.35 |
| 4 | 21.20 | 18.00 | 16.69 | 15.98 | 15.52 | 15.21 | 14.98 | 14.80 | 14.66 |
| 5 | 16.26 | 13.27 | 12.06 | 11.39 | 10.97 | 10.67 | 10.46 | 10.29 | 10.16 |
| 6 | 13.75 | 10.92 | 9.78 | 9.15 | 8.75 | 8.47 | 8.26 | 8.10 | 7.98 |
| 7 | 12.25 | 9.55 | 8.45 | 7.85 | 7.46 | 7.19 | 6.99 | 6.84 | 6.72 |
| 8 | 11.26 | 8.65 | 7.59 | 7.01 | 6.63 | 6.37 | 6.18 | 6.03 | 5.91 |
| 9 | 10.56 | 8.02 | 6.99 | 6.42 | 6.06 | 5.80 | 5.61 | 5.47 | 5.35 |
| 10 | 10.04 | 7.56 | 6.55 | 5.99 | 5.64 | 5.39 | 5.20 | 5.06 | 4.94 |
| 11 | 9.65 | 7.21 | 6.22 | 5.67 | 5.32 | 5.07 | 4.89 | 4.74 | 4.63 |
| 12 | 9.33 | 6.93 | 5.95 | 5.41 | 5.06 | 4.82 | 4.64 | 4.50 | 4.39 |
| 13 | 9.07 | 6.70 | 5.74 | 5.21 | 4.86 | 4.62 | 4.44 | 4.30 | 4.19 |
| 14 | 8.86 | 6.51 | 5.56 | 5.04 | 4.69 | 4.46 | 4.28 | 4.14 | 4.03 |
| 15 | 8.68 | 6.36 | 5.42 | 4.89 | 4.56 | 4.32 | 4.14 | 4.00 | 3.89 |
| 16 | 8.53 | 6.23 | 5.29 | 4.77 | 4.44 | 4.20 | 4.03 | 3.89 | 3.78 |
| 17 | 8.40 | 6.11 | 5.18 | 4.67 | 4.34 | 4.10 | 3.93 | 3.79 | 3.68 |
| 18 | 8.29 | 6.01 | 5.09 | 4.58 | 4.25 | 4.01 | 3.84 | 3.71 | 3.60 |
| 19 | 8.18 | 5.93 | 5.01 | 4.50 | 4.17 | 3.94 | 3.77 | 3.63 | 3.52 |
| 20 | 8.10 | 5.85 | 4.94 | 4.43 | 4.10 | 3.87 | 3.70 | 3.56 | 3.46 |
| 21 | 8.02 | 5.78 | 4.87 | 4.37 | 4.04 | 3.81 | 3.64 | 3.51 | 3.40 |
| 22 | 7.95 | 5.72 | 4.82 | 4.31 | 3.99 | 3.76 | 3.59 | 3.45 | 3.35 |
| 23 | 7.88 | 5.66 | 4.76 | 4.26 | 3.94 | 3.71 | 3.54 | 3.41 | 3.30 |
| 24 | 7.82 | 5.61 | 4.72 | 4.22 | 3.90 | 3.67 | 3.50 | 3.36 | 3.26 |
| 25 | 7.77 | 5.57 | 4.68 | 4.18 | 3.85 | 3.63 | 3.46 | 3.32 | 3.22 |
| 26 | 7.72 | 5.53 | 4.64 | 4.14 | 3.82 | 3.59 | 3.42 | 3.29 | 3.18 |
| 27 | 7.68 | 5.49 | 4.60 | 4.11 | 3.78 | 3.56 | 3.39 | 3.26 | 3.15 |
| 28 | 7.64 | 5.45 | 4.57 | 4.07 | 3.75 | 3.53 | 3.36 | 3.23 | 3.12 |
| 29 | 7.60 | 5.42 | 4.54 | 4.04 | 3.73 | 3.50 | 3.33 | 3.20 | 3.09 |
| 30 | 7.56 | 5.39 | 4.51 | 4.02 | 3.70 | 3.47 | 3.30 | 3.17 | 3.07 |
| 40 | 7.31 | 5.18 | 4.31 | 3.83 | 3.51 | 3.29 | 3.12 | 2.99 | 2.89 |
| 60 | 7.08 | 4.98 | 4.13 | 3.65 | 3.34 | 3.12 | 2.95 | 2.82 | 2.72 |
| 120 | 6.85 | 4.79 | 3.95 | 3.48 | 3.17 | 2.96 | 2.79 | 2.66 | 2.56 |
| ∞ | 6.63 | 4.61 | 3.78 | 3.32 | 3.02 | 2.80 | 2.64 | 2.51 | 2.41 |

*Degrees of freedom for denominator*

# Appendix Table 10 (Continued)

*Degrees of freedom for numerator*

| | 10 | 12 | 15 | 20 | 24 | 30 | 40 | 60 | 120 | ∞ |
|---|---|---|---|---|---|---|---|---|---|---|
| 1 | 6056 | 6106 | 6157 | 6209 | 6235 | 6261 | 6287 | 6313 | 6339 | 6366 |
| 2 | 99.40 | 99.42 | 99.43 | 99.45 | 99.46 | 99.47 | 99.47 | 99.48 | 99.49 | 99.50 |
| 3 | 27.23 | 27.05 | 26.87 | 26.69 | 26.60 | 26.50 | 26.41 | 26.32 | 26.22 | 26.13 |
| 4 | 14.55 | 14.37 | 14.20 | 14.02 | 13.93 | 13.84 | 13.75 | 13.65 | 13.56 | 13.46 |
| 5 | 10.05 | 9.89 | 9.72 | 9.55 | 9.47 | 9.38 | 9.29 | 9.20 | 9.11 | 9.02 |
| 6 | 7.87 | 7.72 | 7.56 | 7.40 | 7.31 | 7.23 | 7.14 | 7.06 | 6.97 | 6.88 |
| 7 | 6.62 | 6.47 | 6.31 | 6.16 | 6.07 | 5.99 | 5.91 | 5.82 | 5.74 | 5.65 |
| 8 | 5.81 | 5.67 | 5.52 | 5.36 | 5.28 | 5.20 | 5.12 | 5.03 | 4.95 | 4.86 |
| 9 | 5.26 | 5.11 | 4.96 | 4.81 | 4.73 | 4.65 | 4.57 | 4.48 | 4.40 | 4.31 |
| 10 | 4.85 | 4.71 | 4.56 | 4.41 | 4.33 | 4.25 | 4.17 | 4.08 | 4.00 | 3.91 |
| 11 | 4.54 | 4.40 | 4.25 | 4.10 | 4.02 | 3.94 | 3.86 | 3.78 | 3.69 | 3.60 |
| 12 | 4.30 | 4.16 | 4.01 | 3.86 | 3.78 | 3.70 | 3.62 | 3.54 | 3.45 | 3.36 |
| 13 | 4.10 | 3.96 | 3.82 | 3.66 | 3.59 | 3.51 | 3.43 | 3.34 | 3.25 | 3.17 |
| 14 | 3.94 | 3.80 | 3.66 | 3.51 | 3.43 | 3.35 | 3.27 | 3.18 | 3.09 | 3.00 |
| 15 | 3.80 | 3.67 | 3.52 | 3.37 | 3.29 | 3.21 | 3.13 | 3.05 | 2.96 | 2.87 |
| 16 | 3.69 | 3.55 | 3.41 | 3.26 | 3.18 | 3.10 | 3.02 | 2.93 | 2.84 | 2.75 |
| 17 | 3.59 | 3.46 | 3.31 | 3.16 | 3.08 | 3.00 | 2.92 | 2.83 | 2.75 | 2.65 |
| 18 | 3.51 | 3.37 | 3.23 | 3.08 | 3.00 | 2.92 | 2.84 | 2.75 | 2.66 | 2.57 |
| 19 | 3.43 | 3.30 | 3.15 | 3.00 | 2.92 | 2.84 | 2.76 | 2.67 | 2.58 | 2.49 |
| 20 | 3.37 | 3.23 | 3.09 | 2.94 | 2.86 | 2.78 | 2.69 | 2.61 | 2.52 | 2.42 |
| 21 | 3.31 | 3.17 | 3.03 | 2.88 | 2.80 | 2.72 | 2.64 | 2.55 | 2.46 | 2.36 |
| 22 | 3.26 | 3.12 | 2.98 | 2.83 | 2.75 | 2.67 | 2.58 | 2.50 | 2.40 | 2.31 |
| 23 | 3.21 | 3.07 | 2.93 | 2.78 | 2.70 | 2.62 | 2.54 | 2.45 | 2.35 | 2.26 |
| 24 | 3.17 | 3.03 | 2.89 | 2.74 | 2.66 | 2.58 | 2.49 | 2.40 | 2.31 | 2.21 |
| 25 | 3.13 | 2.99 | 2.85 | 2.70 | 2.62 | 2.54 | 2.45 | 2.36 | 2.27 | 2.17 |
| 26 | 3.09 | 2.96 | 2.81 | 2.66 | 2.58 | 2.50 | 2.42 | 2.33 | 2.23 | 2.13 |
| 27 | 3.06 | 2.93 | 2.78 | 2.63 | 2.55 | 2.47 | 2.38 | 2.29 | 2.20 | 2.10 |
| 28 | 3.03 | 2.90 | 2.75 | 2.60 | 2.52 | 2.44 | 2.35 | 2.26 | 2.17 | 2.06 |
| 29 | 3.00 | 2.87 | 2.73 | 2.57 | 2.49 | 2.41 | 2.33 | 2.23 | 2.14 | 2.03 |
| 30 | 2.98 | 2.84 | 2.70 | 2.55 | 2.47 | 2.39 | 2.30 | 2.21 | 2.11 | 2.01 |
| 40 | 2.80 | 2.66 | 2.52 | 2.37 | 2.29 | 2.20 | 2.11 | 2.02 | 1.92 | 1.80 |
| 60 | 2.63 | 2.50 | 2.35 | 2.20 | 2.12 | 2.03 | 1.94 | 1.84 | 1.73 | 1.60 |
| 120 | 2.47 | 2.34 | 2.19 | 2.03 | 1.95 | 1.86 | 1.76 | 1.66 | 1.53 | 1.38 |
| ∞ | 2.32 | 2.18 | 2.04 | 1.88 | 1.79 | 1.70 | 1.59 | 1.47 | 1.32 | 1.00 |

*Degrees of freedom for denominator*

*Source:* See Appendix Table 6.

# Appendix Table 11

Values of $d_L$ and $d_U$ for the Durbin-Watson Test

*A. Where $\alpha = .05$*

| | $k = 1$ | | $k = 2$ | | $k = 3$ | | $k = 4$ | | $k = 5$ | |
|---|---|---|---|---|---|---|---|---|---|---|
| $n$ | $d_L$ | $d_U$ | $d_L$ | $d_U$ | $d_L$ | $d_U$ | $d_L$ | $d_U$ | $d_L$ | $d_U$ |
| 15 | 1.08 | 1.36 | 0.95 | 1.54 | 0.82 | 1.75 | 0.69 | 1.97 | 0.56 | 2.21 |
| 16 | 1.10 | 1.37 | 0.98 | 1.54 | 0.86 | 1.73 | 0.74 | 1.93 | 0.62 | 2.15 |
| 17 | 1.13 | 1.38 | 1.02 | 1.54 | 0.90 | 1.71 | 0.78 | 1.90 | 0.67 | 2.10 |
| 18 | 1.16 | 1.39 | 1.05 | 1.53 | 0.93 | 1.69 | 0.82 | 1.87 | 0.71 | 2.06 |
| 19 | 1.18 | 1.40 | 1.08 | 1.53 | 0.97 | 1.68 | 0.86 | 1.85 | 0.75 | 2.02 |
| 20 | 1.20 | 1.41 | 1.10 | 1.54 | 1.00 | 1.68 | 0.90 | 1.83 | 0.79 | 1.99 |
| 21 | 1.22 | 1.42 | 1.13 | 1.54 | 1.03 | 1.67 | 0.93 | 1.81 | 0.83 | 1.96 |
| 22 | 1.24 | 1.43 | 1.15 | 1.54 | 1.05 | 1.66 | 0.96 | 1.80 | 0.86 | 1.94 |
| 23 | 1.26 | 1.44 | 1.17 | 1.54 | 1.08 | 1.66 | 0.99 | 1.79 | 0.90 | 1.92 |
| 24 | 1.27 | 1.45 | 1.19 | 1.55 | 1.10 | 1.66 | 1.01 | 1.78 | 0.93 | 1.90 |
| 25 | 1.29 | 1.45 | 1.21 | 1.55 | 1.12 | 1.66 | 1.04 | 1.77 | 0.95 | 1.89 |
| 26 | 1.30 | 1.46 | 1.22 | 1.55 | 1.14 | 1.65 | 1.06 | 1.76 | 0.98 | 1.88 |
| 27 | 1.32 | 1.47 | 1.24 | 1.56 | 1.16 | 1.65 | 1.08 | 1.76 | 1.01 | 1.86 |
| 28 | 1.33 | 1.48 | 1.26 | 1.56 | 1.18 | 1.65 | 1.10 | 1.75 | 1.03 | 1.85 |
| 29 | 1.34 | 1.48 | 1.27 | 1.56 | 1.20 | 1.65 | 1.12 | 1.74 | 1.05 | 1.84 |
| 30 | 1.35 | 1.49 | 1.28 | 1.57 | 1.21 | 1.65 | 1.14 | 1.74 | 1.07 | 1.83 |
| 31 | 1.36 | 1.50 | 1.30 | 1.57 | 1.23 | 1.65 | 1.16 | 1.74 | 1.09 | 1.83 |
| 32 | 1.37 | 1.50 | 1.31 | 1.57 | 1.24 | 1.65 | 1.18 | 1.73 | 1.11 | 1.82 |
| 33 | 1.38 | 1.51 | 1.32 | 1.58 | 1.26 | 1.65 | 1.19 | 1.73 | 1.13 | 1.81 |
| 34 | 1.39 | 1.51 | 1.33 | 1.58 | 1.27 | 1.65 | 1.21 | 1.73 | 1.15 | 1.81 |
| 35 | 1.40 | 1.52 | 1.34 | 1.58 | 1.28 | 1.65 | 1.22 | 1.73 | 1.16 | 1.80 |
| 36 | 1.41 | 1.52 | 1.35 | 1.59 | 1.29 | 1.65 | 1.24 | 1.73 | 1.18 | 1.80 |
| 37 | 1.42 | 1.53 | 1.36 | 1.59 | 1.31 | 1.66 | 1.25 | 1.72 | 1.19 | 1.80 |
| 38 | 1.43 | 1.54 | 1.37 | 1.59 | 1.32 | 1.66 | 1.26 | 1.72 | 1.21 | 1.79 |
| 39 | 1.43 | 1.54 | 1.38 | 1.60 | 1.33 | 1.66 | 1.27 | 1.72 | 1.22 | 1.79 |
| 40 | 1.44 | 1.54 | 1.39 | 1.60 | 1.34 | 1.66 | 1.29 | 1.72 | 1.23 | 1.79 |
| 45 | 1.48 | 1.57 | 1.43 | 1.62 | 1.38 | 1.67 | 1.34 | 1.72 | 1.29 | 1.78 |
| 50 | 1.50 | 1.59 | 1.46 | 1.63 | 1.42 | 1.67 | 1.38 | 1.72 | 1.34 | 1.77 |
| 55 | 1.53 | 1.60 | 1.49 | 1.64 | 1.45 | 1.68 | 1.41 | 1.72 | 1.38 | 1.77 |
| 60 | 1.55 | 1.62 | 1.51 | 1.65 | 1.48 | 1.69 | 1.44 | 1.73 | 1.41 | 1.77 |
| 65 | 1.57 | 1.63 | 1.54 | 1.66 | 1.50 | 1.70 | 1.47 | 1.73 | 1.44 | 1.77 |
| 70 | 1.58 | 1.64 | 1.55 | 1.67 | 1.52 | 1.70 | 1.49 | 1.74 | 1.46 | 1.77 |
| 75 | 1.60 | 1.65 | 1.57 | 1.68 | 1.54 | 1.71 | 1.51 | 1.74 | 1.49 | 1.77 |
| 80 | 1.61 | 1.66 | 1.59 | 1.69 | 1.56 | 1.72 | 1.53 | 1.74 | 1.51 | 1.77 |
| 85 | 1.62 | 1.67 | 1.60 | 1.70 | 1.57 | 1.72 | 1.55 | 1.75 | 1.52 | 1.77 |
| 90 | 1.63 | 1.68 | 1.61 | 1.70 | 1.59 | 1.73 | 1.57 | 1.75 | 1.54 | 1.78 |
| 95 | 1.64 | 1.69 | 1.62 | 1.71 | 1.60 | 1.73 | 1.58 | 1.75 | 1.56 | 1.78 |
| 100 | 1.65 | 1.69 | 1.63 | 1.72 | 1.61 | 1.74 | 1.59 | 1.76 | 1.57 | 1.78 |

# Appendix Table 11 (Continued)

*B. Where α = .025*

| $n$ | $k=1$ | | $k=2$ | | $k=3$ | | $k=4$ | | $k=5$ | |
|---|---|---|---|---|---|---|---|---|---|---|
| | $d_L$ | $d_U$ | $d_L$ | $d_U$ | $d_L$ | $d_U$ | $d_L$ | $d_U$ | $d_L$ | $d_U$ |
| 15 | 0.95 | 1.23 | 0.83 | 1.40 | 0.71 | 1.61 | 0.59 | 1.84 | 0.48 | 2.09 |
| 16 | 0.98 | 1.24 | 0.86 | 1.40 | 0.75 | 1.59 | 0.64 | 1.80 | 0.53 | 2.03 |
| 17 | 1.01 | 1.25 | 0.90 | 1.40 | 0.79 | 1.58 | 0.68 | 1.77 | 0.57 | 1.98 |
| 18 | 1.03 | 1.26 | 0.93 | 1.40 | 0.82 | 1.56 | 0.72 | 1.74 | 0.62 | 1.93 |
| 19 | 1.06 | 1.28 | 0.96 | 1.41 | 0.86 | 1.55 | 0.76 | 1.72 | 0.66 | 1.90 |
| 20 | 1.08 | 1.28 | 0.99 | 1.41 | 0.89 | 1.55 | 0.79 | 1.70 | 0.70 | 1.87 |
| 21 | 1.10 | 1.30 | 1.01 | 1.41 | 0.92 | 1.54 | 0.83 | 1.69 | 0.73 | 1.84 |
| 22 | 1.12 | 1.31 | 1.04 | 1.42 | 0.95 | 1.54 | 0.86 | 1.68 | 0.77 | 1.82 |
| 23 | 1.14 | 1.32 | 1.06 | 1.42 | 0.97 | 1.54 | 0.89 | 1.67 | 0.80 | 1.80 |
| 24 | 1.16 | 1.33 | 1.08 | 1.43 | 1.00 | 1.54 | 0.91 | 1.66 | 0.83 | 1.79 |
| 25 | 1.18 | 1.34 | 1.10 | 1.43 | 1.02 | 1.54 | 0.94 | 1.65 | 0.86 | 1.77 |
| 26 | 1.19 | 1.35 | 1.12 | 1.44 | 1.04 | 1.54 | 0.96 | 1.65 | 0.88 | 1.76 |
| 27 | 1.21 | 1.36 | 1.13 | 1.44 | 1.06 | 1.54 | 0.99 | 1.64 | 0.91 | 1.75 |
| 28 | 1.22 | 1.37 | 1.15 | 1.45 | 1.08 | 1.54 | 1.01 | 1.64 | 0.93 | 1.74 |
| 29 | 1.24 | 1.38 | 1.17 | 1.45 | 1.10 | 1.54 | 1.03 | 1.63 | 0.96 | 1.73 |
| 30 | 1.25 | 1.38 | 1.18 | 1.46 | 1.12 | 1.54 | 1.05 | 1.63 | 0.98 | 1.73 |
| 31 | 1.26 | 1.39 | 1.20 | 1.47 | 1.13 | 1.55 | 1.07 | 1.63 | 1.00 | 1.72 |
| 32 | 1.27 | 1.40 | 1.21 | 1.47 | 1.15 | 1.55 | 1.08 | 1.63 | 1.02 | 1.71 |
| 33 | 1.28 | 1.41 | 1.22 | 1.48 | 1.16 | 1.55 | 1.10 | 1.63 | 1.04 | 1.71 |
| 34 | 1.29 | 1.41 | 1.24 | 1.48 | 1.17 | 1.55 | 1.12 | 1.63 | 1.06 | 1.70 |
| 35 | 1.30 | 1.42 | 1.25 | 1.48 | 1.19 | 1.55 | 1.13 | 1.63 | 1.07 | 1.70 |
| 36 | 1.31 | 1.43 | 1.26 | 1.49 | 1.20 | 1.56 | 1.15 | 1.63 | 1.09 | 1.70 |
| 37 | 1.32 | 1.43 | 1.27 | 1.49 | 1.21 | 1.56 | 1.16 | 1.62 | 1.10 | 1.70 |
| 38 | 1.33 | 1.44 | 1.28 | 1.50 | 1.23 | 1.56 | 1.17 | 1.62 | 1.12 | 1.70 |
| 39 | 1.34 | 1.44 | 1.29 | 1.50 | 1.24 | 1.56 | 1.19 | 1.63 | 1.13 | 1.69 |
| 40 | 1.35 | 1.45 | 1.30 | 1.51 | 1.25 | 1.57 | 1.20 | 1.63 | 1.15 | 1.69 |
| 45 | 1.39 | 1.48 | 1.34 | 1.53 | 1.30 | 1.58 | 1.25 | 1.63 | 1.21 | 1.69 |
| 50 | 1.42 | 1.50 | 1.38 | 1.54 | 1.34 | 1.59 | 1.30 | 1.64 | 1.26 | 1.69 |
| 55 | 1.45 | 1.52 | 1.41 | 1.56 | 1.37 | 1.60 | 1.33 | 1.64 | 1.30 | 1.69 |
| 60 | 1.47 | 1.54 | 1.44 | 1.57 | 1.40 | 1.61 | 1.37 | 1.65 | 1.33 | 1.69 |
| 65 | 1.49 | 1.55 | 1.46 | 1.59 | 1.43 | 1.62 | 1.40 | 1.66 | 1.36 | 1.69 |
| 70 | 1.51 | 1.57 | 1.48 | 1.60 | 1.45 | 1.63 | 1.42 | 1.66 | 1.39 | 1.70 |
| 75 | 1.53 | 1.58 | 1.50 | 1.61 | 1.47 | 1.64 | 1.45 | 1.67 | 1.42 | 1.70 |
| 80 | 1.54 | 1.59 | 1.52 | 1.62 | 1.49 | 1.65 | 1.47 | 1.67 | 1.44 | 1.70 |
| 85 | 1.56 | 1.60 | 1.53 | 1.63 | 1.51 | 1.65 | 1.49 | 1.68 | 1.46 | 1.71 |
| 90 | 1.57 | 1.61 | 1.55 | 1.64 | 1.53 | 1.66 | 1.50 | 1.69 | 1.48 | 1.71 |
| 95 | 1.58 | 1.62 | 1.56 | 1.65 | 1.54 | 1.67 | 1.52 | 1.69 | 1.50 | 1.71 |
| 100 | 1.59 | 1.63 | 1.57 | 1.65 | 1.55 | 1.67 | 1.53 | 1.70 | 1.51 | 1.72 |

# Appendix Table 11 (Continued)

*C. Where $\alpha = .01$*

| n | k = 1 | | k = 2 | | k = 3 | | k = 4 | | k = 5 | |
|---|---|---|---|---|---|---|---|---|---|---|
| | $d_L$ | $d_U$ | $d_L$ | $d_U$ | $d_L$ | $d_U$ | $d_L$ | $d_U$ | $d_L$ | $d_U$ |
| 15 | 0.81 | 1.07 | 0.70 | 1.25 | 0.59 | 1.46 | 0.49 | 1.70 | 0.39 | 1.96 |
| 16 | 0.84 | 1.09 | 0.74 | 1.25 | 0.63 | 1.44 | 0.53 | 1.66 | 0.44 | 1.90 |
| 17 | 0.87 | 1.10 | 0.77 | 1.25 | 0.67 | 1.43 | 0.57 | 1.63 | 0.48 | 1.85 |
| 18 | 0.90 | 1.12 | 0.80 | 1.26 | 0.71 | 1.42 | 0.61 | 1.60 | 0.52 | 1.80 |
| 19 | 0.93 | 1.13 | 0.83 | 1.26 | 0.74 | 1.41 | 0.65 | 1.58 | 0.56 | 1.77 |
| 20 | 0.95 | 1.15 | 0.86 | 1.27 | 0.77 | 1.41 | 0.68 | 1.57 | 0.60 | 1.74 |
| 21 | 0.97 | 1.16 | 0.89 | 1.27 | 0.80 | 1.41 | 0.72 | 1.55 | 0.63 | 1.71 |
| 22 | 1.00 | 1.17 | 0.91 | 1.28 | 0.83 | 1.40 | 0.75 | 1.54 | 0.66 | 1.69 |
| 23 | 1.02 | 1.19 | 0.94 | 1.29 | 0.86 | 1.40 | 0.77 | 1.53 | 0.70 | 1.67 |
| 24 | 1.04 | 1.20 | 0.96 | 1.30 | 0.88 | 1.41 | 0.80 | 1.53 | 0.72 | 1.66 |
| 25 | 1.05 | 1.21 | 0.98 | 1.30 | 0.90 | 1.41 | 0.83 | 1.52 | 0.75 | 1.65 |
| 26 | 1.07 | 1.22 | 1.00 | 1.31 | 0.93 | 1.41 | 0.85 | 1.52 | 0.78 | 1.64 |
| 27 | 1.09 | 1.23 | 1.02 | 1.32 | 0.95 | 1.41 | 0.88 | 1.51 | 0.81 | 1.63 |
| 28 | 1.10 | 1.24 | 1.04 | 1.32 | 0.97 | 1.41 | 0.90 | 1.51 | 0.83 | 1.62 |
| 29 | 1.12 | 1.25 | 1.05 | 1.33 | 0.99 | 1.42 | 0.92 | 1.51 | 0.85 | 1.61 |
| 30 | 1.13 | 1.26 | 1.07 | 1.34 | 1.01 | 1.42 | 0.94 | 1.51 | 0.88 | 1.61 |
| 31 | 1.15 | 1.27 | 1.08 | 1.34 | 1.02 | 1.42 | 0.96 | 1.51 | 0.90 | 1.60 |
| 32 | 1.16 | 1.28 | 1.10 | 1.35 | 1.04 | 1.43 | 0.98 | 1.51 | 0.92 | 1.60 |
| 33 | 1.17 | 1.29 | 1.11 | 1.36 | 1.05 | 1.43 | 1.00 | 1.51 | 0.94 | 1.59 |
| 34 | 1.18 | 1.30 | 1.13 | 1.36 | 1.07 | 1.43 | 1.01 | 1.51 | 0.95 | 1.59 |
| 35 | 1.19 | 1.31 | 1.14 | 1.37 | 1.08 | 1.44 | 1.03 | 1.51 | 0.97 | 1.59 |
| 36 | 1.21 | 1.32 | 1.15 | 1.38 | 1.10 | 1.44 | 1.04 | 1.51 | 0.99 | 1.59 |
| 37 | 1.22 | 1.32 | 1.16 | 1.38 | 1.11 | 1.45 | 1.06 | 1.51 | 1.00 | 1.59 |
| 38 | 1.23 | 1.33 | 1.18 | 1.39 | 1.12 | 1.45 | 1.07 | 1.52 | 1.02 | 1.58 |
| 39 | 1.24 | 1.34 | 1.19 | 1.39 | 1.14 | 1.45 | 1.09 | 1.52 | 1.03 | 1.58 |
| 40 | 1.25 | 1.34 | 1.20 | 1.40 | 1.15 | 1.46 | 1.10 | 1.52 | 1.05 | 1.58 |
| 45 | 1.29 | 1.38 | 1.24 | 1.42 | 1.20 | 1.48 | 1.16 | 1.53 | 1.11 | 1.58 |
| 50 | 1.32 | 1.40 | 1.28 | 1.45 | 1.24 | 1.49 | 1.20 | 1.54 | 1.16 | 1.59 |
| 55 | 1.36 | 1.43 | 1.32 | 1.47 | 1.28 | 1.51 | 1.25 | 1.55 | 1.21 | 1.59 |
| 60 | 1.38 | 1.45 | 1.35 | 1.48 | 1.32 | 1.52 | 1.28 | 1.56 | 1.25 | 1.60 |
| 65 | 1.41 | 1.47 | 1.38 | 1.50 | 1.35 | 1.53 | 1.31 | 1.57 | 1.28 | 1.61 |
| 70 | 1.43 | 1.49 | 1.40 | 1.52 | 1.37 | 1.55 | 1.34 | 1.58 | 1.31 | 1.61 |
| 75 | 1.45 | 1.50 | 1.42 | 1.53 | 1.39 | 1.56 | 1.37 | 1.59 | 1.34 | 1.62 |
| 80 | 1.47 | 1.52 | 1.44 | 1.54 | 1.42 | 1.57 | 1.39 | 1.60 | 1.36 | 1.62 |
| 85 | 1.48 | 1.53 | 1.46 | 1.55 | 1.43 | 1.58 | 1.41 | 1.60 | 1.39 | 1.63 |
| 90 | 1.50 | 1.54 | 1.47 | 1.56 | 1.45 | 1.59 | 1.43 | 1.61 | 1.41 | 1.64 |
| 95 | 1.51 | 1.55 | 1.49 | 1.57 | 1.47 | 1.60 | 1.45 | 1.62 | 1.42 | 1.64 |
| 100 | 1.52 | 1.56 | 1.50 | 1.58 | 1.48 | 1.60 | 1.46 | 1.63 | 1.44 | 1.65 |

*Source:* J. Durbin and G. S Watson, "Testing for Serial Correlation in Least Squares Regression," *Biometrika*, 38, June 1951.

# Appendix Table 12

COMMON LOGARITHMS

| N | 0 | 1 | 2 | 3 | 4 | 5 | 6 | 7 | 8 | 9 |
|---|---|---|---|---|---|---|---|---|---|---|
| 10 | 0000 | 0043 | 0086 | 0128 | 0170 | 0212 | 0253 | 0294 | 0334 | 0374 |
| 11 | 0414 | 0453 | 0492 | 0531 | 0569 | 0607 | 0645 | 0682 | 0719 | 0755 |
| 12 | 0792 | 0828 | 0864 | 0899 | 0934 | 0969 | 1004 | 1038 | 1072 | 1106 |
| 13 | 1139 | 1173 | 1206 | 1239 | 1271 | 1303 | 1335 | 1367 | 1399 | 1430 |
| 14 | 1461 | 1492 | 1523 | 1553 | 1584 | 1614 | 1644 | 1673 | 1703 | 1732 |
| 15 | 1761 | 1790 | 1818 | 1847 | 1875 | 1903 | 1931 | 1959 | 1987 | 2014 |
| 16 | 2041 | 2068 | 2095 | 2122 | 2148 | 2175 | 2201 | 2227 | 2253 | 2279 |
| 17 | 2304 | 2330 | 2355 | 2380 | 2405 | 2430 | 2455 | 2480 | 2504 | 2529 |
| 18 | 2553 | 2577 | 2601 | 2625 | 2648 | 2672 | 2695 | 2718 | 2742 | 2765 |
| 19 | 2788 | 2810 | 2833 | 2856 | 2878 | 2900 | 2923 | 2945 | 2967 | 2989 |
| 20 | 3010 | 3032 | 3054 | 3075 | 3096 | 3118 | 3139 | 3160 | 3181 | 3201 |
| 21 | 3222 | 3243 | 3263 | 3284 | 3304 | 3324 | 3345 | 3365 | 3385 | 3404 |
| 22 | 3424 | 3444 | 3464 | 3483 | 3502 | 3522 | 3541 | 3560 | 3579 | 3598 |
| 23 | 3617 | 3636 | 3655 | 3674 | 3692 | 3711 | 3729 | 3747 | 3766 | 3784 |
| 24 | 3802 | 3826 | 3838 | 3856 | 3874 | 3892 | 3909 | 3927 | 3945 | 3962 |
| 25 | 3979 | 3997 | 4014 | 4031 | 4048 | 4065 | 4082 | 4099 | 4116 | 4133 |
| 26 | 4150 | 4166 | 4183 | 4200 | 4216 | 4232 | 4249 | 4205 | 4281 | 4298 |
| 27 | 4314 | 4330 | 4346 | 4362 | 4378 | 4393 | 4409 | 4425 | 4440 | 4456 |
| 28 | 4472 | 4487 | 4502 | 4518 | 4533 | 4548 | 4564 | 4579 | 4594 | 4609 |
| 29 | 4624 | 4639 | 4654 | 4669 | 4683 | 4698 | 4713 | 4728 | 4742 | 4757 |
| 30 | 4771 | 4786 | 4800 | 4814 | 4829 | 4843 | 4857 | 4871 | 4886 | 4900 |
| 31 | 4914 | 4928 | 4942 | 4955 | 4969 | 4983 | 4997 | 5011 | 5024 | 5038 |
| 32 | 5051 | 5065 | 5079 | 5092 | 5105 | 5119 | 5132 | 5145 | 5159 | 5172 |
| 33 | 5185 | 5198 | 5211 | 5224 | 5237 | 5250 | 5263 | 5276 | 5289 | 5302 |
| 34 | 5315 | 5328 | 5340 | 5353 | 5366 | 5378 | 5391 | 5403 | 5416 | 5423 |
| 35 | 5441 | 5453 | 5465 | 5478 | 5490 | 5502 | 5514 | 5527 | 5539 | 5551 |
| 36 | 5563 | 5575 | 5587 | 5599 | 5611 | 5623 | 5635 | 5647 | 5658 | 5670 |
| 37 | 5682 | 5694 | 5705 | 5717 | 5729 | 5740 | 5752 | 5763 | 5775 | 5786 |
| 38 | 5798 | 5809 | 5821 | 5832 | 5843 | 5855 | 5866 | 5877 | 5888 | 5899 |
| 39 | 5911 | 5922 | 5933 | 5944 | 5955 | 5966 | 5977 | 5988 | 5999 | 6010 |
| 40 | 6021 | 6031 | 6042 | 6053 | 6064 | 6075 | 6085 | 6096 | 6107 | 6117 |
| 41 | 6128 | 6138 | 6149 | 6160 | 6170 | 6180 | 6191 | 6201 | 6212 | 6222 |
| 42 | 6232 | 6243 | 6253 | 6263 | 6274 | 6284 | 6294 | 6304 | 6314 | 6325 |
| 43 | 6335 | 6345 | 6355 | 6365 | 6375 | 6385 | 6395 | 6405 | 6415 | 6425 |
| 44 | 6435 | 6444 | 6454 | 6464 | 6474 | 6484 | 6493 | 6503 | 6513 | 6522 |
| N | 0 | 1 | 2 | 3 | 4 | 5 | 6 | 7 | 8 | 9 |

# Appendix Table 12 (Continued)

| N | 0 | 1 | 2 | 3 | 4 | 5 | 6 | 7 | 8 | 9 |
|---|---|---|---|---|---|---|---|---|---|---|
| 45 | 6532 | 6542 | 6551 | 6561 | 6571 | 6580 | 6590 | 6599 | 6609 | 6618 |
| 46 | 6628 | 6637 | 6646 | 6656 | 6665 | 6675 | 6684 | 6693 | 6702 | 6712 |
| 47 | 6721 | 6730 | 6739 | 6749 | 6758 | 6767 | 6776 | 6785 | 6794 | 6803 |
| 48 | 6812 | 6821 | 6830 | 6839 | 6848 | 6857 | 6866 | 6875 | 6884 | 6893 |
| 49 | 6902 | 6911 | 6920 | 6928 | 6937 | 6946 | 6955 | 6964 | 6972 | 6981 |
| 50 | 6990 | 6998 | 7007 | 7016 | 7024 | 7033 | 7042 | 7050 | 7059 | 7067 |
| 51 | 7076 | 7084 | 7093 | 7101 | 7110 | 7118 | 7126 | 7135 | 7143 | 7152 |
| 52 | 7160 | 7168 | 7177 | 7185 | 7193 | 7202 | 7210 | 7218 | 7226 | 7235 |
| 53 | 7243 | 7251 | 7259 | 7267 | 7275 | 7284 | 7292 | 7300 | 7308 | 7316 |
| 54 | 7324 | 7332 | 7340 | 7348 | 7356 | 7364 | 7372 | 7380 | 7388 | 7396 |
| 55 | 7404 | 7412 | 7419 | 7427 | 7435 | 7443 | 7451 | 7459 | 7466 | 7474 |
| 56 | 7482 | 7490 | 7497 | 7505 | 7513 | 7520 | 7528 | 7536 | 7543 | 7551 |
| 57 | 7559 | 7566 | 7574 | 7582 | 7589 | 7597 | 7604 | 7612 | 7619 | 7627 |
| 58 | 7634 | 7642 | 7649 | 7657 | 7664 | 7672 | 7679 | 7686 | 7694 | 7701 |
| 59 | 7709 | 7716 | 7723 | 7731 | 7738 | 7745 | 7752 | 7760 | 7767 | 7774 |
| 60 | 7782 | 7789 | 7796 | 7803 | 7810 | 7818 | 7825 | 7832 | 7839 | 7846 |
| 61 | 7853 | 7860 | 7868 | 7875 | 7882 | 7889 | 7896 | 7903 | 7910 | 7917 |
| 62 | 7924 | 7931 | 7938 | 7945 | 7952 | 7959 | 7966 | 7973 | 7980 | 7987 |
| 63 | 7993 | 8000 | 8007 | 8014 | 8021 | 8028 | 8035 | 8041 | 8048 | 8055 |
| 64 | 8062 | 8069 | 8075 | 8082 | 8089 | 8096 | 8102 | 8109 | 8116 | 8122 |
| 65 | 8129 | 8136 | 8142 | 8149 | 8156 | 8162 | 8169 | 8176 | 8182 | 8189 |
| 66 | 8195 | 8202 | 8209 | 8215 | 8222 | 8228 | 8235 | 8241 | 8248 | 8254 |
| 67 | 8261 | 8267 | 8274 | 8280 | 8287 | 8293 | 8299 | 8306 | 8312 | 8319 |
| 68 | 8325 | 8331 | 8338 | 8344 | 8351 | 8357 | 8363 | 8370 | 8376 | 8382 |
| 69 | 8388 | 8395 | 8401 | 8407 | 8414 | 8420 | 8426 | 8432 | 8439 | 8445 |
| 70 | 8451 | 8457 | 8463 | 8470 | 8476 | 8482 | 8488 | 8494 | 8500 | 8506 |
| 71 | 8513 | 8519 | 8525 | 8531 | 8537 | 8543 | 8549 | 8555 | 8561 | 8567 |
| 72 | 8573 | 8579 | 8585 | 8591 | 8597 | 8603 | 8609 | 8615 | 8621 | 8627 |
| 73 | 8633 | 8639 | 8645 | 8651 | 8637 | 8663 | 8669 | 8675 | 8681 | 8686 |
| 74 | 8692 | 8698 | 8704 | 8710 | 8716 | 8722 | 8727 | 8733 | 8739 | 8745 |
| 75 | 8751 | 8756 | 8762 | 8768 | 8774 | 8779 | 8785 | 8791 | 8797 | 8802 |
| 76 | 8808 | 8814 | 8820 | 8825 | 8831 | 8837 | 8842 | 8848 | 8854 | 8859 |
| 77 | 8865 | 8871 | 8876 | 8882 | 8887 | 8893 | 8899 | 8904 | 8910 | 8915 |
| 78 | 8921 | 8927 | 8932 | 8938 | 8943 | 8949 | 8954 | 8960 | 8965 | 8971 |
| 79 | 8976 | 8982 | 8987 | 8993 | 8998 | 9004 | 9009 | 9015 | 9020 | 9025 |
| N | 0 | 1 | 2 | 3 | 4 | 5 | 6 | 7 | 8 | 9 |

# Appendix Table 12 (Continued)

| N | 0 | 1 | 2 | 3 | 4 | 5 | 6 | 7 | 8 | 9 |
|---|---|---|---|---|---|---|---|---|---|---|
| 80 | 9031 | 9036 | 9042 | 9047 | 9053 | 9058 | 9063 | 9069 | 9074 | 9079 |
| 81 | 9085 | 9090 | 9096 | 9101 | 9106 | 9112 | 9117 | 9122 | 9128 | 9133 |
| 82 | 9138 | 9143 | 9149 | 9154 | 9159 | 9165 | 9170 | 9175 | 9180 | 9186 |
| 83 | 9191 | 9190 | 9201 | 9206 | 9212 | 9217 | 9222 | 9227 | 9232 | 9238 |
| 84 | 9243 | 9248 | 9253 | 9258 | 9263 | 9269 | 9274 | 9279 | 9284 | 9289 |
| 85 | 9294 | 9299 | 9304 | 9309 | 9315 | 9320 | 9325 | 9330 | 9335 | 9340 |
| 86 | 9345 | 9350 | 9355 | 9360 | 9365 | 9370 | 9375 | 9380 | 9385 | 9390 |
| 87 | 9395 | 9400 | 9405 | 9410 | 9415 | 9420 | 9425 | 9430 | 9435 | 9440 |
| 88 | 9445 | 9450 | 9455 | 9460 | 9465 | 9469 | 9474 | 9479 | 9484 | 9489 |
| 89 | 9494 | 9499 | 9504 | 9509 | 9513 | 9518 | 9523 | 9528 | 9533 | 9538 |
| 90 | 9542 | 9547 | 9552 | 9557 | 9562 | 9566 | 9571 | 9576 | 9581 | 9586 |
| 91 | 9590 | 9595 | 9600 | 9605 | 9609 | 9614 | 9619 | 9624 | 9628 | 9633 |
| 92 | 9638 | 9643 | 9647 | 9652 | 9657 | 9661 | 9666 | 9671 | 9675 | 9680 |
| 93 | 9685 | 9689 | 9694 | 9699 | 9703 | 9708 | 9713 | 9717 | 9722 | 9727 |
| 94 | 9731 | 9736 | 9741 | 9745 | 9750 | 9754 | 9759 | 9763 | 9768 | 9773 |
| 95 | 9777 | 9782 | 9786 | 9791 | 9795 | 9800 | 9805 | 9808 | 9814 | 9818 |
| 96 | 9823 | 9827 | 9832 | 9836 | 9841 | 9845 | 9850 | 9854 | 9859 | 9863 |
| 97 | 9868 | 9872 | 9877 | 9881 | 9886 | 9890 | 9894 | 9899 | 9903 | 9908 |
| 98 | 9912 | 9917 | 9921 | 9926 | 9930 | 9934 | 9939 | 9943 | 9948 | 9952 |
| 99 | 9956 | 9961 | 9965 | 9969 | 9974 | 9978 | 9983 | 9987 | 9991 | 9996 |

| N | 0 | 1 | 2 | 3 | 4 | 5 | 6 | 7 | 8 | 9 |
|---|---|---|---|---|---|---|---|---|---|---|

*Source:* Adapted from National Bureau of Standards, *Tables of 10ˣ*, Applied Mathematics Series 27, U.S. Department of Commerce, 1953.

# Answers to Odd-Numbered Exercises

## Chapter 1

**1.1** (a) The population consists of statements concerning each of the company's stations as to whether or not it is open 24 hours a day.
(b) The elements or units in the population are the service stations. A list of the company's stations can be obtained from the company.
(c) Finite.
(d) Qualitative.

**1.3** (a) No. It would be a sample of the opinions of those on picket lines.
(b) Finite.
(c) Qualitative.
(d) Teachers on picket lines may be less likely to believe that the strike would be settled on unfavorable terms.

**1.5** (a) Correlation may not imply causation. Even if the neighboring community had not allowed bars to stay open after midnight, the crime rate might have increased.
(b) It would be difficult, but possible. Some communities might allow longer hours for bars, and their crime rates could be compared with communities that maintained existing hours for bars. But it would be difficult to randomize such an experiment properly, because communities are not likely to allow their laws to be dictated by the experimental design.
(c) Perhaps. For example, one might see whether communities that lengthened the hours for bars in recent years have experienced greater increases in crime rates than those that have not lengthened them. However, it is always possible that confounding is present.

**1.7** If there are 10 class intervals, they can be 0.1810 and under 0.1814 inches, . . . , 0.1846 and under 0.1850 inches. The class marks of the first and last class intervals are 0.1812 and 0.1848, respectively.

**1.9**

| Amount of insurance | Number of males |
|---|---|
| Under $40,000 | _____ |
| $40,000 and under $80,000 | _____ |
| $80,000 and under $120,000 | _____ |
| . . . | |
| $360,000 and under $400,000 | _____ |

The class mark in each interval would be \$20,000, \$60,000,..., \$340,000, and \$380,000.

**1.11** (a)

| Score | Number of clerks |
|---|---|
| 40 and under 50 | 2 |
| 50 and under 60 | 4 |
| 60 and under 70 | 6 |
| 70 and under 80 | 16 |
| 80 and under 90 | 7 |
| 90 and under 100 | 5 |

(b)

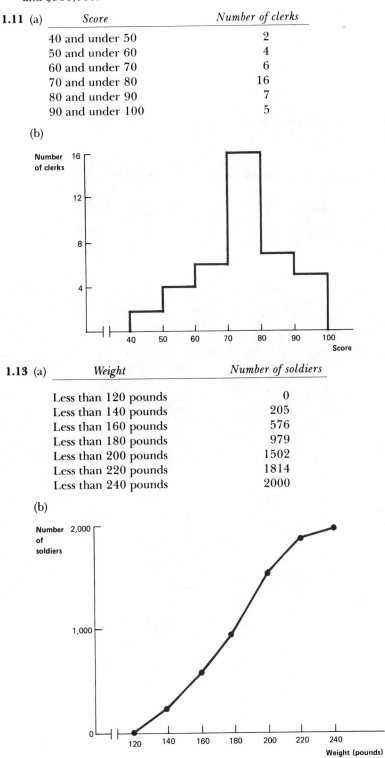

**1.13** (a)

| Weight | Number of soldiers |
|---|---|
| Less than 120 pounds | 0 |
| Less than 140 pounds | 205 |
| Less than 160 pounds | 576 |
| Less than 180 pounds | 979 |
| Less than 200 pounds | 1502 |
| Less than 220 pounds | 1814 |
| Less than 240 pounds | 2000 |

(b)

**1.15**

(a)                                    (b)

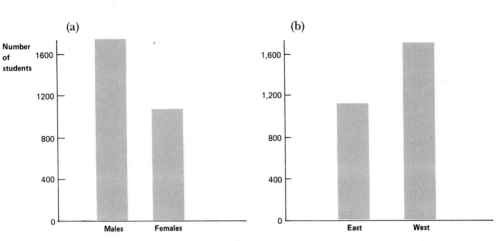

**1.17** 0.25
      0.25

**1.19** No, because the smallest deposit may not be at the lower limit of the low-est class interval, and the largest deposit may not be at the upper limit of the highest class interval.

**1.21** (a)

| Profit rate (percent) | Number of Industries |
|---|---|
| 2.0 and under 3.5 | 3 |
| 3.5 and under 5.0 | 7 |
| 5.0 and under 6.5 | 5 |
| 6.5 and under 8.0 | 2 |
| 8.0 and under 9.5 | 1 |
| Total | 18 |

(b)

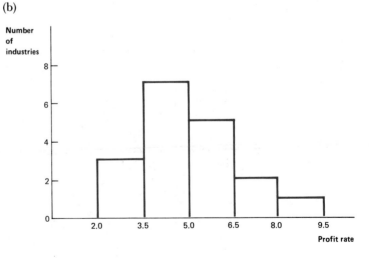

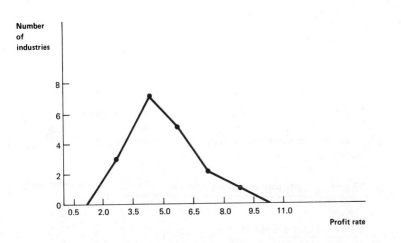

**1.23** (a) They are −$1.00 to −$0.75, −$0.75 to −$0.50, and so forth.
 (b) Yes.
 (c) The fourth class interval: −$0.25 and under $0.00
 (d)

| Error (dollars) | Number of customers |
|---|---|
| Less than −1.00 | 0 |
| Less than −0.75 | 1 |
| Less than −0.50 | 3 |
| Less than −0.25 | 7 |
| Less than   0.00 | 37 |
| Less than +0.25 | 43 |
| Less than +0.50 | 45 |
| Less than +0.75 | 47 |
| Less than +1.00 | 49 |
| Less than +1.25 | 50 |

 (e) The ogive is as follows:

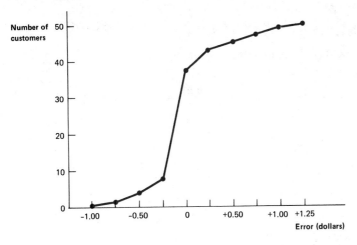

# Chapter 2

**2.1** (a) $\dfrac{28.3}{18} = 1.572$

(b) $(1.56 + 1.68) \div 2 = 1.62$

(c) If one is interested in only these 18 firms, it is the population. If these firms are only part of those one is interested in, it is a sample.

(d) If this is the population, they are parameters; if it is a sample, they are statistics.

**2.3** $\dfrac{\$6.05(4{,}000) + \$7.39(8{,}000)}{12{,}000} = \dfrac{83{,}320}{12{,}000} = \$6.943$

**2.5** (a) $\dfrac{29(.5) + 32(2) + 29(4) + 10(6)}{100} = \$2.545$

(b) $\left(\dfrac{100/2 - 29}{32}\right)\$2 + \$1 = \$2.312$

**2.7** (a) 98 cents.

(b) Yes. To the left.

(c) The mode.

**2.9** (a) $45.5/9 = 5.06$ percent

(b) $41.9/9 = 4.66$ percent

(c) The mean profit rate for durable manufacturing industries was about 4/10 of a percentage point higher than for nondurable manufacturing industries in 1974.

(d) The median profit rate for durable manufacturing industries was 4.8 percent. The median profit rate for nondurable manufacturing industries was 4.8 percent. There was no difference between the medians in 1974.

**2.11** $\displaystyle\sum_{i=1}^{5} X_i = 10.1$

$\displaystyle\sum_{i=1}^{5} X_i^2 = 3.24 + 3.61 + 4.41 + 5.29 + 4.00 = 20.55$

Using the shortcut in equation (2.11),

$$s^2 = \frac{20.55 - \dfrac{1}{5}(10.1)^2}{4} = \frac{20.55 - 20.402}{4} = 0.037$$

$s = 0.19$ pounds

**2.13** (a) $\displaystyle\sum_{i=1}^{5} X_i = 150.11$      $s^2 = \dfrac{4506.6027 - \dfrac{1}{5}(150.11)^2}{4} = .00007$

$\displaystyle\sum_{i=1}^{5} X_i^2 = 4506.6027$      $s\ = .0084$ inches

(b) $\sum\limits_{i=1}^{5} X_i = .11$          $s^2 = \dfrac{.0027 - \dfrac{1}{5}(.11)^2}{4} = .00007$

$\sum\limits_{i=1}^{5} X_i^2 = .0027$          $s = .0084$ inches

(c) They are equal because the standard deviation remains the same if all observations are decreased by the same amount. (Note that the short-cut in equation (2.11) is used in both (a) and (b) above.)

**2.15** $\sum\limits_{j=1}^{k} f_j X_j' = 254.5$

$\sum\limits_{j=1}^{k} f_j X_j'^2 = 29(.25) + 32(4) + 29(16) + 10(36) = 959.25$

$$s^2 = \dfrac{959.25 - \dfrac{1}{100}(254.5)^2}{99} = \dfrac{959.25 - 647.7025}{99} = \dfrac{311.5475}{99} = 3.1469$$

$s = \$1.77$

If the 100 workers were the population, 100 would have to be substituted for 99. Thus, $s^2 = 3.1155$, and $s = \$1.765$.

**2.17** Firm A's rate of return is 1.2 standard deviations above the mean in its industry. Firm B's rate of return is 1.0 standard deviations above the mean in its industry. Thus, relative to other firms in its industry, firm A seems more profitable than does firm B, when the latter is compared to other firms in its industry.

**2.19** Statements (b) and (c) are true.

**2.21** These figures do not prove that civilians were more likely to suffer amputations. There were many more civilians than military personnel during the war. To estimate the likelihood that a civilian would suffer an amputation, the 120,000 would have to be divided by the number of civilians. To estimate the likelihood that a member of the military would suffer an amputation, the 18,000 would have to be divided by the number of military personnel.

**2.23** No. Perhaps divorce is more common among relatively young people who have not had a chance to have many children. Other possible explanations could be given too. Correlation does not prove causation.

**2.25** No. The difference between the means is only about $200. This is a very small difference when compared to the variation about each average. It might have been more meaningful to present the standard deviation (or some other measure of dispersion) of income of farms in each state as well as averages.

**2.27** No. The pitfall here is that the variation about the average is neglected.

**2.29** Yes, because the frequency distribution of income tax payments is likely to be skewed to the right.

**2.31** (a) Zero.
(b) If the frequency distribution is symmetrical, the median equals the mean, so the mean must also equal 3. Thus, the sum equals 1,000(3), or 3,000.

**2.33** 5.5 percent. 4.5 percent. 6.3 percent. No.

**2.35** (a) Using equation (2.12), but substituting N for $(n-1)$ in the denominator because the entire population is given, the standard deviation equals

$$\sqrt{\frac{2.6^2 + 5.8^2 + 2.8^2 + 5.4^2 + 4.8^2 + 6.8^2 + 7.6^2 + 4.0^2 + 2.1^2 - (9)(4.66)^2}{9}}$$

$$= \sqrt{\frac{(224.85 - 195.4404)}{9}} = \sqrt{\frac{29.4095}{9}} = \sqrt{3.27} = 1.8 \text{ percentage points.}$$

(b) The standard deviation equals

$$\sqrt{\frac{3.9^2 + 4.3^2 + 5.6^2 + 4.0^2 + 4.1^2 + 5.4^2 + 4.8^2 + 8.4^2 + 5.0^2 - (9)(5.06)^2}{9}}$$

$$= \sqrt{\frac{(245.63 - 230.4324)}{9}} = \sqrt{\frac{15.1976}{9}} = \sqrt{1.69} = 1.3 \text{ percentage points.}$$

(c) The standard deviation equals

$$\sqrt{\frac{224.85 + 245.63 - (18)(4.86)^2}{18}} = \sqrt{\frac{470.48 - (18)(23.6196)}{18}}$$

$$= \sqrt{\frac{470.48 - 425.1528}{18}} = \sqrt{\frac{45.3272}{18}} = \sqrt{2.52} = 1.6 \text{ percentage points.}$$

**2.37** Since we know from Exercise 2.32 that the mean is $20, the standard deviation (if the data in Exercise 2.32 are the entire population) equals

$$\sqrt{\frac{60(10 - 20)^2 + 30(30 - 20)^2 + 10(50 - 20)^2}{100}} = \sqrt{180} = \$13.42.$$

If we calculate the standard deviation in cents, we have

$$\sqrt{\frac{60(1000 - 2000)^2 + 30(3000 - 2000)^2 + 10(5000 - 2000)^2}{100}}$$

$$= \sqrt{1,800,000}$$

$$= 1,342 \text{ cents.}$$

The ratio of the latter standard deviation to the former is 100, since each observation expressed in cents is 100 times the same observation expressed in dollars.

**2.39** (a) Not necessarily, as explained in (b).

(b) Yes, it is consistent with them. Not unless the female applicants were better qualified than the male applicants, or unless admission was deliberately reduced in majors that have high female application rates.

**2.41**

```
MTB > set in c2
DATA > 73 65 74 46 71 60 73 62 54 73
DATA > 72 72 47 46 73 76 60 41 48 71
DATA > 70 73 64 75 61 57 73 63 70 39
DATA > end
MTB > histogram c2

Histogram of C2  N = 30

Midpoint  Count
      40      2  **
      45      3  ***
      50      1  *
      55      2  **
      60      4  ****
      65      3  ***
      70      6  ******
      75      9  *********

MTB > describe c2

             N    MEAN   MEDIAN  TRMEAN  STDEV  SEMEAN
C2          30   63.40   67.50   64.27   11.27    2.06

            MIN     MAX      Q1      Q3
C2        39.00   76.00   56.25   73.00
```

# Chapter 3

**3.1**

| heads | heads | heads |     | tails | tails | heads |
| heads | heads | tails |     | tails | heads | tails |
| heads | tails | heads |     | heads | tails | tails |
| tails | heads | heads |     | tails | tails | tails |

A probability of 1/8 should be attached to each point in the sample space.

**3.3** (a) 1/6    (b) 1/3    (c) zero    (d) 1/2    (e) 1/3

**3.5** (a) 8/36    (b) 2/36

**3.7** 0.25

**3.9** $0.75 + 0.25 - 0.1875 = 0.8125$

**3.11** (a) No, because $P$ ($A$ and $B$) does not equal zero.
(b) No. If $A$ and $B$ were statistically independent, $P(A$ and $B)$ would equal $P(A)$ times $P(B)$.

**3.13** (a) From Figure 3.1 it is clear that there are five points in the sample space where the sum equals six: (5,1), (4,2), (3,3), (2,4), (1,5). Thus, the probability is 5/36.
(b) From Figure 3.1 it is clear that there are ten points in the sample space where the sum is less than six: (1,1), (1,2), (1,3), (1,4), (2,1), (2,2), (2,3), (3,1), (3,2) (4,1). Thus, the probability equals 10/36.

(c) Since $P(7 \text{ or more}) = 1 - P(6) - P(\text{less than } 6)$, it follows from (a) and (b) that $P(7 \text{ or more}) = 1 - 5/36 - 10/36 = 21/36$.

**3.15** $2/3(2/3) = 4/9$

**3.17** (a) $0.4(0.4)(0.6) = .096$
(b) $3(.096) + 0.4^3 = .352$

**3.19** $1/3$

**3.21** $(.97)^4 = .89$

**3.23** $2/3$

**3.25** A great many statisticians do not regard subjective probabilities as useless.

**3.27** $\dfrac{0.7(0.6)}{0.7(0.6) + 0.2(0.4)} = \dfrac{.42}{.42 + .08} = .84.$

**3.29** $\dfrac{.25(.1)}{.25(.1) + .65(.9)} = \dfrac{.025}{.025 + .585} = .041.$

**3.31** (a)

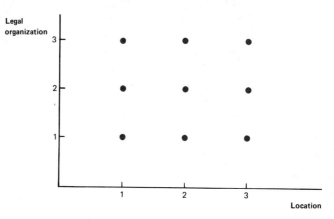

(b)  (i) The subset consists of the three points where location equals 1.
(ii) The subset consists of the three points where legal organization equals 3.
(iii) The subset consists of the point with an *x*-coordinate of 3 and a *y*-coordinate of 1.
(iv) The subset consists of the six points where the *y*-coordinate is 1 or 3.
(v) The subset consists of the five points where the *y*-coordinate is 1 or the *x*-coordinate is 3, or both.

(c)  (i) It buys a Michigan proprietorship, a Michigan partnership, or a Michigan corporation.
(ii) It buys a Michigan corporation, an Illinois corporation, or a New York corporation.
(iii) It buys a Michigan proprietorship, an Illinois proprietorship, a New York proprietorship, a Michigan corporation, an Illinois corporation, or a New York Corporation.

    (d)   (i) Two events must occur: 1) Bona Fide must buy a proprietor-
         ship; 2) Bona Fide must buy a New York firm.
      (ii) Two events must occur: 1) Bona Fide must buy a corporation; 2)
         Bona Fide must buy a Michigan firm.

**3.33** The probability that he will get an A in either subject (or both) is $0.20 +
0.25 - 0.05 = 0.40$. Thus the probability that he will get an A in neither
subject is $1 - 0.40 = 0.60$.

**3.35** (a) .70
    (b) .40

**3.37** (a) $400 \div 1400$
    (b) $1200 \div 4900$
    (c) No. For example, 300 out of 1,100 motors received by plant 1 come
    from supplier III, whereas 400 out of 1,400 motors received by plant 2
    come from supplier III.
    (d)  (i)  $100 \div 900$
       (ii)  $100 \div 1100$
      (iii)  $400 \div 2100$
      (iv)  $300 \div 2500$
    (e) $P(\text{defective}|\text{supplier I}) = P(\text{defective and supplier I}) \div P(\text{supplier I}) =$
    $0.1 \div (800/4900) = 0.1 \div 0.163 = .61$.
    (f) $P(\text{supplier I}|\text{defective}) = P(\text{defective and supplier I}) \div P(\text{defective}) =$
    $0.1 \div .20 = 0.50$.

**3.39** (a) Let $S$ stand for Jones doing well and $N$ stand for his not doing well.
    Let $s$ stand for the consulting firm's prediction that he will do well and
    $n$ stand for the consulting firm's prediction that he will not do well.

$$P(S|s) = \frac{P(s|S)P(S)}{P(s|S)P(S) + P(s|N)P(N)} = \frac{(0.9)(.75)}{(0.9)(.75) + (0.2)(.25)}$$

$$= \frac{.675}{.675 + .050} = \frac{.675}{.725} = 0.93.$$

    (b) $P(N|n) = \dfrac{P(n|N)P(N)}{P(n|N)P(N) + P(n|S)P(S)}$

$$= \frac{(0.8)(.25)}{(0.8)(.25) + (0.1)(.75)} = \frac{.200}{.200 + .075} = \frac{.200}{.275} = .73.$$

**3.41** Since $\dbinom{n}{x} = \dfrac{n!}{(n-x)!x!}$
    it follows that

$$\binom{n}{n-x} = \frac{n!}{[n-(n-x)]!(n-x)!} = \frac{n!}{x!(n-x)!} = \binom{n}{x}.$$

**3.43** $8(7)(6)(5)(4)(3)(2)(1) = 40{,}320$.

**3.45** $6(5)(4) \div 3(2)(1) = 20$ years.

**3.47** There are $5! \div (3!2!)$ different pairs of men that can be included with each
of the two women. Thus, the answer is

$$(2)\left(\frac{5!}{3!2!}\right) = \frac{(2)(5)(4)}{(2)(1)} = 20.$$

**3.49** $5(4)(3)(2)(1) = 120$ hours.

**3.51** $2^8 = 256$

$$\frac{1}{256} = .0039$$

# Chapter 4

**4.1** (a) $P(x) = 1/3$, for $x = 7, 8, 9$.
(b) $1/3 \quad 2/3$

**4.3** None is a probability distribution. In the case of (a) and (b) $\Sigma P(x) \neq 1$. In the case of (c), $P(x)$ is not always nonnegative.

**4.5** (a) Yes.
(b) 0, 1, 2, 3, or 4
(c)

| Number of bicycles | Probability |
|---|---|
| 0 | 1/5 |
| 1 | 1/5 |
| 2 | 1/5 |
| 3 | 1/5 |
| 4 | 1/5 |

(d)

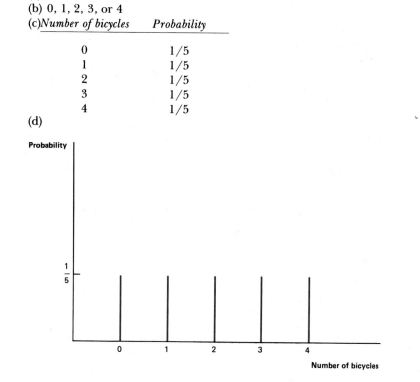

$P(x) = 1/5$, for $x = 0, 1, 2, 3, 4$.

(e) 1/5
3/5
4/5
1
(f) Yes
0, \$20, \$50, \$90

| Income | Probability |
|---|---|
| 0 | 2/5 |
| \$20 | 1/5 |
| 50 | 1/5 |
| 90 | 1/5 |

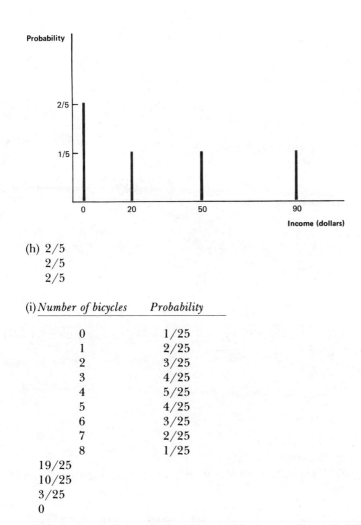

(h) 2/5
2/5
2/5

(i) 

| Number of bicycles | Probability |
|---|---|
| 0 | 1/25 |
| 1 | 2/25 |
| 2 | 3/25 |
| 3 | 4/25 |
| 4 | 5/25 |
| 5 | 4/25 |
| 6 | 3/25 |
| 7 | 2/25 |
| 8 | 1/25 |

19/25
10/25
3/25
0

(j) The salesman can make \$100 or more in the two-day period if he sells six or more bicycles in this period. As shown in (i) the probability of this occurring is 6/25.

(k)

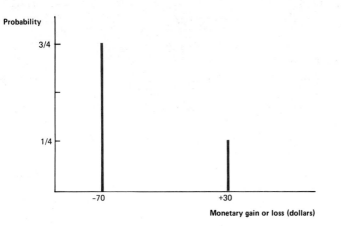

**4.7** (a) 28
(b) −6

**4.9** (a) Discrete
(b) The last digit in a randomly selected two-digit number would be expected to have this probability distribution.
(c) $E(X) = 0(1/10) + 1(1/10) + \ldots + 9(1/10) = 4.5$
$\sigma^2(X) = (0 - 4.5)^2(1/10) + (1 - 4.5)^2(1/10) + (2 - 4.5)^2(1/10) +$
$(3 - 4.5)^2(1/10) + (4 - 4.5)^2(1/10) + (5 - 4.5)^2(1/10) +$
$(6 - 4.5)^2(1/10) + (7 - 4.5)^2(1/10) + (8 - 4.5)^2(1/10) +$
$(9 - 4.5)^2(1/10) = 8.25$
$\sigma(X) = 2.87$

**4.11** (a) The expected value equals $(1/5)(0) + (1/5)(1) + (1/5)(2) + (1/5)(3) + (1/5)(4) = 2$.
(b) Yes.
(c) $\frac{1}{5}(0 - 2)^2 + \frac{1}{5}(1 - 2)^2 + \frac{1}{5}(2 - 2)^2 + \frac{1}{5}(3 - 2)^2 + \frac{1}{5}(4 - 2)^2 = \frac{1}{5}(10) = 2(\text{bicycles})^2$
(d) $\sqrt{2} = 1.414$ bicycles.

**4.13** (a) $\$20(10/16) + \$50(5/16) + \$100(1/16) = \$34.375$
(b) $\sqrt{(20 - 34.375)^2(10/16) + (50 - 34.375)^2(5/16) + (100 - 34.375)^2 1/16}$
$= \sqrt{129.15 + 76.29 + 269.16} = \sqrt{474.60} = \$21.79$
(c) No.
(d) No.

**4.15** 5
5
Yes.

**4.17** (a) $(.20)^{12} = .0000$
(b) $.0155 + .0033 + .0005 + .0001 = .0194$. See Appendix Table 1.
(c) $.0687$
(d) $.0687 + .2062 = .2749$
(e) The mean equals 12 (.2), or 2.4. Thus, the answer is $.0687 + .2062 + .2835 = .5584$.

**4.19** (a) Yes. No.
(b) $\$5,000(.0038) + \$2,000(.9962) = \$2,011$.
(c) $\sqrt{(5,000 - 2,011)^2(.0038) + (2,000 - 2,011)^2(.9962)}$
$= \sqrt{(8,934,121)(.0038) + (121)(.9962)} = \sqrt{34,070} = \$185$.

**4.21** (a) From Appendix Table 1, $.6302 + .2985 = .9287$
(b) $.3874 + .3874 = .7748$
(c) $.2316 + .3679 = .5995$

**4.23**

```
MTB > pdf;
SUBC> binomial n=21,p=.4.
BINOMIAL WITH N = 21 P = 0.400000
     K            P( X = K )
     0             0.0000
     1             0.0003
     2             0.0020
     3             0.0086
     4             0.0259
     5             0.0588
     6             0.1045
     7             0.1493
     8             0.1742
     9             0.1677
    10             0.1342
    11             0.0895
    12             0.0497
    13             0.0229
    14             0.0087
    15             0.0027
    16             0.0007
    17             0.0001
    18             0.0000
```

**4.25** None. All are known constants, not quantities to be determined by chance.

**4.27** The expected gain is $(.993)(\$25) + (.007)(-\$975) = \$18$. Thus, the standard deviation of the gain equals

$$\sqrt{(25-18)^2(.993) + (-975-18)^2(.007)} = \sqrt{48.657 + 6,902.343}$$

$$= \sqrt{6,951} = \$83.37.$$

**4.29** 1/9

**4.31** The expected value of this gamble is $(.9)(\$5,000) + (.1)(-\$10,000) = \$3,500$. Thus, he should purchase the chairs since the expected value exceeds zero.

**4.33** (a) Since $n = 20$ and $\Pi = .20$, Appendix Table 1 indicates that this probability equals $.0005 + .0001 + .0000 = .0006$.

(b) According to Appendix Table 1 this probability equals $.0115 + .0576 + .1369 = 0.206$.

(c) According to Appendix Table 1 the probability that the number expressing dissatisfaction is more than 20 percent of the sample equals

$$1 - (.0115 + .0576 + .1369 + .2054 + .2182) = 1 - .6296 = .3704.$$

Thus, the expected value of the gamble to the personnel director is $(.3704)(-\$100) + (.6296)(\$100) = \$25.92$. This is not a fair bet since the expected value does not equal zero.

(d) The expected number equals $n\Pi$, or $(20)(.2) = 4$ employees. The standard deviation equals

$$\sqrt{n\Pi(1-\Pi)} = \sqrt{20(.2)(.8)} = \sqrt{3.2}, \text{ or } 1.79 \text{ employees.}$$

(e) This upper bound is $1/9$. The true probability that the number will be more than three standard deviations (that is, 3 times 1.79, or 5.37) from the mean (which equals 4) is the probability that the number exceeds 9.37. According to Appendix Table 1 this probability equals $.0020 + .0005 + .0001$, or $.0026$.

# Chapter 5

**5.1** (a) $.5 - .4893 = .0107$
(b) $.5 - .4987 = .0013$
(c) $.5 - .2580 = .2420$
(d) $.4772 - .3413 = .1359$
(e) $.3413 + .4772 = .8185$

**5.3** No. This is true only if the variable's mean equals zero.

**5.5** (a) 0.52
(b) 0.59
(c) 0.23
(d) 0.86
(e) 0.23

**5.7** (a) $.5000 - .3413 = .1587$
(b) $.5000 - .4332 = .0668$
(c) $.1915 + .1915 = .3830$
(d) $.4938 - .3413 = .1525$

**5.9** According to Appendix Table 2, the point on the standard normal distribution that is exceeded with a probability of .05 is 1.64. The weight that corresponds to this point on the standard normal distribution is $\mu + 1.64\sigma$, or $170 + (1.64)(20)$, or 202.8 pounds.

**5.11** The number of times he hits the target is a binomial random variable with mean equal to $(50)(1/3)$, or 16.67 and a standard deviation equal to

$$\sqrt{(50)(1/3)(2/3)}, \text{ or } \sqrt{11.11} = 3.333.$$

We want the probability that this variable is 9 or less, which can be approximated by the probability that the standard normal variable is less than $(9\frac{1}{2} - 16.67) \div 3.333 = -7.17 \div 3.333$, or $-2.15$. Using Appendix Table 2, we find that this probability equals .0158.

**5.13** (a) $Pr\{Z < (7.5 - 10) \div 2.828\} = Pr\{Z < -.88\} = .1894$
(b) $Pr\{Z > (20.5 - 10) \div 2.828\} = Pr\{Z > 3.71\}$, which is less than .001.
(c) $Pr\{(10.5 - 10) \div 2.828 < Z < (18.5 - 10) \div 2.828\}$

$$= Pr\{.18 < Z < 3.01\} = .4987 - .0714 = .4273$$

(d) $Pr\{(8.5 - 10) \div 2.828 < Z < (11.5 - 10) \div 2.828\}$

$$= Pr\{-.53 < Z < .53\} = .2019 + .2019 = .4038$$

**5.15** Let $X$ be the number of pedals in such a carton that will not survive more than ten years of normal wear. Obviously, $X$ has the binomial distribution with mean $(n\Pi)$ equal to $(1/3)(300) = 100$ and with standard deviation equal to

$$\sqrt{300(1/3)(2/3)} = \sqrt{66.67}, \text{ or } 8.165.$$

The probability that between 84½ and 115½ pedals will not survive equals $Pr\{84\ 1/2 < X < 115\ 1/2\}$.

The point on the standard normal distribution corresponding to 84 1/2 is

$$\frac{84\ 1/2 - 100}{8.165} = \frac{-15.5}{8.165} = -1.90.$$

The point on the standard normal distribution corresponding to 115 1/2 is

$$\frac{115\ 1/2 - 100}{8.165} = \frac{15.5}{8.165} = +1.90.$$

Using Appendix Table 2, the probability that a standard normal variable lies between $-1.90$ and $+1.90$ is $.4713 + .4713 = .9426$. Thus, the probability that either less than 85 or more than 115 pedals will not survive is $1 - .9426$, or $.0574$. (Of course, this is only an approximation.)

**5.17** The normal distribution.

**5.19** 3.0

**5.21** Using Appendix Table 1, the binomial probability is .3585. Using Appendix Table 3, the Poisson probability is .3679 since $\mu = (20)(.05) = 1$

**5.23** Since the standard deviation is 1.732, the mean must be $(1.732)^2$, or 3.0. Using Appendix Table 3, we find that the probability of no accidents, given that the mean is 3.0, equals .0498.

**5.25** (a) $P(2) = .0842$
(b) $P(3) = .1404$
(c) $P(0) = .0067$

**5.27** $\sigma = \sqrt{3} = 1.732$

$Pr\{1.268 < X < 4.732\} = P(2) + P(3) + P(4) = .2240 + .2240 + .1680 = .6160$

Thus, the desired probability equals $1 - .6160 = .3840$.

**5.29** (a) $1 - (.0025 + .0149 + .0446) = 1 - .0620 = .9380$
(b) .0892

**5.31** (a) $n\Pi = 20(.25) = 5$

$\sqrt{n\Pi(1 - \Pi)} = \sqrt{20(.25)(.75)} = \sqrt{3.75} = 1.9365$

$Pr\{(3.5 - 5) \div 1.9365 < Z < (7.5 - 5) \div 1.9365\}$

$= Pr\{-.77 < Z < 1.29\} = .2794 + .4015 = .6809$

(b) $.1897 + .2023 + .1686 + .1124 = .6730$
(c) The difference is only .0079.

**5.33** (a) Since the standard deviation is .02 inches, the point on the standard normal distribution corresponding to a width that exceeds the mean by .03 inches is 1.5. Using Appendix Table 2, the probability that the standard normal variable exceeds 1.5 is .0668.

(b) The point on the standard normal distribution that corresponds to a width that falls short of the mean by .05 inches is $-2.5$. Using Appendix Table 2, the probability that the standard normal variable is less than $-2.5$ is .0062.

(c) The point on the standard normal distribution that corresponds to a width that is .015 inches below the mean is $-0.75$. The point on the standard normal distribution that corresponds to a width that is .015 inches above the mean is $0.75$. Using Appendix Table 2, the probability that the standard normal variable lies between $-0.75$ and $0.75$ is .2734 + .2734 = .5468.

(d) 1/2

(e) Since the mean is .01 inches greater than the design, all pedals of widths greater than .01 inches less than the mean exceeded the design. The point on the standard normal distribution corresponding to .01 inches less than the mean is $-1/2$ since the standard deviation equals .02 inches. Using Appendix Table 2, the probability that the standard normal variable exceeds $-1/2$ equals .1915 + .5000, or .6915.

(f) Since the average width equaled the design, if a pedal was .04 inches wider than the design, its width was .04 inches above the mean. Since the standard deviation was .02 inches, the point on the standard normal distribution corresponding to .04 inches above the mean is 2.0. Using Appendix Table 2, the probability that the standard normal variable exceeds 2.0 is .0228.

(g) Since the average width was .01 inches greater than the design, if a pedal was .04 inches wider than the design, its width was .03 inches above the mean. The point on the standard normal distribution corresponding to .03 inches above the mean is 1.5. Using Appendix Table 2, the probability that the standard normal variable exceeds 1.5 is .0668.

**5.35** If $\Pi = .001$, the probability that none of the sample is defective is $(.999)^n$. If $n$ is at least 20, the Poisson approximation should be useful. According to Appendix Table 3, the probability that $X = 0$ (that is, that there are no defectives) equals .9512 (which is very close to .95) when $\mu = .05$. Since $\mu = n\Pi$, $n$ must equal 50 because $\Pi = .001$. Thus, 50 motors should be examined from each day's output.

**5.37**

```
MTB > pdf;
SUBC> poisson 1.45.

    POISSON WITH MEAN = 1.450
         K                  P(X = K)
         0                  0.2346
         1                  0.3401
         2                  0.2466
         3                  0.1192
         4                  0.0432
         5                  0.0125
         6                  0.0030
         7                  0.0006
         8                  0.0001
         9                  0.0000
```

**5.39** $1 - \left(\dfrac{34}{35}\right)^{25} = .52$

**5.41** (a) $e^{-(3)(1)} - 0 = .05$
(b) $e^{-(3)(2)} - 0 = .002$

**5.43** (a) ½ ÷ (2 − 1) = ½ minute
(b) ½
(c) 1 ÷ (2 − 1) = 1 minute
(d) (½)² ÷ (1 − ½) = ¼ ÷ ½ = ½ persons

# Chapter 6

**6.1** (a) *AB, AC, AD, AE, BC, BD, BE, CD, CE, DE*. Of course, this assumes that sampling is without replacement.
(b) *CDE, BDE, BCE, BCD, ADE, ACE, ACD, ABE, ABD, ABC*.
(c) 6/10
  3/10

**6.5** No. Not every guest has the same probability of filling out a question-naire. For example, guests that have complaints are more likely than others to do so.

**6.7** (a) Yes.
(b) No.
(c) The probability that each voter will be included in the sample is un-known and uncontrolled.
(d) He might choose a random sample of voters in each part of his district and ask each person in the sample whether he should vote for or against the proposed legislation.

**6.9** (a) No.
(b) The population of Philadelphia residents with children attending the public schools.
(c) Philadelphia residents with no children currently in the public schools may tend to have different opinions on this score than those with children in the public schools.

**6.11** (a) 2/7 from stratum *A*, and 5/7 from stratum *B*.
(b) 2/3 from stratum *A*, and 1/3 from stratum *B*.
(c) The allocation scheme in (a).

**6.13** Yes. There is some evidence that the candidate named first is at an ad-vantage.

**6.15** The standard deviation of the sample mean is $100 \div \sqrt{100}$, or 10 hours. Thus, if the sample mean differs by more than 15 hours from the popula-tion mean, it differs by more than 1.5 of its standard deviations from its mean. Since the sample size is large, the sample mean is approximately normally distributed. The probability of a normal variable lying more than 1.5 standard deviations from its mean is (2)(.0668), or .1336 ac-cording to Appendix Table 2.

**6.17** (a) $\sigma_{\bar{x}} = \dfrac{\sigma}{\sqrt{n}} \sqrt{\dfrac{N-n}{N-1}}$

$= \dfrac{\sigma}{\sqrt{n}} \sqrt{\left(1 - \dfrac{n}{N}\right) \div \left(1 - \dfrac{1}{N}\right)}$

If $N$ is large, $(1 - 1/N)$ is approximately equal to one. Thus,

$$\sigma_{\bar{x}} \doteq \frac{\sigma}{\sqrt{n}} \sqrt{1 - \frac{n}{N}}.$$

(b) $\dfrac{2}{\sqrt{16}} \sqrt{1 - \dfrac{16}{100}} = \dfrac{2}{4} \sqrt{.84} = (1/2)(.9165) = .458$

$\dfrac{2}{\sqrt{16}} \sqrt{1 - \dfrac{16}{10,000}} = \dfrac{2}{4} \sqrt{.9984} = (1/2)(.9992) = .500$

**6.19** (a) $950,000

(b) $310,000 \div \sqrt{16} = 77,500$

(c) No. The central limit theorem assumes that the sample size is at least 30. But if the distribution of the sales of the 3,000 gas stations is close to normal, the sampling distribution of the sample mean will be close to normal, even though the sample size is only 16.

**6.21** (a) If $\bar{X} < 78$, $(\bar{X} - 80) \div 0.9 < (78 - 80) \div 0.9 = -2.22$. The probability that a standard normal variable is less than $-2.22$ equals $0.5 - 0.4868 = 0.0132$.

(b) Since $(78 - 80) \div 9/7 = -1.56$, and since the probability that a standard normal variable is less than $-1.56$ equals $.0594$, the desired probability is $.0594$.

(c) Because the standard error of the sample mean is inversely related to the sample size.

**6.23** (a) $\sigma_{\bar{x}} = \dfrac{10}{\sqrt{50}} \sqrt{\dfrac{85 - 50}{85 - 1}} = 10 \sqrt{\dfrac{35}{4200}} = 10 \sqrt{.008333} = .913.$

Since $(120 - 115) \div .913 = 5.48$, this probability is less than $.001$.

(b) If there had been 170 students,

$$\sigma_{\bar{x}} = \frac{10}{\sqrt{50}} \sqrt{\frac{170 - 50}{170 - 1}} = 10 \sqrt{\frac{120}{8450}} = 10 \sqrt{.0142} = 1.19.$$

Since $(120 - 115) \div 1.19 = 4.20$, this probability is less than $.001$. Thus, the doubling of the class size would have little effect on the answer to (a).

**6.25** In this case, $\Pi = \dfrac{1,000}{3,000} = 1/3.$

If the sample proportion differs from the population proportion by more than $.02$, the number in the sample having relatives in the United States must be other than 32, 33, 34, or 35. Using the normal approximation to the binomial distribution, the probability that this is the case equals:

$$1 - Pr \left\{ \frac{31.5 - 33\,1/3}{\sqrt{200/9}} < Z < \frac{35.5 - 33\,1/3}{\sqrt{200/9}} \right\}$$

$$= 1 - Pr \left\{ -\frac{1.83}{4.71} < Z < \frac{2.17}{4.71} \right\}$$

$$= 1 - Pr \{-0.39 < Z < 0.46\} = 1 - (.1517 + .1772) = 0.6711.$$

**6.27** (a) It can number the farms on the list from 1 to 3,000, and it can use a table of random numbers to draw the sample.

(b) No, it is a systematic sample.
Yes.

(c) Yes. Holding sample size constant, the sampling error would tend to be smaller if stratification of this sort occurred.

(d) If proportional allocation is used, the number sampled in each stratum is proportional to the total number in the stratum. Since the total sample size is 100, 1/30 of the farms in each stratum will be chosen. Thus the number chosen from each stratum is as follows:

| Farm size (acres) | Number of farms in sample |
|---|---|
| 0–50 | 1,000/30 = 33 |
| 51–100 | 500/30 = 17 |
| 101–150 | 500/30 = 17 |
| 151–200 | 400/30 = 13 |
| 201–250 | 400/30 = 13 |
| Over 250 | 200/30 = 7 |
| Total | 100 |

(e) The product of number of farms and standard deviation in each stratum is shown in column (3) below:

| Farm size (acres) | (1) Number of farms | (2) Standard deviation | (3) (1) × (2) | Column (3) as percent of total | Number in sample |
|---|---|---|---|---|---|
| 0–50 | 1,000 | 10 | 10,000 | 24 | 24 |
| 51–100 | 500 | 10 | 5,000 | 12 | 12 |
| 101–150 | 500 | 10 | 5,000 | 12 | 12 |
| 151–200 | 400 | 15 | 6,000 | 14 | 14 |
| 201–250 | 400 | 15 | 6,000 | 14 | 14 |
| Over 250 | 200 | 50 | 10,000 | 24 | 24 |
| Total | | | 42,000 | | |

In the next column, this product in each stratum is divided by the sum of these products (42,000). Finally, the result is multiplied by 100 to get the sample size in each stratum.

(f) No, because the farms contained in a particular cluster are not close together (unless people with names beginning with the same initial live close together, which is unlikely).

**6.29** Suppose that the median is based on a sample size of $n_1$, and the mean is based on a sample size of $n_2$. If

$$\sqrt{\frac{\pi}{2}} \frac{\sigma}{\sqrt{n_1}} = \frac{\sigma}{\sqrt{n_2}},$$

it follows that $n_1/n_2 = \pi/2$. Since $\pi = 3.1416$, $\pi/2 = 1.57$. Thus, $n_1$ must be 1.57 times $n_2$.

**6.31** (a) Its sampling distribution is normal with a mean of 3.00 inches and a standard deviation of .05 inches.

(b) The point on the standard normal distribution corresponding to 3.01 inches is $(3.01 - 3.00) \div .05 = 0.20$. According to Appendix Table 2, the probability that the standard normal variable will exceed 0.20 is .4207.

(c) 100

**6.35** Draw 100 two-digit random numbers, each corresponding to a day. If a day's random number is 00 to 09, the firm sold zero cars on this day. If the random number is 10 to 24, it sold one car. If the random number is 25 to 54, it sold two cars. If the random number is 55 to 74, it sold three cars. If the random number is 75 to 89, it sold four cars. If the random number is 90 to 99, it sold five cars.

**6.37**

```
MTB > random 30 observations into c3;
SUBC> normal mu=15 sigma=.2.
MTB > histogram c3

Histogram of C3   N = 30

Midpoint  Count
   14.5      1   *
   14.6      0
   14.7      1   *
   14.8      5   *****
   14.9      3   ***
   15.0      4   ****
   15.1      8   ********
   15.2      4   ****
   15.3      3   ***
   15.4      0
   15.5      0
   15.6      1   *

MTB > describe c3

           N     MEAN   MEDIAN   TRMEAN   STDEV   SEMEAN
C3        30   15.037   15.065   15.038   0.220    0.040
          MIN      MAX       Q1       Q3
C3     14.544   15.587   14.898   15.205

MTB > print c3
C3
14.9789 14.8148 14.8301 14.9360 14.7707 14.9999 14.9201
15.0970 15.0787 14.9697 15.5871 15.2427 14.6679 15.2882
15.2043 14.8100 15.1244 15.2214 15.0854 15.1281 15.1372
15.1287 14.7722 14.5444 15.2770 15.0521 14.9555 15.3416
15.2090 14.9482
```

# Chapter 7

**7.1** (a) Estimate. Point estimate.
(b) Estimator.
(c) Estimate. Interval estimate.

**7.3** The sample mean is 4.448 thousands of hours.

**7.5** (a) 8/60, or .13
(b) A sample of this size need not be unreliable; whether it is unreliable or not depends on the required accuracy. The sample percentage is not a biased estimate of the population percentage.

**7.7** (a) .01
(b) .005
(c) .025

**7.9**  (a) $\overline{X} - 1.64 \dfrac{s}{\sqrt{n}} < \mu < \overline{X} + 1.64 \dfrac{s}{\sqrt{n}}$

$810 - 1.64 \dfrac{85}{9.487} < \mu < 810 + 164 \dfrac{85}{9.487}$

$810 - 14.69 < \mu < 810 + 14.69$

$\$795.31 < \mu < \$824.69$

(b) $810 - 1.96(8.96) < \mu < 810 + 1.96(8.96)$

$810 - 17.56 < \mu < 810 + 17.56$

$\$792.44 < \mu < \$827.56$

(c) $810 - 2.576(8.96) < \mu < 810 + 2.576(8.96)$

$810 - 23.08 < \mu < 810 + 23.08$

$\$786.92 < \mu < \$833.08$

**7.11**  (a) $8.4 - 1.64 \dfrac{(1.8)}{\sqrt{90}} < \mu < 8.4 + 1.64 \dfrac{(1.8)}{\sqrt{90}}$

$8.4 - 1.64(.19) < \mu < 8.4 + 1.64(.19)$

$8.4 - .31 < \mu < 8.4 + .31$

$8.1 \text{ years} < \mu < 8.7 \text{ years}$

(b) $8.4 - 2.33(.19) < \mu < 8.4 + 2.33(.19)$

$8.4 - .44 < \mu < 8.4 + .44$

$8.0 \text{ years} < \mu < 8.8 \text{ years}$

**7.13**  (a) $31.15 - 1.96 \dfrac{(.08)}{\sqrt{80}} < \mu < 31.15 + 1.96 \dfrac{(.08)}{\sqrt{80}}$

$31.15 - 1.96(.00895) < \mu < 31.15 + 1.96(.00895)$

$31.15 - .018 < \mu < 31.15 + .018$

$31.13 \text{ ounces} < \mu < 31.17 \text{ ounces}$

(b) No. The sample size is greater than 30, so we can be sure that the sample mean is approximately normally distributed whether or not the population is normal.

**7.15**  (a) 4,376 to 4,624 hours.

(b) The width of such a confidence interval equals $(2)(1.64)(\sigma/\sqrt{n})$. If $\sigma = 500$ and $n = 36$, the width equals $(2)(1.64)(500/6)$, or $1,640/6$, or 273.33 hours. Yet the width of the confidence interval he gives is 300 hours (that is, $4800 - 4500$). Thus, there is a contradiction. He has made a mistake somewhere.

(c) The width of a 95 percent confidence interval is $(2)(1.96)(\sigma/\sqrt{n})$, whereas the width of a 90 percent confidence interval is $(2)(1.64)(\sigma/\sqrt{n})$. Thus, the ratio of the former to the latter is $1.96/1.64$, or 1.20. Thus, he is right.

**7.17** If $s$ is (approximately) distributed normally with mean equal to $\sigma$ and standard deviation equal to $\sigma \div \sqrt{2n}$, then $(s - \sigma) \div \sigma / \sqrt{2n}$ has (approximately) the standard normal distribution. Thus,

$$Pr\left\{-z_{\alpha/2} < \frac{s - \sigma}{\sigma/\sqrt{2n}} < z_{\alpha/2}\right\} = 1 - \alpha$$

$$Pr\left\{\frac{-z_{\alpha/2}}{\sqrt{2n}} < \frac{s}{\sigma} - 1 < \frac{z_{\alpha/2}}{\sqrt{2n}}\right\} = 1 - \alpha$$

$$Pr\left\{\frac{-z_{\alpha/2}}{\sqrt{2n}} + 1 < \frac{s}{\sigma} < \frac{z_{\alpha/2}}{\sqrt{2n}} + 1\right\} = 1 - \alpha$$

$$Pr\left\{\frac{1}{1 + \dfrac{z_{\alpha/2}}{\sqrt{2n}}} < \frac{\sigma}{s} < \frac{1}{1 - \dfrac{z_{\alpha/2}}{\sqrt{2n}}}\right\} = 1 - \alpha$$

$$Pr\left\{\frac{s}{1 + \dfrac{z_{\alpha/2}}{\sqrt{2n}}} < \sigma < \frac{s}{1 - \dfrac{z_{\alpha/2}}{\sqrt{2n}}}\right\} = 1 - \alpha.$$

**7.19** $-0.735$ to $-.265$ points.

**7.21** (a) $.36 - 1.64 \sqrt{\dfrac{(.36)(.64)}{100}} < \Pi < .36 + 1.64 \sqrt{\dfrac{(.36)(.64)}{100}}$

$.36 - 1.64(.048) < \Pi < .36 + 1.64(.048)$

$.36 - .08 < \Pi < .36 + .08$

$.28 < \Pi < .44$

(b) $.36 - 1.96(.048) < \Pi < .36 + 1.96(.048)$

$.36 - .09 < \Pi < .36 + .09$

$.27 < \Pi < .45$

**7.23** (a) Based on Appendix Table 7a,

$0.24 < \Pi < 0.57.$

(b) Based on Appendix Table 7b,

$0.21 < \Pi < 0.62.$

**7.25** (a) $58.15 - 56.35 - 1.96 \sqrt{\dfrac{3.42^2}{50} + \dfrac{4.13^2}{50}} < \mu_1 - \mu_2 < 58.15 - 56.35 +$

$1.96 \sqrt{\dfrac{3.42^2}{50} + \dfrac{4.13^2}{50}}$

$1.80 - 1.96 \sqrt{\dfrac{11.6964 + 17.0659}{50}} < \mu_1 - \mu_2 < 1.80 +$

$$1.96 \sqrt{\frac{11.6964 + 17.0659}{50}}$$

$$1.80 - 1.96\,(.758) < \mu_1 - \mu_2 < 1.80 + 1.96\,(.758)$$

$$1.80 - 1.49 < \mu_1 - \mu_2 < 1.80 + 1.49$$

$$\$0.31 < \mu_1 - \mu_2 < \$3.29$$

(b) No. Each sample size is large enough so that the central limit theorem tells us that the sample mean will be normally distributed.

**7.27** (a) $n = \left[\dfrac{(1.96)(15,000)}{5,000}\right]^2 = 5.88^2 = 35$

(b) People may not be willing to give such data, or they may lie.

(c) Because the population that is being sampled is not the one that the investigator really wants to sample.

**7.29** $n = \left[\dfrac{(1.96)(400)}{20}\right]^2 = \left[\dfrac{784}{20}\right]^2 = 39.2^2 = 1,537$

**7.31** Yes, because part or all of any apparent difference between the half-pieces treated with the chlorinating agent and those that were untreated could be due to the difference in the machines on which they were evaluated.

**7.33** $n = \left(\dfrac{2.576}{.02}\right)^2 (.4)(.6) = 128.8^2\,(.24) = 3,981$

**7.35** (a) $\Sigma X_i = 67$

$\Sigma X_i^2 = 143$

$$s^2 = \frac{143 - \dfrac{67^2}{36}}{35} = \frac{143 - 124.7}{35} = \frac{18.3}{35} = .523$$

$s = .72$

$$1.86 - 1.96\left(\frac{.72}{6}\right)\sqrt{\frac{60-36}{60-1}} < \mu < 1.86 + 1.96\left(\frac{.72}{6}\right)\sqrt{\frac{60-36}{60-1}}$$

$$1.86 - 1.96(.12)(.64) < \mu < 1.86 + 1.96(.12)(.64)$$

$$1.86 - .15 < \mu < 1.86 + .15$$

$$1.7 \text{ persons} < \mu < 2.0 \text{ persons}$$

(b) No. The central limit theorem says that the sample mean is normally distributed if the sample size exceeds 30.

**7.37** (a) $.59 - .52 - 1.64\sqrt{\dfrac{(.59)(.41)}{200} + \dfrac{(.52)(.48)}{200}} < \Pi_1 - \Pi_2 < .59 - .52 +$

$1.64\sqrt{\dfrac{(.59)(.41)}{200} + \dfrac{(.52)(.48)}{200}}$

$0.07 - 1.64\sqrt{\dfrac{.2419 + .2496}{200}} < \Pi_1 - \Pi_2 < 0.07 +$

$1.64\sqrt{\dfrac{.2419 + .2496}{200}}$

$$0.07 - 1.64(.0496) < \Pi_1 - \Pi_2 < 0.07 + 1.64(.0496)$$

$$0.07 - .08 < \Pi_1 - \Pi_2 < 0.07 + .08$$

$$-0.01 < \Pi_1 - \Pi_2 < 0.15$$

(b) $\quad 0.07 - 2.576(.0496) < \Pi_1 - \Pi_2 < 0.07 + 2.576(.0496)$

$$0.07 - .13 < \Pi_1 - \Pi_2 < 0.07 + .13$$

$$-0.06 < \Pi_1 - \Pi_2 < 0.20$$

**7.39** (a) .262 to .418

(b) The confidence interval is

$$\frac{8.7}{1 + \dfrac{1.64}{\sqrt{200}}} < \sigma < \frac{8.7}{1 - \dfrac{1.64}{\sqrt{200}}}$$

$$\frac{8.7}{1 + \dfrac{1.64}{14.14}} < \sigma < \frac{8.7}{1 - \dfrac{1.64}{14.14}}$$

$$\frac{8.7}{1.116} < \sigma < \frac{8.7}{.884}$$

$$7.80 < \sigma < 9.84.$$

Thus, the answer is $7.80 to $9.84.

**7.41** $n = \left(\dfrac{(1.96)(60)}{10}\right)^2 = 11.76^2 = 138$

**7.43** The desired confidence interval is 19.46 to 26.94 years. The number of observations is 10, the sample mean is 23.2 years, the sample standard deviation is 5.22 years, and the standard error of the mean is 1.65 years.

# Chapter 8

**8.1** (a) The alternative hypothesis is that the applicant is not qualified for the job.

(b) The consequence of a Type I error is that the bank turns away a qualified applicant. The consequence of a Type II error is that the bank accepts an unqualified applicant as being qualified.

(c) The applicant can be given tests, and his record and references can be examined.

**8.3** (a) No. With a fixed sample size, the probability of Type I error can be reduced only by increasing the probability of Type II error. Thus, these probabilities should be chosen to reflect the relative costs of Type I and Type II errors.

(b) No.

**8.5** (a) Reject the hypothesis that $\mu = 12$ if $(\bar{x} - 12) \div .09/\sqrt{60} < -1.64$. In other words, reject $H_0$ if $\bar{x} < 11.98$.

(b) If $\mu = 11.9$, the probability that $\overline{X}$ is greater than or equal to 11.98 equals the probability that a standard normal variable is greater than

or equal to $(11.98 - 11.9) \div .09/\sqrt{60} = .08 \div .0116 = 6.90$. This
probability is less than .001.

(c) Since $\bar{x} = 11.95$, the hypothesis that $\mu = 12$ should be rejected.

**8.7** (a) Reject the hypothesis that $\mu = 120$ if

$$(\bar{x} - 120) \div \left[ \frac{10.9}{\sqrt{36}} \sqrt{\frac{(60 - 36)}{(60 - 1)}} \right] > 2.05.$$

In other words, reject $H_0$ if $\bar{x} > 120 + 2.38 = 122.38$.

(b) If $\mu = 121$, the probability that $\overline{X} \leq 122.38$ equals the probability
that a standard normal variable is less than or equal to $(122.38 - 121)$
$\div 1.159 = 1.19$. This probability is .883.

(c) Since $\bar{x} = 122.8$, the hypothesis that $\mu = 120$ should be rejected.

**8.9** (a) The test procedure in the text is: reject $H_0$ if $(\overline{X} - 50) \div 10/\sqrt{64} <$
$-2.33$. In other words, $H_0$ is not rejected if $\overline{X} \geq 47.09$ years. Thus, if
$\mu = 47$, the probability of a Type II error equals the probability that $Z$
$\geq (47.09 - 47)/1.25$, or that $Z \geq .07$; this probability equals .47. If
$\mu = 48$, the probability of a Type II error equals the probability that
$Z \geq (47.09 - 48)/1.25$, or that $Z \geq -.73$; this probability equals .77.
If $\mu = 49$, the probability of a Type II error equals the probability that
$Z \geq (47.09 - 49)/1.25$, or that $Z \geq -1.53$; this probability equals
.94. If $\mu = 50$, the probability of a Type II error equals the probability
that $Z \geq (47.09 - 50)/1.25$, or that $Z \geq -2.33$; this probability
equals .99.

(b)

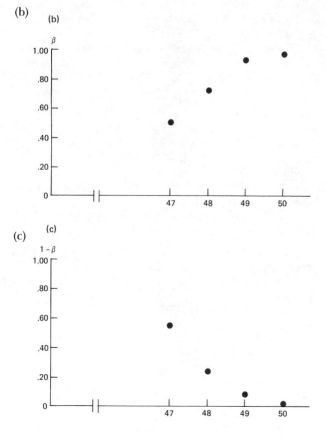

(c)

**8.11** (a) Reject the hypothesis that the teller is performing adequately if one or more out of twenty randomly chosen transactions per day contain an error.

(b) $1 - (.99)^{20} = 1 - (.904)^2 = 1 - .82 = .18$

(c) .3585

**8.13** (a) The null hypothesis is that $\Pi = .40$. It should be rejected if $(p - .40) \div \sqrt{[.4(.6)]/120} < -2.576$ or if $(p - .40) \div \sqrt{[.4(.6)]/120} > 2.576$. In other words, $H_0$ should be rejected if $p < .285$ or if $p > .515$.

(b) This probability is approximately equal to the probability that a standard normal variable's value lies between $(.285 - .42) \div \sqrt{(.42)(.58)/120}$ and $(.515 - .42) \div \sqrt{(.42)(.58)/120}$, which is the probability that $-3.00 < Z < 2.11$. This probability equals .98.

(c) Do not reject the null hypothesis that no change has occurred (and $\Pi = .40$).

(d) Since $z = (.43 - .40) \div .0447 = 0.67$, the $p$-value equals $2(.2514)$ or .5028.

**8.15** (a) If the percentage is the same as last year (81 percent), the probability that 317 or less would be filed on time is approximately equal to the probability that a standard normal variable is less than $(.7925 - .81) \div \sqrt{(.81)(.19)/400} = -0.89$. This probability is .19. Thus, the Service cannot be reasonably sure (if "reasonably sure" is defined as the probability being less than .19).

(b) No.

(c) The test in the text is: Reject the hypothesis that $\Pi = .81$ if $(p - .81) \div \sqrt{(.81)(.19)/400} < -1.64$. In other words, this hypothesis should be rejected if $p < .778$. Thus, if $\Pi = .79$, the probability of not rejecting the hypothesis that $\Pi = .81$ is approximately equal to the probability that a standard normal variable is greater than or equal to $(.778 - .79) \div \sqrt{(.79)(.21)/400} = -.012 \div .020 = -0.60$. This probability is .73.

**8.17** (a) The null hypothesis (that $\Pi = .50$) should be rejected if $(p - .50) \div \sqrt{[(.5)(.5)]/100}$ is greater than 1.64 or less than $-1.64$. In other words, it should be rejected if $p > .582$ or if $p < .418$. Since the observed value of $p$ is .34, it should be rejected.

(b) The null hypothesis should be rejected if $p < .50 - 1.28\sqrt{[(.5)(.5)]/100}$; that is, if $p < .436$. Since the observed value of $p$ is .34, it should be rejected.

(c) Since $z = (.34 - .50) \div \sqrt{[(.5)(.5)]/100} = -3.2$, the $p$-value for the two-tailed test in (a) equals $2(.0013)$, or .0026.

**8.19** (a) (i) The null hypothesis is that the population mean score of the engineering graduates equals that of the business graduates. The alternative hypothesis is that these population mean scores are not equal.

(ii) Reject the null hypothesis if $(\bar{x}_1 - \bar{x}_2) \div \sqrt{(100/100) + (100/100)}$ is greater than $z_{\alpha/2}$ or less than $-z_{\alpha/2}$. Otherwise do not reject the null hypothesis.

(iii) Since $(\bar{x}_1 - \bar{x}_2) \div \sqrt{2} = (80 - 78) \div 1.414 = 1.414$, it follows that it does not exceed $z_{.05}$, which equals 1.64. Thus the null hypothesis should not be rejected.

(iv) Since $z_{.025} = 1.96$, the null hypothesis should not be rejected.

(v) Since $z_{.005} = 2.576$, the null hypothesis should not be rejected.

(b) (i) The null hypothesis is that the population mean score of the engineering graduates equals that of the business graduates. The alternative hypothesis is that the former exceeds the latter.

(ii) Reject the null hypothesis if $(\bar{x}_1 - \bar{x}_2) \div \sqrt{(100/100) + (100/100)}$

$> z_\alpha$, where the scores of the engineering graduates are population 1 and those of the business graduates are population 2. Otherwise do not reject the null hypothesis.

(iii) From above we know that $(\bar{x}_1 - \bar{x}_2) \div \sqrt{(100/100) + (100/100)}$ $= 1.414$. Since $z_{.10} = 1.28$, the null hypothesis should be rejected.

(iv) Since $z_{.05} = 1.64$, which exceeds 1.414, the null hypothesis should not be rejected.

(v) Since $z_{.01} = 2.33$, which exceeds 1.414, the null hypothesis should not be rejected.

(c) (i) The null hypothesis is that the population mean score of the engineering graduates equals that of the business graduates. The alternative hypothesis is that the latter exceeds the former.

(ii) Reject the null hypothesis if $(\bar{x}_1 - \bar{x}_2) \div \sqrt{(100/100) + (100/100)}$ $< -z_\alpha$, where the scores of the engineering graduates are population 1 and those of the business graduates are population 2. Otherwise do not reject the null hypothesis.

(iii) Since 1.414 is not less than $-1.28$, the null hypothesis should not be rejected.

(iv) Since 1.414 is not less than $-1.64$, the null hypothesis should not be rejected.

(v) Since 1.414 is not less than $-2.33$, the null hypothesis should not be rejected.

**8.21** (a) $z = (60.8 - 58.4) \div \sqrt{(9.9^2/100) + (8.7^2/100)} = 2.4 \div \sqrt{1.737} = 1.82$. Since $z_{.05} = 1.64$, the evidence does indicate that men are better than women at this task.

(b) No.

**8.23** (a) $p = (71 + 56) \div 200 = .635$
$z = (.71 - .56) \div \sqrt{(.635)(.365)/50} = .15/\sqrt{.004636} = 2.21$
Since $z_{.01} = 2.33$, the evidence does not indicate that the null hypothesis (that there is no regional difference) should be rejected.

(b) Since $z_{.025} = 1.96$, the evidence indicates that the null hypothesis (that there is no regional difference) should be rejected.

(c) Since $z = 2.21$, the $p$-value equals 2 (.0136) = .0272.

**8.25** (a) $t = (11.3 - 8.9) \div \sqrt{15.625(1/6)} = 2.4/1.614 = 1.49$,

since $s^2 = \dfrac{11(4.1^2) + 11(3.8^2)}{22} = \dfrac{11(16.81) + 11(14.44)}{22} = 15.625$.

Since the value of $t$ is less than $t_{.025}$ (which is 2.074), the data seem to be consistent with the hypothesis that the average effect is the same in both groups.

(b) No.

(c) We are assuming that the weight loss in each group is normally distributed and that the variance is equal in the two groups.

**8.27** (a) Since $6\,\sigma/\sqrt{n} = .0030$, $\sqrt{n} = 6\,(.0015)/.0030 = 3$. Thus, $n = 9$.

(b) No, not as we have described the workings of the control chart. But procedures can be adopted to detect such runs above or below the mean.

(c) No. This seems to be an unusually long run above the mean.

**8.29** (a) Since $n = 36$, $\bar{x} = \$15,561$, and $s = \$9,010$,

$$t = \frac{15,561 - 15,000}{9,010 \div 6} = \frac{561}{1502} = 0.37.$$

Since $t_{.025}$ equals about 2.03, the null hypothesis should not be rejected since the observed value of $t$ is not greater than 2.03.

(b) Since $t_{.005}$ equals about 2.73, the null hypothesis should not be rejected since the observed value of $t$ is not greater than 2.73.

**8.31** (a) Since $s^2 = \dfrac{(35)(9,010)^2 + (8)(5,624)^2}{36 + 9 - 2}$

$$= 71,961,361.$$

Thus,

$$t = \dfrac{15,561 - 15,078}{\sqrt{71,961,361 \left(\dfrac{1}{36} + \dfrac{1}{9}\right)}}$$

$$= \dfrac{483}{\sqrt{(71,961,361)(.1389)}} = \dfrac{483}{\sqrt{9,995,433}} = \dfrac{483}{3162} = .153.$$

Since $t_{.025}$ equals about 2.02, the null hypothesis should not be rejected since the observed value of $t$ is not greater than 2.02.

(b) Since $t_{.005}$ equals about 2.70, the null hypothesis should not be rejected since the observed value of $t$ is not greater than 2.70.

**8.33** (a) and (b) Since $(3)(.0004 \div \sqrt{4}) = .0006$ inches, the upper and lower control limits are at $.1020 + .0006$ inches and at $.1020 - .0006$ inches:

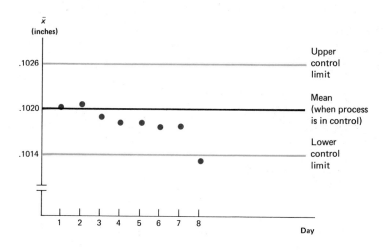

On the eighth day, the firm should reject the null hypothesis.

**8.35** (a) $t = (2.3 - 2.0) \div 1.4/\sqrt{16} = .3/.35 = .86$.
Since this value of $t$ is less than $t_{.05}$ (which is 1.753), there is no reason to reject the hypothesis that $\mu = \$2.0$ million.

(b) $z = (2.3 - 2.0) \div 1.4/\sqrt{160} = .3/.111 = 2.70$.
Since this value of $z$ is greater than $z_{.05}$ (which is 1.64), one should reject the hypothesis that $\mu = \$2.0$ million.

**8.37** (a) Reject $H_0$ if $(p - .40) \div \sqrt{(.4)(.6)/200}$ is greater than 2.576 or less than $-2.576$. In other words, do not reject $H_0$ if $.311 < p < .489$. Since $p = .445$, the difference is not statistically significant.

(b) The probability of not rejecting the hypothesis that $\Pi = 0.4$ when $\Pi = .38$ is approximately equal to the probability that $(.311 - .38) \div \sqrt{(.38)(.62)/200} \le Z \le (.489 - .38) \div \sqrt{(.38)(.62)/200}$. Since this is the probability that $-2.01 \le Z \le 3.18$, it equals .98. The probability of not rejecting the hypothesis that $\Pi = 0.4$ when $\Pi = .40$ is .99. The probability of not rejecting the hypothesis that $\Pi = 0.4$ when $\Pi = .42$ is approximately equal to the probability that $(.311 - .42) \div \sqrt{(.42)(.58)/200} \leq Z \leq (.489 - .42) \div \sqrt{(.42)(.58)/200}$. Since this is the probability that $-3.12 \leq Z \leq 1.98$, it equals 0.98.

(c)

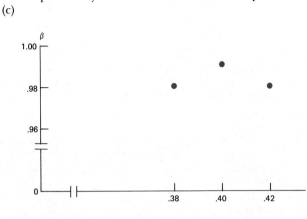

**8.39** (a) $p = (121 + 109) \div 400 = .575$
$z = (.605 - .545) \div \sqrt{(.575)(.425)/100} = .06/.0494 = 1.21$
Since $z_{.05} = 1.64$, the evidence does not indicate that the null hypothesis (that the proportion is the same) should be rejected.

(b) The results are the same as in (a).

**8.41** (a) Reject the null hypothesis if $(\bar{x} - 28) \div 6/\sqrt{100}$ is greater than 1.96 or less than $-1.96$; otherwise, do not reject the null hypothesis. In other words, reject the null hypothesis if $\bar{x} > 29.176$ miles per gallon or if $\bar{x} < 26.824$ miles per gallon; otherwise, do not reject the null hypothesis.

(b) Since $\bar{x}$ is less than 26.824 miles per gallon, the company should reject the hypothesis that the population mean is 28 miles per gallon, based on the test given in Section 8.4.

(c) Reject the null hypothesis if $(\bar{x} - 28) \div 6/\sqrt{100} < -1.64$; that is, if $\bar{x} < 27.016$ miles per gallon. Otherwise, do not reject the null hypothesis.

(d) Yes, because the sample mean is less than 27.016 miles per gallon.

**8.43** (a) 4 percent
(b) 7 and 8
(c) .25
.10
.05

**8.45** According to the $p$-value shown in the printout, one cannot reject the null hypothesis (that the mean was 20 years) if the significance level is set at 0.05. However, if the significance level is set at .085 or higher, the null hypothesis should be rejected, as indicated by the $p$-value.

# Chapter 9

**9.1** (a) 11.0705
(b) 18.3070
(c) 31.4104

**9.3**  Since the mean equals $v$, and the standard deviation equals $\sqrt{2v}$, the coefficient of variation equals $\sqrt{2v} \div v$, or $\sqrt{2/v}$, where $v$ is the number of degrees of freedom. If $\sqrt{2/v}$ equals 1, $v$ must equal 2.

**9.5**  (a) It has the $\chi^2$ distribution with 15 degrees of freedom.
(b) 15
(c) 30

**9.7**  The proportion containing falsified information is $(6 + 8 + 9 + 12) \div 800 = .04375$. Thus, the expected number containing falsified information at each branch is 8.75, and the expected number containing no falsified information at each branch is 191.25. Thus,

$$\sum \frac{(f - e)^2}{e} = \frac{(6 - 8.75)^2}{8.75} + \frac{(194 - 191.25)^2}{191.25} + \frac{(8 - 8.75)^2}{8.75}$$

$$+ \frac{(192 - 191.25)^2}{191.25} + \frac{(9 - 8.75)^2}{8.75} + \frac{(191 - 191.25)^2}{191.25} + \frac{(12 - 8.75)^2}{8.75}$$

$$+ \frac{(188 - 191.25)^2}{191.25}$$

$$= (1/8.75)(7.5625 + .5625 + .0625 + 10.5625)$$

$$+ (1/191.25)(7.5625 + .5625 + .0625 + 10.5625)$$

$$= (.1143 + .0052)(18.7500)$$

$$= (.1195)(18.7500) = 2.2406.$$

Since $\chi^2_{.05} = 7.81473$, the null hypothesis (that the proportion was the same at each branch) should not be rejected.

**9.9**  (a) Summing up the data for all plants, 65 of the 320 items are too small. Thus, if the null hypothesis is true, the expected number of items that are too small in each sample would be 16.25. Thus,

$$\sum \frac{(f - e)^2}{e} = \frac{(9 - 16.25)^2}{16.25} + \frac{(15 - 16.25)^2}{16.25} + \frac{(20 - 16.25)^2}{16.25}$$

$$+ \frac{(21 - 16.25)^2}{16.25} + \frac{(71 - 63.75)^2}{63.75} + \frac{(65 - 63.75)^2}{63.75}$$

$$+ \frac{(60 - 63.75)^2}{63.75} + \frac{(59 - 63.75)^2}{63.75}$$

$$= \frac{1}{16.25}(52.5625 + 1.5625 + 14.0625 + 22.5625)$$

$$+ \frac{1}{63.75}(52.5625 + 1.5625 + 14.0625 + 22.5625)$$

$$= (.06154 + .01569)(90.75) = (.07723)(90.75) = 7.01.$$

Since $\chi^2_{.05} = 7.81473$, the observed differences can be attributed to chance.
(b) Yes.

**9.11**  (a) Summing up the data for all districts, 212 voters support an increase. Thus, if the null hypothesis is true, the expected number in each sample that would support an increase is 53. Thus,

$$\sum \frac{(f-e)^2}{e} = \frac{1}{53}\{(38-53)^2 + (43-53)^2 + (78-53)^2$$
$$+ (53-53)^2\} + \frac{1}{47}\{(62-47)^2 + (57-47)^2$$
$$+ (22-47)^2 + (47-47)^2\}$$
$$= (.01887 + .02128)(225 + 100 + 625)$$
$$= (.04015)(950) = 38.14.$$

Since $\chi^2_{.10} = 6.25139$, it does not appear that the differences can be attributed to chance.

(b) and (c) The conclusion is the same as in (a).

**9.13** The expected frequency of each number is 16 2/3. Thus,

$$\sum \frac{(f-e)^2}{e} = \frac{1}{16\ 2/3}\left\{\left(\frac{5}{3}\right)^2 + \left(\frac{4}{3}\right)^2 + \left(\frac{10}{3}\right)^2 + \left(\frac{1}{3}\right)^2 + \left(\frac{11}{3}\right)^2 + \left(\frac{1}{3}\right)^2\right\}$$

$$= \frac{1}{16\ 2/3}\left(\frac{25 + 16 + 100 + 1 + 121 + 1}{9}\right)$$

$$= 264/150 = 1.76.$$

Since $\chi^2_{.05} = 11.0705$, the null hypothesis (that the die is true) should not be rejected.

**9.15** (a) The actual and expected frequencies are:

|          | A   | B    | C    | D    | F   |
|----------|-----|------|------|------|-----|
| Actual   | 5   | 22   | 50   | 27   | 6   |
| Expected | 8.8 | 27.5 | 37.4 | 27.5 | 8.8 |

Thus,

$$\sum \frac{(f-e)^2}{e} = \frac{(5-8.8)^2}{8.8} + \frac{(22-27.5)^2}{27.5} + \frac{(50-37.4)^2}{37.4}$$

$$+ \frac{(27-27.5)^2}{27.5} + \frac{(6-8.8)^2}{8.8}$$

$$= 1.64 + 1.10 + 4.24 + 0.01 + 0.89 = 7.88.$$

Since $\chi^2_{.05} = 9.48773$, the data seem to be consistent with her claim.

(b) The conclusion is the same as in (a).

**9.17** (a) $(n-1)\dfrac{s^2}{\sigma_0^2} = \dfrac{(29)(10)}{20} = 14.5$

Since $\chi^2_{.025} = 45.7222$ and $\chi^2_{.975} = 16.0471$, the null hypothesis (that $\sigma^2 = 20$) should be rejected since 14.5 is less than 16.0471.

(b) The confidence interval is

$$\frac{(29)(10)}{\chi^2_{.01}} < \sigma^2 < \frac{(29)(10)}{\chi^2_{.99}},$$

or

$$\frac{290}{49.5879} < \sigma^2 < \frac{290}{14.2565},$$

or

$$5.85 < \sigma^2 < 20.34.$$

**9.19** (a) $\dfrac{17(.000009)}{30.1910} < \sigma^2 < \dfrac{17(.000009)}{7.56418}$

$.00000507 < \sigma^2 < .00002023$

$.0023 < \sigma < .0045$

(b) $\dfrac{17(.000009)}{27.5871} < \sigma^2 < \dfrac{17(.000009)}{8.67176}$

$.00000555 < \sigma^2 < .00001764$

$.0024 < \sigma < .0042$

(c) We assume that the population is normal.

**9.21** (a) $\dfrac{n}{2} - z_{.025} \sqrt{\dfrac{n}{4}} = 15 - 1.96 \sqrt{7.5} = 15 - 1.96(2.74) = 9.63$

$\dfrac{n}{2} + z_{.025} \sqrt{\dfrac{n}{4}} = 15 + 1.96 \sqrt{7.5} = 15 + 1.96(2.74) = 20.37$

Since $x$ (which equals 19) does not exceed 20.37 or fall below 9.63, we cannot reject the hypothesis that there is no difference between the alloys.

(b) No.

**9.23** (a) $R_1 = 96$

$$\dfrac{U - E_u}{\sigma_u} = \dfrac{11(11) + \dfrac{11(12)}{2} - 96 - \dfrac{11(11)}{2}}{\sqrt{11(11)(23)/12}} = \dfrac{121 + 66 - 96 - 60.5}{\sqrt{2783/12}}$$

$$= \dfrac{30.5}{\sqrt{231.92}} = 2.00.$$

Since $(U - E_u) \div \sigma_u$ is greater than 1.96, the data do not seem to be consistent with the psychologist's claim.

(b) Since $(U - E_u) \div \sigma_u$ is less than 2.576, the psychologist's claim is not rejected.

**9.25** (a) The sample consists of matched pairs, and the sign test can be used (as in Example 9.3). The number of persons that made fewer errors under the new procedure than the old was nine. Since $n$ is not large, the binomial distribution must be used. If the null hypothesis is true (and there is no difference between the old and new procedures), the probability is .5 that a person will make fewer errors under the new procedure than the old. If this is the case, Appendix Table 1 shows that the probability of nine or more people making fewer errors with the new technique than the old is $.0098 + .0010 = .0108$. Thus, the sign test results in a rejection of the null hypothesis.

(b) The $p$-value for a one-tailed test is .0108.

**9.27** (a) The sequence is

$$\underline{E}\ \underline{OO}\ \underline{EEE}\ \underline{OO}\ \underline{E}\ \underline{O}\ \underline{E}\ \underline{O}\ \underline{EE}\ \underline{O}\ \underline{E}\ \underline{O}\ \underline{E}$$

$$\underline{O}\ \underline{E}\ \underline{O}\ \underline{E}\ \underline{OO}\ \underline{E}\ \underline{O}\ \underline{E}\ \underline{O}\ \underline{EE}\ \underline{OO}\ \underline{E}\ \underline{O}\ \underline{EE}$$

$$\underline{O}\ \underline{E}\ \underline{O}\ \underline{E}\ \underline{O}\ \underline{E}\ \underline{O}\ \underline{E}\ \underline{O}\ \underline{E}\ \underline{OO}\ \underline{E}\ \underline{O}\ \underline{E}\ \underline{O}$$

$$\underline{EE}\ \underline{OO}\ \underline{E}\ \underline{O}\ \underline{E}\ \underline{OO}\ \underline{E}\ \underline{O}\ \underline{EEE}\ \underline{O}$$

Thus, there are 52 runs.

(b) Since $n_1$ (the number of even numbers, including zeroes) equals 34 and $n_2$ (the number of odd numbers) equals 33,

$$E_r = \frac{2(34)(33)}{67} + 1 = \frac{2244}{67} + 1 = 33.49 + 1 = 34.49$$

$$\sigma_r = \sqrt{\frac{2(34)(33)[2(34)(33) - 67]}{67^2(66)}} = \sqrt{\frac{2244(2244 - 67)}{4489(66)}} = 4.06.$$

We should reject the null hypothesis (of randomness) if $(r - 34.49) \div 4.06 > 1.96$, where $r$ is the number of runs. Since the number of runs is 52, the null hypothesis should be rejected.

(c) The sequence is

$$\underline{AAAA}\ \underline{BBBBBB}\ \underline{AAAA}\ \underline{BBBBB}\ \underline{AAAAAAA}$$

$$\underline{BBBBBB}\ \underline{A}\ \underline{B}\ \underline{AA}\ \underline{BBBBB}\ \underline{AAAAA}$$

$$\underline{BBBB}\ \underline{AAAAAAA}\ \underline{BBBBBB}\ \underline{AAA}\ \underline{B}$$

Thus, there are 16 runs.

(d) Since $n_1$ (the number of $A$s) equals 33 and $n_2$ (the number of $B$s) equals 34,

$$E_r = \frac{2(33)(34)}{67} + 1 = 34.49$$

$$\sigma_r = \sqrt{\frac{2(33)(34)[2(33)(34) - 67]}{67^2(66)}} = 4.06.$$

We reject the null hypothesis (of randomness) if $(r - 34.49) \div 4.06$ is less than $-2.576$ or greater than $2.576$. In other words, we reject it if $r$ is less than 24.03 or greater than 44.95. Since the number of runs equals 16, the null hypothesis should be rejected.

**9.29** (a) The number of runs is 15.

$$E_r = \frac{2(13)(11)}{24} + 1 = \frac{286}{24} + 1 = 12.92$$

$$\sigma_r = \sqrt{\frac{2(13)(11)[2(13)(11) - 13 - 11]}{24^2(23)}} = \frac{1}{24}\sqrt{\frac{286(286 - 24)}{23}}$$

$$= \frac{1}{24}\sqrt{\frac{74{,}932}{23}} = \frac{1}{24}\sqrt{3257.91} = 57.08/24 = 2.38$$

Thus, $E_r - 2.33\sigma_r = 7.37$ and $E_r + 2.33\sigma_r = 18.47$. Since the number of runs is between 7.37 and 18.47, there is no evidence, based on the runs test, of a departure from randomness.

(b) No.

(c) Since $\dfrac{15 - 12.92}{2.38} = 0.87$, the $p$-value of a two-tailed test is $2(.1922) = .3844$.

**9.31**

```
MTB > set into c8
DATA> 6.3 7.2 5.2 3.5 4.8 5.7 5.2 6.4 5.9 6.1 6.3 6.8
DATA> end
MTB > set into c9
DATA> 7.4 6.5 7.4 8.2 5.6 5.4 6.3 7.4 6.2 7.3 4.6 5.9
DATA> end
MTB> mann whitney c8 c9

Mann-Whitney Confidence Interval and Test

C8      N = 12    MEDIAN =              6.0000
C9      N = 12    MEDIAN =              6.4000
POINT ESTIMATE FOR ETA1-ETA2 IS      -0.7000
95.4 PCT C.I. FOR ETA1-ETA2 IS  (    -1.50,     0.20)
W =      122.5
TEST OF ETA1 = ETA2 VS. ETA1 N.E. ETA2 IS
SIGNIFICANT AT 0.1190

CANNOT REJECT AT ALPHA = 0.05
```

(a) According to the above Minitab output, the null hypotheses that the median times are equal should not be rejected.

(b) The $p$-value, according to the above printout, is .119, which means that the null hypothesis should be rejected only if $\alpha = .119$ or more.

**9.33** Summing up the data for all types of cars, 162 of the 600 cars stalled. Thus, if the null hypothesis is true, the expected number of cars in each sample that would stall is 27. Thus,

$$
\begin{aligned}
\sum \frac{(f-e)^2}{e} = \frac{1}{27} &\{(31-27)^2 + (22-27)^2 + (29-27)^2 \\
&+ (33-27)^2 + (23-27)^2 + (24-27)^2\} \\
+ \frac{1}{73} &\{(69-73)^2 + (78-73)^2 + (71-73)^2 \\
&+ (67-73)^2 + (77-73)^2 + (76-73)^2\} \\
&= (.03704 + .01370)(16 + 25 + 4 + 36 + 16 + 9) \\
&= (.05074)(106) = 5.38.
\end{aligned}
$$

Since $\chi^2_{.10} = 9.23635$, there is a (considerably) better than 10 percent chance that the observed differences could have been due to chance.

**9.35** (a) The expected frequencies are shown below in parentheses:

|  | California | Illinois | New York | Total |
|---|---|---|---|---|
| First cover | $81 \left( \dfrac{129{,}200}{1200} = 107.67 \right)$ | 60 (107.67) | 182 (107.67) | 323 |
| Second cover | $78 \left( \dfrac{106{,}400}{1200} = 88.67 \right)$ | 93 (88.67) | 95 (88.67) | 266 |
| Third cover | $241 \left( \dfrac{244{,}400}{1200} = 203.67 \right)$ | 247 (203.67) | 123 (203.67) | 611 |
| Total | 400 | 400 | 400 | 1200 |

$$\sum \frac{(f-e)^2}{e} = \frac{1}{107.67}\{(81-107.67)^2 + (60-107.67)^2 + (182-107.67)^2\}$$

$$+ \frac{1}{88.67}\{(78-88.67)^2 + (93-88.67)^2 + (95-88.67)^2\}$$

$$+ \frac{1}{203.67}\{(241-203.67)^2 + (247-203.67)^2 + (123-203.67)^2\}$$

$$= (.00929)(8508.67) + (.01128)(172.67) + (.00491)(9778.67)$$

$$= 79.05 + 1.95 + 48.01 = 129.01$$

Since $\chi^2_{.05} = 9.48773$, the data indicate that there are regional differences.

(b) Since $\chi^2_{.01} = 13.2767$, the data indicate that there are regional differences.

**9.37** (a) $\dfrac{(n-1)s^2}{\sigma_0^2} = \dfrac{14(64)}{100} = 8.96$

Since $\chi^2_{.05} = 23.6848$, there is no reason to reject the hypothesis that $\sigma = 10$.

(b) Since $\chi^2_{.99} = 4.66043$, there is no reason to reject the hypothesis that $\sigma = 10$.

**9.39** (a) There are 15 depositors in the sample with incomes exceeding $15,000.

$$\frac{n}{2} - z_{\alpha/2}\sqrt{\frac{n}{4}} = 18 - (1.96)(3) = 12.12$$

$$\frac{n}{2} + z_{\alpha/2}\sqrt{\frac{n}{4}} = 18 + (1.96)(3) = 23.88$$

Since 15 is not less than 12.12 or greater than 23.88, we should not reject the null hypothesis that the median income equals $15,000.

(b) $\dfrac{n}{2} - 2.576\sqrt{\dfrac{n}{4}} = 18 - (2.576)(3) = 10.27$

$$\frac{n}{2} + 2.576\sqrt{\frac{n}{4}} = 18 + (2.576)(3) = 25.73$$

Since 15 is not less than 10.27 or greater than 25.73, we should not reject the null hypothesis that the median income equals $15,000.

(c) Yes.

No. The present results test a hypothesis about the median whereas Exercise 8.29 tests a hypothesis about the mean.

No. The test in Exercise 8.29 assumes normality whereas this test does not. However, since $n > 30$, the test in Exercise 8.29 should be dependable even if the population is not normal.

**9.41**

```
MTB > read into c13,c14,c15
DATA> 20 12 19
DATA> 24 28 21
DATA> 16 20 20
DATA> 60 60 60
DATA> end
      4 ROWS READ
MTB > chisquare c13,c14,c15

Expected counts are printed below observed counts

              C13        C14        C15      Total
     1         20         12         19         51
              17.0       17.0       17.0

     2         24         28         21         73
              24.3       24.3       24.3

     3         16         20         20         56
              18.7       18.7       18.7

     4         60         60         60        180
              60.0       60.0       60.0

Total        120        120        120        360

ChiSq =   0.53  +  1.47  +  0.24  +
          0.00  +  0.55  +  0.46  +
          0.38  +  0.10  +  0.10  +
          0.00  +  0.00  +  0.00  = 3.82
df = 6
```

Because there are 6 degrees of freedom, Appendix Table 8 indicates that $\chi^2_{.05} = 12.5916$. Thus, since 3.82, the observed value of $\chi^2$, is less than 12.5916, we should not reject the null hypothesis of independence.

# Chapter 10

**10.1** No. There is no way to compare this percentage with what would have occurred if a similar class had used another textbook. In other words, there is no control group.

**10.3** One possible explanation is that, in experiments without controls, there may be a tendency to use the surgical technique on patients that are in relatively good physical condition. This biases the results.

**10.5** (a) No, because it may perform no better than a placebo.
(b) As pointed out in Exercise 10.4, placebos can make people feel better.

**10.7** (a) It would not tell anything about how these men would have performed in the absence of vitamin supplementation.
(b) Soldiers with prior use of vitamin supplements all receive the vitamins, not the placebo. This can bias the results. It is better to allocate the soldiers at random to the two groups.

(c) It would be better not to tell the soldiers whether they are receiving the vitamins or the placebo.

(d) (i) Yes.

(ii) The blocks are the platoons, and the treatments are the vitamins or the placebo.

(iii) No.

(iv) Not applicable.

(v) Yes.

**10.9** (a) .05

(b) .01

**10.11** (a) 2.75

(b) 4.30

**10.13** (a) Using the formulas in Appendix 10.2, we find that

$$BSS = \frac{1}{4}(74^2 + 87^2 + 106^2 + 106^2) - \frac{1}{16}(373)^2$$

$$= 8{,}879.25 - 8{,}695.5625 = 183.6875.$$

$$TSS = 324 + 400 + 361 + 289 + 484 + 441 + 576 + 400$$

$$+ 625 + 729 + 676 + 784 + 841 + 784 + 576 + 625$$

$$- \frac{1}{16}(373)^2$$

$$= 8{,}915 - 8{,}695.5625 = 219.4375.$$

$$WSS = 219.4375 - 183.6875 = 35.75.$$

$$F = \frac{BSS/3}{WSS/12} = \frac{61.229}{2.979} = 20.55.$$

According to Appendix Table 9, $F_{.05}$ equals 3.49. Since the observed value of $F$ exceeds 3.49, we should reject the null hypothesis that the mean number of miles per gallon is the same for all four firms' cars.

(b) The table is as follows:

| Source of variation | Sum of squares | Degrees of freedom | Mean square | F |
|---|---|---|---|---|
| Between groups | 183.6875 | 3 | 61.229 | 20.55 |
| Within groups | 35.7500 | 12 | 2.979 | |
| Total | 219.4375 | 15 | | |

(c) Let $\mu_1$ be the mean for this firm, $\mu_2$ be the mean for its U.S. competitor, $\mu_3$ be the mean for the German firm, and $\mu_4$ be the mean for the Japanese firm. Then

$$\sqrt{3.49\,(1.726)}\,\sqrt{\frac{(3)(2)}{4}} = 3.95.$$

Consequently,

$$-8.0 - 3.95 < \mu_1 - \mu_4 < -8.0 + 3.95$$

$$\boxed{-11.95 < \mu_1 - \mu_4 < -4.05}$$

$$18.5 - 21.75 - 3.95 < \mu_1 - \mu_2 < 18.5 - 21.75 + 3.95$$

$$-7.20 < \mu_1 - \mu_2 < 0.70$$

$$18.5 - 26.5 - 3.95 < \mu_1 - \mu_3 < 18.5 - 26.5 + 3.95$$

$$-11.95 < \mu_1 - \mu_3 < -4.05$$

$$21.75 - 26.5 - 3.95 < \mu_2 - \mu_3 < 21.75 - 26.5 + 3.95$$

$$-8.70 < \mu_2 - \mu_3 < -0.80$$

$$21.75 - 26.5 - 3.95 < \mu_2 - \mu_4 < 21.75 - 26.5 + 3.95$$

$$-8.70 < \mu_2 - \mu_4 < -0.80$$

$$26.5 - 26.5 - 3.95 < \mu_3 - \mu_4 < 26.5 - 26.5 + 3.95$$

$$-3.95 < \mu_3 - \mu_4 < 3.95$$

**10.15** (a) Method B seems fastest, and method D seems slowest.
(b) Using Appendix 10.2,

$$BSS = \frac{1}{4}(78^2 + 63^2 + 79^2 + 86^2) - \frac{1}{16}(306^2) = \frac{23{,}690}{4} - \frac{93{,}636}{16}$$

$$= 5922.5 - 5852.25 = 70.25$$

$$TSS = 5952 - 5852.25 = 99.75$$

$$WSS = 99.75 - 70.25 = 29.50$$

| Source of variation | Sum of squares | Degrees of freedom | Mean square | F |
|---|---|---|---|---|
| Between groups | 70.25 | 3 | 23.42 | 9.52 |
| Within groups | 29.50 | 12 | 2.46 | |
| Total | 99.75 | 15 | | |

(c) Since $F_{.01} = 5.95$, the answer seems to be no.
(d) Let $\mu_1$ be the mean for method $A$, $\mu_2$ be the mean for method $B$, $\mu_3$ be the mean for method $C$, and $\mu_4$ be the mean for method $D$.

$$3.75 - \sqrt{(5.95)(2.46)\frac{(2)(3)}{4}} < \mu_1 - \mu_2 < 3.75 +$$

$$\sqrt{(5.95)(2.46)\frac{(2)(3)}{4}}$$

$$-.94 < \mu_1 - \mu_2 < 8.44$$

$$-4.94 < \mu_1 - \mu_3 < 4.44$$

$$-6.69 < \mu_1 - \mu_4 < 2.69$$

$$-8.69 < \mu_2 - \mu_3 < 0.69$$

$$-10.44 < \mu_2 - \mu_4 < -1.06$$

$$-6.44 < \mu_3 - \mu_4 < 2.94$$

(e) There would be advantages if the skill and experience of the workers were controlled.

(f) No. Even though this method is *faster,* others may be *cheaper* or *better.*

**10.17** (a) Brand *B* seems to wear best; brand *C* seems to wear most poorly.

(b) They seem to wear best in Type II cars and worst in Type I cars.

(c) Using Appendix 10.2,

$$BSS = \frac{1}{5}(140^2 + 150^2 + 130^2) - \frac{1}{15}(420^2) = \frac{59,000}{5} - \frac{176,400}{15}$$

$$= 11,800 - 11,760 = 40.$$

$$RSS = \frac{1}{3}(79^2 + 89^2 + 83^2 + 81^2 + 88^2) - 11,760 = \frac{35,356}{3}$$

$$- 11,760$$

$$= 11,785.33 - 11,760 = 25.33$$

$$TSS = 11,830 - 11,760 = 70$$

$$WSS = 70 - 40 - 25.33 = 4.67$$

| Source of variation | Sum of squares | Degrees of freedom | Mean square | F |
|---|---|---|---|---|
| Treatments (tires) | 40.00 | 2 | 20.00 | 34 |
| Blocks (cars) | 25.33 | 4 | 6.33 | 11 |
| Error | 4.67 | 8 | 0.58 | |
| Total | 70.00 | 14 | | |

(d) Since $F_{.05} = 4.46$, the differences are statistically significant at the .05 probability level.

(e) Since $F_{.05} = 3.84$, the differences are statistically significant at the .05 probability level.

(f) Let $\mu_1$ be the mean for brand *A*, $\mu_2$ be the mean for brand *B*, and $\mu_3$ be the mean for brand *C*.

$$-2.0 - \sqrt{(4.46)(0.58)\frac{(2)(2)}{5}} < \mu_1 - \mu_2 < -2.0 +$$

$$\sqrt{(4.46)(0.58)\frac{(2)(2)}{5}}$$

$$-3.4 < \mu_1 - \mu_2 < -0.6$$

$$0.6 < \mu_1 - \mu_3 < 3.4$$

$$2.6 < \mu_2 - \mu_3 < 5.4$$

**10.19**

```
MTB > read into c18,c19,c20
DATA> 69 1 1
DATA> 88 2 1
DATA> 63 3 1
DATA> 74 4 1
DATA> 73 1 2
DATA> 85 2 2
DATA> 60 3 2
DATA> 71 4 2
DATA> 75 1 3
DATA> 86 2 3
DATA> 59 3 3
DATA> 69 4 3
DATA> end
     12 ROWS READ
MTB > twoway analysis, c18,c19,c20

ANALYSIS OF VARIANCE C18

SOURCE    DF        SS        MS
C19        3     998.00    332.67
C20        2       4.17      2.08
ERROR      6      40.50      6.75
TOTAL     11    1042.67
```

The value of F is $2.08/6.75 = 0.31$, which is less than $F_{.05}$ with 2 and 6 degrees of freedom (5.14). Thus, the null hypothesis that the raw materials produce equivalent results should not be rejected.

**10.21** (a) Catalyst 3. Catalyst 1.
   (b) Plant $C$.
   (c) Using Appendix 10.2,

$$BSS = \frac{1}{8}(294^2 + 309^2 + 318^2 + 310^2) - \frac{1}{32}(1231)^2$$

$$= \frac{379,141}{8} - \frac{1,515,361}{32}$$

$$= 47,392.62 - 47,355.03 = 37.59$$

$$RSS = \frac{1}{4}(156^2 + 153^2 + 161^2 + 160^2 + 159^2 + 150^2 + 147^2$$

$$+ 145^2) - 47,355.03$$

$$= \frac{189,681}{4} - 47,355.03 = 47,420.25 - 47,355.03 = 65.22$$

$$TSS = 47,493 - 47,355.03 = 137.97$$

$$WSS = 137.97 - 37.59 - 65.22 = 35.16$$

| Source of variation | Sum of squares | Degrees of freedom | Mean square | F |
|---|---|---|---|---|
| Treatments (catalysts) | 37.59 | 3 | 12.53 | 7.50 |
| Blocks (plants) | 65.22 | 7 | 9.32 | 5.58 |
| Error | 35.16 | 21 | 1.67 | |
| Total | 137.97 | 31 | | |

(d) Since $F_{.01} = 4.87$, the answer seems to be no.

(e) Since $F_{.01} = 3.64$, the answer seems to be no.

(f) Let $\mu_1$ be the mean for catalyst $A$, $\mu_2$ be the mean for catalyst $B$, $\mu_3$ be the mean for catalyst $C$, and $\mu_4$ be the mean for catalyst $D$.

$$-1.875 - \sqrt{(4.87)(1.67)\frac{(2)(3)}{8}} < \mu_1 - \mu_2 < -1.875 +$$

$$\sqrt{(4.87)(1.67)\frac{(2)(3)}{8}}$$

$$-4.3 < \mu_1 - \mu_2 < 0.6$$

$$-5.5 < \mu_1 - \mu_3 < -0.5$$

$$-4.5 < \mu_1 - \mu_4 < 0.5$$

$$-3.6 < \mu_2 - \mu_3 < 1.3$$

$$-2.6 < \mu_2 - \mu_4 < 2.3$$

$$-1.5 < \mu_3 - \mu_4 < 3.5$$

**10.23** (a) It seems highest in Singapore and lowest in the United States.

(b) $BSS = 15[(15.1 - 14.9)^2 + (16.2 - 14.9)^2 + (13.4 - 14.9)^2]$
$= 15(3.98) = 59.7$

$WSS = 14[3.2^2 + 4.0^2 + 3.1^2] = 14(35.85) = 501.9$

| Source of variation | Sum of squares | Degrees of freedom | Mean square | F |
|---|---|---|---|---|
| Between groups | 59.7 | 2 | 29.85 | 2.50 |
| Within groups | 501.9 | 42 | 11.95 | |
| Total | 561.6 | 44 | | |

(c) Since $F_{.05}$ is about 3.23, the answer seems to be yes.

(d) Output per hour of labor is assumed to be normally distributed in each plant, and the variance is assumed to be the same in each of these plants.

(e) Let $\mu_1$ be the mean in Germany, $\mu_2$ be the mean in Singapore, and $\mu_3$ be the mean in the United States.

$$-1.1 - \sqrt{(3.23)(11.95)\frac{(2)(2)}{15}} < \mu_1 - \mu_2 < -1.1 +$$

$$\sqrt{(3.23)(11.95)\frac{(2)(2)}{15}}$$

$$-4.3 < \mu_1 - \mu_2 < 2.1$$

$$-1.5 < \mu_1 - \mu_3 < 4.9$$

$$-0.4 < \mu_2 - \mu_3 < 6.0$$

(f) They are extremely crude because no allowance is made for international differences in the quantity and quality of equipment, raw materials, and management.

**10.25** The observed value of F is 1.32, which is less than $F_{.05}$ with 4 and 20 degrees of freedom (2.87). Thus, we should not reject the null hypothesis that there are no differences among the typewriters in this regard.

# Chapter 11

**11.1** (a)

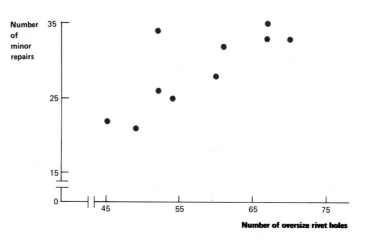

(b) Let $Y$ be the number of minor repairs, $X$ be the number of oversize rivet holes. $\Sigma X = 577$; $\Sigma Y = 289$; $\Sigma X^2 = 33{,}949$; $\Sigma XY = 16{,}987$; $n = 10$.

$$b = \frac{10(16{,}987) - (577)(289)}{10(33{,}949) - 577^2} = \frac{169{,}870 - 166{,}753}{339{,}490 - 332{,}929} = \frac{3117}{6561} = 0.475$$

$$a = 28.9 - (.475)(57.7) = 28.9 - 27.4 = 1.5$$

The regression line is $\hat{Y} = 1.5 + .475X$.

(c) $1.5 + .475(50)$. Thus, the answer is 25.2 repairs.

(d) $1.5 + .475(70)$. Thus, the answer is 34.7 repairs.

**11.3** (a)

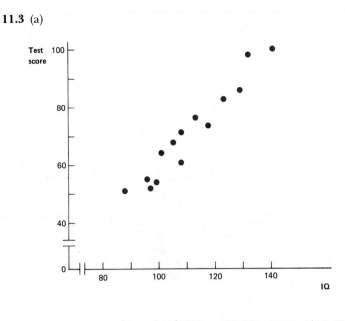

(b) $\Sigma X = 1560$; $\Sigma Y = 989$; $\Sigma X^2 = 176{,}928$; $\Sigma XY = 113{,}355$; $n = 14$.

$$b = \frac{(14)(113{,}355) - (1560)(989)}{14(176{,}928) - 1560^2} = \frac{1{,}586{,}970 - 1{,}542{,}840}{2{,}476{,}992 - 2{,}433{,}600}$$

$$= \frac{44{,}130}{43{,}392} = 1.017$$

$$a = 70.64 - (1.017)(111.43) = 70.64 - 113.32 = -42.68$$

The regression line is $\hat{Y} = -42.68 + 1.017X$.

(c) $-42.68 + (1.017)(130) = -42.68 + 132.21 = 89.53$

(d) No. For values of IQ above 140, the relationship must become closer to a horizontal line.

**11.5** (a)

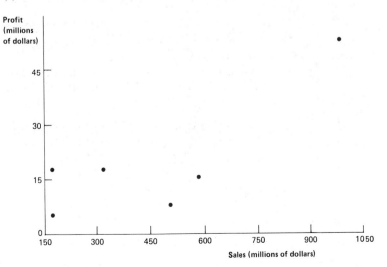

(b) Profits.

(c) The sample regression line is $\hat{Y} = 2.52 + 0.0387X$, where $Y$ is profit and $X$ is sales.

**11.7** (a) $\Sigma X = 158$; $\Sigma Y = 28$; $\Sigma X^2 = 3912$; $\Sigma XY = 653$; $n = 8$; $\Sigma Y^2 = 112$.

$$b = \frac{8(653) - (158)(28)}{8(3912) - 158^2} = \frac{5224 - 4424}{31{,}296 - 24{,}964} = \frac{800}{6332} = 0.1263$$

$$a = 3.5 - (0.1263)(19.75) = 3.5 - 2.494 = 1.006$$

The regression line is $\hat{Y} = 1.006 + 0.1263X$.

(b) $s_e = \sqrt{\dfrac{112 - 1.006(28) - (.1263)(653)}{6}}$

$$= \sqrt{\frac{112 - 28.168 - 82.474}{6}}$$

$$= \sqrt{\frac{1.358}{6}} = \sqrt{.2263} = .476 \text{ thousands of dollars}$$

(c) $1.006 + .1263(20) \pm 1.943(.476) \sqrt{.125 + \dfrac{.25^2}{3912 - 158(19.75)}}$

$$1.006 + 2.526 \pm .925 \sqrt{.125 + \frac{.0625}{791.5}}$$

$$3.532 \pm (.925)(.354)$$

$$3.532 \pm .327$$

Thus, the confidence interval is 3.205 to 3.859 thousand of dollars.

(d) $3.532 \pm .925 \sqrt{1.125}$

$$3.532 \pm .981$$

Thus, the confidence interval is 2.551 to 4.513 thousands of dollars.

**11.9** (a) $s_e = \sqrt{\dfrac{1.1117 + .030(1.83) - .04427(12.9643)}{13}}$

$$= \sqrt{\frac{1.1117 + .0549 - .5739}{13}} = \sqrt{\frac{.5927}{13}} = \sqrt{.04559}$$

$$= .214 \text{ percentage points.}$$

(b) $0.236 \pm 0.694(.214) \sqrt{\dfrac{16}{15} + \dfrac{(6 - 3.431)^2}{327.5882 - (51.46)(3.431)}}$

$$.236 \pm .149 \sqrt{1.067 + \frac{6.600}{151.029}}$$

$$.236 \pm .149 \sqrt{1.111}$$

$$.236 \pm .157$$

Thus, the confidence interval is 0.079 to .393 percent.

**11.11** (a)

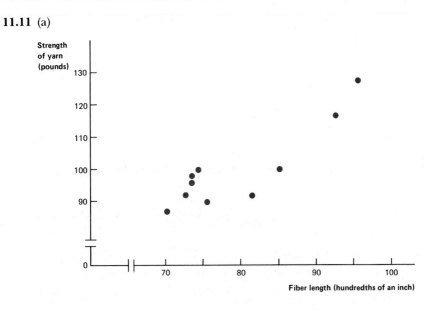

(b) Direct.

Yes, since one would expect longer fibers to be associated with greater strength.

It is hard to tell, but it may be linear.

(c)

| X | Y | $X^2$ | $Y^2$ | XY |
|---|---|---|---|---|
| 85 | 99 | 7,225 | 9,801 | 8,415 |
| 82 | 93 | 6,724 | 8,649 | 7,626 |
| 75 | 99 | 5,625 | 9,801 | 7,425 |
| 74 | 97 | 5,476 | 9,409 | 7,178 |
| 76 | 90 | 5,776 | 8,100 | 6,840 |
| 74 | 96 | 5,476 | 9,216 | 7,104 |
| 73 | 93 | 5,329 | 8,649 | 6,789 |
| 96 | 130 | 9,216 | 16,900 | 12,480 |
| 93 | 118 | 8,649 | 13,924 | 10,974 |
| 70 | 88 | 4,900 | 7,744 | 6,160 |

Total  798     1,003   64,396   102,193   80,991

Mean  79.8    100.3

$$b = \frac{10(80,991) - (798)(1003)}{10(64,396) - 798^2} = \frac{9516}{7156} = 1.330.$$

$$a = 100.3 - (1.33)(79.8) = 100.3 - 106.1 = -5.8.$$

(d) The sample regression is

$$\hat{Y} = -5.8 + 1.33X.$$

$$\hat{Y} = -5.8 + (1.33)(80) = -5.8 + 106.4 = 100.6 \text{ pounds.}$$

$$\hat{Y} = -5.8 + (1.33)(90) = -5.8 + 119.7 = 113.9 \text{ pounds.}$$

(e) $s_e = \sqrt{\dfrac{102{,}193 + (5.818)(1003) - (1.3298)(80{,}991)}{8}}$

$= \sqrt{40.8} = 6.387$ pounds

(Note that $a$ and $b$ are taken out to more decimal places than in (c).) The standard deviation of the conditional probability distribution of the strength of the yarn is estimated to be 6.387 pounds.

(f) The confidence interval is

$$-5.818 + 1.3298(80) \pm (1.860)(6.387)\sqrt{\dfrac{1}{10} + \dfrac{(80 - 79.8)^2}{715.6}},$$

or

$$100.566 \pm (11.8798)\sqrt{.10 + .000056},$$

or

$$100.566 \pm 3.758.$$

That is, it is 96.8 to 104.3 pounds.

(g) The confidence interval is

$$100.566 \pm 11.8798\sqrt{1.100056},$$

or

$$100.566 \pm 12.461.$$

That is, it is 88.1 to 113.0 pounds. This confidence interval is wider than that in (f) because the sampling error in predicting the strength of a particular piece of yarn is greater than the sampling error in estimating the conditional mean strength.

**11.13** (a) $\Sigma X = 190;\ \Sigma Y = 451;\ \Sigma X^2 = 5700;\ \Sigma XY = 11390;\ \Sigma Y^2 = 27{,}503;$ $n = 8.$

$r^2 = \dfrac{[8(11390) - (190)(451)]^2}{[8(5700) - 190^2][8(27{,}503) - 451^2]}$

$= \dfrac{(91{,}120 - 85{,}690)^2}{(45{,}600 - 36{,}100)(220{,}024 - 203{,}401)}$

$= \dfrac{5430^2}{9500(16{,}623)} = \dfrac{29{,}484{,}900}{157{,}918{,}500} = .1867$

$r = 0.43$

(b) $t = \dfrac{.43}{\sqrt{.81/6}} = \dfrac{.43}{\sqrt{.135}} = \dfrac{.43}{.37} = 1.16$

Since $t_{.025} = 2.447$, we cannot reject the hypothesis that the population correlation coefficient is zero.

**11.15** (a) Using equation (11.12b),

$$r^2 = \dfrac{1.5(289) + .475(16{,}987) - \dfrac{1}{10}(289)^2}{8593 - \dfrac{1}{10}(289^2)}$$

$$= \frac{433.5 + 8068.82 - 8352.1}{240.9} = \frac{150.22}{240.99} = 0.62$$

(b) $s_b = \dfrac{3.366}{\sqrt{33,949 - 577(57.7)}} = \dfrac{3.366}{\sqrt{33,949 - 33,292.9}} = 0.131$

$b \div s_b = .475/.131 = 3.63$

Since $t_{.025} = 2.306$, the null hypothesis that the true regression coefficient is zero should be rejected.

**11.17** (a) $s_b = \dfrac{1.1726}{\sqrt{3,145,400 - 5240(524)}} = \dfrac{1.1726}{\sqrt{3,145,400 - 2,745,760}}$

$= .001855$

(b) $.02424 \pm 1.86 \,(.001855)$
$.02424 \pm .00345$

The confidence interval is .02079 to .02769.

**11.19** (a) $r^2 = \dfrac{9{,}516^2}{(7{,}156)(15{,}921)} = .7948$. Thus, $r = .89$.

(b) 79.48 percent of the variation can be explained.

**11.21** $s_b = \dfrac{6.387}{\sqrt{715.6}} = .239$

Thus, $b \div s_b = 1.330/.239 = 5.565$. Since $t_{.01} = 2.896$, the null hypothesis (that $B = 0$) should be rejected because $b \div s_b$ exceeds 2.896.
No. Correlation does not prove causation.

**11.23** (a) One might expect the standard deviation of the amount saved to increase as income increases.
(b) At higher levels of output, the standard deviation of the error term may be greater than at lower levels of output.

**11.25** (a)

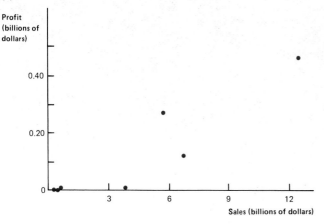

(b) $\Sigma X = 30.0$; $\Sigma Y = 0.94$; $\Sigma X^2 = 248.72$; $\Sigma XY = 8.307$; $n = 7$; $\Sigma Y^2 = .3030$

$$b = \frac{7(8.307) - (30)(0.94)}{7(248.72) - 30^2} = \frac{58.149 - 28.200}{1741.04 - 900} = \frac{29.949}{841.04} = .0356$$

$$a = .134 - (.0356)(4.286) = .134 - .153 = -.019$$

The regression line is $\hat{Y} = -0.019 + .0356X$.

(c) $-0.019 + .0356(2)$. Thus, the answer is about .05 billions of dollars.

(d) No. Prices and costs will be different in 1990 than in 1980.

**11.27** This estimate involves extrapolation of the regression line beyond the range of the data. For reasons given in section 11.15, this is a very hazardous procedure.

**11.29** (a) $r^2 = \dfrac{363{,}450^2}{(52{,}689)(3{,}305{,}500)} = (6.898)(.10995)$

$$= .7584. \text{ Thus, } r = .87.$$

(b) 75.84 percent of the variation is explained.

**11.31** Let the number of television receivers be C35 and the number of telephones be C34.

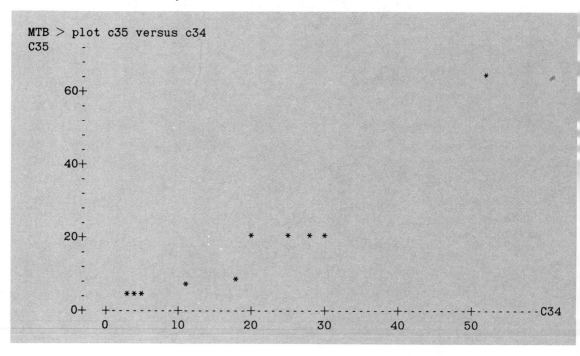

# Chapter 12

**12.1** (a) Let $X_1$ be the number of family members, $X_2$ be family income, and $Y$ be the annual amount spent on clothing.

$$\Sigma(X_{1i} - \overline{X}_1)^2 = 82 - \frac{1}{10}(26^2) = 82 - 67.6 = 14.4$$

$$\Sigma(X_{2i} - \overline{X}_2)^2 = 11{,}244 - \frac{1}{10}(314^2) = 11{,}244 - 9859.6 = 1384.4$$

$$\Sigma(Y_i - \overline{Y})^2 = 71.13 - \frac{1}{10}(23.9^2) = 71.13 - 57.121 = 14.009$$

$$\Sigma(X_{1i} - \overline{X}_1)(Y_i - \overline{Y}) = 71.3 - \frac{1}{10}(26)(23.9) = 71.3 - 62.14 = 9.16$$

$$\Sigma(X_{2i} - \overline{X}_2)(Y_i - \overline{Y}) = 875 - \frac{1}{10}(314)(23.9) = 875 - 750.46 = 124.54$$

$$\Sigma(X_{1i} - \overline{X}_1)(X_{2i} - \overline{X}_2) = 855 - \frac{1}{10}(26)(314) = 855 - 816.4 = 38.6$$

$$b_1 = \frac{(1384.4)(9.16) - (38.6)(124.54)}{(14.4)(1384.4) - 38.6^2} = \frac{12{,}681.104 - 4{,}807.244}{19{,}935.36 - 1{,}489.96}$$

$$= \frac{7{,}873.86}{18{,}445.4} = .427$$

$$b_2 = \frac{(14.4)(124.54) - (38.6)(9.16)}{(14.4)(1384.4) - 38.6^2} = \frac{1793.376 - 353.576}{18{,}445.4}$$

$$= \frac{1{,}439.8}{18{,}445.4} = .078$$

$$a = 2.39 - (.427)(2.6) - (.078)(31.4) = 2.39 - 1.1102 - 2.4492$$
$$= -1.169$$

Thus, the multiple regression is $\hat{Y} = -1.169 + .427X_1 + .078X_2$

(b) Annual clothing expenditure increases by .427 thousands of dollars.
(c) Annual clothing expenditure increases by .078 thousands of dollars.
(d) No. As family income increases, one might expect that the extra clothing expenditure resulting from an extra thousand dollars of income might decrease eventually.
(e) $-1.169 + .427(2) + .078(25) = 1.635$ thousands of dollars.
(f) $-1.169 + .427(4) + .078(35) = 3.269$ thousands of dollars.
(g) No. Minneapolis families may spend more on clothing because the climate is colder than in San Francisco.

**12.3** (a) $.29 \pm (1.717)(.11)$, or .10 to .48.
(b) $.67 \pm (2.074)(.23)$, or .19 to 1.15.
(c) Yes. Divide the regression coefficient (Coef) by the $t$-ratio.

**12.5** $\dfrac{.427(9.16) + .078(124.54)}{14.009} = \dfrac{13.625}{14.009} = .97$

This means that 97 percent of the variation in the dependent variable can be explained by the regression equation.

**12.7** (a) $\overline{R}^2 = 1 - \left(\dfrac{n-1}{n-k-1}\right)(1-R^2)$

$$= 1 - \left(\dfrac{32-1}{32-4-1}\right)0.37 = 1 - \left(\dfrac{31}{27}\right).37 = 1 - .42 = .58$$

(b) $\overline{R}^2$ is an unbiased estimate of the population multiple coefficient of determination.

**12.9** (a) The complete table is

| Sum of squares | Degrees of freedom | Mean square | F |
|---|---|---|---|
| 126 | 3 | 42 | 2.83 |
| 416 | 28 | 14.86 | |
| 542 | 31 | | |

(b) Three.
(c) 32.
(d) 42.
(e) 14.86.
(f) 14.86.
(g) Since $F_{.05} = 2.95$, there is no reason to reject the null hypothesis that the true regression coefficients are all zero.

**12.11** $\Sigma(\hat{Y}_i - \overline{Y})^2 = .831(1.54) + .00214(477.8) = 2.302$
$\Sigma(Y_i - \hat{Y})^2 = 2.322 - 2.302 = .020$
The analysis of variance table is

| Source of variation | Sum of squares | Degrees of freedom | Mean square | F |
|---|---|---|---|---|
| Explained by regression | 2.302 | 2 | 1.151 | 523 |
| Unexplained by regression | .020 | 9 | .0022 | |
| Total | 2.322 | 11 | | |

Since $F_{.01} = 8.02$, we must reject the null hypothesis that the true regression coefficients are all zero.

**12.13** (a) $\hat{Y} = 4.659 - 0.062C2 + 0.096C4 + 0.070C8$.
(b) $.0963 \pm (2.16)(.0154)$. In other words, the confidence interval is .063 to .130.
(c) $-0.062 \pm (1.771)(.009)$. In other words, the confidence interval is $-.078$ to $-.046$.
(d) The $t$-ratio for the regression coefficient of C8 equals 7.20. Since $t_{.025} = 2.16$, the null hypothesis (that the population regression coefficient of C8 equals zero) should be rejected since the value of the $t$-ratio exceeds 2.16.
(e) 0.87. It means that the regression equation explains 87 percent of the variation in the dependent variable.
(f) With 3 and 13 degrees of freedom, $F_{.05}$ equals 3.41. Thus the null hypothesis should be rejected, since the observed $F$-value is 29.01.

**12.15** The dependent variable is the amount saved by the family. There are three independent variables, two of which are dummy variables. The first independent variable is the family's income; the second is a dummy variable that is zero if the family is white and one if the family is non-white; and the third is a dummy variable that is 0 if the family is headed by a male and 1 if the family is headed by a female. We assume that the effect of extra income on extra savings is the same regardless of whether a family is white or nonwhite and regardless of whether it is headed by a male or a female.

**12.17** (a) Let $X_{1i}$ be 0 if the $i$th person is male and 1 if the $i$th person is female. Let $X_2$ be the test score and $Y$ be productivity.

$$\Sigma(X_{1i} - \overline{X}_1)^2 = 8 - \frac{1}{15}(8^2) = 8 - \frac{64}{15} = 8 - 4.267 = 3.733$$

$$\Sigma(X_{2i} - \overline{X}_2)^2 = 72,996 - \frac{1}{15}(1036^2) = 72,996 - \frac{1,073,296}{15} = 1443$$

$$\Sigma(X_{1i} - \overline{X}_1)(Y_i - \overline{Y}) = 397 - \frac{1}{15}(8)(749) = 397 - \frac{5992}{15} = -2.467$$

$$\Sigma(X_{2i} - \overline{X}_2)(Y_i - \overline{Y}) = 53,096 - \frac{1}{15}(1036)(749) = 53,096 - \frac{775,964}{15}$$

$$= 1,365$$

$$\Sigma(X_{1i} - \overline{X}_1)(X_{2i} - \overline{X}_2) = 519 - \frac{1}{15}(8)(1036) = 519 - \frac{8288}{15} = -33.53$$

$$b_1 = \frac{-(1443)(2.467) + (33.53)(1365)}{(3.733)(1443) - (-33.53)^2} = \frac{-3,559.88 + 45,768.45}{5,386.72 - 1,124.26}$$

$$= \frac{42,208.57}{4,262.46} = 9.902$$

$$b_2 = \frac{(3.733)(1365) - (33.53)(2.467)}{4,262.46} = \frac{5,095.54 - 82.72}{4,262.46}$$

$$= \frac{5,012.82}{4,262.46} = 1.176$$

$$a = \frac{749}{15} - 9.902\left(\frac{8}{15}\right) - 1.176\left(\frac{1036}{15}\right) = 49.93 - (9.902)(.533)$$

$$- (1.176)(69.067)$$

$$= 49.93 - 5.28 - 81.22 = -36.57.$$

Thus, the multiple regression is: $\hat{Y} = -36.57 + 9.902X_1 + 1.176X_2$.
(b) $-9.902$
(c) $11.76$
(d) $-36.57 + 9.902(0) + 1.176(82) = -36.57 + 96.43 = 59.86$
(e) $-36.57 + 9.902(1) + 1.176(76) = -36.57 + 9.90 + 89.38 = 62.71$

**12.19** Studies have indicated that other relevant variables are total city revenue, the racial composition of the population, and the ratio of black to white median income.

**12.21** (a) Let $X_2 = 1$ if the year is 1987 and 0 if it is 1985. Express $Y$ in units of 10.

| | $Y$ | $X_1$ | $X_2$ | $Y^2$ | $X_1^2$ | $X_2^2$ | $X_1Y$ | $X_2Y$ | $X_1X_2$ |
|---|---|---|---|---|---|---|---|---|---|
| | 10 | 5 | 0 | 100 | 25 | 0 | 50 | 0 | 0 |
| | 12 | 7 | 0 | 144 | 49 | 0 | 84 | 0 | 0 |
| | 14 | 8 | 0 | 196 | 64 | 0 | 112 | 0 | 0 |
| | 17 | 9 | 0 | 289 | 81 | 0 | 153 | 0 | 0 |
| | 11 | 5 | 1 | 121 | 25 | 1 | 55 | 11 | 5 |
| | 16 | 8 | 1 | 256 | 64 | 1 | 128 | 16 | 8 |
| | 20 | 9 | 1 | 400 | 81 | 1 | 180 | 20 | 9 |
| | 12 | 6 | 1 | 144 | 36 | 1 | 72 | 12 | 6 |
| | 13 | 7 | 1 | 169 | 49 | 1 | 91 | 13 | 7 |
| | 15 | 7 | 1 | 225 | 49 | 1 | 105 | 15 | 7 |
| Total | 140 | 71 | 6 | 2,044 | 523 | 6 | 1,030 | 87 | 42 |
| Mean | 14.0 | 7.1 | 0.6 | | | | | | |

$$b_1 = \frac{[6 - (6)(.6)][1030 - (71)(14)] - [42 - (71)(0.6)][87 - (6)(14)]}{[523 - (71)(7.1)][6 - (6)(.6)] - [42 - (71)(0.6)]^2}$$

$$= \frac{(2.4)(36) - (-0.6)(3)}{(18.9)(2.4) - (0.6)^2} = \frac{86.4 + 1.8}{45.36 - .36} = \frac{88.2}{45} = 1.96.$$

$$b_2 = \frac{[523 - (71)(7.1)][87 - (14)(6)] - [42 - (71)(0.6)][1030 - (71)(14)]}{45}$$

$$= \frac{(18.9)(3) - (-0.6)(36)}{45} = \frac{56.7 + 21.6}{45} = \frac{78.3}{45} = 1.74.$$

$$a = 14.0 - (1.96)(7.1) - (1.74)(0.6) = 14.0 - 13.916 - 1.044$$

$$= -.96.$$

Thus, the regression equation is

$$\hat{Y} = -.96 + 1.96X_1 + 1.74X_2.$$

But it should be recalled that $Y$ is measured in units of 10 in this equation. To put the equation into the original units $a$, $b_1$ and $b_2$ must be multiplied by 10. Thus, the regression equation is

$$\hat{Y} = -9.6 + 19.6X_1 + 17.4X_2.$$

(b) 17.4 major crimes per million people.
19.6 major crimes per million people.

(b) Yes, because these results assume that the effect of a one-percentage-point increase in the unemployment rate on the crime rate is the same in both years.
Yes. One can compute two separate regressions of the crime rate on the unemployment rate, one for 1985 and one for 1987. Then one can test whether the slopes are equal. However, such a test will be quite weak because of the small number of observations.

**12.23** Since $n = 40$ and $k = 2$, Appendix Table 11 shows that, for $\alpha = .01$, $d_L = 1.20$ and $d_u = 1.40$. Since the alternative hypothesis is that there is negative serial correlation, we should reject the null hypothesis (of no serial correlation) if $d > 4 - 1.20 = 2.80$, and not reject the null hypothesis if $d < 4 - 1.40 = 2.60$. Since $d = 2.71$, the test is inconclusive.

**12.25** (a) Using the methods described in previous sections, you might see whether there was evidence of departures from linearity or from homoscedasticity. Also, you might test for serial correlation of the error terms.

(b) It might enable the railroad to predict its costs more accurately and to estimate the marginal costs of its switching and delivery services.

(c) It would be important to have $R^2$, the standard error of estimate, the Durbin-Watson test statistic, and the standard errors of the regression coefficients.

**12.27** (a) The coefficient of determination between $X_1$ and $X_2$ is

$$\frac{38.6^2}{14.4(1384.4)} = \frac{1489.96}{19,935.36} = .07.$$

Since the coefficient of determination is quite low, multicollinearity does not seem to be a problem.

(b) Rather than use annual family income as an independent variable, family income per family member might be used.

**12.29** (a)

| Year | Deviation of sales from the multiple-regression equation | Year | Deviation of sales from the multiple-regression equation |
|------|------|------|------|
| 1987 | 0.65 | 1979 | 0.07 |
| 1986 | 0.09 | 1978 | −0.83 |
| 1985 | −0.09 | 1977 | −0.27 |
| 1984 | −0.53 | 1976 | 0.44 |
| 1983 | 0.35 | 1975 | −0.11 |
| 1982 | 0.02 | 1974 | 0.99 |
| 1981 | −0.27 | 1973 | −0.11 |
| 1980 | −0.38 | | |

(b) $d = [(.56)^2 + (.18)^2 + (.44)^2$

$\qquad + (-.88)^2 + (.33)^2$

$\qquad + (.29)^2 + (.11)^2 + (-.45)^2$

$\qquad + (.90)^2 + (-.56)^2 + (-.71)^2$

$\qquad + (.55)^2 + (-1.10)^2$

$\qquad + (1.10)^2]$

$\qquad \div [.65^2 + .09^2 + .09^2 + .53^2 + .35^2 + .02^2$

$\qquad + .27^2 + .38^2 + .07^2 + .83^2 + .27^2 + .44^2$

$\qquad + .11^2 + .99^2 + .11^2]$

$\quad = 6.0718 \div 3.0244 = 2.01$

Since $n = 15$ and $k = 2$, Appendix Table 11 shows that, for $\alpha = .025$, $d_L = 0.83$ and $d_u = 1.40$. The null hypothesis (of no serial correlation) should be rejected if $d < 0.83$ or if $d > 4 - 0.83 = 3.17$. It should not be rejected if $1.40 < d < 4 - 1.40 = 2.60$. Since $d$ lies between 1.40 and 2.60, the null hypothesis should not be rejected.

**12.31**

```
MTB > read into c44,c45,c46
DATA> 18 105 16
DATA> 17 108 17
DATA> 20 110 17
DATA> 22 109 18
DATA> 24 110 17
DATA> 26 120 21
DATA> 27 125 20
DATA> 19 111 17
DATA> 18 85 9
DATA> 20 96 11
DATA> 21 94 10
DATA> 24 130 26
DATA> 30 134 21
DATA> 28 103 11
DATA> 25 102 12
DATA> 24 89 8
DATA> 22 92 12
DATA> 20 95 13
DATA> 19 100 14
DATA> 17 91 9
DATA> end
      20 ROWS READ
MTB > regress c46 on 2 predictors in c44 and c45

The regression equation is
C46 = −18.2 − 0.339 C44 + 0.385 C45

Predictor        Coef      Stdev    t-ratio
Constant      −18.181      2.586      −7.03
C44           −0.3394      0.1047     −3.24
C45           0.38516      0.02909    13.24

s = 1.419    R-sq = 92.3%    R-sq(adj) = 91.4%

Analysis of Variance

SOURCE        DF        SS         MS
Regression     2     410.72     205.36
Error         17      34.23       2.01
Total         19     444.95
```

# Chapter 13

**13.1** Trends are not useless, since it is interesting and important to recognize and take account of long-run movements over time of relevant variables. Trends are useful in predicting the future and understanding the past.

**13.3** No. You cannot be sure that a trend will continue into the future.

**13.5** (a) Let $t = 0$ when the year is 1960.

| $Y$ | $t$ | $Y^2$ | $t^2$ | $Y_t$ |
|---|---|---|---|---|
| 1.5 | 0 | 2.25 | 0 | 0 |
| 1.6 | 1 | 2.56 | 1 | 1.6 |
| 1.6 | 2 | 2.56 | 4 | 3.2 |
| 1.7 | 3 | 2.89 | 9 | 5.1 |
| 1.9 | 4 | 3.61 | 16 | 7.6 |
| 2.1 | 5 | 4.41 | 25 | 10.5 |
| 2.2 | 6 | 4.84 | 36 | 13.2 |
| 2.5 | 7 | 6.25 | 49 | 17.5 |
| 2.7 | 8 | 7.29 | 64 | 21.6 |
| 2.9 | 9 | 8.41 | 81 | 26.1 |
| 3.0 | 10 | 9.00 | 100 | 30.0 |
| 3.0 | 11 | 9.00 | 121 | 33.0 |
| 3.3 | 12 | 10.89 | 144 | 39.6 |
| 3.9 | 13 | 15.21 | 169 | 50.7 |
| 5.3 | 14 | 28.09 | 196 | 74.2 |
| 5.7 | 15 | 32.49 | 225 | 85.5 |

| Total | 44.9 | 120 | 149.75 | 1,240 | 419.4 |
|---|---|---|---|---|---|
| Mean | 2.806 | 7.5 | | | |

$$b = \frac{16(419.4) - (44.9)(120)}{16(1,240) - 120^2} = .243.$$

$$a = 2.806 - (.243)(7.5) = 2.806 - 1.823 = 0.983.$$

Thus, the equation for the trend line is

$$Y_t = 0.983 + 0.243t,$$

where the origin is set at 1960.

(b) If the origin is set at 1968,

$$a = 0.983 + (.243)(8) = 0.983 + 1.944 = 2.927.$$

Thus, the equation is $Y_t = 2.927 + 0.243t$, where the origin is set at 1968.

If the origin is set at 1965,

$$a = .983 + (.243)(5) = .983 + 1.215 = 2.198.$$

Thus, the equation is $Y_t = 2.198 + 0.243t$, where the origin is set at 1965.

(c) Let $t = 0$ when the year is 1960:

| Y | $u = \log Y$ | t | $u^2$ | $t^2$ | ut |
|---|---|---|---|---|---|
| 1.5 | .1761 | 0 | .0310112 | 0 | 0.0000 |
| 1.6 | .2041 | 1 | .0416568 | 1 | .2041 |
| 1.6 | .2041 | 2 | .0416568 | 4 | .4082 |
| 1.7 | .2304 | 3 | .0530842 | 9 | .6912 |
| 1.9 | .2788 | 4 | .0777294 | 16 | 1.1152 |
| 2.1 | .3222 | 5 | .1038128 | 25 | 1.6110 |
| 2.2 | .3424 | 6 | .1172378 | 36 | 2.0544 |
| 2.5 | .3979 | 7 | .1583244 | 49 | 2.7853 |
| 2.7 | .4314 | 8 | .1861060 | 64 | 3.4512 |
| 2.9 | .4624 | 9 | .2138138 | 81 | 4.1616 |
| 3.0 | .4771 | 10 | .2276244 | 100 | 4.7710 |
| 3.0 | .4771 | 11 | .2276244 | 121 | 5.2481 |
| 3.3 | .5185 | 12 | .2688422 | 144 | 6.2220 |
| 3.9 | .5911 | 13 | .3493992 | 169 | 7.6843 |
| 5.3 | .7243 | 14 | .5246105 | 196 | 10.1402 |
| 5.7 | .7559 | 15 | .5713848 | 225 | 11.3385 |
| Total | 6.5938 | 120 | 3.1939187 | 1,240 | 61.8863 |
| Mean | .4121 | 7.5 | | | |

$$b = \frac{16(61.8863) - (6.5938)(120)}{16(1,240) - 120^2}$$

$$= .036567.$$

$$a = .4121 - (7.5)(.036567) = .4121 - .2743 = .1378.$$

Thus, the equation for the trend line is

$$\log Y_t = .1378 + .03657t,$$

where the origin is set at 1960.

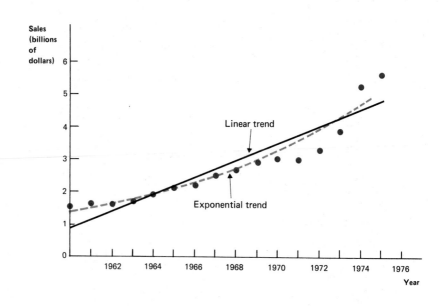

(d) The exponential trend seems to fit better.

(e) The one based on the exponential trend line.

(f) The one based on the exponential trend line.

**13.7** (a) and (b)

| Year | Moving sum 3 year | Moving sum 5 year | Moving average 3 year | Moving average 5 year | Year | Moving sum 3 year | Moving sum 5 year | Moving average 3 year | Moving average 5 year |
|------|------|------|------|------|------|------|------|------|------|
| 1962 | 8.34 | — | 2.78 | — | 1973 | 7.09 | 11.89 | 2.36 | 2.38 |
| 1963 | 8.57 | 14.22 | 2.86 | 2.84 | 1974 | 6.96 | 11.66 | 2.32 | 2.33 |
| 1964 | 8.75 | 14.38 | 2.92 | 2.88 | 1975 | 6.89 | 11.50 | 2.30 | 2.30 |
| 1965 | 8.78 | 14.56 | 2.93 | 2.91 | 1976 | 6.84 | 11.41 | 2.28 | 2.28 |
| 1966 | 8.72 | 14.52 | 2.91 | 2.90 | 1977 | 6.79 | 11.36 | 2.26 | 2.27 |
| 1967 | 8.64 | 14.29 | 2.88 | 2.86 | 1978 | 6.79 | 11.44 | 2.26 | 2.29 |
| 1968 | 8.48 | 14.02 | 2.83 | 2.80 | 1979 | 6.90 | 11.62 | 2.30 | 2.32 |
| 1969 | 8.21 | 13.62 | 2.74 | 2.72 | 1980 | 7.10 | 11.93 | 2.37 | 2.39 |
| 1970 | 7.88 | 13.14 | 2.63 | 2.63 | 1981 | 7.41 | 12.33 | 2.47 | 2.47 |
| 1971 | 7.57 | 12.65 | 2.52 | 2.53 | 1982 | 7.68 | — | 2.56 | — |
| 1972 | 7.27 | 12.23 | 2.42 | 2.45 | | | | | |

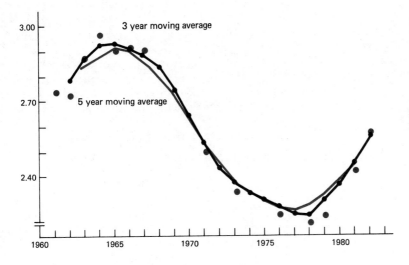

**13.9** (a) $S_0 = 1.5$.

$S_1 = (.2)(1.6) + (.8)(1.5) = .32 + 1.2 = 1.520$.

$S_2 = (.2)(1.6) + (.8)(1.52) = .32 + 1.216 = 1.536$.

$S_3 = (.2)(1.7) + (.8)(1.536) = .34 + 1.2288 = 1.569$.

$S_4 = (.2)(1.9) + (.8)(1.569) = .38 + 1.2552 = 1.635$.

(b) $S_0 = 1.5$.

$S_1 = (.1)(1.6) + (.9)(1.5) = .16 + 1.35 = 1.510$.

$S_2 = (.1)(1.6) + (.9)(1.51) = .16 + 1.359 = 1.519$.

$S_3 = (.1)(1.7) + (.9)(1.519) = .17 + 1.367 = 1.537$.

$S_4 = (.1)(1.9) + (.9)(1.537) = .19 + 1.383 = 1.573$.

**13.11** (a)

| Year | Time series | Year | Time series |
|---|---|---|---|
| 1972 | 1.00 | 1980 | .4(10.00) + .6(6.26) = 7.76 |
| 1973 | .4(2.00) + .6(1.00) = 1.40 | 1981 | .4(12.00) + .6(7.76) = 9.46 |
| 1974 | .4(4.00) + .6(1.40) = 2.44 | 1982 | .4(14.00) + .6(9.46) = 11.28 |
| 1975 | .4(6.00) + .6(2.44) = 3.86 | 1983 | .4(16.00) + .6(11.28) = 13.17 |
| 1976 | .4(5.00) + .6(3.86) = 4.32 | 1984 | .4(13.00) + .6(13.17) = 13.10 |
| 1977 | .4(7.00) + .6(4.32) = 5.39 | 1985 | .4(19.00) + .6(13.10) = 15.46 |
| 1978 | .4(8.00) + .6(5.39) = 6.43 | 1986 | .4(20.00) + .6(15.46) = 17.28 |
| 1979 | .4(6.00) + .6(6.43) = 6.26 | | |

(b)

| Year | Time series | Year | Time series |
|---|---|---|---|
| 1972 | 1.00 | 1980 | .2(10.00) + .8(4.96) = 5.97 |
| 1973 | .2(2.00) + .8(1.00) = 1.20 | 1981 | .2(12.00) + .8(5.97) = 7.18 |
| 1974 | .2(4.00) + .8(1.20) = 1.76 | 1982 | .2(14.00) + .8(7.18) = 8.54 |
| 1975 | .2(6.00) + .8(1.76) = 2.61 | 1983 | .2(16.00) + .8(8.54) = 10.03 |
| 1976 | .2(5.00) + .8(2.61) = 3.09 | 1984 | .2(13.00) + .8(10.03) = 10.62 |
| 1977 | .2(7.00) + .8(3.09) = 3.87 | 1985 | .2(19.00) + .8(10.62) = 12.30 |
| 1978 | .2(8.00) + .8(3.87) = 4.70 | 1986 | .2(20.00) + .8(12.30) = 13.84 |
| 1979 | .2(6.00) + .8(4.70) = 4.96 | | |

**13.13** (a) One factor influencing the competitiveness of U.S. goods in world markets is the extent of U.S. research and development activities.

(b) The percent of GNP devoted to research and development decreased from the early 1960s to the late 1970s, but rose during the early 1980s.

(c) Such a trend certainly isn't obvious from these data.

**13.15** Deseasonalized sales are as follows:

| | | |
|---|---|---|
| January | 2.5 ÷ .97 = | $2.58 million |
| February | 2.4 ÷ .96 = | 2.50 million |
| March | 2.7 ÷ .97 = | 2.78 million |
| April | 2.9 ÷ .98 = | 2.96 million |
| May | 3.0 ÷ .99 = | 3.03 million |
| June | 3.1 ÷ 1.00 = | 3.10 million |
| July | 3.2 ÷ 1.01 = | 3.17 million |
| August | 3.1 ÷ 1.03 = | 3.01 million |
| September | 3.2 ÷ 1.03 = | 3.11 million |
| October | 3.1 ÷ 1.03 = | 3.01 million |
| November | 3.0 ÷ 1.02 = | 2.94 million |
| December | 2.9 ÷ 1.01 = | 2.87 million |

**13.17** (a)

| Month | 12-month Moving sum | 12-month Moving average | Centered 12-month Moving average | Actual rate as a ratio of centered moving average |
|---|---|---|---|---|
| June 1979 | 69.6 | 5.800 | | |
| July | 70.1 | 5.842 | 5.821 | .962 |
| August | 70.4 | 5.867 | 5.854 | 1.008 |
| September | 70.9 | 5.908 | 5.888 | .985 |
| October | 72.0 | 6.000 | 5.954 | .991 |
| November | 73.9 | 6.158 | 6.079 | .971 |
| December | 75.8 | 6.317 | 6.238 | .962 |
| January | 78.0 | 6.500 | 6.409 | .983 |
| February | 79.8 | 6.650 | 6.575 | .943 |
| March | 81.5 | 6.792 | 6.721 | .937 |
| April | 83.1 | 6.925 | 6.858 | 1.006 |
| May | 84.7 | 7.058 | 6.992 | 1.073 |
| June | 86.0 | 7.167 | 7.112 | 1.055 |
| July | 87.1 | 7.258 | 7.212 | 1.082 |
| August | 88.3 | 7.358 | 7.308 | 1.054 |
| September | 89.3 | 7.442 | 7.400 | 1.014 |
| October | 89.7 | 7.475 | 7.458 | 1.006 |
| November | 89.7 | 7.475 | 7.475 | 1.003 |
| December | 89.6 | 7.467 | 7.471 | .977 |
| January | 89.0 | 7.417 | 7.442 | .994 |
| February | 88.6 | 7.383 | 7.400 | 1.000 |
| March | 88.7 | 7.392 | 7.388 | .988 |
| April | 89.2 | 7.433 | 7.412 | .985 |
| May | 90.0 | 7.500 | 7.466 | 1.005 |
| June | 91.5 | 7.625 | 7.562 | .979 |
| July 1981 | | | | |

Multiplying the ratios by 100, the median for each month is

| | | | | | |
|---|---|---|---|---|---|
| January | 98.8 | May | 103.9 | September | 100.0 |
| February | 97.2 | June | 101.7 | October | 99.8 |
| March | 96.2 | July | 102.2 | November | 98.7 |
| April | 99.6 | August | 103.1 | December | 97.0 |

Since the sum of the medians is 1198.2, each of these medians must be multiplied by 1200 ÷ 1198.2 or 1.0015. The result is

| | | | | | |
|---|---|---|---|---|---|
| January | 98.9 | May | 104.1 | September | 100.2 |
| February | 97.3 | June | 101.9 | October | 99.9 |
| March | 96.3 | July | 102.4 | November | 98.8 |
| April | 99.7 | August | 103.3 | December | 97.1 |

(Because of rounding errors, the indexes do not sum precisely to 1200.)

(b) The seasonal variation during the period 1979–81 may be somewhat different from that during the period on which the government's seasonal index is based. (Note that none of the monthly indexes departs very greatly from 100.0.)

**13.19** (a) The deseasonalized value of sales in January is 4.0 ÷ 1.10, or 3.64 millions of dollars. The deseasonalized value of sales in February is 3.8 ÷ 1.01, or 3.76 millions of dollars. Thus, when seasonal variation is taken into account, February's sales were better, not worse, than January's.

(b) Presumably, he multiplied January's sales by 12. But January's sales tend to be raised by seasonal variation. (Recall that its seasonal index is 110.) Thus, his estimate may be too high. Also, he should take account of trend and cyclical variation.

**13.21** Given that the trend value for February 1989 is predicted to be $4 million, the adjusted figure is 4(.96) = $3.84 millions.

**13.23** Changes in investment spending often are responsible for changes in gross national product. In other words, they often play a key role in influencing business conditions.

**13.25** (a) Yes.
(b) Yes.
(c) Since recessions are periods when national output falls, one would expect unemployment to be relatively high in such periods.

**13.27** (a) 0.983 + (.243)(25) = 0.983 + 6.075 = 7.058 billions of dollars.
(b) .1378 + (.036567)(25) = .1378 + .9142 = 1.052. The antilog is 11.27 billions of dollars.
(c) The results are not close to one another at all. Specifically, they differ by over 50 percent.

**13.29**

| Month | Forecasted trend value of sales (millions of dollars) | Seasonal index | Forecasted sales (millions of dollars) |
|---|---|---|---|
| January | 114.9 | 103 | 118.3 |
| February | 116.2 | 80 | 93.0 |
| March | 117.5 | 75 | 88.1 |
| April | 118.8 | 103 | 122.4 |
| May | 120.1 | 101 | 121.3 |
| June | 121.4 | 104 | 126.3 |
| July | 122.7 | 120 | 147.2 |
| August | 124.0 | 139 | 172.4 |
| September | 125.3 | 121 | 151.6 |
| October | 126.6 | 101 | 127.9 |
| November | 127.9 | 75 | 95.9 |
| December | 129.2 | 78 | 100.8 |

**13.31** (a) to (c)

| Year | Moving total 3-year | Moving total 5-year | Moving total 7-year | Moving average 3-year | Moving average 5-year | Moving average 7-year |
|------|--------|--------|--------|--------|--------|--------|
| 1948 | 268 |     |     | 89.3 |      |      |
| 1949 | 263 | 449 |     | 87.7 | 89.8 |      |
| 1950 | 262 | 442 | 629 | 87.3 | 88.4 | 89.9 |
| 1951 | 269 | 442 | 617 | 89.7 | 88.4 | 88.1 |
| 1952 | 271 | 444 | 616 | 90.3 | 88.8 | 88.0 |
| 1953 | 263 | 445 | 625 | 87.7 | 89.0 | 89.3 |
| 1954 | 266 | 444 | 621 | 88.7 | 88.8 | 88.7 |
| 1955 | 264 | 442 | 606 | 88.0 | 88.4 | 86.6 |
| 1956 | 267 | 426 | 600 | 89.0 | 85.2 | 85.7 |
| 1957 | 252 | 425 | 588 | 84.0 | 85.0 | 84.0 |
| 1958 | 244 | 414 | 583 | 81.3 | 82.8 | 83.3 |
| 1959 | 238 | 402 | 573 | 79.3 | 80.4 | 81.9 |
| 1960 | 240 | 397 | 566 | 80.0 | 79.4 | 80.9 |
| 1961 | 239 | 404 | 566 | 79.7 | 80.8 | 80.9 |
| 1962 | 242 | 408 | 581 | 80.7 | 81.6 | 83.0 |
| 1963 | 250 | 419 | 595 | 83.3 | 83.8 | 85.0 |
| 1964 | 260 | 437 | 608 | 86.7 | 87.4 | 86.9 |
| 1965 | 273 | 449 | 624 | 91.0 | 89.8 | 89.1 |
| 1966 | 280 | 460 | 639 | 93.3 | 92.0 | 91.3 |
| 1967 | 283 | 470 | 644 | 94.3 | 94.0 | 92.0 |
| 1968 | 283 | 467 | 643 | 94.3 | 93.4 | 91.9 |
| 1969 | 278 | 456 | 642 | 92.7 | 91.2 | 91.7 |
| 1970 | 269 | 453 | 642 | 89.7 | 90.6 | 91.7 |
| 1971 | 263 | 455 | 640 | 87.7 | 91.0 | 91.4 |
| 1972 | 271 | 450 |     | 90.3 | 90.0 |      |
| 1973 | 277 |     |     | 92.3 |      |      |

(d)

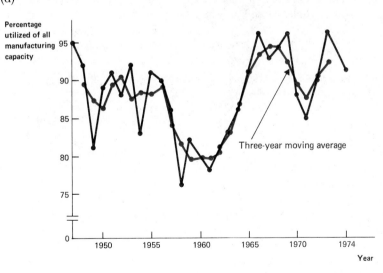

No.

(e) No.

**13.33** (a) Let $t' = 0$ for 1971. Thus, if $Y_{t'}$, is civilian employment in year $t'$,

$$\Sigma Y_{t'} = 1715 \qquad \Sigma t'^2 = 770 \qquad n = 21$$

$$\Sigma t' = 0 \qquad \Sigma Y_{t'} t' = 1369$$

$$b = 1.78 \qquad a = 81.7$$

Thus, the trend line is $Y_{t'} = 81.7 + 1.78t'$.

(b) 1.78 million persons.

(c) $Y_t = 63.9 + 1.78t$

(d) $Y_t = 88.82 + 1.78t$

(e)

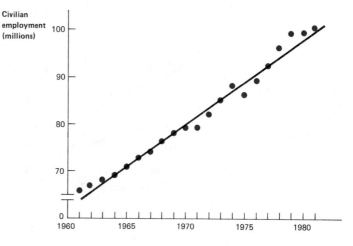

A linear trend does not seem to be a bad approximation, but a curvilinear trend may be better.

(f) The forecast would have been $81.7 + 1.78(13) = 104.84$. Thus the forecasting error would have been only about 0.16 millions, or less than 2/10 of 1 percent.

**13.35** (a) Let $u = t'^2$, where $t' = 0$ in 1971, and let $Y$ equal civilian employment (in millions of persons).

$$\Sigma Y = 1715 \qquad \Sigma t'^2 = 770 \qquad \Sigma Yt' = 1369$$

$$\Sigma t' = 0 \qquad \Sigma u^2 = 50{,}666 \qquad \Sigma Yu = 63{,}577$$

$$\Sigma u = 770 \qquad \Sigma t'u = 0 \qquad n = 21$$

$$\Sigma(t' - \bar{t}')^2 = 770$$

$$\Sigma(u - \bar{u})^2 = 22{,}433$$

$$\Sigma(Y - \bar{Y})(t' - \bar{t}') = 1369$$

$$\Sigma(Y - \bar{Y})(u - \bar{u}) = 63{,}577 - \frac{1715(770)}{21} = 63{,}577 - \frac{1{,}320{,}550}{21} = 694$$

$$\Sigma(t' - \bar{t}')(u - \bar{u}) = 0$$

$$b_1 = \frac{22{,}433(1369)}{770(22{,}433)} = 1.78$$

$$b_2 = \frac{770(694)}{770(22433)} = .031$$

$$a = 81.67 - 1.78(0) - .031(36.67) = 81.67 - 1.14 = 80.53$$

Thus, the trend is $Y_{t'} = 80.53 + 1.78t' + .031t'^2$.

(b) The values of $Y_t$ are calculated and plotted below. The trend seems to fit well.

| Year | $t'$ | $1.78t'$ | $.031t'^2$ | $Y_{t'}$ | Year | $t'$ | $1.78t'$ | $.031t'^2$ | $Y_{t'}$ |
|------|------|----------|-----------|----------|------|------|----------|-----------|----------|
| 1961 | −10 | −17.80 | 3.10 | 65.8 | 1972 | 1 | 1.78 | 0.03 | 82.3 |
| 1962 | − 9 | −16.02 | 2.51 | 67.0 | 1973 | 2 | 3.56 | 0.12 | 84.2 |
| 1963 | − 8 | −14.24 | 1.98 | 68.3 | 1974 | 3 | 5.34 | 0.28 | 86.2 |
| 1964 | − 7 | −12.46 | 1.52 | 69.6 | 1975 | 4 | 7.12 | 0.50 | 88.2 |
| 1965 | − 6 | −10.68 | 1.12 | 71.0 | 1976 | 5 | 8.90 | 0.78 | 90.2 |
| 1966 | − 5 | − 8.90 | 0.78 | 72.4 | 1977 | 6 | 10.68 | 1.12 | 92.3 |
| 1967 | − 4 | − 7.12 | 0.50 | 73.9 | 1978 | 7 | 12.46 | 1.52 | 94.5 |
| 1968 | − 3 | − 5.34 | 0.28 | 75.5 | 1979 | 8 | 14.24 | 1.98 | 96.8 |
| 1969 | − 2 | − 3.56 | 0.12 | 77.1 | 1980 | 9 | 16.02 | 2.51 | 99.1 |
| 1970 | − 1 | − 1.78 | 0.03 | 78.8 | 1981 | 10 | 17.80 | 3.10 | 101.4 |
| 1971 | 0 | 0 | 0 | 80.5 | | | | | |

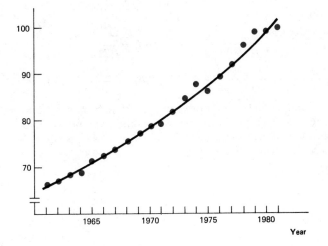

The trend fits well.

**13.37**  $1313 + 111(15) = 2{,}978$ billions of dollars.

**13.39**  $81.7 + 1.78(14) = 81.7 + 24.9 = 106.6$ millions of persons.

**13.41** (a)

| Year | Deviation from trend | Year | Deviation from trend |
|------|---------------------|------|---------------------|
| 1960 | $1.5 - 0.983 = 0.517$ | 1968 | $2.7 - 2.927 = -.227$ |
| 1961 | $1.6 - 1.226 = \ \ .374$ | 1969 | $2.9 - 3.170 = -.270$ |
| 1962 | $1.6 - 1.469 = \ \ .131$ | 1970 | $3.0 - 3.413 = -.413$ |
| 1963 | $1.7 - 1.712 = -.012$ | 1971 | $3.0 - 3.656 = -.656$ |
| 1964 | $1.9 - 1.955 = -.055$ | 1972 | $3.3 - 3.899 = -.599$ |
| 1965 | $2.1 - 2.198 = -.098$ | 1973 | $3.9 - 4.142 = -.242$ |
| 1966 | $2.2 - 2.441 = -.241$ | 1974 | $5.3 - 4.385 = \ \ .915$ |
| 1967 | $2.5 - 2.684 = -.184$ | 1975 | $5.7 - 4.628 = 1.072$ |

(b) In general, the deviations seem to reflect the nonlinearity of the trend rather than the cyclical variation. The trough in 1971 does not correspond to a business cycle trough for the economy as a whole. The downward movement from 1960 to 1971 and the upward movement from 1971 to 1975 do not correspond to a recession and expansion for the economy as a whole.

# Chapter 14

**14.1** (a) The base period for all of these index numbers is 1910–14. The given period for each of them is the year shown at the left end of the row in which the number is located. For example, the given period for all index numbers in the last row is 1974.

(b) A more rapid rate of price increase occurred during 1970–74 than during 1950–70.

(c) A more rapid annual rate of price increase occurred during 1970–74 than during 1930–50.

(d) Tobacco.

**14.3** (a) $\dfrac{505 + 520 + 622}{360 + 480 + 610}(100) = \dfrac{164{,}700}{1450} = 113.6$

(b) $\dfrac{2.1(505) + 1.7(520) + 0.9(622)}{2.1(360) + 1.7(480) + 0.9(610)}(100) = \dfrac{250{,}430}{2121} = 118.1$

(c) $\dfrac{2.0(505) + 1.6(520) + 0.7(622)}{2.0(360) + 1.6(480) + 0.7(610)}(100) = \dfrac{227{,}740}{1915} = 118.9$

(d) The simple unweighted price index suffers from very serious disadvantages described in the text. The other two indexes indicate that the price of the store's television sets increased by between 18 and 19 percent during 1985 to 1987.

**14.5** This index equals

$$\sqrt{\dfrac{\displaystyle\sum_{i=1}^{m} Q_{0i}P_{1i}}{\displaystyle\sum_{i=1}^{m} Q_{0i}P_{0i}} \times \dfrac{\displaystyle\sum_{i=1}^{m} Q_{1i}P_{1i}}{\displaystyle\sum_{i=1}^{m} Q_{1i}P_{0i}}}\,(100)$$

**14.7** For 1950, the index is

$$\dfrac{.08(9.5) + .09(1.2)}{.08(7.2) + .09(1.0)}(100) = \dfrac{86.8}{.666} = 130.3.$$

For 1960, the index is

$$\dfrac{.08(14.8) + .09(1.1)}{.08(7.2) + .09(1.0)}(100) = \dfrac{128.3}{.666} = 192.6.$$

For 1970, the index is

$$\dfrac{.08(21.7) + .09(.6)}{.08(7.2) + .09(1.0)}(100) = \dfrac{179.0}{.666} = 268.8.$$

**14.9** $\dfrac{360(2.0) + (480)(1.6) + (610)(0.7)}{360(2.1) + 480(1.7) + 610(0.9)}(100) = \dfrac{191{,}500}{2{,}121} = 90.3$

**14.11** $\dfrac{173(21,304) + 58(8,243)}{173(18,256) + 58(9,433)}(100) = \dfrac{416,368,600}{3,705,402} = 112.4$

**14.13** No. This would be true only if the sales of its drug division equaled the sales of its chemical division in 1986. Then the 1987 sales of its drug division would be (approximately, not exactly) 11 percent higher than those of its chemical division.

**14.15** (a) No.
(b) No.
(c) Yes.

**14.17** These prices are only a partial and incomplete measure of the costs. There is also the time involved in making the trip. Further, one should consider the amount of discomfort involved. Also, the value of the dollar has fallen.

**14.19** (a) The Consumer Price Index measures changes in prices paid by consumers. The Wholesale Price Index (now the Producer Price Index) measures changes in prices paid at the wholesale level. The price deflator for the gross national product measures changes in the prices of all final goods and services produced.
(b) The price deflator for the gross national product.
(c) The price deflator for the gross national product, because it avoids some double-counting found in the Producer Price Index and it looks only at currently produced goods and services (unlike the Consumer Price Index). However, each of these price indexes is very useful in its own way.
(d) No, because the Consumer Price Index was 116.3 in 1970 whereas the price deflator for the gross national product was 91.36 in 1970.

**14.21** No, because the Consumer Price Index measures *changes over time* in prices, not *differences among regions*. Also, the CPI may not be appropriate for this couple.

**14.23** (a) If 1970 is the given year, this index is

$\dfrac{1000(200) + 500(300) + 2000(20)}{1000(100) + 500(200) + 2000(10)}(100) = 177.$

If 1987 is the given year, this index is

$\dfrac{1000(300) + 500(400) + 2000(20)}{1000(100) + 500(200) + 2000(10)}(100) = 245.$

(b) $\dfrac{2,500(300) + 1,500(400) + 6,100(20)}{2,500(100) + 1,500(200) + 6,100(10)}(100) = 241.$

# Chapter 15

**15.1** (a) No.
(b) No.
(c) Yes. When monetary gain is between $40,000 and $60,000, the utility function has the shape that signifies risk aversion.
(d) No. When monetary gain is between $60,000 and $70,000, the utility function has the shape that signifies risk loving.

**15.3** (a) Let $U(5000)$ equal zero and $U(10,000)$ equal one. Then $U(7,500) = 0.5(0) + 0.5(1) = 0.5$. Also, $1 = 0.5(0.5) + 0.5U(12,500)$, which means that $U(12,500) = 1.5$. The four points are shown below:

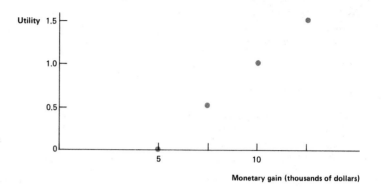

Monetary gain (thousands of dollars)

(b) Indifferent to risk. All four points lie along a straight line.

**15.5** (a) Let 0 equal the utility of losing $90,000, and let 1 equal the utility of gaining $100,000:

$$U(10,000) = 0.5\, U(100,000) + 0.5U(-90,000)$$

$$= (0.5)(1) + (0.5)(0)$$

$$= 0.5.$$

The three points are shown below:

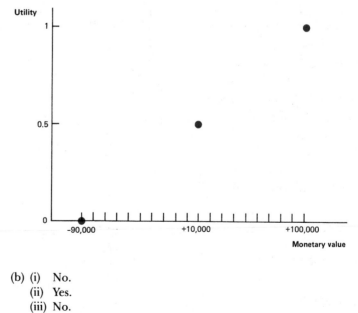

Monetary value

(b) (i)  No.
   (ii)  Yes.
   (iii)  No.
(c)  No.
(d)  No.

**15.7** (a) The firm can inquire carefully about his performance on previous jobs, contact his former employers, give him tests of various kinds, and so forth.

(b) $0.5(100,000) + 0.5(0) - 10,000 = \$40,000$.

**15.9** (a) The situation is shown in the following table:

|  | Profit | | Opportunity loss | |
|---|---|---|---|---|
|  |  | Do not |  | Do not |
| *State of nature* | *Merge* | *merge* | *Merge* | *merge* |
| Drilling activities are successful | $20 million | 0 | 0 | $20 million |
| Drilling activities are not successful | −$10 million | 0 | $10 million | 0 |

The expected opportunity loss is $0.4(0) + 0.6(\$10 \text{ million}) = \$6$ million.

(b) $0.4(\$20 \text{ million}) + 0.6(0) = \$8$ million.

(c) $6 million.

**15.11** (a) The posterior probability that the new bicycle will be a major success is

$$\frac{(0.3)(0.3)}{(0.3)(0.3) + (0.5)(0.4) + (0.2)(0.3)} = \frac{.09}{.09 + .20 + .06} = .26.$$

The posterior probability that the new bicycle will be no better (and no worse) than the existing line is

$$\frac{(0.5)(0.4)}{(0.3)(0.3) + (0.5)(0.4) + (0.2)(0.3)} = \frac{.20}{.09 + .20 + .06} = .57.$$

The posterior probability that the new bicycle will be a flop is

$$\frac{(0.2)(0.3)}{(0.3)(0.3) + (0.5)(0.4) + (0.2)(0.3)} = \frac{.06}{.09 + .20 + .06} = .17.$$

(b)

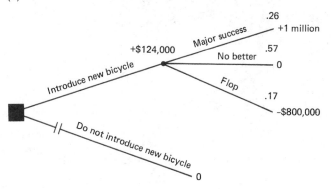

(c) Yes.

(d) The expected value of perfect information is

$(.26) (\$1 \text{ million}) + (.57)(0) + (.17)(0) - \$124,000 = \$136,000.$

(e) The expected value of perfect information is

$(.30) (\$1 \text{ million}) + (.40)(0) + (.30)(0) - \$60,000 = \$240,000.$

The prior expected value of perfect information is $104,000 greater than the posterior expected value of perfect information. Because the posterior probabilities differ from the prior probabilities.

(f)

|  | Opportunity loss | |
| State of Nature | Introduce new bicycle | Do not introduce new bicycle |
| --- | --- | --- |
| New bicycle is major success | 0 | $1 million |
| New bicycle is no better and no worse | 0 | 0 |
| New bicycle is flop | 800,000 | 0 |

(g) If Acme introduces the new bicycle, the expected opportunity loss is (.3)(0) + (.4)(0) + (.3)($800,000)) = $240,000. If Acme does not introduce the new bicycle, the expected opportunity loss is (.3)($1 million) + (.4)(0) + (.3)(0) = $300,000. Thus Acme should introduce the new bicycle.

The previous paragraph shows that the expected opportunity loss corresponding to the best action equals $240,000. (e) shows that the expected value of perfect information equals $240,000. Thus, the two are equal.

(h) If Acme introduces the new bicycle, the expected opportunity loss is (.17)($800,000) = $136,000. If Acme does not introduce the new bicycle, the expected opportunity loss is (.26) ($1 million) = $260,000. Thus, Acme should introduce the new bicycle.

The previous paragraph shows that the expected opportunity loss corresponding to the best action equals $136,000. (d) shows that the expected value of perfect information equals $136,000. Thus, the two are equal.

**15.13** (a) He should expand.

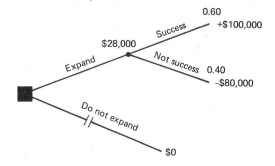

(b) The fork on the left is a decision fork because the owner decides whether or not to expand. The fork on the right is a chance fork because "chance" decides whether the expansion will prove successful.

(c) No.

(d) 4/9

**15.15** (a) The posterior probability that the advertising campaign will be successful is 0.6(0.5) ÷ [0.6(0.5) + 0.4(0.5)] = 0.6. Thus, the expected increase in profit if the firm builds the warehouse is 0.6 ($1 million) + 0.4 (−$0.5 million) = $0.4 million. The firm should build the new warehouse.

(b) The posterior probability that the advertising campaign will be successful is 0.4(0.5) ÷ [0.4(0.5) + 0.6(0.5)] = 0.4. Thus, the expected increase in profit if the firm builds the warehouse is 0.4 ($1 million) + 0.6 (−$0.5 million) = $0.1 million. The firm should build the new warehouse.

**15.17** $0.5(\$1\text{ million}) + 0.5(0) - \$250{,}000 = \$250{,}000$
$0.4(\$1\text{ million}) + 0.6(0) - \$100{,}000 = \$300{,}000$

**15.19** (a) Yes.
(b) Yes. The proportion of potential buyers that would buy the new product. Annual profit $= -1{,}000{,}000 + 1{,}000{,}000(50)p$, where $p$ is the proportion of potential buyers that would buy the new product. (We assume each buyer purchases one set.)
(c) If $E(p) > .02$, the expected value of annual profit is greater than zero. In fact, $E(p) = .0197$. Thus, the firm should not introduce the new product.

# Chapter 16

**16.1** (a)

(b)

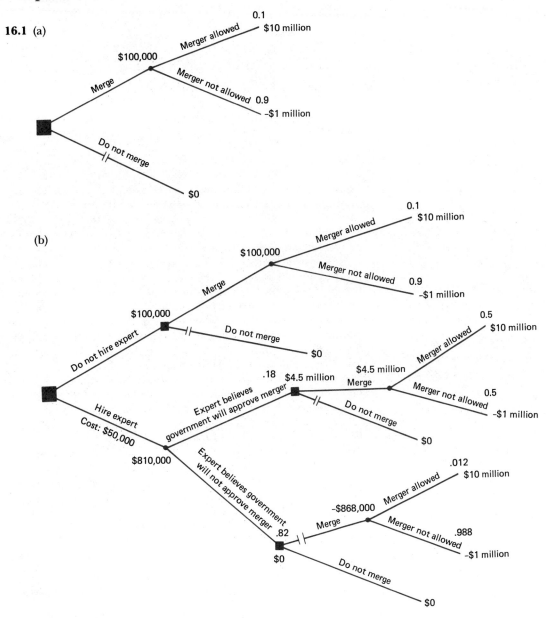

(c)  Yes.
(d)  Yes.
(e)  No.

**16.3** (a)  The opportunity loss for each action and each state of nature is as follows:

| State of nature | Expand plant | Do not expand plant |
|---|---|---|
| Advertising campaign is successful | 0 | $5 million |
| Advertising campaign is not successful | $2 million | 0 |

(b)

| | Strategy | | | |
|---|---|---|---|---|
| Survey outcome | 1 | 2 | 3 | 4 |
| Campaign will be successful | Expand plant | Expand plant | Do not expand plant | Do not expand plant |
| Campaign will not be successful | Do not expand plant | Expand plant | Do not expand plant | Expand plant |

No.

(c)  Consider strategy 1. If the campaign is successful the expected opportunity loss is .8(0) + .2($5 million) = $1 million. If the campaign is not successful, the expected opportunity loss is .2($2 million) + .8(0) = $0.4 million. Thus, the expected opportunity loss is .5($1 million) + .5($0.4 million) = $0.7 million.

Consider strategy 2. If the campaign is successful, the expected opportunity loss is zero. If the campaign is not successful, the expected opportunity loss is $2 million. Thus, the expected opportunity loss is .5(0) + .5($2 million) = $1 million.

Consider strategy 3. If the campaign is successful, the expected opportunity loss is $5 million. If the campaign is not successful, the expected opportunity loss is zero. Thus, the expected opportunity loss is .5($5 million) + .5(0) = $2.5 million.

Consider strategy 4. If the campaign is successful, the expected opportunity loss is .8($5 million) + .2(0) = $4 million. If the campaign is not successful, the expected opportunity loss is .8($2 million) + .2(0) = $1.6 million. Thus, the expected opportunity loss is .5($4 million) + .5($1.6 million) = $2.8 million.

(d)  With strategy 1, the expected opportunity loss equals $P$($1 million) + $(1 - P)$($0.4 million). With strategy 2, the expected opportunity loss equals $(1 - P)$($2 million). With strategy 3, the expected opportunity loss equals $P$($5 million). With strategy 4, the expected opportunity loss equals $P$($4 million) + $(1 - P)$($1.6 million).

(e)  Clearly, strategy 1 dominates strategy 4. Yes.

**16.5** (a)  No.
(b)  The firm should have reached a decision after the 41st item was tested.
(c)  Reject the null hypothesis that the proportion of defectives is .01.

**16.7** (a)

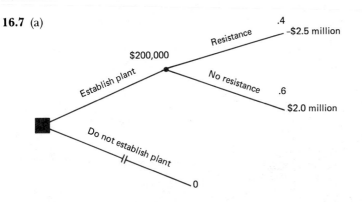

(b) Let $R$ stand for resistance and $N$ stand for no resistance. Let $r$ stand for the survey's indicating resistance and $n$ stand for its indicating no resistance.

$$P(n) = P(R)P(n|R) + P(N)P(n|N)$$

$$= (0.4)(0.2) + (0.6)(0.7) = .50.$$

$$P(r) = P(R)P(r|R) + P(N)P(r|N)$$

$$= (0.4)(0.8) + (0.6)(0.3) = 0.50.$$

(c) The posterior probability of resistance equals

$$\frac{(0.2)(0.4)}{(0.2)(0.4) + (0.7)(0.6)} = \frac{.08}{.50} = .16.$$

The posterior probablity of no resistance equals

$$\frac{(0.7)(0.6)}{(0.2)(0.4) + (0.7)(0.6)} = \frac{.42}{.50} = .84.$$

Thus, the subsequent branches of the decision tree are as follows:

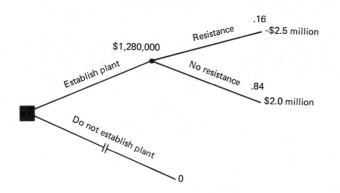

(d) The posterior probability of resistance equals

$$\frac{(0.8)(0.4)}{(0.8)(0.4) + (0.3)(0.6)} = \frac{.32}{.32 + .18} = .64.$$

The posterior probability of no resistance equals

$$\frac{(0.3)(0.6)}{(0.8)(0.4) + (0.3)(0.6)} = \frac{.18}{.50} = .36.$$

Thus, the subsequent branches of the decision tree are as follows:

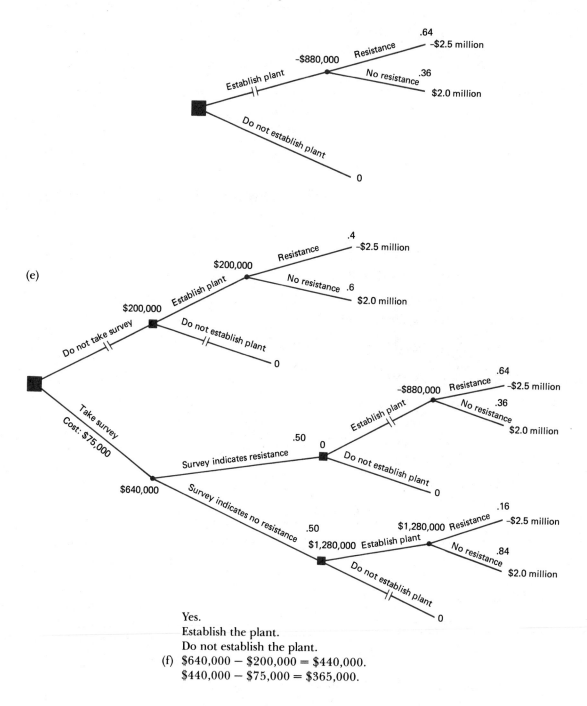

(e)

Yes.
Establish the plant.
Do not establish the plant.
(f) $640,000 − $200,000 = $440,000.
$440,000 − $75,000 = $365,000.

(g) There are four possible strategies, as indicated in the table below:

| Survey outcome | Strategy 1 | Strategy 2 | Strategy 3 | Strategy 4 |
|---|---|---|---|---|
| Indicates substantial resistance | Do not establish plant | Do not establish plant | Establish plant | Establish plant |
| Indicates no resistance | Establish plant | Do not establish plant | Establish plant | Do not establish plant |

(h) The opportunity loss for each state of nature, given each action, is as follows:

| State of Nature | Establish plant | Do not establish plant |
|---|---|---|
| Resistance | $2.5 million | 0 |
| No resistance | 0 | $2.0 million |

First consider strategy 1 in the answer to (g). If there really is resistance, the probability that the survey will indicate resistance is 0.8. Thus, the conditional expected opportunity loss is $(0.8)(0) + (0.2)(\$2.5 \text{ million})$ = $0.5 million. If there really is no resistance, the probability that the survey will indicate resistance is 0.3. Thus, the conditional expected opportunity loss is $(0.7)(0) + (0.3)(\$2.0 \text{ million}) = \$0.6$ million. Since there is a 0.4 prior probability of resistance, the expected opportunity loss equals $(0.4)(\$0.5 \text{ million}) + (0.6)(\$0.6 \text{ million}) = \$560,000$.

Second, consider strategy 2 in the answer to (g). If there really is resistance, the conditional expected opportunity loss is $(0.8)(0) + (0.2)(0) = 0$. If there really is no resistance, the conditional expected opportunity loss is $(0.7)(\$2.0 \text{ million}) + (0.3)(\$2.0 \text{ million}) = \$2.0$ million. Thus, the expected opportunity loss equals $(0.4)(0) + (0.6)(\$2.0 \text{ million}) = \$1.2$ million.

Third, consider strategy 3 in the answer to (g). If there really is resistance, the conditional expected opportunity loss is $(0.8)(\$2.5 \text{ million}) + (0.2)(\$2.5 \text{ million}) = \$2.5$ million. If there really is no resistance, the conditional expected opportunity loss is $(0.7)(0) + (0.3)(0) = 0$. Thus, the expected opportunity loss equals $(0.4)(\$2.5 \text{ million}) + (0.6)(0) = \$1$ million.

Fourth, consider strategy 4 in the answer to (g). If there really is resistance, the conditional expected opportunity loss is $(0.8)(\$2.5 \text{ million}) + (0.2)(0) = \$2.0$ million. If there really is no resistance, the conditional expected opportunity loss is $(0.7)(\$2.0 \text{ million}) + (0.3)(0) = \$1.4$ million. Thus, the expected opportunity loss equals $(0.4)(\$2.0 \text{ million}) + (0.6)(\$1.4 \text{ million}) = \$1.64$ million.

(i) Yes. As pointed out in the text, normal-form analysis and extensive-form analysis will always lead to the choice of the same strategy.

(j) The expected opportunity loss equals $P (\$0.6 \text{ million}) + (1 - P)(\$0.5$ million).

(k) For strategy 2 in the answer to (g), the expected loss equals $P(\$2.0$ million). For strategy 3, it equals $(1 - P)(\$2.5 \text{ million})$. For strategy 4, it equals $P(\$1.4 \text{ million}) + (1 - P)(\$2.0 \text{ million})$.

Clearly strategy 4 is dominated by strategy 1.

(l)  Yes. The strategy in (j) is not as good as strategy 2 if $P < .263$. To see this, note that

$$P(\$2.0 \text{ million}) < P(\$0.6 \text{ million}) + (1 - P)(\$0.5 \text{ million})$$

if

$$P(\$1.9 \text{ million}) < \$0.5 \text{ million};$$

that is,

if $P < 0.5/1.9 = .263$.

Also, the strategy in (j) is not as good as strategy 3 if $P > .769$. To see this, note that

$$(1 - P)(\$2.5 \text{ million}) < P(\$0.6 \text{ million}) + (1 - P)(\$0.5 \text{ million})$$

if

$$\$2.0 \text{ million} < P(\$2.6 \text{ million});$$

that is,

if $P > 2.0/2.6 = .769$.

**16.9** (a)  $0.5(0.75) \div [0.5(0.75) + 0.5(0.25)] = \dfrac{.375}{.375 + .125} = .75$

(b)  $0.5(0.75) \div [0.5(0.75) + 0.5(0.25)] = .75$

(c)  $0.5(0.75)(0.8) \div [0.5(0.75)(0.8) + 0.5(0.25)(0.2)] = \dfrac{.3}{.3 + .025} = .92.$

(d)  $0.5(0.75)(0.8) \div [0.5(0.75)(0.8) + 0.5(0.25)(0.2)] = .92.$

(e)  $0.5(0.75)(0.2) \div [0.5(0.75)(0.2) + 0.5(0.25)(0.8)] = \dfrac{.075}{.075 + .1} = .43.$

(f)  $0.5(0.25)(0.8) \div [0.5(0.25)(0.8) + 0.5(0.75)(0.2)] = \dfrac{.1}{.1 + .075} = .57.$

# Answers to *Getting Down to Cases*

## Does the Production Process Meet the Tolerances? (Chapter 1)

**1** (a) Many frequency distributions could be admissible answers. One possible frequency distribution is as follows:

| *Inches between hole centers* | *Number of die stampings* |
|:---:|:---:|
| 2.996 | 1 |
| 2.997 | 2 |
| 2.998 | 1 |
| 2.999 | 3 |
| 3.000 | 5 |
| 3.001 | 7 |
| 3.002 | 4 |
| 3.003 | 4 |
| 3.004 | 7 |
| 3.005 | 5 |
| 3.006 | 7 |
| 3.007 | 2 |
| 3.008 | 1 |

(b) Based on the above distribution, 15 of the 49 die stampings (that is, 31 percent) did not meet the tolerances.

(c) The die stampings may not all result from the same productive process; they may be a composite of more than one distinct population. Based on this information alone, it is not possible to determine the actual reasons.

**2** (a) Yes. The die stampings based on the use of die *B* all meet the tolerances, whereas many of those based on the use of die *A* do not. The mean distance between holes is greater for those based on die *A* than for those based on die *B*.

(b) Die *A* should be repaired or replaced since a very large percentage of the stampings based on its use (60 percent in this sample) do not meet the tolerances.

## The Reliability of the Apollo Space Mission (Chapter 3)

(a) $(.99)^5 = .9510$

(b) $1 - (.01)^2 = .9999$

(c) $(.99)^4 [1 - (.01)^2] = (.9606)(.9999) = .9605$.

(d) The probability is higher in (c) than in (a) because of the smaller chance that the engine will fail to function properly.

(e) It is expensive to add redundant components, and it adds to the weight of the system.

## Quality Control in the Manufacture of Railway-Car Side Frames (Chapter 4)

(a) Since $n = 10$ and $\Pi = .20$ the probability that $x = 0, 1,$ or 2 equals .1074 + .2684 + .3020, or .6778. Thus, the probability that the process would be stopped is $1 - .6778 = .3222$.

(b) Since $n = 10$ and $\Pi = .40$ the probability that $x = 0, 1,$ or 2 equals .0060 + .0403 + .1209, or .1672. Thus, the probability that the productive process would be stopped is $1 - .1672$, or .8328.

(c) If 20 percent are defective there is a 32 percent probability that the firm will stop the productive process (even though this seems to have been the normal defective rate in the past). If the percent defective jumps to 40 percent, there is a 17 percent probability that the productive process will not be stopped (even though this seems to be a very high defective rate). Based on these considerations, the adequacy of this sampling plan seems to be questionable.

(d) According to Appendix Table 1, since $n = 5$ and $\Pi = .40$, the probability distribution of the number of defectives is

| Number of defectives | Probability | Payment (dollars) |
| --- | --- | --- |
| 0 | .0778 | 0 |
| 1 | .2592 | 100 |
| 2 | .3456 | 200 |
| 3 | .2304 | 300 |
| 4 | .0768 | 400 |
| 5 | .0102 | 500 |

Thus, the graph is

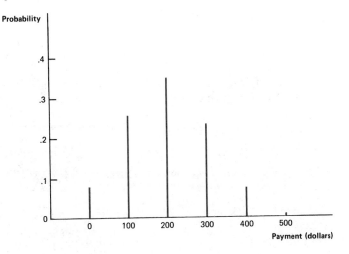

(e) The expected value equals

$(.0778)(0) + (.2592)(\$100) + (.3456)(\$200) + (.2304)(\$300)$
$+ (.0768)(\$400) + (.0102)(\$500) = \$25.92 + \$69.12 + \$69.12$
$+ \$30.72 + \$5.10 = \$200.$

(f) The variance equals

$(.0778)(0 - 200)^2 + (.2592)(100 - 200)^2 + (.3456)(200 - 200)^2 +$
$(.2304)(300 - 200)^2 + (.0768)(400 - 200)^2 + (.0102)(500 - 200)^2$
$= (.0778)(40,000) + (.2592)(10,000) + (.3456)(0) + (.2304)(10,000)$
$+ (.0768)(40,000) + (.0102)(90,000) = 3112 + 2592 + 2304 + 3072$
$+ 918 = 11,998.$

Thus, the standard deviation equals about 110 dollars. (Actually, the variance equals 12,000; the slight difference is due to rounding errors.)

# A Trunking Problem in the Telephone Industry (Chapter 5)

(a) and (b) This is a situation where there are 2,000 Bernoulli trials with $\Pi = 1/30$. Thus, the number of subscribers wanting a trunkline to $B$ has the binomial distribution with $n = 2,000$ and $\Pi = 1/30$. Using the normal approximation, the probability that $L$ or more subscribers want a trunkline to $B$ is the probability that a normal variable with mean of $(2,000)(1/30) = 66.667$ and a standard deviation of $\sqrt{2,000(1/30)(29/30)} = 8.027$ exceeds $(L - 1/2)$. (Because of the continuity correction, $(L - 1/2)$, rather than $L$, is correct.) If this probability is to equal .01, $(L - 1/2)$ must correspond to 2.33 on the standard normal distribution. That is,

$$(L - 1/2 - 66.667) \div 8.027 = 2.33$$

and

$$L = (2.33)(8.027) + 67.167 = 85.9.$$

Thus, the probability that about 85.9 or more subscribers will want a trunkline to *B* is approximately equal to .01. Consequently, about 85 or 86 trunklines should be installed. (This solution is given in W. Feller, *Probability Theory and Applications* [New York: Wiley, 1950], pp. 143 – 144. For further discussion, see this reference.)

# The Effect of a New Enzyme on a Pharmaceutical Manufacturing Process (Chapter 7)

(a) Since $\bar{x} = 1.268$ and $s = .228$, the confidence interval is

$$1.268 - 1.64 \left( \frac{.228}{6} \right) < \mu < 1.268 + 1.64 \left( \frac{.228}{6} \right)$$

$$1.268 - 1.64(.038) < \mu < 1.268 + 1.64(.038)$$

$$1.268 - .062 < \mu < 1.268 + .062$$

$$1.206 < \mu < 1.330.$$

Thus, the answer is 1.206 to 1.330.

(b) The confidence interval is

$$1.268 - 1.96(0.38) < \mu < 1.268 + 1.96(.038)$$

$$1.268 - .074 < \mu < 1.268 + .074$$

$$1.194 < \mu < 1.342.$$

Thus, the answer is 1.194 to 1.342.

(c) The confidence interval is

$$1.268 - 2.576(.038) < \mu < 1.268 + 2.576(.038)$$

$$1.268 - .098 < \mu < 1.268 + .098$$

$$1.170 < \mu < 1.366.$$

Thus, the answer is 1.170 to 1.366.

(d) It is assumed that the sample is random.

(e) 1.268

(f) Yes.

(g) Yes, since the 99 percent confidence interval is entirely above one.

(h) The mistake lies in comparing the difference between the sample mean and 1.00 with the sample standard deviation. It makes more sense to compare this difference with the standard error of the sample mean, as we shall see in Chapter 8.

# Testing for Normality at the American Stove Company (Chapter 9)

(a) and (b)

| Upper limit | $\dfrac{\text{Upper limit} - .8314}{.0059}$ | Area under standard normal curve to the left of point in column 2 | Theoretical proportion in class interval | Theoretical frequency |
|---|---|---|---|---|
| .8215 | −1.68 | .0465 | .0465 | 6.7 |
| .8245 | −1.17 | .1210 | .0745 | 10.8 |
| .8275 | −0.66 | .2546 | .1336 | 19.4 |
| .8305 | −0.15 | .4404 | .1858 | 26.9 |
| .8335 | 0.36 | .6406 | .2002 | 29.0 |
| .8365 | 0.86 | .8051 | .1645 | 23.9 |
| .8395 | 1.37 | .9147 | .1096 | 15.9 |
| .8425 | 1.88 | .9699 | .0552⎤ | 12.4 |
| ∞ | ∞ | 1.0000 | .0301⎦ | |

The second column in the table expresses the upper limit of each class interval (shown in column 1) as a deviation from the sample mean divided by the sample standard deviation. Using Appendix Table 2, we find the probability that the standard normal variable is less than each number in column 2; this probability is shown in column 3. Column 4 shows the probability that a normal variable (with mean equal to the sample mean and standard deviation equal to the sample standard deviation) will fall in each class interval. To obtain these numbers, we subtract the figure in column 3 for the previous class interval from the figure in column 3 for this class interval. Finally, multiplying the numbers in column 4 by 145, we get the theoretical frequencies in each class interval, shown in column 5. (The last two class intervals are combined so that the theoretical frequency will exceed 5.)

(c)

$$\sum \frac{(f-e)^2}{e} = \frac{(9-6.7)^2}{6.7} + \frac{(5-10.8)^2}{10.8} + \frac{(14-19.4)^2}{19.4} + \frac{(21-26.9)^2}{26.9}$$

$$+ \frac{(55-29.0)^2}{29.0} + \frac{(23-23.9)^2}{23.9} + \frac{(7-15.9)^2}{15.9} +$$

$$\frac{(11-12.4)^2}{12.4}$$

$$= 0.79 + 3.11 + 1.50 + 1.29 + 23.31 + 0.03 + 4.98 + 0.16 =$$

35.17.

(d) and (e) Since there are two parameters estimated from the sample, there are $8 - 3$, or 5 degrees of freedom. Thus, $\chi^2_{.05} = 11.0705$. Since the observed value (35.17) exceeds 11.0705, the null hypothesis (of normality) should be rejected.

(f) The evidence indicates that the heights of these metal pieces are not normally distributed. In particular, there were far more pieces with heights of .8305 and under .8335 inches than would be expected if they were normally distributed.

## Pig Iron and Lime Consumption in the Production of Steel (Chapter 11)

(a) The scatter diagram is shown below.
(b) $\hat{Y} = 93.74 + 2.40X$
(c) $r = 0.505$
(d) There seems to be a direct relationship between the percentage of pig iron in a cast of steel and the lime consumption (in hundred weights) per cast. For each percentage point increase in the former, the latter increases by 2.40 hundred weights. This regression explains about 25 percent of the variation in lime consumption per cast.

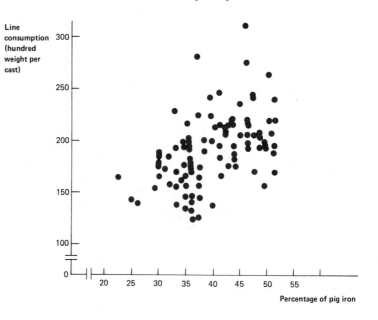

## Changes in Employment in Five Nations (Chapter 13)

(a) A 10-year period is too short to estimate a long-term trend. However, it does appear that during this period (1975–84), employment grew much more rapidly in the United States and Japan than in France, Germany, and Great Britain.
(b) Yes. Although the pattern varies from country to country, there tend to be broad similarities among many countries. For example, employment fell in most of these countries between 1981 and 1982.
(c) Yes. Although the pattern differs somewhat from country to country, recessions tend to spread from one country to another. For example, in 1982 and 1983, the unemployment rate was relatively high in all of these countries.
(d) Clearly, employment did not grow as rapidly in the major European countries as in the United States and Japan. Also, unemployment rates at the end of this period were higher there than in the United States and Japan. These figures alone cannot tell us why these differences occurred.

## Should Cutler-Hammer Purchase an Option? (Chapter 15)

(a)

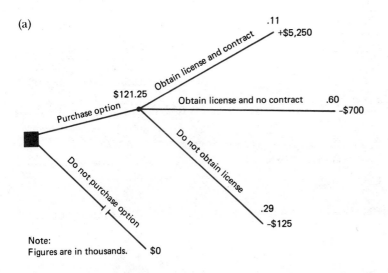

(b) Yes.
(c) Yes. The expected monetary value of this course of action is $49,800, which is considerably less than if it purchased the option ($121,250, according to part (a) above). But there is a zero chance of loss, in contrast to a .89 probability of loss if it purchased the option.

*Note:* Figures are in thousands.

# Answers to *Statistics in Context*

## Japan, W. E. Deming, and the Use of Histograms (Chapter 2)

**1** No. It is possible that the firm's supplier was buying metal from only one firm, but that the metal was very heterogeneous.

**2** The standard deviation of the hardness numbers of the panels might be used. This standard deviation might be compared with what experts in the field would expect.

**3** There is a gap at 0.999 centimeters and a peak at 1.000 centimeters. There sometimes are gaps in histograms, particularly if there is a small number of observations. But the combination of the gap and the adjacent peak (and the fact that there are 500 observations) certainly is suspicious.

**4** No. Undoubtedly, there could have been other causes.

**5** Lower. If the mean is higher than last month (and the standard deviation is the same), the proportion below 1 centimeter is probably lower than last month. However, this assumes that a rod is defective only if its diameter is less than 1 centimeter. If it is also defective if its diameter exceeds some amount (say 1.007 centimeters), the proportion defective for this reason is likely to increase.

**6** Higher. If the standard deviation is higher in December than November (and the mean remains at 1.003 centimeters), the proportion below 1 centimeter is probably higher in December than in November. Moreover, if a rod is also defective if its diameter is too large (above 1.007 centimeters, say), the proportion defective for this reason too is likely to be higher in December than in November.

## Playing the Slots at the Sands (Chapter 6)

**1** The amount of space that a slot machine of each type takes up, the amount of revenue per machine from each type of machine, the effect of the mix of types of machines on the amount of revenue per machine from each type of machine, and the cost per machine of obtaining and maintaining a machine of each type are relevant factors.

**2** If we assume that the 400 people are a random sample (which may or may not be true), the number of players preferring the $0.25 machines is a binomial variable with mean equal to 400 (.60) = 240 and standard deviation equal to $\sqrt{400(.60)(.40)} = 9.80$. The probability that such a variable exceeds 300 is approximately equal to the probability that a standard normal variable exceeds $(300.5 - 240)/9.80$, or 6.17. According to Appendix Table 2, this probability is less than .001.

**3** No. From Table 1, it appears that the net revenue per machine is highest for the $1.00 machines. There is no sampling error, since Table 1 refers to all of the Sands' machines.

**4** It is a sample of people staying at the Sands, not of people playing the slot machines there. The people who stay at the Sands may have quite different characteristics from those playing there.

**5** No.

**6** No. You need to know the standard deviation of the amount that players who prefer $0.25 machines are willing to spend.

## Acceptance Sampling at the Pentagon and Dow Chemical (Chapter 8)

**1** The null hypothesis is that the proportion defective is 4 percent. The alternative hypothesis is that the proportion defective exceeds 4 percent. This is a one-tailed test.

**2** $\beta = .75$.

**3** On page 309, we assumed that the sample size was large enough so that the normal approximation to the binomial distribution can be used. Also, the test statistic used there was $(p - \pi_0) \div \sqrt{\pi_0(1 - \pi_0)/n}$, whereas in Table 1 it is the number defective.

**4** The cost of a Type I error is the expense of stopping the clerical process to determine why so many invoices contain errors when in fact the error rate does not exceed .25 percent. The cost of a Type II error is the expense of tolerating an error rate in excess of .25 percent.

**5** No. It could be done by classifying the invoices (or error opportunities) into groups, by the potential costliness of errors in them, and treating them separately.

**6** No, because the test procedure is not supposed to reduce completely the probability of an incorrect decision. It controls the probability that batches with a relatively high error rate will be accepted, but it does not reduce this probability to zero.

## Industrial Experimentation in Cotton Textiles (Chapter 10)

**1** No. It depends on a variety of factors (for example, the cost of such a flyer), not just on the mean number of breaks per 100 pounds of material.

**2** No.

**3** Yes. The analysis of variance assumes that the standard deviation in each population is the same. If this is not the case, the analysis may be misleading.

**4** Yes. Cottons A and B can be regarded as treatments, and high, medium, and low number of turns can be regarded as blocks. Of course, if the experiment is viewed this way, a two-way analysis of variance should be used.

**5** For the nine observations where there is a low number of turns with cotton A,

$$\Sigma x_i = 401 \qquad \Sigma x_i^2 = 20{,}487.$$

Thus, $(n-1)s^2 = \Sigma x_i^2 - \dfrac{(\Sigma x_i)^2}{n} = 20{,}487 - \dfrac{160{,}801}{9} = 2620.2$

To test whether $\sigma = 10$, we calculate $(n-1)s^2/\sigma_0^2$, which equals $2620.2/100 = 26.202$. If $\alpha = .05$, $\chi^2_{.025} = 17.5346$, since there are 8 degrees of freedom. Since $(n-1)s^2/\sigma_0^2 > 17.5346$, the null hypothesis that $\sigma = 10$ should be rejected.

**6** Assuming that the number of breaks per unit-length of warp is normally distributed, the confidence interval is

$$\frac{2620.2}{17.5346} < \sigma^2 < \frac{2620.2}{2.17973}$$

Thus, $149.4 < \sigma^2 < 1202.1$, or $12.2 < \sigma < 34.7$.

## How to Estimate the Mileage per Gallon of a New Car (Chapter 12)

**1** No. To get $\hat{Y}_i$, the sum of the other terms on the right-hand side of equation (12.15) must be added to $0.415X_{4i}$.

**2** No. The data are not a time series.

**3** Calculate the correlation between all pairs of independent variables, as well as $R^2$ for each of the independent variables on all or some of the others.

**4** The interpretation differs. In equation (12.15), the regression coefficient estimates the effect of a diesel engine when the EPA estimate is held constant. In equation (12.14), the regression coefficient estimates the effect of a diesel engine when the EPA estimate is *not* held constant.

**5** The standard error of estimate is a measure of goodness of fit of the regression, but it cannot prove the existence or direction of causation.

**6** No. One cannot be sure that this will be the case, although it obviously seems likely that a better-fitting regression will predict better than a poorer-fitting regression.

## The Index of Industrial Production and Seasonal Variation (Chapter 14)

**1** A curvilinear regression might provide a better fit to the data.

**2** No. It is a coincident series, which means it tends to go down at the peak and up at the trough of the business cycle.

**3** $[147.6 \div 137.1] \times 100 = 107.7$.

**4** Yes, because climatic conditions have a considerable effect on construction.

**5** They meant that the seasonal index for December output was likely to decline.

**6** It would be better to add up the value-added of all firms, since there would be double counting if sales were added up.

## Decision Making at Maxwell House Concerning the Quick-Strip Can (Chapter 16)

**1** It occurs because the cost of coffee in the new can was about 0.7 cents higher than in the old container.

**2** It is assumed to be linear, since the firm is assumed to maximize expected monetary value.

**3** Such prior probabilities should reflect the experience and best judgment of the relevant experts. However, as pointed out on page 679, there is the danger that such probabilities may be biased for personal or other reasons. In some cases, it may be difficult for someone without considerable expertise to judge the reasonableness of such probabilities.

**4** Yes, because the expected profit if it did not raise its price would have been $4,104,000, which is less than $5,058,500, the expected profit if it did raise its price.

**5** No. It would adopt the new can, regardless of the outcome.

**6** The decision maker's utility function would have to be estimated, and utilities would be used rather than monetary values, as explained in Chapter 15.

# Index

Does a vehicle ^coming from south affect the percentange of vehicles that go west? why or why not? (Hint use statistical independence)

4 pts

① $P(A_2 \text{ and } B4) \overset{?}{=} P(A_2) \cdot P(B4)$

0.03 $\overset{?}{=} (0.18)(0.15)$

0.03 $\neq 0.027$

② $P(A_2/B4) \overset{?}{=} P(A_2)$

$P(A_2/B4) = \frac{P(A_2 \text{ and } B4)}{P(B4)} = \frac{0.03}{0.15} =$

$P(A_2) = 0.18$

$0.2 \neq 0.18$

Events are not statistically independent

2. The winner of the Michigan Super Lotto this week would get $3.2 Million. The probability for a one dollar ticket to win, as we know, is $1/7,059,052$ If you buy a one dollar ticket:

a) what is the probability distribution of the amount that you will win or lose?

5pts

| $X_i$ | $P(X_i)$ |
|---|---|
| 3,200,000 | $1/7,059,052$ |
| -1 | $1 - 1/7,059,052$ |

b) what is your expected net gain?

5pts

$E(x) = \sum_{i=1}^{n} X_i P(X_i) = (3,200,000)(1/7,059,052) + (-1)(1 - 1/7,059,052)$

$= 0.453 - 0.999 = -0.546$

The net expected gain is -0.546 cents

In order to est the mean scores a random et 38 exam scores,
A sample mean is found.
a) point estimate for the pop mean.

$$\bar{x} = 74.3.$$

b) construct .98 confidence int. $\sigma = 14$.

$$\bar{x} - z\alpha/2 \frac{\sigma}{\sqrt{n}} < \mu < \bar{x} + z\alpha/2 \frac{\sigma}{\sqrt{n}}$$

$$74.3 - 2.33 \frac{14}{\sqrt{38}} < \mu < 74.3 + 2.33 \frac{14}{\sqrt{38}}$$

$$74.3 - 5.29 < \mu < 74.3 + 5.29.$$

$$69.01 < \mu < 79.59.$$

c) construct .8 confidence int. for the
mean of all " "   $\sigma = 14$.

$$\bar{x} - z\alpha/2 \frac{\sigma}{\sqrt{n}} < \mu < \bar{x} - z\alpha/2 \frac{\sigma}{\sqrt{n}}$$

$$74.3 - 1.28 \frac{14}{\sqrt{38}} < \mu < 74.3 + 1.28 \frac{14}{\sqrt{38}}.$$

$$74.3 - 2.9 < \mu < 74.3 + 2.9$$

$$71.4 < \mu < 77.2.$$

$1-\alpha = .8$
$\alpha = .2$
$\frac{\alpha}{2} = .1$

Where do you attribute the diff. in your answers
in b) and c)

can be attrib to the level of significance $\alpha$.
A smaller $\alpha$ results in a "wider" confidence
interval.

# GLOSSARY OF SYMBOLS
(Continued from inside the front cover)

| Symbol | Meaning* | Symbol | Meaning* |
|---|---|---|---|
| $P_w$ | Probability that a newly arrived customer will have to wait (Appendix 5.1) | $S$ | Seasonal variation (13.2) |
| $P_{oi}$ | Price of the $i$th commodity in the base year (14.2) | $S_t$ | Smoothed value of time series at time $t$ (13.7) |
| $P_{1i}$ | Price of the $i$th commodity in year 1 (14.2) | $s$ | Sample standard deviation (2.5) |
| $p$ | Sample proportion (6.7) | $s^2$ | Sample variance (2.5) |
| $p_1, p_2$ | Sample proportions from populations 1 and 2 (7.7) | $s_1, s_2$ | Sample standard deviations of populations 1 and 2 (7.7) |
| $Pr\{a<X<b\}$ | Probability that random variable $X$ lies between $a$ and $b$ (7.5) | $s^2_j$ | Sample variance from $j$th population (10.5) |
| $Pr\{A\}$ | Probability of event $A$ (5.9) | $s_e$ | Sample standard error of estimate (11.8) |
| $Q_i$ | Dummy variable for $i$th quarter (13.10) | $s_b$ | Standard error of sample regression coefficient (11.13) |
| $Q_{oi}$ | Quantity of $i$th commodity in base period (14.3) | $s_{b1}, s_{b2}$ | Standard errors of $b_1$ and $b_2$ (12.5) |
| $Q_{1i}$ | Quantity of $i$th commodity in year 1 (14.3) | $s_E$ | Square root of error mean square (10.9) |
| $Q_{wi}$ | Quantity of $i$th commodity in year $W$ (14.4) | $s_w$ | Square root of within-group mean square (10.7) |
| $R$ | Sample multiple correlation coefficient (12.6) | $\sigma$ | Population standard deviation (2.5) |
| $R_1$ | Sum of ranks in Mann-Whitney test (9.11) | $\sigma^2$ | Population variance (2.5) |
| $R_n$ | Critical limit in sequential sampling plan (16.7) | $\sigma_0^2$ | Population variance if null hypothesis is true (9.8) |
| $R^2$ | Sample multiple coefficient of determination (12.6) | $\sigma(X)$ | Standard deviation of random variable $X$ (4.5) |
| $RSS$ | Block sum of squares (10.8) | $\sigma^2(X)$ | Variance of random variable $X$ (4.5) |
| $r$ | Sample correlation coefficient (11.11) | $\sigma_{\bar{x}}$ | Standard deviation of sampling distribution of sample mean (6.8) |
| $r^2$ | Sample coefficient of determination (11.10) | $\sigma_1, \sigma_2$ | Standard deviations of populations 1 and 2 (8.6) |
| $r$ | Number of rows (9.5) | $\sigma_e$ | Population standard error of estimate (11.8) |
| $r$ | Number of population proportions being compared (9.4) | $\sigma_b$ | Population standard error of $b$ (11.13) |
| $r_r$ | Sample rank correlation coefficient (Appendix 11.2) | $\sigma_p$ | Standard deviation of the prior distribution of the sample mean (Appendix 7.1) |
| $\rho$ | Population correlation coefficient (11.12) | $\sigma_r$ | Standard deviation of number of runs (9.12) |
| $S$ | Sample space (3.2) | $\sigma_u$ | Standard deviation of $U$ in Mann-Whitney test (9.11) |

*Number in parenthesis indicates the section where the symbol is introduced.